WORLD RAILWAY ACCIDENTS 1916-30

Francis Voisey

ISBN: 978-1-9162011-6-3

This book is dedicated to Signalman Brian Stone,
who first awakened my lifelong interest in railways way back in 1966.

ACKNOWLEDGMENTS

Thanks to the staff of the Exeter Reference Library and the University of Exeter Library for their assistance while I was researching this book, to my friend June Nightingale for encouraging me to go ahead and get it finished when my enthusiasm was waning a little, and to my publishing team at KINDLE BOOK PUBLISHING, without whose guidance and assistance it is doubtful if I would have been able to navigate the mysteries of getting books published.

CONTENTS

INTRODUCTION

This book is the fourth of what is hoped will be a series detailing railway accidents worldwide. A good many books have been written over the years about railway accidents, some concentrating on single disasters, others covering accidents in a given country or geographical area. Accidents in countries such as Great Britain, Germany, the Netherlands, the United States, Australia, New Zealand and so on have been documented comprehensively in one or more books but no attempt has been made previously to detail every significant accident worldwide. This is clearly an ambitious project, but three volumes dealing with the earlier years having been completed. This book will attempt to detail the accidents which occurred between 1916 and 1930. Previous volumes have covered the periods 1811-80, 1881-1900 and 1901-15. Further volumes are planned to cover the remainder of the 20th century and the opening years of the 21st century. As before, an attempt will be made to include all accidents involving five or more fatalities to train passengers. A number of less serious accidents which have been considered of interest have also been included. The accidents have been grouped into chapters according to their primary cause and are dealt with as far as possible chronologically within each chapter.

The gathering of information about accidents in distant parts of the world has not been easy, though it is much easier nowadays than when this project began in the days before the internet. Outside of Great Britain and Ireland few countries seem to have had a formal body charged with investigating all serious railway accidents until recent years, though one or two major disasters such as the Bussey Bridge failure in the United States and the Baptiste Creek collision in Canada were the subject of full formal investigations even in the early days. Otherwise reliance has largely had to be placed on contemporary newspaper reports and previously published works on the subject, though the author has tried as far as possible to keep to primary sources. If the book appears to be biased towards accidents which occurred in Great Britain and the United States this is probably partly because these are the countries about which information on accidents has been easiest to obtain, though these were also in any case the countries where many of the most serious accidents did occur.

Contemporary spellings of place names have been used where appropriate and for the many American accidents, terms such as railroad, engineer, caboose and so on have been used instead of railway, driver, brake van etc. The 24-hour clock was gradually coming into use, beginning with Italy from 1893, and the 12-hour clock and 24-hour clocks will both be used depending upon the custom in the country concerned.

By the beginning of the 1930s railways had spread to almost all parts of the world, though some countries only had very small systems or perhaps just one line.

The names and the boundaries of countries have undergone many changes over the years. Considerable changes occurred following the end of the First World War and the Treaty of Versailles in 1919 and there have of course been many name changes since, plus the splitting up of the former Soviet Union and Yugoslavia, the separation of India into India, Pakistan, and Bangladesh and so on. The names of some other countries and of numerous places have changed. The names of locations which were in use at the rime an accident occurred have been given, together with present names where appropriate, as for example Ürsküb (now Skopje). Some attempt has been made to use local names where these differ from the English names, for example Wien rather than Vienna.

CHAPTER 1

DERAILMENTS: TRACK DEFECTS

The tracks over which trains run are obviously a vital part of the railway and any problems with the track are likely to lead to the train being derailed. By 1916 the technology involved in laying railway track had much advanced since the early days, but train speeds and weights had also much increased. While accidents occurred from time to time due to failures to keep the track in a good state of maintenance, insufficient allowance being made for hot weather expanding the rails, damage caused to the track from floodwaters and so on. Also included in this chapter, though they were strictly not due to defects in the track, are accidents due to trains having become derailed on open points. Broken rails were another hazard which caused mishaps and worse from time to time.

An additional 19th century derailment has recently come to light, not included in the 1881-1900 volume. On Saturday 2nd October 1897 what was stated to be 'the most serious railway accident that had ever occurred in Japan' happened between Oyama and Yakameta. The line had been damaged by floods and a passenger train was precipitated into a river. Of the 40 passengers on the train seven were killed and 28 injured. [1]

Mexico was in the throes of a civil war between 1910 and 1920 but the derailment of a military train carrying General Luis Gutiérrez and his troops on the night of Thursday 8th June 1916 was apparently the result of a defective (broken?) rail rather than of sabotage. The engine and three coaches of the train left the track on a curve near Saltillo, Mexico. The driver, fireman and ten soldiers were killed, and 17 Carranza soldiers were seriously injured. The general himself escaped with slight bruises. [2]

A northbound passenger train of the Ferrocarril Central Mexicano (Mexican Central Railway) was derailed near Ortiz, Mexico, in Chihuahua province, on Sunday 10th September 1916. Meagre reports of the accident reaching the American press from

---

[1] *The Indianapolis Journal*, Indianapolis, Indiana, United States, 21st October 1897

[2] *The Anaconda Standard*, Anaconda, Montana, United States, 12th June 1916

Ciudad Juárez stated that the accident had occurred about 75 miles south of Chihuahua City and was due to spreading rails. The train had been crowded with civilians and soldiers. Several cars overturned and 22 passengers died; half of the dead were Carranza troops en route from Torreon to Chihuahua City. [3] [4]

A Mexican troop train was, again, involved in a derailment apparently due to track defects on Sunday 6th May 1917, this time near Queretaro, Mexico, on the Ferrocarriles Nacionales de México (Mexican National Railway). A passenger train for Nuevo Laredo left Mexico City early on Saturday 5th, accompanied by a military escort. From Queretaro the passenger train was preceded by a special military train carrying about 600 Carranza soldiers, the passenger train following close behind. The two trains had only gone a short way when the whole of the military train was derailed due to the bad condition of the track, its coaches overturning into a ditch. Fourteen passengers were killed outright, including four women and three children. The women and children were families travelling with their soldier husbands and fathers. A hundred or more other soldiers were seriously injured, and a horrific scene was said to have met those riding in the ordinary passenger train which arrived at the scene a little later. The passenger on this train who related the details of the accident upon his arrival in Texas stated that while those on the second train had rushed forward to help the victims, many of the uninjured soldiers from the troop train had just stood around after the accident looking at their less fortunate comrades but doing nothing to assist. It was not until Monday that the injured and the bodies of the dead were taken to Queretaro and the line cleared. The ordinary passenger train did not finally reach Nuevo Laredo until Tuesday evening.[5] The same day as the Queretaro accident bandits dynamited a train near Oarneros, Mexico, in an attempt to steal $50,000 from an express car, killing 30 soldiers. [6] [7]

On 14th January 1918 westbound (geographically northbound) Houston and Texas Central Railway train No. 17 from Houston to Dallas, named *The Owl,* [8]was in charge of Conductor Morgan with Engineer Hyndman. The train crossed eastbound passenger train No. 10 at Calvert, where it received orders to wait at Hammond, Texas, United States, until 3.45 a.m. for extra eastbound freight train No. 465. It left Calvert on the single track at 3.24 a.m., 14 minutes late, and arrived at Hammond at about 3.50 a.m., finding extra 465, with Conductor Benge and Engineer Elms, waiting in the siding for it there, with engine No. 465 standing 250 feet west of the facing points leading into the siding. Engine No. 104 of train No. 17, its tender and the leading bogie of the following baggage car passed safely over the points, but the switch was defective and the rear bogie of the baggage car and all the wheels of the passenger train on the left-hand side were diverted to the left. The chair car which was the third vehicle behind the engine was thrown against engine No. 465 and the superstructure of the chair car above the window sills was torn away. The train was completely derailed except the rear truck (bogie) of the last

---

[3] *The Salt Lake City Tribune*, Salt Lake City, Utah, United States, 13th September 1916

[4] *The Plano Star-Courier*, Plano, Texas, United States, 22nd September 1916

[5] *Laredo Weekly Times*, Laredo, Texas, United States, 13th May 1917

[6] *Brownsville Herald*, Brownsville, Texas, United States, 9th May 1917

[7] *San Antonio Light*, San Antonio, Texas, United States, 9th May 1917

[8] *Victoria Daily Advocate*, Victoria, Texas, United States, 14th January 1918

Pullman sleeping car. All the coaches were of steel construction, but the heavier four sleeper cars crushed the lighter coaches ahead of them, while passengers were scalded by steam from the damaged freight locomotive or possibly from the broken heating pipes of the day coaches. Seventeen passengers died and ten were injured. Rescuers worked by torchlight to release the victims and the injured were loaded onto an express car and rushed off to Waco. [9]

Engineer Hyndman of train No. 17 stated that he had approached Hammond travelling at 35 or 40 miles an hour and that the switch target indicating the position of the points had shown that they were closed for the main line. He thought that his locomotive had escaped being derailed because it was on a curve and its wheels had good flanges. After the accident he had examined the points and found that the bridle bar (stretcher bar) of the switch was bent as if it had been struck by something dragging from a train, and his observation was confirmed by Conductor Benge and Engineer Elms of extra 465. Before the arrival of train 465 the last train to use the dead end siding at Hammond had been eastbound extra 617, which had backed into it to allow following train No. 5 to pass. Conductor Stagell and Brakeman Mckee had failed to see anything wrong with the switch when they had reset it after their train had left the siding again but on arrival at Calvert, Conductor Stagell had discovered a brake beam hanging down from the 19th car, this no doubt being what had damaged the switch.

The Interstate Commerce Commission blamed Conductor Stagell and Brakeman McKee of extra 617 and Engineer Elms and Brakeman Bimsky of extra 465 for having failed to notice the defective points. They were all experienced men, and each claimed to have examined the switch, making it inexplicable that none of them had seen anything wrong. [10]

Heavy rain during a storm led to flooding and the washing away of embankment at Hochstetten, between Kirn and Martinstein, Germany, on the Bad Münster – Saarbrücken line, on Wednesday 16th January 1918. Leave train D 243, crowded with troops, was derailed, and the engine, baggage van and three coaches fell into the Nahe River. Thirty-eight soldiers died and 25 were injured. [11] It was impossible to carry any rescue work until the level of the water had gone down. [12]

We go to the Netherlands for the next accident; perhaps appropriately for the superstitious the accident occurred on Friday 13th September 1918, near Weesp, on the Hollandsche Ijzeren Spoorweg-Maatschappij (HSM). This was by far the worst in the history of the country up to that time and would remain so for the next 44 years. Heavy rain was again a factor. The train involved in the disaster was No. 102, and it consisted of portions from Zwolle and Enschede, which were combined at Amersfoort, while further along the line another carriage was added to the train at Hilversum, so that at the point of the derailment it consisted of locomotive N 520 and 11 carriages.

---

[9] *Athens Daily Herald*, Athens, Georgia, United States, 14th January 1918
[10] Interstate Commerce Commission report
[11] *Eisenbahn-Katastrophen in Deutschland* by Hans Joachim Ritzau, pg. 72 (in German)
[12] *Eisenbahnunfälle in Europa* by Erich Preuß, pgs. 16 and 68 (in German)

On the Amsterdam side of Weesp Station the railway climbed on an embankment to a substantial metal truss bridge over the Merwedekanaal. It was on this approach embankment that the accident occurred. The train approached the bridge at about 10.20 a.m., travelling at a speed of 55 to 60 km/hr. As its engine was about 30 metres from the bridge the engine crew noticed their machine suddenly jump and tip over to the right. The locomotive came to a stand entangled with the metalwork of the bridge superstructure. Adjoining the bridge was blokpost (signal box) CB and the signalman, his wife and a platelayer watched in bewilderment as the engine left the track and hit the bridge, while some of the eleven carriages following plunged down the seven-metre-high embankment. The leading baggage van stopped close to the bridge behind the engine tender, the next three carriages slid down the embankment and partially telescoped one another. The next five vehicles also went down the embankment and some of them ended up alongside of and crashed into the front three. The rear two vehicles remained on the track. A gas cylinder in one of the front carriages also exploded and set it alight but the fire was quickly extinguished using sand. A worse disaster was thus averted but even so the casualty list was a heavy one, with 41 dead and 42 injured.

None of the several railwaymen on the train was killed or injured in the accident so they were all available to protect the train and begin the rescue work. By chance Professor H. Treup, a surgeon, was a passenger on the train, as were four nuns from Apeldoorn, and an infantry regiment under Captain Van Andel was also in the vicinity and rendered aid. The alarm was sounded by the Weesp stationmaster at 10.45. A relief train arrived from Naardem-Bussum at 11.40, followed by a breakdown train from Amsterdam Rietlanden and later a Red Cross train from Amsterdam Centraal station. Both non-injured and injured passengers were taken by train to Amsterdam, and some were also conveyed there by boat via the Merwedekanaal. The line could not be reopened for almost a week, meanwhile trains between Amsterdam and Amersfoort were diverted via Breukelen and Utrecht. After reopening there was a 5 km/hr speed restriction at the accident point while the embankment and bridge were repaired. Normal working could finally be resumed on 2nd December.

The investigation into the cause of the disaster began the next day, presided over by Dr C. Lely. The conclusion of the investigation was that a prolonged period of rainfall (100 mm of rain had fallen in the ten days preceding the accident) had saturated the embankment and impenetrable layers within it had prevented the excess water from escaping. There had also been a leakage of water from the adjoining Merwedekanaal. The level of groundwater in the embankment had risen above ground level and the passing of the train and resulting vibration had resulted in a collapse and displacement of the semi-liquid matter comprising the earthwork and had derailed the train. In the opinion of the investigating committee the accident could not be blamed on negligence or lack of maintenance by the company. The HSM therefore stated that it could not accept any claims for compensation in connection with the accident but after a lengthy judicial procedure a court in Amsterdam decided otherwise in 1924.

The solution to the accident, it was pronounced, was simple; the installation of drainage openings to prevent the accumulation of excess water and the broadening of the foot of the embankment to increase its strength. The investigating body also considered the condition of 21 other embankments at different locations which were five or more metres high and decided that five of them were also at risk of a similar collapse and needed attention. [13]

The worst railway accident in the history of the Netherlands up until 1962 was followed in less than three weeks, on 1st October 1918, by the worst railway accident up to the present time (June 2021) in Swedish history, near Getå. The accident occurred exactly three years after the line had been opened between Åby and Nyköping. At the location of the accident the railway ran along the hillside, with a rock face on one side and a steep embankment falling to the shore of Bråviken on the other. A geological fault which existed in the area had been known of before the line was built and geological surveys had been undertaken. The spring of 1918 had been unusually dry, the level of Bråviken had dropped and the ground had dried out and begun to crack. September 1918, on the other hand, was unusually wet, and the ground became saturated. The railway line had been formed by cutting into the hillside, and the embankment beneath it consisted of layers of colloidal clay and gravel. The bearing capacity of the clay diminished as the ground got heavier and on the day of the accident a layer of gravel began to slip against a layer of clay and the embankment beneath the railway line began to collapse.

Train No. 422 had left Malmö at 7 a.m. for Stockholm. Engines were changed at Mjölby, and from there the train was hauled by engine No. F1200 with Driver Axel Wahlström and Fireman Carl-Erik Carlsson. In charge of the train was Chief Conductor Karl Nordqkvist, together with Conductors Karl Johan Ström and Carl J.V. Löfstedt. Beyond Mjölby a vehicle was taken off at Linköping, and from there the train consisted of ten vehicles, with an unbraked goods waggon at its rear. After Linköping the train passed through Norrköping, then Åby, and arrived at a halt at Getå, which served Prins Carls villa just to the south at Fridheim and thus often saw Royal patronage. The accident occurred on the far (Stockholm) side of the halt, between there and the next station at Krokek.

A train had passed through Getå safely at 6.26 p.m. After that, a milk cart had passed the site of the embankment failure at 6.50 p.m., its driver seeing nothing wrong. Between two and five minutes later, however, the embankment gave way. The train had left Åby at 6.54 p.m. and Porter Svensson attempted to inform Kroket of its departure at 6.57 p.m. Shortly afterwards Station Clerk Johansson or possibly Stationmaster Edén at Kroket enquired of Åby as to why they had not received the ‘Tåg ut’ (equivalent to ‘train entering section’) signal from Åby for train No. 422 but could get no reply. The collapsed embankment had brought the telegraph wires down. Even if anyone at either station had realised what had happened at this point they had no way of stopping the train.

Travelling at 65 or 70 km/hr, well below the maximum permitted speed of 90 km/hr, the train passed the house of Platelayer A.F. Andersson. Shortly afterwards Driver Wahlström felt the engine give way beneath his feet and a crash, then lost consciousness for several minutes. When he came round, he managed to drag himself under the engine. His fireman was partially buried in coal, lost an arm, and was scalded by hot water and steam from a broken injector. Despite the best efforts of his mate he died shortly afterwards. Behind the engine the vehicles up to the restaurant car went down the embankment and were wrecked. The restaurant car slid partly down the embankment and the last two vehicles remained on the track. Chief Conductor Nordqkvist chanced to be in

[13] *Spoorwegongevallen in Nederland* by R.T. Jongerius, pgs. 97-103 (in Dutch)

the last passenger carriage. He went forward to the engine after the derailment and spoke to the driver, then headed back down the line towards Åby. On his way he met Platelayer Andersson, who was able to notify Åby from a nearby house. Having protected the obstruction with detonators Andersson then went and assisted with the rescue efforts.

Meanwhile 500 metres on the Krokek side of the accident site was Platelayer Eriksson's house, situated at a level crossing which his wife managed. She had lowered the crossing barriers for the expected train, but it did not appear. Then she heard something and told her husband that it sounded as if the train had stopped down the line, whereupon he set off to investigate. As he approached the scene he could smell burning and see a glow, then he saw telegraph poles uprooted and knew for certain that there had been an accident. When he got to the train, he found that it was on fire, spilt engine coals having set the wooden carriages alight. Eriksson met Chief Conductor Nordqkvist and Platelayer Andersson, then hurried back down the line past his house to Getå halt, from where a message could be telephoned to Krokek. Back at the wreck confusion as to what to do prevailed at first after the accident, then the survivors organised themselves and several passengers were highly active in the rescue work until the advancing fire forced them to desist. Some of the trapped victims were regrettably burnt to death before the eyes of the rescuers who were unable to save them. The fire apparently began with the mail van but quickly spread throughout the wreckage. It was rumoured that damaged acetylene gas cylinders had also contributed to the blaze but the inquiry into the disaster could find no evidence of this. Of the approximately 170 people on the train 42 bodies were recovered and five were reported 'missing', their remains possibly consumed by the fire, or perhaps they were not on the train, or were unhurt and had left the accident scene without giving their names. In addition to Fireman Carlsson, Conductor Ström was also killed.

The first help from outside came from Åby, where Stationmaster Callmé converted passenger train No. 418 into a relief train, which was accompanied by men with lights and rescue equipment and carried a physician, Doctor Landelius, who on arrival at the scene was able to assess the injured and prioritise those who needed immediate removal to Norrköping. A second relief train arrived later from there, a third from the opposite direction from Nyköping and finally a fourth train, also from Norrköping. With the line well blocked by the accident passenger traffic was managed by working trains from Åby to a point just short of the slip and from Krokek station to Getå halt on the other side, with passengers and goods going via the parallel road between the two.

The Royal Railway Board inquiry into the accident began on 2nd October, and there was a second inquiry conducted by a geological commission. The Kungliga järnvägsstyrelsen investigation ruled out an escape of gas as a cause of the fire. Nevertheless, gas lighting of carriages was replaced by electric lighting in the 1920s. The geological investigation drilled holes in the embankment and concluded that there had been a prehistoric landslide at the point of derailment which had increased the flow of water at that spot and hence the likelihood of a slide. This exceptional circumstance had probably misled the engineers building the line into overestimating the loading strength of the embankment. Following the clearing of the line the track was realigned a metre closer into the side of the hill and the embankment was reinforced with rocks. The line reopened in December with a 15 km/hr speed restriction.

Another landslide occurred at the same place in 1923, this time without any consequences.[14]

A report received by Reuters news agency in Amsterdam quoting sources in Wien, Austria, spoke of the derailment of a double-headed Bucharest – Budapest express (train No. 2) near Piatra, Romania, due to a cloudburst having washed the track away on about Wednesday 9th October 1918. The Căile Ferate Române (Romanian State Railway) train reportedly fell into the Bistrita River and 100 persons were killed and 200 seriously injured.[15] Another report stated that the boiler of one of the engines had exploded and that some of the carriages had caught fire. A goods train had passed the place half an hour earlier without incident.[16]

A train on the Lüneberg – Soltau Kleinbahn (light railway) was derailed between Retmer and Melbeck halts on Sunday 11th May 1919 due to a 'rejection' (verwerfung in German), meaning presumably track defects. Nine people died and 11 were injured.[17]

Fourteen people died in an accident to a Bruxelles – Mons train at Lembecq, Belgium, on Thursday 12th June 1919. Almost all of them were soldiers on leave. The pre-war safety installations which would have rendered such an accident impossible had been removed by order of the occupying Germans and had yet to be re-installed. The derailment occurred on a set of points. Thirty-five more people were injured. A couple of days later the Belgian Minister of Railways attributed the derailment to enemy, clearly meaning German, sabotage, presumably with reference to their removal of the safety devices.[18] Quite why these were removed anyway it is hard to see unless the military wanted them for re-use elsewhere.

At 7 a.m. on the morning of 10th January 1920, a train carrying workmen to the blast furnaces of the Société Métallurgique de Normandie de Colombelles at Caen, France, was derailed on a set of points. Seven died and 41 were injured.[19]

As a Paris d'Orsay – Nantes express of the Chemin de fer de Paris à Orléans was entering Aubrais station, France, at 16:15 on Monday 5th July 1920, a set of faulty electro-pneumatic facing points worked from poste (signal box) A, shifted beneath its sixth carriage.[20] The train broke in two, and the rear coaches, crowded with holidaymakers, were derailed, and overturned. Six or seven passengers died and 20 were seriously injured. The London *Times* assured its readers that 'It is said that no British or

---

[14] *Getå 1918* by Rolph Wegmann (in Swedish)
[15] *Winnipeg Free Press*, Winnipeg, Manitoba, Canada, 16th October 1918
[16] *Eisenbahnunfälle Bind 1* by Ludwig Stockert, pgs. 88-9 (in German)
[17] *Eisenbahn-Katastrophen in Deutschland* by Hans Joachim Ritzau, pg. 92 (in German)
[18] *La Croix*, Paris, France, 15th and 18th July 1919
[19] Wikipedia – Liste des accidents ferroviaires en France au XX siècle
[20] Ibid.

Americans were among the casualties.'[21]

Northbound passenger train No. 2 of the Southern Railway, otherwise, *The Royal Palm Limited*, was derailed just north of New River, Tennessee, United States, at about 2.30 p.m. on Wednesday 6th April 1921. The train consisted of locomotive No. 6492, a combination baggage car and coach, two coaches, eight Pullman cars and a dining car and was in charge of Conductor Dineen with Engineman Miller and Fireman Duncan. The engine and tender were derailed to the left and the first six cars were derailed to the right, four of the all-steel cars being damaged by scraping up against a lineside rock bluff. Two passengers were killed outright, three fatally injured and another 74 passengers, three other persons and a railroad employee were less seriously hurt. Those who lost their lives were F.E. Cook, of Orion, Michigan; E.J. Bussey, of Detroit, Michigan; Harry Sickles, of St Marys, Ohio; Joe Kramer, of Chicago, Illinois, and William Parks, of Soldier's Home, Grand Rapids, Michigan. [22]

The engineer and fireman told the Interstate Commerce Commission (ICC) inquiry that they had whistled for a road crossing immediately before the derailment. They had not been flagged and noticed nothing wrong with the track. Examining his locomotive after the accident the engineer could find nothing wrong with it but looking at the track he discovered that angle bar bolts had been sheared from a joint on the inside of the east rail and the east rail was low.

Section Foreman McDonald told the inquiry that about a week before the derailment he had been notified by Track Supervisor Argo that the track at the curve where the accident occurred was rough and told to carry out repairs. On the morning of the accident he had found the gauge at the point of derailment to be ¼ inch wide and the elevation 3½ inches instead of 4½ or 5 inches. In the early afternoon he was advised of problems at another location in his section, which he had attended to and had then hurried back to the accident point as he had been worried about the heat expanding the rails. He had worked to increase the elevation, had put in 43 new ties, and tamped the ballast with shovels. He had not considered the track unsafe for trains to run over at scheduled speed but when train No. 2 arrived he saw the track slue beneath it and afterwards found that it had slued 11½ or 12 inches.

The ICC investigators concluded that Section Foreman McDonald had not properly tamped under the ties on the outside of the curve, so the outside rail had not been properly supported and had sunk beneath the weight of the locomotive, resulting in a rebound to the left which had caused the inside rail to overturn. The section foreman had been employed since 1882 and was considered by his superiors to be a capable and reliable foreman and one of the best in the district. [23]

A very brief report in a single American newspaper spoke of the derailment of a passenger train from San Louis Potosi to Tampico at Las Cososos, Mexico, due to a washout on Thursday 7th July 1921. Eleven persons lost their lives, and 25 others were

---

[21] *The Times*, London, England, 6th July 1920

[22] *Marysville Evening Tribune*, Marysville, Ohio, United States, 7th April 1921

[23] Interstate Commerce Commission report

injured. [24]

The washing away of the track due to the failure of an impounding dam was the primary cause of a disaster near Empala Gonzales, Mexico, on Tuesday 19th September 1922. A military escort train which was passing at the time was derailed and a northbound passenger train crashed into it. Fourteen persons were killed and 17 injured. It was believed that other bodies might be found in the wreckage. Americans Henry Burton of Atlixco, Mexico and J.D. Lestay, of Texas City, Texas, United States, were said to be amongst those seriously injured. 'All traffic on the Laredo line from Quetaro to Mexico is suspended.' The disaster happened on the Ferrocarriles Nacionales de México (Mexican National Railway). [25] [26]

What was then the British island colony of Ceylon, now Sri Lanka, was subject to severe flooding early in 1923 and on 14th January the mail train from Talaimannar was derailed near Anuradhapura due to the track having been washed away, though without any loss of life. The bad weather continued and on the night of 25th/26th January a severe storm led to flooding and the washing away of the track in the same area. At 1 a.m. on the morning of the 26th of January 1923 the Jaffna portion of the down mail train left the track on the approach to a bridge, the engine and tender, travelling post office, brake van and one coach being submerged in the rapidly rising waters and according to one report six bogie carriages being reduced to matchwood. The accident occurred in a remote area. A special train carrying senior railway officials was sent to the scene. [27]

A report from Colombo on the day of the disaster stated that 16 bodies had been recovered and 12 injured had reached Anuradhapura Hospital. It was feared that a number of other bodies had been carried away and that the final number of dead would be as high as 39. [28]

On 6th March the accident was the subject of a question raised in the British House of Commons, Mr Linfield asking the Under-Secretary of State for the Colonies if he had any information on the number of dead and injured in the accident, whether some of the victims had been prisoners who had been handcuffed at the time of the derailment, and whether there would be an independent inquiry into both derailments by a committee including at least one railway expert, in view of the dissatisfaction prevailing in Ceylon with the committee of inquiry set up by the local government to investigate the accidents. Mr Ormsby-Gore replied that he had no information on the matter but would ask the Governor of Ceylon for a full report. [29]

Note: The two derailments are listed in the wrong order in *Railway Disasters of the World* by Peter Semmens. It was the minor accident which occurred first.

---

[24] *Leesville Vernon Parish Democrat*, Leesville, Louisiana, United States, 21st July 1921
[25] *Ogden Standard Examiner*, Ogden, Utah, United States, 20th September 1922
[26] *The Evening Gazette*, Cedar Rapids, Iowa, United States, 20th September 1922
[27] *The Railway Gazette,* Volume 38, pgs. 102 and 258
[28] *The Times*, London, England, 27th January 1923
[29] Webpage: hansard.parliament.uk/commons/1923-03-06/debates

Presumably because those who lost their lives were only tramps the wreck of a Chicago Milwaukee and St Paul Railway freight train near Fairdale, Illinois, United States, at 5.30 p.m. on Thursday 5th July 1923 was not investigated by the Interstate Commerce Commission. Thirty-five cars of a fast freight train, several of which were carrying livestock, went into the ditch when it was derailed due to a broken rail. A report from Rockford, Illinois, stated that the victims of the accident had been members of a work gang from Omaha who had been working on the line between Leaf Junction and Davis Junction and were being transferred to Chicago. According to this report five men were killed and 18 injured. The men had been riding in a car in the centre of the train and none of the train crew was injured. Some of the injured were cared for in Fairdale village, the more seriously injured being taken to Rockford.

A conflicting report from Chicago stated that railroad officials there had been informed by Superintendent Lollis, of Savannah, Illinois, that a wreck train accompanied by Assistant General Manager C.O. Bradshaw had left Savannah for Fairdale, that none of the local officials knew of any gang of workmen riding on the train and that if they had been they would not have been riding in the middle of it, but provision would have been made for them. The dead and injured must therefore all have been tramps. [30]

Another newspaper named three of the dead as George Henton, of Chicago; Walter Rooves, of Omaha and a man named Livingstone. The other two were unidentified. All the dead were negroes, an additional reason, possibly, why the Interstate Commerce Commission did not bother to look into the accident. At least one of the injured was said to be in a critical condition at the time of the newspaper report. [31]

Eastbound Wabash Railroad passenger train No. 2 from St Louis, Missouri, to Detroit, Michigan, [32] was wrecked at 11.44 p.m. on Sunday 1st June 1924 at Williamsport, Indiana, United States. Westbound freight train second No. 91, in charge of Conductor Nelson with Engineer Bouchard, arrived at Williamsport at 11.21 p.m. and entered the passing track to cross eastbound trains Nos. 50 and 2. The locomotive, 59 cars and caboose of train No. 91 would not fit in the passing track and after train No. 50 had been flagged and had come to a stand on the main track opposite the passing track, the engine of train second No. 91 pulled out onto the main track with four cars and pushed them back into a siding known as the back track. When this had been done it was found that the caboose was still foul of the main track at the east end of the siding and the engine had to be uncoupled from the four cars in the back track and set back on the main part of train second No. 91 to pull it slightly forward. While this was being done Head Brakeman Davis Torrance went forward to the main switch at the west end and reset it for the main line. Conductor Nelson saw the indication of the switch lamp change from red to white, asked the head brakeman if he had locked the switch and was assured that he had. Conductor Nelson saw that the automatic signal protecting the switch was clear. The conductor and brakeman then busied themselves dealing with a hanging brake beam on the first of the four detached cars and were thus engaged when train No. 2 arrived and was derailed at the west switch.

---

[30] *The Burlington Hawkeye*, Burlington, Iowa, United States, 6th July 1923
[31] *Jacksonville Daily Journal*, Jacksonville, Illinois, United States, 6th July 1923
[32] *The Kingston Daily Freeman*, Kingston, New York, 2nd June 1924

Train No. 2 was in charge of Conductor Martin with Engineer Bickel and Fireman Siblisk. It consisted of engine No. 672, a mail car, an express car, a combination car, a coach, a chair car, three sleeping cars and a business car. When it reached the west switch it left the track. As a result of the derailment it split into three parts. The locomotive and three cars came to a stand 3,000 feet beyond the switch; the second section, consisting of the coach and chair car, came to a stand 1,075 feet beyond the switch, having struck the side of engine No. 2210 of the freight train. The third section only ran a short way beyond the switch and the third sleeping car came to a stand alongside engine No. 2210. It was here that most of the fatalities occurred, steam entering the sleeping car due to the check valve and other parts having been torn from the side of the engine. Six passengers and six employees off duty (deadheading on train No. 2?) lost their lives, together with Engineer Bouchard. Forty-four passengers and four railwaymen were injured.

Engineer Bickel told the Interstate Commerce Commission inquiry that the automatic signals had been clear, and he had noticed nothing wrong until the train was derailed. He could not see the indication of the switch lamp from his side of the locomotive.

Fireman Siblisk agreed that the automatic signal had been clear and added that the switch lamp had also been showing clear.

Conductor Martin said that he had examined the switch shortly after the accident and found that it was badly damaged. He had called the attention of Head Brakeman Torrance to the switch lock lying on the ground and the latter stated that he had put the lock in the switch. The conductor was of the opinion that prior to the accident the switch had been closed but not locked. Flagman Hauk of train No. 2, Conductor Smith, who was deadheading on that train, and other railwaymen confirmed at the inquiry that after the accident the switch had been open and the lock on the ground.

Head Brakeman Torrance told the accident investigators that he had flagged train No. 50 and agreed that he had been questioned by Conductor Nelson and had confirmed that he had told Conductor Nelson that he had closed and locked the switch. At first he denied having gone to the switch shortly after the derailment but confronted with Conductor Martin he admitted going to the switch, having done so and given his lantern to the conductor so that the latter could examine it. The head brakeman stated that at no time had he laid the lock on the ground, and he was confident that he had put the lock in the staple after closing the switch, though his statements indicated some doubt in his mind as to whether he had actually snapped the lock shut.

The commission concluded that Head Brakeman Torrance had merely closed the switch and not locked it. In this condition the vibration of the engine and leading vehicles of train No. 2 passing over the connection at high speed had caused the lever to fly out of the notch, whereby the points had opened and derailed the train. The head brakeman had been employed on the railroad since the end of January and had two months' previous experience on the New York Central and about a year's experience on the Chesapeake and Ohio. [33]

A passenger from Moscow to Sebej (or Sebezh) was derailed at 11:30 p.m. on Tuesday 12th May 1925 between Rumiantsevo and Volokolamsk, Russia. Its first-class

[33] Interstate Commerce Commission report

carriage was completely smashed, and all its occupants were killed or injured. All the other carriages but five were more or less badly damaged. A panic reportedly set in amongst the passengers as they scrambled through the windows of overturned coaches in the darkness. A relief train arrived from Moscow after three hours; meanwhile it was alleged that peasants from the surrounding area had congregated and some of them had begun looting the train until stopped by passengers and some militia from Moscow. Many passengers were pinned for a long time in the wreckage. The driver was killed, and the fireman was pinned beneath the engine for six hours, being freed only by the amputation of his leg. A few British, Germans, Latvians and Swedes were among the passengers on the train. Fourteen people died and 30 were injured. The accident was blamed on the condition of the track, 50 per cent of the sleepers between Moscow and the Latvian border being said to be rotten, leading to frequent derailments. [34]

The Sunnyland, a crack express of the St Louis-San Francisco Railway, plunged over an embankment near Victoria, Mississippi, United States, between 6.35 and 6.40 a.m. on Tuesday 27th October 1925. Twenty passengers and one railroad employee died and 117 passengers, eight Pullman employees, three mail clerks and two railwaymen were injured. As far as the railroad was concerned train No. 108 was northbound; it was en route from Florida to Memphis, Tennessee, Kansas City, Missouri, and points west and consisted of locomotive 1060, a refrigerator express car, a combination baggage and mail car, a baggage car, two coaches and seven Pullman sleeping cars. In charge of the train was Conductor J.W. Ryan with Engineer R.B. Herring and Fireman T.H. Young. The engine and tender did not leave the track but almost all the remainder of the train was derailed. The railroad employee killed was Conductor Ryan, who had been riding in the day coach.

A newspaper reporter managed to interview one of the surviving passengers, a Mr W.G. Davis, of Oxford, Mississippi, who related, "We were travelling at 60 miles an hour when the train suddenly gave a lurch and car after car toppled down the embankment. Cars were piled on top of one another. I saw several dead and a number of others who were badly mangled. Passing motorists and farmers living nearby came to the rescue with axes and shovels and extricated those who were trapped. I was only slightly injured and aided in rescuing Miss Olive Webster, a nurse from Oxford, who suffered a broken hip. We were brought to Memphis in a car. The two rear Pullmans in the train apparently did not leave the rails. I saw nothing of the engine. The report was that a split rail had caused the accident." Miss Webster was more badly hurt than Mr Davis perhaps realised for she later died of her injuries in a Memphis hospital. [35]

It was quickly apparent that the derailment had been caused by a broken rail, the result of a transverse fissure, and the most serious accident of its type in the United States since that at Manchester, New York, in August 1911. The Interstate Commerce Commission investigation resulted in a lengthy report, extending to over a hundred pages, which dealt in great detail with the broken rail involved in the disaster and with the general subject of transverse fissures. In the report's concluding remarks it was stated that strictly speaking it was possible that the fissure might have been detected before the accident, but this was not likely. The report went on to mention the recently invented Sperry transverse fissure

---

[34] *The Times*, London, England, 13th, 14th and 15th May 1925

[35] *Waukesha Daily Freeman*, Waukesha, Wisconsin, United States, 27th October 1925

detector, by which such defects might be identified, and their relative sizes measured. The prospect of avoiding or overcoming the formation of transverse fissures did not, however, seem promising. As to the cause of transverse fissures the report stated that 'they are caused by the introduction of internal strains in the head of the rail set up by the action of the wheel pressure on the running surface of the head. The internal strains directly introduced are strains of compression. To resist these strains of compression the metal next below is put into a state of tension. Transverse fissures are formed in that zone of metal which is put into a state of tension.' Rails differed in their ability to endure stresses, so some exhibited transverse fissures and others did not. There was, at the time of the report, no known method of preventing such fissures, though having a means of detecting them was a great accomplishment.

The report ended by suggesting the improvement in the manufacture of rails by steel mills to reduce the incidence of all types of rail fracture, the collection of data by the railroad companies relating to the longevity of rails exhibiting transverse fissures and further research specifically referring to steel rails and into the physics of steels in general. [36]

What was described as unprecedented rain washed away the track at mile 246/16-12, between Pattukkottai and Adirampattinam stations, India, on the Tanjore District Board line of the South Indian Railway on the night of 10th December 1925. As a result at ten minutes after midnight on the morning of Friday 11th December mixed train No. 55 on the metre gauge line ran into the breach, which in the darkness the driver had no chance of seeing in time to pull up. The engine and four bogies capsized, and two of the two upset vehicles were completely telescoped and the other two partly so. Nine passengers lost their lives and 20 were injured. An inquiry concluded that there had been no negligence on the part of the staff and the accident was entirely due to abnormal weather which could not possibly have been foreseen. [37]

Late in the night of Wednesday 17th March 1926 (possibly early on Thursday 18th) five or six cars of an eastbound Canadian Pacific Railway freight train left the rails at Sicamous Lake, British Columbia, Canada. A westbound passenger train was held up two hours by the wreck and on arrival at Vancouver on Thursday evening passengers from the latter expressed fears that the lives of six tramps might have been lost in the derailment. They related that trainmen at the scene had found a man lying half in the water at the base of a rock and gravel fill (embankment), the subsidence of which had caused the train to leave the track. Both the man's legs were broken, and he allegedly told his rescuers that he had been one of a party of eight men beating their way eastwards on the freight train. A search led to the discovery of one of the man's companions, who was only slightly hurt, but there was no trace of the others. The badly injured man was named as D.A. Rogers, address unknown. Neither of the two survivors knew the names of any of the other men. [38]

---

[36] Interstate Commerce Commission report

[37] Report by the Railway Board on Indian Railways for 1925-26.

[38] *The Winnipeg Evening Tribune*, Winnipeg, Manitoba, Canada, 19th March 1926

On Thursday 29th April 1926, a mail train from Astorga to Cácheres was derailed between Hervás and Baños de Montemayor stations, Spain, killing four people outright and injuring a number of others. The derailment occurred at kilometre 58. A relief train left Las Delicias station, Madrid, for the scene of the accident, at 10 a.m., carrying doctors and workmen. At the same time a motor car left Madrid for the scene carrying the director of the railway company, Engineer Moreno and two other senior company officials. The derailment was said to have occurred at a dangerous point on the line, on a severe curve and an embankment.

An initial newspaper report named three of the dead as Dr Jasé Riuz Rodriguez, his daughter Maria Luisa and a civil guard on the train, Ángel López López Castro. A fourth victim, a young person aged about 25, was at that time still unidentified. Civil guard Ángel Rodriguez, who was the wife of the dead doctor, and another woman, Ramada Remedios Puerto, were seriously injured. Some of the other unnamed victims had required the amputation of limbs. [39] One of the injured subsequently succumbed to his or her wounds. The derailment was caused by a broken rail. [40]

The disruption of schedules due to a storm reportedly meant that the only Long Island Railroad service to leave New York Penn station on the early evening of Friday 13th August 1926 was the company's crack train, the so-called *Millionaire Special*, which was well filled with commuters and 'weekenders.' The train was in charge of Conductor Neaves with Engineers William Squires and Jackson with Firemen John Montgomery and Fitzgerald. It consisted of locomotives Nos. 214 and 2, a parlour car, a combination baggage and chair car and four coaches, the vehicles all being of all-steel construction.

The train left Manorville at 5.59 p.m., 16 minutes late, and four miles on was derailed at the west switch of the south siding at Calverton, New York, United States, at 6.05 p.m., while travelling at a speed of between 40 and 55 miles an hour. Both engines and the parlour car were derailed and overturned and the remainder of the train, except the rear coach, also left the rails but stayed upright. The engines struck a lineside factory building. The rescue operation was aided by Colonel M. G. Spink of the 62nd Coast Artillery, who brought 300 men from Centerville, five miles away, while ambulances and commandeered cars rushed the injured to hospitals. A Roman Catholic priest who had been a passenger on the train administered the last rites to several of the victims. [41]

Five passengers were killed, as were Engineer Squires and Fireman Montgomery on the lead (pilot) engine. The passengers who lost their lives were Mrs George Schofield, of Asheville, North Carolina, her children George, aged three and Dorothy Louise, aged one; J. B. Wayne, of Greenport, New York, and stockbroker Harold Fish, of New York City. Mrs Schofield, the wife of an Asheville attorney, died in hospital after being pinned in the wreckage for five hours. She did not lose consciousness before she was released but was not told of the death of her two children. [42] Also taken to hospital were Mr and Mrs Charles Augull, Mrs Schofield's parents, and Mrs Laura Conley, Mrs Schofield's coloured maid. All were expected to recover, though the maid had to undergo the

---

[39] *La Vanguardia*, Barcelona, Spain, 30th April 1926

[40] Federación Castellano Manchega de Amigos del Ferrocarril: Accidentes Ferroviarios en España

[41] *East Liverpool Review-Tribune*, East Liverpool, Ohio, United States, 14th August 1926

[42] *Centralia Evening Sentinel*, Centralia, Illinois, United States, 14th August 1926

amputation of part of her foot. In addition to the Augulls and Mrs Conley another 18 passengers and seven railwaymen were injured.

Engineer Jackson on the train engine told the Interstate Commerce Commission (ICC) investigation that the first warning that he had had of the accident was seeing engine No. 214 jump in the air; immediately afterwards he was thrown out of his cab. Fireman Fitzgerald said he had been working on the fire and had had no anticipation of danger. The first knowledge that Conductor Neaves, Ticket Collector Lautner and Trainman Munson had of trouble was the sudden application of the air brakes.

After the accident, the surviving train crew examined the switch at which the derailment had occurred. They found the switch lever locked for the main track and the target showing clear, but the nut and washer had worked off the end of the crank shaft and were lying on the ground beneath the connecting rod, which in turn had dropped off the end of the crank and had released the points, which were lying partly open. No cotter pin was present in the vicinity and the cotter pin hole at the crank shaft was clogged, indicating that there had been no cotter pin in it for some time.

Section Foreman Dinizo stated that the nut had been placed on the crank shaft about a year previously and that he had last examined the switch on 10th August; he at first alleged that on that day he had used a wrench on the nut, and it had been screwed up to the maximum distance and was tight. There was also a wire nail in the cotter pin hole, bent so that it could not work out. Later he admitted that he was not sure if he had used a wrench on the nut that day. He had gauged the track in the vicinity of the switch about six weeks previously and it had been correct. He was not aware of any tampering with equipment in his section and could not account for the nut, washer and nail having come off.

Supervisor Steers told the inquiry that he had examined the switch at 10.45 p.m. on the day of the accident when he had found it in the condition described by the train crew. He was of the opinion that the nut had been removed from the shaft rather than gradually working off, though he had had no reports of tampering in the area. He had last inspected the switch three weeks earlier, but only by giving it a casual glance. His instructions were to use nails in cotter pin holes in the absence of cotter pins, though the nails were not as secure.

The day after the accident the switch was examined by ICC investigators, who found that the cotter pin hole in the crank shaft was clogged with a gummy mixture of old dirt and grease, which it took a considerable pressure to shift, and that the nut and washer present at the time of the accident were likewise covered with a deposit of old grease and dirt, to the extent that the sides of the hexagonal nut appeared to be circular. There was not the slightest evidence of a wrench having been used as stated by the section foreman and the application of a standard wrench would have removed the deposit from the side of the nut. All indications were that the nut had not been fully screwed onto the crank shaft for a considerable time prior to the accident. There was no evidence of any defect in the locomotives or cars of the train which might have contributed to the accident.

In conclusion the ICC decided that the derailment had resulted from the defective condition of the west switch of the south siding at Calverton, for which Section Foreman Dinizo was primarily responsible. It appeared that there had been no cotter pin or nail in the hole at the end of the crank shaft to prevent the nut working off, and the great weight of evidence indicated that the rather loosely fitted nut had gradually worked off as a result of the pounding effect of the connecting rod eye against the upper surface of the nut due to the vibration from passing trains and the tendency of the nut to work off when the switch was operated. Once the nut had worked off the bottom of the crank shaft there was nothing to prevent the connecting rod from dropping off, leaving the points to move at will. The investigators noted that it was a very old switch stand, and that there were several similar stands on the main line, some of which had nails in place of cotter pins, several of which had no nuts, the connecting rod being held in place only by a washer and cotter pin. In one of the latter cases the ends of the cotter pin had not been spread, so that there was nothing to prevent it from working out or being easily pushed out by hand. The commission stated that, 'The existence of such conditions indicates laxness in the maintenance of switches as well as in the supervision and inspection of the same' and that, 'Immediate steps should be taken to correct this situation.'

It was noted that Section Foreman Dinizo had begun as a section labourer on 1st August 1913, had been promoted to a section leading labourer on 16th March 1923 and to section foreman on 1st January 1924. He was unable to read or write, nor could he understand English to any degree. It had been difficult to examine him, even with the aid of an interpreter, due to his apparent lack of understanding of what was being asked. [43]

Torrential rain during a typhoon led to a swollen river washing away the track and derailing a train on the San'yo main line between Aki-nakano and Kaidaichi, on the shores of the Inland Sea in Hiroshima Prefecture, Japan, on 23rd September 1926. The train involved was a limited express from Tokyo to Shimonoseki, one of the fastest trains in the country, which proved a connection via the Straits of Korea with Harbin and the Trans-Manchurian Railway. [44] Thirty-four people died and 39 were injured in the accident, which occurred at 3.30 a.m. Three sleeping cars were wrecked. Although it might be considered that the accident was a result of factors outside of the railway's control the official report into it concluded that there had been insufficient inspection of the line given that washouts might be expected in connection with the severe weather conditions at the time. There was also heavy criticism from the public of the use of wooden carriages in the train. All steel carriages were introduced the following year. [45]

A brief Reuter message from Belgrade spoke of the overturning of a coach of a passenger train on a set of points as it was leaving the station at Csóka (Hungarian spelling – Serbian spelling Čoka), on the line between Zenta and Szegodin in Yugoslavia on about Friday 4th May 1928. The passengers attempted to jump from the coach as they felt it going over but some of them were caught and crushed. Five passengers died and 18 were injured. Whether the points were defective or were moved under the train, or whether perhaps something dragging from the train caught in them, was not stated in the report and was probably not immediately clear. [46]

---

[43] Interstate Commerce Commission report

[44] *Northern Star*, Lismore, New South Wales, Australia, 24th September 1926

[45] Webpage: Japan Railway and Transport Review No. 33: Japanese Railway Safety and the Technology of the Day.

[46] *The Times*, London, England, 5th May 1928

At 5.10 p.m. on Friday 24th August 1928, during the evening rush hour, an Interborough Rapid Transit subway train was derailed on a set of facing points at Times Square, New York, United States, which according to a newspaper was 'the busiest spot in the world.' The packed eighth car of the train 'jumped the points,' skidded sideways for a hundred feet and then collided with one of the massive steel pillars supporting the street above. A newspaper reporter described what happened next. 'It sheared off the pillar, tore loose from the forward seven cars, split itself in two, and while part of it hurled forward, tossing passengers against stanchions onto the track, under the wheels of the car, against the sides of the tunnel, and piling them up in masses against what had left of the car floor. Some 200 feet below the stanchion, it finally stopped – its force spent, its ghastly work done.'

As for the occupants of the car, 'A piercing shriek went up. Men fought but hardly could strike, so closely were they packed one against another. Women fainted but they could not fall, so firmly were they wedged in their places. There was a hiss and a curl of blue smoke, as short-circuited wiring burnt out. Everywhere there was terror. Men in the rear cars and those ahead, began smashing windows, climbing through them, dragging others after them. One man pulled two women he believed dead through a window and left them beside the track, to learn later they had but fainted. Another scrambled through an open window to find that two others had been dragged with him, clinging to his coat tails. And all the while the injured were writhing and groaning in pain.'

Meanwhile, 'Up on the station itself, where hundreds had been waiting for the next train, there had been a sudden crunching and crashing of twisting, breaking steel. Then came the curl of blue smoke and with it a terrifying shriek – the frightened shriek of 2,000 people. The crowd surged in the direction from which the terrifying sounds and smoke had come, after the manner of crowds. Then it turned and fought towards exits. It was but a few minutes until the clang of gongs and the piecing shrill of whistles announced the coming of police. Squad after squad of police, and company after company of firemen massed in Times Square and crowded through the throngs into the tubes beneath ground. A dozen ambulances, and then another dozen, whirled up. Rescue groups were at work in a twinkling. Five priests and to each one of them a score of police, worked through the panic-stricken crowd, quieting them, the priests stopping to minister here and there to those they thought might die. In a moment, it seemed, ten blocks around Times Square were roped off. On the sidewalk a first-aid station was in operation. In less than half an hour the procession of stretchers began coming up and through the exits. On this one man writhed in pain; on another a woman cried; a third was silent, its occupants unconscious or dead. One by one they were placed in ambulances and carried away to a half dozen hospitals where approximately 100 doctors were ready besides operating tables and in emergency rooms to ply their skill.'

Mayor Warren and Police Commissioner Warren arrived at the scene, the former in his limousine.

The stories of some of the passengers also appeared in the newspaper, such as that of Peter Mollitor, of Brooklyn. '"I felt a terrible impact and heard a crash. I thought I had better get out," Mollitor said. There was no laughing as Pete, a cab driver, told how he happened to have a monkey wrench in his pocket. He smashed two windows with his wrench, tossed a girl out and then leaped out himself. He never saw the girl again and in this city of millions probably never will.'

The reporter also interviewed a child, 'Alice Braithwaite, seven year old, was here from her home in Bridgeport, Conn., and with her parents, Mr and Mrs Edgar Braithwaite, was enroute to a friend's home. "I was very frightened. Oh, it was awful," Alice said.'

Some of the adult passengers were clearly also severely traumatised. 'Typical of the survivors who were brought to the surface was Mrs Julia Smith. She came out crying hysterically. She waved a check for a large amount of money at a policeman. "I'll give you this check if you'll only give me a nickel," she almost screamed. The policeman reached in his pocket for the coin, waving aside the check. "But how will I ever get home?" she asked after taking the coin from the policeman. "I'll never ride in the subway again as long as I live. It was horrible. I cannot bear to think of it." [47]

Fourteen passengers were killed more or less outright in the accident and another 99 injured, three of whom subsequently also died. The Interborough Rapid Transit had advertised itself as 'the safest railway in the world.'

Four separate inquiries were immediately opened into the accident. By the following morning Maintenance Foreman William Baldwin had been arrested on a charge of homicide for permitting the train to pass over a defective switch. An investigation disclosed that the events leading up to the disaster had begun when Towerman Henry King in his cabin in the Seventh Avenue tunnel south of 40th Street had attempted to operate the switch to put an empty train into the storage track, prior to it being returned northwards. The set of facing points (switch) concerned lay only 85 feet from the south end of the platform at Times Square. The switch had failed to work properly. The towerman had then contacted Baldwin, who had been at 34th Street and Seventh Avenue, and the foreman had hurried to the scene with his men. Examining the switch Foreman Baldwin could see nothing wrong with it but the protecting signal in its rear was being held at danger. The foreman sent his assistant, Joseph Carr, to the tower to press a button which overrode the automatic train stop which would otherwise have applied the brakes on a train which had passed the signal at danger. With Carr deactivating the automatic train stop the empty train was then successfully moved into the little used storage track. As the problem was apparently solved the points were not spiked closed, nor was a speed restriction imposed over them. It was alleged by the New York Transit Commission that Baldwin had then signalled to Motorman Frank McCormick of the passenger train that he could proceed and for some reason had operated the switch from the track, perhaps inadvertently. Baldwin denied that he had done so. He was released on $10,000 bail. [48]

The arrest of Baldwin was not the end of the matter, for on 4th October he was released, having been reportedly exonerated by the investigation, and Towerman King was arrested instead. It transpired that King was not his real name, and that 'he was an ex-convict, living under a false name, using a false address, and repeating falsehood after falsehood during the investigation. Formal charges of manslaughter in the second degree were filed against him. He was finger-printed and taken to jail pending a hearing.' [49] King had formerly been Harry C. Stocksdale, of Baltimore, Maryland. He had been convicted

[47] *South Haven Daily Tribune*, South Haven, Michigan, United States, 25th August 1928
[48] *Charleston Daily Mail*, Charleston, West Virginia, United States, 26th August 1928
[49] *Chester Times*, Chester, Pennsylvania, United States, 5th October 1928

there of stabbing a man and had subsequently changed his name. [50]

Eastbound passenger train No. 1 of the Peoria Terminal Company was derailed near Bismark Siding, Illinois, United States, at about 6.43 a.m. on 20th February 1929. The train was carrying about 1,000 miners to the Crescent coal mines. Six miners died in the accident and 91 miners, and one railwayman were injured. The railway was only used to carry miners to and from work. At one time it had been in the hands of the receiver.

The train was in charge of Conductor Hulting with Engineer F.O. Fisher and Fireman Bowers. It consisted of Chicago Rock Island and Pacific Railroad engine No. 2030 and eight wooden coaches and was running at between 12 and 20 miles an hour when the rear six coaches left the track, four of them falling down an eight-foot-high embankment. Upon seeing what had happened Engineer Fisher detached his engine from the remaining coaches of the train and ran forward to Hollis, from where help was summoned, doctors, nurses and ambulances being sent from Peoria and from the Martinville state hospital. Meanwhile persons living nearby had taken a number of the injured miners to Peoria in their cars, a well-intentioned move which unfortunately led to considerable confusion, with miners' wives frantically running from coach to coach searching for their relatives who were eventually found to have already been taken to hospital.

The miners who lost their lives were Cecil Walker, a union official; George Jones; George Wilkinson; William Brown, a man tentatively named as Warner and Charles Luthy, who died in hospital. In addition an elderly miner named Jim Clark fell across one of the stoves provided in the coaches for heating and his clothes caught alight. A fellow miner rolled him in the snow until the flames were extinguished, but physicians thought that he would probably die. [51]

Engineer Fisher told the Interstate Commerce Commission investigators that the first he had known of anything being wrong was when he had looked back and seen two of the coaches buckling, when he had made an emergency brake application. Fireman Bowers thought, however, that the brakes had applied from the rear of the train before his engineer had applied them. Conductor Hulting and Brakemen Miller and Marshall said that they had known nothing until the accident had occurred. A hurried examination of the track after the derailment had revealed a shattered rail.

Section Foreman DeRose stated that he had been a section foreman since 1907 but had only been in charge of the section involved in the accident for one day because the regular section foreman was sick. He had not passed over the section of track on that day but had given the track walker the usual instructions to do so. He said that broken rails were a considerable problem, especially in cold weather, when they averaged one a week.

Track Walker Farano said that he inspected the track every day and had done so on the day prior to the accident and found nothing wrong.

The investigation concluded that the derailment had resulted from a defective piece of track which included a broken rail and splice bar, and that a defective joint had been allowed to exist for a considerable time. [52]

---

[50] *The Escanaba Daily Press*, Escanaba, Michigan, United States, 5th October 1928
[51] *The Ada Evening News*, Ada, Oklahoma, United States, 20th February 1929
[52] Interstate Commerce Commission report

Bessarabian soldiers returning to duty after Easter leave were the principal victims of the derailment of a fast train from Chişinău and Jassy (now Iaşi) to Bucharest at Boboc station, Romania, on the evening of Thursday 4th April 1929. The train left the track at a set of points which were half open, the locomotive taking one route and the carriages another. Three coaches were derailed, and 29 passengers were reportedly killed (some accounts spoke of only 24 deaths) and 59 more injured. Breakdown gangs from Bucharest worked all night on the wreckage, their job made more difficult by a violent snowstorm. The pointsman who had been on duty and the station agent both disappeared after the accident, doubtless fearing retribution. A report on the disaster added that this was the latest of several recent accidents involving points in Romania, including one to the Simplon express the previous autumn, this being, no doubt, a reference to the collision at Resca on 26th October 1928. [53] [54] The safety record of Romanian railways was not particularly good at this period; in 1929 there were said to have been 325 collisions, 1,555 derailments and 1,895 other accidents, with 500 deaths, a lot for a relatively small country. [55]

Early on the morning of Wednesday 19th June 1929 workman's train No. 1804 left Ghent for Haine-Saint-Pierre. It was booked to run as an omnibus service, stopping at all stations, as far as Grammont from where it ran direct to its destination. On this day it had passed Sotteghem (Zottegem) and was within 350 metres of Viane-Moerbeke station, Belgium, when it became derailed, due reportedly to a deficiency of ballast which caused the track to subside. The derailment happened on a curve and at that moment train No. 1871 from Manage (a place between Charleroi and Mons) was approaching from the opposite direction. Due to the curve the driver of this train was unable to see the obstruction until the last moment and although he then braked hard he could not avoid colliding with the derailed carriages of the first train, which were fouling the opposite track. The engine of the Ghent train was buried in the ballast, and it was described as a miracle that its driver and fireman escaped unhurt. Not so lucky was the third-class carriage marshalled second in the train for it was described as having been literally flatted by the leading van, and it was in this coach that all the casualties occurred apart from the guard in the van. The accident occurred at 05:00.

Farmers working in neighbouring fields who heard the cries of the victims rushed to help and were soon joined by Red Cross personnel and staff from Grammont hospital. The vicar of Moerbeke also went to the scene. Nine passengers were killed outright; their bodies were at first taken to the second-class waiting room of the station, to be later removed to Grammont hospital. Ten of the 27 or so injured were women; the condition of eight of the injured was said to be serious.

Mme Théophile Somers, the wife of a crossing keeper, who was one of those involved in the rescue, declared, "I was attracted by the cries; people were running around and fleeing to the adjoining fields. Some of the witnesses of the accident were also screaming and running away at the horrible sight."

[53] *The Times*, London, England, 6th April 1929
[54] *East Liverpool Review*, East Liverpool, Ohio, United States, 5th April 1929
[55] *The Railway Gazette*, Vol 51, p. 490

Driver Victor Verbeken, in charge of the engine of train No. 1804, stated, "When I reached level crossing No. 47, I felt my engine tilting over for a few moments. Suddenly there was quite a violent shock and I realised that the track had given way beneath the locomotive, and it was derailed. I rushed to the brake which immediately came into action. The train began to slow down but with that train No. 1871 was into us."

Driver Gérard Désiré, of train No. 1871, for his part said, "We had just left Viane and had gone about 500 metres when I saw train No. 1804 and it looked as if it had left the rails. The engine of the other train was only a few metres away when I saw that she was coming into us. I applied the brake, but it was too late and my engine overturned."

Work began on clearing the line as soon as the initial investigation into the accident had been completed. [56]

Queen Elisabeth visited the site of the accident during the afternoon and then went to see the injured in Grammont hospital, where they were also visited by M. Jaspar, the Prime Minister.

Repairs to the track, which at this point ran on an embankment, had been in progress for some days, and it would seem that due to an error of judgement it had been left in too weakened a condition. [57]

A washout caused the derailment of a Guadalajara – Mexico passenger train on probably Thursday 12th September 1929, between Prieto and Lena, Mexico, 70 miles north-west of Mexico City. Seven cars, including an armoured escort car, overturned. Five of the escorting soldiers died but there were no casualties among the passengers. [58] [59]

Early on the evening of Saturday 30th November 1929 a party of 488 excursionists, including sailors from the *Utah*, *Oklahoma*, and *Florida* battleships who were on weekend leave, boarded a ferry at Norfolk, Virginia, United States, and crossed Chesapeake Bay to Cape Charles, from where they boarded a Pennsylvania Railroad train bound for New York. The train never reached its destination, for 15 minutes after midnight, on Sunday 1st December, it was derailed at Onley, Virginia, United States, resulting in the deaths of eight passengers and one railroad employee and injuries to 30 passengers, a news agent, and a railwayman.

V.V. Kemp, a passenger from Norfolk, related his experiences of the accident. "My wife and I were sitting in a seat facing Mrs Wilde and a sailor when we first knew something was wrong. The car fell over on its side and went plunging along the ground. Everything seemed to be crashing at once. Dirt, rocks flew at the broken windows and glass struck me in the face and eyes. Women and men were screaming and being thrown all over the car and against each other. I went blind, still holding to Marge {his wife} and to the car. I don't know what happened next, but suddenly I came to again and I saw Marge lying there with her head all cut and her arm half off bleeding and looking like she was dead. Mrs Wilde was crying and telling me to do something for Marge; that she was dying. I didn't

---

[56] *Le Populaire*, Paris, France, 20th June 1929
[57] *The Times*, London, England, 20th June 1929
[58] *The Port Arthur News*, Port Arthur, Texas, United States, 13th September 1929
[59] *Arizona Silver Belt*, Miami, Arizona, United States, 13th September 1929

know what to do." Fortunately, a sailor appeared and tied a handkerchief around Mrs Kemp's arm, stemming the bleeding and saving her life according to a doctor.

After the accidents calls had gone out to all the physicians and surgeons in the area, and 'fast automobiles' rushed 19 of the more badly hurt to the hospital at Nassawadox, 12 miles from Onley. All of the dead passengers but one were women. They were Mrs Annie Abrams, Julia Coogan, M.E. Diggs, and Miss Grace Raby, all from Norfolk, and three unidentified negresses. The only male passenger to lose his life was Ralph Smith, aged 19, of the U.S.S. *Utah*. The railwayman who died was Marion Callaway, of Delmar, Delaware. [60]

The train was running as northbound extra No. 5087. It was in charge of Conductor Cordrey with Engineer Northam and Fireman Hudson and consisted of locomotive No. 5087 and ten all-steel coaches. It had departed from Cape Charles at 11.23 p.m. on the Saturday and was running at between 45 and 55 miles an hour when it left the track.

The Interstate Commerce Commission investigators were told by Engineer Northam that the first he had known that anything was wrong was when the air brakes had been applied from the rear of the train (the engine, tender and leading two cars were not derailed). He had noticed no unusual motion of the engine such as might have resulted from bad track, a low joint, or a broken rail. Fireman Hudson gave similar evidence.

Track Foreman Hancock stated that the accident had occurred about midway through the section he was in charge of. He had a crew of five labourers and a track walker. The track foreman said that he had inspected the track about a week earlier and found no evidence of defective rails or flaws.

Track Walker Finney said that he had inspected the track at about 3 p.m. the previous day looking for broken rails and bad joints but had found none.

District Engineer Lewis submitted a written statement to the inquiry in which he wrote that examination had revealed a broken rail on the east side of the northbound track, the result of a transverse fissure covering three-quarters of the head. There were several broken pieces of rail lying about the track, examination of which showed more transverse fissures. There was also a break in the west rail of the northbound track, but this was apparently a result of the accident. The alignment, surface and general condition of the track were good. The break in the east rail was exactly midway between two ties. Mr Lewis considered that it was this break which had caused the derailment.

The commission agreed that the broken rail had caused the accident and noted that the derailment had features similar to others due to transverse fissures. The question of whether present wheel loads and speeds had exhausted the margin of resistance of rails to a point from which there was no escape or whether means could be discovered to eliminate this dangerous type of fracture demanded serious attention. [61]

A military special train carrying reservists who had been for their annual training with the colours at Valdahon, and in which they were returning to depots at Neubrisach, Belfort and Colmar, met with an accident at Laissey, France, about 20 kilometres from

[60] *The Salisbury Times*, Salisbury, Maryland, United States, 2nd December 1929

[61] Interstate Commerce Commission report

Besançon, at 09.10 on Friday 11th April 1930. The soldiers on the train belonged to the 152nd infantry and the 4th battalion of *chasseurs alpins*. At the exit from a short tunnel, 200 metres from Laissey station, the train was running at 50 kilometres an hour when it entered a portion of the line which was under repair. At a point where the track was insecure the engine suddenly left the rails and crashed into some rocks bordering the line. The two leading coaches were derailed in turn, telescoped into one another, and struck the rocks. Ironically, these were the only two vehicles of the train which were carrying passengers, the ten remaining were loaded with equipment and with horses. [62]

Assistance quickly came from Laissey station and a little later from Besançon. Six bodies and 38 injured, four very seriously, were extricated from the wreckage, the injured being taken to hospital in Besançon. One of the injured died either before or soon after reaching hospital and the condition of a second was considered to be hopeless. The names of six of the dead were given as Clotaire Gauthereau, of Bezons; Marcel Sauzet, of Paris; René Paris, of Milly; Lucien Schwendemann, of Mulhouse; Pierre Fournier, of Mâcon and Claude Gallot, of Longwy. [63] Driver Billaut and Fireman Chatet, both from Besançon, escaped unhurt.

It was reported that the work on the track was protected only by a small red flag stuck in the ground 130 metres from the start of the work. An investigation into the accident was opened. [64]

In late June 1930 very severe weather was affecting the province of Ontario, Canada, with excessive rain leading to much flooding. The abnormal weather was to lead to two bad railway accidents in two days.

On Thursday 26th June Canadian National Railway train No. 4 from Winnipeg to Toronto was running at about 30 miles an hour west of Capreol, 17 minutes late, the time having been mostly lost as the result of speed restrictions (slow orders) issued because of possible track damage related to the flooding. There was, however, no speed restriction at the point where the railway ran on an embankment alongside the Vermilion River. Trains had previously negotiated the embankment at 6 p.m. and 8 p.m.. When train No. 4 reached the spot at 10.30 p.m. its engine, tender and leading express car passed over the point of danger safely, but the following vehicles left the track and a tourist sleeping car and day coach slipped off the embankment and into the flooded river. In all seven of the nine coaches of the train were derailed. After the accident Engineer Virtue uncoupled his locomotive and the front express car from the derailed part of the train and proceeded to Capreol to summon help.

Four or possibly five passengers died in the accident. Most of the newspaper reports spoke of only four dead, all of them children, identified as Lenore, aged seven and Arthur, aged four, the children of Mr and Mrs Gilbert, of Fairmount, Saskatchewan; Helen Nedderburn, aged three, travelling with her mother from Saskatoon, Saskatchewan, to Toronto, and Gisley Brussiere, aged seven months, child of Mr and Mrs Brussiere, who were bound for Sturgeon Falls, Ontario, from Elsas. Mr Brussiere managed to save his two other children as they were floating away from the coach. His wife seized the

[62] *Le Grand écho du Nord de la France*, Lille, France, 12th April 1930
[63] *Le Matin*, Paris, France, 12th April 1930
[64] *Le Grand écho du Nord de la France*, Lille, France, 12th April 1930.

baby and held on to it desperately as she was helped from the submerged car, not realising that it had already drowned. [65] [66] It was stated that the bodies of other victims might have floated away in the river and a recent author states that there was at least a fifth fatality, a man who drowned in his bed. [67] There were about 35 others injured, most of them seriously. Conductor Hamilton, in charge of the train, was among those hurt. [68]

The fact that more lives were not lost was probably due mainly to the rescue efforts of the passengers from the rear of the train, plus a good deal of luck. 'As soon as the train stopped passengers in the Pullman cars pulled on what clothes they had and set to rescuing their less fortunate fellow passengers in the coaches ahead. Some of the men plunged into the river up to their necks, and stood there for half an hour, in numbing cold, helping entrapped passengers out of broken windows. The heroism of the men passengers aroused the praise of those who owed their lives to their assistance. "Not a man lost his head," one aged lady said. "Even those in the car with us remained cool. They lived up nobly to the motto of 'women and children first,' for not a man tried to save himself till every woman and child in the cars were safely out. The large preponderance of women casualties was due to the shock which proved harder on feminine nerves, and to the fact that the majority of passengers in the tourist coach were women.

'Every railroad accident has its stories of individual heroism, and the Capreol wreck is no exception. In this case the outstanding heroine was undoubtedly Mrs Hayes of Hornepayne. She was travelling to Parry Sound with her three children. Without waiting for outside help, or without losing her head for an instant, this brave mother broke through one of the windows, and with a child under each arm and an infant gripped firmly in her teeth, struggled to shore, a distance of about 15 feet.

'Many of the passengers had extremely narrow escapes. One man crawled out of a window on the wrong side of the car and was unable to get to the shore. He dared not enter the car again. He could not swim, and the cold water had so numbed his muscles that he was unable to pull himself up over the car. All that he could do was to cling to the top of the car, and {he} was finally rescued in so weakened a condition that in another minute he would have had to let go.' [69]

The inquest on the accident found that it was due to a waterlogged roadbed, one rail of which had sagged, derailing most of the train. [70]

A second train, No. 3, was derailed in the same area on either 26th or 27th June, without any serious consequences. Then on Friday 27th June there was a third derailment, which proved to be the most serious of the three. This time most of the victims were what were variously described as 'trespassers,' 'transients,' 'hoboes' and 'migrant workers,' travelling about the country looking for work. The Great Depression, which had begun with the Wall Street Crash the previous year, had affected Canada as elsewhere, and there were 15, 18 or possibly as many as 25 unauthorised passengers riding on westbound Canadian National Railway freight train No. 401 when it encountered a 'sink hole,'

---

[65] *Winnipeg Free Press*, Winnipeg, Manitoba, Canada, 28th June 1930
[66] *Medicine Hat News*, Medicine Hat, Alberta, Canada, 3rd July 1930
[67] *Wreck!* by Hugh H. Halliday, p. 161
[68] *The Lethbridge Herald*, Lethbridge, Alberta, Canada, 27th June 1930
[69] *Winnipeg Free Press*, Winnipeg, Manitoba, Canada, 28th June 1930
[70] *Wreck!* by Hugh H. Halliday, p. 162

another waterlogged portion of embankment, near Crerar, Ontario, at 4.15 a.m. This time the engine was also involved when the track gave way beneath it and Fireman Albert McLeod was killed outright, together with eight of the trespassers, who had been riding in one of two telescoped cattle cars in the middle of the train. Only a few of the cars were derailed and none were upset. Four uninjured trespassers who had been riding in car at the rear of the train went forward to the engine, where they found Engineer John McDonald enveloped in scalding steam and frantically attempting to amputate his own foot trapped in the wreckage with a knife. [71] The trespassers disarmed him and after fifteen minutes were able to free him, but the engineer was extensively scalded and died later. Meanwhile Brakeman Clifford Boivin had gone back two miles along the line to Spindal Siding and was able to stop a following express and prevent a worse disaster. It was said that when a relief train arrived at the scene later the railwaymen refused at first to give the surviving trespassers a lift to Capreol, telling them to walk there, though eventually they relented. This second disaster was dealt with at the same inquest as that on the accident of the day before and was ruled to have been due to the abnormal weather, with no one to blame. [72]

Early on Wednesday 29th October 1930 the international express of the Chemin de fer de Paris à Orléans, which had left Geneva, Switzerland, at 13.35 the previous afternoon for Bordeaux, was derailed between Négrondes and Ligueux stations, France, about 25 kilometres short of Périgueux. The train was running at close to 95 kilometres an hour down a falling gradient when it left the rails at about 04:00. The two locomotives of the double-headed express were derailed and dragged most of the 12 coaches of the train after them; only the last three remained on the rails intact. Three first and second-class and one third-class coach were telescoped into one another. Immediately after the accident Conductor Guichard, although badly hurt, got down from his van and made his way along the line to Négrondes station to raise the alarm and from there messages were sent to Limoges and Périgueux and a relief train was despatched to the scene. [73] Meanwhile, the keeper of a level crossing near the accident scene, who had doubtless been awakened by the noise of the derailment, ran to Ligueux village and awakened the curé (parish priest), who sounded a tocsin (alarm signal) on the church bell to rouse the villagers. For a while though, the uninjured passengers were the only ones who could do anything for their less fortunate fellow travellers. [74] In the almost total darkness they worked with torches made from burning newspapers to free the dead and injured from the wreckage. About 20 of the injured were immediately sent to the Francheville clinic in Périgueux but it was necessary to use cutting equipment to reach other victims trapped in the steel coaches.

The bodies of those who had died were taken to a room at Ligueux station which had been converted into a *chapelle ardente*. By the evening, after much effort, 17 dead had been recovered: 15 passengers and Drivers Prens and Vendredat. As well as the passengers and the conductor the injured reportedly included Firemen Goursenneau, Barem and Douisonas – did one or both engines have two firemen or was one of them

[71] *Winnipeg Free Press*, Winnipeg, Manitoba, Canada, 28th June 1930
[72] *Wreck!* by Hugh H. Halliday, pgs. 163-4
[73] *Le Petit Journal*, Paris, France, 30th October 1930
[74] *The Times*, London, England, 30th October 1930

'riding on the cushions'?

The passengers on the train included 210 Romanian emigrants, on their way to a new life in South America. Only one of them was hurt and they continued their journey to Bordeaux and the New World on another train.

It was initially stated that the cause of the derailment was unclear and that express train No. 75 from Paris to Agen had passed the spot only 20 minutes earlier without its driver having noticed anything wrong. It was thought probable though that the heavy rain of the previous few days had weakened the track and the sleepers on the left side of the side had subsided beneath the engines of the heavy train. Another initial theory was that a rail had broken beneath the previous train, which had also been double-headed. Various senior officials, including the director general of the company and M. Epinet, the chief engineer, took a train to Limoges and then continued by car to Ligueux to open an inquiry into the accident.

It soon became apparent that contrary to the first report the driver of train No. 75 had in fact noticed a bump at the point of the accident and that he had reported this on arrival at Périgueux. Notice had then been sent to both Négrondes and Ligueux but had come too late to stop the Geneva express. Had train No. 75 itself become derailed an even worse disaster would probably have occurred as the Geneva express would then have run into the wreckage.

Colonel Lesme, a military commissioner from Evreux, and his wife were passengers on the Geneva express. Mme Lesme told a newspaper reporter, "We were woken up with a start by a terrible noise and could feel our carriage rolling along the ballast. The next moment there was a tremendous crash, our coach went down an embankment and we were all thrown against one another and covered in debris of all kinds. My husband, whose forehead was bleeding profusely, climbed out through the broken window, and assisted me out. The two locomotives which had been pulling our train were lying overturned with steam pouring from them, illuminated by the ghostly red glow of the engine fireboxes." [75]

It appears that the derailment was definitely due to track subsidence following heavy rain, the initial alternative theory of a broken rail having been discarded. [76]

---

[75] *Le Petit Journal*, Paris, France, 30th October 1930

[76] *L'Homme libre*, Paris, France, 30th September 1930

CHAPTER 2

DERAILMENTS - BROKEN TYRES, BROKEN AXLES AND OTHER VEHICLE DEFECTS

One of the frequent causes of derailments of trains other than problems with the track was defects in the trains themselves, such as broken tyres, wheels, and axles. Such accidents were not as frequent by the twentieth century as in the early days of railways, but they still happened quite often and were to a large degree unavoidable.

A rather similar accident to that at Weedon, England, in 1915, caused a major disaster in Germany towards the end of the First World War. On Tuesday 30th July 1918, the broken piston rod of the locomotive of goods train 6641 damaged the opposite running line between Gurkow and Zantoch on the Berlin – Schneidemühl main line just as D Zug (express train) 22 was passing on the opposite track. The passenger train was derailed and five of its carriages were thrown against waggons of the freight train. Fire broke out in the wreckage and 42 people lost their lives and 25 were injured to a greater or lesser degree. The drivers and firemen of the two trains escaped unhurt [77] [78]

A broken axle caused the derailment of an Ungarisch Staatsbanen (Hungarian State Railway) troop train near Rakos, Hungary, on the outskirts of Budapest, at about 11 p.m. on possibly Sunday 4th November 1918 (another source states that the accident took place on the 6th). Three carriages overturned and the front part of the train was run into by the rear part. Sixty passengers died and between 150 and 180 were injured. [79] [80]

An armistice between Austria-Hungary and the Allied powers came into effect on 4th November 1918 and the following day, Monday 5th November 1918, so-called

---

[77] *Eisenbahnunfälle Bind 1* by Ludwig Stockert, pgs, 88-9 (in German)
[78] *Eisenbahn-Katastrophen in Deutschland* by Hans Joachim Ritzau, pg. 134 (in German)
[79] *The Times*, London, England, c8th November 1918
[80] *Eisenbahnunfälle Bind 1* by Ludwig Stockert pg. 89 (in German)

Heimkehrerzug (homecoming train) No. 259, which was presumably carrying defeated troops, was en route Pragerhof (now Pragersko in Slovenia) to Csáktornya (now Čakovec in Croatia), when it was involved in a major catastrophe between Sternthal and Pettau, Austria (now Ptuj in Slovenia). As the train was running between the two places an overheated journal at the leading end of its fourth coach led to a broken axle and the derailing of the vehicle. At a level crossing 100 metres further on the train then separated into two, the portions came together again, and 10 or 12 carriages were telescoped into one another and wrecked. Sixty-three passengers and a conductor died, 96 passengers were seriously injured and 25 slightly hurt in this effectively unavoidable accident on the Südbahn. [81]

Paris – Toulouse express No. 55, which was composed of a locomotive and 14 vehicles and had left the capital at 10:00, was derailed in open country well away from any habitation, while running at 80 k/ph, at 18:15 on the evening of Saturday 14th May 1921. The accident happened between Vigeois and Uzerche, France. One coach ended up across the track and seven more piled up behind it. After a moment's panic the passengers pulled themselves together and began the rescue work. The engine and three vehicles remained on the track, but the driver quickly realised that he had lost part of his train and stopped and alerted the staff at Brive and Limoges stations. The bodies of three dead soldiers were soon recovered from the wreckage and a woman passenger died en route to hospital. According to the initial report about 15 other passengers were injured. [82]

A second report attributed the derailment to the breakage of one of the axles of the fourth vehicle of the train and stated that in addition to the four dead 24 passengers had been hurt, six seriously. [83]

A series of headlines in an English newspaper on the morning of Tuesday 4th November 1924 read 'Train Smash in Lancashire,' 'Express Derailed,' and 'Many Killed and Injured.' The newspaper report went on to report that the 4.40 p.m. train from Liverpool to Lytham and Blackpool had been wrecked at Moss Side, between Kirkham and Lytham, on the London Midland and Scottish Railway, at about 5.20, (actually at 5.46) adding that the exact location of the accident was Warton signal box, about a mile from Lytham, England. The train was derailed a short distance on the Kirkham side of the box and the locomotive demolished the cabin. Signalman Hornby was thrown over a wooden fence and into a lineside brook and although quite badly injured was lucky to escape with his life. The engine and tender were overturned and ended up facing in the direction from which they had come; the two leading carriages were also overturned, and it was in these that the casualties occurred.

According to the newspaper the presence of mind of some of the passengers averted a worse disaster. Three men and a woman climbed out of the rear coaches and running down the line were able to signal to the driver of an oncoming up train in time for him to be able to pull up short of the obstruction.

---

[81] Ibid, 89-90

[82] *Le Petit Parisien*, Paris, France, 16th May 1921

[83] Ibid, 17th May 1921

A special train took some of the injured to Blackpool and others were removed by ambulances and private motor cars. Breakdown trains were called from Preston and later from Manchester and Liverpool (see below) and police and ambulances from Blackpool, St Anne's, Lytham, and Fleetwood, together with stationmasters and their staff from the Blackpool stations. The initial newspaper report stated that four passengers had lost their lives; Miss Emma Pickup, aged 48, of Lytham; sisters Annie and Isa Greenwood, aged 29 and 21, respectively, of Blackpool, and Thomas Hartley, of Blackpool. Driver William Crookes of the express was also killed, while 11 passengers and Fireman Frank Livingstone were said to have been injured. Meanwhile work was proceeding in the darkness on searching the wreckage for further victims, using torches and acetylene lamps borrowed from a neighbouring aeroplane works. [84]

The following day the full magnitude of the accident had become clearer, and the number of passengers known to have died had reached 12. (With the deaths of two of the injured it eventually rose to 14.) Over 30 passengers were injured, as well as Fireman Livingstone and Guard Bullock.

'Daylight first revealed the horrible character of the accident. The locomotive, which had been lifted by the crane on to the metals, was scraped all along its side, evidently by the brickwork of the signal cabin it had demolished on leaving the track, and bricks were scattered for a distance of 30 yards over the adjoining field. The top of the engine cab was buckled, and turf stuck out of the chimney stack. The track had begun to be torn up some distance on the Kirkham side of the signal box, and the bogie of the engine was found 15 yards short of the box. The flanged tire of one of the bogie wheels had apparently come off 200 yards short of the signal cabin, and the discovery of this tire this morning started a new theory as to the cause of the disaster.'

As to the victims, the paper noted, 'One very sad feature of the accident is that many of the dead are working women weavers at a Kirkham mill, who entered the train there at the end of their day's work and expected to be home in a quarter of an hour. Kirkham is the first stop the 4.40 p.m. Liverpool express makes on its way to Blackpool. Six of them were in one compartment, and one survivor found that her three companions who were brought out of the train after her were all dead. It will be noted that the dead include two pairs of sisters who were employees at Kirkham.

'The first two coaches after the engine, which collapsed, presented a terrible sight. A magazine lay on a blood-smeared and broken window. In another compartment the passengers had left the scattered cards of a pack with which they had been playing. In this coach also the roof was bloodstained and there were visible portions of flesh. Although some of the coaches took fire it does not appear that fire was the cause of death or injury. The Lytham fire brigade, which was early on the scene, put out the flames with chemical extinguishers and water from an adjoining brook, but several of the uninjured survivors speak of escaping gas being one of the first horrors menacing them when the train overturned, and their first thought was to break the windows to get air if not to escape.

'Many persons have spoken of the bravery of Commander Creame, of whom Mr. A. Tarbuck, of Clifton-street, Lytham, says: - "If ever there was a brave man it was Commander Creame, for he never made the slightest complaint and directed his own treatment. Both legs were nearly severed, and all he asked for was a cup of tea and asked

---

[84] *The Times*, London, England, 4th November 1924

to be remembered to his people at Blackpool." He died at midnight in Lytham Hospital, whither he had been taken after urging that others should be attended to before himself.

'Single-track working of trains from Preston to Blackpool was established by 1.30 this afternoon, and the other line was restored, and normal running resumed to-night.' [85]

The comprehensive Ministry of Transport inquiry into the accident was conducted by Colonel J.W. Pringle. It stated that the train had consisted of 4-4-0 steam locomotive No. 1105 and tender, a composite coach, a third-class van, a composite coach and a third-class van, all gas lit and non-corridor. The train was fitted throughout with the continuous vacuum brake. The brakes had been tested before leaving Liverpool Exchange, and again after three coaches had been slipped at Midge Hall (for Burnley according to the report).

The report noted that correspondence had suggested excessive speed as one of the causes of the accident. There was no evidence to support this suggestion. Only over the short section between Ormskirk and Burscough Junction did the train timings require a speed in excess of 50 miles an hour. The average speed for the express over the whole journey was 35.6 miles an hour, and between Kirkham and Lytham 39 miles an hour. There was no reason to suppose that the speed approaching Warton box had been more than about 50 miles an hour, and the permanent way was certainly of sufficient strength to justify such a speed and higher.

The railwaymen in charge of the train could throw little light on the circumstances attending it. Driver Crookes had succumbed to his injuries shortly after the accident and Fireman Livingstone could only recall that the engine had been "running nicely" until near Warton. After passing the Warton down distant he had been looking out for the Lytham Goods Junction distant when he had noticed sparks flying from the leading end of the engine. He had been about to alert his driver when the engine left the rails, and he could remember nothing more. He could not recall if steam had been shut off or the continuous brakes applied.

Goods Guard Bullock, in charge of the train, told the inquiry that the continuous brakes had been tested by the rear guard (riding in one of the coaches which were slipped) before the train left Liverpool. He had many times travelled around the curves in the area of Warton box faster than on the night of the disaster. Riding in what after Midge Hall had become the rear vehicle of the train, he had known of nothing wrong until he was suddenly thrown to the floor of his van and knocked semi-conscious. He had not noticed any application of the continuous brake before the accident and said that it was not usually applied at that point.

Assistant Motive Power Superintendent Housley stated that he had examined the engine on the night of the accident and had found the vacuum brake fully applied and the regulator closed. He had concluded that Driver Crookes had not had sufficient warning of anything wrong to close the regulator and apply the brake and not to reverse his engine.

Colonel Pringle remarked that it was clear that the engine crew had not realised that any of the engine wheels were off the rails and had therefore not made any attempt to stop the train, until the engine had struck the rail crossing the down track to the Gas Works siding at Warton.

After the derailment one of the engine wheels had been discovered to be missing a

---

[85] Ibid, 5th November 1924

tyre. And a careful examination of the track revealed unmistakeable marks of derailed wheels from a point 352 yards on the Preston side of the Main Dyke Road bridge (the spot near Warton box where the engine became totally derailed). From that point there were various marks and indications of derailment all the way to the bridge, while the tyre of the left-hand leading bogie wheel of the locomotive was found lying in a field east of the railway, 50 yards from the down track and 120 yards from the initial mark of derailment. The fractured surface of the tyre showed a large defect in the centre of the tread.

Colonel Pringle reconstructed the circumstances of the derailment. He believed that the tyre had possibly broken some distance before the first mark on the track. The broken tyre had gradually opened out until it had become loose enough to slip off the wheel centre. The tyre had then become completely detached from the wheel and flown off into the field. The engine had continued, running at perhaps 50 miles an hour, around the curve for some 300 yards with its two leading wheels off the rails, unnoticed by the engine crew. Then the obstruction caused by the rail crossing and the adjoining cast iron girder of the Main Dyke Road bridge had led to the final derailment.

The faulty tyre was identified as having been on of 34 manufactured at Horwich on 27th February 1920 and had been put into service in May 1920. The wheel to which the tyre was attached had run 101,067 miles before the accident. The 30 tyres of the batch which remained in use were all withdrawn and broken up, but none showed any defects. Laboratory examination of the tyre showed a defect which had been introduced during its manufacture and which could not have been detected.

Going on to a general review of tyre failures the report noted that the average number of steam locomotives in service in 1924 was about 17,000. The average number of failures of engine or tender wheels over the previous four years was 17.3, or about one per 1,000 locomotives. In terms of engine miles there was one failure per 27 million miles. It was also comparatively rare for a failed tyre to leave a wheel, and the inspecting officer was unable to trace a tyre failure causing the death or injury to a passenger since 1893. Great improvements had thus been effected in the manufacture, design, and material of wheel tyres. Colonel Pringle considered, however, that there should be further consideration and research into the shrinkage of tyres onto wheels and into the detection of flaws in steel, perhaps using X-ray apparatus.

Regarding the fire, it was clear that it had originated from the signal box stove and although it had eventually involved two compartments of one of the carriages it had been quickly extinguished by the fire brigade. It was fortunate, though, that escaping gas from two of the carriages had not been set alight.

No salvage tools or ambulance equipment had been carried in the two brake compartments of the train, it having been the policy until recently to only carry such items in corridor express trains. Some tools had been available in a nearby ganger's (permanent way) hut, and it was fortunate that many more had chanced to be available close at hand due to the proximity of the Lytham St Anne's Corporation works. The destruction of Warton box had cut the telephone and telegraph connections and the first telephone message from Lytham had had to be transmitted via three signal boxes to Blackpool Central station, thence by road to Blackpool Talbot Road station before it could be transmitted to the control office in Manchester. (No mobile phones in 1924!) A further problem had arisen when the Preston steam crane had been rendered useless by the bursting of a tube at Kirkham and the Liverpool and Newton Heath (Manchester) steam cranes had had to be ordered out. [86]

Fast train No. 11 of the Chemin de fer de Paris à Orléans, from Paris d'Orsay to Bordeaux, was derailed at 15.20 on Sunday 20th June 1926, between Noizay and Vernou, France, and close to Vouvray, while running at 75 kilometres an hour. The engine and tender reportedly left the track the leading van crashed into them and the first passenger carriage, a first and second class composite, slid past them and fell down an embankment several metres high. The third-class carriage which was the third vehicle of the train telescoped the baggage van. Two other carriages also left the rails.

The first relief train from Tours reached the scene at 16.50. Five passengers were killed outright; M. Alfred Tassard (or Stossard), principal customs inspector, of Paris; M. Blanvillain, a merchant of Saint-Symphorien, near Tors; a man believed to be M. Legrand, of Chinon, and two unidentified women. A sixth passenger, M. Maurice, of Billy, was dead on arrival at Saint-Gatien hospital. About 30 people were injured, four of them seriously. The injured included Conductor Hivert (or Yvert) and Driver Bresset. [87] [88]

The derailment was provisionally attributed to the breakage of an axle of the locomotive, due to a 'hot-box.'

While the wreckage was being cleared away on the day after the accident the body of a woman was discovered, bringing the total number of dead up to seven. [89]

Running downhill at probably 30 miles an hour – a newspaper stated a mile a minute or 60 miles an hour, doubtless a considerable exaggeration – an eastbound Chicago Burlington and Quincy Railroad fast freight train was derailed in the heart of Wyanet, Illinois, United States, at 10.57 a.m. on the morning of Sunday 22nd August 1926. Thirteen cars in the middle of the train, most of them refrigeration cars, left the track and four were demolished. The derailment occurred at the west switch of the eastbound passing track. Seven trespassers riding on tank car GATX 20072 were killed and four others were hurt. The injured men were taken to hospital and were said to be near death, though probably, being fit young men, they all survived, as the official report on the accident does not mention any more fatalities. A tank car loaded with molasses, presumably the one on which the trespassers had placed themselves, burst and flooded the tracks with sticky fluid.

This time, although only tramps were involved, the Interstate Commerce Commission did condescend to investigate the accident. Its report stated that the train involved, No. 70, was in charge of Conductor Weidenhamer with Engineer Mohlenhoff. It consisted of locomotive No. 4953, 55 cars and a caboose. None of the train crew was aware of any problem until the train brakes were applied in emergency.

Examination showed that first marks of the derailment were on the crossing planks at Locust Street, these having apparently been made by column bolts striking the crossing

---

[86] Ministry of Transport report
[87] *Le Petit Parisien*, Paris, France, 21st June 1926
[88] *Le Peuple*, Brussels, Belgium, 21st June 1926
[89] Ibid, 22nd June 1926

timbers and ties. Inspection of tank waggon GATX 20072 showed that the bottom arch bar on the right side of the forward truck was broken just behind the journal box. It had an old flaw. There was no evidence of any other defect. It was supposed that when the arch bar broke the body strap sagged and permitted the column bolts to strike the ties, causing the derailment.

The commission simply concluded that the derailment had resulted from a broken arch bar and made no recommendations. [90]

The wreck was apparently the fourth to occur at the spot within two years. [91]

An unlucky accident on the London Midland and Scottish Railway on Friday 19th November 1926 near Parkgate and Rawmarsh, station, England, cost nine lives. Three trains were involved in the accident, the 10.10 a.m. express passenger train from York to Bristol running on the up fast line, the 7.20 a.m. special mineral train from Westhouses to Royston, running on the down slow line and the 8 a.m. special empty waggon train from Normanton to Toton on the up slow line.

The York express consisted of 4-4-0 engine No. 387, a third-class coach, a first and third-class composite coach, a third-class brake composite, a first and third-class composite and a third-class brake composite, in charge of Passenger Guard Frederick Peaker with Driver James Collingwood and Fireman Alfred Cooke. The Westhouses mineral train consisted of 2-4-0 engine No. 270 piloting 0-6-0 engine No. 3980, 90 loaded waggons of coal and a brake van, in charge of Goods Guard Frederick Smith with Driver Ernest Stainforth and Fireman Charlesworth on the pilot engine and Driver Charles Ellis and Fireman Harry Bosworth on the train engine. The Normanton empty waggon train consisted of engines Nos 2537 piloting engine No. 303, 81 empty waggons and a brake van, in charge of Goods Guard Hales with Driver Colin Griffiths and his (unnamed) fireman and Driver Arthur Birkenshaw and Fireman Joseph Freeman on the train engine.

As the express was passing the mineral train several waggons of the latter were derailed and the sides of the carriages of the express were ripped out. Nine passengers were killed outright or fatally injured and 25 passengers and the guard of the express were hurt. Several trucks of the empty waggon train were also derailed as a result of the accident.

The first persons to give evidence at the Ministry of Transport inquiry into the accident were two miners who had witnessed the accident. John Edwards was working on the roof of a chapel about 170 yards from the scene of the collision and probably had the best view of it. He spoke of seeing the empty waggon and mineral trains, the latter being divided not far behind the engine. He had heard the engines of the up empty train whistling, then the down mineral train had stopped, the gap between the two portions of it had diminished to nothing and they had collided. Just then the up passenger train had appeared and collided with the wreckage.

Miner Harry Mitchell, standing on some waste ground about 80 yards from the collision site, said that the first thing that he had noticed was a grinding noise made by the mineral train. He had seen, he thought, the first waggon behind the engines sagging

---

[90] Interstate Commerce Commission report

[91] *The Daily News*, Frederick, Maryland, United States, 23rd August 1926

and then the fourth or fifth waggon from the engines burst outwards. At that point he had made his way towards the railway and as he was getting over the boundary fence he had heard the noise of the collision.

Signalman Henry King in Rawmarsh South cabin stated that the empty mineral train had arrived at his up slow starting signal at 11.11 a.m. and had been accepted by Parkgate Junction cabin at 11.12, at which time he had also received the "Train entering Section" signal for the down mineral train. The York express had passed at 11.14 a.m. Immediately afterwards he had heard an unusual noise as if something had struck the signal wires and looking towards Parkgate Junction had seen that several waggons were off the rails. He had not at that point realised that there had been a collision between the express and the waggons of the mineral train.

Signalman Benjamin Hambleton at Parkgate Junction had booked the "Train entering Section" signal for the down mineral train at 11.13 a.m. and for the up empty waggon train and the up express at 11.14. He said that he had seen nothing wrong with the down mineral train as it had passed him and certainly no division. The rattling of his levers had been his first warning of the accident.

Driver Griffith of the empty waggon train said that it had been stopped at the Rawmarsh South up slow starting signal for half a minute. Just after his train had restarted he had seen the down mineral train approaching and as the pilot engine of that train had passed him he had seen that the fourth or fifth waggon from the engine had collapsed and he had whistled and stopped his train. He had seen the express passing on the up fast line, and that its rear carriages were badly damaged.

Driver Birkenshaw of the same train stated that he had also seen the waggons of the mineral train collapse and some of them strike his own train, while Fireman Freeman and Goods Guard Hales gave similar evidence.

Driver Stainforth of the mineral train testified that the Rawmarsh South distant signal had been at danger, and he had reduced speed to about five miles an hour before he could see that the home signal was off. As he had re-applied steam he had been alerted by the popping whistle of an engine on the up slow line and looking around had seen that a waggon sixth or seventh behind his engine was off the line. He had applied the brake gently, whereupon the waggons at the head of his train had immediately begun to pile up. Fireman Charlesworth corroborated his driver's evidence.

Driver Ellis of the train engine of the mineral train stated that he had heard the whistles of the empty waggon train and his fireman had cried, "There is something wrong with our train." Looking round he had seen some waggons of his train off the line and had shouted to Driver Stainforth to take his brake off. The piling up of the waggons of his train had, he thought, still been going on as the express had passed them.

Fireman Bosworth generally confirmed his driver's evidence and said that he had jumped off the engine as the waggons of his train had struck the tender and had run across the up fast line directly in front of the express.

Goods Guard Smith said that he had taken over the train from Goods Guard Arthur Morris at Masborough Sorting Sidings North, where the train had been watered and examined. The first that he had known of the accident, he went on, was seeing the Rawmarsh South up fast starting signals collapse (struck by the derailed waggons). He had then heard the waggons of his train buffering up, had been thrown to his knees, and getting up had applied his brake as the waggons rebounded..

Driver Collingwood and Fireman Cooke of the express stated that they had heard the thud as the waggons of the mineral train had struck theirs. Fireman Cooke said he had looked back and seen the damage, had shouted, "Whoa," and Driver Collingwood had immediately applied the brakes.

Guard Peaker of the express said that his train had passed Swinton Junction on time. Just after it had passed Rawmarsh South cabin he had been hit on the head by what he had thought was a piece of glass. He was unable to speak further as to the nature of the obstruction struck by his train.

Sub-ganger Harry Scott stated that the down slow line had been in good condition when he had examined it on the day before the accident. He had seen the derailed waggon strike the train on the up slow line and had made his way to Parkgate Junction to tell the signalman.

Carriage and Waggon District Foreman Robert Wilson of Sheffield detailed his examination of the wrecked waggons after the collision.

Waggon examiners Frederick Jackson at Clay Cross and Edward Veal and Herbert Vickers of Masborough spoke of examining the waggons of the mineral train and of finding nothing wrong with them.

Reviewing the events which had led up to the accident, Colonel J.W. Pringle, the Ministry of Transport inspecting officer, stated that from deposits of coal in the four-foot of the down slow line and the wreckage the course of the initial derailment could be traced. One hundred and twenty yards from the demolished Rawmarsh South signal post the side door of Hickleton Main Colliery private owner waggon No. 1264 had been found, and it was evidently this side door which had struck the engines of the empty waggon train. It was the striking of his engine by the side door which had prompted Driver Griffiths to pop his whistle, whereupon Driver Stainforth had applied his brake. In doing so he had checked his train more sharply than he had intended, causing the waggons to pile up. The derailed waggons had then struck and demolished the Rawmarsh South up starting signal as the express was passing and the coaches of the express had been struck by the signal post and possibly also by the derailed waggons.

In the conclusions to his report Colonel Pringle wrote that the originating cause of the accident had been the defective condition of the Hickleton Main Colliery waggon, marshalled probably fifth behind this engines. This vehicle had a pre-existing breakage in the eye-end of the rear intermediate draw-bar, perished and ineffective rubber springing, at least one rotten timber sole-bar and corroded through rods. The immediate cause of the accident had been the sudden check to the mineral train which had caused the waggons to pile up, leading to the demolition of the signal post which had been struck by the passenger train.

He considered that the breakage of the intermediate draw-bar had probably occurred after the last examination of the waggon by the Hickleton Colliery Company on 21st April 1926. The defect was not such as could be discovered during a routine examination in sidings or on running lines and he did not therefore place any responsibility on any of the three waggon examiners. The action of Driver Stainforth in applying his brake was, he thought, normal and necessary, though he had misjudged the effect of the application upon a heavy unfitted train pulled by two engines moving at 8 miles an hour. However, there had been danger to the up empty waggon train and thus a special reason to quickly stop his train and Stainforth could not have anticipated the entirely fortuitous arrival of the express train. With regard to the exceptional circumstances Driver Stainforth's misjudgement was excusable.

In general the inspecting officer considered the circumstances of the accident to have been exceptional and some of them, such as the existence of the signal post, very fortuitous. He considered, though, that further precautions must be taken to maintain a high standard of waggon maintenance, as ordinary checks by carriage and waggon examiners were insufficient. He recommended that all waggons should be subject to careful and detailed examination at least once in every ten years, which would necessitate the re-registration of private owner waggons, which otherwise would not be permitted to run. [92]

The inquest jury on the six passengers who were killed outright in the accident returned a verdict that their deaths had been due to multiple injuries sustained through their being passengers in the express train which came into contact with the signal post, knocked down by the derailed mineral wagons, and that no blame was to be attached to any railway servant. [93] The inquest on the remaining three victims doubtless reached a similar conclusion.

On probably Thursday 24th February 1927 the Ferrocarriles Nacionales de México (Mexican National Railway) suffered a major disaster between Chocoy and Manuel stations, 44 and 50 miles respectively north west of Tampico, Mexico. [94] A Tampico – Monterey passenger train was derailed, apparently due to a broken axle on the engine tender. The engine and tender, baggage cars and two coaches left the rails and overturned, and a Pullman car was derailed and damaged. At the time of the report from Mexico City it was stated that the bodies of three adults and a child had been recovered from the shattered wooden coaches and that the bodies of 36 other victims were still trapped in the wreckage. [95] Between 70 and 100 passengers were said to have been injured. 'Most of the victims were working men and their families.' [96]

At 02:00 on Friday 3rd June 1927, Chemins de fer de Paris à Lyon et à la Méditerranée (PLM) train No. 1415, an express which had left Paris at 20:30 the previous evening for Nimes, collided with the derailed waggons of a goods train at Bessay-sur-Allier, France, while travelling at 95 kilometres an hour. The express engine was derailed and overturned, the leading van and following third-class carriage were damaged and the next two carriages were damaged to a lesser degree. Driver Henri Chardet and eight passengers (several of them railwaymen travelling as passengers?) were killed and Fireman Pierre Laugié, Head Conductor Louis Rondepierre and about 20 passengers were injured. [97]

---

[92] Board of Trade report
[93] *The Times*, London, England, 3rd December 1926
[94] Webpage: Proceedings of the United States National Museum: Some Upper Cretaceous Shells of the Rudistid Group from Tamaulipas, Mexico.
[95] *The Wisconsin State Journal*, Madison, Wisconsin, United States, 25th February 1927
[96] *The Bee*, Danville, Virginia, United States, 25th February 1927
[97] *La Peuple*, Paris, France, 4th June 1927

An investigation showed that the sequence of events leading to the derailment of the parcels train had begun at Varenne-sur-Allier when a vehicle had lost one of the spacer rods which served to hold the guard plates. Between Varenne and Ferté Hautelrive the vehicle had then lost a spring. At this point it had left the rails, swinging first to the left and then to the right. Driver Giron, becoming aware that there was a problem with his train, had whistled to his conductor to apply the brakes and at the same time had shut off steam. The jerk caused by the brake application had thrown the derailed waggon and the following one across the track. [98]

The Bessay signalman, M. Chérasse, seeing the danger, had attempted to replace his signal to danger against the express, but was unable to do so, probably because the derailed waggons had fallen on the signal wire. The express collided with the derailed waggons shortly afterwards. Immediately after the accident Head Conductor Chambon of the goods had gone back the line and was able to stop a following goods train and thus prevent a worse disaster. None of the railwaymen involved was in any way to blame for the accident. [99]

Fourteen people died and twenty were injured in an accident late in the evening of Thursday 18th April 1929 at an unstated location between Iquique and Antofagasta, Chile. The engine of a passenger train reportedly lost its wheels in some unexplained manner and the carriages of the train piled up behind it. A relief train with physicians and Red Cross equipment was despatched from Baquedano station, Santiago. [100]

A special Canadian National Railway train carrying the circus of Al G. Barnes from Newcastle to Charlottetown became derailed at 7 a.m. on Sunday 20th July 1930 at Canaan, New Brunswick, Canada. The long train consisted of 29 cars and nine of them behind the 17th left the rails: five flat cars and four passenger cars. Four people were killed outright, all had been riding on one of the flat cars. Those who lost their lives were Frank Finnigan and Albert Johnson, prop men; J. MacFarland, waiter and a one-armed man believed to be James Stephens, who was not a circus employee but probably a trespasser. Another 17 people were injured, one probably fatally. [101] The derailment was ascribed to a failed truck arch on one of the cars. [102]

[98] *La Presse*, Paris, France, 4th June 1927
[99] *Le Petit Parisien*, Paris, France, 5th June 1927
[100] *Evening News-Journal*, Clovis, New Mexico, United States, 19th April 1929
[101] *Manitoba Free Press*, Winnipeg, Manitoba, Canada, 21st July 1930
[102] *The Lethbridge Herald*, Lethbridge, Alberta, Canada, 21st July 1930

CHAPTER 3

OTHER DERAILMENTS

Other chapters in this book cover derailments caused by track defects, defects with the trains themselves and those caused deliberately by acts of vandalism or sabotage. This chapter will deal with derailments due to other causes such as excessive speed, collisions with animals and other accidental obstructions, traction shocks and so on, together with those for which no cause was given.

The 11.30 a.m. express passenger train from Launceston to Hobart was derailed two miles north of Campania, Tasmania, Australia, just after 4 p.m. on Tuesday 15th February 1916. The train consisted of a Garrett engine, four coaches and a guard's van and was in charge of Guard William Williams, with Driver Philip Goodchild and Fireman Douglas Bessel. It was carrying about 200 passengers. Among them was a party of 50 delegates of the Young Women's Christian Association (YWCA) from Sydney, Adelaide, Ballarat, Bendigo, and Geelong on their way to attend a conference at Browns River. About the same number of passengers were soldiers returning from leave to Claremont Camp (a military training camp opened in October 1914 near Hobart to cater for the large number of men wishing to enlist to fight for King and Empire) [103] while others were young men going to the same camp to enlist.

The engine of the express left the rails on a curve, dragging three carriages after it, and the leading carriage was smashed to pieces. Four passengers were killed outright, and a dozen more seriously injured. Those passengers who died at the scene were Mrs Olive Howell, president of the Geelong branch of the YWCA; Miss Margaret Barrass, a Bendigo delegate to the YWCA conference; Gunner William Orr, and James Quinn, a labourer. [104]

Fireman Bessel was thrown from the engine over a wire fence into a paddock and escaped with a sprained wrist. Driver Goodchild also managed to struggle out of the engine and run clear without sustaining serious injury but fearful of the engine boiler exploding he then returned to its steam-filled cab to manipulate the steam cocks and in doing so was

---

[103] Webpage – Tasmanian Government: Centenary of ANZAC, Tasmania Remembers 2014-2018
[104] *The Mercury*, Hobart, Tasmania, Australia, 17th February 1916

fearfully scalded. A witness to the accident made his way as fast as he could on a motorcycle to Campania station, from where the stationmaster telephoned news of the accident to Hobart at 4.20 p.m. and a relief train was despatched to the scene at 5.05 p.m.

Surviving passengers described the accident to a newspaper reporter. 'Mr Frank Smith, a representative of the Vacuum Oil Co, who had come over from Melbourne on business, told a *Mercury* reporter of the remarkable escape of himself and six or seven other men in the same compartment. "We were in the third carriage from the front," he said. "The engine jumped the rails, and went clear over the bank, and the first carriage twisted round across the track, and was splintered to atoms. The second car was smashed to pieces, and the third car had half the front torn away. We had a wonderful escape. The roof of the carriage was torn off and the wheels came through the floor. It was the biggest wonder in the world that we were not all killed, but not one of us was injured."

'Gunner Green, of the Artillery, who was returning to camp, perhaps saw more of the actual accident than most of the passengers, as he chanced to be looking out of the window at the time. His story is best told in his own words, "I was in the last carriage, with other soldiers. As we approached Campania, I was looking out of the carriage window and saw what looked like a cloud of dust under the engine. Then I saw the engine and first carriage disappear followed by the third carriage on the opposite side. The engine is upside down, buried about three or four feet, with her wheels in the air. My friends and I got off with nothing worse than a severe shaking. We were in a terrible plight as regards the treatment of the injured. It was about two hours and forty minutes before the relief train arrived with the doctors. Unfortunately, none of the passengers knew much about first-aid, and those of us who escaped unhurt had to patch up the injured as best we could. Practically everybody in the first two carriages was injured."' [105]

By Saturday 19th the number of fatalities had increased to seven with the deaths of Driver Goodchild, labourer John Castles, who had been on his way, like James Quinn, to enlist; and 70-year-old John Roughton, a boot warehouseman. [106]

The inquest on the victims of the accident was quickly opened and at its resumption on 22nd February Fireman Bessel testified that Driver Goodchild had been perfectly normal and that the train had not been running any faster than it should have been when it left the track. According to him the train had been travelling at 25 miles an hour at the moment of derailment, his driver having previously applied the brakes. The fireman was followed by Dr Gregory Sprott, who had attended the victims and gave medical evidence, then by Guard Williams. The guard said that his train had left Conara Junction six minutes late and Colebrook five minutes late, at 3.42 instead of 3.37. He had said nothing to the driver about making up time. At the 29½ milepost the speed of the train had increased, and he had applied the brakes. He put the speed of the train as it entered the fatal curve at 35 miles an hour. His opinion was that something had gone wrong and caused Driver Goodchild to lose control. He timed the accident at 4.02 or 4.03 p.m. Guard Williams said that he had remarked to Permanent Way Inspector William Paynter, who was riding in the van with him, that the train was going too fast to negotiate the curve. The guard also spoke of seeing blue steam or smoke coming from the front of the locomotive which the driver appeared to be preoccupied with. Guard Williams stated that

[105] Ibid, 16th February 1916
[106] Ibid, 19th February 1916

it was not his duty to control the speed of the train. [107]

The inquest continued, with the testimony of various witnesses and experts and numerous adjournments, until 9th March, when the jury brought in a verdict that 'the said railway accident was caused by the overturning of the engine running at an excessive speed, but from the evidence before us we cannot ascribe any direct cause for the said excessive speed.' As a Hobart newspaper commented in an editorial the verdict of the jury was not really a very satisfactory one, but the best that could be expected in the circumstances. Charles Baird, a passenger on the train who had spoken to Driver Goodchild after the accident said that the driver had blamed the derailment on a broken steampipe, but Fireman Bessel denied that there had been any such problem and why Driver Goodchild had failed to control the speed of his train must remain a mystery. [108]

Soldiers, mostly from southern Italy and especially from the Puglia region, who were going home on leave from the front, were the victims of an accident at Cortona, Italy, early on the morning of 26th February 1916. The train from Florence to Rome, carrying 450 troops, was entering the station at about 01:45 when Driver Giuseppe Bartolini felt and heard that something was wrong and brought his engine to a stand as soon as he could, though not before the train was badly derailed, its carriages overturning and being broken up. The leading luggage van and coaches of the train collided with the tender of the engine which had ended up sideways across the track while other carriages fell down an embankment. Twenty-two soldiers died in the accident, together with Fireman Giuseppe Chiappi. Driver Bartolini was unhurt. The derailment was believed to have been caused by a piece of metal which had fallen from an earlier train and become jammed in a set of points, but a judicial investigation was ordered to determine its exact cause or to rule out sabotage. [109] [110]

A major catastrophe on the Ferrocarril Interoceánico de México between Dehesa and San Miguel, Mexico, on 12th November 1916, was reportedly due to excessive speed by the engineer over a dangerous part of the road. All but five of the cars of the passenger train were said to have rolled over a cliff. The accident, near Jalapa, killed 159 people and left many others more or less seriously injured. A widow and her seven children were among the dead. 'No Americans were on the train.' [111] [112]

Double-headed passenger and mail train No. 34a from Trieste to Wien, travelling at 50 kilometres an hour, ran into a landslide at Km 383.5 between Trifail, (now Trbovlje) and Sagor (now Zagorje ob Savi; both places are now in Slovenia) Styria, Austria, at 1.42 a.m. on Tuesday 16th January 1917. Both locomotives and the leading vehicle of the train were derailed. Immediately afterwards, probably precipitated by the shock of the derailment, a much larger landslide, estimated at 1,800 square metres of soil and rock,

---

[107] Ibid, 23rd February 1916
[108] Ibid, 24th, 25th and 26th February and 1st, 2nd, 3rd, 4th, 6th, 8th, 9th and 10th March 1916
[109] Webpage – Cento ani fa, le strage de la stazione di Carmucia – Valdichiana Oggi
[110] *Goshen Daily Democrat*, Goshen, Indiana, United States, 29th February 1916
[111] *The Sedalia Democrat*, Sedalia, Missouri, United States, 20th November 1916
[112] *Oak Creek Times*, Oak Creek, Colorado, United States, 23rd November 1916

occurred, and engulfed the whole train, part of which was reportedly buried, and the remainder thrown into the River Sava. Eleven people died, 26 were seriously and eight slightly injured. [113] According to one report further deaths occurred when a relief train sent to the scene ran into the debris. [114] There had been heavy rain in the days preceding the accident and the area was also prone to earthquakes.

The colony of Newfoundland suffered a serious accident early on the morning of Monday 5th February 1917. Westbound Reid-Newfoundland Railway express train No. 1, in charge of Conductor Lush, had left St Johns at 12.30 p.m. on Sunday 4th for Port-aux-Basques and had changed engines at Clarenville, from where it was taken forward at about 9.05 p.m. by Driver Arnold Webber, with engine No. 115. All went well until the train was about 10 miles east of Glenwood. The train was rounding a right-hand curve on an embankment about 10ft high, and despite its designation as an express was only moving at 8 or 10 miles an hour when the accident happened at 3.30 a.m. Suddenly Driver Webber felt a jolt as the tender left the track. He applied the brake in emergency and brought his engine to a stand, still on the rails, then looking back he saw that the tender, baggage car and following carriages had left the track. The second-class carriage was overturned onto its side and by the time that Driver Webber had got back to it this carriage was already on fire. The passengers who were able to had already got out and Conductor Lush and a brakesman were calling to the other passengers to "Come this way." [115]

Conductor Lush told the subsequent inquiry into the accident that he had been riding in the first-class carriage at the rear of the train when he had felt it stopping and a slight bump. Jumping out, he had gone forward past the dining car, which was off the road and leaning over. Attached to the diner was the second-class carriage, which was on her side about 15 feet from the track and afire. He and a baggageman kicked open the door at the east end of the carriage while the brakesman went to the west end and did the same thing. As soon as they had kicked open the east end door he had sung out and some of the passengers came out that way, he was not sure if any escaped from the west end. He had then gone to the west end where the driver had joined the brakesman. There had only been the four of them – himself, the baggageman, the brakesman and the driver – engaged in the rescue at that time. He complained that they had received no help from the mail clerks Knight and Thistle in the baggage car or from the first-class passengers. After three or four minutes everything had gone quiet in the second-class coach, and they had concentrated on preventing the fire from spreading to the remainder of the train.

Conductor Lush said that 15 passengers had escaped from the second-class carriage, two of whom had been injured. The bodies of eight persons who had not been able to get out in time were discovered after the fire had been extinguished. They were, Joseph Watts, of Carbonear; Mr and Mrs Moses Rodway, of Muscle Harbour Arm, Placentia Bay; Phillip Hounsell, of Pound Cove; Thomas Stapleton and John Jacobsen, of St Johns; and Ambrose and Simon Rodgers, of St Johns Within. Mr and Mrs Rodway had only been married a month. John Jacobsen was a Norwegian seaman staying at the Seaman's

[113] *Eisenbahnunfälle Bind I* by Ludwig Stockert, pg. 89 (in German)
[114] *The Times*, London, England, 22nd and 23rd January 1917
[115] *St John's Evening Telegraph*, St John's, Newfoundland, 5th, 7th, 8th and 13th February 1917

Home in St Johns and en route to North Sydney, while Joseph Watts was going with four others to Nova Scotia to collect a newly-purchased schooner.

The conductor swore that everything possible had been done to rescue the passengers who had lost their lives, but the time had been too short. Only about three minutes had elapsed between the initial jolt of the derailment and things going quiet, indicating that those left in the coach had either suffocated or burnt to death. The second-class coach was a Colonist car with double windows and a wooden screen. He did not think that the windows would have been big enough for any of the passengers to squeeze through them. The car was steam heated but was lit by three oil lamps suspended from the ceiling and of the same type as used in other cars. The lamps had been lit at the time of the accident and their upsetting as the car overturned was clearly the cause of the fire. [116]

The cause of the derailment was not at all clear. When daylight came after the accident Driver Webber discovered a broken rail, but this he found *ahead* of the point where the tender had first left the track, implying that it must have been a result of the derailment and not its cause. Driver Webber said that there had been a little roughness in the road but not sufficient to be worth reporting. There had been nothing wrong with his engine when he examined it after the derailment except that the drawbar had snapped between it and the tender. This had happened as a result of the derailment and had prevented his engine following the tender down the embankment. Driver Webber gave evidence at considerable length at the inquiry but was not really able to throw any additional light on why the accident had happened. [117]

Italy suffered a bad accident on the late evening of Tuesday 7th August 1917 when train No. 74, which had left Genoa at 20:10 for Milan, was derailed as it was leaving the new station at Arquata Scrivia, opened only the previous year. Three coaches overturned and three others were wrecked. The casualty list was high, with 40 reportedly killed (other sources say 34) and another 60 injured. Some of those who had been hurt were taken to a local hospital while others were transported in rescue trains to hospitals in Genoa and Novi Ligure. Among those killed were Driver Pietro Bosetti, Fireman Edmondo Sevo and an American passenger named Rene Hammond, while those injured included at least two foreigners, Swiss and Chinese. The derailment was due to excessive speed. [118] [119]

The Midland Railway suffered an unlucky accident on Saturday 19th January 1918 just north of Little Salkeld station in Long Meg (or Throstle Hill) cutting, England. The train involved was the 8.50 a.m. from London St Pancras to Glasgow, 1.45 p.m. from Leeds, in charge of Head Guard Calder, with Driver Whitworth and Fireman Costello. It consisted of Compound engine No. 1010, Midland & Glasgow and South Western Joint (M & G & SW Jt) bogie van, four M & G & SW Jt third class coaches, a Midland Railway dining car, three M & G & SW Jt first-class coaches, and a Midland and North British Joint composite coach and brake van. Driver John Whitworth said that after ditching the pilot

---

[116] Ibid, 13th February 1917
[117] Ibid, 16th February 1917
[118] Wikipedia – Incidente ferroviario de Arquata Scrivia
[119] *Brandon Daily Sun*, Brandon, Manitoba, Canada, 9th August 1917

locomotive at Ais Gill he had had clear signals at Appleby and Little Salkeld. Approaching the scene of the accident he had been looking out as a matter of habit as it was previously the location of the Long Meg Sidings down distant signal (the cabin had been done away with about 1915). It was 3.57 p.m., still broad daylight, and the line was then clear. He had looked away for no more than half a minute to adjust the controls, and when he looked out again a mound of earth was covering the rails directly in front of him. He had immediately applied the brakes as his engine struck the obstruction and mud and water filled the cab. The engine was lifted into the air, came down onto the ballast, ran 180 yards and stopped. Fireman Costello, busy with his work, had noticed nothing wrong before the collision.

The whole of the train except the last van was derailed and the two leading vehicles were broken up, the others sustaining less damage. Three boys and three women were killed outright, and 38 passengers injured. Two of the dead boys were sons of a Mr Whiting of St Albans, on their way to what must have been a boarding school in Carlisle – what a distance to send your children to school. Several days after the accident there were fears for the lives of Nurse Catherine Johnston of Hamilton and Miss Henry, of Carlisle, both of whom were still in a critical condition, but it was Mr John Barclay, a Glasgow merchant, who died in hospital a week after the accident to become the seventh fatality.

Giving evidence at the Board of Trade inquiry into the accident, held by Colonel J.W. Pringle, Foreman Platelayer George Spreadbury said that he had held his position for the last 37 years. On the day of the accident he had walked past the spot where the landslide occurred at 3.20 and 3.35 p.m., the second time thus only just over 20 minutes before the accident. There had been no sign then of any slippage of the cutting. The drains had been full of water and the ground had been very wet but there had been no water escaping from the cutting face or running over the edge of the cutting. He had been about a mile further north when he had heard the express coming, then the sound of the derailment. He remembered a small slip a little further north in 1911 but nothing else.

Sub-inspector Hook stated that he had walked along the top of the cutting on 8th January, when there had been no signs of settlement.

Permanent Way Superintendent Hunt said that the line had been built between 1870 and 1875, and the existence of a masonry toe wall erected then was evidence that the location had given problems. There was a 6-inch pipe-drain in the cess at the bottom of the cutting and a ditch at the top of it. The 1911 slip had been very small.

Colonel Pringle stated that he believed that the weather conditions in the days before the slip had had a very significant effect in causing it. There had been first a severe frost, then a considerable snowfall, then a sudden thaw. Then on the afternoon of the accident there had been heavy showers of rain. The inspecting officer theorised that the frost had interfered with the normal drainage of the melting snow, as a result of which water had collected in an unusual quantity at the base of the cutting till the area had become waterlogged and the weight of it caused a slip. The fact that the ground east of the cutting, which had formerly been pasture, had been ploughed up in the past year for turnips, might have contributed to the interference with the drainage.

The inspecting officer concluded that the slip must have been sudden and that no blame should be attached to either the permanent way men or the driver of the express.

He went on to note that a number of the coaches of the express had been gas lit but there had been no fire, the first action of Driver Whitworth, Guard Wilce (there were three guards in all on the train) and Traffic Inspector Thomas having been to ensure that one did not break out. After the accident, a local train from Carlisle had reached the scene at 5.05 p.m. and had returned to Carlisle with the uninjured and some of the injured passengers at 6 p.m. A relief train from Carlisle had arrived with staff and equipment from Fusehill military hospital at 6.45 while doctors had arrived earlier from Langwathby and on a ballast train from Appleby.

A brief newspaper report mentioned the derailment of a train in Cordoba province, Argentina, on 29th January 1918. Five persons were killed and 15 injured. [120]

Very little information has come to light regarding a major disaster at Pizzo Calabro, Italy, on 19th February 1918. It apparently involved the derailment of a train carrying soldiers home on leave from the front and resulted in 42 deaths. [121]

A preoccupation with the war news meant that the derailment of four carriages of a Montdore-Paris mail train on Tuesday 16th July 1918 received little coverage in the French press. The accident occurred between Theillay and Salbris, France, on the Orleans – Vierzon line of the Chemin de fer de Paris à Orléans. Fifteen people died in the accident and 20 were seriously and 50 slightly injured. [122] [123]

The inexperience of the motorman in charge of a Brooklyn Rapid Transit train on 1st November 1918 was demonstrated when he approached the tower at the junction at Franklin Avenue. His train was not showing any headcode to inform Towerman Peter Gorman which way it was to go, so Gorman left the road set for straight ahead for Fulton Street. A strike was in progress, train schedules were disrupted, and the towerman could not therefore be sure which way the train was to go in the absence of its headcode. The train ran through the junction onto the Fulton Street line but then stopped beyond the junction. It was in fact a Brighton Beach train, but instead of stopping short of the junction, the signals at which clearly showed that the route was set for Fulton Street, it had overshot it. A discussion then followed between the towerman, the motorman and the conductor. Towerman Gorman was not happy about the train setting back over the junction in the wrong direction so instead it went forward to the next mains crossover, was crossed to the other line, returned to the junction, was crossed back onto the proper line via another crossover on the Manhattan side of the junction and finally proceeded down the Brighton Beach branch, now ten or fifteen minutes late.

This little episode was, however, merely the prelude to a major catastrophe, which then and now (June 2021) is the third worst railroad accident in United States history. In

---

[120] *The Times*, London, England, 30th January 1918

[121] Webpage – Giornata del ferroviare : a Pizzo cerimonia in memoria del 42 soldati morti nel 1918 in un incidente.

[122] *The Times*, London, England, 18th July 1918

[123] *La Presse*, Paris, France, 17th July 1918

charge of the train was Motorman Anthony E. Lewis, whose original surname was Luciano. Lewis was not a qualified motorman and although he had had supposedly 80 hours' training, he had never driven a train on the Brighton Beach branch. However, the Brooklyn Rapid Transit (BLT) company were desperately short of motorman as the regular motormen, members of the Brotherhood of Locomotive Engineers, had gone out on strike for the reinstatement of 22 fellow workers who had been discharged and whom the National War Labour board had ordered the company to reinstate. The previous day Lewis had worked as a train dispatcher at the Culver Depot and when he had learned that the BRT was short of six motormen to work trains on the Brighton Beach line he had volunteered to act as one. A non-union man, he was therefore basically a scab. [124] He said that he wanted to earn more money for his family.

The train which Lewis was driving was jammed with passengers as there were fewer trains and delayed services because of the strike; there were estimated to have been 900 passengers on it. It was said that every train which had left Brooklyn Bridge had been jammed with delayed residents trying to get home and that they had almost fought one another to get on the train. Having lost time in the mix-up at Franklin Avenue, Lewis now decided to run hard to try and regain some time. The Brighton Branch was mostly straight but being unfamiliar with it, he failed to realise that at Malbone Street there was a very sharp curve, restricted to six miles an hour and clearly indicated by two warning signs. Instead of reducing the train's speed to a crawl Lewis had attempted to take the curve while running at 30 miles an hour at least. The second car of the train left the track at the entrance to Malbone Street Tunnel, followed by the third car, then the first and the fourth. In the tunnel there was a concrete wall containing iron uprights on the right-hand side, with a row of steel pillars separating the two tracks on the left-hand side. The cars of the train were scraped against the concrete wall. The first and fifth cars, which were steel, suffered little harm but the three flimsy middle cars of wooden construction and 35 years old, were very severely damaged. Fire also broke out in the wreckage. The death toll, initially given as over 100, was probably 92. About 200 were injured.

Police reserves from a dozen stations were rushed to the scene and they ordered out all the ambulances in Brooklyn. Manhattan hospitals were asked for assistance and the fire department worked to put out the fire and attend to the injured. The location of the accident made the rescue work more difficult and the bodies of the dead and injured had to be carried up ladders provided by the firemen. More than 100 members of the women's motor corps of America lent assistance and all kinds of vehicles were called into service to transport the injured and dead to hospitals and morgues. [125]

After the accident, the BRT agreed to reinstate the discharged men and the strike was called off.

Motorman Lewis survived the accident more or less unhurt and fled the scene in terror. He was arrested on a charge of homicide at his home the next day. Guard Samuel Russof and Conductor Michael Turner were also arrested on the same charge. The same day BRT President Timothy S. Williams, apparently unrepentant, asserted that although Lewis had only worked previously as a motor switchman, he was a trained man capable

---

[124] *The New York Times*, New York, New York, United States, 5th November 1918

[125] *Bluefield Daily Telegraph*, Bluefield, West Virginia, United States, 2nd November 1918

of running a train. [126]

President Williams, BRT Vice Presidents John Dempsey and John Hallock, Division Superintendent Blewitt and BRT Chief Engineer William Menden were also subsequently arrested. Motorman Lewis and Superintendent Blewitt were acquitted in April 1919. [127] The others were eventually all acquitted as well, or the charges against them were not pursued. The last defendant was John Dempsey, as the jury could not agree at his first trial. He was acquitted at the end of his second trial on 14th April 1920. [128]

The derailment of an Ohio Electric Railway interurban car at London, Ohio, United States, on Monday 13th January 1919 was a strange affair, since it was an excessive speed derailment at a very low speed.

Train No. 22 from Dayton to Columbus, Ohio, consisted of motor car No. 263, in charge of Conductor Willard with Motorman Henry Lahrmer. It passed the station at London and left the rails on a curve with a radius of only 75 feet at the junction of Main and High Streets, was thrown over on its right side on the south side of High Street and its roof and interior were sheared by a steel trolley wire pole. Five passengers lost their lives and nine passengers, and two railwaymen were injured. Those who lost their lives were J.N. Ascheliman, a farmer of Wauseon, Ohio; Mrs Kinkead, of Columbus, Ohio; Frances Reeder, a child; another unnamed Reeder child, both of London, Ohio, and an unidentified negress. [129]

Motorman Lahrmer told the Interstate Commerce Commission (ICC) inquiry that he had taken over the car at Springfield from Motorman McClurg, who had told him that the car was not braking evenly and to be on the lookout. He (Lahrmer) had inspected the car and had found that the piston travel was excessive and had adjusted it. He had not noticed the condition of the brake shoes, tried the air, noticed the amount of piston travel, tried the sander, or looked in the sand box. While he was adjusting the brakes the regular inspector had appeared and told him that he was not doing it correctly, to which he (Lahrmer) had retorted that he had been making adjustments for four years without any problems and considered himself competent to do the job. The trip from Springfield had been normal until he had attempted to stop at London station. There the wheels had begun to slide, and he had been unable to release the brakes to stop the sliding, so had applied power. By that time, the car was almost on the curve. The wheels continued to slide, and he had applied sand but had not used the hand brake. He had been talking to Conductor Willard and when both had realised that the car was going too fast to take the curve both had attempted to alight but had been thrown off first.

Motorman McClurg testified that the brakes on car No. 263 had been working unevenly on the day of the accident and it had slid at points where the rails were slippery, but he had had no serious trouble controlling it. When he had handed over the car to Motorman Lahrmer he had mentioned the problem and his colleague had replied that he would adjust the brakes. The sand box had been two thirds full.

---

[126] *Boston Evening Globe*, Boston, Massachusetts, 2nd November 1918
[127] *The New York Times*, New York, New York, United States, 6th April 1919
[128] *The Bridgeport Telegram*, Bridgeport, Connecticut, United States, 15th April 1920
[129] *Sandusky Star-Journal*, Sandusky, Ohio, 14th January 1919

Track Inspectors Green and Johnson said that the car had been in good condition on 11th January and Inspector Morris spoke of Motorman Lahrmer telling him on the day of the accident that he would not ask any barn man to adjust the brakes as he could do a better job himself.

Master Mechanic Foote testified that he had examined the car after the accident and had found it to be in good condition apart from a broken brake rod, the result, he considered, of the accident.

Conductor Willard put the speed of the train approaching 5th Street, where the train usually braked to stop at the station, at 14 or 16 miles an hour. At 4th Street the speed was still excessive, 14 or 15 miles an hour, and the wheels had then begun to slide. A little later, convinced that the car would derail on the curve, he had jumped or been thrown off. He said that he had not opened the conductor's valve as he had considered that that would be futile with the wheels sliding.

The ICC concluded that Motorman Lahrmer was responsible for the accident because he had failed to observe the speed restriction of 10 miles an hour through the town of London and had failed to have the car under control approaching a station. Conductor Willard was, according to Rule No. 234, equally responsible for allowing him motorman to operate the train at an unsafe speed.

Following the accident the company made 5th Street a safety stop for all eastbound trains. [130]

According to the author's notes a train was wrecked, presumably derailed, between Tiflis (now Tbilisi) and Batum (now Batumi), Georgia, at an unknown date in January 1919, with the loss of 200 lives.

A train carrying demobilised British troops was derailed on the evening of Thursday 29th May 1919 between Marchienne-au-Pont and Luttre stations on the Brussels – Charleroi line in Belgium. Thee coaches overturned. Seven soldiers were killed and about 20 seriously injured. [131]

Train No. 513, which had left Pau at 16.56, became derailed on a set of points at Artix station, France, on the Chemins de fer du Midi, on Saturday 23rd August 1919. The 13th carriage of the train then struck a large German goods waggon in an adjacent siding, cutting it in two. Five people were killed outright and more than 20 injured. The bodies of those killed were taken to Artix town hall and numerous cars transported the injured to Pau, Ortez and Dax. [132] [133] Seven people eventually died in the accident.

The passengers on the train included 33 officers and 24 men of the Royal Irish Regiment who were returning to Köln (Cologne) after a pilgrimage to Lourdes. 'The

[130] Interstate Commerce Commission report

[131] *The Times*, London, England, 31st May 1919

[132] *Le Petit Journal*, Paris, France, 25th August 1919

[133] *Le Petit Parisien*, Paris, France, 26th August 1919

British soldiers were not injured, and they performed fine rescue work.' [134]

A North Western Railway mixed train was derailed when it ran into a landslide between Uchhri and Chhab stations, India, on Monday 1st September 1919. Nineteen passengers died, 44 were injured and there was considerable damage to rolling stock. An investigation found that the patrolling arrangements were satisfactory and that no one was to blame for the disaster. [135]

A brief press report spoke of the death of at least 60 second-class passengers in the derailment of a Mexico City – Laredo train near Yurecurao, between Venegas and Saltillo, Mexico, on Sunday 5th October 1919. It was feared that the number of dead would rise to 75. [136] The first-class passengers escaped with more or less serious injuries. [137]

On about 18th October 1919 a train was derailed between Ujda and Oran, Algeria. Twelve passengers were killed and many injured. [138]

A mail train from Vladivostok to Harbin was 'wrecked', presumably derailed, at Pogranichnaya, RSFSR (since January 1918 and until the formation of the Union of Soviet Socialist Republics in January 1923 Russia was officially the Russian Soviet Federative Socialist Republic, or RSFSR), on about Saturday 23rd October 1920. One hundred persons were said to have been killed, 'including several actors and actresses of the Dolins Company, the best-known theatrical enterprise in the Far East.' [139] [140]

A passenger train appears to have become derailed between Panbult and Piet Retief, South Africa, in November 1920, with the loss of 13 lives, but it has proved impossible to trace any further details of this accident and the original reference has unfortunately been mislaid.

As a result of the blockage of the line between Vigeois and Uzerche, France, by the accident which happened on the evening of Saturday 14th May 1921 (see previous chapter) trains were diverted from Limoges via Périgeux, re-joining the usual line to Toulouse at Brive. Unfortunately, a train using this diversionary was then involved in a rather more serious accident the following day, Sunday 15th May. Train No. 71, which had left Paris at 22.26 on Saturday evening, arrived at the little station at La Bachellerie, France, at about 05:00. Its driver, who was perfectly familiar with the normal line through Vigeois, seemed, in the words of the newspaper, to have had an unfortunate lack

---

[134] *The Brisbane Courier*, Brisbane, Queensland, Australia, 26th August 1919
[135] Railways in India, Administration Report 1919-20
[136] *The Bryan Daily Eagle*, Bryan, Texas, United States, 6th October 1919
[137] *The Washington Daily Post*, Washington, D.C., United States, 8th October 1919
[138] *The Times*, London, England, 21st October 1919
[139] Ibid, 25th October 1920
[140] The Singapore Free Press and Mercantile Advertiser, Singapore, 27th October 1920

of caution in running over the alternative route, and attempted to negotiate the sharp curve which preceded the station at much too high a speed, allegedly 75 or 80 kilometres an hour. Predictably the train failed to take the curve. The coupling at the rear of the locomotive snapped and it continued down the track for 200 metres. The leading van was swung around, and the following coaches piled up against it. One of the carriages demolished the stationmaster's office, crushing all the furniture. The other coaches formed an indescribable pile of wreckage. By the time of the report the bodies of Conductor Bourdet and seven passengers had been recovered and there were said to be 50 passengers injured, no less than 20 of them seriously. [141] The rescue work had been delayed by the remoteness of the station but the uninjured passengers and doctors on the train did what they could and a Dr Gardette, who lived three kilometres away, was soon on the scene. [142] One of the injured passengers died the day after the accident. [143] Clearly the driver, who does not seem to have been named in any of the newspaper accounts, was not sufficiently familiar with the diversionary route and should have asked for a pilot.

On the afternoon of Saturday 25th June 1921 disaster struck Chemins de fer du Nord express No. 328 from Tourcoing to Paris, which had departed from Lille at 13:40. At 15:30 three vehicles and a van at the rear of the train left the track near Beaucourt-Hamel station, France, at a point where the line bordered the River Ancre. The vehicles remained coupled to the front of the train for some distance before eventually plunging down an embankment. The accident happened, as the *Times* correspondent in Paris wrote, at a location 'well-known to many Englishmen as the scene of some of the fiercest fighting in which British divisions advanced foot by foot to enlarge the salient created by the battle of the Somme in 1916.' The accident occurred only five years later, and 'The villages hereabouts are barely traceable amid the wilderness of shell craters and remnants of woods, but a few hut colonies have been erected, and the main double track, which here ran through the old no-man's-land, was reconstructed soon after the Armistice.' [144]

Paris – Lille express train No. 311 coming from the opposite direction was able to pull up short of the derailed coaches and its passengers and crew assisted in the rescue work. Relief trains were also later sent from Arras and Amiens. The bodies of 25 people were taken from the wreckage and 24 very seriously injured passengers were taken to Arras; two of them died on the journey. Most of the victims, many of whom were women and children, were from Lille and the northern mining area. One of those injured was 33-year-old Conductor Demay of Lille, who had been in charge of the train and had been riding in the rear van. He was not badly hurt. [145] News of the disaster was telephoned to Paris and M. Le Trocquer, the Minister of Public Works, and various officials of the Nord company went to the scene.

By the following day two more of the injured had died, bringing the death toll to 29. The possibility that the derailment had resulted from sabotage was quickly ruled out. It was clear that the rear luggage van had been the first to leave the track as the train was crossing a small concrete bridge over the River Ancre and it was theorised that the rear

---

[141] *Le Petit Parisien*, Paris, France, 16th May 1921
[142] *La Dépêche*, Toulouse, France, 16th May 1921
[143] *Le Petit Journal*, Paris, France, 17th May 1921
[144] *The Times*, London, England, 27th June 1921
[145] Le Grand écho du Nord de la France, Lille, France, 27th June 1921

van had been too light to be attached to an express, which at the time of the derailment was travelling at 100 kilometres an hour. Possible contributory factors were slight track defects, due to the poor condition of the subsoil, and heat distortion of the rails. [146]

Between February 1919 and October 1920 the little known Polish-Soviet War was fought between the Second Polish Republic and Soviet Russia. A ceasefire was followed by the Peace of Riga in March 1921 [147] After the peace treaty had been signed Polish soldiers who had been captured by the Russians were repatriated, and it was a train of such repatriates which ran off the rails near Baranowicze, Poland (now Baranavichy in Belarus) on about Monday 28th August 1921. The carriages were reportedly completely demolished and 13 died and 20 were injured. [148] [149]

The French accidents continued with a disaster on Saturday 10th September 1921. The Chemins de fer de Paris à Lyon et à la Méditerranée (PLM) express which had departed from Strasbourg at 12.20 and was due to arrive at Lyons at 22.55, was wrecked at Les Ėchets, France. The train consisted of an engine and tender, a baggage van, a mail van, a light four-wheeled van, a number of third-class carriages, two first- and second-class composite carriages, a first-class carriage, a restaurant car, and a baggage car bringing up the rear. It was crowded with soldiers on leave from Alsace-Lorraine and the French Army of Occupation on the Rhine.

Between Bourg and Sathonay, the main line, which had been double track before the First World War, was now reduced to a single line with crossing loops at the principal stations, the second track having been removed in 1917 for military use in anticipation of a French advance. The speed of trains through the points at the crossing places was limited to 20 kilometres an hour. The tachometer of the engine indicated a speed of 85 kilometres an hour at the time of the accident; there was no need to look any further for the cause of the derailment.

In defence of the driver it must be said that the situation at Les Ėchets was difficult. Really the connection back onto the single line was unsafe to negotiate at a speed greater than 20 kilometres an hour, but a little further on the line climbed sharply and drivers needed to preserve the speed gained on the descent from Mionnay in order to retain sufficient momentum to ascend the incline. As a result the drivers of almost all the expresses had got into the habit of taking the connection at speeds varying between 40 and 50 kilometres an hour. Previously they had managed to do so without having an accident, though probably with a rough ride for the passengers. On the day of the disaster the driver had approached faster than usual, perhaps because the train was running behind time. He certainly applied the brakes as the application was felt by surviving passengers who made statements to this effect, but it was too little too late. The engine and leading van were diverted onto a siding serving a goods shed, but in a manner difficult to explain, by centrifugal force, after ploughing up the ballast and tearing up sleepers, they almost regained the main line. The couplings behind the front van broke and the carriages

[146] *The Times*, London, England, 28th June 1921
[147] Wikipedia – Polish-Soviet War
[148] *The Times*, London, England, 30th August 1921
[149] *La Lanterne*, Paris, France, 29th August 1921

following were telescoped into one another and some struck the stone goods shed, while the locomotive and van came to a stand 160 metres from the initial point of derailment and right in front of the station.

The leading third-class coach was split in two and its left-hand half became embedded in the wall of the goods shed, while the following third-class carriage, pushed by the remainder of the train, was crushed into the wreckage in front. Another coach broke into two and its front half ran past the goods shed to fall to pieces some 20 metres further on, while the second half joined the general mass of wreckage. The bodies of the front third-class coaches separated from their underframes. The bogie third-class carriages which came next did not, but their front and rear ends were crushed. In the darkness the little vehicles at the rear of the express, which the momentum had carried forward to the locomotive, formed the impression of an intact train standing in front of a heap of wreckage. This impression was so convincing that when the Les Ėchets stationmaster emerged from his office he believed at first that the train had simply broken down, not realising that a major disaster had happened. The cries and groans of the injured soon enlightened him. [150]

The accident was a very serious one. Men women and children in the leading coaches were crushed to death or terribly injured, and even in the rear of the train the passengers were thrown against one another with great violence, and some badly hurt. The Paris correspondent of a London newspaper wrote that, 'Lieutenant Dubois, of the 13th Regiment of *Tirailleurs Indigènes*, who was travelling in one of the rear coaches, says that the train was travelling at 80 kilometres an hour when there was a violent shock, which threw him against the partition of his compartment. The train, however, continued its progress for another 50 yards. He then jumped onto the permanent way and was horrified by the spectacle which the three leading carriages presented. They were completely demolished, and agonizing cries from the wounded buried under the debris filled the air.

'Relief parties were hurriedly organized, and medical aid was brought by train and motor-car from Lyons. The work of rescue continued until daylight, but several of the injured who were pinned under the wreckage died before they could be extricated. A dozen condemned prisoners who were in the train in charge of gendarmes were released from their handcuffs, and they helped in the work of rescue. They first saved two of their guards, who were crushed under the wreckage, and afterwards gave devoted service to the doctors in attending the injured.' [151] The prisoners were soldiers convicted of disciplinary offences who were on their way to the French North African colonies via Marseilles to serve in the *Bataillons d'Infanterie Légère d'Afrique*. [152]

M. Durand, deputy judge, and M. Blanc, interim prosecutor, arrived at the scene at 03:00 and questioned Joseph Coquand, the driver of the express, who was then arrested. Fireman Nicholas was also questioned but stated that he had been fully occupied with his firing. [153]

---

[150] *La Petit Journal*, Paris, France, 12th September 1921
[151] *The Times*, London, England, 12th September 1921
[152] Wikipedia – Battalions of Light Infantry of Africa
[153] *La Petit Journal*, Paris, France, 12th September 1921

By Monday, the number of dead was put at 38, with 60 injured. [154]

The arrest of Driver Coquand brought forth strong protests, especially from Pierre Semard, Secretary General of the Fédération des Cheminots, the railwaymen's union, and after a few days he was released on the orders of M. Le Trocquer, the Minister of Transport, pending his trial on charges of manslaughter and having caused injuries. The usual two investigations were opened, one by M. Maison, of the Ministry of Public Works, and the other by the Trévoux public prosecutor and the matter was also twice raised in parliament.

Driver Coquand was tried in March 1922. The defence lawyers pressed for an acquittal. They failed to get one, but Driver Coquand was only punished by a symbolic suspended fine of 50 francs and the Minister of Transport did not appeal against the sentence. The prosecution had conceded that the case had substantial mitigating factors, for example the dangers of running an intensive train service over the single track had been recognised two years before the disaster and there had been pressure to redouble the line, but the company had pleaded that the necessary rails were not obtainable. The order to begin the redoubling had in fact been finally received only two days before the accident.

It also emerged during the investigations that Driver Coquand had properly reduced speed for the single to double line points at the four previous stations, and that the light of the signal which should have warned him to reduce speed for the connection at Les Ėchets had gone out. Various other defects in the operation of the line were highlighted and it was argued that the responsibility for the disaster must be shared by the higher officials of the company.

In a footnote to the disaster in December 1922 M. Le Trocquer, who was still Minister of Transport, criticised the management of the PLM in parliament.[155]

By the beginning of October 1921 the White Army forces in Russia had been defeated by the Red Army (Bolsheviks) and several train loads of White Army troops in southern Russia commanded by General Pyotr Wrangel were passing through Yugoslavia on their way to the Crimea. On about 1st October one of these trains was involved in an accident, probably a derailment, near Ragusa, Yugoslavia, in which many were killed or injured, according to a report from Bari, Italy. [156]

On Wednesday 12th October 1921, the Badajoz – Madrid – Madrid mail train was derailed on the approach to Villagonzalo station, Spain. A violent storm had caused a stream to become a swollen torrent. The train was only travelling slowly at the time, but by an unlucky chance the third-class coach placed second behind the engine was telescoped into the leading mail van. The coach was crowded with passengers, mostly soldiers of the 75th Segovia regiment who were travelling to Melilla as reinforcements of an expeditionary battalion of their regiment which was fighting in Morocco. (Spain was involved at the time in a war with the Berber tribes of the Rif mountains of Morocco,

---

[154] *The Times*, London, England, 13th September 1921
[155] Wikipedia – Accident ferroviaire des Échets
[156] *The Times*, London, England, 3rd October 1921

which had begun in 1920 and would last until 1927. [157]) The remainder of the train, consisting of another third-class carriage, a second-class carriage, a first-class carriage, and a rear van was undamaged. Eleven of the occupants of the third-class carriage, three of them soldiers, lost their lives, as did a postal worker riding in the mail van. Thirty-one other people were seriously injured and 45 were slightly hurt. [158]

On the evening of Sunday 8th January 1922, a passenger train 'fell down a slope' on the main line between Kosice (Koške) and Oderberg in Czechoslovakia. Five people died and 22 were injured. [159]

Winslow Junction, New Jersey, United States, on the Atlantic City Railroad main line from Camden to Atlantic City, was the scene of a serious derailment on Sunday 2nd July 1922. At this point a branch diverged from the main line to the right towards Cape May via a sharp curve. The junction with the main line was controlled by WA tower and fully signalled. The automatic block system was in operation over the line with home and distant signals of the enclosed disk type showing red, yellow, and green for stop, caution, and proceed, while the semaphore signals at WA tower were merely route signals. Situated 742 feet north of the Winslow Junction points was automatic home signal 25, above which were placed semaphore route signals 48 and 47, while signal 25 was preceded by automatic distant signal 25-X, 3,131 feet from the junction points. When the route was set for the main line the top semaphore arm, signal 48, displayed at night green, semaphore signal 47, below it, displayed red and signals 25 and 25-X were green. When the route was set for the Camp May branch signal 48 was red, signals 47 and 25 below it were green and the distant signal, 25-X, was yellow.

The train involved in the accident was southbound passenger train No. 33, the *Midnight Flyer*, which consisted of a coach, a Pullman parlor car, three coaches and a combination passage and baggage car, in charge of Conductor Nace with Engineer Walter Westcott and Fireman William Sautter. It left Camden at 10.45 p.m., five minutes late; passed Williamstown Junction, 7¾ miles from Winslow Junction, at 11.21 p.m., 12 minutes late, and travelling at about 70 miles an hour was diverted onto the Cape May branch and became derailed on the curve at 11.28 p.m. The engine, tender, and first four cars of the train went down an embankment and the engine, tender and first coach were damaged and the parlor car was very badly damaged. Engineer Westcott, Fireman Sautter, Conductor Nace, three passengers and a Pullman porter were killed, and 84 passengers and five railwaymen were injured.

At about 11.10 p.m. a local freight train for the Cape May branch, No. 491, had finished its work at Winslow Junction and was standing on the Hay's branch, north of signal 25-X. Conductor Archut of this train telephoned WA tower and arranged with Towerman John DeWalt that his train would follow train No. 33, which was due to pass Winslow Junction at 11.17. A southbound train then passed, which the conductor assumed was train No. 33, and he saw that the signals had been set for the Cape May branch, with the distant 25-X at

---

[157] Wikipedia – Rif War

[158] Federación Castellano Manchega de Amigos del Ferrocarril: Accidentes Ferroviarios en España

[159] *The Times*, London, England, 10th January 1922

yellow, but his train did not start. He was about to go forward from the caboose to the engine to find why his train had not started when Flagman Bacon pointed out that another train was approaching. The first train had in fact not been train No. 33 but extra 120, a train of empty coaching stock bound for Atlantic City.

Towerman DeWalt told the Interstate Commerce Commission (ICC) inquiry that having spoken to Conductor Archut he had been busy as train No. 120 had passed but having been informed, he said, that train No. 33 was two minutes late and as train No. 120 had passed at 11.18, one minute late on the passing time of train No. 33, he had assumed that it must be that train. He had replaced signal 48, reset the switch for the Cape May branch and cleared signal 47. He had then reported to the dispatcher that train No. 33 had passed, and the latter had replied, "OK." Then he had been engaged in other work and had not seen the two headlights of train No. 33 until it was passing his tower. [160]

According to a newspaper report Towerman DeWalt had advised his dispatcher, "My God, I have sent No. 33 on the wrong track," and then fainted. [161]

Dispatcher Eagen stated that he had told Towerman DeWalt that train No. 33 was six minutes late, not two. When the towerman had reported No. 33 passing at 11.18 he had realised that it must be extra No. 120 and had told the towerman, "That is extra 120," but he had not pursued the matter further. Copy Operator Keefe in the same office verified what the dispatcher had said.

The ICC found that the accident had been due to the failure of Engineer Westcott to obey signal indications. As he had died in the accident his failure must remain unexplained. The signalling was rather old and there was no approach locking to prevent the towerman from resetting the route after the train had passed the distant signal, but he strongly denied having done so and there was no evidence that he had. He had contributed to the accident only by mistaking extra No. 120 for train No. 33, but this was a mistake which should only have caused delay.

The ICC stated yet again that the accident pointed for the need for automatic train control. [162]

Six died and 18 were injured in a train wreck near Mexico City, Mexico, on Tuesday 30th January 1923, according to a United Press report.

A double-headed passenger train of the Denver and Rio Grande Western Railroad met with an accident on Wednesday 2nd May 1923. Eastbound passenger train No. 2 was in charge of Conductor Turner and consisted of engines No. 779 and 790, a baggage car, two coaches, three Pullman sleeping cars, a dining car and four Pullman sleeping cars, the vehicles being of all-steel construction except the fourth, fifth, seventh and eight cars which had steel underframes. It was derailed at 10.53 p.m. while running at an estimated speed of between 30 and 40 miles an hour. Engineer Joe Westbrook and Fireman A. Anderson (who was killed in the accident) were on the pilot engine and Engineer Fred

[160] Interstate Commerce Commission report

[161] *The Daily Times*, New Philadelphia, Ohio, United States, 3rd July 1922

[162] Interstate Commerce Commission report

Rader (who also lost his life) and Fireman William Gillis were on the train engine. At the moment of the derailment the train was travelling on a low fill (embankment) about three miles beyond Grassy, Utah, United States. In the derailment locomotive No. 779 and its tender went off to the left and came to rest north of the embankment, bottom up, while locomotive No. 790 went off on the opposite side, turned completely over and came to rest on the south side of the embankment. The third and fifth cars of the train came to rest on the south slope of the embankment; the remaining cars remained on the top of the fill and the last four were not derailed. In addition to the two enginemen four passengers and a trespasser were killed and 62 passengers, nine railwaymen and three other persons were injured. A special hospital train took 32 of the worst injured to Salt Lake City. [163]

Engineer Westbrook told the Interstate Commerce Commission (ICC) inquiry that he had made an 8 or 10 pound application of the air brakes to steady the train around the curve approaching Grassy, had allowed the train to drift down the grade past the station, then made another brake application for the first curve beyond the station. He had then released the brakes to allow the train to drift around a second curve when the parting of the train applied them in emergency. He estimated the speed at that point as 30 or 35 miles an hour. He had not noticed any roughness of the track, nor any rocking or rolling of the engine. Coal and water had been taken at Soldier Summit, 65 miles back, and there had been at least half a tank of water and 10 tons of coal left at Grassy.

Fireman Gillies of the train engine said that he had noticed nothing wrong until he was getting on his seat box immediately before the derailment, when he saw the forward end of the engine begin to rise and he noticed a streak of fire. He found Engineer Rader's watch, which had been broken in the accident, stopped at 10.52½.

Travelling Engineer Cowan, who had been riding on the train, and Master Mechanic Perkins, who inspected the engines afterwards, could not explain the derailment. Likewise Roadmaster Quinn, who had ridden over the line in an inspection car on the day of the accident, Section Foreman Demas, who had inspected the track three days earlier, and an un-named track walker, who had been over the line a few hours before the accident, could see nothing wrong with the permanent way.

Post-accident inspection of the track showed two low joints in the south rail about three rails lengths short of the first mark of derailment and two rail lengths apart. These joints showed signs of having worked in the ballast. Otherwise the surface and alignment of the track appeared good, and the gauge was correct.

The ICC investigators surmised from the marks on the track that the rear tender truck of the pilot engine had been the first part of the train to leave the track. This truck had then pulled the front of the train engine against the north rail with sufficient force to break it, causing the general derailment.

As part of the investigation a tender engine of the type involved in the derailment, carrying railroad officials and commission investigators, carrying 8 tons of coal and with 22 inches of water in the tender, was run through Grassy at the speed said to have been attained by engines 779 and 790. It was noted that there was a considerable surging of the water in the tender.

The ICC concluded that the accident had been caused by the rocking of the tender of

---

[163] *The Salmon Herald*, Salmon, Idaho, United States, 9th May 1923

the first engine, due to the lateral surge of the water, the unevenness of the track and the speed at which the train was being operated. It was believed that the train had been running faster that the train staff believed, as the 2.8 miles from Grassy to the point of derailment had been covered in 3½ minutes, giving an average speed of 48 miles an hour, and probably higher when the train was drifting down the 1.1 per cent (1 in 91) grade after the brake application for the Grassy curve. [164]

One of the aspects of the Treaty of Versailles which followed the First World War was the imposition of massive reparations on Germany, mostly to be paid to France, over whose soil the war had been fought. The sum of the reparations decided in 1919 was somewhat reduced in 1921 but even then was huge, and the payments crippled the German economy. By the end of 1922 Germany was regularly defaulting on its reparation payments, and in January 1923 France (and Belgium) began a military occupation of the Ruhr, the industrial heartland of Germany, with the objective of extracting the reparations themselves. The occupation was opposed by both passive resistance and civil disobedience by the German population, and the French occupying troops killed 130 German civilians. The occupation lasted until August 1925. [165] It was against the background of the occupation of the Ruhr that a major disaster apparently occurred at St Goar, Germany, within the occupied area, on Wednesday 9th May 1923. According to German sources a French troop train was derailed and fell into the River Rhine, killing at least 29 people, and possibly the whole complement of the train being drowned. However, following the accident the French authorities sealed off the whole area, using Spahi (North-African colonial) troops and imposed a total news blackout. [166] The French press stated that they knew nothing of the accident, implying that the German reports were false, while mentioning the near escape of a train from crossing a bridge damaged by sabotage near Osterfeld, in the same area, the following day. [167]

Just after 2 p.m. on Monday 25th June 1923, a train of the Brooklyn Manhattan Transit (BMT) company became derailed in Brooklyn, New York, United States, and its two wooden cars fell from the elevated railway structure to the street below. Seven passengers died and about 80 were injured. Six of those who died were women. A newspaper reporter described the scene shortly after the accident:

'Carrying with it a mesh of electric trolley wires, the train hurled to the pavement amid spurts of blue flame and crackling electric wire. As screams of the pinioned passengers, many of them transfixed by jagged sections of broken wood, rose above the crash, flames shot from underneath the cars and drove back pedestrians who ran to the scene.

'The driver of a delivery truck, the radiator of which had been caught under the falling train, rose white and trembling to his feet, burst into tears, and staggered away. Two women, shaking as if with palsy, their faces bleeding from cuts, climbed out of the wreck and walked after him.'

---

[164] Interstate Commerce Commission report
[165] Wikipedia – Occupation of the Ruhr
[166] *Eisenbahn-Katastrophen in Deutschland* by Hans Joachim Ritzau, pg. 73 (in German)
[167] *L'Ouest Éclair*, Caen, France, 13th May 1923

The motorman of the train, Edward Parcell, escaped unhurt. After being questioned for over an hour by Police Inspector Coughlin, he was arrested on a charge of homicide, which given that the cause of the derailment was not at all clear and that there was no indication that it was the motorman's fault, seems a completely unjustified knee-jerk reaction on the part of the police. The motorman said that about 20 feet before his train had reached a trailing switch between the tracks he had felt a sag and a bump. "I was coasting at the time. I put on the emergency brake, and it did not hold. The switch had been green and had given me a clear headway, so I looked back to discover what was the matter. Then suddenly I heard the ties and guard rails splinter, and the car went over the side of the structure in a nose dive. A few feet from the tracks I felt it catch and then both cars went over on their sides. That saved me. They pulled me out of my box."

Another train had passed through the crossover without any problem four minutes earlier and Switchman John Mundone declared that the switch had been green and in good condition when he had examined it at 10 a.m. An unofficial theory was that an axle of one of the cars had broken and had ripped up the outer rail.

Major Hylan, District Attorney Dodd, chairman of the BMT board of directors Gerhard Dahl, and the police all reportedly started inquiries into the accident. The mayor, and the district attorney, having examined the accident site, declared that the derailment had been due to 'faulty construction.' The mayor stated, "I am convinced that the accident was caused by a defective truck leaving the tracks and crashing through a rotten guard rail, which, if good, should have prevented the fall to the pavement. The present guard rails are thirty years old. The cars are from 35 to 40 years old." Mr Dodd declared that the guard rails were nothing but a mass of matchwood and had long outlived their usefulness. [168]

The Interstate Commerce Commission did not investigate subway accidents and it has not been possible to discover the results of any formal investigation to determine the real cause of the derailment.

Towards the end of the first week of July 1923, railway operations on the North Island Main Trunk of New Zealand were disrupted by severe weather with heavy rain which had led to flooding at Frankton Junction, south of Auckland, while on Wednesday, 4th July, a northbound express ran into a landslide at Kakahi, though without serious consequences. Concerns regarding the safety of the line led to a delay in the despatch of the southbound express the following evening, but eventually it left Auckland for Wellington at 11 p.m., four hours late. As a result of the delay some passengers abandoned their journeys. [169] Normally a goods train preceded the express from Te Kuiti to Taumarunui and acted as a sort of pilot which would have detected any obstructions but due to the disorganisation of traffic it did not run that night. By about 6 a.m. on Friday 6th July 1923 the express had reached Ongarue, some 30 miles short of Kakahi, and it was here that disaster struck, when the train ran into another landslide, resulting in the worst railway accident in New Zealand up to that date.

The Ongarue stationmaster said that he heard the train pass through his station, then shortly afterwards a terrific splintering crash. He had hurried down the line to find an

---

[168] *The Philadelphia Inquirer*, Philadelphia, Pennsylvania, United States, 26th June 1923

[169] *Daily Standard*, Brisbane, Queensland, Australia, 7th July 1923

appalling scene, with wreckage piled high over the engine and terrible screams from the injured. [170]

A passenger who was unhurt and assisted with the rescue work gave a detailed account of his experiences to a newspaper reporter:

'Mr A.M.C. Croucher, who did great work in assisting the injured, has arrived at Wellington. He tells a moving story. He is convinced that he owes his life to the interposition of Providence, for just before the smash he had gone to the lavatory to bathe his eyes, which were smarting, and was leaving the lavatory when the crash came. The lights went out, but Mr. Croucher had a pocket torch, which proved invaluable in succouring the disabled. He tried to open the carriage door, but it was jammed, and as there was a burst of flame it looked as if the occupants of the car were doomed to be roasted alive. Fortunately, the flames, which came from the gasometer, quickly died down. A young man broke a window, crawled through, and they kicked the carriage door in. Mr Croucher held the door open, and said, "Come along, women and children first," and about 15 adults and one child passed out of the carriage. When all were out Mr. Croucher tried to open the door into other compartments, but could only open it about 15 in. When he got a glance inside, an agonizing scene was revealed. The iron and woodwork of the carriage was bent and twisted, and inextricably mixed with mangled human bodies. He could see that herculean efforts would have to be put forth before these unfortunate people could be extricated. A man's voice called, "Take this boy," and a little fellow about 7 years was passed out. Mr. Croucher returned to the carriage door and called out, "Is anybody else able to get out," and a voice replied, "No, we are all pinned down." Going outside, Mr. Croucher tried to pull away the broken side of the car, at the same time calling for assistance, but every time he moved a piece of woodwork, one of the victims called out, "Oh, don't move that." By this time others had arrived on the scene, and everybody was calling for the guard, but the people in the rear portion of the train did not know that an accident had occurred. Then the guard, presumably having been unable to get a reply to his signals, came along to see why the train was not going ahead. When told of the accident, he returned to the van for an axe, a saw, and a crowbar. With these rescue operations, were facilitated.

'Seeing that the rescue work was well in hand, Mr. Croucher proceeded to see how the front part of the train had fared. He found the engine, the postal van, and another car had been derailed, and were lying on their sides in the mud. The engine whistle was going full blast, steam was hissing in clouds from, the safety valve, and the door of the firebox had been wrenched off, exposing the glare of the fiery furnace. Mr. Croucher asked if the driver and fireman got out all right, and the driver, who was standing by said, "Yes, but the fireman is badly scalded, and I got an arm scalded when helping him." The engine driver was in a greatly distressed state, but he joined in the rescue work, and though suffering from his scalded arm, worked like a Trojan. Asked if the slip put the engine over, the driver said, "No; we picked up that big boulder and dragged it 50 or 60 yards, ripping up the sleepers and the line." The driver pointed to the rock and estimated the weight at about three tons. The rock, which was covered with clay, was the same colour as the bank, and could not be seen until one was close up to it. It was just round a curve in the line, and right in a cutting. Mr. Croucher returned to where the rescue operations were in progress. By this time, the side of the car had been taken off, and willing helpers

[170] Ibid, 9th July 1923

proceeded to get out. all the injured. Cushions and mattresses from the train were laid along the banks on either side of the train in readiness to enable the doctors to attend to the injured with dispatch. Splendid fortitude was displayed, and scarce a groan was heard as the injured were moved from the wrecked carriages. Mr. Croucher called stretcher bearers to put Fireman Campbell into the relief van. He said to Campbell, "We will try not to hurt you." Campbell replied, "Go on, sir. I'm a British bulldog," and not a sound escaped him as his badly scalded body was lifted from the ground, placed on the stretcher, and borne to the van. The work of carrying the injured required great endurance, as the stretchers had to be taken through the slip knee deep in mud and slush. Everyone kept at it though, and would not give in, notwithstanding that they were nearly ready to drop from fatigue. Mr. Croucher's journey from Auckland occupied 32 hours, and five different trains were employed in making the trip.' [171]

The story of the brave fireman, enduring his sufferings in silence, was rather contradicted further down the same report by the account of Driver Alec Stewart, '"On reaching the top of the rise near the scene of the accident I shut off steam, and the train proceeded slowly down the slope under its own weight. My first glimpse of the obstruction was on rounding the sharp bend, and it was impossible to pull up the train, which struck the rock before I could apply the brakes. I consider the train would have got through the limited debris but for the boulder, which the cowcatcher carried along about 50 yards before the engine was derailed. A violent lurch followed, and the postal and other cars and the engine were flung against the bank. I was thrown violently from side to side of the cab, which was filled with steam, and it was impossible to see. I called my fireman, Campbell, but there was no answer. I groped a way through the steam, and found the fireman jammed between the engine and the bank and screaming with pain. The postal men and I made unsuccessful efforts to reach him, being stopped by the furiously escaping steam. I plunged my arm through the cloud of steam to reach my comrade but was unsuccessful. Eventually Campbell succeeded in crawling out of the cab and into the postal van terribly scalded and in great agony. He implored us to pull off his clothing, and compliance caused awful pain. His screams were heartrending. In addition to the scalding Campbell's shoulder was injured. My injuries were scalded arms and severe bruising, but not serious."'

'One man, who was being extricated from the wreckage, said, "Give her another turn, digger, and I'll be 'jake'." Another turn was given, but as the wood was lifted the man expired. Another extricated passenger, after drinking a cup of tea, ejaculated, "God bless you," and fell back, dead. A returned soldier's chest was crushed, but he declined a drink of brandy, saying, "Give it to others. They need it more. I'm about all in." He expired before he reached the hospital. A pathetic case was that of Ronald Ace, a seven year old Whangarei boy, whose leg was broken, and who was otherwise injured. He was found lying covered with debris. When lifted out he said, "Mummie will be here soon." Quite a number of those killed were either going or returning from holiday. Among them were Misses Gordon and Kelly, both of whom were jammed between seats and scalded, but were rescued by the Maori footballers. All the doors of the women's special car were jammed and had to be prised open. Two occupants of the telescoped carriage were friends. One was killed and the other had both legs broken. An old man terribly injured was found lying on top of his dead son. A woman when she saw that the car was crashing

[171] *The Register*, Adelaide, South Australia, Australia, 9th July 1923

kept off the debris with her feet, and thereby saved the lives of herself and daughter.' [172]

In all 17 passengers died in the accident and 28 passengers, the driver and Fireman Bill Campbell were injured. Guard Hobson was unhurt. Only one woman passenger lost her life. She was Mrs M.W. McDonald, of Auckland, the wife of a tram conductor, who had been going to visit her sister's husband, who was lying at death's door. He died the same day as the accident.

Bad though it was, the accident could have been worse but for several lucky circumstances. For example, the light postal van which was normally placed behind the engine was not available due to the disruption to traffic and a heavier luggage van had been substituted. This had withstood the shock much better than the usual van would have and the three postal officials riding in it were unhurt. Again there could have been a much more serious fire from the escaping gas from ruptured cylinders. A passenger attempted to extinguish the fire which broke out by throwing mud onto the burning wreckage, then his efforts were dwarfed by a further small slip which smothered the flames. Members of the *Scandal Theatrical Company* had a narrow escape. The actors had a reserved car, which was usually placed in front of the train next the postal van, but it was not in the usual position on this occasion. The company had joined the train 60 miles before the scene of the accident. All escaped unhurt. A party of Maori footballers, returning from Australia, had similar good fortune. Their carriage was in the rear. It was supposed to have been marshalled near the front, and the alteration on leaving Auckland caused some murmuring. It is stated that at Frankton Junction, halfway to Ongarue, an attempt to move the carriage, to the front, was made, but the couplings did not match, and the coach remained in the fourth position, next the three wrecked cars, and only a severe jolt was felt by its occupants. There was also the possibility that the train might have fallen down an embankment into the river.

At the opening of the inquiry into the accident on 20th July railway officials gave evidence that every precaution had been taken to prevent a disaster. The ganger in charge of the section of the line concerned stated that he had inspected the line every day and had passed over the spot four times on the previous day, when there had been no sign of danger. The slip was so small that it would not have wrecked the train but for the huge boulder, weighing a ton, falling on the line. [173]

On the second day of the inquiry of the inquiry the district engineer stated that the fact that the door of the engine smoke-box was broken away indicated that the boulder had fallen as the train was passing, while Fireman Campbell, speaking from his hospital bed, confirmed that he had heard a loud report as the boulder struck the smoke-box. [174]

The coroner's inquest concluded on 26th July with a verdict of "Accidental death." The coroner raised the question of the suitability of the engine headlights (although as there had been a curve immediately preceding the site of the landslide a better headlight would not have illuminated the obstruction any sooner). The railway engineer stated that the department had been working on an improvement to the headlights for several years

---

[172] Ibid.

[173] *Northern Star*, Lismore, New South Wales, Australia, 21st July 1923

[174] *Daily Mail*, Brisbane, Queensland, Australia, 23rd July 1923

without making any progress. [175]

The Railway Board inquiry reopened on 7th August. The Chief Traffic Manager, Herbert Burton, stated that the journey between Auckland and Wellington occupied about 18 hours. The maximum speed at different points varied between 25 and 35 miles an hour, at Ongarue it was 35. He stressed that 'safety first' was the general principle of the railway department and safety was not sacrificed for speed. Between 50,000 and 60,000 trains had passed the point of the accident safely. He would not recommend abandoning the heavy, powerful engines used to run lighter trains or think that the present trains were unwieldy.

The Railway Department Chief Mechanical Engineer, Francis McLean, said that he considered suggestions to run wires alongside the lines to indicate slips to be impractical. The line was regularly patrolled and the speed on sharp curves was limited to 25 miles an hour. There had been indication of a tendency for a slip when he had visited Ongarue the day after the disaster. A channel had caused the face of the cutting to be undermined and had precipitated the slip. [176]

Dr Henderson, a mining geologist, gave evidence on the 11th of August, the final day for the taking of evidence. He said that in his opinion no inspection by the ganger could have enabled him to predict a slip. The batter (slope of the cutting) was about one in one. He would not consider the place dangerous. [177]

The board of inquiry, after considering the evidence, concluded, like the coroner at the inquest on the victims, that the accident was unavoidable, and that no railway official was to blame for it. The rescue operation was also praised. The board recommended increasing the number of tools carried on trains to deal with an emergency, an increase in the number of lamps carried on trains, and a special examination of cuttings in the locality with a view to making them more secure. [178] On the question of the rolling stock the board considered that cars should be materially strengthened to withstand the shocks of collisions and prevent telescoping, and that the use of stronger cars on the fast mail trains should be especially considered, with a view to the economics of using steel cars. The chairs at present used easily broke from shocks, and ones with steel frames and more effective fastenings to the floors should be substituted. [179]

On Thursday 23rd August 1923, an express train from Vilna (then in Poland but now Vilnius in Lithuania) to Warsaw was derailed at a location variously stated as near Lida and (probably more precisely) as near Rozhanka, Poland. Both Lida and Rozhanka were at the date of the accident in Poland and are now part of Belarus. (Some newspaper reports gave the starting point of the train as Wien, Austria, [180] but this is clearly incorrect as Lida lies between Vilna and Warsaw and north of Warsaw whereas Wien is south of Warsaw.) A large proportion of the passengers on the train were Jewish emigrants, bound for the United States. (The mass exodus of Jews to North America and elsewhere after

---

[175] *The Sun*, Sydney, New South Wales, Australia, 26th July 1923
[176] *The Register*, Adelaide, South Australia, Australia, 8th August 1923
[177] *The Telegraph*, Brisbane, Australia, 13th August 1923
[178] *The Register*, Adelaide, South Australia, Australia, 31st August 1923
[179] *The Examiner*, Launceston, Tasmania, Australia, 1st September 1923
[180] *Nebraska State Journal*, Lincoln, Nebraska, United States, 24th August 1923

the First World War had its origins in the persecution which they had suffered during the war. 600,000 Jews whose loyalty was suspect were deported by the Czarist government and then after the 1917 October revolution 100,000 Jews were killed in anti-Bolshevik campaigns. [181])

Following the early morning derailment the train crashed into and demolished a dam and the ensuing flood of water swelled the casualty list. The flooding also made the rescue operation much more difficult and a relief train from Lida was delayed by the flooded track. Over 50 people reportedly died in the accident and about 100 were injured. [182]

A despatch received in Riga from Moscow stated that on Saturday 8th September 1923, 82 people had been killed and 152 injured in the derailment of an express train at Omsk, USSR. [183]

A passenger train from Walk to Reval was derailed near Dorpat, Estonia, on the morning of Tuesday 3rd June 1924. Six carriages were reportedly destroyed, and 12 persons died and 36 were injured, six seriously. [184] [185]

After the Polish – Soviet War had ended in October 1920 the new eastern border of Poland remained insecure, with armed bands of saboteurs crossing the border on a daily basis. To deal with the problem, in August 1924 it was decided to create a new militarised border police force. [186] On Tuesday 9th September 1924 an armoured train carrying members of this force, destined for the defence of the Eastern border, was derailed at an unspecified location near the frontier with Lithuania. A lieutenant, a sergeant and seven privates were killed, and six soldiers were injured. [187] Given the circumstances it is possible that the derailment was the result of an act of sabotage, but the brief Reuter report does not say so.

Early on the morning of Friday 12th September 1924 a third-class Tokyo express was derailed when it ran into a landslide near Kaiki, Japan. The landslide, not far from Hiroshima, had been caused by heavy rains related to a typhoon which had swept the south of the country. Nine persons lost their lives and 20 were injured. Telegraph and telephone communication between Osaka and Nagasaki was interrupted by the typhoon but 'no serious damage was done to shipping.' [188] [189]

---

[181] Webpage: My Jewish Learning – World War 1 and the Jews
[182] *The Philadelphia Inquirer*, Philadelphia, Pennsylvania, United States, 24th August 1923
[183] *Nebraska State Journal*, Lincoln, Nebraska, United States, 11th September 1923
[184] *The Times*, London, England, 5th June 1924
[185] *Le Journal*, Paris, France, 5th June 1924
[186] Wikipedia – Border Protection Corps
[187] *The Times*, London, England, 11th September 1924
[188] *The Kalgoorlie Miner*, Kalgoorlie, Western Australia, Australia, 13th September 1924
[189] *The Times*, London, England, 15th September 1924

On Wednesday 8th October 1924, a luxury Paris-Rome express which had left the French capital at 17:00 the previous evening was derailed at Santa Margherita station, Italy. The engine, leading van and a dining car overturned and the driver, fireman, the head conductor and either two or three passengers were killed. Driver Puccioni Emilio and Fireman Conti Marcello were from Pisa and Head Conductor Pignotti Collatino was from Genoa. Professor Dott Padoa of Florence and Lady Wallis, consort of the Minister of Columbia in Rome, lost their lives as they were eating breakfast in the dining car. A third passenger possibly died later of his or her injuries. The derailment was initially attributed to a defective set of points, but an investigation found that the express had been running too fast in an attempt to make up time, and travelling at 80 km/h at a point where the speed was restricted to only 40 km/h. [190] [191]

A passenger train derailed near Magny Saint Médard, France, on the Dijon – Champlitte line on Saturday 17th January 1925. The engine was derailed and overturned, and the leading coach was precipitated down an embankment into a stream. Five people lost their lives; two of them were residents of Magny Saint Médard, two were initially unidentified passengers and the fifth was a mail clerk named Michouillet, who had been working his leave day in place of a colleague. Three passengers were injured. A relief train was sent from Dijon at 23:00 and returned with the passengers and the driver and fireman, who were unhurt.

The derailment occurred at a sharp curve where the line fell to a bridge across the Albanne stream. There was a severe speed restriction around the curve but when the driver attempted to reduce speed the brakes failed to work properly. An initial investigation found that the driver was not at fault but was investigating if the brakes of the engine had been properly checked before it left the depot. [192]

A major disaster in Spain on Thursday 9th April 1925 was the result of a series of errors by the motorman of an electric train. The day was Holy Thursday (Maundy Thursday) and the train for Tarrasa left the station at Los Planas, a popular resort for trippers, at 18:30 crowded with passengers, either 600 or 700 according to different accounts, in a three-coach train designed to only accommodate 300. As the train left Los Planas it reached a point where there was a voltage change in the catenary system, but Motorman Juan Martinez, despite having been with the railway for 12 years, apparently failed to make the change properly. As a result there was a short circuit, and a small fire broke out. By this time, the train had entered a tunnel and the passengers, who were beginning to be affected by smoke from the fire, began to panic as the train came to a stand in the tunnel. They tried to open the carriage doors, while the railwaymen on board attempted to calm them. After some delay the motorman restarted the train, still unaware of the initial problem, and possibly in an effort to regain some of the time lost in the stop, he accelerated the train more than he should have done. The short circuit had left the rear car without power and without its brakes, while the line hereabouts began to fall very steeply at 40‰ (1 in 25).The result was that the train began to run away. Approaching

[190] Webpage: Stazioni Liguria Levante – Märklinfan Club Italia
[191] *Le Rappel*, Paris, France, 9th and 10th October 1924
[192] *La Dépêche de Brest*, Brest, France, 19th January 1925

Sarrià station, just beyond another tunnel, the motorman made, it would seem, a final, fatal error. Alarmed by the increasing speed of the train he applied the brakes violently, and the resulting combination of high speed on a curve and a traction shock derailed the train, the leading coach crashing first against the posts supporting the overhead catenary and then against the wall of an electricity sub-station. Immediately after the derailment the fire broke out again and spread to two of the coaches of the train. Some of the passengers who escaped from the wreckage reportedly made their way into the sub-station and were prevented with the greatest difficulty from smashing the equipment. Worst damaged was car No. 502, the leading third-class coach, while the middle car, No. 501, also third-class, was only slightly less damaged. The rear, first-class car, No. 402, did not suffer so badly. The motorman and 24 passengers lost their lives, 46 passengers were badly hurt enough to require hospital treatment and about others were 70 more less badly hurt. [193] [194] [195]

One of the outcomes of the Treaty of Versailles after the First World War was the creation of the so-called Polish Corridor, a narrow strip of land separating East Prussia from the remainder of Germany and granting Poland access to the Baltic Sea. Running through the corridor was the main railway line between Königsberg in East Prussia and Berlin, then the German capital. Passenger trains passing over this line were worked by Polish train crews and had their carriage doors locked while the passengers were, as it were, in transit through Poland. At 1 a.m. on the morning of Friday 1st May 1925, a train passing through the corridor was derailed between Swaroschin and Preussich Stargard, Poland, with disastrous results.

The train involved was D 4, the same train which was derailed near Schneidemühl, Germany, in an act of sabotage in January 1920 (see Sabotage chapter). The locomotive of the train left the track on a curve just short of Preussich Stargard and fell down a 25ft embankment, followed by the leading luggage van and five coaches. The luggage van was wrecked, the second vehicle of the train, a first-class compartment coach, was telescoped into the following coach, and it was in these two coaches that most of the fatalities occurred, though two other coaches were badly damaged. The sleeping car and first-class compartment coach at the rear of the train kept the rails and were undamaged.

Twenty-five bodies were recovered from the wreckage and at least 20 were seriously injured, one of whom died shortly afterwards. Apart from a Polish customs officer all the dead were Germans, including a trade union secretary from Danzig and a two-year-old child. Two women, one of them a train attendant, lost both legs. Fortunately, there was no fire.

The cause of the derailment was shrouded in controversy. Early official Polish reports alleged that it was the result of a Communist outrage. However only 20 minutes before the derailment a train from Marienburg to Schneidemühl had passed the spot safely, and it was believed that there would have been insufficient time for saboteurs to have loosened a rail in the interval. Another possibility was that recent heavy rain had caused

---

[193] *La Vanguardia*, Barcelona, Spain, 11th April 1925

[194] Federación Castellano Manchega de Amigos del Ferrocarril: Accidentes Ferroviarios en España

[195] *The Times*, London, England, 11th and 13th April 1925

the embankment to subside. German engineers declared that the derailment had resulted from poor maintenance of the track by the Polish administration. [196]

A week after the accident the death toll had risen to 29. Meanwhile the controversy over the cause of the accident continued. The Polish authorities stated that fishplates had been removed from a section of the line and the rails had been shifted sideways three inches with the aid of a cart jack found concealed nearby under grass and twigs, causing the derailment of the train moving at 80 kilometres an hour. A reward of 50,000 zlotys was offered for the discovery of the perpetrators. For its part, the German government continued to allege that the track had fallen into a state of disrepair and that rotten sleepers had caused the accident, adding that a bridge across the River Vistula at Dirschau damaged by floods the previous year had never been repaired. [197]

In an attempt to settle the matter an inquiry was launched into the accident, but its verdict was hardly satisfactory. It ruled that the derailment had not been due to subsidence, while the burden of proving that it had resulted from an act of sabotage rested with the Poles. Possibly high speed over track not in the best condition had caused the train to leave the rails.

A serious accident was reported on the railway between Sofia and Kustendil at Gorna Banya, Bulgaria, on about Tuesday 26th May 1925. About ten people lost their lives and 20 were injured in the derailment of a passenger train, possibly because a set of points was moved under a train, as a signalman was blamed for it. The Sofia correspondent to *The Times* in London, remarked that, 'It is impossible to say whether any political significance should be attached to the accident, but the pointsman has been arrested.' [198]

In the early hours of Tuesday 16th June 1925, a major disaster happened at Rockport Crossing, rather more than three miles west of Hackettstown, New Jersey, United States, on the line of the Delaware Lackawanna and Western Railroad. Early newspaper reports stated, wrongly, that a rail had been struck and shattered by a stroke of lightning an hour before the accident. [199] The train involved was a special carrying native Germans who had emigrated to the United States some years earlier and had prospered in the corn belt. They were on a trip back to their homeland, together with their children and grandchildren, on a special from Chicago to Hoboken, New Jersey, from where they were to cross the Atlantic to Bremen, Germany, on the S.S. *Republic*. The Germans were travelling as a group under a special-rate plan known as a 'Neuman's excursion'. Most had reserved cabins on the ship, but some were travelling as steerage passengers; [200] more cheaply but travelling in cramped conditions with little privacy, scant security for their belongings and poor sanitary conditions and food.

Eastbound passenger extra No. 1104, as the train was properly known, was in charge of Conductor Banker with Engineer Fred Loomis and Fireman Carl Haern. [201] It consisted

---

[196] *The Times*, 2nd and 4th May, 1925
[197] Ibid, 6th and 8th May 1925
[198] *The Times*, London, England, 28th May 1925
[199] *The Kingston Daily Freeman*, Kingston, New York, United States, 16th June 1925
[200] Ibid
[201] *The Commercial Tribune*, Cincinnati, Ohio, United States, 17th June 1925

of locomotive No. 1104, two coaches and five Pullman sleeping cars, all of all-steel construction. It passed Washington station at 2.17 a.m. and became derailed at 2.25 a.m. while travelling at about 50 miles an hour. The engine and first five coaches of the train left the track, and the two leading coaches were very badly damaged. First reports put the number of dead at 12, later it was said to have risen to 34 and the final figure was 45 passengers, a Pullman porter and four railwaymen, with another 23 injured. The driver and fireman were among the dead.

Despite their ordeal 125 of the survivors continued their journey to Hoboken to catch their ship on another train. One of the survivors, who was also active in the rescue work, was interviewed by a reporter. 'To Mrs E.B. Haaker, of Park Ridge, Ill, a trained nurse who was on the train, many owe their lives. With her first-aid kit, Mrs Haaker set up an impromptu hospital in the observation car.

'Mrs Haaker was travelling with her husband and a friend, Frederick Oberheld, of Chicago. All three were awakened by the crash and by the screams of members of the train crew who ran through the rear cars with terribly scalded hands, screaming, "For God's sake, somebody help me." Still in her night clothing, Mrs Haaker dressed the man's hands while her husband and Oberheld hurried ahead, improvised stretchers, and brought many of the injured to the observation car. Sheets were torn up for use as bandages and the emergency hospital force worked for hours to relieve the agony of the scalded.

'"I could hear, it seemed, a hundred persons screaming," said Mrs Haaker, describing the crash, "and, through their voices, the roar of escaping steam. When we came near the engine the sight was overwhelming. The first two day coaches were telescoped and were ground together in {a} mass of wood and steel, and some passengers in the first coach had been thrown bodily into the tender. The third coach was lying squarely on top of the engine. The engine's steam pipes had broken and a fog of steam was pouring into the car perched on top. We could see hands waving and then falling back into that car. Those in the tender who were unable to get out because of the car on top of them were lying close against the coal in the hope of avoiding the blistering steam. We could do very little until the flow of steam slackened. When we pulled them out those who were still conscious screamed with the pain. I saw at least 12 dead, and at least 25 of those treated were so badly scalded that they will probably die."

'Mrs Anniee Jollie, 55, of Chicago, told how she was hurled from her berth amid a shower of glass. "I saw a poor woman running towards me distractedly," said Mrs Jollie. "Her hand, held high, was nearly severed just above the wrist. "Get the scissors and cut this off quickly," she gasped. Instead I tore my skirt and with a piece of wreckage made a tourniquet, stopped the flow of blood, and led her to a country doctor who was doing fine service. "Farm people scurried about with food and country doctors came. The news seemed to spread like magic. I saw woman dig a child out of the wreckage and I helped farm boys dig out a woman with bare hands."'[202]

The Interstate Commerce Commission (ICC) report into the disaster outlined the real cause of it. Lightning had not broken a rail, as the initial newspaper reports had suggested, but rather the cloudburst which had accompanied the thunderstorm had washed a quantity of dirt and small stones onto the track. The obstruction had been

---

[202] *The Morning Herald*, Hagerstown, Maryland, United States, 17th June 1925

sufficient to throw the leading pair of engine truck (bogie) wheels off the line, and in this condition the engine had continued for 198 feet until it had come into contact with the rail of a crossover frog and the whole engine had left the track, taking most of the remainder of the train with it.

Flagman Judge told the ICC investigation that it had been raining very hard at Binghampton, 131 miles from Hackettstown, but the rain had stopped at Milford, 30 miles further on, and there was no further precipitation, though at Washington he had seen lightning in the distance. He had noticed no brake application before the derailment.

Mr Joseph Snyder, who lived near the scene of the accident, said that there had been a severe thunderstorm and a cloudburst about 20 minutes before the accident. A lightning strike had set fire to a lumber yard at Rockport, and he had stayed up for a while to watch the blaze. As he was about to go back to sleep the special train arrived and he had witnessed the derailment. Afterwards he had found five or six inches of dirt over the rails at the crossing.

Section Foreman Smith said that he had reached the accident scene at about 6 a.m. and had cleared dirt and rocks from the track, the soil three or four inches deep on the eastbound track and less on the westbound. Where he lived at Port Murray, three miles distant, there had been considerable rain, heaviest at he thought about 1 a.m., but he had not considered it severe enough to warrant an urgent examination of the track. He had passed the spot with a track walker at about 4 o'clock the previous afternoon, when all had been in order.

Roadmaster Sexton stated that he had not been aware of any heavy rain at Port Murray, but he agreed that there were clear signs of very heavy rain at Rockport. He said that he had been in the area for about 25 years and that this was the first time that he had experienced any trouble at Rockport. He added that a similar level crossing only 600 feet further west had been unaffected by the heavy rain that night.

Principal Assistant Engineer Neafe said that the obstruction had been cleared by the time that he had reached the scene, but he was of the opinion that the material had come from the highway itself rather than from the ditches paralleling the highway. He considered the track immediately west of the crossing to be perfectly safe for a speed of 70 miles an hour.

Master Mechanic Root stated that there was no indication of a boiler explosion but that practically all the appurtenances on the top of the boiler must have been torn off, permitting the steam to immediately escape. He could find nothing wrong with the engine which might have caused the accident.

Operator Linaberry at Washington said that the headlight of engine No. 1104 had been burning properly. It had begun raining there at 11 p.m. and was still raining when extra 1104 passed, but not sufficiently for him to issue a slow order.

The ICC concluded that unusually heavy rainfall at Rockport had washed earth and rocks from the highway onto the crossing, covering the rails and filling the flangeways between the crossing planks and the rails such as to derail the forward pair of engine truck wheels. There had been no problem at the crossing in the previous 25 years and no one on the railroad was aware that there had been such a severe local storm such as to endanger the safe passage of trains. The large number of fatalities had been due to the chance that two cars of the train had come to rest upon or against the damaged engine boiler. [203] Altogether a very unlucky accident.

Nine persons reportedly died and 32 were injured when a Chita – Moscow express was derailed on the Trans-Siberian Railway at Taiga (now Tayga), Russia, about 100 miles east of Novonikolayevsk (renamed Novosibirsk 1926). The accident occurred on Monday 29th June 1925. [204]

Express train No. 3689 of the Chemin de fer de Paris à Orléans, which left Le Mans at 22:25 for Tours, was derailed just before midnight on Wednesday 29th July 1925 at Saint-Antoine-de-Rocher, France, 20 kilometres short of its destination. An early report stated that the train had left the rails at the points where the line went from single to double track and appeared to be due to excessive speed. A reporter wrote that approaching the scene of the accident the engine of the train could be seen lying on its side, and the body of the driver, who had tried to jump at the last moment, but had been crushed between the engine and its tender. The engine went down an embankment and the sudden stop telescoped the following four carriages and two vans. By the time of the report, in addition to Driver Emile Gorse it was known that 18 passengers had lost their lives. Fireman Eugène Bodard, Conductor Henry Bonneau and 17 passengers were identified as having been injured.

After the accident, the uninjured passengers and railwaymen began the rescue work, assisted by local residents, while the single member of staff on duty at the little station at that hour ensured that the obstruction was protected and telephoned to Tours for assistance. A relief train was soon sent to the scene, along with a 50-ton breakdown crane. M. Grimaud, the prefect of Indre et Loire, and the state prosecutor were soon on the scene and were followed by M. Pierre Laval, the Minister of Public Works, and other officials.

The investigation into the accident noted that the single line approached Saint-Antoine on a gently falling gradient and described a curve with a radius of 800 metres. It was on this curve that the line became double track, the new line diverging on a curve with a radius of 250 metres. It was therefore evident that trains must approach this spot at moderate speed. Trains normally took the connection at 30 kilometres an hour. Train No. 3689 was travelling at 93 kilometres an hour according to its tachometer, as was also evidenced by the fact that the track had been shifted from left to right by five or six centimetres. It appeared that the wheel of a coach had mounted a rail at the points, resulting in the initial derailment. Almost immediately there had been a second derailment on the new curve and this time the engine had left the rails, had continued along the line for 100 metres, carried by its own momentum, and had then gone down the embankment and the following vehicles had piled up against it. Only the last carriage, a Paris à Lyon et à la Méditerranée vehicle, and the rear van, had remained on the line. [205]

Another French newspaper, describing the circumstances of the accident, pointed out that there was a signal de ralentissement at the previous station, Neuilly Pont Pierre, to

[203] Interstate Commerce Commission report
[204] *The Commercial Tribune*, Cincinnati, Ohio, United States, 1st July 1925
[205] *Le Figaro*, Paris, France, 31st July 1925

warn drivers that they must reduce speed for the transition to double track. [206] The accident seemed therefore to be a simple case of a driver failing to reduce speed despite ample warning, for reasons which must remain a mystery as he was no longer alive to explain himself. The investigation, however, found that other factors had played a role. Newspaper correspondents who had visited the scene had commented on the poor state of the permanent way, as well as upon the faulty composition of the train. [207] The investigators agreed, finding that due to a lack of ballast the sleepers were simply resting on the ground, making the whole track unstable. As for the make-up of the train, the rear vehicle was three times heavier than those in front of it, so that when the engine stopped suddenly after the derailment the rear vehicle had crushed those preceding it. [208]

The Times correspondent in Paris wrote that, 'Four accidents have taken place at the same spot within the last three years. The catastrophe adds another bad mark to the deplorable record of the Orleans Railway, which already has ten or more accidents to its discredit since the beginning of this year.' [209]

A day or so after the disaster M. Jean Montigny, deputy for the Sarthe department, wrote to the Minister of Public Works stating that he had noticed rotten sleepers in the track a few kilometres from Saint-Antoine-du-Rocher and making various other complaints regarding the railway company. He called upon the government to prosecute the company directors for culpable homicide and to consider whether the state could not take over the railway. [210]

Barely a fortnight after the Saint-Antoine-du-Rocher disaster another French train was derailed on a set of points, again apparently due to excessive speed and again with disastrous consequences. This time a Paris – Boulogne express of the Chemins de fer du Nord was involved. On Thursday 13th August 1925 express No. 313 left Paris Nord at 14:28; it came to grief at the entrance to Amiens station just under two hours later at 16:24. The train reportedly consisted mostly of old third-class carriages, usually reserved for racecourse traffic, and it was unusually full, most of the passengers being holidaymakers bound for the seaside. It was a relief train to the 'less democratic' main express, which probably did not cater for third-class passengers.

The train was due to stop at Amiens but approached the station at high speed and the wheels of the engine apparently left the rails at the points. The tender was then derailed, as were 11 of the 14 vehicles of the train. 'The guard's van behind the engine was reduced to splinters. The first passenger coach toppled over, and the third coach telescoped into the second, while the fifth coach was piled upon the coach in front, badly damaged by the following vehicle.

'Out of the mass of smashed timber and tangled ironwork could be heard the cries of the injured. The horror of the scene was increased when the fourth, fifth and sixth coaches caught fire. First aid parties were quickly organized from the staff and people in the station and the fire brigade did splendid work in putting out the fire caused by the

---

206 *La Petit Parisien*, Paris, France, 31st July 1925
207 *The Times*, London, England, 31st July 1925
208 *La Petit Parisien*, Paris, France, 31st July 1925
209 *The Times*, London, England, 31st July 1925
210 Ibid, 3rd August 1925

explosion of the gas cylinder.

'A number of mangled bodies were found in the debris of the guard's van and another carriage and were brought out amidst great excitement. There was a painful scene when the body of a young man was being carried away. A woman who was his wife made frantic efforts to reach the body and had to be led away in a fainting condition. The bodies of the dead were removed to the mortuary, and motor cars were requisitioned to convey the injured to hospital.' [211]

The number of dead was initially thought to be eight, with over 60 injured. In contrast to the Saint-Antoine accident Driver Alfred Verhaeghe survived, as did Fireman Taquet, but Conductor Henri Vitaz was one of those who lost their lives.

Considerable concern was reportedly felt in Paris regarding the safety of M. Aristide Briand, a socialist politician who had several times been Prime Minister and at the time of the accident had been returning from a trip to England. The boat train on which he was returning from Boulogne was not involved in the accident but was still on the Boulogne side of Amiens station and hence was considerably delayed by it.

M. Pierre Laval, the Minister of Public Works, arrived at the accident site during the evening and visited the injured in hospital. He arrived back at Paris Nord station 20 minutes after midnight, when he made a statement regarding the accident, describing the scene and the rescue work. Asked as to the cause of the accident, he stated it was too early to say for certain, but it was clear that the train had been travelling at too high a speed whilst a piece of broken rail had been found embedded in the locomotive. He added that Driver Verhaeghe was one of the best Nord drivers. After the accident, the driver had been found wandering about the line in a dazed condition. Asked if he thought the driver had lost his mind M. Laval replied that no one was sure and added that he had asked the investigating judge, who had arrested the driver, to release him, at least for the present, and the driver had been taken to an Amiens hospital and placed under observation. [212]

Some of the witnesses alleged that the speed of the train had risen to as high as 128 kilometres an hour on the journey from Paris and the tachometer indicated that it had still been going at 93 kilometres an hour at the moment of derailment.

The following morning Driver Verhaeghe had recovered somewhat and made a statement to the investigating judge, in which he maintained that he had been running under clear signals and had been justified in approaching Amiens station at the speed that he had. He said that the brakes of the train had worked normally through the journey but admitted that they had failed to "bite" when he had applied them just before the derailment. Meanwhile a provisional report from the Ministry of Public Works attributed the derailment to excessive speed, stating that the train had been running at twice the regulation speed of 40 kilometres an hour as it entered the station. [213]

By the following day, two of the injured had died and the day after, the death of M. Phillipe Macquet increased the number of those who had lost their lives to 12,[214] while

---

[211] Ibid, 14th August 1925

[212] *Le Petit Parisien*, Paris, France, 14th August 1925

[213] *The Times*, London, England, 15th August 1925

[214] *Le Figaro*, Paris, France, 16th August 1925

the number of injured had risen to 160. The final death toll was 14.

Also on Friday 14th the Nord company suffered another bad accident, this time a collision at St Denis, just outside of Paris, in which four people died, due to a train overrunning signals, while there was a less serious derailment at Pontoise, due to a signalman shifting the points beneath a train. *The Times* Paris correspondent wrote that, 'The frequent occurrence of such disasters is beginning to disturb even the French public, which is so much accustomed to railway accidents that it usually displays towards them an indifference which would be inconceivable in England. But even a Latin race has been shaken out of this indifference by a realization of the appalling death-roll of the last six weeks. Since July 1, 35 people have been killed, bringing the total for the year so far to up to 43. Thus in seven and a half months more persons have been killed on the railways than during the whole of 1924 (39) or 1923 (27). There have been 13 railway accidents since July 28.' On the same day it was reported that there were suggestions that Driver Verhaeghe was epileptic, though there was no mention of this in his medical records. The driver continued to insist that he had not been going too fast and had braked normally. [215] A medical examination of the driver was then carried out and suggested that he was suffering from some disease of the central nervous system, possibly meningitis. [216]

Driver Verhaeghe appeared in court at Amiens on 22nd December.[217] Despite the efforts of his union the court found him guilty of causing the disaster and on 30th December he was sentenced to a year's imprisonment and a fine of 100 francs. [218]

A passenger train derailed on the Pleskau line, about 160 kilometres from Leningrad, USSR, on about Thursday 15th October 1925. Six people were killed and 15 injured, six seriously. [219]

Seven people died and 40 were injured when a train 'turned over' on the Longitudinal railway at Vallenar, Chile, on Wednesday 23rd December 1925. 'Details of the accident are lacking.' [220]

A passenger train on the Rede Sul Mineira was derailed on 27th March 1926 in Minas Geraes state, Brazil, near Pedro Grande. An engine and a second-class coach rolled down a 200 ft embankment. Five people reportedly lost their lives and eight were injured. [221] [222]

On Wednesday 12th May 1926, a goods train travelling from Peking (now Beijing) to Hankow, China, was derailed at an unstated location. More than 50 people who had been riding on the tops of the cars were killed and a large number were injured. 'The wreck

---

215 *The Times*, London, England, 17th August 1925
216 Ibid, 19th August 1925
217 *Le Petit Journal*, Paris, France, 23rd December 1925
218 *Journal des Débats* Politiques et Littéraires 31st December 1925
219 *Journal de Lausanne*, Lausanne, Switzerland, 16th October 1925
220 *Daily Lake Region*, Eustis, Florida, United States, 24th December 1925
221 *Winnipeg Free Press*, Winnipeg, Manitoba, Canada, 29th March 1926
222 *L'Humanité*, Paris, France, 29th March 1926

held up an express on which the Extra-Territoriality Commissioners were travelling to Hankow until a special was dispatched.' [223] (The commission referred to was probably an American one which had been attempting to carry out an investigation since the beginning of April but had had problems due to the lack of transport available. [224])

Initially it was reported that 15 persons had died, and many had been injured in the derailment of a passenger train from Pachuca to Ixmiquilpan, Mexico, near the latter place, on Sunday 16th May 1926. It was said that the train fell into a canyon and that almost every coach was wrecked. [225] A follow-up report stated that the train had been carrying troops and that ten men had been killed and 30 injured. A military relief train had been sent to the scene. [226]

South Africa suffered a major disaster on Wednesday 9th June 1926, at Salt River, a suburb of Cape Town. Salt River was an important junction and the train involved was a suburban one which was probably well filled, as these trains usually were packed for the various stations as far as the naval port at Simonstown. During the course of the accident the train became divided in two and the accident was made much more severe because the three rear coaches struck the stanchions of a road overbridge. Sixteen persons died, including Sir Malcolm Searle (Judge President of the Cape Provincial Division of the Supreme Court of South Africa), and there were 211 injured. The victims were terribly mangled, arms, legs and feet being picked up in the debris. [227]

The derailment was attributed at the time to the fact that the train had become separated, but the mere division of a train into two should not, of course, have been enough in itself to cause it to leave the tracks. A commission of inquiry investigated the accident. It commended the condition of the permanent way and the general operation of traffic and found that the speed of the train involved in the accident had been well within the regulation limits. The derailment, it found, had occurred not merely because the train had broken in two but because a coupling had fallen from a coach and lodged in a set of points. It was impossible to determine who was directly responsible for the coupling having dropped from the train, but it was considered that the guard should have become aware of the problem prior to the accident. The commission recommended the abolition of screw couplings and other auxiliary couplings on suburban trains. [228]

France suffered yet another major disaster on 3rd July 1926, this time on the Chemins de fer de l'État, near Achères. The accident happened on the *grande ligne* – literally the big line – from Le Havre to Paris St Lazare, about a kilometre on the Le Havre side of Achères station at a junction known as des Ambassadeurs. *Rapide* (express) No. 158, which had departed from Le Havre at 17:00, was due in Paris at 20:08, and became

---

[223] *Maryborough Chronicle*, Wide Bay and Burnett Advertiser, Queensland, Australia, 15th May 1926

[224] www.history.state.gov – The American Commission on Extraterritoriality in China

[225] *The Commercial Tribune*, Cincinnati, Ohio, United States, 17th May 1925

[226] *The Salt Lake Tribune*, Salt Lake City, Utah, United States, 18th May 1926

[227] *The Advertiser*, Adelaide, South Australia, Australia, 11th June 1926

[228] *The Advocate*, Burnie, Tasmania, Australia, 12th July 1926

derailed at 19:50 about 23 kilometres short of its destination. On the engine were Driver Sondach and Fireman Cheminade, accompanied by *Chef Méchanicien* – literally Chief Driver, probably a travelling inspector – Mausel. Unknown to engine crew the main line through Achères had been closed at 17:00 for emergency repairs to a set of points and the signalman at des Ambassadeurs had set the junction there to divert the express to a branch, the *voie de ceinture*. A kilometre in the rear of the junction, the diversion was indicated to the driver of an approaching train by a signal known as a *rond vert,* which indicated that speed must be reduced to 30 kilometres an hour at the junction points, but as at Saint-Antoine-de-Rocher and Amiens the driver failed to reduce speed and arrived at des Ambassadeurs still travelling at 90 kilometres an hour, as if his train was continuing along the main line. [229] Inevitably, at the speed at which it was running, the train was derailed. The Pacific class engine, after leaving the track, ploughed along the ground for some distance and then overturned. The leading luggage van and two second-class coaches following were dragged off the line and telescoped one another while the fourth and fifth vehicles in the train passed alongside of those in front and some way beyond the engine before they overturned and stopped. The rear vehicles in the train suffered a severe shock but did not overturn. It was initially stated that 18 people had died and 73 injured, five of them being in a critical condition. M. Mausel and Conductor Menoux were amongst the fatalities, while Driver Sondach was seriously injured. A violent thunderstorm was raging at the time of the accident, with torrential rain.

As soon as news of the disaster reached Achères a relief train was sent from there while two senior officials of the state company who chanced to be on the train and had escaped injury went on to St Lazaire station, presumably by road, and two more relief trains were sent from Paris. Many of the injured were taken to hospital at Saint-Germain-en-Laye, others were taken to Paris by train and to the Breaujon Hospital. [230] Fifty-ton breakdown cranes went sent from Mantes and Batignolles to clear the wreckage.

The investigation into the disaster was hampered by the fact that M. Mausel had lost his life and Fireman Cheminade, who miraculously escaped unhurt, stated that he had been firing and knew nothing as to cause of the accident. Driver Sondach, who was seriously injured, testified from his bed that neither he nor M. Mausel had been able to see the signals through the torrential rain, but he added that moments before the derailment M. Mausel had assured him that everything was all right. The driver and fireman of a Cherbourg train which had been diverted in the same way a few minutes previously testified that they had been able to see the signals and they had reduced speed accordingly. Possibly the storm had been less intense a little earlier. Much was made of the fact that the driver had not been told of the diversion when the train had stopped at Rouen, but that of course was what the signals were for. It was proved that the signals themselves had been working properly and had not been affected, as had been suggested, by the electrical storm. [231] [232]

The final number of dead in the accident was probably 20.

It does not appear that any charges were ever brought against Driver Sondach, probably because he was seriously injured (his left arm was amputated), the severe weather

---

[229] *Le Petit Journal*, Paris, France, 5th July 1926
[230] *The Times*, London, England, 5th July 1925
[231] *The Chicago Tribune*, Chicago, United States, 6th July 1926
[232] *Le Ouest Éclair*, Caen, France, 6th July 1926

conditions at the time of the accident, and, perhaps most importantly, the fact that he shared the responsibility for it with the deceased M. Mausel, who had been his superior.

A very brief message from the Central News Agency in Bucharest spoke of many people being killed and injured in a train wreck at Ploesti, Romania, on about Thursday 15th July 1926. [233]

Failure to protect work being carried out on a set of points led to a bad derailment at Langenbach, Germany, on the line between München and Landshut, at 9.28 a.m. on Friday 13th August 1926. Twelve lives were lost and 27 were injured. Because of the maintenance work a speed restriction of 45 miles an hour was in force through the points as opposed to the normal maximum permissible speed of 80 kilometres an hour. BP (accelerated passenger) train 858 from Regensburg to München normally ran through Langenbach without stopping. The foreman in charge of the maintenance work failed to carry out any protection with the result that train 858 approached at normal speed, the clamp temporarily securing the points gave way, and the points opened beneath the train. [234] [235]

The mail train which had left Barcelona for Valencia at 21:35 the previous evening was derailed between Ametlia and Ampolla, Spain, at 2 a.m. on Wednesday 1st September 1926.

The train was an hour behind time and running at high speed when the engine ran into a landslide and derailed. Except for the leading van and a sleeping car placed second in the train, the coaches of the train then crashed into one another. A reporter said that it was impossible to describe the horrors of the scene.

On receipt of news of the disaster in Tarrogona a relief train with Dr Luis Soler and staff was despatched to the scene followed by a second train at 06:00. Another train was sent from Tortosa with Mayor Joaquin Bau, the municipal police chief and more doctors and a third relief train was sent from Tarragona later. An early report stated that 18 persons had died, five were critically injured, 35 were seriously hurt and 90 slightly so. The more badly hurt were taken in one of the relief trains to Tortoga and lodged in one of several hospitals, while another train took the 90 slightly injured passengers to Valencia. The only railwayman to lose his life was a ticket collector who was crushed in the van in which he was riding. He had been making his first trip after having been off on a month's leave. The remainder of the train crew, the postal staff and the civil guards were unhurt.

A seriously injured woman with a baby in her arms tried to take a few steps towards the station but collapsed in a dying state. The body of another woman was found hanging out of a coach window, supposedly killed as she tried to jump from the train.

The passengers on the train who should have reached Valencia at 05:00 arrived there at 3 in the afternoon.

---

[233] *East Liverpool Review-Tribune*, East Liverpool, Ohio, United States, 16th July 1926

[234] Eisenbahn-Katastrophen in Deutschland by Hans Joachim Ritzau, pg. 40 (in German)

[235] Wikipedia – Eisenbahnunfall von Langenbach

The ground which had given way was said to have become very unstable following heavy rain and the vibration set up when the train passed at speed was sufficient to precipitate the slip. [236]

The day after the accident Driver Francisco Fenoyosa was questioned regarding the disaster. He was physically unhurt but clearly greatly affected by his experiences. He said that he had taken charge of the engine at 00:20 at Tarragona. It was a very dark night and there was a tremendous storm raging with torrential rain and although his train was already behind time he was driving very cautiously despite this meaning further delay. He added that at the point where the accident had occurred the schedule demanded a speed of between 45 and 50 miles an hour but because of the bad weather he had gone at a lower speed. All had gone well until they reached kilometre 219, between Ametlia and Ampolla stations, when he and his fireman had realised that a mass of earth and rock had fallen from the right-hand side and was blocking the line. He had made a full brake application. The engine had jumped, run 50 metres off the road and then embedded itself in the bank on the left-hand side of the track. The third and following coaches of the train were telescoped into one another. The coaches of the train remained on the track. The fireman was thrown to the side of the tender by the violence of the shock, and he had found himself on the ground, though he did not know how he got there. It was an experience that he would never forget, he said.

After the accident, the police escort of the train had made its way to Amerlia, from where help could be summoned from Tortosa. Meanwhile the rain was still falling in torrents. By the scream of the victims the driver and his fireman had realised the extent of the disaster and that there must be many victims. By the faint light of a lantern they could make out the sad scene. The saddest thing that he saw was when they discovered the body of a woman clutching her young child. The little creature, which was apparently unharmed, was crying endlessly for its mother, and when they approached the child made a tremendous effort to free itself from the mass of wreckage. When they looked again it was dead. They were somewhat relieved when at 5 a.m., before daybreak, the first relief train arrived from Tortosa with doctors and rescuers. By dawn the last of the bodies of those who had died had been recovered. [237]

There was confusion at first over the exact number of deaths following the derailment of a Denver and Rio Grande Western Railroad express train on Sunday 5th September 1926, but it was believed that between 23 and 26 people had lost their lives, so it was clearly a major disaster. The train had come off the rails on a curve nine miles below Granite, half a mile west of Waco, Colorado, United States. Its Mogul engine and first five coaches plunged down an embankment into the Arkansas River. After the accident, the crews of two nearby construction camps hurried to the scene, rescued the injured and retrieved the bodies of the dead from the cars of the train, which were partially submerged in the water, and administered first-aid.

'By the time a relief train from Salida, carrying doctors and nurses, arrived, cries from the wounded rent the air along the banks of the river. Half-dazed survivors huddled about camp fires in an effort to thaw out after being precipitated into the stream. Motor tourists

[236] *La Vanguardia*, Barcelona, Spain, 2nd September 1926
[237] Ibid, 3rd September 1926

passing the scene of the wreck also rendered what assistance they could before the arrival of the relief train, which brought many of the dead and 41 injured to this city {Salida, Colorado}. Pathetic scenes were enacted as mothers, husbands and wives searched for loved ones in the wreckage. Other survivors, apparently dazed, ran about unable to tell their names for several hours. Many of the dead were imprisoned in the wreckage of coaches while rescue workers, equipped with acetylene torches, brought many wounded from their imprisonment within the twisted steel coaches. According to the Rio Grande Western officials, the train was on time and was proceeding at its normal speed or slightly less when the accident occurred {but see below}. The giant engine that pulled a score to death with it in its plunge was returning from its maiden trip over the route after a successful run two days before to Salt Lake City from Denver. It was one of the largest locomotives in use on any American railroad with a length of 106 feet and equipped with the most modern of engineering devices.' [238]

The Interstate Commerce Commission (ICC) report on the accident stated that the train was eastbound passenger No. 2, consisting of locomotive No. 1604, a baggage car, a smoking car, a coach, four Pullman cars, two dining cars, and another five Pullman cars. It was in charge of Conductor Cope, with Engineer Harpending and Fireman Willingham. Also riding on the engine was Road Foreman of Equipment G.M. Lillis. The engineer and the road foreman were both killed in the derailment, together, according to the ICC report, with 27 passengers and an off duty railroad employee, while 45 passengers, three railroad employees on duty and three more off duty, two news agents and a Pullman porter were injured.

Fireman Willingham told the investigators that the road foreman had taken charge of the engine at Tennessee Pass and had appeared in general to drive the train properly, though he had run past the water tank at Malta. The fireman did not recall his engineer saying anything to the road foreman about braking, though the engineer had remarked to him (the fireman) that the road foreman was certainly running fast. Approaching the curve where the derailment occurred the road foreman had applied the brakes lightly for seven or eight seconds, then applied them harder. The brakes had worked properly but had been applied too late to reduce the speed below 45 miles an hour as the train reached the 11 degrees 30 minutes curve. The fireman said that he had been sure that the train could not make it around the bend but that there had been nothing that he could do except hold tight. He added that the air brakes had been tested at Tennessee Pass and had worked properly at all points.

The evidence of Conductor Cope was not very helpful. He said that after dealing with passengers who had got on at Malta he had gone to one of the dining cars to see the steward and had been there when the derailment happened. He had noticed two brake applications and did not think that the train's speed had exceeded 30 miles an hour. He put the time of the accident at 11.15 a.m.. The ICC remarked that in that case the train must have taken 27 minutes to reach Waco from Malta, whereas the scheduled running time between the two points was only 20 minutes. The investigators believed that the derailment had occurred at about 11.05 a.m.

The statements of Head Brakeman Angel and Flagman Grout more or less agreed with that of the conductor.

---

[238] *The Havre Daily News-Promoter*, Havre, Montana, United States, 6th September 1926

Trainmaster Lynch said that he could see nothing to account for the accident other than excessive speed. He had been riding on the train but had been talking to his son, who had been away for the summer, and hence had not been paying close attention to the running of the train.

Section Foreman Gillon and Roadmaster Chambers stated that they had examined the track after the derailment. It had been in good condition, and they could see nothing which could have thrown the train off the line.

Divisional Engineer Darby spoke of having found several rails torn from the track and broken, but he was sure that this was a result of the accident and there was no evidence of a broken rail having caused it. The rails were 15 years old but showed no perceptible wear and in his opinion the derailment was due to excessive speed.

The report noted that engine No. 1604 had only recently been delivered to the company. It was a 4-8-2 type and somewhat heavier than those previously used on passenger trains. The total weight of the loaded engine and tender was 710,310 pounds (355 tons). There was no sign of there having been anything wrong with the engine or with it having struck any obstruction on the line.

As to the time at which the derailment had occurred, the ICC noted that this was disputed. The most reliable source must be the time recorded by the train dispatcher. According to the latter Trainmaster Lynch had reported the accident at 11.10 a.m. Lynch would have had to crawl out of the car in which he had been riding, climb over another car and up to the track, look over the situation, instruct one of the brakemen to flag the rear of the train and then run almost 2,800 feet to a lineside telephone, all of which would have taken him at least five minutes. It was believed, therefore, that the accident had occurred at about 11.05 a.m., which if correct would indicate that train No. 2 had run at an average speed of 38.8 miles an hour from Malta.

Continuing, the report noted that the time-keeping of passenger trains on the railroad had not been satisfactory and about two months previously the general superintendent had issued a bulletin on the subject and stated that he would begin issuing a bulletin showing the proportion of trains which had kept scheduled times during the previous 24 hours. This had resulted in his subordinates pressuring employees to do better. A check of the train sheets covering the 56 miles between Malta and Salida disclosed that during August it was an almost daily event for a train to make up 30 minutes of time between the two points and on occasion 40 or 45 minutes. It was true that the schedules were relatively easy, but the stretch included 25 curves with radii varying between 8 and 14 degrees and innumerable curves of less than 8 degrees. On 17th August, the superintendent of the Pueblo and Denver division had issued a bulletin to the effect that engineers were taking curves too fast. No equivalent bulletin had been issued for the Salida division although the average speeds there were somewhat higher. It must be a matter of conjecture, the report stated, to what degree the taking over of engine No. 1604 by the road foreman had been done with a view to regaining time.

Under the timetable rules the speed of trains on all sharp curves was limited to 30 miles an hour. According to the recommended practice of the American Railway Engineering Association the theoretical elevation on a curve of 11 degrees 30 minutes was 6.81 inches, whereas the actual elevation was only 4 inches, meaning that that prescribed speed could not be much exceeded without there being a danger of the engine overturning.

The commission concluded that from the best available evidence the train had been running at 40 to 45 miles an hour as it entered the curve. The brake application had come too late to make any difference and the derailment had resulted from excessive speed. Had the more experienced Engineer Harpending been driving the locomotive it was unlikely that the accident would have occurred.

The penultimate paragraph of the ICC report remarked that, 'It is well recognized that in mountain territory where there are many heavy grades and sharp curves it is desirable to keep down the amount of elevation on curves in order to facilitate the movement of heavy trains when ascending grades. This requirement, however, has a very necessary corollary which is even more important and that is that trains capable of moving at high rates of speed shall not exceed the limits beyond which they cannot go with any reasonable degree of safety, in view of the limited elevation provided. If the desire to expedite the handling of heavy tonnage trains is paramount, resulting in low elevation on curves, then it is necessary very materially to restrict the speed of passenger trains when rounding such curves. But if passenger trains are to be operated at speed, then the elevation provided on these curves should be increased accordingly, keeping in mind the necessity for providing an adequate margin of safety between the authorized speed and the speed beyond which trains cannot be operated with any degree of safety. This is a matter which has been referred to on previous occasions; it is one of primary importance and should not be overlooked in the desire to make up lost time.' [239]

Eight men died and 11 were seriously injured in the derailment of an Insular Lumber logging train at Fabrica in Occidental Negroes province in the Philippines, on Thursday 16th December 1926. [240]

On Saturday 20th August 1927, the 10.51 a.m. Southern Railway express from London Charing Cross to Margate was derailed between Bearsted and Hollingbourne, England, it was believed as a result of track subsidence. Six passengers were taken to hospital suffering from shock. A relatively minor accident, therefore, but it was to be a forewarning of a much more serious one.

Only four days after the Bearsted derailment the 5 p.m. Southern express from London Cannon Street to Deal met with disaster on the approach to Sevenoaks Tubs Hill station when it was derailed at Riverhead, as it approached the Shoreham Lane road overbridge. The engine and leading two coaches passed through the bridge but the third coach smashed into the piers of the structure and was demolished. The engine continued along the line for some distance before overturning 600 yards short of Sevenoaks station. Thirteen passengers lost their lives, 20 passengers and the fireman of the train were seriously injured, and 38 passengers and two railwaymen were slightly hurt or suffered from shock.

People living near the scene at first mistook the sound of the crash for a peal of thunder but were soon alerted by the whistling of the express engine and news of the accident quickly spread. 'It was at once realized that the accident was a very bad one, and

---

[239] Interstate Commerce Commission report

[240] *The Laredo Daily Times*, Laredo, Texas, United States, 16th October 1926

it was early ascertained that several people had been killed and many injured. Motor-cars were rushed from all parts of the district, and the injured were taken to hospitals at Sevenoaks and Tonbridge.'

'A local resident who arrived soon after the accident said people were scrambling out from all parts of the wreckage. One man in a top hat was climbing up the bank, with a white handkerchief folded round his forehead, and helping to pull up a woman who had also apparently been injured. Mr S. Francis, of Riverhead, said in conversation that he had just returned from his allotment, when his nephew told him there had been an accident. This was about 5.30. "I didn't hear the crash," he said, "but I rushed to the spot, and saw a number of coaches leaning over in the vicinity of the bridge. I got in under the bridge for some distance and saw a number of people lying about, who were apparently dead. I assisted to remove them. There also appeared to be a very great number of injured. These were quickly removed to hospital in private motor-cars, and in other conveyances that came from all parts of the district."'

A newspaper reporter also managed to interview a several passengers from the train. 'Mr. F.S. Upton, a well-known Folkestone business man, said: - "I was in the coach immediately behind the Pullman. We were about 200 yards on the London side of the bridge, when the engine began bumping. I immediately put my feet up on the carriage seat because I thought there was something coming. The bumping became more violent, and the six of us who were in the compartment were thrown in a heap on the floor. When we came to a stop I tried to open the door, but it was jammed. I forced it open and went into the corridor and looking out of the window I saw that the Pullman coach immediately in front of us lay with the front portion at the top of the bridge completely blocking the opening. I scrambled up the bank and reached the top of the bridge, and on looking over saw an appalling sight. Apparently the whole bodywork of two coaches had been completely ripped off the bogeys. The dead and injured were lying about in all directions. The Pullman coach had evidently struck the bridge and stopped the back portion of the train going through. My impression is that the engine and first coach left the rails, and they got through the bridge with a certain amount of knocking about, and the third coach hit the stonework of the bridge and so brought the disaster."

'Mr M.C. Frye, who is a member of the committee of the Folkestone Lawn Tennis Tournament, said: - "I was in the first coach immediately behind the guard's van. We were going at fair speed just before we reached the bridge. The first thing I felt was a bump. We then went on for about 50 yards, when there was another frightful bump, which splintered the glass in the carriage and threw us in all directions. Our coach then apparently left the metals. When the big bump occurred the corridor side of our coach hit the bridge and the whole of the corridor was carried away. The engine and two front coaches must then have run on for the best part of 100 yards, being off the metals all the time. The engine then turned over on its side and buried its nose in the embankment, bringing the train to a standstill. So far as I could see the two front coaches remained coupled to the engine and the coach immediately behind us turned broadside across the metals. All in our compartment were cut by splintered glass and received a severe shaking."'

'A graphic story of the accident was told last night by Mr Barry O'Brien, theatrical manager, who was travelling in the train to see one of his companies in *Lady Luck*. He said his chauffeur put his luggage in the Pullman car, but as there were so many people in it he went to another carriage at rear of the train. That act probably saved his life, because it was the passengers in the Pullman car who suffered most. "I settled down to go through a play, and the first intimation of anything wrong was the sudden rocking of the train, which was travelling at top speed. Another male passenger opposite me got up and looked out of the window but seeing nothing unusual sat down again. Then came a sudden grinding and tearing noise, as though the train had left the metals. We all looked at each other in the compartment, but before we had time to wonder what was going to happen next there was a terrific, deafening crash, and we were all thrown from our seats.

'Then the piercing shrieks of frightened women and children reached us above the din, and we picked ourselves up and looked out. We were amazed to see the engine on its side and two coaches, one the Pullman I had left, completely smashed. At once we jumped down on to the track and rendered what service we could to those who were lying injured in the debris."

'Mr O'Brien added that there were several doctors in the train, and they at once set about doing whatever they could for the injured. Nurses, doctors, ambulances, and police seemed to appear as from nowhere. They worked hard in getting the injured away, and the male passengers who were unhurt gave a hand in clearing the debris of the smashed coaches to get at the imprisoned passengers. Mr O'Brien suddenly collapsed, and when he recovered found himself on an ambulance. He was taken to the Holmesdale Cottage Hospital, where he soon recovered and returned to help.' [241]

The Ministry of Transport inquiry into the accident was conducted by Colonel J.W. Pringle. In his report he described in detail the site of the accident, the composition of the train, and the details of the derailment itself, then detailed the evidence of the different witnesses.

Four signalman gave evidence but those at Knockholt, Polhill and Dunton Green could only say that the train had been running at its usual speed, Signalman John Sears at Sevenoaks No. 1 said that as the express had approached his box he had heard two loud reports and knew that something serious had happened. The track circuit indicators for the up as well as the down lines had gone to 'Occupied' so he knew that the opposite running line must be fouled, and he had sent the 'Obstruction Danger' signal to Dunton Green and Sevenoaks No. 2 boxes at 5.30 p.m. He did not remember any heavy rain since he had come on duty at 2 p.m.

Much more significant was the evidence of Driver Frederick Buss. He stated that he had left Cannon Street station half a minute late with "River" (K) class tank engine No. A 800 (*River Cray*). He had had clear signals all the way to Sevenoaks, though a distant signal at Orpington had almost checked his train before it was cleared. Because of the gradients the speed of his train had dropped to 35 miles an hour at Knockholt. Approaching Polhill Tunnel, just beyond Knockholt and now on a falling gradient, he had told his fireman to shut the front and open the back damper, because they had 200 lbs of steam and a full boiler of water. Driver Buss said that he had anticipated that this would check any tendency of the engine to roll on leaving the tunnel. The engine ran steadily through the tunnel and on to Dunton Green station at a speed he estimated as 55 to 57 miles an hour. Steam was still applied halfway between Dunton Green and Sevenoaks when he had heard a knocking noise and had immediately closed the regulator. The noise

[241] *The Times*, London, England, 25th and 26th August 1927

had stopped, then started again, and he had made a full brake application, but he had not felt the continuous brake taking hold until his train was near the Shoreham Road overbridge. He had noticed the top of the engine cab scraping the bridge pier as they passed beneath it. He had never made any complaint as to the rough riding or rolling of the engines. He thought the engines good for the work but liable to roll at speeds above 50 miles an hour. Rolling sometimes occurred as the engines were going through crossings and took some time to die down, though in less than a minute. The rolling, he considered, was more pronounced on tank than on tender engine, perhaps because of the movement of water in the tanks. When the engine began rolling he attempted to steady it by closing the regulator to reduce speed.

Fireman Henry Barter said that he had fired on a 'River' class engine for the first time on the day of the accident. He had heard drivers say that the engines were powerful and efficient, and an easier job for the firemen than other classes, but also that the engines rolled somewhat at high speed, though not dangerously so, and that it was a different kind of movement from the usual swaying of tender engines. To his recollection his driver had told him to reverse the dampers when they were halfway through Polhill Tunnel, and he had done so. He had had no idea of anything being wrong until they were within 100 or 150 yards of the catch points, when he had heard a crack in front of the engine as if something had broken. On hearing the sound Driver Buss had closed the regulator and applied the brakes, first partially and then fully. He could not give any opinion as to the speed of the train at any point.

Guard William Bailey put the speed of the express through Dunton Green at between 55 and 60 miles an hour, which was no faster than usual. He had had no idea that anything was wrong until the continuous brake was applied.

Mr Barry O'Brien, the passenger who had been interviewed by the newspaper reporter, also gave evidence at the inquiry. He told the inspecting officer that he had been dozing in the rear carriage but one when he had felt a quite unusual sensation, similar to that on a switchback line when one was suddenly switched in another direction.

Mr F.S. Upton, whose interview with a reporter is reproduced above, was a season ticket holder who also gave evidence at the inquiry. He spoke of having noticed severe swaying on previous journeys and said that on the day of the accident there had been very severe swaying or rolling, as though they were being checked in their forward direction and at the same time rolling from side to side. He had anticipated an accident, and eight or ten seconds later had realised that they were off the road.

Locomotive Inspector E.A. Turner was a frequent passenger on the train. He remembered the carriage throwing from side to side halfway between Dunton Green and Sevenoaks. The knocking had died out, but a few seconds later the crash had come. The previous June a driver had complained of serious rolling with a 'River' class engine. He had ridden the engine with no problem between Folkestone and Tonbridge, but on the return journey there had been serious rolling between Saltwood Tunnel and Cheriton Junction. The engine was taken out of service. He considered that the condition of the road was partially to blame.

The crew of the 4.30 p.m. down slow train from London Charing Cross to Sevenoaks, the previous train over the line to the doomed express, were Driver T. Knight and Fireman S.J. Osborne. They had been in charge of a class C tender engine and had noticed no problems.

Driver W.F. Gooding, on the other hand, had driven 'River' class engines, and said that they rode well at speeds below 50 miles an hour but tended to have a nasty roll at higher speeds and he had heard other drivers speak of unpleasant oscillation with this class of engine. The oscillation was not peculiar to curves; he had experienced it while travelling at high speed on a straight road between Tonbridge and Ashford. In his experience the rolling was worst on the Chatham line at Buckland Crossing, on the curve between Dunton Green and Sevenoaks and on the straight line between Tudeley Intermediate box and Paddock Wood.

Driver F. Donaldson agreed that the rolling began at speeds between 50 and 60 miles an hour. He considered that the rolling with K-class engines was due partly to high speed and partly to problems with the track. He had always been able to check the rolling by closing the regulator without needing to apply the brake.

Locomotive Running Superintendent A.D. Jones stated that the first 'River' class engine had been put into service in August 1917 and complaints about its rolling had begun in April 1925. Engine No. A 800 had been brought into use in July 1926 and the only complaint made regarding it had reference to the permanent way between Cheriton Junction and Shorncliffe. Complaints had also been made regarding the running of several other locomotives of the 'River' class. He was of the opinion that the class of engines was perfectly safe running on a road in thorough order and that weak spots in the track would accentuate the rolling of heavy tank engines.

Permanent Way Inspector Thomas Vousden stated that he had examined the track after the derailment and had found the down track to be 1¼ inches out of gauge 135 feet on the London side of bridge 194 (outside of the Sevenoaks No. 1 down distant signal) and the curve to be out of line all the way through, in some places by as much as three inches. He had received no complaints from his men that it was difficult to keep the curves in question to line, gauge, and level.

Ganger J.E. Warn agreed that there was some wideness to gauge, and that the road had been slewed towards the outside of the curve. He had walked the length at 4 p.m. on 24th August and had then been satisfied that it was good for express services.

Sub-ganger H. Boakes said that he had walked past the spot at 5 p.m. on his way home after finishing work and had looked at it out of habit and seen nothing wrong.

Other witnesses, including Mr. R.E.L. Maunsell, the Chief Mechanical Engineer, and Mr George Ellson, Chief Engineer, also gave evidence at the inquiry.

Reviewing the evidence of the various witnesses, Colonel Pringle remarked that the accident was one of those of which the cause was difficult to elucidate and the evidence taken at the inquiry was not sufficient to decisively prove the cause of the derailment. From the evidence of the crew of the 4.30 p.m. slow train the track was adequate for low-speed traffic immediately before the derailment. From the evidence of various witnesses the train was probably travelling at 60 miles an hour when it left the track. Various persons travelling on the express had reported unusually severe rolling and oscillation, or jolting and lurching, just before the derailment. Driver Buss had reported hearing a knocking noise prior to the derailment. Various drivers had reported the tendency of the K class engines to roll. There had been no problems with the engines on the Central (ex-London Brighton and South Coast) section but several when they were running on the

Eastern (ex-South Eastern and Chatham) section and Mr Maunsell had given evidence of design improvements made to deal with the problem. Mr Ellson stated that although the speed of the trains had not increased there had been a material increase in their weight over the previous seven years, putting more stress on the track. A heavy rainstorm on the morning of the 24th might have affected the track. The general conclusion of the various staff of the permanent way department was that there had been nothing wrong with the track which could have caused the derailment.

Colonel Pringle stated that he had thoroughly inspected engine No. A 800 at Ashford Works in its damaged state and had concluded that no radical defect in its design or condition had existed.

Examination of the permanent way revealed that the rails were 20 years old but in good condition and of adequate strength. The sleepers were in fair condition. The ballast was a mixture of Kentish ragstone and Dungeness beach ballast and was inferior in rigidity and holding capacity to that used on first-class roads. The cess on the inside of the curve was too high and there was no drainage from the top to the bottom ballast except through the bank. There were inequalities in superelevation over a considerable length likely to cause or intensify the rolling of an engine and there were irregularities of gauge. The inspecting officer wrote that 'though the strength of the rails, fastenings, and of the sleepers were adequate, the drainage, the quality of the ballast, and in particular of the bottom ballast, the gauge and the level of the rails could not be described as first class. The road had not, I think, a sufficiently firm foundation to carry such heavy loads at high speeds as had during recent years passed over it, with the result that it had gradually been knocked down. It is possible that the heavy rainfall had caused the road to go down rapidly, as stated by Mr. Messer. It appears that the permanent way staff either had not realised the extent to which the road had suffered under this traffic and wet weather or were unable to maintain it. The provision of a check rail, when the track was laid in 1907, on a curve of so large a radius as 54 chains is some proof that great reliance was not at that time, even with the lighter loads carried, placed upon the stability and strength of the foundation of the track.

To obtain more information on the behaviour of 'River' class engines at high speeds trials were carried out on the ex-Great Northern main line of the London and North Eastern Railway over the 7 miles between Huntingdon and St Neots. Two engines were used, Nos 803 and 890, and it was found that they ran very steadily at high speeds, even in one case with engine No. 890 at slightly over 83 miles an hour. At the request of Sir Herbert Walker, the General Manager of the Southern Railway, further trials were then carried out on the Western (ex-London and South Western) section of that company, over the 7½ miles between Woking and Walton-on-Thames. The results were not so impressive here, with engines Nos. 803 and 890 both rolling at the same places and severely at higher speeds (the maximum speed attained was 77¾ miles an hour). It was clear that the Southern Railway track was markedly inferior to that on the London and North Eastern.

In conclusion Colonel Pringle stated that the derailment must be attributed to the rolling movement of engine No. 800 while travelling at perhaps 60 miles an hour. The heavy 'River' class engines were more susceptible to weaknesses in the foundations and irregularities in the track than tender engines because they were more lightly sprung, and the more resilient springs reacted more readily to track inequalities. The 500 gallon side tanks had the effect of making the rolling movements slower and easier but more difficult to damp out, while increased speed made the rolling more liable and it more likely that its amplitude would increase until it became dangerous.

The insufficiency of hard and clean ballast foundation, the lack of proper drainage and the irregularities in the level and gauge of the rails on the curve between Dunton Green and Sevenoaks had been sufficient to set up serious rolling and lateral motion in a tank engine travelling at high speed. The heavy rainfall on the morning of the accident might also very well have caused the track to go down rapidly under traffic.

In his final remarks the inspecting officer stated that:

(a) The necessity for permanently strengthening the road bed of some sections of the main line, with heavier locomotives pulling heavier trains at the same or increased speeds, did not appear to have been fully appreciated.

(b) The harsh springing of engines, though it might reduce the extent and therefore the danger of some of the movement at high speed, caused rough riding and damage to the track and was uneconomical in the long term.

(c) Sufficient attention did not appear to have been paid, at any rate on certain sections, to instructions issued by the Chief Engineer regarding the amount of superelevation on curves.

(d) More frequent footplate riding by permanent way officers and inspectors, especially in bad weather, would keep them better informed of problems with the track.

(e) There should be a speed limit on "River" class engines, they ran more steadily when travelling bogie first and when fitted with laminated bearing springs on the pony and bearing trucks. [242]

At the conclusion of the inquest upon the victims of the disaster the jury returned a verdict of 'Accidental death', exonerating Driver Buss and Fireman Barter from all blame. [243]

In the aftermath of the disaster the whole of the 'River' class locomotives were rebuilt as tender engines, the highly-strung Mr Ellson suffered a nervous breakdown, and the Southern Railway permanent way was improved with the use of rock ballast from Meldon Quarry throughout the system. [244]

No. 30 down mixed train on the Bengal and North Western Railway was derailed at about 20:30 on Sunday 22nd January 1928 between Haiya Ghat and Kishanpur stations, India. An empty road van left the track at Bridge No. 11-A at mile 8/8, struck the parapet wall and was thrown back against the following third-class bogie carriage, which was wrecked. The engine became detached from the van and ran 80 feet beyond the wreck while the remainder of the train escaped harm. Eight persons died, 15 were injured and 10,000 rupees damage was done to rolling stock. An investigation by the senior government inspector concluded merely that the empty road van marshalled between the

[242] Ministry of Transport report
[243] *The Times*, London, England, 4th November 1927
[244] *Historic Railway Disasters* by O.S. Nock, pgs 107-9

engine and heavy bogie passenger stock had become derailed, without saying why. [245]

A passenger train was derailed near the Cape Town suburb of Fish Hoek, South Africa, on the evening of Tuesday 24th January 1928. The engine and two third-class coaches left the rails and were smashed.[246] Five 'natives' and one European were killed and 25 'natives' and one European were injured. Railway officials declared that there was no apparent cause for accident, but 'an engineer' (the train driver?) attributed it to improperly fastened fishplates.

A reporter wrote that, 'The dying natives showed remarkable fortitude, One, who was offered water, said: "Me finished, baas; help my brothers." Although all were frightfully injured, there were no screams.'

The Minister for Railways, Mr. C.W. Malan, expressed the Government's sympathy with the relatives of those killed and injured in the House of Assembly and promised the Opposition that he would consider the appointment of outside experts on the inquiry board. [247]

The report of the inquiry has not been traced but it seems likely that the real cause of the accident was not track defects but excessive speed. [248]

Morning coffee was just being served to passengers on the mail train from Johannesburg to Cape Town when it was derailed in the Hex River Pass, a mountainous region of South Africa 140 miles north of Cape Town and near Worcester, at 6 o'clock on the morning of Friday 27th April 1928. The train rolled over, and seven coaches caught fire. The driver, fireman, and seven passengers were killed outright. The engine had left the rails at the sharpest point of the curve over a culvert and buried itself in the steep slope. The fire had broken out in the cook's galley from the upsetting of an oven and spread to the dining saloon. In what was described as an incredibly short space of time only the steel frames of the carriages were left. Everything on the train, including passenger's personal luggage, was destroyed.

A newspaper wrote, 'The public has been much disturbed during the past two years by the long series of derailments and minor accidents. Although it is difficult to assign any common cause to these, the opinion is universally expressed that there must be something wrong with the railways. An exhaustive enquiry is demanded into the general question of the frequency of accidents. The original death roll of five Europeans was increased tonight by the discovery of the charred remains of three others among the wreckage of the coaches, which were completely destroyed by fire.'

In the Assembly General Smuts, the leader of the opposition South African Party, demanded an impartial inquiry into the accident, and the Minister for Railways and the Minster for Justice, who both expressed the deepest sympathy with the relatives of the

---

[245] Report by the Railway Board on Indian Railways for 1927-28

[246] *Longreach Leader*, Queensland, Australia, 27th January 1928

[247] *Daily Advertiser*, Wagga Wagga, New South Wales, Australia, 27th January 1928

[248] Webpage: Fish Hoek Fossickings – Fish Hoek 100 and beyond

victims, promised such an inquiry. [249] [250]

On 1st May the *Cape Times* was expected to publish numerous letters supporting the journal's demand that there should be a strong Inquiry Commission appointed to investigate the general poor safety record of the railway, this commission being separate to the normal department inquiry and statutory inquiry held into just the Hex River Pass accident. [251] Given the fact that the derailment had occurred on the sharpest point of a curve it is tempting to suggest that it was another of those due to excessive speed, though it is not known if this was the conclusion of the inquiries. The accident happened close to the site of the September 1914 derailment of a troop train.

After several days of heavy rain a serious landslide blocked the railway line between San Benedetto and Grottamare, Italy, at a point where it runs along the coast of the Adriatic Sea. The fallen earth covered the railway line for a distance of about 150 metres and to a depth of from three to six metres and also involved two houses. On the evening of Wednesday 9th May 1928 a train ran into the obstruction and was derailed. Reports as to the number of casualties varied, some stating that the bodies of six dead had been recovered and that there were 18 injured, while another source stated that there were 11 dead and that some of the injured might not recover. Some of the victims might have been the occupants of the two houses rather than people on the train. [252] [253]

München – Dortmund express D 47 became derailed at 02:21 on Sunday 10th June 1928 at the west exit from the station at Siegelsdorf, Germany. A track gang had been working in the area until midday on Saturday. Thirteen trains subsequently passed through the area before the one which met with the accident. As a result of the accident 25 passengers and Driver Gustav Imhof lost their lives, five passengers were seriously injured, 20 were slightly injured and 103 suffered from shock. Scalding steam from a broken boiler feed pipe killed most of the victims.

The derailment caused something of a sensation, not least because it was one of a series of similar derailments which had begun with the front axle of the leading bogie of the engine. Bahnmeister (Permanent Way Inspector) Stuhlfath was blamed for the accident, prosecuted, and sentenced to three months imprisonment for having ordered that a speed restriction signal be removed as unnecessary given the condition of the track. Rottenführer (Ganger) Volland and Strekengeher (Track Walker) Geller were also prosecuted but acquitted as being less responsible. [254]

Chemins de fer de l'État express train No. 519, which had left Paris Montparnasse for Brest at 21:15 the previous evening, was derailed at the entrance to Le Mans station, France, at 01:15 on the morning of Thursday 2nd August 1928. The engine of the train

---

249 *Evening News*, Sydney, New South Wales, Australia, 28th April 1928
250 *The News*, Adelaide, South Australia, Australia, 30th April 1928
251 *The Times*, London, England, 1st May 1928
252 Ibid, 11th May 1928
253 *Le Matin*, Paris, France, 11th May 1928
254 *Von Siegelsdorf nach Aitrang* by H.J. Ritzau, pgs. 9-24 (in German)

scraped along the side of the platform and came to a stand some metres from the entrance to the station, overturned on its left-hand side into an inspection pit. Behind the engine tender telescoping occurred between the leading baggage van and the following three postal vans and the leading passenger coach, a second and third-class composite, mounted the rear postal van. Several of the following coaches were derailed but not damaged. Cries of pain came from the wreckage, and they were begging for help, but it was impossible to quickly release them. The rescue work began and involved the station staff, passengers and two army doctors who worked to shift the mass of wood and metal beneath which the victims lay. Doctors Lelong, Maisons, Neau and Oudiette arrived at the station soon afterwards. Five postal workers, MM. Fernand Villeroy, Louis Gaussérès, Marcel Venisse, Louis Aury and Gaston Letienne lost their lives, together with one passenger, M Hervé Kerlouet. Seventeen others were injured, including one postal worker, a luggage guard, and Fireman Gaston Dhommée. It transpired that one of the dead postal workers, Marcel Venisse, had almost lost his life a few years earlier in a derailment at Rennes, when he had been trapped in a van which caught fire and was half-asphyxiated.

Marcel Bernar, prefet of Sarthe, Arsène Le Feuvre, the mayor of Le Mans, and public prosecutor Vetillard were immediately advised of the accident and made their way to the station to initiate an investigation. The initial inquiry suggested that the train had been running too fast but as the record of the speed recorder was not available (see below) it was not possible to easily determine the speed at the moment of derailment.

The railwaymen worked to clear the line. Other tracks through the station were not affected so there was no disruption to traffic. The uninjured passengers continued their journey on a special train at 02:20. The prefet and the mayor visited the injured victims in hospital at 05:00.

On the evening of the day of the accident Driver Jean-Françoise Uguen, aged 45, was arrested charged with causing death and injury through negligence by entering the station at excessive speed.

A reporter managed to interview Driver Uguen, who related, "All of a sudden I felt my engine 'dancing' and I realised that it was derailed. I applied the brakes and then...after that I don't know. I don't understand," he added, "I don't know what happened."

Fireman Dhommée, stood close to his driver, declared, "I thought I was lost, and then I found myself here. I don't know anything, but them..." He was referring to the dead postal workers. [255]

A little later Driver Uguen and Fireman Dhommée alleged that the brakes of their train, which was reportedly carrying an exceptionally heavy load of passengers, had failed to respond when they were applied. Meanwhile Driver Uguen's union, La Fédération des Cheminots, protested at his arrest and demanded his release. The postal workers' union, La Fédération Postale, for their part, protested at the frequent accidents to which their members were exposed and the use of obsolete material; 80 of the 418 vehicles in use were in need of replacement but remained in use for reasons of economy. [256]

---

[255] *Le Petit Parisien*, Paris, France, 3rd August 1928

[256] *Le Peuple*, Paris, France, 4th August 1928

At the end of November Driver Uguen had his final interrogation from M. Hébert, the *juge d'instruction* (examining magistrate) and his file was handed to the *parquet* (public prosecutor) for consideration. [257]

Driver Uguen stood trial in January 1929. The circumstances of the accident were reviewed in detail in a newspaper report on the proceedings. It was stated that the train had been 44 tonnes over its maximum weight. Drawn by a *Pacific* class engine it had been due at Le Mans at 00:53 and had arrived at 01:15. It had crossed bridge bourge-Belé and at Poste No. 1 had been directed to platform 6. When he had felt that the train was derailed Driver Uguen had tried not to pull up too suddenly for fear of telescoping the following vehicles, but this had been frustrated when the front of the engine had fallen into an engine inspection pit, when the train had been pulled up very suddenly. The driver insisted that he had strictly obeyed the signal de ralentissement and had reduced speed to the required 30 kilometres an hour. He attributed the derailment to track defects, to the fact that the train was overloaded by 44 tons and to its faulty composition, with a van weighing 12 tons followed by one weighing 37 tons. Against this, it was stated that there were numerous witnesses who said the train was going too fast and the damage to the train bore this out. The question of the speed of the train should have been easily settled by reference to the Flaman speed recorder. However, in the chaos which had followed the accident M. Martin, a traction inspector, had gone to the engine to seize the vital evidence and unfortunately in his haste as he opened the recorder he had torn the paper, and the 8 centimetre fragment which recorded the speed of the train over the last kilometre or so into Le Mans had disappeared.

Giving evidence, M. Benoist, in charge of Le Mans station, stated that had the engine not fallen into the inspection pit no telescoping would have occurred. Asked if it was normal to have engine inspection pits on reception lines, he said that it was.

Bridges and Track Engineer Maatuszek stated that the track had been in a perfect condition. The train was partly composed of very heavy carriages and had in fact weighed 579 tonnes, which was still within the permitted 15% excess of the normal maximum load of 520 tonnes. His opinion was that the engine must have been running at more than 30 kilometres an hour when it fell into the inspection pit. Pressed by M. André Berthon, for the defence, he would not care to say exactly how fast he thought the train had been travelling but repeated that he was sure it had been more than 30 kilometres an hour. M. Matuszek was of the opinion that if Driver Uguen had made a normal brake application he should have stopped 100 metres short of the inspection pit. M Berthon suggested that the loss of the readings from the Flamen recorder was extremely regrettable. Indeed, M. Matuszek replied, as it reduces us to theorizing.

Signalmen Geffot, Dehoux and Daudibon in No. 1 signal box stated that they had been surprised at how fast the train had come into the station.

Traction Inspector Martin explained exactly how the vital part of the speed recording had been lost. After the accident he had gone to the locomotive and torn off the paper strip with his fingers. Going to a first-class carriage to examine it, he had found that part of it was missing. He had returned to the engine but had found it impossible to remove the last part of the recording, even with Driver Labbé helping him by holding a light and lending him a knife. He had gone to get a tool and when he returned a little later with

---

[257] *Le Ouest Éclair*, Caen, France, 24th November 1928

Head Driver Giroud and other witnesses the 'famous fragment' (as the newspaper referred to it), had disappeared.

Fireman Dhommée, who had been on the tender as the train reached the station, stated that he had not been under the impression that the train's speed had been any higher than usual.

Several other witnesses gave evidence, without adding anything of importance.

After the testimony of all the witnesses had been taken M. Berthon spoke in favour of Driver Uguen. He criticised the layout at Le Mans which made access to the station difficult and the weight and composition of the train, with heavy and light vehicles intermingled and a wooden postal van marshalled between steel vans. Turning to the charge against Driver Uguen, the disappearance of the Flaman recording meant, he maintained, that it was impossible to be certain that the speed of the train had been excessive, whilst the evidence of Control Inspector Yokel appeared to show that the train was only travelling at 33 kilometres an hour at the moment of the accident. There was, therefore, some doubt as to the excessive speed of the train which must benefit the defendant and he pressed for an acquittal. [258]

The judgement of the court was given on 1st February. It declared that the composition of the train had been in accordance with regulations, there had been no defect in the Westinghouse brakes, nor was there any fault with the signalling or the points. In short, the derailment could only be attributed to the train coming into the station at too high a speed. According to the observations of M. Benoit the train first left the rails at points No. 55. Once it had left the rails its speed would have been quickly reduced as it ploughed along the ballast and further reduced by the derailed coaches striking three vehicles standing in an adjacent siding, by the action of the brakes and by its negotiating the reverse curves between points 45 and 55. That its speed upon entering the station was too high was also borne out by the evidence of the signalmen in No. 1 signal box and by some of the passengers on the train. According to the signalmen the speed of the train when it passed them, where it should not have exceeded 30 kilometres an hour, had been well over 45 and probably close to 60 kilometres an hour.

In conclusion the court found that Driver Uguen, who had an unblemished record of 20 years' service, was guilty of reckless driving. He was sentenced to six days' imprisonment and a 50 francs fine. The Chemins de fer de l'Ėtat was declared to be civilly liable. [259]

Fate, a newspaper reporter wrote, had played a part in the wrecking of a northbound Illinois Central Railroad express, *The Chickasaw*, on Monday 6th August 1928. The railroad's service cards of eternal vigilance, cement-like roadbeds and nth degree efficiency had all come to nothing. [260]

The sequence of events which led to the accident began with the departure from the yard at Mounds, Illinois, United States, of northbound freight No. 1522, consisting of locomotive No. 1522, 18 cars and a caboose, in charge of Conductor Hess and Engineer

---

[258] Ibid, 26th January 1929

[259] Ibid, 2nd February 1929

[260] *The Daily Independent*, Murphysboro, Illinois, United States, 7th August 1928

Rochte. The train left at 12.15 a.m. and had only travelled a short distance when, unnoticed by the train crew, a 12 inch cast-iron soil pipe fell from the tenth car and fouled the southbound track.

Over two hours after the departure of the freight train, southbound passenger train No. 203, in charge of Conductor Walker with Engineer Joest and Fireman Roy, passed Mounds yard, travelling at a speed of about 50 or 55 miles an hour, struck the pipe and knocked it onto the northbound track. The train stopped at Mounds station, some distance south of the yard, where Engineer Joest found marks on the pilot of the locomotive, showing that it had struck something. However, he did not report the occurrence there, but waited until his train had reached Cairo Junction, five miles further south, where he told his fireman to warn the enginemen of northbound train No. 16, *The Chickasaw.* This was done, and train No. 16, in charge of Conductor Ed Whalin with Engineer Abel and Fireman Palmer, arrived at Mounds at 3.09 a.m., travelling at a reduced speed of 15 miles an hour after the warning at Cairo Junction. Despite this, in the darkness the obstruction was only seen at about an engine length's distance, was struck again, and knocked against the southbound track, forcing it out of position. Train No. 16 was brought to a halt but by then southbound train No. 3, the *Chicago and New Orleans Flyer*, was close at hand, travelling at high speed. With no time to warn it, it ran onto the distorted track, was derailed, and sideswiped the coaches of train No. 16.

Six passengers, a Pullman porter and an express messenger died in this complex accident and 137 passengers, a Pullman porter and six railroad employees were injured. Two of the passengers who lost their lives were Joe Sedinch and Mrs Bud Snider, members of a vaudeville team known as *Bud Snider and Company.* Mr Snider was among the injured.

Speaking to a reporter, Conductor Whalin said, "'It was the most terrible thing I have ever experience. Our train had been warned that the pipe was lying across our track, but we struck it 2,000 feet before we had expected it. Though we were able to stop immediately because we were going at a slow speed, evidently, when we struck the pipe it hit a rail on the other track throwing it out of line. Half a minute later, and before we could flag it, No. 3 came roaring down the right of way, sixty miles an hour. The engine was derailed and ploughed into the Pullmans of No. 16, ripping through the entire side of a number of cars, taking before it the whole tier of sleeping passengers. There in the darkness, made more dense by the smoke and dust, the screams of agony from the injured and the moans of the dying were awful.'" [261]

The crew of freight train No. 1522 told the Interstate Commerce Commission (ICC) that there had been no rough riding as the train had left Mounds yard. Conductor Hess said that he had inspected the train as it left the yard and Flagman Burris related that he had walked over the top of the car loaded with pipes. Neither had noted anything wrong.

Engineer Joest explained that having inspected his engine a second time at Cairo Junction he had become apprehensive and had asked the roundhouse foreman to notify the dispatcher. He had not said anything at Mounds as he had not been sure that his train had hit anything and if it had, it had posed no danger to his train so he had not supposed that it would to train No. 3. Conductor Walker, Baggageman Richardson and Flagman Arvin had been unaware of anything before train No. 203 reached Cairo Junction.

[261] *The Kokomo Daily Tribune*, Kokomo, Indiana, United States, 6th August 1928

Engineer Abel of train No. 16 said that from the warning he had been given at Cairo Yard, he had understood that the obstruction was near the lead track switch at Mounds. He had slowed to 15 miles an hour after passing Mounds station and Fireman Palmer had shouted, "There it is now!" when they were still about 1,500 feet short of that switch. He had made an emergency brake application and stopped his train. Train No. 3 then been very close and became derailed about eight seconds after he had applied the brakes.

Engineer Banks of train No. 3 testified that his first intimation of anything wrong had been when his engine had reared up in the air and then swerved to the left to collide with the cars of train No. 16. Fireman Alstat, Conductor Reader and Flagman Thompson of train No. 3 could add nothing of importance.

Supervisor Thetford and Superintendent Shaw gave evidence as to finding the pipe after the accident, and that it had been clear that it was the pipe being thrown against the track which had knocked it out of position.

Inbound Car Inspectors Knight and Yosick and Outbound Car Inspectors Billingsley and Schweiger had all checked train No. 1522, all the inspections being undertaken from the ground and none disclosing a problem. Master Mechanic Seely stated that inspectors were not required to climb onto the tops of cars to determine if they were loaded according to American Railway Association rules as this would result in delays. They were, however, instructed to pay particular attention to the lading to determine that it had not shifted.

Roundhouse Foreman Remacle at Cairo Junction stated that on being informed of the obstruction he had at once notified Dispatcher Wilson of its location. The latter said that he had tried to contact the operator, the switchman's shanty, and the yard office at Mounds at 3 a.m., but could not get any answer until 3.05 a.m., when Yard Clerk Nelms had answered. The latter immediately contacted Yardmaster Palmer, who attempted to take the yard engine to inspect the track, but by the time that several switches had been aligned it was too late. Operator Green at Mounds had overheard the conversation between the dispatcher and the yard clerk but misunderstood the location of the obstruction and hence had made no attempt to flag train No. 16.

The ICC noted that the pipe had been loaded onto car IC 209125, a high-side fixed-one gondola. The billing indicated that there were 69 pipes but on arrival of the train at East St Louis one was missing. It was apparent that some of the lading had been shifting back and forth and there were no wires or stakes to hold the lading in position.

In their conclusions the ICC investigators remarked that if Engineer Joest had reported the obstruction to the Mounds operator there would have been ample time for it to have been removed. Again the crew of train No. 16 would probably have seen the obstruction earlier had they not been expecting to find it 970 feet further north. The commissioners considered that the faulty lading ought to have been discovered had a proper inspection been made at Mounds. [262]

The Insular Lumber company of the Philippines, which had suffered a serious derailment of one of its trains in December 1926 (see above), suffered another one on Saturday 15th September 1928, this time near Cadiz in Negros Occidental Province.

[262] Interstate Commerce Commission report

Sixteen labourers died and 15 were injured when the engine of the train left the tracks and 10 loaded cars of logs fell down an embankment, possibly after the train had run away. [263]

An international express from Timişoara to Paris, France was derailed near Prusnila station, Romania, not far from Turnu Severin, probably late on Tuesday 15th January 1929. A few minutes later a Timişoara – Bucharest express arrived at the scene and running past the signals set at danger collided with the derailed train. Fire then broke out in the wreckage. Fourteen people died and 40 more were seriously injured. The investigation into the disaster concluded that the driver of the Bucharest express had probably been intoxicated. The cause of the initial derailment was not stated. [264]

Six persons died and nine were seriously injured in an accident on the Trans-Siberian Railway between Irkutsk and Chita, USSR, on about Tuesday 23rd April 1929. The train apparently ran into a landslide 'caused by recent frequent earthquakes,' and was derailed. [265]

Train D 23 became derailed at Buir, Germany, on the Köln – Aachen line, on Sunday 25th August 1929. Fourteen died and 43 were injured. [266] Because of bridge strengthening work between Buir and Sindorf to cope with heavier trains there was single-line working in operation between those points. The first train to which this working applied was delayed train D 23. When this train made its booked stop at Düren at 07:50-07:51, it was cautioned by the clerk, who was standing in for the fahrdienstleiter (signalman), who was on leave, and advised of the single-line working and a speed restriction between Buir and Sindorf of 30 kilometres an hour at the site of the bridgework. The notice given to the driver was the same as that issued to the driver of the previous train, which specified that the bridge in question was on the far side of Buir. However, the clerk had failed to observe that the worksite had now changed to a bridge on the Düren side of Buir, and so this was not mentioned in the caution issued to the driver. The train therefore approached Buir at full speed, its driver presumably assuming that he did not need to reduce speed until he reached the station, and he also apparently failed to observe the warning position of the Einfahrsignal (home signal), which should have warned him to reduce speed. The train therefore ran onto the facing crossover onto the opposite running line at almost 90 kilometres an hour and the engine and eight, mostly wooden, coaches were derailed. [267]

The driver and clerk appeared in court in connection with the accident in the summer of 1930. The driver was acquitted, and the clerk was sentenced to six months' suspended imprisonment with probation. [268]

A report from Moscow stated that a passenger train from there had derailed near

---

[263] San Antonio Express, San Antonio, Texas, United States, 18th September 1928

[264] *The Times*, Hammond, Indiana, United States, 16th January 1929

[265] *The Times*, London, England, 24th April 1929

[266] *Eisenbahn-Katastrophen in Deutschland* by Hans Joachim Ritzau, pg. 87 (in German)

[267] *Von Siegersdorf nach Aitrang* by H. J. Ritzau, pg. 96 (in German)

[268] Wikipedia – Eisenbahnunfall von Buir

Zuevka (between Kursy and Kharkoff) in the USSR on Monday 23rd September 1929. Six carriages were wrecked, and 30 people were killed (some reports spoke of 46 deaths). [269] [270]

Semi-direct passenger train No. 1202 left Bruxelles Quartier Léopold (now Bruxelles Luxembourg) station, Belgium, at 05:42 on Tuesday 10th December 1929 for Arlon. The train was crowded with workmen on their way to factories in Namur and the surrounding area. Just after leaving Gembloux station, between which place and Namur the lines falls sharply, the brakes of its engine it was at first reported had failed. The speed of the train gradually increased, and by the time that it had reached Rismes, a couple of kilometres from Namur, it was allegedly going very fast. On arriving at the maze of tracks at Namur station the train had been reportedly travelling at 100 kilometres an hour and its engine was derailed and overturned, taking the leading three carriages with it. In the words of a French newspaper these vehicles were 'reduced to crumbs.' The remaining vehicles of the train stayed on the track and were undamaged. The derailment occurred in front of the signal box controlling the junction of the lines to Bruxelles, Charleroi and Tirlemont.

As the accident had happened at a major station help was quickly at hand from station staff, the Red Cross and doctors from local clinics and the military hospital. About 75 passengers were injured. [271]

An early report stated that ten bodies had been recovered from the wreckage, including those of two women who had not at that time been identified, and ten very seriously injured people had been taken to the Sainte-Elisabeth clinic, not far from the station. Among the dead were the conductor of the train Zacharie Bouillon, and Fireman René Dewilde. The body of the latter was trapped between the engine and the tender, and it took three hours' work to free him with cutting torches. It was believed that there were six or seven more bodies buried in the wreckage. No further bodies were, it seems, recovered though.

A surviving passenger told his story: 'To begin with train was running exactly as usual. We had stopped at Gembloux and were running through Rismes and on the steep descent which leads to Namur when I got the impression that something unusual was happening. Several passengers were getting up to put on their coats; the train was going faster and faster; we were entering the station when there was a tremendous shock; everyone was thrown one upon the other; there were terrible cries from all parts. I sat there for a moment, then pulling myself together I left the wrecked compartment. A terrible spectacle presented itself: the train was reduced to a mass of scrap metal and crushed coaches.'

M. Lippens, the Minister of Railways, and King Albert I visited the injured in hospital during the afternoon.

M. Jarrigion, general secretary of the Belgian railway union, stated that it was too early to be sure of the cause of the accident, but it appeared that the brakes of the train had failed, adding that if the Rodolauwse safety apparatus had been in use the driver

[269] Ibid, 25th September 1929
[270] *Evening News Journal*, Clovis, New Mexico, United States, 24th September 1929
[271] *The Times*, London, England, 11th December 1929

would probably have been alerted to the problem in time to prevent the accident. [272]

Another newspaper, published the same day, stated that contrary to what had first been believed, the speed of the train had not been excessive, but that instead of taking track 4 into Namur station, the engine had become derailed on a set of facing points, so that it had tried to take tracks 4 and 5 at the same time and had overturned on its right side. Driver Correman had been thrown from the engine as it went over and had picked himself up from the ground and run towards the station like a madman, doubtless suffering from shock. [273]

The driver was taken to the Sainte-Elisabeth, where he was interviewed by representatives of the Namur parquet.

"Is it true that the accident was caused by the brakes failing?"

"No, they were working properly."

"What do you think, then, was the cause of the accident?"

"When we reached the crossing the bogies of the engine jumped the rails and it overturned."

"What did you do to prevent the accident."

"I pushed the brake valve to the fifth position to stop the train as quickly as possible, but it was already too late."

"How fast was the train going at that moment?"

"I don't know. I was told after the accident that we had not been going more than 50 kilometres an hour."

"Is it correct, as has been said, that you jumped from the engine when you realised that an accident was inevitable?"

"No, I was thrown from the locomotive as it overturned." [274]

On 13th November it was announced that experts had concluded that the accident had been caused by a sudden violent application of the brakes when the train was travelling at 90 kilometres an hour, in other words, a traction shock. The driver had presumably realised that had allowed his train to reach too high a speed and had taken drastic action to put matters right, but with disastrous results. [275]

Twenty-two people died and 28 were injured when an Irkutsk – Leningrad express was derailed near Cherepok station, USSR, not far from Leningrad, on Sunday 29th June 1930. [276] The accident was reportedly the result of a signalman's error. [277]

A train on the narrow-gauge electric railway between Porto San Giorgo and

---

[272] *Le Peuple*, Paris, France, 11th December 1929
[273] *L'Homme Libre*, Paris, France, 11th November 1929
[274] *Le Petit Journal*, Paris, France, 12th November 1929
[275] *Comœdia*, Paris, France, 13th November 1929
[276] *Manitoba Free Press*, Winnipeg, Manitoba, Canada, 1st July 1930
[277] The Railway Gazette, Vol 53, pg. 42

Amandola, Italy, was derailed at 08:35 on Wednesday 26th November 1930 at Fermo Santa Lucia. The line was steeply graded, and as electric car No. 21 was descending an incline and negotiating a curve it left the rails, broke through the parapet at the side of the line and fell about ten metres into the street below. Initial reports stated that seven people had died and eleven were injured. [278] [279] The number of dead later rose to eight. The motorman of the train survived but was seriously hurt and paralysed. Among the victims was a 22-year-old girl who had been on her way to Ancona to purchase a wedding dress and who was due to be married a few days later. [280]

[278] *Le Temps*, Paris, France, 27th November 1930
[279] *The Times*, London, England, 27th November 1930
[280] Webpage: Cronache fermane.il – 87 anni fa il deragliamento del trenino: una targa ne ricorda le vittime

CHAPTER 4

REAR COLLISIONS

The risks associated with one train running into the back of another had been low in the formative years of the railways when trains were few and train speeds low, but as the years progressed train speeds and traffic intensity often grew faster than advances in brake power, rolling stock construction and signalling and as a result by the early twentieth century serious rear collisions had become common. This chapter is therefore a long one, even though a great number of less serious accidents have not been included.

An accident on Sunday 20th February 1916, takes us back to the United States, this time to the West Coast and South Cheney, Washington. The line here was owned by the Spokane Portland and Seattle Railroad but two Northern Pacific Railroad trains were concerned in the accident, that company's trains having been diverted over Spokane Portland and Seattle tracks for several days because of washouts on their own route. The trains were Northern Pacific Nos 42 and 2, running over the foreign company's line as trains second and third No. 4. The trains rejoined the Northern Pacific route at Marshall Junction but as this was not a registering station on the Spokane Portland and Seattle all Northern Pacific trains had to stop for orders at South Cheney, 8.5 miles short of Marshall Junction. The Spokane line was single track and worked by train order and timetable without a block system or automatic signals; trains were spaced 10 minutes apart at telegraph stations.

Train second No. 4 (Northern Pacific train No. 42) was the *Mississippi Valley Limited* express from Seattle, Washington, to St Louis, Missouri. It consisted of Northern Pacific locomotive No. 2106, two baggage cars, a coach, a chair car, a tourist sleeping car, a dining car, two Pullman sleeping cars and a deadhead coach, and was in charge of Conductor La Violette and Engineer Smith. It left Pasco, Washington, at 3.20 a.m., an hour and five minutes late, with four train orders, collected eight more orders at Snake River, three more at Benge, and three more at Lemont. Most of these were slow orders related to reductions of speed at different points due to the dangers of washouts, landslides and so on. At 7.15 a.m. the train reached Amber; it left there at 7.20 and stopped for orders at South Cheney at 7.40.

Train third No. 4 (Northern Pacific train No. 2) was the *North Coast Limited* from Seattle, Washington, to Chicago, Illinois. It consisted of locomotive No. 2211 and nine cars, with Conductor Wilkins with Engineer Gander and Fireman Wallace, and left Pasco at 4 a.m., one hour and 45 minutes late. Receiving the same orders as train second No. 4, it left Amber at 7.30 and it ran into the rear of second No. 4 at 7.43. In the collision engine No. 2211 penetrated three feet beyond the vestibule of the deadhead coach, while the forward end of that coach was pushed forward and telescoped the Pullman sleeping car *Crookston*, killing five passengers within that car. Seventeen other passengers and five railwaymen were injured. [281]

The passengers who died were Professor Elton Fulmer, a state chemist of Pullman, Washington; I.J. Minnick, deputy grain inspector, of Spokane; J.J. White, of Spokane; Lee M. Conroy, Northern Pacific travelling passenger agent, of Spokane; and B.J. Berkey, of Portland, Oregon. Professor Fulmer was carrying an accident insurance card which he had obtained just before he started out, providing a pay-out of $25,000 should he die in an accident during his journey. [282] The victims had all been either still asleep or had been dressing. Others who had already dressed and gone forward to the dining car escaped injury. The accident was an almost exact parallel to one which occurred at nearby Cheney on the Northern Pacific line back in January 1911 detailed in the previous volume of this series.[283]

Flagman Lude of train second No. 4 testified at the Interstate Commerce Commission inquiry that he had thrown down red fusees at various points, each time the train slowed down; at Snake River, at Farmington, before the tunnel approaching Kahlotus, at Washtucna, at Benge, at Lamont and at Amber. He had not gone back very far at most of these points. At South Cheney it had been foggy and after half a minute he had begun to walk back with torpedoes, a fusee and a red flag. There was no sound of an approaching train but after he had gone about 600 feet he had seen the glow of a headlight, had lit a fusee and begun to run towards it. There had been no answer from the engineer of the oncoming train and about 200 feet further on he had to jump out of its way. He had known that there was another section following his own but had seen nothing of it since leaving Pasco.

Conductor La Violette stated that his train had stopped at Amber at 7.15, the block signal being in the stop position, and he was told by the operator that it was to hold his train for the required 10 minutes behind the previous one. His train then went on to South Cheney, after taking water, at 7.20, and at South Cheney it was very foggy. He saw the flagman set off back when they stopped and went into the station for his orders, having no reason to believe that his train would be stopped there very long. There did not seem any reason to speak to his flagman about making a special effort to protect his train. Coming out of the office a couple of minutes later he saw third No. 4 approaching and his flagman running back towards it waving a red fusee. He had had Flagman Lude for about a month and had considered him competent.

Engineer Smith said that after leaving Amber he had slowed down at a point where train order No. 46 had warned of a broken rail (which had in fact already been replaced)

---

[281] Interstate Commerce Commission report

[282] *The Salt Lake Tribune*, Salt Lake City, Utah, United States, 21st February 1916

[283] *The Ogden Standard*, Ogden, Utah, 21st February 1916

and then nearer South Cheney he had reduced speed after seeing a yellow fusee left by the previous train. He did not think that they were stopped at South Cheney over three minutes and had not whistled out the flagman as it did not seem necessary in view of the short stop and the fact that he was expecting to be ordered into the siding at the west switch.

Head Brakeman Ball of train second No. 4 testified that the train had not exceeded 30 miles an hour between Amber and South Cheney and had slowed up twice. He had gone into the office at South Cheney with the conductor and agreed that the train had not stopped more than three minutes. He said that it was customary for the flagman to go back at every stop, especially when it was foggy.

Engineer Gander of third No. 4 stated that he had passed Amber without stopping at 7.26, running at 13 or 14 miles an hour. Between there and South Cheney he had slowed up at a point where the track was in poor condition and at the location of the supposed broken rail. At no point had he seen a fusee on the track until he saw the one waved by the flagman immediately before the collision. He testified that he could not cover the distance between Amber and South Cheney in less than 15 minutes, though the time sheets indicated that on the day of the accident he had taken only 13.

Fireman Wallace said that he did not know what time their train passed Amber. He had not seen any fusees at any point. He thought the flagman was 400 feet from the rear of the front train.

Conductor Wilkins did not time the train at Amber either and said that he had seen no fusees. He did not think that his train had been travelling at an unusual speed.

Operator Thompson at Amber testified that the first section of No. 4 had passed at 7.10 and the second section had arrived at 7.15 so he had held it until 7.20. When the third section appeared Agent Nedwed, who had been at the station but not on duty, had looked at the clock, seen that it was just 7.30, and told him to clear the block signal and let the third section go. Operator Thompson was only working his third day; asked where he had learnt about the 10 minute rule he said that he had been told of it verbally by the agent and in conversation with a brakeman.

Midway between Amber and South Cheney was a non-telegraph station at Mock. Here Spokane Portland and Seattle Railroad freight train extra No. 356 was waiting on the side track and Engineer Winsla put the passing times of the three sections of train No. 4 as 7.16, 7.27 and 7.34. Fireman Robinson did not note the passing time of the second train but agreed with the times of his engineer with regards to the other two. Both men thought that third No. 4 was running rather faster than the two trains which had preceded it.

Superintendent Votaw of the Spokane Portland and Seattle stated that since November of the previous year almost all eastbound Northern Pacific freight trains had run over his company's lines and Northern Pacific employees should therefore be well familiar with the tracks between Pasco and Marshall Junction. Superintendent Tyler of the Northern Pacific concurred with this statement.

The Interstate Commerce Commission struggled somewhat, as was often the case, to reconcile the conflicting statements as to times and actions but finally decided that the basic cause of the accident was the failure of Engineer Gander to abide by train order No. 34 restricting his speed to 30 miles an hour between Amber and South Cheney. Even if his own statement were correct he had run at an average of 40 miles an hour between those points, while if the statements of the operator and agent at Amber were correct it was 50 miles an hour. He had also approached South Cheney at a much higher speed than the 12 or 15 miles an hour that he alleged, despite the thick fog, as evidenced by the telescoping of the Pullman sleeping car. The fact that Flagman Lude had dropped off lighted fusees at various points indicated that he had been fully alive to his responsibilities, however, his failure to drop one off approaching South Cheney had contributed to the accident. It was possible that Operator Thompson at Amber had also contributed to the collision by failing to hold third No. 4 for the full 10 minutes but the conflict in testimony made it impossible to say.

The accident once again pointed to the inadequacy of the time interval system and the need for the adoption of some form of block system to ensure that trains were kept apart.

Two days after the South Cheney collision there was an accident on the far side of the United States, in the New England state of Connecticut. The accident happened two miles east of Milford, on the quadruple track of the New York New Haven and Hartford Railroad. Three trains were involved, two passenger and one freight.

Westbound passenger train No. 79, the *Connecticut River Special*, consisted of engine No. 1346, a combination baggage and smoker car, two coaches, the Pullman cars *Wrentham*, *Sangatuck* and *Napoleon* and a coach, in charge of Conductor W.T. Bray with Engineer John Kennedy and Fireman Charles Courts. [284] The Pullman cars had steel underframes and the other vehicles were of all-steel construction. This train passed Tower 73 at 11.19 a.m. on Tuesday 22nd February 1916 and came to a stand on track 3 at 11.21 a.m. with its rear 450 feet west of automatic signal 65.23, due to a burst air hose causing the brakes to apply. After it had been standing there three minutes it was run into by local passenger train No. 5, in charge of an unnamed conductor with Engineer William Curtis and Fireman Edward McGinnis. Immediately after the collision the boiler of engine No. 824 exploded and a moment later the wreckage was run into by westbound extra freight train No. 5003, running on track 1. Engineer Curtis, Fireman McGinnis, ex-railwaymen Harry Sweeney, who was riding without authority on the engine of the local train, Flagman George Tourtellotte of train No. 79, Pullman Porter Joseph Frye and five passengers, two of them railwaymen not on duty, were killed and 235 passengers, nine railwaymen, 15 railwaymen off duty, four Pullman employees and three persons carried under contract were injured.

Engineer Spaulding of extra No. 1003 told the Interstate Commerce Commission (ICC) inquiry that his train had been running at 25 or 30 miles an hour. He saw the rear of train No. 79 and its flagman (George Tourtellotte) at about the same moment; the flagman was going back at a fast walk. He had whistled. As he passed the rear of train No. 79 he saw that train surge forward and the flagman was thrown across the track in front of his locomotive. He had applied his brakes in emergency and almost at once felt his train strike the wreckage. He added that according to his understanding enginemen did not necessarily need to apply the brakes on sighting a distant signal at caution as long as their train was under control so that they could stop at the home signal but at this location the engineman should have shut off steam at the distant signal (65.93) in order to stop at the home signal (65.23).

[284] *Boston Post*, Boston, Massachusetts, United States, 23rd February 1916

Fireman Kipputt of train No. 5003 said that he had been busy firing just before the collision. He suggested that smoke drifting across the tracks might have obscured the view of the signals from Engineer Curtis.

Section Foreman Heath stated that it was common for passenger trains to pass yellow signals at a high rate of speed, adding that some engineers seemed to be more cautious than others.

Signal Engineer Morrison testified that the signal spacing and the rules meant that engineers did not need to reduce speed immediately upon receiving a caution signal, but simply had to have their trains under control so as to be able to stop at the home signal. He said that since the automatic signals had been installed they had performed 7,132,660 operations. Over this period there had been 250 right-side failures and only eight wrong-side failures, all of which had been fully looked into. Signals 65.93 and 65.23 were of course carefully examined after the accident and no fault was found with them which could have caused a wrong-side failure whereby the distant signal was clear when the home signal was at danger.

The ICC concluded that the accident had resulted from the failure of Engineer Curtis to obey signals, for reasons which could not be determined as he had lost his life in the accident. It noted that the New Haven company was looking into the question of automatic train stop devices, but that most of the 4,000 plans for such a system had been rejected out of hand as they did not indicate that they were suitable for use on an electrified line such as the New Haven. One system was still under trial. [285]

A serious accident occurred on the Paris-Brest main line of the Chemins de fer de l'État at La Loupe, France, on Saturday 11th March 1916. Omnibus train B6 had left Brest for Chartres at 19:35 and was due to arrive at its destination at 00:18. On reaching La Loupe the omnibus train had to attach some vehicles and it was while it was performing this operation that it was run into by a goods train which normally overtook it at the previous station. There was a thick fog at the time. The two rear carriages of the omnibus train, which contained a detachment of soldiers, were smashed and according to one source seven of the soldiers were killed outright and 50 injured. Another source gave the exact number of injured as 53, of whom six were very seriously hurt. These six were taken by car to the mixed hospital (one which catered for both public and private patients) in Chartres. The less seriously hurt were transported to the public hospital in La Loupe or cared for in private houses. As soon as news of the accident reached Paris M. Legrain, the director of the network, and M. Touchard, the Chief Medical Officer of the Chemins de fer de l'Etat went by a special train to the scene, accompanied by the chief permanent way and locomotive engineers. [286]

In contrast to accidents in the United States on unsignalled lines such as those described just above at Orient and South Cheney, that on the New York Central Railroad near Amherst, Ohio, at 3.18 a.m. on 29th March 1916 happened on a multi-track main line with interlocking towers between which was automatic signalling. Amherst tower was

[285] Interstate Commerce Commission report
[286] *Le Petit Parisien*, Paris, France, 13th January 1916

located about a mile west of the station of the same name. East of the tower there were four tracks, west of it only three; the northernmost line, No. 3, being used by westbound trains and the middle and southernmost, Nos. 2 and 1, by fast and slow eastbound trains, respectively. The tower contained a 40 lever Saxby and Farmer lever frame which operated the switches and their protecting signals and there was approach and route locking to prevent the signalman setting up conflicting routes or reversing the points in front of an oncoming train. Signal and track indicators showed the towerman the positions of the signals and of the trains and annunciators warned him of the approach of a train. The signals themselves were oil lit, using long-burning signal lamps, and at night showed red (stop), green (caution) and white (clear) aspects. Located 1,178 feet west of the tower were its home interlocking signals for eastbound track No. 2, with the top blade signal 40, for track No. 2, beneath which was signal 39 for a movement from track No. 2 to track No. 1. 5,449 and 7689 west of signal 40 were automatic signals S7.2 and S9.2 respectively, each with two arms, the upper a stop signal and the lower a distant signal. On the approach to Amherst of an eastbound train and its passing signal S9.2 an approach indicator was displayed in the tower and the approach annunciator sounded. Provided the line was clear from signal S9.2 to beyond the interlocking the home and distant arms of signal S9.2 and the home arm of signal S7.2 would all be clear, but the distant arm at signal S7.2 would remain at caution until signal 40 was cleared.

Three trains were involved in the accident: the first and second sections of eastbound passenger No. 86 on track No. 2 and westbound passenger train No. 25 on track No. 3. Train first No. 86 from Chicago, Illinois, to Pittsburgh, Pennsylvania, was in charge of Conductor Bunnell with Engineer Leonard and Fireman Robinson; train second No. 86, from Toledo, Ohio, to Cleveland, Ohio, was in charge of Conductor Keller with Engineer Herman Hess and Fireman Turner; and train No. 25 was the *Twentieth Century Limited* from New York, New York, to Chicago, Illinois, with Conductor Burke and Engineer Robertson.

On duty in Amherst tower that night was Towerman Albert Ernst, who told the Interstate Commerce Commission (ICC) inquiry into the accident that the last train on eastbound track No. 2 before those involved in the collision had been an extra at 2.05 a.m. He had then been reading until the annunciator had sounded for train first No. 86 at about 3.10 a.m. whereupon he pulled the lever to clear signal 40 and saw, he said, that its indicator showed it to have come off. He had then sat down again and a few minutes later the dispatcher rang him to ask if train first No. 86 had passed and about the same moment he heard the engine of first No. 86 whistle. The track and signal indicators for signal 40 had then both been horizontal and the latch of the lever for signal 40 had not been quite properly over in the frame. He had replaced the lever and pulled it again and signal 40 had cleared. About a minute later the annunciator had sounded for westbound train No. 25, he had cleared his signal for it, and it had passed his tower about 3.17½. He had known nothing of any collision until he saw a flagman coming back with a lighted fusee. It had been very foggy.

Engineer Leonard of first No. 86 testified that it had first become foggy at Ceylon, 15 miles from Amherst, and gradually got thicker; at Amherst, the signals were only visible from 200 feet away. Signal S7.2 had been at caution, he had whistled, drifted down to signal 40, whistled again and stopped. The signal had cleared after half a minute, he had called in the flagman and restarted but his train had stalled. He took in the slack, restarted, and had moved six or seven car lengths when his train was run into by second

No. 86. A few seconds after the shock of the first collision train No. 25 appeared out of the fog and there was the shock of a second collision. He thought it safe to run at 60 or 70 miles an hour through fog depending entirely upon the signals. He had never experienced a false-clear indication (wrong-side signal failure).

Fireman Robinson agreed with his engineer's testimony on just about every point.

Conductor Frank Bunnell said that he had first noticed fog at Bay Bridge, 33 miles from Amherst. He saw the flagman alight as soon as his train stopped, go back a car's length, return when he was recalled and set off again when the train had stalled. When the train had again restarted he (the conductor) had ridden a short distance when he heard another train approaching and anticipating a collision he had jumped off and clambered up the bank. It had then been so foggy that he could not see the lineside fence. He did not think that the flagman could have got more than a car length and a half back the second time. No orders had ever been given to him regarding slow running in fog and trains usually ran at the same rate regardless of the weather conditions.

Flagman Beach of first No. 86 said that second No. 86 had been in Toledo station when they had left there but he did not think it would leave less than ten minutes after them. It had been more or less foggy all the way from Bay Ridge. He had thrown off a five-minute fusee when the train had first braked approaching Amherst. When it had stopped he had at once gone back, was quickly recalled, then the train had stopped again, and he went back a second time. When he had heard the second train coming after he had gone back about a car length he had lit a red fusee and started to run; he thought he was about 10 car lengths back when second No. 86 passed him, with its engine steaming and running at about 50 miles an hour. The rear lights of his train had been burning but the fog was so thick that he could only see the fusee that he had thrown off for about three car lengths.

Engineer Hess of second No. 86 said that he had first noticed fog about 20 miles from Amherst, west of Huron, and it had increased in thickness as his train continued eastwards. According to him all the signals at Amherst had been clear except the distant signal for track 1 at S7.2. The first warning he had seen was a burning fusee on track 2 about two car lengths from the rear of the train in front while at the same moment he saw the flagman, and he had immediately made an emergency brake application. He agreed that he had been travelling at that point at about 50 miles an hour. He said that he had been running a little faster than scheduled, hoping to make up time and not get into Cleveland late. He was certain in his own mind that he had not missed any signals and that they had all been clear.

Fireman Turner had noticed the fog at Sandusky, 28 miles from Amherst, and said that it became thick after Vermilion, 6 miles from the accident point. He had not been able to make out any signals beyond Vermilion and was lost as to his whereabouts approaching Amherst. He saw the burning fusee about three car lengths from the rear of the first train. They had been running as fast as they could, which was the normal procedure in foggy weather as long as they had a clear block.

Conductor John Kellar of second No. 86 said that he had had no warning of anything being wrong until the brakes had gone on in emergency.

Flagman Gates of second No. 86 stated that after the collision he had gone back to protect his train and confirmed that 15 minutes after the collision both blades of signal

S7.2 had been horizontal, the stop signal being red and the distant signal green.

Engineer Robertson of train No. 25 stated that it had been clear at Cleveland and only became foggy about a mile and a half east of Amherst. The signals at Amherst had been clear and he had had no warning of an obstruction on the track until his train had run into it. He agreed that fog made no difference to his speed as long as he was sure of the signal indications. In 35 years', experience he had never passed a clear distant signal and then found a stop signal at danger. [287]

The results of the double collision were very serious. It was believed that only a few passengers died in the initial collision, but this threw the last two cars of first No. 86 foul of the westbound track where they were struck by train No. 25. All but two of the 28 deaths (27 passengers and one railroad employee) took place in the rear car, which was of steel construction. It was theorised that in the second collision the steel underframe of the wooden club car which was the penultimate vehicle of the first train had been forced lengthways through the steel car, causing the casualties. As well as those who lost their lives about 40 passengers and 3 railroad employees were injured. Travelling on the Twentieth Century Limited were movie stars Mary Pickford and Ella Hall, who being woken by the crash quickly dressed and went to help in the rescue work.

As often happens after such disasters various wrong theories as to how it had occurred circulated. In this case it was none other than New York Central Superintendent A.S. Ingalls who, after a hurried investigation, stated the following afternoon that Towerman Ernst had caused the accident. It so happened that Mrs Ernst had given birth on Sunday night, 26th March, and it was alleged that the towerman, having had little sleep from caring for his wife and new-born baby, had dozed off and let second No. 86 into the back of first No. 86. [288] Towerman Ernst vehemently denied having fallen asleep and the Interstate Commerce Commission found that the fault lay elsewhere.

The commission noted that by the towerman's own admission he had not pulled lever 40 properly the first time (due to tiredness?) and that signal 40 had therefore not cleared and caused the unnecessary stoppage of first No. 86. This, however, the commission noted, had no direct bearing on the cause of the accident, for trains were often stopped out of course for one reason or another (a passenger pulling the communication cord, animals on the line, brakes binding, engine failure and so on) and automatic signals S9.2 and S7.2 were provided to protect trains stopped in such circumstances.

Engineer Hess had testified that signals S9.2 and S7.2 had been clear but as train first No. 86 was occupying the track in the rear of signal 40 this would have meant the false clear operation of signal S7.2 and its corresponding distant signal S9.2. All the signals had been working correctly before the accident, signal S7.2 was shown to have been correctly at danger after the collision and the signals had worked correctly since without needing any adjustments or repairs. The likelihood of such a temporary wrong-side signal failure which had corrected itself and had not recurred was, the commission decided, extremely remote, and it was much more likely that Engineer Hess had failed to see them or misread them because of the fog and the speed at which he had been travelling. Time card Rule 551 was to the effect that enginemen must control the speed of their trains according to the weather conditions prevailing and it was apparent that in this instance

[287] Interstate Commerce Commission report.
[288] *Centralia Evening Sentinel*, Centralia, Illinois, 29th March 1916

the rule had not been observed and the high speed of the train through the fog and the short distance at which signals could be observed was one of the proximate causes of the collision. Not only train second No. 86 but all three of the trains involved in the accident had been running fast to make up time despite the fog and the evidence of almost all the employees indicated that the presence of fog made no difference to the speed at which they ran. There was no evidence that the officials in charge were making any checks to ensure that Rule 551 was being complied with and it seemed that they were well aware that trains continued to run just as fast during foggy conditions as in clear weather.

The accident again demonstrated, the report of the commission continued, that even competent and experienced enginemen were not infallible and that even a modern and complete block system could not provide adequate protection against disastrous collisions. There was therefore a need for an automatic train-stopping device and many such devices had in fact been devised and several brought to a comparatively high state of development. If the railroads were to fulfil their obligations to the travelling public steps should be at once taken to avail themselves of such devices and assist in their further development. [289]

The New York New Haven and Hartford Railroad suffered another rear collision on 17th April 1916 at Bradford, Rhode Island, United States. The line here was double track; the northern track, No. 1 was used by westbound trains and the southern, No. 2 by eastbound trains. The line was worked by timetable, train orders and a manual block system. Towers were provided at each block station, usually about 4 miles apart, working switches, stop and distant signals and communicating with one another by means of bock instruments and bells. The system was therefore similar to that normally used in Great Britain at this date and there were no intermediate automatic signals. Track circuits extended from just beyond the most advanced starting signals to the distant signals. The stop signals showed red and green and the distant signals yellow and green. Bradford tower had 24 levers and worked to Wood River Junction to the east and Westerley to the west.

On the day of the accident westbound passenger train No. 633 from Boston, Massachusetts, to New London, Connecticut, was in charge of Conductor George French, Engineer Weatherbee and Fireman Ross and consisted of locomotive No. 1106, a baggage car, a postal car, the Pullman cars *Capita, Phillipston, Trotwood, Lesno, Rushville* and *Mahomet*, a club car and a day coach. Only five miles out of Boston it was noticed that the engine was not steaming properly and as it continued its journey it also continued to lose time for the same reason. It reached Bradford at 7.25 p.m. and stopped just east of the tower. Conductor French and Engineer Weatherbee went into the tower and asked the operator to put them through to the dispatcher for instructions as to what they were to do, given that their train was losing time badly due to poor steaming. The dispatcher told them to take the train to the station platform, discharge its passengers to be collected by the following train No. 25 and then put their train into a siding. Returning to their train, they had just brought it to the platform when it was run into by train No. 25.

The *Gilt Edge Express*, otherwise train No. 25, ran from Boston, Massachusetts, to New York, New York. It was in charge of Conductor Read with Engineer Charles

[289] Interstate Commerce Commission report

Mansfield and Fireman King and consisted of locomotive No. 1338, two Pullman cars, three coaches, a baggage car, an express car, three baggage cars and Pullman cars *Moorefield, Glenford* and *Rutherglen*. This train left Boston on time at 5.34 p.m., passed Wood River Junction five minutes late at 7.27 p.m., passed the Bradford distant signal, which was allegedly clear, was braked on passing the home signal at danger and collided with the rear of train No. 633 at a speed of 8 or 10 miles an hour. The rear coach of train No. 633 was telescoped a distance of 20 feet and fire then broke out in the wreckage, spreading to envelop the whole train, the station building and the freight house, all of which were destroyed. Five train passengers died and 17 were injured. [290]

The passengers who lost their lives in the accident were Thomas Boardman, of Westerly, Rhode Island, a dye broker; Miss Jeanette Clark, daughter of the president of the Westerly mill of the American Thread Company and described as a 'society girl'; Mrs Oscar Martell, of Southbridge, Massachusetts; William Barber, a Westerly carpenter and a woman believed to be Mrs Margaret Bliss, of Wickford, Rhode Island. The passengers for New York, New London and New Haven had all had time to alight from the train before the collision and only the local passengers for Westerly, Stonington, Mystic and so forth remained to suffer in the accident. Those who died were trapped in the wreckage and escaping gas caught alight at once from contact with the contents of the firebox of the colliding engine. The fire soon became intense and later involved three barrels of gasoline stored on the platform of the freight house, which exploded. Volunteer firemen from the Bradford Dye Association Works brought out their fire-fighting apparatus but had to lay over a quarter mile of hose and the volume of water at low pressure which could be obtained was wholly inadequate to deal with the blaze. Further assistance was provided by a motor fire truck from Westerly which was hurried to the scene over eight miles of rough country roads, while medical assistance was provided from Westerly, Ashaway, Shannock, Stonington, Providence, and New London and two relief trains were sent from New London. [291]

Giving evidence at the subsequent Interstate Commerce Commission inquiry, Engineer Weatherbee told of noticing his engine steaming badly at Forest Hills and of losing time onwards to Bradford. There he and Conductor French went to the tower. Towerman McManus was there but first trick Towerman McClusky then arrived and it was he who had contacted the dispatcher and set the road for their train to go into the siding. He and the conductor had then returned to the station and had just drawn it forward to the platform when it was run into. He had seen the distant and two home signals in the clear position; the weather had been slightly hazy but not such as to interfere with observing signals. Five minutes had elapsed between first stopping and returning to his engine, pulling down to the station took a minute and they had been stopped half a minute when they were run into.

Fireman Ross described the weather as hazy but not foggy. He said he saw the distant and home signals at clear and two minutes after they had stopped he noticed the second home signal revert to danger and the dwarf signal cleared for the siding.

Conductor French said that when he was returning to his train after following his

---

[290] Ibid.

[291] *Boston Evening Globe*, Boston, Massachusetts, United States, 18th April 1916

engineman to the tower he had observed the red and white lights of his flagman, about 20 car lengths from the rear of his train; there was no fog between him and the flagman. His train then moved forward, and he was opening the vestibule door to let the passengers out when the collision occurred, extinguishing all the lights of the train. Fire broke out after about five minutes about the middle of the rear coach and near the front of engine No. 1338. Fire extinguishers were brought from both trains. Two of them failed to work and the fire had soon taken too strong a hold for them to have any effect anyway.

Flagman Coombs of the first train stated that he had started back not more than half a minute after his train came to a stand. He saw signal 21, at the west end of the station, was at danger and had not reached a bridge east of the station when he saw the conductor signal for the train to draw forward. When he first saw train No. 25 he was about 170 feet east of the bridge and between it and home signal 23. He immediately waved his red lantern at the engine, lit a fusee and waved it across the track. The train did not stop, and he had had to step aside. He threw the burning fusee, the red lantern and the white lantern at the engine as it went by. Fire was flying from the brakes of the train as it passed him. It had been hazy or foggy. He did not put down any torpedoes as he thought a fusee was better. He did not think that he had gone back far enough to provide full protection as per Rule 99, but it was as far as he could get walking, though he could have gone back further had he run. Bradford was not a regular stop for his train. He fully understood the importance of Rule 99.

Engineer Mansfield testified that he had left Providence, Rhode Island, about four minutes late. The weather had been misty all evening, worse in some places than others, but not unusual in that area and he had had no difficulty in seeing signals until he reached the home signal at Bradford. Approaching Bradford, he had seen its distant in the clear position, had shouted to his fireman, "High ball," and the latter had looked and made some reply. Because of the river the fog had been thicker at the home signal, and he had not seen it until the last moment, when he saw it was at stop. He had immediately made an emergency brake application and applied sand. At the same moment that he saw the home signal he had seen the flagman and the rear markers of train No. 633. He had expected to pull up in time but struck the train in front at 8 or 10 miles an hour. His brakes had been in excellent condition and had worked properly at Woodlawn, just beyond Providence. The rails had been slippery, but he thought the sand had prevented his engine slipping. He had not seen the lanterns or fusee of the flagman.

Fireman King stated that he had just finished attending to the fire when his mate had called out, "High ball," and he had looked out and seen the distant off. He had then begun raking the fire when he felt the brakes being applied and he looked out and saw the light of what he took to be the home signal, the flagman and the rear markers of train No. 633. He did not notice any fusee or lantern thrown at the engine, nor any fog.

Flagman Camp of train No. 25 went back after his train had stopped and met Flagman Coombs. They found Coombs' red lantern just east of the east end of the bridge and there was a fusee sticking in the end of a tie close by. Ten minutes after the collision he saw the distant showing a yellow light. There was no fog and the home signal, showing a red light and the rear markers of train No. 25 were plainly visible from the distant signal.

Towerman McManus said he had come on duty at 7 p.m. The distant signal, 24 and home signals 21 and 23 were then all in the normal position. When train No. 633 had entered the block at Wood River Junction he had cleared the signals. When the train passed the distant and first home signals he had replaced them and unlocked the block to permit train No. 25 to enter the section at Wood River Junction. He could not have replaced the home signal unless the distant signal had been replaced, nor released the block to Wood River Junction unless both signals were normal. Train No. 633 stopped and its conductor and engineer came to his tower for instructions. The weather had been a little hazy, but he could see signal 20 about 3,000 feet west of his tower.

Towerman John McClusky said that after Towerman McManus had taken over he had gone to the station and had been there when train No. 633 had stopped at the tower. Signal 21 was clear for the train and wondering why the train had stopped, he had returned to the tower, where Towerman McManus had asked him to take over. He had done so and had telegraphed the dispatcher for instructions and then restored signals 20 and 21 to normal and set the road for the siding. Seeing train No. 25 approaching he had shouted to McManus, "Twenty-five has run the boards," whereupon McManus had gone to a window and shown a red lantern to the train. Levers 23 and 24 working the first home and distant signals had been normal; had they not been, the locking would have prevented him setting the road for the siding. The weather had been hazy but not foggy and he could see the eastbound block signal 2,400 feet from the tower. [292]

By chance Signal Inspector Nicholson was riding on train No. 25. Alighting after the collision he was met by someone whom he later learnt was the tower operator who told him to, "Go back and look at the signal." He said that he had done so, running part of the way, and found it in the stop position with the lamp burning. Accompanied by the flagman of train No. 25 he then went back to the distant signal and found its blade in the horizontal position with a yellow light displayed. There was a light fog, but not one likely to interfere with observing signals. Back in the tower he found levers 23 and 24, working the first home and distant signals, in the normal position and told the towerman to secure them so that they could not be moved.

Signal Repairman Anderson stated that he had examined, cleaned, and adjusted the signalling apparatus on 15th April as part of a routine check and all had been working correctly. Distant signal 24 was worked electrically and there had been one failure in the previous year, due to a Western Union telegraph wire breaking and falling across the signal wires between the home and distant signals. The problem had been at once noted and fixed. The only other wrong-side failure of a distant signal in his area had been at Wood River Junction the previous year due to water getting into a circuit breaker.

Signal Engineer Morrison stated that with a train standing in the position of No. 633 the distant signal was both electrically and mechanically locked. Unless the home and distant signals levers were normal Bradford could not have given permission to Wood River Junction for train No. 25 to approach. If the distant signal had been clear for train No. 25 it could only have gone to clear after the release had been given and he could think of no way in which this could have happened other than deliberate tampering. He listed three wrong-side failures, two of signal 5 due to damage to the wire carrier causing the arm to stick off but which would have prevented the towerman from releasing the block, and the failure of the distant signal described by Signal Repairman Anderson, due to a steam shovel falling on a telegraph wire. He considered the signalling system as safe as the more modern signalling at the western end of the line, just not capable of handling

[292] Interstate Commerce Commission report

as much traffic. Estimates for a more modern system were being prepared and tests were being made on various systems of automatic train stops.

The Interstate Commerce Commission report noted that despite the engineer and fireman of train No. 25 stating that the Bradford distant signal had been clear, it had been found to be at caution by a signal inspector and flagman immediately after the accident and various tests had shown nothing wrong with it. It must be concluded therefore that the accident was due to the failure of Flagman Coombs to properly protect his train and of Engineer Mansfield to obey signal indications.

Rule 99 stated most positively that block and interlocking signals would not relieve flagmen from the duty of protecting their trains. Train No. 633 had stopped at 7.25 and was struck at 7.32. Flagman Coombs therefore had seven minutes to protect his train but only got back 1,120 feet and by his own admission did not start off back at once. Had he used all the time at his disposal he could have covered 2,200 feet on the basis of subsequent tests.

Investigations and tests showed that the signalling was working normally and the possibility of the distant signal showing a false clear was exceedingly remote. There was evidence of some fog but not sufficient to interfere with the viewing of signals and in any case, if the signal was obscured by fog the speed of the train should have been reduced. Moreover, the rear of train No. 633 was 1,400 feet west of the home signal. At the speed at which Engineer Mansfield stated he was running an emergency brake application at the home signal should have enabled him to stop within this distance. He claimed that he could not pull up due to the bad condition of the rails and the wheels sliding.

The cars of train No. 633 had been of wooden construction and gas lit. Had they been of steel construction they would not have been destroyed by fire and the lives of some of the passengers who were caught in them and cremated might have been saved. [293]

The coroner's inquest placed the responsibility for the accident squarely on Engineer Mansfield. [294] The engineer's trial in October lasted several days but on Saturday 21st he was found not guilty. It took the jury eight hours to reach their verdict. [295]

On the night of Wednesday 30th August 1916 the down Lahore-Calcutta mail train ran into the rear of a goods train standing in Kalanour station, India, on the North Western Railway line between Ambala and Saharanpur. Two European and seven Indian passengers were killed, and five European and 20 Indian passengers were injured. Most of the casualties occurred in the two leading vehicles of the mail train, which were wrecked, while the brake van and five waggons of the goods train were 'completely smashed.' From the extent of the damage it was estimated that the mail train must have been travelling at at least 30 miles an hour at the moment of collision.

A departmental inquiry found that the driver of the mail train was at fault, as he had run past all the signals which were set at danger. He was subsequently prosecuted and sentenced to six months' rigorous imprisonment. [296]

---

293 Ibid.

294 *Norwich Bulletin*, Norwich, Connecticut, United States, 17th May 1916

295 *The Daily News*, Frederick, Maryland, United States, 23rd October 1916

296 Railways in India: Administration Report 1916-17

Preußisch Staatsbahnen Vorzug (relief train?) D24 was standing at Jahnsfelde block post, Germany, at 4 a.m. on Sunday 8th October 1916 when it was run into by Eydtkuhnen (now Chernyshevskoye in Russia) to Berlin Hauptzug (main train) D24. The first train had come to a stand due to an engine problem and a signalman had irregularly cleared it back and accepted the second train, which was following ten minutes behind. The collision was a violent one and 12 passengers died, 4 were seriously injured and 11 were slightly hurt. [297]

Early on the morning of Sunday 15th October 1916 two stock trains were in collision near Smithfield, Nebraska, United States, on the Chicago Burlington and Quincy Railroad. Traffic over the single line involved in the accident was very light and only one train operated over it at night, No. 156, due to leave Curtis, Nebraska, at 10.45 p.m., and which ran only once a week, on Saturdays. On Saturday 14th October though this train was run in three sections, and it was the second and third sections which collided. The line was operated by telegraph and train orders, supplemented by a manual block system for trains running in the same direction, which was absolute for passenger trains and light engines but permissive for freight trains. At night Curtis was the only open telegraph office in the 229½ miles between Sterling, Colorado and Holdrege, Nebraska.

Train second No. 156 left Curtis at 12.50 a.m. on Sunday, 2 hours and 5 minutes late on the schedule of the first section, with engines Nos. 1232 and 1236, 44 cars of stock and a caboose. It was in charge of Conductor Sawyer with Engineers Ward and Hansen. Twenty minutes were lost at Eustis taking coal and water and another 20 minutes at Elwood dealing with a hot box and detaching the helper engine, No. 1236. The train then passed Smithfield but had to stop 2½ miles further on because of another hot box.

Train third No. 156 had originated at Holyoke, as eastbound extra No. 1226, consisting of engine No. 1226 and consisted from that point of only two cars of stock and a caboose, but another 19 cars were taken on at Grant, Nebraska. Conductor Albro, Engineer Patten and Fireman Reiber accompanied the train. Arriving at Curtis at 12.40 a.m. its crew were given order No. 67, reading: 'Engs. 1227, 1232 and 1236 run as first, second and third 156 Curtis to Holdrege,' together with a permissive card warning to the crew to expect a train in the block between Curtis and Holdrege. The train left Curtis as third No. 156 at 1.01 a.m., 11 minutes after second No. 156. Having seen a fusee thrown down by second No. 156 it waited ten minutes at Moorfield, the first station out, and there was another ten-minute stop at Eustis for the same reason. Another half hour was lost at Eustis taking coal and water and trying to fix the electric headlight. As the headlight could not be fixed a white lantern was substituted, and the train proceeded until it collided with second No. 156 at 4.10 a.m., while travelling at 8 or 10 miles an hour. In the collision the caboose was telescoped for almost its entire length by the stock car in front of it. The caboose was crowded with stock caretakers (drovers), 11 of whom were killed and 15 injured.

Engineer Ward told the Interstate Commerce Commission inquiry into the accident that when Head Brakeman Cougher, who was riding on the engine, told him that there was another hot box he had stopped the train and whistled out the flagman. There had

---

[297] *Eisenbahnunfälle Bind 1* by Ludwig Stockert, pg. 34 (in German)

been a drizzling rain, but he could easily see the green markers at the rear of his train and considered that the markers could have been seen for 40 or 50 or perhaps 60 rods (330 yards) from a following train. He was not immediately aware that his train had been run into. He thought his train had been stopped four or five minutes when the crash occurred.

Engineer Ward's fireman, named Harriman, thought they had only been stopped 2 or 2½ minutes and said that visibility had been variable.

Conductor Sawyer stated that no fusee was thrown off when the train slowed up and stopped beyond Smithfield as it should have been in accordance with Rule 99. He did not know why. With 31 stockmen riding in the caboose it had been difficult to get out of its rear door, but he had followed the flagman out of the front door. He had assumed that the flagman had gone back at once but having gone three of four car lengths up the train his flagman had shouted that the hot box was on his (the flagman's) side. The conductor said that he told the flagman to go back and protect. The conductor thought that the train had been stopped about two minutes and said that visibility had been good despite a little rain.

Flagman Talbot agreed that he had at first gone forward when the train stopped beyond Smithfield, he said he had done so because he thought he could save time by telling the conductor where the hot box was. He then started to go back but the collision occurred before he reached the caboose. He knew that he should have gone back at once but did not think it would have made any difference in this instance as there had been no time, though he admitted that it would have been better if he had thrown out a fusee earlier.

Engineman Patten of third No. 156 said that he had seen the second section in front at Moorfield and had been flagged at Eustis. Unable to fix the electric headlight at Eustis he had substituted a white lantern. He had expected to see the second section at Elwood where he knew the helper engine would be taken off but had not seen it there. He had seen the rear markers of the second section 20 or 25 car lengths away and thinking that it was moving initially only made a service brake application. Realising that it was stationary, he then made an emergency brake application and opened the sanders. It seemed at first that he would be able to pull up, but the driving wheels had begun to slide, and the train had surged forward. About 100 feet away he had seen a red light being waved from the caboose, being shown, apparently, by one of the stockmen. Had a fusee been thrown off or if someone had waved a red lantern earlier he thought he could have stopped in time despite the weather conditions.

Fireman Reiber was firing when Engineer Patten braked and although he then looked he could see nothing at first on account of the glare from the fire.

The Interstate Commerce Commission placed the responsibility for the accident on Conductor Sawyer and Flagman Talbot for having failed to protect their train. They rejected the flagman's assertion that he would not, in any case, have had sufficient time to go back far enough and stated that there could be no excuse whatsoever for such utter negligence in the performance of his duties by an experienced flagman. While the conductor knew that his flagman had protected the train properly at other points he should still have checked definitely that he was doing so again.

Given the fact that train No. 156 was the only night-time train over the division, and then only once a week, and that the daytime service consisted of only one passenger train each way daily and a freight train each way three or four times a week the commission was satisfied that the method of operation was sufficient for the safe running of trains provided that the rules were properly enforced and obeyed. [298]

At the inquest on the victims, held at Smithfield on the 16th and 23rd, one of the stockmen testified that when they had got to Curtis they had gone into the office and asked for a coach to be attached to the train, as there was not enough room in the caboose for them all to sit down, but rough language had been used and they had been told that they could not have one. The jury returned a verdict blaming the conductor and flagman for failing to protect their train, and the Burlington company for failing to provide proper accommodation for the stockmen, running an engine without a headlight, and not providing agents at night to keep trainmen informed of their relative positions to one another. [299]

Five or six people died and about the same number were injured in a rear collision on the Canadian Pacific Railway at De Beaujeu, Quebec, Canada, just west of Saint-Polycarpe, on Wednesday 27th December 1916, when a Toronto-Montreal express ran into the rear of a local train standing on a siding at 6.40 p.m. A semaphore signal was at danger against the express but was presumably not seen by the engineer of the express in conditions of poor visibility, with thick fog and a heavy fall of sleet. Those injured included Conductor Hilton of the local train and Canadian Pacific Bridge and Building Master Malley. Although the rear cars of the local train were demolished the express was little damaged and it later continued to Montreal with the dead and injured. [300] [301]

Wartime censorship was possibly the reason for only very brief accounts appearing in the French press of a major disaster at Roncherolles-en-Bray, France, 4 kilometres from Serqueux on the Amiens – Rouen line, while although the victims were British soldiers no mention of it at all appears in the British press. The passenger train concerned in the accident had left Le Havre at 18:30 on Tuesday 13th February 1917 and was packed with 1,100 soldiers who were returning from leave. At about 21:00 it collided with three waggons which had become detached from a train carrying supplies from England. The force of the collision was sufficient to completely crush five old wooden carriages in the middle of the leave train. At least 26 soldiers lost their lives, and they were buried in a military funeral on 16th January at Bois-Guillaume. Many others were injured. [302] [303] [304]

Towards the end of February 1917, on the 27th of the month, there was another serious rear collision in the United States, with the familiar features of fog, an engineer overrunning signals and the negligence of a flagman. This accident happened on the Pennsylvania Railroad, at Mount Union, Pennsylvania. The line here was quadruple track with automatic signals, in addition to which there were towers at approximately five-mile intervals controlling crossovers between the running lines. The tracks were numbered 1

---

[298] Interstate Commerce Commission report
[299] *The Nebraska State Journal*, Lincoln, Nebraska, United States, 25th October 1916
[300] *Winnipeg Free Press*, Winnipeg, Manitoba, Canada, 28th December 1916
[301] *Logansport Daily Tribune*, Logansport, Indiana, United States, 28th December 1916
[302] *Le Figaro*, Paris, France, 15th February 1917
[303] *Le Matin*, Paris, France, 19th February 1917
[304] *Journal de Rouen*, Rouen, France, 15th and 17th February 1917

to 4 from the south, tracks 1 and 2 being for eastbound trains, passenger trains usually running on track 1 and freight trains on track 2. MU tower was a mile west of Mount Union station.

Eastbound passenger train No. 6 consisted of engine No. 748, two mail cars, a combination car, a coach and four Pullman sleeper cars with Conductor Stewart, Engineer Gearhart and Fireman Parks. It left Altoona, Pennsylvania, at 10.45 p.m. and at Tyrone, 16 miles from Altoona, and again at Huntington, 19 miles further on, there was trouble with brakes binding. Approaching Mount Union, it was foggy and unsure of his exact location Engineer Gearhart braked a little early for the station. When he released the brakes those on one car stuck again, so it took longer than usual to cover the mile from MU tower to the station. Then when the work at the station was completed the train would not restart after several attempts, again due to the brakes sticking. Fireman Parks had just identified the sticking brake as being on the first car, had released it and had just got out from under the car when the train was run into by eastbound freight train PG 20.

The freight train consisted of locomotive No. 614, 35 loaded cars and a caboose, with Conductor Fagan, Engineer Thomas Cook and Fireman Thomas. It had left Altoona at 11.59 on track 2, was diverted to track 1 at FR tower, to track 2 at SC tower and back to track I at PG tower because of a slow freight train on track 2. Passing MU tower at 12.06 a.m. it failed to reduce speed at the green signal on signal bridge 1912, ran past the red signal on signal bridge 1904 and struck the rear of the passenger train 276 feet beyond. The force of the collision drove the passenger car forward 193 feet and its rear vehicle, the Pullman car *Bellwood*, was almost completely telescoped between the engine of the freight train and the next Pullman car, *Bruceville*. Nineteen passengers and a Pullman porter in the *Bellwood* were killed, while in the *Bruceville* only one passenger was injured, and some did not even wake up. [305]

It was reported that after the accident it was the women passengers who remained calm while the men were panicking. 'Immediately after the crash men ran wildly about sleeping coach No. 13, according to Mrs Arthur Pritchard, of Ithaca, N.Y. The number, Mrs Pritchard said, had proved lucky, since not a single occupant of the car had been injured. It was the third coach from the end of the train. "I was lying my berth and was about dozing off when I felt a jolt," declared Mrs Pritchard. There was a rumbling noise after the crash. I looked out of my berth and saw men jumping out of their berths and running up and down the car.

'"In a short time, the train crew came through our car and quieted everybody, telling us there was no further danger and we could take our time about dressing. The men then went to the assistance of the women and helped them out. Of course, everybody hurried out with clothes in their hands as soon as they learned what had happened, but there was no further excitement. In fact, a death-like stillness prevailed, everybody realizing that some had been killed. On account of the condition of the wrecked car it was impossible to help anybody in the debris. The men passengers in the train did what they could to pull the wreckage away, but they did not succeed in helping anybody. They kept at it with lamps until the wrecking crew arrived. Residents of Mt Union were very kind to us. They brought us hot coffee and threw open their homes to us."' [306]

---

[305] Interstate Commerce Commission report

[306] *The Philadelphia Inquirer*, Philadelphia, Pennsylvania, United States, 28th February 1917

Conductor Stewart told the Interstate Commerce Commission (ICC) inquiry into the disaster that his train had stopped at about 12.01 a.m. He thought that it had been stationary about six minutes before the collision.

Flagman Samuel Jacobs said that he had not noticed the previous problems with the brakes except perhaps at Huntingdon. He did not know what time they stopped at Mount Union, but it was 12.02 when he climbed down to the track. He had helped unload and load the passengers and after about three minutes the train was ready to restart but the head brakeman then came back and told him they were having trouble with the brakes. He was about signal bridge 1904 when the westbound Mount Union passed him at 12.03 or 12.04. Train PG 20 passed him when he was 20 or 30 feet beyond the signal bridge. He gave it stop signals, which were not acknowledged, and had a fusee, but did not stop to light it. The commissioners pointed out that although he claimed to have been walking back fast for 4½ or 5 minutes he had only covered 300 feet.

Brakeman Folk did not think that half a minute elapsed between the flagman starting to go back and the collision.

Engineman Cook of freight train PG 20 stated that the automatic signals on bridge 1912 east of MU tower for tracks 1 and 2 had both distinctly been white. He had called his signal 'white' and Fireman Thomas had queried if it had not been a green caution signal. By then it was too late to look at the signal again as with the fog it could only be seen close-up. He had not ignored what his fireman said but thought that his vision might have been temporarily interfered with from working on the fire and he was confident that the signal had been clear. With the noise on the engine he could not hear what the head brakeman said. He saw the signal on bridge 1904 was red but with the fog he was then only 5 car lengths from it. He shut off steam, opened the sanders and made an emergency brake application but travelling at 40 miles an hour he could not pull up in time. He had not seen the flagman or received any warning before he saw the red signal.

Fireman Thomas said that both he and Head Brakeman Barr had called the signal green, but Engineer Cook had insisted it was white and did not apply the brakes. The fireman then wondered if his sight had been affected by looking at the fire. He saw the red signal at about the same time as his engineer and the flagman near the signal bridge.

Conductor Fagan put the speed of his train at 48 miles an hour a few moments before the shock of the collision.

Signalman Clemens (as he is referred to in the ICC report, rather than the more usual towerman or operator) said that the passenger train had passed him at 11.53 and the freight train at 12.06. With the shifting fog he could at times see both signals on bridge 1912 and at times neither of them. One of the signals had been clear when the freight train passed it, but he could not tell which. The other signal had been obscured by smoke from the freight train engine. He was not aware of any recent trouble with the signals.

Tests conducted on the brakes of the first car of the passenger train showed that there was in fact nothing wrong with them, and it was concluded that the problems had been caused by a 'slight irregularity in their manipulation' with subsequent attempts to release them merely aggravating the problem instead of solving it.

The ICC concluded that the cause of the disaster was the failure of Engineer Cook to correctly observe the caution indication shown by the signal on bridge 1912, despite the fact that the two other men in the cab had seen it. His colour vision had been tested on New Year's Eve 1915 and found to be correct. He was clearly someone who did not lack self-confidence and was so sure that he was right that he was willing to take an unnecessary risk. Had he assumed that he was right but reduced speed just in case the fireman was right then no harm would have come about other than losing a couple of minutes with a freight train. As it turned out the fireman had been correct and by not acting on the side of caution he had caused a disastrous collision.

The commission added that a very material contributing cause of the accident was the failure of Flagman Jacobs to properly protect his train. By the best evidence he had had at least four minutes between the time that he was told that his train would be delayed and the freight train passing him but had only managed to get back 500 feet. The commission concluded that this lapse from duty was inexcusable, especially in view of the dense fog, and that the flagman must bear his full share of responsibility for the collision.

The vague and indefinite character of the flagging rule was also criticised. Rule 99 of the American Railway Association Standard Code read: 'When a train stops or is delayed, under circumstances in which it may be overtaken by another train, the flagman must go back immediately with stop signals a sufficient distance to insure full protection. When recalled he may return to his train, first placing two torpedoes on the rail when conditions require it. When conditions require it a fusee must be used.'

The commission noted that the rule was full of phrases such as 'when conditions require it' which left matters to the judgement and moral responsibility of the flagman. While conceding that to some degree matters must rely on the judgement of the flagman the commission thought that the rule should be amplified to suit the varied operating conditions of various railroads. It was also noted that on lines with automatic signalling these tended to be relied upon and the flagging of trains neglected.

Turning to the signals, the ICC noted that these had been installed in 1913 and their electrical and mechanical features represented the highest development in signalling. They were semaphores giving three indications in the upper quadrant. There was no overlapping so that only one danger and one caution aspect was displayed behind a train. The signals were electrically lit, showing red for danger, green for caution and white for clear, but a change to the more usual yellow for caution and green for clear had already been authorised. Tests made after the accident had shown no indication of a tendency for the signal on gantry 1912 to wrongfully show clear instead of caution.

The report concluded with considerations as to the strengthening of rolling stock so as to prevent telescoping, the danger of running trains at full speed in foggy conditions and the need for the further development of a system of automatic train protection. [307]

On 12th March, the coroner's jury at the inquest on the victims at Huntingdon returned a verdict that, 'We find that the collision was caused by the neglect of duty of Engineer A.T. Cook of the freight train and Flagman Samuel K. Jacobs of the express train. We recommend that they be held in the sum of $2,000 bail each to answer before the grand jury of Huntingdon county, at the ensuing term of criminal court, on the charge of involuntary manslaughter.' Cook and Jacobs were then arrested, furnished bail, and were

---

[307] Interstate Commerce Commission report

released. [308]

On 10th April Superintendent N.W. Smith announced that the engineer and flagman had been dismissed.[309]

No report of any trial of the two railroad men has been discovered.

According to the tachograph passenger train Südbahn passenger train No. 31 approached the Einfahrsignal (home signal) at Divača, Austria (now in Slovenia), travelling at 52 miles an hour and it was still travelling at 40 kilometres an hour when it passed the signal at "Halt" and collided with goods train No. 811 standing in the station on Sunday 8th April 1917. The engine and 11 vehicles were derailed, and 15 vehicles were considerably damaged. Ten 'officers' (railwayman?) were killed, one person was seriously and 17 were slightly injured. [310]

Two eastbound Chicago Burlington and Quincy Railroad freight trains were involved in a rear collision at Earlville, Illinois, United States, late on the evening of Sunday 16th September 1917. The trains involved were extra No. 5006, consisting of engine No. 5006, 60 cars of livestock, a wooden coach for the stockmen and a caboose, in charge of Conductor John Fisher with Engineer Eskdahl; and extra No. 5307, consisting of engine No. 5307, 67 cars and a caboose, in charge of Conductor Carolan with Engineer A. C. Carlson and Fireman J. F. Long.

Train No. 5307 was in front as far as Mendota, where it was overtaken by train No. 5006. Train No. 5006 left Mendota at 10.33 p.m. and at 10.55 p.m. came to a stand at an automatic signal at Earlville, where another extra train was busy shunting. At about 11.05 p.m. it was run into by train No. 5307. The caboose and coach of the stock car were telescoped into one another and demolished, several other cars of the stock train were damaged and or derailed, engine No. 5307 was derailed, and ten cars of the colliding freight were demolished by buckling. Seven stockmen were killed, and five stockmen and Engineer Carlson were badly enough hurt to be taken to hospital.

Engineer Carlson told the Interstate Commerce Commission (ICC) that train No. 5006 had been in sight when he left Mendota and all the automatic signals had been clear to the next station at Meriden, by which time the rear lights of train No. 5006 had disappeared. He had not called out any of the signals to his fireman nor to Head Brakeman G.J. Hickman, who was also riding on the engine. He had not spoken to his fireman because the latter was new and having trouble with his fire, nor to the brakeman because he was an experienced man who was familiar with his duties. He remembered whistling for one road crossing, then admitted that he had fallen asleep, not waking until he saw the rear lights of train No. 5006 about 10 cars ahead, when he had applied the brakes in emergency. Had he been flagged or awoken a little earlier he could have stopped in time. He said that he had had a bad cold, for which he had taken a day off sick, and had taken some pills for his cold he had obtained from a drug store before he set out. He did not know what the pills had contained but thought that they must have made him drowsy (c.f.

---

[308] *Altoona Mirror*, Altoona, Pennsylvania, United States, 12th March 1917
[309] *The Daily News*, Frederick, Maryland, United States, 10th April 1917
[310] *Eisenbahnünfalle Bind I* by Ludwig Stockert, pg. 35 (in German)

the Ivanhoe collision of 1918, below).

Head Brakeman Hickman said that three distant signals between Mendoza and Meriden had been at caution. but the home signals had been clear. He had not drawn the engineer's attention to the caution signals. Approaching Earlville he had seen a flagman about ten car lengths from the rear of the stock train with red and white lights and although Engineer Carlson had not acknowledged the flagman he (Hickman) had presumed that Carlson had seen him and would pull up. It was only when it had become clear that there was going to be a collision that he had called out to the engineer and fireman that they were going to hit and to jump. At that time, his train was still travelling at 20 miles an hour and still working steam.

Fireman Long stated that he had been making his first trip over the line and he knew nothing of the automatic signals. He had not known that there was anything wrong until the head brakeman had shouted for him to jump.

Conductor Fisher said that the electric headlight of train No. 5307 had been visible continuously since they had left Mendoza. Brakeman E.W. Johnson, riding in the caboose with him, had remarked that train No. 5307 was close behind them but had not dropped off a fusee or made any effort to protect their train until it had stopped, when he had immediately alighted and begun running back, getting back perhaps 25 or 30 car lengths. When he (Fisher) had realised that train No. 5307 was not stopping he had gone into the coach and warned the stockmen to alight as quickly as they could.

Brakeman Johnson testified that it had not occurred to him to throw off a fusee although there was another train close behind them because he had not expected his train to have to stop, believing that the extra ahead would pull into the siding out of their way. He alleged that he had got off as soon as his train had stopped and gone back with lights, fusees and torpedoes, but he had not lit a fusee nor put down any torpedoes, because he was depending on the automatic signals. He had got back 25 car lengths when train No. 5307 passed him and admitted that had he hurried he could have got back 40 or 60 car lengths. At the last moment he had also thought of putting down a torpedo but by then the train had been too close.

The ICC concluded that the direct cause of the accident was the failure of Engineer Carlson to observe and obey signals. Contributory causes were the failure of Brakeman Johnson to properly protect his train, the failure of Conductor Fisher to see that he did so and the failure of Head Brakeman Hickman to draw the attention of Engineer Carlson to the need to stop his train. The ICC was also critical of the failure of the crews of either train to call out the signal aspects to one another, of Engineer Eskdahl to whistle out a flagman and of Engineer Eskdahl, Conductor Carolan and Brakeman Hickman to treat an unlit signal as a stop signal. [311]

The inquest on the stockmen found Engineer Carlson, Fireman Long and Brakeman Hickman all criminally negligent and guilty of manslaughter. [312] It has not been possible to find a record of a trial of any of the men, so the charges were probably later dropped.

Two trains collided on the Madrid-Irún main line in Spain at 12.45 a.m. on Sunday

[311] Interstate Commerce Commission report

[312] *The Earlville Leader*, Earlville, Illinois, United States, 20th September 1917

23rd September 1917, the accident occurring between Matapozuelos and Pozaldez. In a situation reminiscent of the 1913 Ais Gill disaster in England, Santander – Madrid mixed train No. 84 came to a stand in the section because of low steam pressure in its engine, supposedly due to the poor quality of the coal being used, less good-quality English coal being employed because it had become too expensive due to wartime inflation. The train had already had to stop eight times for the same reason between Santander and Palencia. The collision occurred because train No. 26, the Irún – Madrid mail, which was also badly delayed, was wrongly allowed to leave Matapozuelos before the mixed train had reached Pozaldez. The third-class carriage at the rear of mixed train, which had been added at Valladolid, was also not carrying any tail lights. The rear vehicles of the mixed train were telescoped by the locomotive of the mail train. According to the usual procedure the passenger carriages had been marshalled behind the freight waggons. Thirteen passengers died in the accident and 37 were injured. [313] [314]

Early on Monday 24th September 1917 a troop train, No. 858, on the Südbahn was admitted into the section between Rakek and Planina stations, Austria. It reached Planina safely, and the arrival track was clear, but the Einfahrsignal (home signal) remained stubbornly at red, so its driver stopped in the rear of it and waited for it to clear. Meanwhile Rakek had goods train 812 waiting to follow the troop train. After the expected interval of time had elapsed and Planina had not reported the troop train as having arrived the Rakek fahrdienstleiter (signalman) telephoned his colleague and asked if it was there. The reply was, "Nein, obwohl Einfahr frei." ('No, but the road is clear for it.') The goods train was therefore stopped at Rakek.

A little later Rakek advised Planina, "Zug 812 hier," ('Train 812 here.' That is, 'I am waiting for you to clear back the troop train.') and the Planina fahrdienstleiter replied, "Weiter hahren lassen," ('Let her come on') although the troop train was still stopped at his Einfahrsignal. He could not explain afterwards why he had done so but said he thought that he must have confused the two trains. Train No. 812 was then allowed to leave Rakek. Its driver found the Planina Vorsignal (distant signal) at clear, then suddenly saw the tail lamp of the troop train. At about the same moment the driver of the troop train realised what was happening and hurriedly opened his regulator wide to try and pull away. The sudden shock was too much for the couplings and the troop train became divided into four parts, the rearmost three coaches running back down the incline to collide with the oncoming engine. Twenty-seven soldiers were killed outright, two were fatally injured, 16 soldiers and two railwaymen were seriously injured, and 23 soldiers and ten railwaymen were slightly hurt. [315]

A tragic accident in Germany on 16th October 1917 led to the death of 25 children. It happened at Schönhausen (Elbe) on the Berlin – Lehrte line of the Preußisch Staatsbahnen. The children were among 250 who were returning home to Essen on a holiday special, train No. 2240. Preceding the passenger train was a westbound freight

---

313 *The Times*, London, England, 24th September 1917

314 Webpage – Federación Castellano Manchega de amigos del ferrocarril: Accidentes ferroviarios en España (in Spanish)

315 *Eisenbahnünfalle Bind I* by Ludwig Stockert, pgs. 25-26 (in German)

train. The driver of the latter found the Einfahrsignal (home signal) at Schönhausen clear for him but pulled up anyway as he was due to stop there. The driver of the following children's special overran the stop signal at the preceding station and when his train approached Schönhausen he saw the Einfahrsignal still standing clear for the freight train. Misled by this, he only braked at the last moment when he saw the tail lamp of the freight train in front of him. The collision therefore occurred at quite a high speed and several of the carriages of the passenger train were smashed. In addition to the children a conductor lost his life and 15 children, and an escort were injured. The driver and fireman of the colliding train jumped and escaped injury. [316]

By the end 1917 the United States had been drawn into the First World War and thus an accident on 12th November 1917 involved two troop trains. The accident occurred near Cotopaxi, Colorado, on the Denver and Rio Grande Railroad single-track main line between Salida and Pueblo. Four trains were indirectly involved, the first being a freight train, No. 1134, and the second eastbound passenger train No. 16, which on the day of the accident was running as first No. 16, the following troop trains being run as second and third Nos. 16.

The events which led to the accident began with the need to remove some bad order (defective) cars from the rear of the freight train. The engine of train first No. 16 was busy removing these when the leading troop train, second No. 16, came up behind it and stopped at about 4.50 a.m. Train second No. 16 consisted of locomotive No. 1001, six baggage cars, nine Pullman sleeping cars, six Pullman tourist cars and a caboose, and was in charge of Conductor Schmelzer and Engineer Greenman. It came to a stand rather more than half a mile short of Cotopaxi station and was stood there for about an hour. It had just got moving again, had covered about 500 feet and was moving at 4 miles an hour when it was run into by third No. 16.

Third No. 16 consisted of locomotive No. 1177, three baggage cars, five Pullman sleeping cars, another sleeping car, five Pullman tourist cars, eight box cars of 'government supplies' (munitions) and a caboose. It was in charge of Conductor Kemp with Engineer Bowsher and Fireman Burgess. This train left Salida at 3 a.m. and was held by the dispatcher at Swissvale, the last office before Cotopaxi, which was open at night, until he could find out what was delaying first and second Nos. 16, Cotopaxi itself not being a night telegraph office. The train left Swissvale at 4.50 a.m. and ran into the rear of second No. 16 at 5.33 while running at 10 or 15 miles an hour. At the rear of train second No. 16 was a combination mail and baggage car used as a hospital car and this forced the caboose against the Pullman car ahead of it, which was telescoped for half its length. Three soldiers were killed outright, two more died of their injuries, and 18 soldiers and five railroad men were injured. [317]

The soldiers on the trains were mainly infantrymen from Salt Lake City, Utah. The day after the accident schools were closed and businesses suspended in Salida, Colorado, while the whole town assembled to honour three of the soldiers. [318]

Conductor Schmelzer stated at the Interstate Commerce Commission inquiry that

---

[316] *Eisenbahn-Katastrophen in Deutschland* by Hans Joachim Ritzau, pg. 91(in German)
[317] Interstate Commerce Commission report
[318] *Nebraska State Journal*, Lincoln, Nebraska, United States, 14th November 1917

when his train had stopped at Cotopaxi he had told his flagman to go back, and the latter had set off at once. He (the conductor) had then gone forward to the station and at 5.20 after getting a Complete order had come out to give his engineer the signal to proceed. Trees and obstructions prevented him signalling directly to his engineer, so he got the engineer of a train on the passing track to whistle a proceed signal. His engineer then in turn whistled in the flagman. Conductor Schmelzer admitted it was not in accordance with Rule 99 to call in the flagman before the train started when there was a passenger train following but said he assumed that his train would stop again on a straight stretch at the station where there was a good view from a following train.

Flagman Lewis of train second No. 16 stated that he had alighted before his train had quite come to a stand with his lantern, two red fusees and he thought three torpedoes. When he got back about eight or ten telegraph poles he found that he only had one torpedo, so he put that on the rail and lit a fire. From that point he could see back about a mile but not the rear of his train because of a rock bluff and cutting. After about an hour he was called in. He had then lit a ten-minute fusee, stuck it in a tie (sleeper) and run back to his train, stopping to pick up the torpedo he had put down. It took him about five minutes to get back to his train and he had just climbed into the caboose when he had turned around and seen train third No. 3 approaching. About 40 minutes after the accident he had gone back with a military officer and another man, and they had found the half-burned fusee. He said that he had picked up the torpedo as he had not been expecting another train for some time and thought that if he had left and it was exploded later by a following train it would cause unnecessary delay. He had taken a written examination of the rules on 10th July and had read a circular letter to the effect that every possible precaution must be taken with the movement of troop trains.

Engineer Bowsher of train third No. 16 said that when he arrived at Swissvale at about 3.15 a.m. the operator there told him that first and second No. 16 had not yet passed Cotopaxi and he was held at Swissvale until 4.50 a.m., when his conductor gave him an order to meet westbound passenger train No. 15 at Fernleaf, the same place as first and second No. 16, and he had then proceeded. Approaching Cotopaxi, his train had been travelling at 20 miles an hour and he had seen no fusee and had no warning of second No. 16 until the headlight of his engine had shone on the windows of its cars, when he had made an emergency brake application and jumped. Conductor Kemp, Fireman Burgess and Flagman Porter of third No. 16 all testified to either having seen no fusee or only an unburnt one.

The commission concluded that the accident had been caused by the failure of Flagman Lewis to properly protect his train by failing to ensure that he had enough torpedoes, and by picking up the one torpedo that he had put down. As for the fusee which he alleged he had lit and which he said he had found half burnt after the collision the evidence of all the other witnesses was that it had not been burnt at all but just broken, while no one on third No. 16 had seen any fusees burning. Flagman Lewis had also failed to remain out when he knew that train third No. 16 was overdue, in accordance with Rule 99. His failures represented inexcusable negligence on his part.

Contributing to the accident was the action of Engineer Greenham in calling his flagman in before his train was in such a position that there was a clear view of it for a quarter mile. [319]

On the night of Monday 3rd December 1917 a broken coupling caused a special train of the Preußisch Staatsbahnen, carrying Italian prisoners of war, to become divided at Heeßen, Germany, between Ahlen and Hamm. A signalman presumably failed to notice that the train had been incomplete when it passed him and D-Zug No. 22 from Berlin to Köln was allowed to enter the still occupied section. In the collision which followed the wreckage of the special train fouled the opposite running line where it was run into by a goods train. Thirty-five of the POWs and one of the military escort lost their lives and 45 POWs were injured but the passengers on the express all escaped injury. [320]

On 11th December 1917, the driver of train D 253 (train number also given as D421) from Herbesthal to Köln overran a stop signal at Gürzerach on the Aachen – Köln and collided with leave train No. 5 carrying troops from Laon on the front, destroying its last two carriages, at Düren station, Germany. Nineteen died and 36 were injured [321] (another source states that 23 died and 30 were injured). [322] [323]

A disastrous collision on the Louisville and Nashville Railroad at Shepherdsville, Kentucky, United States, on Thursday 20th December 1917, highlighted once again the primitive operating methods of some American railroads at this relatively late date. The main line of the Louisville and Nashville southwards from Louisville to Bowling Green, Kentucky, was, like many American lines, largely single track, but the first stretch of slightly under 30 miles as far south as Lebanon Junction was double track. It was worked by time interval with train orders, there being no block system. Intervals of ten minutes were kept at open telegraph offices by the train order signals, which were normally kept at danger. The enginemen of approaching trains were required to sound four short blasts on their whistles, whereby if there were no orders and the previous train had been gone ten minutes the signal would be cleared; it being replaced to danger when the rear of the train had passed 200 feet beyond it.

On the day of the accident southbound local train No. 41 from Louisville to Springfield, Kentucky consisted of engine No. 18 and three wooden cars in charge of Conductor Campbell with Engineer Keyer and Fireman Masden. This train ran on the main line for 22 miles before leaving it at Bardstown Junction. It left Louisville on time at 4.35 p.m., but time was lost en route due to the heavy pre-Christmas traffic. Behind it was passenger train No. 7 from Cincinnati, Ohio, to Montgomery, Alabama and New Orleans, Louisiana; nine cars hauled by engine No. 230 and in charge of Conductor Ogle with Engineer Wolfenberger and Fireman Gossom.

At Brooks station, 5 miles north of Shepherdsville, Conductor Campbell was told by the dispatcher to shunt his train at Shepherdsville out of the way of train No. 7 if he could

---

319 Interstate Commerce Commission report
320 Ibid, pgs. 37-38
321 *Eisenbahn-Katastrophen in Deutschland* by Hans Joachim Ritzau, pg. 79 (in German)
322 *Eisenbahnunfälle Bind 1* by Ludwig Stockert, pgs. 38 (in German)
323 *Eisenbahnunfälle in Europa* by Erich Preuß, pg. 67 (in German)

not reach Bardstown Junction on time. The train left Brooks at 5.13, seven minutes late, and arrived at Shepherdsville at 5.24, six minutes late. The station work at Shepherdsville having been completed, Conductor Campbell instructed Engineer Keyer to move the train forward to the south passing track switch, 400 feet beyond the station, so that it could be set back into the siding. The train duly pulled forward, the passing switch was opened, and the train was just on the point of setting back when it was run into by train No. 7 at 5.30 p.m. Train No. 41 was driven forward no less than 800 feet by the force of the impact; its rear car was completely telescoped and the rear compartment of the compartment car next to it was crushed. The engine and baggage car were driven forward another 150 feet. Forty-four passengers died in the accident and 52 people were injured. Conductor Campbell and Flagman Greenwall, who were standing on the platform between the compartment car and the rear coach, were also killed. [324]

Given its magnitude the accident does not appear to have received a great deal of press coverage, probably because of the amount of space devoted to war news, but the fate of two child passengers was mentioned in several papers. 'Two children whose identity had not been determined early today were found in snow drifts into which they had been thrown and were being cared for last night by Shepherdsville citizens.

'One, a girl of about eleven years, was apparently unhurt, but had suffered such a violent shock as to lose her memory, at least temporarily. The other child, a little girl about four years old was asleep when picked up from the snow drift into which she had been pitched by the impact of the collision. Her rescuers decided not to awaken her after it was found that she was unhurt, and the child slept the night through.' [325]

The first witness to give evidence at the Interstate Commerce Commission inquiry into the accident was Train Dispatcher Sams, who testified that he had instructed Operator Sanders at Brooks to tell Conductor Campbell to shunt at Shepherdsville if his train was unlikely to reach Bardstown Junction on time, and had told the conductor himself the same thing over the telephone at Shepherdsville. He said that it was permitted by the rules to allow train No. 41 to proceed to Shepherdsville close in front of train No. 7, the conductor being told that the fast train was close behind so he could avoid delaying it. He had expected the conductor to protect against the following train.

Agent Thompson at Shepherdsville said that the stop of the local train there was somewhat prolonged due to the heavy traffic. He saw train No. 7 approaching at 35 or 40 miles an hour. It did not whistle, and its brakes were not applied.

Operator Jesse Weatherford at Shepherdsville said that after assisting the agent in unloading the mail, baggage and express he had returned to the telegraph office and met Conductor Campbell, who told him that he was going to shunt the local for train No. 7. He then saw train No. 7 coming and running to the office, threw the train order signal to danger before train No. 7 had passed it and emerged with a lantern to signal to the train. Like Agent Thompson he said that he did not hear train No. 7 whistle and it did not seem to be braking. He had not replaced the train order signal earlier because the rear of the local train had still been less than 200 feet from it when it was stopped at the station. It had been a clear night, with no fog, and the markers on train No. 41 and the switch light were easily visible.

[324] Interstate Commerce Commission report

[325] *Kokomo Daily Tribune*, Kokomo, Indiana, United States, 21st December 1917

Engineer Keyer said that he did not know that train No. 7 was close behind them, though he knew it was late (it should have left Louisville at 4 p.m., in front of the local, but was an hour and 53 minutes late leaving there). As he was leaving Shepherdsville a porter told him that he was to shunt there for the fast train, this being the first he knew of his needing to do so. He did not know what protection was being provided but assumed that it was being taken care of. The night was clear.

Porter Chase of the local train said that he had asked Conductor Campbell at Gap in Knob, just north of Shepherdsville, if the engineer knew they were to shunt at Shepherdsville for train No. 7. The conductor said that the operator there would tell them where the fast train was. When they reached Shepherdsville, the conductor instructed him to tell the engineer to shunt there and he had done so. As he was operating the switch to let train No. 41 into the siding he saw train No. 7 approaching at high speed. Neither the conductor nor the flagman had gone back to flag No. 7, and he did not know where they were.

Engineer Wolfenberger of train No. 7 testified that he had had orders to run an hour and 30 minutes late from Louisville to Bowling Green but was actually an hour and 53 minutes late starting. He had had to stop soon afterwards at Oak Street and then at the Southern Railway crossing for other trains and had passed Brooks at about 5.24. He knew the local train was in front of him. He had seen the Shepherdsville train order signal at green, had whistled four short, the signal had not changed so he whistled again when it had changed to red. Having previously made a slight brake application he had then made an emergency application, but he had been expecting the signal to be cleared for him by the operator. He estimated his speed on approaching Shepherdsville at 40 to 45 miles an hour, reduced to 20 or 25 miles an hour at the moment of collision. He understood the rule that his train must approach the train order signal prepared to stop and that he must see the signal change from red to green, and that any other signals were improper and required him to stop. The rear markers of train No. 41 he had not seen until his train had been within 300 feet of it because he had been looking at the train order signal and the markers had been obscured by steam from the engine of the local train. He had seen no flag and no fusees, had heard no torpedoes and had received no signal from the Shepherdsville operator. It had not occurred to him that the fact that the train order signal had not been cleared for him was because the local train was in front of him.

Fireman Gossom corroborated his engineer's account. He said that they saw no flagmen and they ran over no torpedoes. He knew that train No. 41 was running ahead of them.

Operator Morrison, first trick operator at Shepherdsville, said he had not been on duty at the time of the accident but had gone to the station to get a newspaper. As he was leaving the station again he had heard train No. 7 whistle, and saw it coming in at a speed he estimated as 40 to 45 miles an hour. There was no evidence of its brakes being applied, nor did he see anyone protecting the local train.

Engineer Johnson, who was also not on duty, was standing on the platform at Shepherdsville. The markers of train No. 41 had been burning brightly and there was nothing to interfere with the view other than that it was smoky. He said that train No. 7 had been braking and put its speed at 20 or 25 miles an hour.

The Interstate Commerce Commission concluded that the direct cause of the accident was the failure of the conductor and flagman of train No. 41 to protect their train and that their failure was inexcusable. A major contributing cause was the failure of Engineer Wolfenberger to observe the train order signal at Shepherdsville and stop before passing it as required by rule.

The commission stated, however, that a large measure of responsibility for the accident must rest with the Louisville and Nashville operating officers for their failure to provide for a proper means of spacing trains in the area. There were 44 scheduled trains in each direction daily, too high a traffic density to be handled by a time interval system. Rules 221 (a) and (d) of the company, which required trains to whistle train order signals off when 600 feet from them, and if the signals were not cleared to stop short of them, were, the commission stated, grossly inadequate if not positively unworkable. For example, the schedule speed of train No. 7 between Brooks and Shepherdsville was 50 miles an hour, and to be sure of stopping at the Shepherdsville train order signal the engineer would need to reduce speed materially much further away than 600 feet. Nor, in the absence of a block system, was the method of operation whereby crews were informed verbally of where a following train might overtake them a safe one, and in fact Dispatcher Sams had violated Rule 103, which directed that messages regarding the movement of trains must be in writing. The Louisville and Nashville had repeatedly stated in its annual reports to the commission that the line between Louisville and Bardstown Junction was worked by the block system, but this was patently not true. There had been some progress in installing automatic block signals in the previous 5 years, mostly on single line sections, but there were clearly long sections of its main lines carrying heavy traffic without adequate protection. [326]

A Canadian Pacific Railway troop train bound for Vancouver, British Columbia, was run into from behind by a Montreal to Pointe Fortune local train at Dorval, Quebec, Canada, at about 6.40 p.m. on the evening of Friday 4th January 1918. Reports as to exactly how the accident happened are a little confusing but it would seem that the troop train was booked to take the siding at Dorval to allow the local train to pass it but ran past the switch and was stopped on the main line at the time of the accident. An initial report spoke of the deaths of Privates Thomas Kelly, Hunter, Mackie, Clarke, and Lalonde, with Privates Arthur and Sherren 'missing' presumed dead, and the injury of 13 soldiers, two porters and a waiter. The injured were taken to the Royal Victoria Hospital in Montreal. The local train was reportedly travelling at high speed and had run through signals.

The troop train was carrying 250 men. It included a number of colonist cars. The rear car of the troop train had been vacated shortly before the accident. In the collision it was pushed up on top of the cars ahead and it was the middle cars of the train which were telescoped and in which the fatalities and injuries occurred. Several of the victims were trapped in the wreckage and it was some time before they could be freed. When news of the accident reached Montreal ambulances and doctors were rushed to Westmount station, to which a special train took the injured and the bodies of those killed.

Most of the soldiers on the train were returned soldiers who had seen active duty in France, though some had been sent to England and then returned to Canada for duty there. One of the dead, Tom Kelly, had been at the front three times and only back three

[326] Interstate Commerce Commission report

months.[327]

The inquest on the victims of the accident was held the day after it occurred, by which time the bodies of the two 'missing' soldiers had been discovered in the wreckage, bringing the number of dead up to seven. As was often the case the inquest heard conflicting evidence as to what had happened. Engineer Gabriel David of the Pointe Fortune train testified that the automatic block signals had been 'with' him and alleged that there had been previous wrong-side failures of the signals, naming fellow employees who would support this claim. His fireman, Irwin Palmer, was in Montreal hospital recovering from injuries received in the accident and was not expected to be well enough to give evidence for another week.

Signalling Supervisor G.S. Taylor testified, however, that the signals had been examined after the accident and found to be working normally.

Conductor Thomas Markey of the troop train said that he had sent his brakeman back to flag the Pointe Fortune train at Dorval, intending to shunt the troop train into the siding, but the local train had been on them before there was time to do so. He went back to look at the protecting signal after the accident and it was then at 'danger.' He did not think it could have been changed after the collision. The troop train had been five hours late.

Brakeman McDonald corroborated the evidence of his conductor and said that he had only got back seven or eight car lengths before the collision. Afterwards he had gone to Engineer David and asked him, "What's the matter with you?"

The engineer had replied, "Why didn't you flag?" Nothing was said about the signals being out of order.

Baggageman George Sauve of the local train stated that he had heard Engineer David state that the signal had been 'with him.'[328]

The signal maintainer having admitted that it was possible, if unlikely, that a particle of frozen moisture might have caused a wrong-side signal failure without leaving any trace, the inquest jury decided that the block signals were not altogether reliable and returned a verdict clearing the train crews of blame for the accident.[329]

A collision occurred on about Monday 7th January 1918 (exact date unclear) at Karpovo station, in the Ukraine, about 20 miles from Odessa. A passenger train from Kharkoff bound for Odessa collided with some waggons loaded with sugar for the military stores there. The engine of the passenger train was smashed, several carriages, mostly occupied by soldiers, were derailed, and overturned and their occupants were pinned beneath them. About 10 passengers were killed and 100 injured. According to a newspaper correspondent disorder and anarchy reigned after the accident. 'The revolting feature was that while the wounded were lying in urgent need of aid, people sufficiently callous and ravenous were found to rob the helpless, seizing anything they could lay their hands on. From the trucks of the goods train bags of sugar fell out. Passengers and soldiers fell to looting the precious article, filling anything they could find with sugar mixed with mud. Next a cry was raised among the soldiers that the railway servants had

---

327 *Manitoba Free Press*, Winnipeg, Manitoba, Canada, 5th January 1918

328 Ibid, 7th January 1918

329 *Lethbridge Herald*, Lethbridge, Alberta, Canada, 11th January 1918

maliciously caused the accident, and a howling crowd of soldiers attacked them with intent to lynch them. Only the timely appearance of militiamen saved them.'

An inquiry subsequently found that the goods train had accidentally become divided, and its forward portion had proceeded, the train crew presumably having been unaware that part of their train had been left behind. [330]

A severe blizzard led to a collision on the Kaiserslautern – Homburg line of the Pfälzer Eisenbahn at Bruchmŭhlbach, Germany, at 11.30 p.m. on Monday 7th January 1918. The snow obscured the driver's view of an Einfahrsignal (home signal) at danger and a leave train ran into a goods train. Twelve died and 87 were injured. [331] Another source gives 21 fatalities. [332]

An express and a leave train collided at Ölingen, Germany, on Wednesday 16th January 1918. The driver of the express had failed to observe the Ausfahrsignal (starting signal) at Bohmte station, three kilometres in the rear of Ölingen, and had entered the occupied section. Thirty-one people died and 66 were injured. [333] [334]

Preußisch Staatsbahnen D-Zug 252 collided with leave train No. 3 at Köhn-Ehrenfeld, Germany, on 8th February 1918. Six soldiers on the leave train and a railwaywoman on the express died and 20 soldiers were injured. [335] The Einfahrsignal had been cleared prematurely. [336]

The Southern Railway of the United States featured in the next serious rear collision to occur near Frost station, South Carolina, on Monday 25th February 1918. Two eastbound trains were concerned in the accident, local train No. 18 from Greenville, South Carolina, to Columbia, South Carolina, and train No. 42, also bound for Columbia from Ashville, North Carolina. The former train consisted of engine No. 942 and four cars with Conductor Meredith, Engineer Nicely and Fireman Robinson. The latter train had engine No. 1200 with five cars and was in charge of Conductor Gelston with Engineer Long and Fireman Stricklan.

The line on which the accident took place was single track and worked by timetable and train orders. There was no block system but according to the rules passenger trains were spaced ten minutes apart at open telegraph stations. Train No. 18 passed Bookman, the last open telegraph office, 8 miles short of Frost, at 12.47 p.m., and left Frost station at 1.59 p.m. Half a mile east of Frost station the train came to a halt again, a leak in the brake pipe having caused the air brakes to apply. Six or seven minutes were consumed

---

[330] *The Times*, London, England, 15th January 1918

[331] *Eisenbahnunfälle Bind 1* by Ludwig Stockert, pg. 38 (in German)

[332] *Historische Eisenbahn-Katastrophen* by Bernhard Püschel, pg. 113 (in German)

[333] *Eisenbahnünfalle in Europa* by Erich Preuß, pg. 68.(in German)

[334] Katastrophen der deutschen Bahnen Teil II: Chronik 1845-1992, by H.J. Ritzau, pg. 43 (in German)

[335] *Eisenbahn-Katastrophen in Deutschland* by Hans Joachim Ritzau, pg. 79 (in German)

[336] *Eisenbahnunfälle Bind 1* by Ludwig Stockert, pg. 39 (in German)

fixing the problem, the train was restarted but stopped again after only covering 50 feet for the flagman who had gone back to rejoin his train. It then restarted a third time but after travelling 300 feet and having attained a speed of 10 miles an hour it was run into by train No. 42 at about 2.13 p.m. In the collision engine No. 1200 telescoped about 6 feet of the rear coach of the local train, while the front end of the rear car in turn telescoped two thirds of the length of the next car. Apart from the leading car of train No. 42 all the cars of both trains were of wooden construction, which doubtless increased the amount of destruction. Twelve passengers and one person carried under contract died and 108 passengers and four railroad employees were injured.

Engineer Nicely told the Interstate Commerce Commission (ICC) inquiry that when the local train stopped short after leaving Frost he thought at first that an air hose had burst but instead discovered a bad leak in the brake pipe under the middle of the leading car. Ascertaining what was wrong he had returned to his engine, whistled out the flagman and then assisted in repairing the leak, which was done by plugging the pipe. This had occupied six or seven minutes. Before he got back on his engine the conductor had signalled for him to call in the flagman, which he did. He had immediately got a signal to go and had started at once without waiting for the flagman but had only moved an engine's length when he was signalled to stop again. He did so and after two minutes received another signal to proceed and had moved two train lengths when train No. 42 ran into them. He said that he was familiar with Rule 99 requiring the flagman to remain out if a passenger train was due within 10 minutes, but that his conductor had signalled him to wait for the flagman.

Conductor Meridith said that the flagman had started back as soon as the train stopped. When the repairs had been completed he told his engineman to call the flagman in. The train then restarted but the flagman had been so close to it that he had stopped the train again so that the flagman could rejoin it, assuming that he had put down torpedoes. He knew that the flagman should not have got back onto the train in such circumstances and that from the short interval between his being whistled in and his appearing close to the rear of the train he could not have gone back very far but said that the flagman had gone right around the curve.

Flagman Locklier stated that as soon as the local train had stopped he had begun to run back to protect his train before the engineer had signalled him to do so. He had not counted the telegraph poles but had put down two torpedoes east of Frost station at what he estimated was half a mile back. After that, he had come back to a position eight or nine telegraph poles from the rear of his train waiting to be called in. When he was, there was no sign of train No. 42, so he did not put down a further torpedo. He had re-boarded his train and it was under way before he saw train No. 42 approaching. He knew that they were running on the other train's time and that he had violated the flagging rule by coming in with a passenger train due in 10 minutes.

Fireman Robinson of the local train (referred to in the ICC report as 'Coloured Fireman Robinson', as if his race had anything to do with the accident) said that when his engineer had whistled in the flagman he (the engineer) had clearly had misgivings about doing so as he had remarked, "We ought to go ahead and leave the flagman." Fireman Robinson said he did not hear train No. 42 whistle, or any torpedoes explode.

Engineer Long of train No. 42 said that when checking the register at Alston for opposing trains he had not noticed if the local train (which joined the main line there) was ahead of him. He did not hear any torpedoes explode, did not see a flagman and had no warning that the local train was in front till his fireman called out. At that moment he was running at about 45 miles an hour.

Fireman Stricklan stated that he saw section men at work and then the rear of train No. 18 150 feet away. They exploded no torpedoes and were not flagged, though he saw the flagman at the rear of the local train.

Conductor Gelston of train No. 42 said that he had also checked the train register at Alston but had not noticed if train No. 18 was in front. He did not hear any torpedoes explode, nor did Baggagemaster Lawson, Trainman Harris and Porter Thompson of train No. 42, Section Foreman Derrick or Trainmaster Bishop of train No. 18. The latter added that he did not think that the flagman had gone back far enough to ensure full protection.

The commission concluded that the direct cause of the accident was the failure of Flagman Locklier to properly protect his train in accordance with Rule 99. From his own admission he had only gone back just over 700 yards as opposed to the three-quarters of a mile required by Rule 99. He had then come back to a point only 300 yards from the rear of his train as opposed to the half mile dictated by the rule. Furthermore, he had not remained out, as required by Rule 99, and there were doubts as to his claim of having laid two detonators near Frost as no one on either train had heard them explode, nor had the section foreman, nor could any trace of them be discovered afterwards.

Conductor Meridith was at fault for signalling his engineer to stop a second time for the flagman. Had he not done so the collision would probably have been avoided.

The accident, the commission underlined, once again pointed to the inadequacy of the time interval system. Between Alston and Columbia there were 14 trains each way daily plus an average of four extras. Twice daily passenger trains followed one another at an interval of ten minutes. Under the time interval system trains were kept ten minutes apart at open telegraph offices but the long distances between open offices provided ample opportunity for fast trains to overtake slower ones between them and showed the necessity for some sort of block system. [337]

At the inquest on the victims of the accident the coroner's jury came to the same conclusion as the ICC, returning a verdict that, 'We find that the deceased came to their death by train No. 42, running from Spartanburg to Columbia, running into train No 18, running from Greenville to Columbia, near Frost, through the rules of caution not being thoroughly carried out by the crew of train No 18.' The jury had some difficulty in coming to a verdict but the one that they finally agreed on thus stopped short of finding a verdict of manslaughter against Flagman Locklier. [338]

The series of rear collisions in the United States continued on 22nd June 1918 with a major catastrophe at Ivanhoe, Indiana, on the Michigan Central Railroad. This time it was not the flagman who was at fault but the engine crew.

The Michigan Central line at Ivanhoe was double track and equipped with semaphore three position upper quadrant signals, displaying red, yellow, and green lights, which

[337] Interstate Commerce Commission report

[338] *The Wilmington Dispatch*, Wilmington, North Carolina, United States, 26th February 1918.

were placed about a mile apart. At Ivanhoe itself, however, there was a signal tower, controlling the level crossing with the Elgin Joliet and Eastern Railroad line, crossovers and a connection to the parallel Gary and Western line. Protecting the rail/rail crossing and the connection to the Gary and Western was a home signal located 950 feet east of the crossing, while automatic signal 2581, 5,360 feet east of the home signal, served as a distant signal for it. The next automatic signal to the east, 2571, was 6,200 feet further back. The approach of westbound trains was announced by telephone from Tolleston tower, about three miles to the east.

The first train involved in the accident was westbound extra 7826, a Hagenback Wallace circus train, consisting of engine No. 7826, 7 stock cars, 14 flat cars, 4 sleeping cars and a caboose, in charge of Conductor R.W. Johnson with Engineer Gasper and Fireman Phillips. The train had left Michigan City at 2.30 a.m. and was to take the connection to the Gary and Western at Ivanhoe. Automatic signal 2581 was therefore yellow and it approached the tower and crossing at 25 miles an hour. Just as it was nearing the crossing the conductor signalled his engineer to stop, the flagman at the rear of the train having noticed a blazing hot box on a flat car 10 or 11 cars from the rear of the train.

Behind the circus train was extra 8485, consisting of engine No. 8485 and 21 Pullman cars. This train had left Michigan City at 2.57 a.m., en route from Detroit, Michigan, to Chicago, Illinois. It was in charge of Conductor L. Johnson with Engineer Alonzo Sargent and Fireman Gustav Klauss. All the evidence indicated that automatic signal 2581 was at danger and automatic signal 2571 was at caution but extra 8485 overran both of them, ran past the flagman of extra 7826 and collided with the rear of the circus train at 3.57 a.m. The wreckage immediately caught fire after the collision and the four sleeping cars were entirely destroyed. Sixty-seven passengers and one railroad employee were killed according to the Interstate Commerce Commission, and 127 passengers were injured. The circus stated after the disaster that in addition to the identified dead 61 people were missing, many of them itinerant roustabouts hired in Michigan City the day before, and it seems likely that in all, 86 must have died. [339]

The circus lost a number of its performers in the accident, including strongman Arthur Dierickx; clown Eddie Devie; animal trainer Millie Jewell; bareback rider Louise Cottrell and aerial artist Jennie Ward Taylor. The railroad employee who lost his life was Trainmaster Fred Whipple. [340] Despite the loss of its performers and of much equipment the circus performed again as scheduled at Beloit, Wisconsin, having missed only two performances due to the disaster, a good example of the circus tradition that 'The show must go on.'

Giving evidence at the Interstate Commerce Commission (ICC) inquiry Engineer Gasper testified that he was slowing to 8 miles an hour for the connection at Ivanhoe tower when he saw a stop signal being given with a fusee from the caboose. About two minutes later he was looking back, trying to find out what was wrong, when he saw a headlight coming around the curve, but it might have been something on the parallel Gary and Western line. Then he saw the flagman on the ground with his red and white lanterns, saw the latter light a fusee and then wave it violently at the approaching train.

[339] *No Performances Today* by Warren A. Reeder, pg. 9

[340] *Defiance Crescent-News*, Defiance, Ohio, United States, 24th June 1918

He then saw the fusee apparently thrown in the air. Shortly afterwards he felt a slight shock and realised that there had probably been a collision, though it did not seem to be a serious one.

Conductor R.W. Johnson testified to seeing signal 2581 go from caution to danger as his train passed it. As his train had rounded the curve east of Ivanhoe, he had spotted the hot box and had lit a fusee and waved it to the engineer. As they slowed down the flagman had dropped off, taking with him a fusee, lanterns and torpedoes. The conductor said that he had then gone forward to deal with the hot box and had only gone five or six lengths when he had seen the headlight of another train coming around the curve. He saw the flagman waving a fusee at the oncoming train and thought that he had gone back 25 car lengths.

Flagman Oscar Timm of extra 7826 stated that he had seen the hot box and had notified the conductor who lit a fusee to display to the engineer. He had got down as soon as the train stopped and begun to walk back. He could see the smoke of the following train before he saw its headlight. When he realised it had run past signals he had begun to run. Lighting the fusee he had run across the track to try and attract the fireman's attention, re-crossed to the engineer's side, shouted, and finally thrown the fusee at the engine cab. His signals had not been answered, and the train had not reduced speed. He thought it was travelling at 25 miles an hour. He had not put down torpedoes as he was trying to get back as far as possible; he thought he had got back about 700 feet.

Conductor L. Johnson stated that the first he knew of the collision was feeling a shock which he had initially attributed to a burst air hose. After the accident he had asked Engineer Sargent how it had happened and the latter had replied, "I was dozing, otherwise asleep." He had heard his engineer whistle at Tolleston but not after that. He had observed no signals between Michigan City and Ivanhoe.

Brakeman Jackson of extra 7826 said that he had noticed no brake application before the collision. After the accident, the fireman told him that the engineer had been asleep.

Engineer Sargent was arrested after the accident and refused to make any statement to the commissioners on the advice of his counsel. He had, however, already made a statement to the railroad officials, in which he stated that he had been called at 8 p.m. on 21st June to work train 7826. He had been up since 5 a.m. that morning, deadheading from his home at Jackson on train No. 41 and had had little or no sleep. Believing that he would have no further chance to eat until the next morning he had had two heavy meals before going out. He had been travelling at 25 miles an hour when he saw signal 2571 at caution but he had not applied the brakes, expecting that the next signal would clear. The wind had been blowing in hard on his side and he had closed the window to make the cab more comfortable. It had clearly been made too comfortable for he had then dozed off and had not realised anything was wrong until within 75 or 90 feet of the rear of the freight train. Even then he had not realised it was so close and had initially only made a service brake application, changing it to an emergency application only at the very last moment. He thought that he had closed the throttle, though he was not sure. He did not see a fusee. He was in perfect physical and mental condition and had suffered from no illness in the previous 25 or 30 years. As far as he was aware there was nothing wrong with his train's brakes or other equipment or with the signals; the accident had been entirely due to his falling asleep.

Fireman Klauss, who had also been arrested, refused to make any statement to the commissioners.

The Interstate Commerce Commission came to the rather obvious conclusion in view of Engineer Sargent's statement that the accident was due to his falling asleep and thus running past the signals and failing to see the frantic warnings from the flagman. In the absence of any statement from Fireman Klauss as to what he had been doing approaching Ivanhoe it was impossible to decide if his actions had contributed to the disaster. Examination of the signals after the accident and the statement of Conductor R.W. Johnson demonstrated conclusively that the signals had been showing their proper indications.

Listing all the accidents which had occurred since July 1911 due to engineers overrunning signals, some, as at Ivanhoe, in clear weather, the commission repeated yet again that automatic train stop devices had reached a state of development which warranted extensive trials of them. It was also noted that automatic signalling meant that trains could run too close together for flagmen to have time to go back far enough to properly protect their trains, increasing the need for some sort of automatic train control.

The sleeping cars of the circus train were rebuilt wooden Pullmans, lit with oil lamps, no other form of lighting being permitted. The construction of the cars was such as to be unable to withstand an impact even less than that at Ivanhoe. Had the cars been steel they would not have burned. The large number of berths in the cars also resulted in more casualties than would have occurred had they been ordinary sleeping cars. The exact origin of the fire could not be ascertained but it probably originated from the oil lamps and oil headlight of the colliding train as there was no contact with the hot engine coals.

Engineer Sargent finally stood trial on 15th April 1919. Evidence was given that he had been kept waiting at Dowagiac, Michigan 17 hours before taking his train out and had been exhausted, while Flagman Timm had laid no torpedoes. After being out over four hours the jury were unable to agree on a verdict. There was then talk of another trial but eventually on 9th June the case against the engineer was dropped. [341] [342]

At 3.30 a.m. on Sunday 30th June 1918 hospital train No. 840, consisting of an engine and 32 vehicles, was stopped 200 metres in the rear of the Einfahrsignal (home signal) at Hrastnigg, Austria (now Hrastnik in Slovenia) on the Südbahn, when it was run into from behind. The colliding train was leave train No. 38, consisting of an engine and 19 vehicles and travelling at 40 kilometres an hour. The rear nine vehicles of the hospital train were wrecked and crushed into one another but as chance would have it there were only four people in them, one of whom was killed and two seriously injured. The engine and leading six carriages of the leave train were derailed and broken gas cylinders then set them ablaze. Five soldiers and two guards in this train lost their lives and 21 soldiers were seriously injured. The principal cause of the accident was a mistake by the fahrdienstleiter (signalman) at Trifail (Trbovlje) who had permitted the leave train to enter the Trifail-Hrastnigg section before the hospital train had been reported back to him. Contributory factors were the excessive speed of the leave train and the fact that the driver of the hospital train had pulled his train up well short of the Einfahrsignal, so that

---

[341] *The Fort Wayne News and Sentinel*, Fort Wayne, Indiana, United States, 16th April 1919

[342] *No Performances Today* by Warren A. Reeder, pgs. 118-120

its rear was only 20 metres in advance of the Vorsignal (distant signal). [343]

In the early hours of 11th September 1918 (to be precise at 4.53 a.m.) goods train 7592 of the Preußisch Staatsbahnen was stopped at Schneidemühl block post, Germany, when it was run into from behind by special passenger train 2244, carrying children bound for München-Gladbach. The rear vehicles of the goods train were badly damaged, and the leading carriage of the passenger train was driven into the guard's van and shattered. Thirty-three children, a guide and the rear conductor of the goods train were killed, two children were seriously injured, and 15 children and a woman were slightly hurt. [344]

The major disasters in Germany continued with one which occurred at about 10 p.m. on 22nd September 1918 at Dresden-Neustadt on the quadruple track Leipzig-Dresden main line of the Sächsischen Staatsbahnen. The sequence of events which led up to the disaster began with the stoppage of passenger train No. 1513 from Döbeln at the Dresden-Neustadt Einfahrsignal (home signal) due to problems with its engine. Following this passenger train was delayed D-Zug (fast train) 196 from Berlin, the driver of which pulled up his train in turn at the preceding stop signal at Block post 30. The last carriage of train 196 came to a stand on a bridge over Risaer Platz. The train was still standing in this position when it was rammed from behind by the engine of D-Zug 13, bound for Leipzig, the driver of which had run through the signals at Pieschen and entered the occupied section, braking only at the last moment. The colliding train was still travelling at considerable speed and the rear four carriages of train 196 were completely wrecked, the rear two being telescoped, while the wreckage caught fire following the collision. The locomotive of the colliding train, in contrast, was little damaged.

According to one source 33 passengers in the Berlin train died, 35 were seriously hurt and 29 slightly so. In the colliding train only a few passengers were slightly hurt. [345] Other sources state that there were 38 deaths and 118 passengers injured.

On 10th November, Driver Ferdinand Schneider and Fireman Gustav Becher of the colliding train were sentenced at Dresden to eight months' imprisonment for having caused the accident. [346] Tests conducted after the accident indicated that Driver Schneider suffered from red-green colour blindness, though he presumably must have been able to distinguish the indications of signals previously, if with difficulty. [347]

Südbahn goods train No. 847 passed Marien station late on Monday 20th October 1918, 12 minutes late, and was stopped at the Einfahrsignal (home signal) at the next station at Kapfenberg, Austria, for an hour. Early on Tuesday 21st express passenger train No. 5 from Wien to Laibach was in the section from Kindberg to Marien but Kapfenberg had still not reported back the goods train. The Marien fahrdienstleiter (signalman) enquired of his colleague at Kapfenberg on the telephone as to the goods. "I must have

---

343 *Eisenbahnunfälle Bind 1* by Ludwig Stockert, pg. 40 (in German)
344 Ibid, pg. 41
345 Ibid.
346 *Eisenbahnunfälle in Sachsen* by Claus Schlegel, pg. 121 (in German)
347 *Eisenbahnunfälle in Europa* by Erich Preuß, pgs. 71-4 (in German)

dozed off and failed to report it back," was the reply, and without further thought the Kapfenberg signalman placed his instrument at 'Freie Fahrt' (Line Clear) for train No. 5. With train No. 5 then on its way from Marien the Kapfenberg signalman first dealt with leave train No. 32c going towards Marien before he cleared his signals for train No. 5. When he cleared his Einfahrsignal goods train No. 847 started away, but by that time train No. 5 was already between the Vorsignal (distant signal) and the Einfahrsignal signal, and the collision followed shortly afterwards. The leading five vehicles of the express (brake van, saloon, mail van and two passenger coaches) were badly damaged, and the first three were telescoped and caught fire. Six waggons of the goods train were badly damaged and three slightly. The leave train then ploughed into the wreckage and a brake van, and two carriages of that train were also damaged. Thirteen passengers lost their lives in the accident, 36 were seriously injured and 70 slightly. A Südbahn director was among those who died. Significant fog and track curvature prevented the tail lamp of the goods train being seen for more than a short distance. At the time of the accident the Marien fahrdienstleiter had been on duty for 19 hours, his mate at Kapfenberg for 16 hours, doubtless due to wartime exigencies. [348]

On the evening of 26th October 1918 a troop train carrying the 138th United States Field Artillery from Cherbourg to a training camp at Meucon, came to a stand in Gael station, France, probably as a result of engine problems which had already become apparent earlier in the journey. While it was stopped there, it was run into at 8.50 p.m. by a following train carrying the 113th Ammunition Train. The six rear vehicles of the front train – three compartment carriages and three box cars – were completely demolished (some reports spoke of 14 cars telescoped) and there were reportedly 38 fatalities. The collision brought down the telegraph wires on either side of the station and news of the accident had to be conveyed by someone on foot to a location five miles away. It was 1 a.m. before troops and ambulances arrived at the scene of the accident, probably from Camp Coelquidan, 20 miles distant. All the dead and injured had been removed by 03:00. Among those who lost their lives were ten members of the Kentucky National Guard: William Aubrey, Bulford Craig, Norbert Henry, Charles Lucas, Watkins Moss, Walter Neagle, Roy Ogle, Ralph Rose, James Tucker and Garland Wells. The dead were buried in U.S. Government cemetery 18 at Camp Coelquidan.

The aftermath of the accident featured a heroic rescue attempt. Major Short, Captain Freehan, Captain Kavanagh and Private Sheehan worked for an hour in darkness and rain to free a soldier named Aubrey who was trapped, pinned beneath an upturned waggon wheel upon which rested the greater part of three telescoped cars. To rescue the victim the wreckage had to be raised to a very dangerous angle. The four, 'without regard to the personal danger involved, crawled through the wreckage and in a lying and sitting position, worked against odds for three hours and succeeded in rescuing the soldier alive.' One of Aubrey's legs had to be amputated with a penknife by Major Short. [349]

At about 3.40 a.m. on the morning of 1st November 1918, heavy local goods train 7708 (108 axles so about 50 waggons) arrived at Briesen station on the Berlin – Frankfurt

---

[348] *Eisenbahnunfälle Bind 1* by Ludwig Stockert, pgs. 28-29

[349] Webpage – RR Deaths 1917-18: Kentucky National Guard Memorial

(Oder) double-track main line. Although Briesen station was situated on a quite busy main line it had a primitive layout. Goods trains were often stabled in two sidings (tracks 3 and 4) to the north of the running lines but the connections to these sidings were worked by hand points, involving a walk of 225 metres for the pointsman from one end of the station yard to the other and goods trains must often wait at the entrance to the station while this was done. Goods trains usually took 13 minutes to travel from the previous station at Jakobsdorf to Briesen but on the night in question train 7708 arrived in only ten minutes. When he was advised of the train's approach from Jakobsdorf the fahrdienstleiter (signalman) telephoned to the pointsman in Bude II at the Berlin end of the station to operate the points there, a job which took four minutes. As his train was going into the yard its driver failed to realise that it had become divided, and about 40 waggons had been left in the Briesen – Jakobsdorf section. His train was not fitted with automatic brakes, so these did not apply when the train separated into two. The rear conductor was obviously asleep in his van and did not realise either that anything was wrong.

Assistant Pointsman Franz Biedny in endstillwerk (end signal box) Bob at the Jakobsdorf (Frankfurt) end of the station should have noticed that the tail lamp at the end of the goods train was missing and reported the problem to the signalman but at that moment long-distance goods train 7665 to Frankfurt (Oder) passed his box on the line next to the box and the pointsman imagined, like other signalmen elsewhere before and since, that he had seen the tail lamp through the thick fog. He replaced the home signal and cleared the goods train back to Jacobsdorf.

Following the goods train was military leave train 4026. The driver of this train only glimpsed the tail lamp of the detached waggons through the fog too late to stop and the collision was unavoidable. Following the impact some of the waggons ran forward down the 1 in 280 gradient towards Briesen until they were stopped by a shunter using scotches.

Nineteen soldiers and the rear conductor of the goods train were killed outright, 30 soldiers were seriously injured and 13 soldiers and two railwaymen, presumably the driver and fireman of the leave train, were slightly hurt.

Although the railway company considered Assistant Pointsman Biedny responsible for the accident he was acquitted by the court. [350]

On the evening of 5th December 1918 a train loaded with American goods and supplies was stopped close to Meung-sur-Loire station, France, on the Chemin de fer de Paris à Orléans, when it was run into by an express train from Orléans. The engine and leading two carriages of the express climbed up onto and 'smashed to atoms' four waggons of the goods train. Twenty people died and 30 were injured. The Paris correspondent to *The Times* in London added that those killed included four French officers, five American soldiers and a barrister of the Paris Court of Appeal. [351] [352]

Only two days after the Meung-sur-Loire disaster France suffered a considerably

---

[350] *Eisenbahnunfälle in Europa* by Erich Preuβ, pgs. 146-9 (in German)
[351] Wikipedia - Liste des accidents ferroviaires en France au XX siècle
[352] *The Times*, London, England, 7th December 1918.

worse accident at Lothiers, near Luant, also on the Chemin de fer de Paris à Orléans. Whilst news of the accident was not entirely supressed only brief details of it appeared in the French press. The accident involved a freight train and two trains carrying soldiers returning to their regiments from leave. At 04:00 on Saturday 7th December 1918 the freight train was stopped in Lothiers station and the first leave train, which had left Limoges at 23:54 the night before, was stopped behind it. The morning was foggy, and the driver of the second leave train, which had left Limoges ten minutes after the first and was travelling at 70 or perhaps 85 kilometres an hour, with men from Haute-Vienne, Dordogne, Corrèze and Creuse, must have failed to see the signals, though he apparently shut off steam and braked at the last moment.

The sound of the impact and the screams and cries of the injured which followed were apparently enough to wake the whole household of a farmhouse at Moulin-à-Vent, two kilometres away, while a journalist reported that 'A terrible clamour and piercing cries arose immediately, mixed with the din of broken waggons. Seeing these masses of twisted iron and broken and pulverized wood, one can get an idea of the terrible collision.'

After the accident Americans who had arrived at the scene by car on hearing of the accident managed to set up a wire which enabled them to telegraph an airfield at Châteauroux, and a biplane was flown to the scene with medical dressings and supplies, while a relief train was sent from Limoges. Between 700 and 800 uninjured soldiers from the two trains worked for 24 hours to extricate the dead and injured, assisted by Americans from Montierchaume Camp and Red Cross nurses from Argenton. Various accounts gave different figures as to the number killed in the accident, one stating that there were 88 dead, but a careful modern analysis suggested that in fact there were 74 deaths, including several who died later from their injuries. Of these 73 were passengers (soldiers), and one was a railwayman, Conductor Jules Magne. 151 others were injured. The driver and fireman of the colliding train escaped unhurt. The majority of the victims (67 or 70 according to different reports) were buried at a mass funeral in Châteauroux on 9th December. [353] [354]

Fog was blamed for another rear collision in France nine days later, again involving a leave train. At Belleville station, 20 kilometres from Nancy, late in the evening of Monday 16th December 1918, a passenger train from Metz ran into a train carrying soldiers home on leave, the last three carriages of the leave train being completely demolished. Sixty victims were recorded, of whom nine were killed outright, one of them a woman. 'An inquiry has been opened to determine the causes of the accident.' [355] [356]

We must now return to the United States and to the little flag station of Norge, Oklahoma, on the Chickasha sub-division of the St Louis-San Francisco Railway. Like many in the country at this time the line was single track and worked by timetable and train orders without a block system, so a train stopped out of course relied entirely on

---

[353] Webpage -Lothiers, 6 Décembre 1918, une catastrophe ferroviaire

[354] Webpage – La Nouvelle Republique 7/12/2013

[355] *Le Petit Parisien*, Paris, France, 17th December 1918

[356] *The Times*, London, England, 18th December 1918

flagmen for protection against following trains. On Christmas Eve, 24th December 1918, westbound passenger train No. 407 consisted of engine No. 604, a baggage car, two all-wooden coaches and a wooden coach with a steel underframe, in charge of Conductor Lewis with Engineer Quinn and Fireman Lawson. It left Oklahoma City, Oklahoma, at 7.45 p.m. on Christmas Eve, two hours late and by the time that it reached Norge, at 11.55 p.m., it was about three hours late.

Behind the passenger train was a short freight train, extra No. 719, consisting by the time it reached Norge of engine No. 719, four cars and a caboose. This train had left Oklahoma City at 4.45 p.m. and was overtaken by the passenger train at Tuttle. It was in charge of Conductor Sawyer with Engineer Atkins and Fireman Essell.

The freight train reached Norge at 12.40 a.m. on Christmas Day, 25th December, by which time the passenger train should, of course, have been long gone, but it was still there because its engine was low on steam. The result was that the freight train crashed into its rear at an estimated speed of 18 miles an hour, pushing the rear coach into that in front and telescoping it for 30 feet, half of each car being broken up. Five passengers lost their lives and 21 were injured.

Engineer Quinn told the Interstate Commerce Commission (ICC) investigators that when the train reached Norge he had only had 100 lbs of steam and although there were no passengers to drop off or pick up there he decided that he had to stop to get up steam. He said that he did not whistle out the flagman as such but went back and informed the conductor that they would be stopped there until he could get up steam. He had then returned to his engine and about the same time Brakeman Higgins had come up and asked his fireman if he needed any help, whereupon his fireman had replied that he did not and told the brakeman to go back and protect the train. Five minutes later the conductor had also come up to the engine and had remained there until the collision, telling the engineer that the brakeman had gone back to protect. After ten or fifteen minutes unsuccessfully trying to get steam up Engineer Quinn said that he had decided that the only solution was to clean out the fire, and he and the fireman had just done so and were starting to rebuild it when they were run into. Although he did not see him while they were working on the fire he recognised the voice of the flagman and heard the conductor telling the flagman sharply to hurry back and resume protecting the train. This had been five or ten minutes before the accident.

Fireman Lawson had nothing to add to his engineer's statement.

Conductor Lewis testified that his train had been made up of extra equipment and he had told Porter Burns to get torpedoes and fusees which the porter had done. Wondering why the train had not restarted from Norge he had learnt that the engineer needed to get steam up first. He then told Brakeman Higgins to go back and flag, mentioning the fact that they had overtaken freight extra 719 at Tuttle, and he saw the brakeman go back with a red and white lantern, but did not know if he had torpedoes or fusees. The red light was good and bright, and he had never received any complaints about lanterns. He had then gone back to the engine.

Porter Burns confirmed that he had got five torpedoes and five or six fusees at Oklahoma City, which were in his bag at the front of the train, there being none at its rear. The lantern was burning brightly, and the train markers were burning brightly at Amber. When the train had stopped at Norge the brakeman had come up to him in a rush and had asked to borrow his overcoat to go out flagging with as it was bitterly cold.

Brakeman Frank Higgins said that he was familiar with the flagging rules and had never been dismissed for improper flagging. At Norge, the engineer had not whistled him out, but the conductor had told him to go back and protect. He had then gone back to the rear coach, picked up a red lantern and looked for torpedoes and fusees but could not find any. Without looking further, he had then gone back with just red and white lights, stopping briefly at a section house, and had gone back about 16 car lengths. He had stayed there about 25 or 30 minutes, by which time the lights had got so low that he thought that they would be useless to flag a train. In view of this he had returned to the train in search of a fusee. He had found one fusee in the coal box. Hurrying back with it he had got about three car lengths from the rear of his train when the freight train had passed him. He had waved his lanterns at it but received no response, the fusee had been missing its cap and he could not light it. He denied having been up in the baggage car at all during the journey, borrowing the porter's overcoat before he went back to flag or of going up to the locomotive.

Engineer Atkins testified that his brakes had been in good order. He had had no knowledge of the passenger train being ahead until his engine headlight picked out the rear of the train when it was 300 feet away. He had been travelling at about 20 miles an hour and his emergency brake application probably reduced the speed of his train slightly. He thought that he saw the flagman just before the collision but was too busy trying to stop to see what lights the flagman was carrying but got no signal from him, nor were there any torpedoes or fusees. Fireman Essell corroborated his mate's testimony.

Conductor Sawyer said that his train had stopped very suddenly at Norge and going forward he found that there had been a collision. He had heard no torpedoes.

The Interstate Commerce Commission found Conductor Lewis and Brakeman Higgins equally responsible for the accident. The brakeman had been negligent in going back initially without either torpedoes or fusees, while the conductor was to blame for their not being available at the rear of the train as per Rule 415. Notwithstanding the flagman's going back without all his equipment there was no excuse for his returning to the train before being called in, while had the conductor made proper provision for the protection of his train instead of remaining on the engine where he was not needed, the accident would probably have been avoided.

In addition Engineer Quinn's belief that he did not need to whistle out a flagman until the train was delayed more than five minutes was based on an edition of the rule book which had been superseded nine years earlier, indicating that the examination and supervision of employees with regard to the rules could not be very efficient. [357]

An inquest held at Chickasha on 28th December on Lieutenant James H. McLauren, one of the passengers killed, returned a verdict that his death had been the 'result of culpable and felonious negligence and omission of duty on the part of Frank Higgins, the flagman of the passenger train.' Higgins was forthwith charged with second degree manslaughter. [358]

---

[357] Interstate Commerce Commission report

[358] *Joplin News Herald*, Joplin, Missouri, United States, 29th December 1918

Disaster struck the New York Central Railroad at the beginning of 1919, in the form of a rear collision near South Byron, New York, United States, on Sunday 12th January. The trains involved were No. 17, *The Wolverine*, and No. 11, the *South Western Limited*, both westbound and both en route from Syracuse to Buffalo. Train No. 17 consisted of engine No. 3364, a Pullman club car and six Pullman sleeping cars, in charge of Conductor William Starr with Engineer Gibbons and his fireman. Train No. 11 consisted of engine No. 3340, a Pullman club car, six Pullman cars and two day coaches, with Conductor Harry Stewart, Engineer John Friedley and Fireman David Brill.

Train No. 17 had left Rochester, its last booked stop, at 2.54 a.m., passed tower SS 35, six miles east of South Byron, at 3.25, and stopped at tower SS 37 at South Byron at 3.35. There Engineer Gibbons had gone to the tower and requested assistance as his engine was not steaming properly. Engine No. 3088 was just about to be attached to the front of his train when it was run into by train No. 11 at 3.42 a.m. The force of the collision drove the front train forward about 250 feet and the rear sleeping car, the *Canfield*, was forced into the car ahead, the *Croton Falls*, and telescoped it for almost its entire length. The third car from the end of train No. 17 was considerably damaged and derailed and two other cars of that train were slightly damaged. The colliding engine was considerably damaged, and the leading car of train No. 11 was derailed. Twenty-one passengers and a Pullman porter died, and 71 passengers were injured. [359]

Describing the aftermath of the accident a newspaper reporter related that, 'The injured were cared for by emergency trains which carried doctors and nurses to the accident from Batavia. Mr and Mrs Nathan Lichtermann of New York and Flora Doherty, passengers in the rear car, were the only ones who escaped instant death. They are in a Batavia hospital. Miss Doherty suffered two fractured legs. Mrs Lichtermann is more seriously injured than her husband.

'A photographer was immediately engaged to photograph bodies as they were taken from the wreck and all articles of clothing were photographed for identification purposes. South Byron folks were called to the aid of those in distress by the ringing of church bells. Father Mullins stood on a ladder at the side of the wreck and administered the last rites to a dying woman.' [360]

There were four lines at the point of the accident, numbered from north to south 4, 3, 1 and 2, tracks 1 and 2 being used by westbound and eastbound passenger and fast freight trains, respectively. Between the towers (interlockings) the line was provided with automatic block signals consisting of three position upper quadrant semaphores showing night-time indications of red, yellow, and green lights for danger, caution and clear. The automatic block sections were about a mile long. The collision had occurred on track 1, about 900 feet east of the home signal at South Byron worked from Signal Station 37. This signal was on a bridge 60 feet east of the tower, the main arm for track 1 was controlled by lever 55 in the tower. Also controlled from the tower, by lever 56, was the rear home signal 39461, located on a bridge 4317 feet east of the home signal. 4954 feet further back was automatic signal 39361, which acted as a distant signal for the rear home signal, showing caution when the rear home signal was at danger.

Towerman Palmer at South Byron told the Interstate Commerce Commission (ICC)

[359] Interstate Commerce Commission report

[360] *The Brookfield Courier*, Brookfield, New York, United States, 15th January 1919

inquiry that train No. 17 arrived at his tower at 3.35. Its engineman whistled one long and three short and then came to his tower and said that he needed assistance as far as Batavia. There were two helper engines standing at the time in the siding between tracks 3 and 4 so having consulted with the dispatcher he had signalled engine No. 3088 to come out of the middle road, cross the eastbound track and come back to couple up to the front of train No. 17, the whole operation taking about 6½ minutes. Just as the coupling was being made he saw train No. 11 approaching and someone whom he presumed was the flagman standing a few car lengths back from the rear of train No. 17 and swinging his red and white lanterns in the face of the oncoming train. He did not see a fusee or hear any torpedoes and there had been no response to the flagman's signals. When he first saw train No. 11 he had looked at the repeater for signal 39461, which showed it to be at danger.

Engineer Thomas Moynihan of the other helper engine, No. 3120, said that his engine had been on the middle track about three car lengths from the rear of train No. 17. Two or three minutes after that train had stopped he had seen the headlight of another train approaching but paid it no attention until his fireman had remarked that he thought it was on track 1 rather than track 4, the other westbound line. At this point he had seen the flagman, apparently 15 or 20 car lengths away and going back slowly. He had seen the flagman give a couple of swings with the white lantern but not the red one. He had seen no fusees.

Fireman Arthur Jasper of helper 3120 put the flagman 12 car lengths back and saw him swing his lantern once. He saw no fusees and heard no torpedoes.

Flagman McMahon of helper 3120 said that he had been standing at the door of his shanty. He estimated that the flagman had been 15 car lengths back. The flagman's lanterns were swinging, but only as much as they would with the normal movement of his arms as he walked along and not enough to be giving stop signals. He saw no fusees.

Engineer Gibbons of train No. 17 said that he had whistled his flagman out as soon as he had stopped. He had seen nothing of him. All the South Byron signals had been clear and their lights burning brightly.

Conductor Starr of train No. 17 agreed that his engineman had immediately whistled the flagman out. After the accident, the flagman had said that he had gone back 20 or 25 car lengths and had put down a torpedo and lit a fusee when he saw train No. 11 coming. He (the conductor) had not heard any torpedo exploding.

Flagman Thomas Groves testified that he had gone back at once when his train stopped and had begun to run when he saw the lights of train No. 11, stopping only to put down two torpedoes about 20 car lengths back. He had then continued to run back, lighting a red fusee as he did so and dropping it between the rails and then swinging both lanterns. He estimated that he had been about 25 car lengths back when train No. 11 passed him. Its engineer had not responded to his signals but after the engine had passed him he had noticed that steam was shut off and there was fire flying from the wheels. He had continued to go back till he was in the rear of signal 39461, where he was joined by the flagman of train No. 11, and they found that the signal was at danger. The rear markers of his train had been burning brightly. He later corrected his statement in several respects including saying that train No. 11 had been about a quarter mile away when he had lit the fusee.

Engineer Friedley of train No. 11 alleged that the rear home signal, 39461, had been green. After passing that signal he said that he had seen the flagman's lanterns, apparently on the ground and only five or six car lengths from the rear of train No. 17. When he had first seen the lanterns, he had not been sure whether they were on tracks 1 or 3 but he had shut off steam and whistled. He had not seen the flagman or any sign of a fusee. It was only after passing the lanterns when he was looking for the next block signal that he first saw the rear markers of the train ahead. He had at once applied his brakes in emergency and at the same time ran over a torpedo. He had got up at 10 a.m. on the Saturday so had not had any rest for almost 18 hours at the time of the collision but he had not felt tired.

Fireman Brill stated that he had called signal 39361 as clear (it should of course have been at caution) after Engineer Friedley had first done so and he distinctly remembered the engineer calling the rear home signal as clear while he had been busy firing. He had felt the air brakes come on and as he had been getting onto his seat to look out of the window the collision had occurred. He did not hear a torpedo nor see the reflection of a fusee.

Conductor Stewart estimated that there had only no more than ten seconds between his feeling the brakes going on and the collision. He saw no glare of a fusee when he first got off his train.

As might be expected the signals were very carefully examined after the accident. Nothing was found to be wrong with them and the Interstate Commerce Commission pointed out that they had also been working correctly after the collision. In view of this and the other evidence it was believed that Engineer Friedley had failed to see them, nor had he brought his train under proper control when he said he had seen flagman's lanterns and his failure to see the rear markers of the train in front at this time was inexplicable. In fact, given that Engineer Friedley had failed to apply the brakes until his engine exploded a torpedo there was a strong suspicion that he had briefly fallen asleep.

It was also questionable, the commission's report continued, if Flagman Groves had in fact gone back at once to protect his train and if he had really gone back as far as he alleged. The flagman's statement that he had lit a fusee and shown it to the approaching train was also disputed.

The commission concluded that the accident had been principally caused by the failure of Engineer Friedley to observe and obey the automatic block signals, while a contributory cause was the failure of Flagman Groves to go back far enough to protect his train and to display a lighted fusee.

Asked by the commission what could be done to prevent such collisions Signal Engineer William Elliott had replied, automatic train control, adding that the New York Central had been looking into the matter for some years but there were objections, such as the fact that it would take control of the train away from the engineer and that it would be expensive. In view of the cost a decision had been made to install automatic signalling on those lines not so equipped rather than installing automatic train control on lines already having automatic signalling. Mechanical train stops as used on underground lines would suffer difficulties in open country from snow and ice, from ballast being deposited on them and from the variations in train speeds.

The commission was not impressed by Mr Elliott's arguments and pointed out that arguments against automatic train control other than cost had been known for many years and the problems were difficult but not insurmountable. Automatic train control was not intended to take away control of the train from the engineer under normal conditions but only to intervene should he fail to respond to adverse signals through inadvertence or incapacity. The commission added, 'The startling recurrence of accidents resulting from the failure of enginemen to observe and heed signal indications calls attention with great emphasis to the fact that the sense of responsibility for the safety of their trains which is almost universally felt by enginemen is not always effective with present safeguards to prevent collisions.'

Regarding the telescoping of the rear two cars of train No. 17 the commission noted that the circumstances had been almost identical to those of the Mount Union collision of February 1917 (qv) and it repeated the comments made in that report regarding the inadequate strength of the rolling stock to resist the effects of a collision and to prevent disastrous telescoping, noting that, 'In the present case the inadequacy of the construction of the end car which was telescoped is clearly shown. It possessed but little strength in a comparative sense against telescoping, while its horrible shock-absorbing feature was the mass of humanity forced from all parts of its length into its extreme end. Types of construction have been proposed and cars are in service which offer greater resistance than the car which was telescoped and in which all the occupants were killed. The features of adequate strength in end construction and collapsibility of vestibules as the means of lessening the dangers of collisions demand serious consideration.' [361]

The day after the South Byron disaster another rear collision occurred in very similar circumstances near Fort Washington, Pennsylvania, United States, on the Philadelphia and Reading Railroad. Again, an engineer overran automatic signals and again there was a contributory failure on the part of a flagman.

Fort Washington station lay on what was known as the Bethlehem Branch of the railroad. The line here was double track and was provided with automatic block signals of an enclosed disc design, with each post being provided with two discs one above the other acting as home and distant signals. A stop indication was given by a red disk over a green disk, a caution indication by a white disk over a green disk and a clear indication by two white disks. At night lights displayed the same indications. From the south the signals for northbound trains were 309 at Oreland, 311 just south of Camp Hill station and 5,718 feet from signal 309, and 313, rather less than half a mile south of Fort Washington station and 3,756 feet from signal 311.

On Monday 13th January 1919 freight extra 1508, consisting of engine No. 1508 and 24 cars, in charge of Conductor Russell and Engineer Love, arrived at Fort Washington at 5.50 p.m. A waggon was dropped off and the train was ready to proceed at 5.55 when a coupling pulled out from the fourth vehicle from the front of the train. The damaged car was then taken to Ambler, a mile and a half away, and left there, and the front of the train returned to Fort Washington for the remainder of the train. While these operations were in progress train No. 381, the Doylestown local, had arrived at Fort Washington and pulled up behind the freight train. This train consisted of engine No. 587 and eight wooden coaches, in charge of Conductor Barndt with Engineer Rouland. Having pulled up at signal 313, which was showing a stop indication, the train had then drawn

[361] Interstate Commerce Commission report

northwards as far as the rear of the freight train, stopping behind it at 6.17 p.m., with its rear 413 feet north of signal 313. While still standing there at 6.33 p.m. it was run into by train No. 319.

Train No. 319 was the Scranton express, consisting of 4-4-2 engine No. 339, a Central of New Jersey Railroad combination car, a Central of New Jersey Railroad coach, the Pullman parlor car *Brant Brach* and a Philadelphia and Reading coach. This train had left Philadelphia for Scranton at 6 p.m. in charge of Conductor Fawthorp with Engineer Philemon Githens and Fireman Frank Coulton. Stopping at Columbia Avenue and Wayne Junction it had passed Oreland at 6.31 p.m., three minutes late, passed a flagman about 1,000 feet south of signal 313, passed signal 313, and striking the local train had telescoped its rear coach about 45 feet, demolishing it. The next coach of the local train and the colliding locomotive were damaged. Thirteen passengers and an off-duty railroad employee died and 22 were injured. [362]

The Doylestown train had left Philadelphia at 5.30 p.m. and had picked up scores of munitions workers at Midvale Steel and other plants along the line. [363]

Engineer Love and Conductor Russell of the freight train told the Interstate Commerce Commission investigators that signals 311 and 313 had both shown double white clear indications, which of course were their proper indications at that time.

Flagman Jessop of the freight train said that he had gone back 600 feet to protect his train, had been called in, then had gone back again when the coupling broke, this time 800 feet, and had flagged and stopped the Doylestown local.

Engineer Rouland of local train No. 381 testified that he had received a white over green indication at signal 311, had stopped at Camp Hill station, had stopped at signal 313, which was displaying red over green, then proceeding cautiously had been stopped by the freight train flagman.

Conductor Barndt of the local train said that he had gone forward to the front of his train to find the cause of the delay, had returned to the rear, and seeing the express approaching had waved a white lantern at it. His flagman had gone back to the Trenton cut-off bridge (the overbridge where the Pennsylvania Railroad line crossed the Reading tracks).

Flagman Powers of the local train said he had not known why his train had stopped but he had set off back at once. He had stood for three or five minutes at a concrete bridge north of the Trenton cut-off from where he could see both the rear markers of his train and signal 313 standing at danger. When the Scranton express had come around the corner at 30 or 35 miles an hour he had flagged it, but he had received no response until, when half the train had passed him, he saw that its brakes were applied.

Engineer Githens of the Scranton train stated that his brakes had worked properly at his two previous station stops. According to his testimony signal 311 had been showing two white lights. As he was approaching the Trenton cut-off he had been travelling at 50 or 60 miles an hour. When he saw the flagman waving a stop signal he had made an emergency brake application, but the brakes had not seemed to take hold properly. After passing the flagman he saw both signal 313 at danger and the marker lights of the local train but he was unable to stop before striking it. He stated that five or six weeks earlier

---

[362] Ibid.
[363] *The Evening Tribune*, Albert Lea, Minnesota, United States, 14th January 1918

he had also had a clear indication at signal 311 and a stop indication at signal 313 which he had told someone about on his arrival at Bethlehem.

Fireman Coulton was too badly injured to be interviewed but he had previously told railroad officials that he had been firing and had not seen signal 311. His attention had been attracted after passing Camp Hill by the engineer whistling in response to the flagman and applying the brakes, when he had looked out and seen signal 313 and the rear of the local train.

Conductor Fawthorp thought that his train was travelling at 40 miles an hour when the brakes were applied and 20 miles an hour at the time of the collision. He did not see either signal 311 or 313.

Engineer Yeakel and Fireman Knapp of train 387, the Lansdale local, which was the next train to come along, half an hour after the collision, stated that signals 311 and 313 had been displaying the correct caution and danger indications to their train, white over green and red over green.

Various examinations and tests were carried out after the collision to determine if the brakes of the express were working properly, how far a train would take to stop after seeing the flagman and if the signals were working correctly. No evidence was found of any problem with the signals and in particular nothing that would have made signal 311 show clear when it should have been at caution. The commission's investigators concluded that Engineer Githens must have mistaken or misinterpreted the signal. With regard to the engineer's statement of a previous failure Signal Supervisor Steele explained that this probably related to an incident on 25th November of the previous year when signal 313 had failed to clear after the passage of a train. This had thus been a 'right-side' failure as although the signal had wrongly been displaying danger there was in fact no train in advance of it. The primary cause of the accident was therefore the failure of Engineer Githens to observe signal 311 at caution and so bring his train under control so as to stop at signal 313. Why he had failed to observe the signal it was futile to speculate. Given the physical characteristics of the line, with curvature and the view obscured by trees and the overbridge, combined with the speed of the train, it could not be expected that the engineer would have seen the flagman early enough to be able to stop short of the local train.

As for Flagman Powers, according to the evidence he had had 15 minutes in which to go back and protect his train and yet he had only covered 1,500 feet, a distance which could be walked easily in five minutes, while according to his own evidence he had stood in that location for three or five minutes. He had therefore had ample time to provide full and proper protection, had been derelict in his duty and was partly responsible for the accident.

In a final remark the commission noted that had the cars of the local train been of steel rather than wooden construction the number of deaths and injuries would have been materially reduced. [364]

The rear portion of a hospital train on the Chemins de fer de l'Est was left in the section near Mauvages, France, due to the breakage of a coupling. It was then run into by

[364] Interstate Commerce Commission report

a following mixed train, at the head of which were two carriages carrying 54 soldiers on leave. Fire appears to have broken out following the collision, for the newspaper reports speak of the discovery of the charred bodies of the conductor, driver, and fireman, while 13 of the soldiers died. The two coaches were thrown down a 12-foot-high embankment and all the other 41 soldiers were injured.

The following day at Theil-sur-Vanne on the Sens-Troyes line a passenger train ran into a shunting goods train. A soldier lost his life, and a brakeman was seriously injured. Finally, in Belgium at about the same time a train carrying returning German soldiers was derailed at Kinkempois and a number of them were injured. [365]

At 05:00 on the morning of Tuesday 4th February 1919 a Paris – Strasbourg express overtook a leave train at Cousanses-au-Bois, between Ernecourt-Loxéville and Lérouville stations on the Chemins de fer de l'Est. Nine soldiers died in the accident and 21 were injured. [366] [367]

Another disaster occurred in France on the Paris – Brest line of the Chemins de fer de l'Ouest at Crissé at 00:38 on Thursday 17th April 1919. A leave train carrying French soldiers stopped near the station, 31 kilometres from Mans, due to its locomotive breaking down. The conductor of the train reportedly went back down the line and placed detonators but a train carrying American soldiers to Brest was running fast downhill and could not be pulled up in time. An initial report spoke of 20 dead; six of them French and 14 Americans, with 45 others hurt. The injured were taken on a rescue train to Mans. The condition of a number of them was said to be serious, and a follow-up report stated that the number of dead had increased to 33. 'An inquiry has been opened to determine responsibility for the accident.' [368] [369]

An American newspaper added that, 'Among the dead is a ten-year-old French boy who had been adopted as a mascot by the American soldiers.' [370] How can soldiers adopt a child as a mascot?

Thirteen of the dead American soldiers were buried at Mans on 20th April, their bodies carried in 13 Red Cross motor ambulances covered with flowers and flags from the American hospitals to the cemetery. The route was lined with American troops while the cortege was followed by American commanders with their staffs as well as the French district commander and the prefect of the department of the Sarthe. [371]

The Paris – Strasbourg express due at Lunéville at 20:30 was considerably delayed on the evening of Thursday 14th August 1919. At 23:00 it was still a kilometre from the station at Blainville-le-Grande, France, instead of at Morteau, and stopped at a signal.

---

[365] *La Croix*, Paris, France, 24th January 1919
[366] Wikipedia – Liste des accidents ferroviaires en France au XX siècle
[367] *The Times*, London, England, 6th February 1919
[368] *Le Figaro*, Paris, France, 18th April 1919
[369] *The Times*, London, England, 19th and 21st April 1919
[370] *Muskogee Times-Democrat*, Muskogee, Oklahoma, United States, 19th April 1919
[371] *Joplin News Herald*, Joplin, Missouri, United States, 21st April 1919

Before the signal cleared it was run into from behind by a leave train from Metz, travelling at 40 kilometres an hour. The van at the rear of the express was smashed but no one on that train was much hurt. Unfortunately the same was not the true for the crowded troop train. Six were killed outright and 34 were injured, many seriously; two more died soon after arrival at hospital in Lunéville. The accident was blamed on 'the faulty working of the signals.' [372]

Express train 145 which left Paris d'Orsay at 21:54 on 4th September 1919 for Toulouse came to a stand between Castelneau-d'Estretefonds and St Jory stations, France, at about 23:00, when it was run into from behind by express train 121 from Bordeaux to Cette. The Bordeaux express was double-headed, and a French newspaper spoke of the last three vehicles of the first train being annihilated; one of them was fortunately a goods vehicle, the others were a second-class and a first-class carriage. A fourth vehicle, another first-class carriage, was half destroyed.

When the accident occurred the mayor and curé (village priest) of the village of St Jory sounded the tocsin (alarm bell) and the whole population of the village rushed to the scene. In the darkness it was difficult to distinguish anything, but the air was filled with cries and groans and the noise of the steam escaping from the boiler of the damaged locomotive. Near the leading engine of the colliding train lay the body of its driver, who had been thrown against a wall. Some metres away lay the seriously injured fireman, while uninjured passengers were searching for their relatives and friends.

The rescue operation was speedily undertaken and the injured were taken to the Hôtel Dieu Saint Jacques in Toulouse (which despite its rather misleading name was the main hospital there and not a hotel). Several of them were critically hurt. The mayor of Toulouse, M. Rieux, and his deputy, M. Deltour, visited the injured in hospital where they congratulated a doctor who had been a passenger on the Paris train and was himself badly hurt but had provided the first care for other injured passengers.

By the morning of the next day the bodies of 18 victims had been recovered and Mme Marie Giraud, aged 40, of Bourges, and an unidentified young woman had died of their injuries after arrival at hospital. About 40 others were injured.

An initial investigation disclosed that the Paris express had stopped because its engine had broken down. It had left Paris 13 minutes late and instead of its driver being able to regain time it had lost more at each point along the route (it should have arrived in Toulouse already at 22:25). The driver had finally been forced to stop because of a lack of steam pressure due to poor coal (c.f. Ais Gill, England, September 1913). After the train had been stopped five minutes the head conductor reportedly went back to protect it with detonators and flares, but he failed to stop the Bordeaux express travelling at 80 kilometres an hour, either because he did not go back far enough quickly enough or because the driver of the express had failed to see him. How the Bordeaux train came anyway to be in the same section of a line which presumably was worked on some sort of block system was not explained. [373] [374]

---

[372] *Le Radical*, Paris, France, 17th August 1919

[373] *Le Petit Parisien*, Paris, France, 6th September 1919

[374] *The Times*, London, England, 6th September 1919

Our next accident brings us back to the United States, and to one of the interurban railways which briefly flourished in that country before the automobile reigned supreme. The accident occurred at Dunkirk, New York, on the Buffalo and Lake Erie Traction Company on 4th October 1919. The line of this company ran for just over 91 miles between Buffalo, New York and Erie, Pennsylvania. It was double track, and the trains were electrically propelled trolleys. The line was operated by timetable and train orders. In addition to the longer distance interurbans there were so-called local cars which worked only over the 5 miles between Dunkirk and Fredonia. The longer distance interurbans had worked to a timetable and had 9 stops in the 5 miles; the local trains had 37 stops and their movements were not indicated in the schedules of the interurbans.

On the day of the accident local car No. 440, in charge of Conductor Elmer Fick and Motorman LaBarron, left Dunkirk for Fredonia at 9 p.m. It stopped at the Ennis stop to discharge a passenger and take on two, proceeded a quarter mile to the Corwin Stop, moved off again from there and had travelled a short distance when it was struck by interurban car No. 602 at 9.22 p.m.

Interurban car No. 602, designated in the timetable as train No. 37, was in charge of Conductor Perrin with Motorman Irving Fick. It had left Dunkirk at 9.06 p.m., 5 minutes late, and made its last stop at Country Club, colliding with the local car at Corwin Stop. The local car was driven forward between 150 and 200 feet in the collision and its rear vestibule was destroyed as was the front vestibule of the interurban. Six passengers lost their lives and 25 passengers, and three employees were injured.

Motorman LaBarron told the Interstate Commerce Commission inquiry that as the conductor had rung the bell for him to proceed from Corwin Stop he had seen the reflection of the headlight of the interurban and heard its whistle. He had tried to get his car under way but only managed to move it a car length.

Conductor Fick was too badly injured to say much about the accident other than that his car had been crowded.

Conductor Perrin of train No. 37 said that as far as he knew his motorman was in a normal condition. He thought that the motorman had dimmed the headlight of his car after leaving Dunkirk. He could not estimate the speed of his train before the collision and was not sure if the brakes had been applied by the motorman or as a result of the collision.

Motorman Fick was interviewed at his home by an attorney. He stated that he had applied full power after leaving Country Club but had reduced to half power 800 feet before Corwin Stop. Approaching the site of the accident his vision had been impaired by the light of automobiles coming from the opposite direction and with the water on the windscreen (it was raining) he had been unable to see the local car until just after his train passed Ennis Stop. He had then shut off power, applied the brakes and whistled.

Another motorman, named Woodward, stated that he had worked car No. 602 that day until it was taken over by Motorman Fick and that its brakes and other equipment had worked properly. He was aware that there was a speed restriction through the Dunkirk city limits but said that at present crews used their own discretion with regard to the speeds at which they ran and did not feel prohibited from making up time, while nothing had been said to him recently about going too fast.

The Interstate Commerce Commission blamed the accident on the failure of Motorman Fick to properly control the speed of his car and his failure to obey Rule 5 of the special instructions printed in the timetable, which read: 'Trains will reduce speed to five (5) miles per hour through Dunkirk Subway, and will not exceed a speed of twelve (12) miles an hour through Westfield, Fredonia, Dunkirk and Silver Creek.'

Contributing to the accident, the commission found, had been the absence of a red light or marker at the rear of the local car, and had there been such a light the collision would probably have been prevented. [375]

Irving Fick was tried for manslaughter at the end of December but after deliberating for seven hours the jury were unable to agree on a verdict and they were discharged. [376] At this point it seems likely that the prosecution was abandoned.

Denmark suffered, on Saturday 1st November 1919, a railway disaster of comparable magnitude to its worst ever at Gjentofte 22 years earlier, and again it was a rear collision. The events began as train No. 168 was approaching Vigersley, on the outskirts of København, running some 15 minutes late. A 7-year-old boy was riding with his family in the third-class carriage last but one in the train. In what was probably a moment of boredom he played with the door lock, the door flew open, and he fell out onto the track. Another passenger then pulled the communication cord, and the train came to a stand in the section, just a little short of Vigerslev station, at 20:50. Once the train had stopped two railwaymen named Hoburg and Larsen, having found what had happened, made their way back along the track to search for the boy. It was of course dark, and Larsen had no handlamp, while Hoburg, rather than wasting time to go forward to his van to collect his lantern, seized one of the train's tail lamps to use instead. Having gone back three or four hundred metres they found the missing boy, who had suffered a broken leg and was being assisted back to the train by his 20-year-old brother. Hoburg and Larsen then decided that it would be easier if the train were to set back to them and one or both of then signalled and whistled for the driver to do so. At the front of the train Driver A. Petersen, had found what was wrong, but the brakes having been applied in emergency there was a delay while the brake valve was closed, and the vacuum re-created. There was then a further delay when they saw a train approaching on the opposite running line and signalled for it to stop, lest it should present a danger to the people who had gone back down the line.

Meanwhile Signalman A. Hansen in Vigerslev signal box was wondering why train No. 168 had stopped short of his home signal. Having discovered the cause, he replaced his signals against the train. Matters were complicated for him by the fact that he had a special fire train, No. 9185, approaching from the goods station branch, on its way to tackle a fire at Kjøge, which in coming off the branch had to cross the path of the express. The signalling controls at Vigerslev were such that once the signals had been cleared for an approaching train they could not normally be replaced, and the junction points reset, until the train in question had cleared a track circuit in advance of the points. To get around this, Signalman Hansen had to break two seals and depress two buttons to override the locking. This was justifiable in the circumstances but unfortunately he also

[375] Interstate Commerce Commission report

[376] *Cattaraugus Republican*, Little Valley, New York, United States, 25th December 1919

inadvertently cleared the block back to Brøndbyøster signal box, three kilometres in the rear, and moreover failed to immediately realise what he had done. He was then able to reset the junction points and clear his signals for the fire train. It was only when the latter train had passed him and was being brought to a stand by the crew of train No. 168 that he belatedly realised what he had done. He immediately sprang to the telephone and called Signalman O. Nielsen at Brøndbyøster to tell him that there had been a mistake, but the latter had replied either, "It is passing," or, "It has passed." As soon as Vigerslev had, as it appeared, cleared back train No. 168, Signalman Nielsen had lowered his signals for train No. 8064, a duplicate service for train No. 44, to cope with the extra November Change Over Day traffic. Nielsen did have a handlamp, which, had he turned it to red and displayed it towards train No. 8064 as it was receding, there might have been some chance even then of stopping it, as Traffic Inspector Juel Hansen was looking back from the rear platform of the last vehicle of the train. Unfortunately, Nielsen, who had only been a signalman for six weeks, did not grasp the urgency of the situation and hence did nothing.

As soon as he found that train No. 8064 had passed Brøndbyøster Signalman Hansen left his box and ran towards train No. 168, shouting for them to move forward as there was another train coming up behind them. The train had just begun to move forward when the collision occurred. The engine of the colliding train crushed the rear five carriages of train No. 168, then overturned and fell with them down an embankment, taking the leading three vehicles of its train with it.

Driver A. Høgholt of the colliding train was killed outright and Fireman C. Nielsen was fatally injured, so it was impossible to determine why Driver Høgholt had not seen the tail lamp of train No. 168 earlier, though from the fireman's ante-mortem statement he did see it at the last moment. However, tests conducted afterwards showed that the tail lamps were not as easy to discern as they might be with background street lighting on Vigerslev Road, an electric light on the roof of the rear carriage of the train, possibly misty conditions and so on. It appeared that Driver Høgholt did apply the brakes at the last moment. In addition to the driver and fireman 38 passengers died in the accident and 27 others were injured. [377]

On the evening of Monday 3rd November 1919, the Simplon express of the Chemins de fer de Paris à Lyon et à la Méditerranée (P.L.M.), train L. 33, left Paris Lyon station at 20:40. The *train de luxe* consisted of a locomotive, four sleeping cars, a dining car and two luggage vans. It passed through Pont-sur-Yonne station, France, at about 22:40 but was then stopped by signals at kilometre 104 because train No. 5 for Marseilles, which was front, had to stop in Sens station to have its engine changed. At this point it should have been protected by the Villeperrot signals, but while the Simplon train was still waiting to proceed it was run into by the following Paris – Geneva express, train No. 661, which had departed Paris Lyon ten minutes later.

The colliding train was reportedly running at 96 kilometres an hour at the moment of the collision and the two luggage vans at the rear of the Simplon express were pulverised, while the dining car was pushed forward and telescoped the *Wagons-Lits* sleeping car next to it, bound for Bucharest. The vehicles at the front of the Geneva express suffered

[377] Webpage – Jerbaneartiker på Jernbanen.dk – Vigerslevulykken

even more; the leading postal van was flattened against the tender of the locomotive, the following third-class carriage, containing 67 passengers, was also broken up, as was the first and second-class composite carriage behind it. Of the eleven postal workers in the van six were killed outright according to the first reports and the other five seriously injured.

After the collision, several railwaymen ran to the nearest houses of Villeperrot village, which was close to the line, and were directed to the home of the mayor, M. Montéré, who sounded the tocsin (alarm bell). The locals came running with lanterns, for the night was very dark. The amount of lighting still proving insufficient fires were lit on either side of the tracks, but the rescue work was not that efficient until dawn, though a steam crane arrived at the scene at 02:00.

One of the passengers who escaped without serious injury from the accident was the member of parliament for l'Ain, M. Laurent Derognat. Less fortunate was the chef of the dining car, who suffered a fractured thigh and other injuries, though he was able to relate his story to reporters. He told them, 'I was in the kitchen, and the second service was just over when we pulled up suddenly. I looked out and saw open country. We had been waiting there about a quarter of an hour when there was a fearful crash. I was flung to the ground among debris of every kind and fainted.'

A French newspaper reporter wrote of the corpses of the victims being wrapped in sheets or whatever was available and taken to nearby houses, while surviving passengers, some of them only half-dressed, shivered as they stood around the trackside bonfires, for which there was plenty of wood (presumably from the shattered carriages). At dawn 200 soldiers arrived from Sens and with the help of the breakdown crane the rescue work could then be done quickly. [378]

The seriously injured victims were taken by the engine and undamaged carriages of the Simplon express to Sens; the slightly injured were taken to Montargis, Montereau and some to Paris. Most of those who were killed could not be immediately identified as they had been asleep, undressed, at the time of the collision, and hence not wearing clothes containing wallets and purses which would have aided their identification.

An official statement issued by the P.L.M. railway company and published in the newspapers on the 6th of November blamed the accident on the driver of the Geneva express overrunning signals and stated that 15 passengers, two postal workers (as opposed to the six initially reported) and a conductor for the *Wagons-Lits* company had been killed and there were 60 injured. A newspaper of the 6th of November also stated that there were two drivers, M. Badier and M. Bonzon, in the cab of the locomotive of the Geneva express together with Fireman Moreau. Of these only M. Badier had returned to Paris at this date and he blamed the accident on thick clouds of steam escaping from the engine which had obscure the signals. [379] [380]

On 7th of November it became clear that Driver Bonzon had actually been driving the locomotive of the Geneva express. He stated that he had complained the day before the accident of problems with the engine. Repairs had then been carried out, but these had not been effective in curing the problem and before leaving Paris on the day of the

---

[378] *Le Petit Parisien*, Paris, France, 5th November 1919

[379] Ibid, 6th November 1919

[380] *The Times*, London, 5th and 6th November 1919

accident he had complained to Travelling Inspector Badier, who had accompanied him on the engine for the trip, for reasons which were not stated, of the difficulty of viewing signals. Chief Inspector Rimbaut and the head conductor of his train had also been present when he had made these comments. During the journey he had repeatedly had to close the regulator in order to see the signals. At Pont-sur-Yonne the avertissieur (distant signal) was on a curve and he had to cross the cab to see it. The entrance signal to the station was clear but it was possible that he had confused the white light of a hydraulic crane on track 1 for the Pont-sur-Yonne exit signal. At kilometre 104 he had failed to see the avertissieur and had closed the regulator, then he had seen the semaphore at danger and had made an emergency brake application just before the collision. [381]

In a footnote to the accident *The Times* in London added that one of the passengers who had lost their lives appeared to have been a smuggler of silver coin, for in his trunk were found 6,000 francs in one- and two-franc pieces, weighing almost 66lb, which it was assumed he had been intending to smuggle into Switzerland. [382]

On the evening of Thursday 18th December 1919, a rear collision occurred at Walton, Virginia, United States, on the Norfolk and Western Railway. The main line westwards from Roanoke, Virginia, to Walton divided at the latter point into branches to Bristol, Virginia and Bluefield, West Virginia, both of which were double track from Walton, though the Bristol line became single track at Radford. There were three tracks east of Walton for some miles, the southbound track being used by eastbound passenger trains, the middle track by eastbound freight trains and the northern track by westbound trains. The lines were worked by train orders and an automatic block system and the connections between them at Walton were operated from a 35-lever tower. Opposite the tower was signal 14-L controlling the movement of eastbound trains from the Bristol line, while 759 feet further east signal 8-L protected the connection from eastbound track to the middle track. This signal had two semaphore arms, the upper being a semiautomatic signal controlling movements over the main track and the lower a permissive signal which when cleared allowed trains to proceed with caution past the signal when the upper arm was at danger. The indications of the signal were red over red (stop), red over yellow (proceed with great caution), and green over red (proceed).

At the time of the accident there were two trains standing on the two eastbound tracks at Walton. Immediately east of the crossover, standing in its usual position, was a shuttle passenger train which took passengers from through trains and ran between Walton and Radford on the Bristol line. To the east of the shuttle train on the south line taking water from a tank was eastbound passenger train second No. 4 from Cincinnati, Ohio, to Roanoke, consisting of engine No. 120, six sleeping cars, two steel sleeping cars and three wooden coaches, in charge of Conductor Munsey with Engineer Heslep and Fireman Long.

Train second No. 4 had been standing at the water tank five minutes when it was run into by eastbound passenger train No. 26. The latter consisted of engine No. 106, a mail car, a combination car, three sleeping cars, a dining car and two coaches, all of steel construction, and was in charge of Conductor Davis with Engineer Hooper and Fireman

[381] *Le Temps*, Paris, France, 7th November 1919
[382] *The Times*, London, England, 8th November 1919

Martin. In the collision the rear car of the Cincinnati train was pushed forward and telescoped the preceding car for two thirds of its length, while the rear car was badly damaged and penetrated by the colliding engine. The third car from the end of the Cincinnati train was also quite badly damaged by being pushed forward into the steel coach in front. No damage was done to the colliding train, other than to the front of the engine.

Five passengers died in the accident and 10 passengers, and seven railwaymen were injured. [383] The passengers who lost their lives were Dr Charles Hamlin, of High Point, North Carolina; the Rev Bernard Green, of Nottoway, Virginia; A.N. DeLong, of Roanoke, Virginia; Mrs Will Dinkins, of New Castle, Indiana and an unidentified coloured woman, probably the wife of the Rev Green. Four of the victims were killed outright but Mrs Dinkley lived for about 15 minutes, trapped in the wreckage.

Ten minutes after the accident westbound train No. 3 arrived at Walton with a trained nurse from Lynchburg on board, who temporarily aborted her journey to aid the victims, while physicians from Radford and nearby towns went to the scene. [384]

The first witness to give evidence at the Interstate Commerce Commission (ICC) inquiry was Engineman Heslep of train second No. 4. He testified that he had stopped at Walton tower and picked up an order and then drawn forward to the water tower. He had just whistled the flagman in when the collision occurred. He did not notice any smoke or steam blowing across the track.

Fireman Long on the other hand said that he had seen quite a lot of steam or smoke coming from the engine of the shuttle train which somewhat obscured the view.

Head Brakeman Mallory said that the flagman had been about seven car lengths back when the collision occurred.

Conductor Mouncey stated that when his train had stopped, he had gone to the rear platform to check that the flagman had gone back and had remained there until he saw train No. 26 coming around the curve west of the tower. There was no smoke or steam obscuring his view. Train No. 26 had been running at about 30 miles an hour as it passed the tower and did not shut off steam until within 10 car lengths of his train. He saw the flagman signalling to the train; his signals did not seem to be responded to and the brakes of the approaching train were not applied until it was eight car lengths away. When he saw that a collision was inevitable, he had jumped from his train and run down the embankment. The speed of the train when it struck was still 20 or 25 miles an hour. He thought that the flagman had gone back about 400 feet, to the rear of the shuttle train, and that he could have reached the tower in three minutes. He had had no lantern when he was stood at the rear of his train, having previously lent it to the head brakeman. The rear markers of his train had been burning brightly.

Flagman Pyrtle of train second No. 4 said that he had started to walk back slowly when it pulled up and was six or eight car lengths back when he saw train No. 26 passing the tower, travelling at about 45 miles an hour, and working steam. There was no smoke or steam obscuring his train and its rear markers were burning brightly. The headlight of train No. 26 appeared to be dim. As soon as he saw the train, he had begun to give stop signals and had continued to do so until he had had to jump out of the way. Steam was

383 Interstate Commerce Commission report

384 *Bluefield Daily Telegraph*, Bluefield, West Virginia, United States, 19th January 1919

shut off just before the train reached him and the brakes were applied shortly after it passed him. He had not used a fusee because he thought that the train would have passed him before he could light it and he thought swinging a red lantern was better.

Engineer Hooper of train No. 26 said that he had found that the electric headlight of his locomotive was not working after leaving Pulaski and a message to have someone ready to repair it at East Radford had not been delivered, so to avoid delay he had substituted a lantern for the headlight. Signal 14-L at Walton tower had been clear and he had continued, letting his train drift at a speed he put at 15 miles an hour. Signal 8-L had been showing a yellow caution indication and he had had to work steam to maintain his speed but had then, he said, run into a bank of steam which had obscured his view, but he had not shut off steam. When he came out of the steam, travelling at 20 miles an hour, he saw a red light which turned out to be one of the rear markers of train No. 4. Leaning out of the window he had then seen two red lights and realising that there was a train ahead he had applied the brakes in emergency. He had seen nothing of the flagman either before or after the collision. He could not say why he had not shut off steam or applied the brake when his view was obscured and thought that he would have seen the train in front sooner had he had a proper headlight. This was his third trip with train No. 26, and he had had trouble with the electric headlights on each trip.

Fireman Martin of train No. 26 said he had not seen either signal because he had been looking to see the train order board at the tower. He put the speed of the train at this point at 15 miles an hour. Between the tower and signal 8-L his engineer had begun to work steam and he had begun to shovel down coal. His first knowledge that anything was wrong was when his engineer had braked and called out immediately before the collision.

Conductor Davis put the speed of his train at the tower at 20 miles an hour, said that it had then been increased to 25 miles an hour and was again reduced to 20 miles an hour at the moment of the collision. Head Brakeman Cruise put the final speed of train No. 26 at 20 miles an hour; Rear Brakeman Keesee put it at 25 miles an hour.

Conductor Wright of the shuttle train said that he saw the flagman four car lengths from the rear of his train giving stop signals to train No. 26 and he (the conductor) had given stop signals with a white lantern from the platform of his car, while he had shouted to the crew of the oncoming engine as it passed. He did not notice any smoke or steam blowing across the track.

Operator Harman in Walton tower said that train second No. 4 passed him at 6.36 p.m. When the engine of train No. 26 was passing him, he had cleared signal 8-L, the calling-on signal, to a permissive indication, although he did not then know the whereabouts of train second No. 4, except that it had not passed the next station at Vicker. He thought that train No. 26 was travelling faster than any other train he had seen and would not have been able to stop at signal 8-L had it not been cleared. When he cleared signal 8-L he was expecting train second No. 2 from Bluefield within 10 or 15 minutes but in any case it was the custom to clear the calling on signal at all times and for all purposes, and according to his understanding these calling-on signals had been installed to keep traffic moving as well as to reduce congestion through the interlocking.

The ICC found that the accident had been caused by the failure of Engineer Hooper to operate his train 'with great caution' through the block section east of Walton tower after receiving a permissive signal indication, as required Rule 708, which read in part,

'Engineers and conductors running on permissive block must handle their trains with great caution. Where the view is obscured, speed must be reduced to insure against collision with a train running ahead...' There could be no excuse for his failure to have his train under full control and at any event he should have shut off steam and been prepared to quickly stop when running through the smoke and steam which he had said had obscured his view.

Flagman Pyrtle had contributed to the accident by not going back as far as signal 8-L, which he could easily have done in the time available.

Operator Harman had not exercised good judgement in allowing train No. 26 to pass signal 8-L when he knew there was a passenger train in the section ahead, but in doing so he had only been following the regular practice which had grown up and there were no specific regulations prohibiting him from doing so. The calling-on arm had originally been installed to prevent trains coming to a stand behind trains calling at a coal wharf which had since been moved, and to allow helper engines to couple to the rear of trains without the need for hand signals, and instructions should be issued to prevent its use for allowing passenger trains to enter an occupied section or trains to enter the occupied section behind a passenger train except in emergency.

Had the rear cars of train second No. 4 been of steel rather than wooden construction the results of the accident would have been much less serious. [385]

Bitterly cold weather, with a temperature of -15° Fahrenheit (-26°C) was supposedly a factor in a collision a mile and a half miles east of Corbeil, Ontario, Canada, at 11.15 a.m. on the morning of Sunday 25th January 1920 on the Canadian Pacific Railway. The trains involved were the first and second sections of train No. 1, the transcontinental express. The engine crew of the first section had fought a losing battle in attempting to keep their engine steaming sufficiently until finally the train had stalled on a curve. Flagman T.J. Curley went back to flag his train, though how far back he had gone was disputed. He laid torpedoes on the rails and had left them there when he was recalled by a whistle from Engineer James Davidson. He had almost regained his train when he heard the second section approaching at unreduced speed and ran back again to try and warn it, but his signals were unseen, and the second section was still travelling at 35 or 40 miles an hour when the collision occurred. The engine of the second train ploughed through almost three cars before the train came to a standstill. The observation car at the rear of the first section was described as having been 'crumpled almost to matchwood.'

Neither Engineer Isidore Godin nor Fireman Gleason of the second section had time to jump but they escaped serious injury, although the fireman suffered from shock. The fireman had been busy firing and a newspaper report theorised that the noise of his shovel combined with the noise of the engine had masked the sound of the exploding torpedoes. Standing on the rear platform of the observation car had been Brakeman Kent and Porter Simmons. When they saw the second section approaching the brakeman jumped off and survived, the porter retreated into the car, possibly to warn the passengers, and was killed. In addition to the porter at least seven passengers were killed. All the casualties had been in the last car of the train.

---

[385] Interstate Commerce Commission report

'Most of the passengers were immigrants, and men and women ran up and down the length of the train, eagerly seeking news of this or that one who was missing. Rescuers seeking bodies walked back and forth over what seemed like a door lying almost flat in the snow. Then someone noticed a piece of cloth protruding beyond one edge, and a hurried search revealed an unconscious woman, who is now in the hospital. In another instance a girl was pinned down by part of the tender of the overturned engine. All attempts to lift the great weight were unavailing, until finally one man who had served as a stretcher bearer in France thought of the plan of getting away the frozen earth beneath her. This was done, and she is now also in the hospital and expected to recover.

'One girl, formerly a nurse in a British hospital, who was on her way to join relatives in the West, was washing out a few handkerchiefs in the second last sleeper. The first impact threw her against the wall of the small compartment she was in. After the second she found herself being forced out through the window into the snow. She was not seriously injured.

'Mail Clerk Joseph D Larose, of Ottawa, was on the second section. He believes he owes his life to the fact that two cars of British mail for the West were between his car and the engine, and thus absorbed most of the shock. As it was, he was thrown to the other end of his car, amid a bundle of tumbled letters. He opened his door and found the body of Porter Simmons buried in the snow bank with the exception of his feet.

'C.S. Beall with his wife and their son, the latter aged 20, had left England to take up a ranch which he had recently bought just outside Vancouver. They had had a very rough and slow passage on the steamship *Empress of France*, and after reaching shore breathed freely, thinking their troubles were over. Young Beall was killed by the impact, and his father cut about the head and otherwise injured.' [386]

Charles Murphy, General Manager of the Canadian Pacific western lines, might well have lost his life in the accident but for a fortunate mistake. He had been asleep in his private car at Montreal Central station, which had been intended to attach to the rear of the first section of the train, but the car was forgotten in the station yard until after the first section had departed. [387]

The inquest on one of the victims of the collision, A.R. Tilley, returned a verdict that all the trainmen concerned had carried out their orders and instructions according to the railroad regulations, but recommended that the Canadian Pacific Company adopt the block system. Conductor J.S. Alexander stated at the inquest that the failure of engines to keep up steam was not uncommon when the coal was inferior. He was confident that Flagman Curley had gone back far enough and said that sometimes the explosion of torpedoes could not be heard in severe weather. A flagman would go out half a mile and leave one or two flags out.

Flagman Curley stated that he had gone back 20 poles (110 yards) which he had thought was far enough and could see back another 15 poles (82½ yards) thus well less than the half mile stated by the conductor.

Engineer Godin of the second section stated that he had neither heard the torpedoes nor seen the flagman and he was confident that had the torpedoes been exploded he

[386] *The Brandon Daily Sun*, Brandon, Manitoba, Canada, 27th January 1920
[387] *Medicine Hat News*, Medicine Hat, Alberta, Canada, 29th January 1920

would have heard them. His ears were not covered, and the cab window was broken. He only saw the train ahead when he was about a car length away and had shut off steam and applied the brakes in emergency. His fireman had not seen the other train at all.

Driver Davidson and Fireman John Page of the first section and Conductor Oscar England of the second section also gave evidence at the inquest. [388]

The inquest verdict exonerating all the railwaymen involved was not met with satisfaction by the office of the provincial attorney-general in Toronto, and Crown Attorney McKee, of Nipissing, was asked to report on the advisability of holding a second inquest on another of the victims. [389]

No further inquest appears to have been held but the Attorney General appointed a special commission to investigate the accident, and commissioner Julien Lowden reported at the beginning of March that the accident must be blamed on the railwaymen trusting to luck and failing to obey instructions. [390]

Reading the evidence relating to the accident, it must be concluded that Flagman Curley in particular was negligent in not properly protecting his train and that he had been fortunate not to have found himself facing a manslaughter charge.

At 23:40 on Wednesday 4th February 1920 fast goods train No. 7442 from Bercy was standing on the main track at Dijon-Perrigny station, France, on the Chemins de fer de Paris à Lyon et à la Méditerranée when it was run into from behind by the express which had left Lyon-Perrache at 20:00 and was due to arrive in Paris at 05:40 the following morning. The engine of the express smashed the rear waggon of the goods train before overturning together with its tender to the left-hand side of the track. The three leading carriages of the express, two composites and one first-class, were telescoped into one another to form a mass of wood and iron from which poured the cries of the injured. The driver, fireman, head conductor and two baggage handlers were thrown out or jumped onto the embankment and some of them were seriously injured.

After the accident local residents were soon on the scene, aroused by the tocsin sounded by the mayor, and assisted the railwaymen and uninjured passengers in releasing the injured from the wreckage. The injured received first aid at the mayor's house before being taken to hospital in Dijon. A rescue train also arrived the scene from Dijon three hours after the accident. The bodies of 18 dead were removed from the wreckage and taken to a room at the hospital. About 70 others were injured. While the rescue operation was in progress an individual was caught searching the clothes of the victims. He was arrested and was found to be in possession of a number of wallets. [391]

It took a considerable time to clear the line, even with the aid of a 50-ton breakdown crane sent from Paris. The Simplon express which was following the Lyons express arrived in Paris at 11:00, six hours late, all traffic having to be diverted via Amour and St Jean de Losne. [392]

---

[388] *The Brandon Daily Sun*, Brandon, Manitoba, Canada, 28th January 1920
[389] *The Lethbridge Daily Herald*, Lethbridge, Alberta, Canada, 28th January 1920
[390] *The Banner*, Russell, Manitoba, Canada, 4th March 1920
[391] *Le Petit Parisien*, Paris, France, 6th February 1920
[392] *The Times*, London, England, 7th February 1920

An investigation into the accident quickly placed the responsibility for it on Driver Labonnat of the express. The examining magistrate questioned Driver Fortier, of a train from Chalon-sur-Saône, which was shunted at the previous station at Gevrey, and he declared that the signals there had been standing at danger a quarter hour before the arrival of the express. The crossing keeper at Gevrey said the same thing, and the station staff at Perrigny stated that the signals there had been at danger. [393] Driver Labonnat and his conductor continued to maintain, however, that the signals had been clear. He was placed under arrest after the accident and faced a possible manslaughter charge.

The colliding train was reportedly travelling at over 80 kilometres an hour and Driver Labonnat ran past four signals at caution or danger and a red lamp being waved at him, presumably by the conductor of the goods train. The night had been clear, the line was dead straight and there were three red lights at the rear of the goods train, yet the driver only applied the brakes at the last moment. Again, Driver Labonnat asserted that he had braked earlier, but the officials pointed out that in that case the collision would not have pushed the goods train forward 75 metres or caused the degree of destruction. [394]

An horrific collision occurred at Mewa Nawada station, India, near Moradabad on the Oudh and Rohilkhand Railway, on 24th April 1920. Goods train No. 159 had been stopped for an hour at the outer signal for the station when it was run into from behind by the Allahabad Dehra Dun Express travelling at about 35 miles an hour which had been wrongly admitted into the occupied section, having been issued with an identically numbered train order. The passenger train was crowded but what made the accident much worse was that a fire, fed by broken gas cylinders, broke out in the wreckage. Four carriages caught fire, though several others were separated and pushed clear thanks to the initiative of the Deputy Inspector General of the Government Railway Police, who chanced to be a passenger on the train. An initial report that 50 passengers had died was followed by a second report that there had been 150 deaths, but it was officially stated that a most careful search of the scene indicated that the number of dead could not exceed 45, with 69 injured, many of them only slightly so. It was stated that the dead included 'almost the entire company of three Hindu marriage parties.' There were reports that the police themselves had looted the train; possibly they were merely collecting valuables for safekeeping.

An official inquiry into the disaster blamed the Mewa Nawada assistant stationmaster, who was to be prosecuted for his negligence. [395] [396] [397] [398]

On 9th June 1920 New York Central Railroad passenger train No. 28 from Buffalo to New York consisted of engine No. 3389, an express car, a mail car, a baggage car, two coaches and the Pullman sleeping cars *Wandin* and *Florimond*, in charge of Conductor Flynn and Engineer Hogan. It left Syracuse at 8.15 p.m. and about 20 miles out, at

---

[393] *Le Petit Parisien*, Paris, France, 7th February 1920
[394] *The Times*, London, England, 7th February 1920
[395] Ibid, 27th and 29th April 1920
[396] *The Railway Gazette,* Vol. 32, pgs. 680, 710 and 911
[397] *Cootamundra Herald*, Cootamundra, New South Wales, Australia
[398] *The Pioneer Mail* and *Indian Weekly News*, Madras, India, 27th August 1920

Oneida, Engineer Hogan noticed a problem with a leaking brake pipe. At Rome, 13 miles further on, the engineer found that the problem was a leak at the nipple in the pipe under the front end of the tender. At Utica, 14 miles beyond Rome, the faulty nipple was removed but as the inspector there did not have a replacement he cut some additional threads out of the original nipple and replaced it, which appeared to fix the difficulty, and all went well until the train neared Schenectady, New York, United States, 78 miles further on, where the brake pipe failed again, and the train was brought to a stand.

Following the Buffalo train was Michigan Central Railroad train No. 34, consisting of nine baggage and express cars and a coach, in charge of Conductor Merritt, with Engineer Martin Doyle and Fireman Kopp. This train passed Signal Station 11 at 12.23, an hour and 59 minutes late, and ran into the back of the New York Central train while it was still at a stand having the brake pipe repaired at 12.29 a.m., while still running at a speed of 55 or 60 miles an hour. The front train was driven ahead three car lengths and in some way the Pullman car *Wandin* was thrown over on its side, badly damaged, on the south side of the track, while the *Florimond* car was driven under the third coach from the rear and its superstructure almost demolished. The leading two cars of the colliding train were also demolished. Fourteen passengers and Engineer Doyle lost their lives and 37 passengers, five persons carried under contract and one railroad employee were injured. The passengers who lost their lives had all been in the *Florimond* car.

The line on which the accident happened was quadruple track and equipped with automatic signalling, the three-position upper quadrant signals being spaced about a mile apart. The accident occurred about 2,100 feet beyond automatic signal 16452 and 5.003 feet east of automatic signal 16552. The tracks were numbered 4, 3, 1 and 2 from north to south, the collision occurring on track 2. The weather at the time of the accident was clear.

The Interstate Commerce Commission inquiry heard from Flagman Robinson of train No. 28 that he had thrown off a lighted fusee when he noticed that his train was slowing down. When his train had stopped he saw that the next signal for it was clear and realising that there must be a problem and that train No. 34 was closely following on a faster schedule he had set off back at once, equipped with red and white lanterns, four fusees and eight torpedoes. Looking back to his train he saw that its markers were burning brightly and had gone back 25 car lengths when he had heard train No. 34 approaching and saw the reflection of its headlight. He had put down one or possibly two torpedoes and then lit a fusee and run towards the oncoming train, jumping aside as the engine reached him and throwing the lighted fusee at it. He thought that the engineer shut off steam at that time but saw no brake application. The flagman's statement was verified by Conductor Flynn, Pullman Conductor English and Baggagemaster Hildreth of train No. 28, also deadheading employee L. H. Eckhart on that train. Not all of them heard the torpedoes, but the baggageman apparently did and warned the fireman to get out from under the tender, where he was trying to fix the brake pipe. Two independent witnesses, farmer E.C. King and his daughter, also witnessed the fireman going back and trying to stop train No. 34.

Fireman Kopp of train No. 34 said that they had realised that they were close behind train No. 28 at Amsterdam, 12 miles short of Schenectady, when the automatic signal had only cleared as they approached them. All the succeeding signals had been clear up to what he thought was signal 16552, which Engineer Doyle had called as clear. He was then firing and did not see signal 16452, and half a minute or so later the engineer had called to jump and had applied the brakes in emergency, whereupon he had seen the markers of the train ahead. He had not jumped but had braced himself for the impact. He did not see the flagman, nor a fusee, nor hear a torpedo. He thought that Engineer Doyle was in a normal state both physically and mentally. He admitted that the signal which he had confirmed to his engineer as being clear could have been before signal 16552.

The statements of Conductor Merritt, Brakeman Smith and Flagman Peters of train No. 34 did not really add much to that of the fireman.

The crew of the engine which followed train No. 34 to draw the rear of that train back from the accident site found the automatic signals clear until they reached signal 16552, which was at caution, while signal 16452 was at danger. The signals were therefore working correctly after the collision, as was shown by subsequent tests.

The Interstate Commerce Commission concluded that the accident had been caused by the failure of Engineer Doyle to observe and comply with the automatic signals and the warning signals of the flagman. It was impossible to account for his failure.

It was clear that the trains were only running about four minutes apart and that Flagman Robinson had not had sufficient time to go back far enough to protect his train. While the evidence as to whether he had used torpedoes was conflicting it was clear that he had gone back as far as could reasonably be expected and the warning signal of his fusees had not been observed.

The accident again drew attention, the commission continued, to the need for a system of automatic train control to supplement the automatic block signals. [399]

Engineer Doyle had died on his 50th birthday. [400]

An American interurban railway suffered another bad accident on 3rd July 1920, this time near South Pittston, Pennsylvania, on the Lackawanna and Wyoming Valley Railroad. The double-track line ran for just over 18 miles from Scranton and Wilkes-Barre and was worked by timetable and train orders without block signals, though there were semaphore spacing signals at Avoca, 7 miles south of Scranton, and at Plains, 3 miles north of Wilkes-Barre. Train orders and messages could be received by the train crews at the spacing signals, and by the operator at Pittston, just under 11 miles south of Scranton.

On the day of the accident northbound passenger train No. 6, consisting of two coaches with Conductors Leo Cawley and McGrail and Motorman Moran, left Wilkes-Barre at 6.10 p.m. and Inkerman at 6.28 p.m. 5 minutes late, and at 6.37 was run into from behind by train No. 21. The latter train consisted of a single coach in charge of Conductor Mulroney and Motorman White. It had departed from Wilkes-Barre at 6.25 and collided with train No. 6 at a speed of about 35 miles an hour. The rear of the second car of train No. 6 and the front vestibule of the car of train No. 21 were badly damaged.

Following this initial accident Conductor Mulroney went back and flagged passenger train No. 3, which had left Wilkes-Barre at 6.32 at Yatesville Crossover. This train had two coaches and with Conductors F. Ratchford and Boone and Motorman Dailey. Train

[399] Interstate Commerce Commission report
[400] *The Boston Daily Globe*, Boston Massachusetts, United States, 9th and 10th June 1920

No. 3 then went forward to the rear of train No. 21, picked up the fatally injured Conductor Cawley, returned to Yatesville Crossover, picked up the flagman and having obtained verbal authority from the dispatcher proceeded northwards over the southbound track. Meanwhile the crew of train No. 6 had uncoupled the first car of their train and proceeded to Pittston with its passengers, intending to return for the passengers in the rear car. Shortly after the front car of train No. 6 had started forward, the rear of train No. 21 was run into by a fourth train, No. 4, two coaches coupled with Conductors J. Ratchford and Gabriel and Motorman Schlager, which had left Wilkes-Barre at 6.50, the second collision occurring at 7.5 p.m. Train No. 4 telescoped train No. 21 for the greater part of its length and the front car of train No. 4 was badly damaged, but it was thought that all of the deaths and most of the injuries occurred in train No. 21. In addition to Conductor Cawley, 16 passengers died and 20 passengers and three railroad employees were injured. [401]

Initial newspaper reports of the accident stated that the first train had stopped after colliding with a telegraph pole which had fallen across the tracks after having been struck by lightning. There is no mention of any such incident in the official report on the accident. [402]

Motorman Moran explained to the Interstate Commerce Commission that Conductor Cawley had told him to run slowly after leaving Inkerman as far as Yatesville Crossover so that there was time to collect fares and the distance had been covered at only 4 miles an hour.

Conductor McGrail said that he knew that they were only 9 minutes ahead of train No. 21 but claimed that he had thought that Conductor Cawley would protect the train and that the motorman would whistle a flagman out, though he admitted that as the rear conductor he should have done the flagging. When he heard train No. 4 approaching he had got back about a car length.

Motorman White said that his vision had been limited by a curve and an embankment so that he could not avoid the first collision. After the accident he had told Conductor Mulroney to go back and flag.

Conductor Mulroney, who was slightly hurt in the first collision and seriously injured in the second, remembered going back to flag and returning on train No. 3, but not much else.

Conductor F. Ratchford of train No. 3 said that when train No. 3 returned to Yatesville Crossover and received authority from the dispatcher to use the northbound track he had picked up the flagman (Conductor Boone). He knew that the rear of train No. 21 was then no longer protected but assumed that it did not matter as the dispatcher had arranged for single-line working over the southbound line between Yatesville Crossover and Pittston.

Conductor Boone said that when going back to protect he had taken only flags, but neither fusees nor torpedoes, as he did not believe that there was a need to provide any further protection once his train had crossed over to the other track.

Motorman Dailey of train No. 3 said that he had not given any thought to the need for protection.

---

[401] Interstate Commerce Commission report

[402] *The Philadelphia Inquirer*, Philadelphia, Pennsylvania, United States, 4th July 1920

Motorman Moran of train No. 6 said he had not realised that Conductor Cawley had been injured and thought that he had gone back to protect.

Motorman Schlager testified that he had had no warning of a train ahead and had only seen the rear of train No. 21 from a distance of 200 feet when he had been travelling at 40 or 45 miles an hour, when he had applied the brakes in emergency and jumped.

Dispatcher O'Malley said that he was advised of the first collision by Conductor Maloney at 6.49 p.m. Failing to contact train No. 4 at Wilkes-Barre he had sent a message to the operator at Plains to notify the crew of the next southbound train to contact the dispatcher when they arrived at Yatesville Crossover. He had not kept a written record of this message nor of that to train No. 3 authorising it to use the southbound track between Yatesville Crossover and Pittston, though it was customary to do so. He had said nothing about single line working to either Conductor Mulroney or F. Ratchford.

Operator Brennan at Plains said that the dispatcher had contacted him at 6.59, a minute after train No. 4 had gone by, with instructions to tell the next southbound train to call the dispatcher on arrival at Yatesville Crossover. This seemed strange, as Plains was south of Yatesville Crossover, so a southbound train would have already passed Yatesville before reaching Plains. He thus asked the dispatcher to repeat the location, but not the direction. The dispatcher repeated Yatesville, and so Operator Brennan gave the message as received to Motorman Boland of southbound train No. 1 at 7.02 p.m. Not surprisingly the motorman, having already passed Yatesville, queried the message as well, and the misunderstanding came to light when Operator Brennan informed the dispatcher that he had delivered the message to southbound train No. 1.

The Interstate Commerce Commission investigators concluded that both accidents had resulted from a failure to provide proper flag protection. One of the special rules in the timetable was that the conductor of the first car must carry the train orders and was responsible for the safety of the train. The late Conductor Cawley was thus primarily responsible for the first collision, though as an experienced man Conductor McGrail should not merely have assumed that protection was being proved, nor should Motorman Moran.

Several different employees were jointly responsible for the second collision. Conductor Mulroney had taken no steps regarding protection once his train had been crossed over to the other track. Motorman White knew that Conductor Mulroney had returned to his train and that train No. 3 had proceeded on the southbound track but did nothing about protecting his train. These two were mainly responsible for the second collision. The fact that they had notified the dispatcher of the first collision did not relieve them of the need to provide protection. Conductor Boone did not put down any torpedoes when his train was crossed over to the southbound track and there had been no consultation between the crew of train No. 3 and those of train No. 21 as to the need to protect the latter train, as might be reasonably expected. Likewise, Dispatcher O'Malley had taken no action to protect train No. 21 after being informed of the first collision and there had been a misunderstanding between him and Operator Brennan.

Had an adequate block system been in use the accident would probably not have occurred, the commission concluded. [403]

[403] Interstate Commerce Commission report

Direct train No. 184 for Milan and Turin, with Driver Giovanni Poli and Fireman Giovanni Boccate, left Venice at 23:30 on Friday 8th October 1920, about 40 minutes late due to a technical problem, and had not gone far before it was stopped by signals at the entrance to Mestre station. After a few minutes, the signal cleared, and the train restarted, but had to stop again because the brakes were binding on some of the coaches, and it came to a stand once more on the long bridge over the lagoon between Venice and Mestre, while its crew attempted to rectify the problem. Ten minutes after the departure of train No. 184, direct train No. 619 from Trieste to Rome was allowed to follow it. This train apparently received clear signals at the first two block posts but between the second and third posts its driver caught sight of train No. 184 still standing on the viaduct. Although it was travelling only at moderate speed its engine destroyed the rear coach and forced it halfway through the penultimate one. The locomotive, however, was little damaged. Its driver jumped off before the impact and left the scene to avoid arrest. Twenty-five passengers died and over 30 were injured. The accident was clearly due to a mistake in the signalling and Signalman Damiano of the second block post and Signalman Tantuo of the third block post were arrested. There was also a failure to protect the first train, probably because its crew were preoccupied with trying to rectify the problem and thought that they would soon be on their way again. [404]

The Detroit Jackson and Chicago Railway, part of the Detroit United interurban system, suffered a rear collision on Saturday 18th June 1921. The accident occurred at Warsaw Siding, Michigan, United States. The section of line on which the collision took place was single track, worked by timetable and train order without block signals, and Warsaw was a passing place. On the day of the accident westbound train first No. 169, in charge of Conductor Fletcher with Motorman Root, arrived at Warsaw and entered the siding to meet eastbound train No. 170. The east switch was left open for train second No. 169, which was also to meet train No. 170 at Warsaw.

Train second No. 169, in charge of Conductor Lyon and Motorman Winchell, duly arrived and entered the siding but failed to pull up before it had run into the rear of the first section, while it was still moving at a speed variously estimated at 15 and 40 miles an hour. Car No. 7293, which formed the first section, was of wooden construction with a steel underframe whereas car No. 26, forming the second section, was of all-steel construction. In the collision this fact resulted in the forward car being telescoped for about a third of its length, as well as being pushed through the end of the loop and out onto the single track. Three passengers were killed outright and 22 were injured, two fatally. Two railwaymen were also hurt. The collision happened at about 5.30 p.m.

Motorman Root told the Interstate Commerce Commission (ICC) investigators that he had stopped his train at a telephone booth midway along the siding and had gone to the telephone to speak to the dispatcher, who had then asked to speak to the conductor to give him an order.

Conductor Fletcher said that having taken the order he had started back towards the east switch. It was customary, he said, to leave it open for the following train in such circumstances, though the rules stated that he should then have remained at the switch.

[404] Wikipedia – Incidente ferroviario del ponte sella Laguna di Venezia.

He saw the second section approaching but there was no indication that it was not under control and its motorman did not whistle. Had he had more warning he could have closed the switch, but it was only when the train was close at hand that he had realised, too late, that it was not going to stop in time. He did not know when the brakes were applied but when the car passed him its wheels were sliding and the power seemed to have been reversed.

Motorman Winchell of the second section said that the brakes of his car had been slack throughout the journey and he had had to brake earlier than usual. Coming down the hill into Warsaw at 35 miles an hour he had shut off power and applied the brakes, which had not held as they should. Entering a curve he had reapplied power, as usual, then shut it off and applied the brake, using straight air until he reached a road crossing 1,200 feet short of the point of collision. Then, as the brakes did not seem to be reducing the speed, he had applied them in emergency. Then he had released them, made several service applications and again another emergency application, still without a reduction in speed. At a point 600 or 700 feet east of the beginning of the passing siding he had reversed power. He had not whistled, applied sand, or used the hand brake. He knew that he was to cross train No. 170 at Warsaw and saw the switch open for the siding. His opinion was that the brake rod had broken before the car had started down the hill.

Conductor Lyon stated that he had gone towards the front of the car as they had approached Warsaw, had seen the switch open for the siding and knew that they were not going to stop. The motorman had not given any indication that he could not control the car and he (the conductor) had not noticed any brake application nor felt any reduction in speed.

Joint Inspector Ballard said that the car had been inspected on the morning of the day of the accident and found to be in good condition. A brake test was not part of the inspection.

Foreman Inspector Ritter stated that a running inspection of car No. 26 had been made at Ypsilanti 50 minutes before the accident and it had then been reported to be in good condition. An hour after the collision he had examined the car and had found the front truck brake rod broken but he was of the opinion that this was a result of the collision.

Motorman Schaible said that he had been operating the car until relieved by Motorman Winchell at Dunn at 4.17 p.m. Having been operated all day without adjustment the car brakes were a little slack and he had told his colleague this and that the braking was a little slow. The problem had not been bad enough to need to stop the car to adjust the brakes. After being relieved he had ridden in the car over the 10 miles from Dunn to Ann Arbor and had not noticed Motorman Winchell having any difficulty in controlling it.

The ICC investigation concluded that the direct cause of the accident was the failure of Motorman Winchell to control his car after having been informed that its brakes were slack, and that he had used poor judgement in not taking advantage of the retarding effect of the curve. Had he braked earlier he could have reversed the car earlier if the brakes had not worked well enough, while had he whistled Conductor Fletcher would have had sufficient time to close the switch and thus been able to avert the accident.

The commission noted that since beginning as a motorman in December 1918 Winchell had been suspended or reprimanded five times for operating his car too fast through curves and over switches, for overrunning his order and passing stop signals and for being involved in a rear collision. [405] Not quite an unblemished record over only two and a half years.

On the evening of Monday 28th November 1921 a local train which was stopped in Duffel station, Belgium, to allow a Brussels to Antwerp express to pass, was run into by the latter train. The engine of the express overturned the rear carriage of the stopping train, killing probably five passengers and injuring a number of others. [406]

The Missouri Pacific Railroad suffered a disastrous collision at Sulphur Springs, Missouri, United States on Saturday 5th August 1922.

Northbound local train No. 32 consisted of engine No. 4310, two baggage cars, a mail car, a combined mail and baggage car, two chair cars, three coaches and a chair car, all of wooden construction except the first and third cars, in charge of Conductor J.A. Long with Engineer Gross and Fireman Long. The train had arrived at Riverside, 3½ miles south of Sulphur Springs, at 6.48 p.m., had been run into a siding there to cross southbound train No. 41, the line being single track, and had left Riverside at 7.07 p.m., 2 hours and 11 minutes late, had stopped at Bushburg and Glen Park, and then at Sulphur Springs to take water.

Northbound express No. 4 consisted of engine No. 5312, three baggage cars, a mail car, two baggage cars, a chair car, two coaches, a Pullman sleeping car and a dining car, in charge of Conductor Gregg with Engineer Matt Glenn and Fireman Tinsley. The line was equipped with two position upper quadrant automatic signals, with alternating stop and distant signals, and approaching Riverside Engineer Glenn found distant signal 262-D at caution, which indicated that either the Riverside train order was at danger, that stop signal 262 was at danger, or that both were at danger. When the train reached Riverside station platform the train order signal was at danger and the operator was on the platform with a Form 19 train order (one which could be picked up by the train in passing as opposed to a Form 31 order for which the train must be stopped and the engineer and conductor sign the order). The train order, No. 51, read:

No. 1 engine 5306 wait at Whitehouse until 7.37 a.m., Wickes 7.40 p.m. for No. 4, engine 5312.

Train No. 4 passed Riverside at 7.13 p.m., passed signal 262 at danger, passed distant signal 232-D at caution and ran into the rear of train No. 32 opposite stop signal 232. The last two cars of train No. 32 were demolished and thrown to the bed of a creek and the next two were badly damaged. One pair of wheels of the colliding engine was derailed and the second car of that train was demolished. Thirty-two passengers, one trespasser and Engineer Glenn were killed and 171 passengers, 10 railwaymen, a mail clerk, two newsagents, an express messenger and a Pullman porter were injured. [407]

A passenger in the express, Miss Mabel Steingaubey, 20, who was travelling with her sister Florence, 22, related her experiences to a reporter. "We were coming home from

---

[405] Interstate Commerce Commission report

[406] *The Times*, London, England, 29th and 30th November 1921

[407] Interstate Commerce Commission report

our vacation, and had left Ironton about 5.25 p.m. We were in the second coach behind the baggage car, and I was resting my head on the seat. Florence was sitting across the aisle. Suddenly, without warning, there came a great crash – I did not know what had happened. Something hit me in the mouth – I think it was the seat. For a time everything was confusion. I remember going to the water cooler and washing the blood from my face, then everything seemed black – I don't remember how I got out of the car. I looked for my sister and saw her helping the injured. I can't describe the scene, it was too horrible. How do I feel? I feel 'all in.'" [408]

Conductor Long told the Interstate Commerce Commission (ICC) inquiry that there was no regular point at which train No. 4 overtook his train and that at Riverside the operator, after talking to the dispatcher, had told him to try and make Wickes (4¼ miles north of Sulphur Springs) for train No. 4. His train had been stopped at Sulphur Springs less than a minute when train No. 4 ran into it. He thought that the flagman had gone back at Riverside though he did not think it necessary to comply with Rule No. 99 (regarding the protection of trains) given that train No. 4 had been reported as 20 minutes late at De Soto. He agreed that if the flagman had gone back at Riverside the engineer of train No. 4 would have had warning that there was a train close in front.

Engineer Gross said that he had just been getting down to oil his engine at Sulphur Springs when his train was run into. He had not whistled out the flagman at either Riverside of Sulphur Springs.

Fireman Long said that it was not customary to whistle out the flagman at Riverside.

Flagman Boston of train No. 32 said that he had not done any flagging between Poplar Bluff and Riverside, nor at the latter place, although the train was there longer than usual. It was not customary to do so, he had not been instructed to do so, and the engineer had not whistled him back. He had not known where train No. 4 was. He had lit the markers at the rear of his train but did not know if they had been burning at the time of the collision. He added that it was not the practice to flag trains at stations, though if there were officials on board he made a pretence of doing so. At Sulphur Springs it would have taken him three or four minutes to collect his torpedoes and lantern before going back.

Fireman Tinsley of train No. 4 testified as to collecting the train order at Riverside and handing it to Engineer Glenn, who had read it twice before handing it back to him. He had suffered a head injury in the collision and was unconscious for several hours; he could remember nothing of the immediate period before the collision. He did not think it the practice for enginemen to whistle out the flagman at stations unless there train was likely to be overtaken.

Signal Supervisor Ragland stated that he had tested the signals after the accident and they had been working properly.

The ICC concluded that the accident was due to the failure of Engineer Glenn to observe and obey the automatic signal indications and the failure of the crew of train No. 32 to protect it in the rear as per the rules, for which Flagman Boston, Conductor Long and Engineer Gross were responsible. In view of the fact that Engineer Glenn was killed and Fireman Tinsley had post-traumatic amnesia it was impossible to account for the engineer's mistake. It was possible that he had supposed that the caution aspect of signal

---

[408] *The News-Sentinel*, Fort Wayne, Indiana, United States, 7th August 1922

262-D had related solely to the train order signal being at danger and that in reading train order No. 51 he had missed seeing signal 262 at danger, but he had also failed to reduce speed at signal 232-D. His was a fast train, which usually had a clear road, and he had an order to meet train No. 1 at Wickes, which might have led him to suppose that he had a clear road to there.

Flagman Boston's most important duty was to protect his train, but according to his own statement he had made no attempt to do so at any station. He would not have had time to do so at Sulphur Springs but had he done so at Riverside, Engineer Glenn would have been made aware that there was another train close in front. For not seeing that he did so Conductor Long was equally responsible, while Engineer Gross should have whistled him out. For this lax observation of the rules Missouri Pacific supervisory officials must also be held responsible. [409]

At 4.14 a.m. on Tuesday 31st July 1923 Hamburg – München Vorzug (relief train) D 88 was run into from behind by the main train, D-88, at Kreiensen station, Germany. The two rear coaches and half of third from last coach of the front train were almost completely telescoped into one another. Four hours after the accident there were already known to be 27 dead and 25 injured and by midnight the numbers had risen to 44 dead and 34 injured. The cause of the disaster quickly became evident; the driver of the 11 p.m. train from Hamburg had overrun the Einfahrsignal (home signal). He claimed that at the crucial moment he had got a foreign body in his eye and had thus failed to observe the signal. The first train was not booked to stop at Kreiensen but a problem had developed with its engine and it had stopped there to change locomotives. In the colliding train the luggage van and the mail van were telescoped into one another but eight postal officials in the mail van escaped injury. A woman passenger who had been riding in the rear carriage of the first train survived because she had gone into the station restaurant. Her husband was killed. The final number of dead was 47 or 48, with 45 injured. The guilty driver was tried and found guilty of negligence and sentenced to 18 months' imprisonment. [410]

Due to a signalling error on Thursday 6th September 1923, Berlin – Köln express D 10 ran into the rear of stationary international express D 138 from Leipzig to Amsterdam at Lohnde, Germany. The second train had been admitted to the occupied block section before the first train had cleared it. Eighteen people died and 19 were injured. [411]

A relatively minor level crossing collision, in which no one was hurt, led in turn to a serious collision on the New York Central Railroad near Forsyth, New York, United States, early on the morning of Sunday 9th December 1923. Messrs C.H. Harney, G.B. Maier and W.S. Stratford were on their way by car from Pittsburgh, Pennsylvania, to Buffalo, New York, with Mr Stratford driving. The latter stated at the Interstate Commerce Commission (ICC) investigation into the subsequent accident that it had been

---

409 Interstate Commerce Commission report

410 *Historische Eisenbahn-Katastrophen* by Bernhard Püschel, pgs. 125-26 (in German)

411 *Eisenbahn-Katastrophen in Deutschland* by Hans Joachim Ritzau pg. 92 (in German)

foggy, so he had been driving slowly. Turning the car to the right to negotiate the level crossing, he had had to change down into second gear and at that moment Mr Maier had shouted that there was a train coming, so he had stepped on the accelerator. The sudden jerk forward, combined with the fact that he not turned far enough to the right to begin with, had caused the forward pair of car wheels to go off the left side of the crossing. He had tried to reverse, but the car's rear wheels could not get enough traction on the wet crossing planks. He had then tried to go forward, again without success, and seeing that the train was now very close he had quickly abandoned the vehicle. Mr Stratford said he had not been aware that the road turned at that point and had neither heard the crossing bell nor seen the swinging red light of the wigwag signal. Mr Maier said that he had neither heard the crossing bell nor seen any of the signals warning of the crossing, though he had heard the crossing bell afterwards. Mr Harney said that he had been half asleep and had known nothing until he had felt the car bumping along the track.

Approaching the crossing at the time were three sections of train No. 25, the westbound *Twentieth Century Limited,* which for some reason were running as the second, third and fourth sections of train No. 5. There were four lines at Forsyth, numbered from north to south 3, 1, 2 and 4. The line was operated by timetable, train orders and an automatic signalling system, with two arm, two position, lower quadrant semaphores, indicating red over yellow (stop, then proceed), green over yellow (approach next signal prepared to stop) and green over green (proceed). Signals 61.1 and 62.1 were about 6,300 and 1,050 feet east of the point at which the accident occurred. The trains involved in the accident were running on track 1.

Train second No. 5 was in charge of Conductor Burke with Engineer Cross and Fireman Houser and consisted of engine No. 3288, a club car, three Pullman sleeping cars and an observation car. Engineer Cross said that his train had been travelling at about 60 miles an hour at the crossing. He had whistled for it and then seen the two white lights of an automobile about 100 or 150 feet ahead and had applied the automatic brakes in emergency. The train had then struck the car at about 1.10 a.m. and was brought to a stand. Flagman Davis had immediately gone back to protect it and had stopped train third No. 5 with a burning fusee at a time he put as 1.17 a.m. He stated that he had heard the third section stop at signal 62.1, whistle it off an proceed towards him; after he had flagged it the train had moved forward to the crossing where it stopped again.

Train third No. 5 consisted of engine No. 3287, a club car, seven Pullman sleeping cars and an observation car and was in charge of Conductor O'Donnell with Engineer Bradley and Fireman Batt. The latter said that he had been travelling at 70 miles an hour when he had seen signal 61.1 at caution. He had applied the air brakes, bringing the speed of his train down to 10 miles an hour, applied the brakes again when signal 61.1 did not clear and stopped at it. Proceeding again, he had been stopped by Flagman Davis, had picked him up and gone forward to the crossing, stopping just short of it. He thought that his train had been stopped two minutes when it was run into by fourth No. 5. [412] The collision was a violent one, and the rear car of train third No. 5 was demolished, being telescoped for almost its entire length by the car ahead. Eight passengers and a railwayman lost their lives, and another 35 passengers were injured. The engine of the colliding train was considerably damaged.

[412] Interstate Commerce Commission report

One of the passengers who died was William E. (Wild Bill) Donovan, a veteran baseball player and manager of the New Haven baseball club, who had been on his way to Chicago to attend a baseball conference. [413]

Conductor O'Donnell said that his flagman had thrown off a lighted fusee at signal 61.1. He saw another lighted fusee behind his train at signal 62.1. When his train had stopped at the crossing he had got down from the club car in which he had been riding and could see the reflection of the fusee his flagman was carrying as well as the markers at the rear of his train. Not more than two minutes later his train was run into.

Flagman Ackerman of the third section confirmed that he had thrown off a lighted ten-minute fusee when his train had slowed at signal 61.1 At signal 62.1 he had started back but had only got 1½ car lengths when the train had whistled off at 1.17 a.m. and he had reboarded it. The train had then stopped at the crossing at 1.24 a.m., and he had started back once more, covering about two car lengths when he had heard the fourth section approaching. Lighting a five-minute fusee he had run towards the approaching train but had received no acknowledgement of his warning and the engine had passed him at 48 or 50 miles an hour, still working steam, though it appeared that the brakes were applied immediately afterwards.

Train fourth No. 5 consisted of engine No. 3316, a mail car, a club car, seven Pullman sleeping cars and an observation car, in charge of Conductor Timmons with Engineer Patterson and Fireman Pegler.

Engineer Patterson told the inquiry that he had had clear signals all the way after leaving Buffalo until he reached signal 61.1, which because of the foggy conditions he had only observed when his engine had been almost under it. He had been travelling at 45 to 50 miles an hour and had shut off steam and applied the brakes, then, when he had thought that he had reduced speed sufficiently, he had released the brakes and applied a little steam to keep the smoke from the engine. Sooner than he had expected he had seen signal 62.1 in the stop position about two car lengths ahead. He had not previously seen a fusee. At that point he had made a 10 lb brake application, then when he had seen the flagman he had placed the brake valve in the emergency position. He thought that his train had been travelling at 25 miles an hour when he passed that flagman, it was only after doing so that he had seen the rear of the third section. Asked why he had not made an emergency brake application as soon as he had passed signal 62.1 at danger he replied that he had not wanted to tear the train apart on account of a broken knuckle or a similar cause. The engineer's testimony was different to what he had said to Superintendent Brogan on the night of the accident, when he had stated that he had not applied the brakes at all until he had seen signal 62.1 at danger and the flagman's fusee and had been working steam up to that point.

The ICC noted that Engineer Patterson, who was 57 years old, had not had a physical examination for several years. He was given one after the accident, when it was found that he had a slight leakage to the heart and a systolic blood pressure of 200 (100 to 140 is normal). The examining physician was apparently not much concerned about the high blood pressure and said that he had found no condition which might have contributed to the accident.

Conductor Timmons, who had been riding in the second car, did not recall any brake

[413] *New Britain Herald*, New Britain, Connecticut, United States, 10th December 1923

application before the emergency one. After the collision, afraid that his flagman might have been injured by the sudden stop and unable to protect the train, he had gone to the rear and not seeing the flagman there he had gone back half a mile before overtaking him. He saw Flagman Ackerman, who was then his way back to his own train, a car or two behind train fourth No. 5.

Brakeman Case, Flagman Hopper and Baggagemaster Yeska of train fourth No. 5 said that they had not noticed any earlier brake application either. The flagman stated that the rear markers of his own train were visible from about four or five car lengths, and the stop indication of signal 62.1 about the same distance.

Signal Engineer F.B. Wiegand, in charge of lines west of Buffalo, said that the signal system was modern in every way and had the signal indications been obeyed the collision would not have occurred. He agreed that to prevent enginemen from running past signals an additional measure was needed. An ICC order of 13th June 1922 had required the installation of an automatic train control system between Albany, New York, and Cleveland, Ohio, and bids for the installation of a continuous system were to be returned by 28th December. He considered the continuous system more suitable to lines with automatic signalling, as with it the engineers were constantly informed of the state of the line ahead, though it probably would not have been any more effective than the intermittent system for preventing the Forsyth accident.

The ICC noted that there had been several accidents at the Forsyth crossing, and there were plans for the erection of an overbridge to replace both it and two other level crossings with both the New York Central and the parallel New York Chicago and St Louis Railroad tracks, but disputes over details had delayed matters.

In conclusion the ICC stated that the initial level crossing collision was significant only in that it had resulted in stopping the first of the three sections of train No. 5, so that the following two sections had to stop in turn. All should have been protected by the automatic signalling system and the accident had been caused by the failure of Engineer Patterson to observe and obey the signals. The engineer had stated at the inquiry that with fast trains on receiving a caution signal it was the practice to reduce speed just sufficiently to allow the train in front to get ahead, but on the night of the accident because of the fog he had run through the section quicker than he had expected. At any rate he had made no attempt to stop until he had seen signal 62.1 at stop at close range and had overrun it by 1,000 feet. He had had many years of service, was used to running fast trains, had had adequate rest, and was not suffering from any medical condition which might have affected his driving. There was no explanation or excuse for his failure.

The commission stressed that the collision again forcibly demonstrated the need for automatic train control. It was 18 months since the commission had issued an order requiring the installation of automatic train control on portions of 49 railroads, among them the New York Central, before 1st January 1925. Regardless of this the company had not yet made much progress towards such an installation. The train timings between Buffalo and Westfield required trains to run at high speeds over considerable distances, and it had emerged at the inquiry that engineers continued to run at full speed under caution indications, a practice which was not safe without automatic train control installations to ensure that trains obeyed stop signals. The railroad should revise its rules to require trains running at high speed to reduce speed to a prescribed minimum rate after encountering a caution signal and to approach the next signal prepared to stop. [414]

The 1924 English FA Cup Final was played at Wembley, in north London, on Saturday 26th April, between Birmingham football team Aston Villa, and Newcastle United. The event was marred by a serious accident early that day involving a London Midland and Scottish Railway (LMS) train carrying Aston Villa supporters from Coventry. The trains in the accident were No. 71 up special, 5.30 a.m. from Coventry, bringing the football fans from there, and the 7.15 a.m. up local electric train from Watford. The football train consisted of *Precursor* class 4-4-0 No. 625 (pilot engine) and 4-4-0 *George V* class No. 1725 (train engine), a composite brake van, 12 third-class corridor coaches and a brake van, in charge of Guard Thomas Reading with Driver Gower and Passed Cleaner John Hone on the pilot engine and Driver Sear and Passed Cleaner Pearce on the train engine. The Watford train consisted of a third-class motor car, a first and third-class composite trailer car, two third-class driving trailer cars, a first and third-class composite trailer car and a third-class motor car, in charge of Guard Reynolds with Motorman Edwin Adams.

The excursion (football) train was stationary at the up slow home signal of Euston No. 4 signal box with its rear within a covered way known as Park Street overbridge when it was run into by the electric train. The rear van and penultimate coach of the football train were destroyed, and the front of the electric train was damaged. Five passengers in the football train were killed or fatally injured, 68 passengers were injured, Guard Reading was badly hurt, as was Motorman Adams, who was trapped in his cab for about five hours; Guard Reynolds was less badly hurt. The passengers who lost their lives were Thomas Bradshaw, William Dix, James Gray, Samuel Ives, and Samuel Jones, all from Coventry. [415]

The Ministry of Transport inquiry determined that because of the extra traffic in connection with the cup final the regular man in Euston No. 4 box, Signalman William Hill, had been given the assistance, from 6 a.m. on the day of the accident, of Signalman Arthur Chambers, a relief signalman who normally worked the box on Sundays. From about 7.40 a.m. the two signalmen divided the work between themselves, Signalman Hill taking charge of the carriage and engine lines and the telephones while Signalman Chambers dealt with the main lines. No booking was done in the box. At about 7.50 a.m. there were trains stopped at the up home signals for the box on both the up fast and up slow lines, awaiting a platform at Euston station, with an up Scotch express on the up fast line and the football train on the up slow line, the leading engine of the latter train being stopped only 23 yards from the box . Signalman Chambers twice offered both trains to Euston No. 2 box and twice they were refused. After several minutes Euston No. 2 accepted the Scotch express on the up fast line, Signalman Chambers cleared his home signal for it, watched it pull away, checked that it had a tail lamp, went to the block shelf, and gave the 'Train out of Section' signal to Camden No. 1 box not on the up fast line but on the up slow line. Signalman Harris in that box immediately offered the Watford train on the up slow line and it was accepted. About this time Passed Cleaner Hone had come into the box to carry out Rule 55. As he was signing his name in the train register book he

[414] Interstate Commerce Commission report
[415] *The Times*, London, England, 28th April and 1st May 1924

heard the sound of the collision.

Motorman Adams was in hospital eight weeks before he was able to give evidence. He said that had been stopped briefly at the Camden up slow home signal before it was cleared. He had then set off and had approached the Park Street covered way at about 20 miles an hour. Visibility within the covered way was severely limited due to smoke and steam and his train had been within about 15 yards of the rear coach of the football train before he had been able to make out its tail lamp, when he had made a full emergency brake application and released the dead man's handle.

Colonel J.W. Pringle, the Ministry of Transport inspecting officer, stated that it was clear that the accident had been caused by Signalman Chamber's mistake, as he had himself admitted, and he had not attempted to blame anyone else. Down trains had passed through the covered way at 7.41, 7.43 and 7.51 a.m. and it was clear that the view afforded to Motorman Adams must have been very restricted, so that he was in no way to blame for failing to see the tail lamp of the first train sooner. Driver Gower and Passed Cleaner Hone had also acted correctly in carrying out Rule 55.

There had been talk of a fire breaking out in the wreckage but there was in fact no evidence of such other than a small amount of smouldering caused probably by drippings from the lighted flare lamps used by the rescuers. It was satisfactory that no fire had arisen from the gas cylinders used for train lighting but the dangers of gas as an illuminant had been shown by the recent disaster at Bellinzona. All new LMS rolling stock had electric lighting and existing gas-lit vehicles were being converted to electric lighting if their life expectancy justified it. Since 1912 the proportion of electrically lit rolling stock on the London and North Western (LNW) section of the LMS had increased from 33 to 49 per cent. In addition the frames and bogies of all new rolling stock would be made of steel with body framing of jarrah (non-inflammable) timber.

As for the protection of trains against signalmen's mistakes, 457 track circuits were in use on the LNW section of the LMS, but the installation of track circuits at Euston No. 4 could not be considered a priority as there was an unrestricted view of trains.

The current rails of the up and down slow lines were cross-bonded so that the heavy short circuiting caused by the collision had rendered both lines dead, but staff should be trained to use short-circuiting bars on live rails, so they were not afraid to employ them in an emergency.

Turning to recommendations, Colonel Pringle considered that track circuits might be installed at the up home signals in view of the fact that trains were frequently stopped at them with their rear vehicles inside Park Street covered way. As regards the rescue operations an improved alternative to naphtha flare lamps would be useful.

Referring again to the Swiss Bellinzona disaster, the inspecting officer stated that it was satisfactory that the LMS was moving in the desired direction, that is the replacement of gas with electric lighting.

Motormen, conductors, and guards of electric trains should be practically instructed in the use of short circuiting bars on live rails.

The ambulance and rescue equipment available was adequate but ladders might be carried on electric trains for use as stretchers in an emergency. [416]

Aston Villa lost the cup final, Newcastle United winning 2-0.

In the early hours of Monday 30th June 1924, Chicago Burlington and Quincy Railroad eastbound passenger train No. 2, in charge of Conductor Ogden with Engineer Sullivan and Fireman Stephenson, was running late. It left Galesburg, Illinois, 24 minutes late at 3.19 a.m. and passed Neponset, Illinois at 4.26 a.m., now 41 minutes late. The problem was poor coal and Engineer Sullivan decided to obtain a fresh supply at Buda, Illinois, United States, 6½ miles further on. Buda station was passed at 4.34 a.m., 40 minutes late, and the train was stopped at the coal chute, a mile further on, at about 4.37 a.m., the engineer not being aware that the coal chute men were not on duty at that hour.

Behind train No. 2 was an eastbound mail train, in charge of Conductor Boyers with Engineer J. J. Huber and Fireman W. J. Brechtel. This train passed Galesburg at 3.44 a.m., four minutes late, and Neponset at 4.30 a.m., now only four minutes behind train No. 2. The double track line was provided with two-arm two-position lower quadrant semaphore signals capable of showing at night red, yellow, and green, for stop, caution, and proceed. In addition at stations manually operated calling-on signals were provided, enabling trains to pass automatic signals at danger to take a diverging route or enter a side track. Train No. 8 received a caution indication at the automatic signal a mile west of Buda station and reduced speed. Automatic signal S-116.7, 312 feet west of Buda station, was at danger, but it was provided with a calling-on signal which the Buda operator cleared on the approach of train No. 8. That train continued along the main line, passed the station, passed the flagman of train No. 2 without his having been seen, and collided with the rear of train No. 2 while still running at about 20 miles an hour. The observation sleeping car at the rear of the first train was telescoped for about 20 feet, the car ahead was derailed, and the wooden sleeping car marshalled third from the end of the train was telescoped over two thirds of its length. Six passengers and Pullman Porter J.A. Triplett of train No. 2, along with Fireman Brechtel of train No. 8, were killed and 13 passengers, a Pullman porter, a mail clerk, and Engineer Huber were injured.

At the Interstate Commerce Commission inquiry Engineer Sullivan testified that he had whistled out his flagman as he was stopping at the coal chute and had also whistled for the coal chute man. The latter did not appear despite repeated whistles, and he was then told by the brakeman of train extra 4974, standing on the passing track, that there was no coal chute man on duty at that hour. His fireman had then been occupied trying to work the coal chute when his train was run into.

Conductor E.J. Ogden told the inquiry that he had remarked to his brakeman, W.P. Allard, at either Kewanee or Galva, two or three stations short of Buda, that train No. 8 must be catching them up. He had not been expecting his engineer to stop and when the train had done so, at 4.37, he had got down and begun to walk towards the locomotive to find out why it had done so. Seeing the fireman reaching for the coal chute apron had answered his question and he had turned backwards to make sure that the rear of his train was being protected. The collision had occurred, he thought, only two and a half or three minutes after his train had stopped; barely time, he considered, for his flagman to have gone back far enough. Asked about his flagman dropping off a fusee before his train had

[416] Ministry of Transport report.

fully stopped, he said that they had been taken by surprise by the stop, though he agreed that it was the practice to drop fusees when trains were losing time and encroaching on the time of a following time, as his train had been, and that his flagman should have been on the alert.

Brakeman Holmquist, who had been riding in the car with the conductor, had also gone forward when the train had stopped and turning round had witnessed the collision, which he stated occurred at 3.40 a.m. He had then gone forward and had succeeded in flagging westbound mail train No. 7, thus preventing a double collision.

Flagman Allard agreed that his conductor had mentioned that the mail train was gaining on them but that he had not thrown down any fusees. The stop had been unexpected. He had got down and proceeded back to protect, carrying red and white lanterns, two torpedoes and fusees. After getting two or three car lengths back he had heard a train approaching and had lit a fusee. The headlight of train No. 8 had appeared when it was 10 or 12 passenger car lengths away from the rear of his own train. His stop signals were not answered, and he had thrown down the fusee and got out of the way, the train passing him travelling at 20 or 30 miles an hour, with its brakes not applied. There had been no time to put down torpedoes. Had he run back instead of walked he could have got back 15 or 16 passenger car lengths instead of nine. Had he known the other train was right behind his he would have run. He had not previously thrown down any fusees because he had not though it necessary to do so.

Engineer Huber stated that after leaving Galesburg he had stopped at signal S-127.8, after which he had received a caution signal approaching Neponset, but the home signal had been cleared. Approaching Buda he had reduced speed at the caution signal, prepared to stop at signal S-116.7 but five or six car lengths before that signal the calling-on signal had been cleared and he had released the air brakes and allowed his train to drift towards the station at 8 or 10 miles an hour, expecting to receive instructions to cross over or head in, but none were given. He had continued forward, as he felt at caution, knowing that train No. 2 was ahead, his speed increasing a little on the falling grade. Looking for the automatic signal east of the coal chute he had instead seen the rear of train No. 2 about 80 feet distant. His train had been travelling, he estimated, at 15 or 18 miles an hour and he had applied the brakes in emergency. He had not seen anything of a flagman or a fusee, nor of the Buda operator, nor had his fireman said anything about them. The sun had just been rising but he did not think that it had interfered with his vision. He considered that if the calling-on signal was cleared, his train was not switched to another line and there was no indication from the operator that he was permitted to continue at caution to the next automatic signal, prepared to stop within his range of vision.

Conductor Bart Boyers of train No. 8 agreed with his engineer's statement as to the journey between Galesburg and Buda. Watching for instructions from the Buda operator he saw the latter give what he took to be a proceed signal from the bay window of the station. As freight train No. 4974 was on the passing track he knew that his own train must be routed along the main track. His train had continued, at 12 or 14 miles an hour for half a mile and then a little faster. He was looking ahead but the vehicles of the freight train on the curve, the smoke and the sun shining in his face meant that he saw nothing of train No. 2 before the brakes came on in emergency. He estimated his train's speed at the moment of collision at 18 miles an hour or perhaps a little more. He saw a burning fusee lying against the right rail after the collision. His engineer had seemed to be in a normal physical condition at Burlington.

Head Brakeman Milroy of train No. 8 corroborated the statement of his conductor. He put the speed of his train at the time of the collision at 20 miles an hour and said that he saw the burning fusee after the collision.

Flagman Kramer of train No. 8 agreed with the statements of his colleagues. He saw the Buda operator seated at his desk raise his hand as his train passed. He saw the fusee. With the sun rising he was of the opinion that it would have been difficult for an engineer to observe it but said that if he had been flagging he would have used night signals.

The crew of train extra 4974 also gave evidence. Engineer Metz was not sure whether or not the engineer of train No. 2 had whistled out the flagman. He thought that four or five minutes elapsed between that train stopping and the collision. Fireman Carlson could not recall any whistles. Head Brakeman Hosier, who had been in the coal chute when train No. 2 arrived, was positive that there had been no whistle for the flagman to go back but said that there had been four or five for the coal chute man. Flagman Smith was also in the coal chute when train No. 2 stopped and likewise remembered whistles for the coal chute man but not for the flagman. Conductor Rowe in the caboose said that train No. 2 had passed him at a good speed and train No. 8 five or six minutes later, at a speed he put at 8 or 10 miles an hour. He had heard no whistle for the flagman and had known nothing of the collision for several minutes after it had happened.

Operator A.M. Kermeen stated that train No. 2 had passed at 4.34 a.m. under clear signals, and he had not known that it was going to stop at the coal chute. Train No. 8 had approached at a speed at which it could have pulled up at signal S-116.7 but when it had been three car lengths short of the signal he had cleared the calling-on arm so that the train could drift by without stopping, which it had done at 4.34. He had held up two fingers to the fireman to indicate that train No. 2 was ahead, the latter had nodded, and he had given the same indication to other members of the train crew. He had cleared the signal entirely on his own responsibility, on the understanding that the rules permitted him to do so, though he could not recall anyone in authority saying that he could do so. He had last been examined on the rules in April. He maintained that the train proceeding at caution to the next block signal after having the calling-on signal cleared and receiving an instruction from him was no different to their stopping and then proceeding from signal S-116.7.

Train Rules Examiner S.H. Schultz stressed that the operator was forbidden to use the calling-on signal except when instructed to do so by the dispatcher to avoid delay, but that the clearing of the signal did not relieve the engineman from observing the provisions of Rule 509 A and B (relating to proceeding at caution expecting to find a train ahead, a broken rail, an obstruction and so on)

The ICC concluded that responsibility for the collision was shared by Engineer Huber for failing to obey signal indications and by Flagman Allard for failing to properly protect his train.

The engineer, instead of giving close attention to the track immediately ahead of him as required by the circumstances, had been stood up trying to see over the tops of the freight cars of train No. 4974 to determine the aspect of the next automatic signal, and while doing so had missed seeing the flagman and the fusee. He had also failed to keep his train under proper control, allowing its speed to rise on the falling gradient.

Flagman Allard had disobeyed the provisions of Rule 99, which required a flagman to throw down fusees in situations where his train was in danger of being overtaken, and when his train stopped to immediately go back a sufficient distance, put down two torpedoes and display fusees, as necessary. Although Conductor Ogden had spoken to him about their being on the time of train No. 8 he had not thrown off a fusee before his train stopped, nor had he hurried back after it had stopped, and he had not begun to run even when he had heard train No. 8 approaching and he knew that the view ahead was much obscured. Even assuming that the stop had been sudden and unexpected and there had been no opportunity to throw off a fusee there was no reason why the flagman should not have alighted the moment his train came to a stand and made every effort to get back as quickly as possible.

There was conflicting evidence as to whether or not Engineer Sullivan had whistled the flagman out. It was important that he do so, especially as he was losing time on the following train, and it was an unusual stop.

Operator Kermeen had cleared the calling-on signal without having been given any instructions to do so and without the train being diverted to diverging route. It did not appear that the clearing of the calling-on signal had had any direct effect on the accident but had train No. 8 been brought to a complete stand at signal S-116.7, Flagman Allard would have had time to get a little further back. The officials should take steps to ensure that the use of the calling-on signals was restricted to the purpose for which they were installed. [417]

The inquest jury found that the deaths of six of the victims (two died later) had resulted from the collision and that the accident was 'due to the failure of the engineer of train No. 8 to observe due caution as defined in the rules governing the operation of trains on the Burlington system; and we further find that the system of operating trains at Buda, Ill, does not afford adequate protection to the travelling public in view of the local conditions at Buda, taking into consideration the grade, the curve, and the position of the coal chute with reference to the block signals.' [418]

The evening of Monday 28th July 1924 saw a serious collision in Scotland, at Haymarket, Scotland, in the suburbs of Edinburgh, on the London and North Eastern Railway. The 6.41 p.m. down Inner Circle passenger train from Leith Central via Edinburgh Waverley was held up slightly at the down South platform at Haymarket station to provide a connection for a large party from an Edinburgh hostel who had been spending the day at St Andrews and who had just arrived at Haymarket on a train from there. While it was standing at the platform it was run into from behind at about 6.58 p.m. by the 6.54 p.m. passenger train from Edinburgh Waverley to Kirkliston (Port Edgar). The colliding train was only travelling at 10 miles an hour at the moment of impact but the second coach from the rear of the forward train mounted that in front, crushing several compartments together.

Four passengers were killed outright, one died later, and 54 were injured. The passengers who were killed outright were Miss F.M. Umpleby, of York; Miss Helen Fountain, sister-tutor at Brompton and Lambeth Hospitals, London; Janet Love, aged 17,

---

[417] Interstate Commerce Commission report

[418] *The Bureau County Tribune*, Princeton, Illinois, United States, 4th July 1924

of Edinburgh and Miss Spratley, of Brixton, London. [419]

The four tracks between the west end of Waverley station and Haymarket station were known as the up and down North lines and the up and down South lines. They ran partly in a cutting with retaining walls and partly in tunnel, with the 124 yard Mound Tunnel between Waverley and an intermediate signal box at Princes Street Gardens and the 1,020 yard Haymarket Tunnel between Princes Street Gardens and Haymarket station. The usual semaphore signals were provided and there was ordinary block working between Waverley West and Princes Street Gardens, but between the latter box and Haymarket East box the line was signalled with Sykes Lock and Block. The lock and block working meant that the lever of the down home signal, once reversed, clearing the signal for a train, could not be fully replaced until the train had passed over a treadle (A) in advance of it and could not be again reversed until the train had passed over a second treadle (B) in advance of the starting signal and the lever of that signal had been reversed. The starting signal also had to be released by the Haymarket East signalman plunging his instrument, and the latter could not give a second release until the train for which the starting signal had been cleared had passed over a third treadle (C) in advance of the Haymarket East down home signal. A gong at the Princes Street down South line starting signal was designed to sound to warn the driver of a train passing that signal at danger.

Signalman Thomas Armstrong at Waverley West box told the Ministry of Transport inquiry onto the accident that the working between him and Princes Street Gardens had been normal, and he had known nothing of the accident until told of it by the signalman at Princes Street just after 7.

Relief Signalman George Duncan at Princes Street testified that the 6.45 p.m. Edinburgh to Hyndland train had passed him normally on the down South line running under clear signals. He had duly cleared it back to Waverley West and accepted the Inner Circle train. Haymarket East had yet to clear the Hyndland train and he had 'severely checked' the Inner Circle train at his down South line home signal and stopped it at his down South line starting signal. Haymarket East had then cleared back the Hyndland train and accepted the Inner Circle train, the down South starting signal had been cleared and the Inner Circle train had proceeded. When he had cleared back the Inner Circle train to Waverley West he had been offered and had accepted the Kirkliston train. He admitted that when the Kirkliston train had emerged from Mound Tunnel he had immediately cleared the down South line home signal for it and had not checked it in accordance with of Rule 40 (a). (Readers of the previous volume of this series will perhaps recall that this was exactly the sin of Signalman Sutherland at Mallerstang cabin back in 1913 in the sequence of events leading to the Ais Gill disaster). Relief Signalman Duncan said that he was absolutely certain that his down South line starting signal had been at danger when he had cleared the home signal and the Kirkliston train had appeared to be under control as it had passed his box. At the time, his up South line home and down North line starting signals had been off for other trains. Having checked that the tail lamp of the Kirkliston train was in place he had turned to check a train diagram on the box wall. About two minutes later Haymarket East had offered him an Outer Circle train on the up South line and he had seen that the Kirkliston train was not standing, as he had supposed, at his down starting signal. He had refused the up train and at almost the same moment the Haymarket East signalman had telephoned him and asked what was following the Inner

[419] *The Times*, London, England, 30th July 1924

Circle train. "The ferry," he had replied, and the Haymarket signalman had replied, "Well, it's into the back of the suburban train here." About 35 trains had passed on the down South line between the start of his shift at 2 p.m. and the accident and there had been no problems with the Lock and Block.

Also in Princes Street box at the time of the accident had been Telegraph Linesman Munro. He had been carrying out a routine inspection of the electrical instruments and was not paying any attention to the signalling, nor had he had anything to with the South line Lock and Block instruments. He did not even realise that there had been a collision until the guard of the Kirkliston train had arrived at the box at about 7.12 p.m. After the accident he had found the Sykes instruments for the down South line home and starting signals to be both in the 'Locked' position, which he considered proved that the lever for the down South starting signal could not have been cleared for the Kirkliston train.

Signalman Robert Lauder at Haymarket East said that his down home signals were at the west end of the station platform and down trains were not cleared back until they had left the station. The down Inner Circle train had arrived at the South line platform at 6.56 and had been held two minutes longer than usual for passengers from the Polytechnic special from St Andrews, which had arrived at the North platform at the same time. The Kirkliston train had not been offered to him from Princes Street and he had known nothing of it until he heard the noise of the collision. A down empty stock had been offered and accepted on the North line at 6.55 p.m. and passed him at 6.58, the starting signal for it at Princes Gardens would have been clear when the Kirkliston train passed there. The Princes Street signalman could not have cleared his down South line starting signal until he (Lauder) had plunged his instrument, which he could not do until the Inner Circle train had left the platform and activated the treadle in advance of his down South line home signal.

Signalman David Harper of the double-manned Haymarket East box corroborated Signalman Duncan's statement as to the telephone conversation. He thought that the empty carriage train on the North line and the Kirkliston train had been running nearly parallel. His only experience of failures with the Lock and Block in ten years was when the electric treadle had failed to replace the home signal to danger, in which case the lever controlling the signal would not work. There was no release key for the Lock and Block instruments in either box.

Driver John Hill of the Inner Circle train stated that his train had stood at Haymarket with its brakes off for about four minutes when it was run into. He had not experienced any difficulty in seeing the Princes Street down South starting signal because of smoke and steam issuing from Haymarket Tunnel and did not think it likely that a driver would confuse the down North line starting signal for the down South one. He thought that a driver emerging from Mound Tunnel and finding the down South home signal off would expect to find the starting signal also off.

Fireman Jamieson of the Inner Circle train agreed with his driver regarding the misleading impression given by the premature clearance of the down home signal.

Passenger Guard Andrew Brown of the Inner Circle train said that his train had been stopped half a minute at the Princes Street starter. His train had been about a minute at Haymarket when the porter gave him the tip to start but before he could signal to his driver to do so the stationmaster had come hurriedly down the steps and told him to hold the train for 30 passengers from the St Andrews service. Almost all these passengers had got into the second and third coaches from the rear (which, by chance, were those most damaged in the collision). He thought that fully three minutes had been thus lost and he had just been assisting the last passenger to board and closing the doors when his train had been run into. He considered that the rear of his train would have been difficult for the driver of the Kirkliston train to see because of smoke and steam issuing from the mouth of the tunnel.

Driver James Swan of the Kirkliston train related that his engine had been running tender first, but his view of signals had not been obscured. He had seen the Princes Street Gardens down South line distant was "on" from the east end of Mound Tunnel and had anticipated that the home and starting signals would be against him. Emerging from the tunnel there had been a heavy atmosphere with drizzling rain, but the visibility had been fairly good. Just outside of the tunnel he had found the down South line home signal "off." Nevertheless he had been prepared, he said, to stop at the down starting signal if it were "on" but he was positive that it had been "off." He could not say what the down North line starting signal had been showing. He had not heard the warning gong and doubted if he would have done anyway. Emerging from Haymarket Tunnel smoke and steam had prevented him seeing the rear of the Inner Circle train until he was within about 15 yards of it. He was certain that he had not mistaken the down North line starter at Princes Street for the down South line one.

Fireman John Ross of the colliding train testified that he had been busy with his fire leaving Waverley station but that he had seen the Princes Street starter "off" with the Haymarket East distant signal below it "on" as the train had passed under the second footbridge east of Princes Street box and he had turned to get coal. Reminded that the signal was not visible from that point he said that it must have been the third footbridge, west of the box.

Guard Martin M'Que of the Kirkliston train could say nothing about the events leading up to the collision. Afterwards he had hurried back through the tunnel and reached Princes Street box at 7.12 p.m. Signalman Duncan had said to him, "Your train has run through my starting signal."

Signalman William Bell said that he had relieved Signalman Duncan at 10 p.m., at which time the levers for the down South line home and starting signals were back in the frame and both instruments showed "locked." The lines had been cleared at 11.03 p.m. and at 11.15 p.m. Signal Supervisor Thomson had asked him to test the working of the down South starting signal and he had got a plunge from Haymarket East. The replacement of the starting signal lever then freed the lock on the down home and the instrument associated with that lever had then gone back to its normal position of "Free."

Signal Supervisor Adam Thomson said that he had tested the warning bell (gong) at the Princes Street down South line starter, and it had worked correctly. At about 11.20 p.m. the Lock and Block instruments for the down South line had been used for the first train movement since the collision. They had worked satisfactorily and no adjustment of or work on the instruments had been necessary.

The inspecting officer, Colonel J.W. Pringle, stated that in his mind the record of block signals and evidence established in his mind beyond contention that:

(a) The Inner Circle train had not been cleared back to Princes Gardens on the block bell.

(b) The Kirkliston train had not been offered to or accepted by Haymarket East.

(c) It was impossible for the Princes Street starter to be cleared without a plunge from Haymarket East.

(d) Haymarket East could not have given a release under the Inner Circle train had left the station and passed over treadle (C) in advance of the Haymarket East down South line starting signal.

(e) The Lock and Block apparatus had been tested after the collision and found to be in perfect order, including the action of treadle (A) in reversing the Princes Street down home signal.

Driver Swan had been emphatic in his assertion that the Prince Street starting signal had been "off" and had been unshaken in cross-examination. He had been supported by Fireman Ross, though the latter's evidence was less convincing. If their evidence be accepted there must have been irregular working on the part of the signalmen or a serious failure of the Lock and Block apparatus. That the signal had been half off, was ruled out by the evidence of Driver Swan and that it had jammed off was ruled out by it having worked properly after the collision.

The possibility of Signalman Duncan having used a release key or interfered with the instrument could be ruled out by the fact that the signalmen were not provided with release keys, together with the fact that had the instrument been interfered with to release the down South line starting signal the pulling and replacing of the starting signal lever would have freed the instrument of the down South line home signal, and there was conclusive evidence that the home signal instrument had shown "Locked." The only other possibility was that Signalman Duncan had failed to replace the home and starting signals behind the Inner Circle train and treadle (A) had failed to replace the home signal. In addition Signalman Duncan would then have to have replaced, contrary to normal practice, the starting signal before the home signal behind the Kirkliston train in order for the home signal instrument to show "Locked." Colonel Pringle stated that he was unable to accept this improbable chain of events.

The holding of the Edinburgh and Glasgow Trades holidays on the same week, together with the Fife District Trades and Aberdeen holidays, had, the inspecting officer continued, resulted in the withdrawal of the normal heavy bogie stock used on the suburban services for long-distance traffic and its replacement by older four- and six-wheeled coaches, built between 1891 and 1896. Had more modern rolling stock been in use the telescoping of two of the coaches of the Inner Circle train would probably have been less severe. The vans of the older rolling stock were also not equipped with rescue appliances and ambulance boxes like modern trains. Crowbars had had to be obtained from a surfaceman's hut and saws and hammers from joiners who chanced to be working at Haymarket station.

In conclusion Colonel Pringle remarked that he had been impressed with the manner in which Driver Swan had given his evidence, but he was forced to the conclusion that the driver was mistaken. He could only conclude that the driver, whose engine had been running tender first, had misread the signals. The warning bell had no doubt rung but he (the inspecting officer) was not convinced that its sound would have reached the ear of the driver of a passing train (in which case, it would seem, the bell was of little use).

The responsibility for the accident therefore mainly rested upon Driver Swan for misreading signals. Given the atmosphere at the exit of Haymarket Tunnel Driver Swan could not be blamed for not avoiding the collision upon leaving the tunnel.

The failure of Signalman Duncan to give proper effect to Rule 40 (a) and bring the Kirkliston train almost or completely to a stand before clearing the down South line home signal was to some extent a contributory cause to the accident and some of the responsibility for it therefore rested upon Signalman Duncan.

Colonel Pringle ended his report with several additional remarks:

(a) The treadle bell arrangement at the Princes Street down South starting signal at the east end of Haymarket Tunnel could not in his opinion be relied upon.

(b) The most effective way of preventing such an accident was a system of automatic train control at the Princes Street starting signals whereby the continuous brakes were applied automatically should a train pass a stop signal at danger. Pending such a system the installation of detonator placers at the two signals working in conjunction with them was advisable.

(c) The circumstances of the accident emphasised the importance of General Rule 40 (a).

(d) The action of the Haymarket stationmaster in delaying the Inner Circle train appeared to be contrary to the instructions of the Traffic Superintendent but the train should of course have been fully protected by the Lock and Block signalling.

(e) Although none of the vans involved in the accident had been equipped with the usual salvage and ambulance equipment the arrangements for dealing with the passengers had been adequate and the rescue work had been carried out expeditiously. [420]

Driver Swan was arrested at the beginning of August and released on £40 bail. [421] He stood trial on a charge of "Culpable Homicide" at the end of January 1925. The defence argued that the signals had been clear. The jury returned a unanimous verdict of "Not Guilty," a decision which was applauded in court. [422]

Our next accident happened across the Atlantic and across the United States, in Oakland, California, on the Key Division of the Key System Transit, a 600 volt DC electric railway. The line on which the accident happened ran to the Pier Terminal on San Francisco Bay, from which ferries crossed the bay to the Market Street terminal in San Francisco. San Francisco-Sacramento Railroad trains also used the line between Fortieth Street and Shafter Avenue and the Pier Terminal. The line was double track with automatic stop signals of the three position upper quadrant time together with automatic train stops. The latter engaged a valve on the roof of the train cars, applying the brakes in emergency should a car pass a signal at danger. An overlap of two stop signals and one

---

[420] Ministry of Transport report

[421] *The Times*, London, England, 6th August 1924

[422] Ibid, 31st January 1925

preceding caution signal was provided behind every train enabling trains to be run with a headway of only 45 seconds. Key System trains were meant to operate at a maximum speed of 36 miles an hour and should have been able to stop in a distance of 300 to 350 feet from this speed.

Two westbound trains were involved in the accident on Thursday 4th December 1924. Key System train No. 729, in charge of Conductor Noone with Motorman Compton, was pulled up short of signal 100, which was at danger because of switching operations at the Pier Terminal, when it was run into by San Francisco-Sacramento Railroad train No. 15, in charge of Conductor Ernest Clyde Knoblock with Motorman Brubaker. Train No. 729 consisted of cars 655, 656, 664 and 665 coupled together. In the collision car 1014, the only car of train No. 15, telescoped car 665 of train No. 729 for 18½ feet. Eight passengers and two railroad employees not on duty were killed and 36 passengers, Conductor Knoblock with Motorman Brubaker were injured. [423]

After the accident emergency calls were sent out to all the ambulance companies and hospitals in Oakland and the police emergency hospitals, while a wrecking train was sent from the Key Route shops. Passengers described their experiences to a newspaper reporter. '"We were chatting in the train when suddenly Conductor Clyde Knoblock exclaimed, 'My God, we're in for it now!' I looked out of the front vestibule window and saw a Key train standing directly in front of is. Then there was a crash." Thus N.W. Thompson, of 6046, Lawton avenue, Oakland, a passenger in the "short line" train described what had happened.

'Lying on a cot at the Berkeley General Hospital with a broken arm and fractured collarbone, Thompson believes his escape from death was almost a miracle. He was seated on the first seat of the train. Thrown upward by the impact, he was buried but protected by the metal frame of the seats when the front of the "Short Line" train ploughed through the rear coach of the Key route train.

'"I don't see how it could have happened," he said. "The Key train was standing still. I noticed it a few minutes before the crash. We came out of the little tunnel and were travelling pretty fast. Knoblock, the conductor, or 'Nobby,' as he is known to commuters, was passing through his train and had just said a few words to me. He looked out down the train, saw the train standing ahead.

'"My God, we're in for it now!' His whole body stiffened but he managed to reach for the bell rope. As I went over the seat as the trains crashed I saw 'Nobby' go sailing upward and headfirst down the aisle between two sears. The last I saw of the engineer of the train was when he screamed out something. His arms were moving as if he was putting on his brakes. That was the last I saw of him."

'Earl C Blasingame, aged 24, of 728 East Twentieth street, who was a passenger in the Key train, said the train came to a stop. "I thought nothing of it," he said, "until there was a peculiar grinding sound followed by a terrific crash. I remember shooting through the air and seeing something black behind me. I guess it was part of the seat. My glasses tore a gash in my neck, and I first thought my jugular vein was cut.

"I just said to myself I wasn't going to bleed to death, and I stuffed my handkerchief into the open wound, stopping the flow of blood. Pretty soon my leg began to pain me.

[423] Interstate Commerce Commission report

Somebody pulled me out of the wreckage. I think it was a college student. He held his hand to my neck on the Shattuck train which brought us to Berkeley.

"The doctors tell me I had a close call from the cut. I guess I owe my life to my handkerchief and to that student." Blasingame is also at the Berkeley General Hospital.' [424]

The Interstate Commerce Commission (ICC) set to work to investigate what had gone wrong.

Motorman Compton of train No. 729 told the commissioners that it had been raining a little on the morning of the accident, but visibility had been good. At signal 108 he had seen signal 104 at caution, had shut off power and made a full application of the air brakes. There had been no difficulty in stopping his train short of signal 100. He had immediately whistled out his flagman and released his brakes. About half a minute later his train had been run into. He said that the front of his train had stopped about 130 feet short of signal 100, with its rear a few feet in advance of signal 104. He explained that he had stopped so far back from signal 100 after having made a full brake application and feared that if he had released his brakes before stopping he might have overrun the signal.

Conductor Noone stated that his train had been travelling at 30 or 35 miles an hour. It had come to a stand with its rear just beyond signal 104. As soon as his train had stopped his motorman had whistled out the flagman and he had picked up his flagging equipment and started back, but the collision had occurred before he had even reached the rear of his train and before he had seen train No. 15.

Brakeman Hamma of train No. 729 said that he had been collecting fares in the leading car. He agreed that there had been only 30 seconds between his train stopping and it being struck.

Brakeman Van Dalsen said that he had been collecting tickets in the third car of train No. 729. He too put the interval between his train stopping and the collision at 30 seconds.

Motorman Hobson, deadheading and riding at the front of the third car, said that the flagman had been whistled out at once. Looking back he had seen train No. 15 coming towards them, at the moment two car lengths away and travelling, he thought, at 30 miles an hour. Realising that a collision was inevitable and there was no time to alight he had braced himself for the impact.

Conductor Knoblock of train No. 15 said that his train had consisted of two cars from Concord, but one had been detached at Fortieth Street. His train had been stopped at signal 46 at tower No. 2 to give precedence to train No. 729 and had followed about a minute and a half later. His train had run at a normal speed, and he had seen nothing of train No. 729 until immediately before the collision. Approaching the scene of the accident there had been a service application of the brakes, followed perhaps half a minute later by an emergency application due to the automatic train stop which had thrown him off balance. He put the speed of his train at 35 to 40 miles an hour at the time of the service brake application. After the emergency application the train wheels had seemed to lock, and the train had slid into collision.

Motorman Willis, of the train following No. 15, said that his train had been running at

[424] *Berkeley Daily Gazette*, Berkeley, California, United States, 4th December 1924

about 30 miles an hour and he could see for half a mile. Passing through the subway beneath the Southern Pacific Railroad west of tower No. 2 train No. 15 had been constantly in view, running faster, he thought, than his train could go.

Other motormen operating trains that morning agreed that the rain had not obscured signals and that although the rails were wet they had had no difficulty in controlling their trains and obeying signals.

Towerman Corker at tower No. 2 stated that he had given train No. 729 priority and train No. 15 had left about a minute later. No times were recorded at his tower.

Motorman Brubaker refused to give any evidence at the inquiry on the advice of his counsel as it could be used against him in any pending criminal proceedings.

Electrical Engineer Bell stated that with train No. 729 stopped where it had signal 104, a few feet in its rear, and signal 108, 420 feet in the rear of the train, would have been at danger and signal 112, 840 feet back, would have been at caution. A very fresh mark on the automatic stop arm of signal 108 clearly showed that it had been struck by an automatic stop valve arm and there was a corresponding mark on the valve arm of car No. 1014. Apart from the replacement of the damaged automatic stop arm on pole 104 nothing had needed doing to the signalling, which had functioned properly before and after the accident. Since the installation of the signalling in 1911 there had been approximately 18,000,000 45 degree movements annually, with only four wrong-side failures (false clear signals) during that time, the last in 1916.

Master Mechanic Jackson said that he had examined train No. 15 after the accident and had found no indication of brake failure.

Superintendent Thornton of the Key System said that the trains were being operated in their proper order. All motormen were instructed to obey signals and the responsibility for a rear collision rested with the motorman of the colliding train. Motorman Brubaker had been involved in a previous rear collision on 1st March 1922.

Analysing the accident, the ICC decided that train No. 729 had actually come to a stand 185 feet east of signal 108 having taken 430 or 440 feet to stop. The automatic signalling system and train stop had functioned correctly, and the collision had occurred because train No. 729 had come to a stand in a position which left only the minimum braking distance available, while train No. 15 had been going too fast. Wet rails and the train wheels locking after the emergency brake application had contributed to the collision. The signalling system was designed for trains travelling at a maximum speed of 36 miles an hour while train No. 15 was capable of 55 miles an hour. Motorman Brubaker had failed to comply with the rules requiring him to run at an even speed and to pass a caution signal prepared to stop at the next stop signal. In view of the motorman's refusal to testify no explanation for his failure could be put forward.

The commission recommended that the Key System Transit Company establish a maximum speed for all trains running on the line. [425]

On 17th December, the grand jury investigating the collision adjourned indefinitely without making any recommendations or taking any action, citing insufficient evidence. Jury Foreman R.E. Beach announced after the adjournment that, "The Key collision is a

[425] Interstate Commerce Commission report

closed incident as far as the grand jury is concerned. There is not sufficient evidence to warrant an indictment, nor even to indicate who was to blame for the catastrophe." At this point the ICC and state railroad commission investigations were still in progress. [426]

Over the night of 12th and 13th January 1925 an exceptionally thick and impenetrable fog had descended over the Ruhr area of Germany. Passenger train No. 230, used mainly by workmen and early commuters, was due to leave Herne, Germany, at 7.19 a.m. for Wanne, but on the morning of Saturday 13th January 1925 it was a little late in starting, and it was just setting off at 7.20 when it was run into from behind by Berlin – Köln D-Zug No. 10, due in Herne from Dortmund at 7.25. The rear fourth-class coach of the crowded front train was telescoped and wrecked; 22 men, women and children were killed and 60 injured, most seriously. The engine of the colliding express was derailed but otherwise it was hardly damaged. Its driver and fireman were unhurt and after two sleeping cars had been detached it continued its journey with a fresh engine.

There was no question but that the accident had been due to the driver of the express overrunning the Einfahrsignal. Herne station was equipped with the most modern signalling. There were in fact two stop signals, what in Great Britain would be termed outer and inner home signals, as well as a Vorsignal (distant signal). The inner of the two stop signals was automatically replaced to danger as a train passed it and could not again be cleared as long as a train was in the station. The claim by the driver that the signal had been clear for him was therefore impossible, moreover other witnesses had seen it at danger. The driver had also only just begun his run at Hamm, a few kilometres away, so he could not have been overtired. The official investigation determined that the experienced driver who was familiar with the line had overrun signals which he should have observed. A German author writes that it was too early for any form of automatic train control but incomprehensible given the thick fog that detonators had not been used to protect the obstruction. [427]

Bressana station, Italy, between Pavia and Voghera, was the scene of a serious accident about midnight on Friday 16th October 1925. A passenger train, which was running about ten minutes late, was still standing in the station when a through goods train approached. The signals were against the latter, but neither the driver nor the fireman appears to have observed them; they entered the station at undiminished speed, and ran into the rear of the passenger train. Three carriages were smashed to pieces and 12 passengers, who were mostly peasants and soldiers, were killed and about 30 more were hurt. The engine crew of the goods train jumped to safety just before the collision. [428]

Darkness and thick fog were features of a collision which occurred at about 5.55 a.m. on the morning of Thursday 12th November 1925 on the Pennsylvania Railroad near Monmouth Junction, New Jersey, United States. The line at this point was quadruple track, worked by timetable, train order and automatic block signals. The signals were of

---

[426] *The Bakersfield Californian*, Bakersfield, California, United States, 17th December 1924
[427] *Historische Eisenbahn-Katastrophen* by Bernhard Püschel, pgs. 127-28 (in German)
[428] *The Times*, London, England, 19th October 1925

the upper quadrant type, mounted on bridges above the four tracks, and at night displayed indications of red, yellow, and green indicating danger, caution and proceed. From north to south the tracks were numbered 4, 3, 2 and 1, the accident occurring on track 1, 2¾ miles west of Monmouth Junction. Approaching from the west the track was tangent for a considerable distance and practically level.

On the morning of the accident eastbound passenger train No. 166 from Washington to New York, consisted of engine No. 3775, two mail cars, a combination mail and baggage car, two coaches and six Pullman sleeping cars. It was in charge of Conductor Neal with Engineer Moore and Fireman Boyd. Leaving Trenton, New Jersey, at 5.31 a.m., seven minutes late, it passed Princeton Junction, six miles from Monmouth Junction, at 5.44 a.m., having lost another minute, and stopped at signal bridge 440, the signal being red. It restarted at once, stopped again to enable a burning fusee to be extinguished, started again, and was going at 8 or 10 miles an hour when it was run into from behind by train No. 6.

Train No. 6, from St Louis to New York, the *Mercantile Express*, consisted of engine No. 3884, an express car, a business car, three Pullman sleeping cars, a coach, a combination passenger and baggage car, two mail cars and an express car, in charge of Conductor Hamilton with Engineer T. J. Carroll and Fireman Armstrong. It had left Trenton at 5.39 a.m., nine minutes late, passed Princeton Junction at 5.52 a.m., still nine minutes late, and was travelling at 35 or 40 miles an hour when it struck train No. 166.

In the collision the rear sleeping car of Washington train was telescoped for two thirds of its length by the car ahead of it and practically demolished, while the car which had telescoped it was somewhat damaged. Nine passengers and a Pullman porter in the rear car lost their lives and 20 passengers, 11 mail clerks and four railroad employees were injured.

Engineer Moore of train 166 told the Interstate Commerce Commission (ICC) that he had received a caution signal at signal bridge 448, made a service brake application, then allowed his train to drift to signal bridge 440. He had stopped his train at the second signal bridge, recalled his flagman at once and proceeded, then stopped again when a red fusee was encountered. His fireman had got down to extinguish the fusee, the flagman had again been recalled and his train had been proceeding at about 10 miles an hour when it was run into.

Conductor Neal testified that he had gone forward into the baggage car and had been standing in the doorway of that car with Baggagemaster Sortor when the train stopped a second time. He (the conductor) had no personal knowledge of how his flagman had performed his duties and had not been aware that there had been a collision until told of it by his engineer, when he had gone to the rear and reported what had happened to the dispatcher.

Flagman Cunningham of train No. 166 said that the engineer had given two blasts on his whistle when the train had stopped at signal bridge 440. He had at once given the engineer a proceed signal and thrown off a lighted fusee, which had disappeared from his view after about three car lengths because of the fog. When his train had stopped again, at about only 15 or 20 feet beyond the signal bridge, he had done the same thing. As the train had proceeded again he had thrown off a third fusee, but the head of that one had broken off, so he had thrown off a fourth. All had been five-minute fusees. He said that he had not alighted from his train at any point and had not been aware that it had been sufficiently delayed to be on the time of train No. 6. He maintained that in any case the brief stops had not been long enough for him to put down torpedoes.

Engineer Carroll of train No. 6 stated that after leaving Philadelphia Broad Street his train had stopped at West Philadelphian, North Philadelphia and Trenton, and the air brakes had worked properly each time. Approaching the site of the accident it had been very foggy and he had passed signal bridge 448 without realising it and only seen the stop indication at signal 440 at the last moment, when he had shouted to his fireman, applied the brakes in emergency, shut off steam and attempted to open the sanders. The speed of his train had been reduced from 55 miles an hour to 15 miles an hour when it struck train No. 166. He had only seen one fusee, less than two car lengths from the train in front. He was unable to explain how his train had run about 1,800 feet beyond signal 440 with its brakes applied in emergency.

Fireman Armstrong recalled his engine taking water at the track pans (water troughs), receiving a clear indication at signal bridge 458 and passing Plainsboro station. He had then been firing and had thus missed seeing signal bridge 448. Returning to his seat box to ring the bell for a road crossing he had done some more firing, then returned to the seat box again just in time to see signal bridge 440 at red as his engine was passing under it. He thought that it was he who had first alerted his engineer, and the latter had at once applied the brakes in emergency. He put the speed of his train at 50 to 60 miles an hour at the signal bridge and still at 50 miles an hour at the moment of impact. He saw no fusees.

Flagman Dickens of train No. 6 was riding in the express car at the rear of the train. He said that he was knocked down by the collision, then had difficulty in opening the damaged door. Going back to protect his train he said that he had found a lighted fusee in the centre of the track about six car lengths behind the rear of his own train. He put the speed of his train at about 55 miles an hour, reduced before the collision to 40 miles an hour. Visibility in the fog he put at about 50 feet.

Conductor Hamilton put the speed of his train at 45 to 50 miles an hour, reduced by the braking to 35 to 40. On his train stopping he had gone directly to the front of it, then to a telephone to contact the dispatcher, so he had not seen the fusee at the rear of his train.

Head Brakeman Green estimated that only a few seconds had elapsed between the brake application and the collision, the speed dropping from about 40 to 30 miles an hour. Baggagemaster McKelby thought that train No. 6 had covered 500 feet between the brake application and the collision with the speed having been reduced from 50 or 55 miles an hour to 35 or 40.

Electrical Inspector Rainey confirmed that the signals were working properly after the accident, nothing having been done to them. The signal on bridge 440 for track 1 was correctly showing red and the corresponding signal on bridge 440 was showing yellow.

Acting General Superintendent Davis stated that Engineer Carroll had passed periodical examinations for acuteness of vision, colour sense, and hearing in April 1923, July 1924, and July 1925.

The ICC concluded that the collision had been caused by the failure of Engineer Carroll to obey the indications of the automatic signals in foggy conditions. The failure of Flagman Cunningham to put down torpedoes was a contributing cause. Engineer Carroll admitted having failed to see signal bridge 448 and according to the rules should have reduced speed prepared to stop at signal bridge 440 but had not done so. Both Engineer

Carroll and Fireman Armstrong positively maintained that an emergency brake application had been made immediately upon seeing signal bridge 440 at red, but the statements of other crew members indicated that the brakes had been applied considerably later and had only reduced the speed of the train by about 10 miles an hour before the collision. Had the brakes indeed been applied in emergency at signal bridge 440 the train would have stopped in the 1,859 feet available, or at any rate have been moving very slowly.

Flagman Cunningham had thrown off lighted fusees but had not put down two torpedoes, as required by Rule 99, when his train stopped. Had he done so Engineer Carroll might have been warned in time.

The commission pointed out that the collision again demonstrated the need for an automatic train control device which would intervene to stop a train running past an automatic signal at danger. [429]

Two days after the accident Engineer Carroll was arraigned on ten charges of manslaughter. He pleaded 'Not Guilty' and was held on $10,000 bail for the grand jury. [430]

No report has been found of Engineer Carroll having come to trial, so it appears likely that the grand jury threw out the indictment.

The Public Utilities Commission of New Jersey also investigated the accident, coming to the same conclusions as the ICC. [431]

At least ten persons were known to have been killed and twenty injured on the evening of Monday 29th March 1926 when an express train from Bogotá to Girardot and the coast, carrying passengers and overseas mails, ran into the rear of a freight train standing in El Corso station on the outskirts of Bogotá, Columbia. Two first-class coaches were thrown into a river. [432]

A major disaster occurred in Germany at 10.30 p.m. on Whit Monday, 24th May 1926, at München Ost. Train No. 820 was stopped at the entrance to the station when it was run into by train No. 814. Thirty people died and 353 were injured. [433] The investigation into the disaster determined that the block instrument had been correctly at "Blocked" against the following train from Berchtesgaden and all the seals of the instrument were intact, implying that the signalman could not have overridden the controls and given a release. Nevertheless, the driver of the second train maintained that the Ausfahrsignal (starting signal) at Berg am Laim had been clear for him and was supported by other witnesses. It seemed possible that there had been a rare 'wrong-side' failure of the signalling. [434]

---

[429] Interstate Commerce Commission report
[430] *The Sandusky Star Journal*, Sandusky, Ohio, United States, 14th November 1925
[431] *High Point Enterprise*, High Point, North Carolina, United States, 2nd February 1926
[432] *The Lincoln Star*, Lincoln, Nebraska, United States, 30th March 1926
[433] *Eisenbahn-Katastrophen in Deutschland* by Hans Joachim Ritzau, pg. 40 (in German)
[434] *Von Siegelsdorf nach Aitrang*, by H. J. Ritzau, pg. 44 (in German)

The Pennsylvania Railroad was in trouble again Wednesday 16th June 1926. This time the accident happened a mile east of Gray, Pennsylvania, United States, again on a stretch of quadruple track equipped with automatic three-position upper quadrant signals, installed on signal bridges as at Monmouth Junction. Trains were operated by timetable, rules and block and interlocking signal indications; train orders were not ordinarily used. The accident occurred on track No. 2, the eastbound passenger track.

The first train involved in the accident was eastbound No. 50, the *Washington Express*, which consisted of engine No. 299, a deadhead Pullman observation car and six Pullman sleeping cars, in charge of Conductor C.B. Pennell with Engineer H.M. Kinkead and Fireman C.N. Groves. This train left Pittsburgh, Pennsylvania, at 10.28 p.m., on time, passed DR tower at Derry, about 5½ miles short of Gray, at 11.29 p.m., two minutes early, and came to a stand a mile east of Gray at 11.36 because of a burst air pipe hose. The rear of the train was about 820 feet east of signal bridge 3012. At about 11.44 or 11.45 it was run into by train No. 40.

Train No. 40 was the *Cincinnati Limited.* It consisted of engines Nos. 3850 and 612, a Pullman combined club and baggage car and nine Pullman sleeping cars in charge of Conductor L. Faust with Engineers W.S. Gordon and Edmund A. McConnell and Firemen N.L. Hollingsworth and W.S. McConnell. This train left Pittsburgh at 10.35 p.m., passed DR tower at 11.38, on time, and collided with train No. 50 while running at a speed of between 40 and 50 miles an hour. Engine No. 3850 penetrated the *Mt Union* sleeping car at the rear of train No. 50 by about 16 feet. The penultimate sleeping car of the front train, the *Fieldsboro*, telescoped the car ahead of it, the *Entriken*, over almost its entire length. The *Neshaming*, the car ahead of the *Entriken*, was only slightly damaged. Train No. 40 did not escape either. The two engines were thrown from the track and the leading car, the *Watkins*, was jammed between engine No. 612 and its tender on one side and the third car, the *Finley*, and practically destroyed. The second car, the *Inwood*, went off to the left and was less badly damaged. Eleven passengers, Engineers Gordon and McConnell, the fireman of engine No. 3850 and the baggageman of train No. 40 were killed and 82 passengers and four railroad employees were injured, one of the passengers fatally. The passengers who lost their lives had all been in the *Entriken* and *Watkins* cars. [435]

A local newspaper naturally had much to write about the disaster, of which a small excerpt must suffice. 'Mr and Mrs Raymond Hoelsener, Buffalo, N.Y., bride and bridegroom, who escaped with slight injuries in the railroad wreck near here early today, were so dazed they could give but little description of the accident. The honeymooners said their escape from death was miraculous. They climbed from a car window suffering from bruises and shock and were taken to the Latrobe hospital. They reported the scene appalling. Persons were fighting their way out of coaches and others who had escaped uninjured struggled to get back in the cars to aid the less fortunate, Hoelsener said. Many heroic rescues were enacted. Fatalities would have been tripled had the wreckage caught fire, he said.

'Mrs W.B. Dunwoody, of Wilkinsburg, who, with her two little children, was among the dead, was one of the first identified, a card in her pocket giving the name of a relative to notify in case of accident.' [436]

---

[435] Interstate Commerce Commission report
[436] *Altoona Mirror*, Altoona, Pennsylvania, United States, 17th June 1926

Engineer Kinkead was the first witness at the Interstate Commerce Commission (ICC) inquiry, which was held jointly with the Public Services Commission of Pennsylvania. He said that he had had a normal run from Pittsburgh until his train had been a few car lengths short of signal 3012, the distant signal for BH tower, when the air brakes had gradually come on. He then stopped his train. He did not whistle the flagman out but had expected him to go back anyway. Finding that an air hose at the rear of the tender had burst, his fireman began working to replace the hose. The flagman he saw going back and the conductor came forward to his engine. The hose was replaced, a road test of the brakes was made with the co-operation of the head brakeman at the rear of the train, and he whistled the flagman in once or perhaps twice. He thought that the flagman was a little slow in returning and was thinking of calling him again when the fireman said that the flagman was coming. Looking at his watch at that point, he had seen that it was 11.45. Just as he was replacing his watch in his pocket the collision had occurred.

Fireman Groves corroborated his engineer's evidence. He had seen the flagman's lamps and the burning fusee left when the flagman was called in. He heard train No. 40 before he saw it and had assumed at first that it was on track No. 1. The brakes of train No. 40 had not been applied when he first saw it, they appeared to have been first applied just west of signal 2012 and were fully applied at the time of the collision. He had not heard the noise of any torpedoes exploding. When he realised that there was going to be a collision he had shouted to his engineer. He had not looked at his watch but thought they been stopped about six minutes. After the accident he had gone east with his lantern to protect against westbound trains.

Conductor Pennell's account of the events leading up to the collision agreed with those of his engineer and fireman. He said that he had been standing on the rear platform of the second car and had only seen train No. 40 approaching when it was close at hand. Continuing, he stated, "I heard it as I was looking over the other side and was glancing around and when I looked back I saw the fire, flying from the wheels, and the headlight burning, and I saw the front brakeman jump off the rear end and the flagman jump – the flagman had not reached the rear end yet." He thought that his flagman had had ample time to go back far enough and in his judgement the flagman had done so. He had not heard any torpedoes go off. He considered that if his flagman had laid torpedoes and left a burning fusee then he was justified in returning to his train even if he had heard another train approaching and had been unclear as to which line it had been running on.

Front Brakeman Thomas Beecher of train No. 50 stated that he had been riding in the rear car with Flagman MacDonald. When the train had stopped suddenly they had both alighted and he (Beecher) had set off towards the front, but the conductor had shouted to him that an air hose had come off and to return to the rear end to test the brakes when the pipe was replaced, which he had done. He saw the flagman just under the signal bridge, and saw the flagman light a fusee when he was called in. He could then first hear and then see train No. 40 but had assumed at first that it was on the other line because it was coming in so fast. Then he had heard the torpedoes explode.

Flagman D.S. Macdonald said that he had previously flagged his train when it had stopped at East Liberty. When his train had stopped near Gray he had begun to run back, having picked up some fusees and with torpedoes in his pockets. His progress had been interrupted when he had tripped over a surveyor's stake, at which point he had lit a fusee, but he thought that he had got back about 700 feet from the rear of his train to a spot just east of signal 2012 when he had been called in. When called in he had put down two torpedoes and lit another fusee. Like Brakeman Beecher, when he had heard train No. 2 he had thought that it must be on track No. 1 from its speed, it being difficult to tell on the curve. When he had seen that the train was on his track he had begun swinging his lantern and had continued to do so but his signals had not been answered by the engineer. The fusee had been burning when the engine ran over it, and he had heard the torpedoes explode. He did not notice any brake application. The rear markers of his train were good, and the headlight of engine No. 3850 was burning full strength. He had not noticed the signal aspects before the collision; afterwards, signal 3012 had been red and the preceding signal, 3020, had been yellow. Having put down torpedoes and used fusees he considered that he had given full protection to his train.

Fireman McConnell of engine No. 612 was positive that the signal that he had seen at Gray had been clear. He did not remember seeing any further signals but saw the rear end of train No. 50 and heard the torpedoes explode, at which point his engineer (his brother) had applied the brakes in emergency. He had noticed Engineer Gordon apply the brakes at Copeland and Edgewood and whistle for a crossing before Gray but did not notice him applying the brakes before the collision.

Conductor Faust could add little information. He put the speed of his train at 50 miles an hour at the moment of the collision. He saw a burning fusee behind his train afterwards.

Flagman W.E. Penfield of train No. 40 stated that he had noticed nothing unusual until the emergency brake application at signal 3012 had thrown him down. He had picked himself up and had been starting for the rear of his car when another shock had thrown him down again. He had then gone back to protect his train and noted that signal 3012 had been showing red. Going back just west of Gray station, accompanied by Flagman MacDonald, they had found that signal 3020 was bobbing between yellow and red, which it continued to do for 45 minutes until stopped by a signalman. Both signals were burning properly; at no time was a clear position displayed.

General Foreman H.W. Lehr said that he had examined the brakes of train No. 40 at 3 a.m. He had been unable to find any closed angle cocks, brakes cut out or other irregularities which might have prevented the brakes from operating properly. The engine throttles had been closed or nearly so, the reversing gears had been in the full forward position and the wheels had shown no sign of abnormal braking.

Operator Little at BH tower said that he had cleared his signal for train No. 50 when it had been reported by DR tower at 11.29. The approach indicator had gone out at 11.35, indicating that train No. 50 had passed signal 3020 and had not cleared again. DR tower had reported the passing of train No. 40 at 11.38. He had heard the crash at 11.44, when the approach indicator for track No. 1 had gone out, the advance signal for track No. 4 had gone to danger and the telephone line west had failed.

Signals 3012 and 3020 were carefully observed after the accident to determine if there was any irregular operation. It was established that insufficient tension on a contact spring and somewhat glazed carbon contacts had caused signal 3020 to pump between caution and stop positions. It could not cause the signal to go to clear but would tend to move it from caution to stop. Records for the previous five years showed one failure of signal 3012 and three of signal 3020, all of which had been right-side failures, that is the signal had wrongly displayed "stop." There was nothing to indicate that any condition had existed on the night of the accident to cause a false clear or false caution signal indication.

It was, the ICC stated, impossible to determine why the signal indications had not been observed. The brakes of the leading (pilot) engine had not been applied after the flagman and rear markers of train No. 40 had come into view, or even after the burning fusee had been passed and the torpedoes exploded, while steam was still being used up to the last moment.

Engineer Gordon had been the subject of a post mortem, which had found moderate arteriosclerosis, and an attack of angina was a possibility, though this was not, according to the doctor, anything more than a theory. The engineer was 57 years of age and had passed a physical examination on 11th September 1925. No one had noticed anything wrong with him in the hours before the accident.

The fact that Engineer McConnell on the second (train) engine had not braked until the rear end of the forward train and the fusee had come into view and the torpedoes had exploded might be explained by his inability to see the signals due to smoke and steam from the engine in front, or his confidence in Engineer Gordon.

It was probable that the clear signal seen by Fireman W.S. McConnell at signal bridge 3020 was that for track No. 1.

From the evidence, train No. 50 was standing between seven and nine minutes before it was struck. Flagman MacDonald had had at least four minutes to go back and protect his train. The circumstances demanded that he exercise the utmost diligence. The train had stopped at a point where trains ran at high speed, on a curve with the view obscured by trees and a cutting, and at a point which was barely the emergency braking distance beyond signal 3012, while the flagman knew that train No. 40 was following close behind. The flagman maintained that he had gone back as far as he could, but the point marked by the burned out fusee was only 720 feet from the rear of his train and demonstrations after the accident indicated that he could have gone back considerably further in the time available.

The ICC concluded that the accident had resulted from the failure of Engineer Gordon to control his train as required by the signal indications and the warnings of the flagman. The commissioners accepted the theory that this had been due to the engineer's physical incapacitation or death. Had Engineer McConnell realised the danger a little earlier or Flagman Macdonald gone back a little further the accident might have been averted. The ICC made the usual comments regarding the need for an automatic train control or train stop system.

The commissioners noted that the actions of Flagman MacDonald were apparently in line with those commonly employed by flagmen, and that the provisions of Rule 99 were commonly not being observed or enforced. It noted that the responsible operating officers should give this matter immediate and thorough consideration. [437]

The next rear collision of any magnitude also happened in the United States, this time

[437] Interstate Commerce Commission report

on the Chicago and North Western Railway at Clybourn, Illinois, on the outskirts of Chicago. There were two tracks through Clybourn station, near the western end of the platform on which the accident occurred. A little further west was Wood Street tower, westward from which for several miles there were three tracks. Tracks 1 and 3 were reserved for westbound and eastbound movements respectively, track 2 was normally used by eastbound trains between 11 p.m. and 11 a.m. and westbound trains between 11 a.m. and 11 p.m. Eastbound movements on track 3 at Wood Street tower were controlled by home signal 34, and on the same post as this signal was a calling on signal, 38.

Although track 2 was normally used for most of the day by westbound trains, on the early evening of the day of the accident, Monday 6th September 1926, eastbound passenger train third No. 508 was given authority to use track 2 from Hunting Avenue to Wood Street. This train consisted of engine No. 1635, an express car, a mail car, a baggage car, a smoking car, four coaches, a parlor car and four coaches, in charge of Conductor Koepp, with Engineer Thompson and Fireman Goetke. This train had passed train No. 734 on track 3 at Maplewood, just over a mile from Clybourn, had passed Wood Street tower at 7.03 p.m. and stopped at Clybourn station at 7.04 p.m.

Train No. 734 consisted of engine No. 125, five coaches and a combination passenger and baggage car, in charge of Conductor Wolff, Engineer Louis J. Smith and Fireman Laidley. When it reached Wood Street tower signal 34 had been at danger but calling-on signal 38 had been cleared to caution before the train reached it and the train passed the tower, according to the towerman's record, at 7.05 p.m. Continuing at reduced speed it collided with the rear of train third No. 508 at a speed variously estimated as 5 and 20 miles an hour. Engine No. 125 telescoped the rear end of the last car of the first train and this car in turn telescoped the car in front. Five passengers lost their lives and about 200 were hurt.

Train No. 534 was crowded with passengers on their way back into Chicago after spending the day at the Wisconsin lakes. A newspaper wrote, 'The vacation train {No. 534} had left Elroy, Wis., this morning with three baggage cars, six day coaches and one parlor car. The rush of Labor Day traffic, with so many people returning to the city at the ends of their vacations, was so great that at Lake Geneva two more day coaches were added to the rear end. At Crystal Lake another coach was added. The last three were all constructed of wood. The coaches were filled to capacity, many of the passengers being compelled to stand. Children, being brought back to resume school tomorrow, were seated on parents' laps to make room for others, but still there was a crowd in the aisles. At Clybourn junction the train was halted, and many passengers were getting off when the crash came. A suburban train from Barrington, also crowded, but with evening theatre goers and others, was driven at top speed into the halted vacation train, and the two wooden coaches at the rear of the steel parlor car on the forward train collapsed as though made of crepe bark.' [438]

The reporter appears to have some of his facts wrong, for example, it was not a morning but an evening train from Elroy, while Lake Geneva station was not on the line from Elroy to Clybourn, all three coaches had been added at Crystal Lake, two of them off the branch train from Williams Bay, just beyond Lake Geneva. [439] The account

[438] *The Times*, Harrison, Arkansas, United States, 10th September 1926

[439] *Chicago Genealogical Society,* Vol 48, No. 3: Horror at Clybourn, 1926

nevertheless gives a good idea of the atmosphere of the accident.

At the Interstate Commerce Commission (ICC) inquiry into the accident Conductor Koepp stated that when his train had stopped at Clybourn station he and the head brakeman had helped to unload the passengers from the front of the train while the flagman opened the door of the four rear coaches. This was the normal practice, it not being customary to flag a train making a normal station stop. He considered that in the circumstances of the accident the responsibility lay with the engineer of the following train, who having received a calling-on signal should have been driving prepared to stop within his range of vision.

Flagman Elling testified that the rear markers at the back of his train had been burning brightly when it had stopped at Mayfair tower, 4¾ miles from Clybourn. He had seen train No. 734 pulling away from Maplewood station as his own train passed there and was thus aware that the other train was closely following them. Like his conductor, however, he had supposed that train No. 734 would be driven prepared to stop after passing the calling-on signal and he said that it was not customary to go back and flag a train making a routine station stop. When his train had stopped at Clybourn he had got off with red and white lanterns and assisted the passengers to alight from the rear cars. Perhaps half a minute later he had looked back and seeing the headlight of the following train approaching he had started back, swinging his lanterns as a stop signal. When he had covered two and a half car lengths he had seen that train No. 734 was not going to stop and had had to jump up onto the girder of a highway overbridge out of the way. He had had fusees and torpedoes with him but there had been no time to use them.

Engineer Smith told the inquiry that he had left Maplewood station slowly to give the other train time to get away but that even so, the signal at Wood Street had been at danger. However, when he had been a quarter mile or so short of it the calling-on signal had been cleared and he had allowed his train to drift forward at about 15 miles an hour. His vision had then been obscured by the headlight of a westbound train and he had reduced speed to about 6 miles an hour. Thinking that his train was about to stop dead he had then released the brakes. A few seconds later he had seen the train ahead and had applied the brakes in emergency, opened the sanders, reversed his engine, opened the throttle and the collision had occurred.

In trying to explain his failure Engineer Smith said that the calling-on signal having cleared, he had assumed that the train in front had left Clybourn station. He considered that he had operated his train strictly in accordance with Rule 501-G, which required him, on receiving a calling-on signal, to "Proceed at slow speed prepared to stop short of train or obstruction." His failure to stop he blamed on the brakes not having been recharged after previous applications and their long piston travel, together with the fact that he had been unaware that train third No. 508 was immediately in front. Usually, he said, when the calling-on signal was cleared there was no train ahead and he maintained that the towerman had used the calling-on signal for a purpose other than that for which it was intended. He persisted in refusing to admit that he had violated Rule 501-G, nor would he admit that his train had been travelling at more than 5 miles an hour at the moment of collision, despite the magnitude of the damage which had resulted.

Fireman Laidley of train No. 734, who was described as inexperienced, thought that the speed of his train had been reduced to 25 miles an hour when his engineer had applied the brakes at Wood Street tower. He said that the air brakes had been released and the train allowed to drift forward when the calling-on signal was cleared. About ten car lengths from the scene of the accident the headlight of a westbound train had come into view and his engineer had dimmed the headlight of his own engine, before shortly turning it on fully, when he, the fireman, could see the markers of the train in front and its flagman assisting passengers. As soon as the headlight of engine No. 125 was turned on fully the flagman had jumped to the track and started giving stop signals, while he had called to Engineer Smith. Train third No. 508 had then been about four car lengths away. His engineer had seen it at about the same moment and had applied the brakes, opened the sanders, and they had braced themselves for the collision. Fireman Laidley estimated the speed at collision at 15 to 20 miles an hour.

Conductor Wolff put the speed at 15 miles an hour at Wood Street and the same at the moment of collision, not having noticed any brake application. Ticket Collector Lowell on train No. 734 put the respective speeds at 25 and 20, while Flagman Stallman was unsure about the brake application and put the closing speed at 15 miles an hour.

Towerman McLain put the speed of train No. 734 at between 10 and 15 miles an hour as it had passed his tower and said that the engine had been working steam.

The ICC remarked that Engineer Smith's low estimate of the colliding speed of his train was not supported by any other witness except perhaps Towerman McLain, and from the resulting damage must have been considerably higher, at least 15 miles an hour. Had Engineer Smith complied with the intent and plain meaning of Rule 501-G and reduced his speed such that he could stop his train within his range of vision, the accident would not have occurred.

Towerman McLain was criticised by the commission for clearing the calling-on signal for train No. 734, a move of little value given that the station platform at Clybourn was not long enough to accommodate both trains at once. The train should have been held at signal 34 until train third No. 508 had cleared the interlocking at the east end of the station. [440]

The morning after the accident Engineer Smith was found in a local hospital, 'a quivering, sobbing human wreck, his mind deranged by the tragedy.' It would seem that he must have recovered himself considerably by the time of the ICC investigation. He was placed under arrest after the accident, with detectives assigned to guard his room at the hospital. [441]

At the inquest on the victims of the accident there was discussion of Engineer Smith's physical condition as a mitigating cause for his failure but after seven hours' deliberation the jury found on 17th September that this did not absolve him from gross and criminal negligence, and he was indicted on five charges of manslaughter. The other members of the train crews were all absolved from blame. The jury recommended that all employees operating trains should be subject to regular medical examinations, which Smith, apparently, had not been. [442]

It has not been possible to trace any report of Engineer Smith having been brought to trial.

---

[440] Interstate Commerce Commission report

[441] *Waterloo Evening Courier*, Waterloo, Iowa, United States, 7th September 1926

[442] *The Kokomo Daily Tribune*, Kokomo, Indiana, United States, 18th September 1926

A serious accident occurred in France on Thursday 23rd September 1926, involving a collision between two express trains on the Chemins de fer de Paris à Lyon et à la Méditerranée. Express train No. 112 had left Lyons at 21:30 on Wednesday night and was due in Paris at 06:05 on Thursday morning. Behind it was train No. 106, which had left Lyons at 22:50 and was due in Paris at 06:45. At Vulaines-sur-Seine, a small station near Melun, a sudden stoppage of the first train caused a coupling to fail and the train became divided, unknown to its driver, who continued with the front portion of the express towards Héricy. Some minutes later train No. 106, travelling at 60 kilometres an hour, collided with the detached coaches. The accident happened at 05:30.

A reporter who arrived at the scene at 10 wrote, 'Under the blue sky and blazing sun was a heart-breaking sight. There was a great mass of smashed wreckage, of steel, wood, and glass. Shorts and towels can be seen hanging in a sad tangle. The locomotive of the colliding train is derailed and leaning over slightly to the right. The front of the tender is jammed into the back of the engine, perched at 45 degrees like a monstrous shell which has fatally struck the moving train. A third-class carriage is suspended over a flattened van like an airliner. In front of the locomotive are the vehicles of the front train, half broken, a van, a first-class carriage, and a sleeping car. A crowd has gathered around the station, they have come on foot, on bicycles and by car and are standing looking and commenting on the accident. Fifty workers, mostly soldiers of the Fontainebleau 46th, are working with axes to clear the track. An enormous 50 tonne crane, brand new, has arrived from Paris, and is manoeuvring its big light grey arm, weighed down with chains.'

By 15:00 the bodies of five victims had been recovered. They were Antonin Badain, aged 52, courier agent, of Lyons; Georges Reynard, aged 19, courier agent, of Lyons; Mr Allan Blackie, an Englishman, aged 31, who was a representative of Lunn's travel agency in London and was returning from the south of France; Mr Eugene Montgomery, aged 24, at first also identified as an Englishman but actually an American from Kansas City, possibly a student, and M. Charles Vuidard, aged 35, the conductor of train No. 106.

Traction Engineer Germans stated that the initial investigation had placed the blame for the accident on Driver Georges Rodot of train No. 106 who he said had failed to obey the closed disc located a kilometre from Vulaines station. In addition, he had received a slow order at Champagne-sur-Seine. On the other hand the conductor of train No. 112 had not protected his train by putting down detonators 500 metres in its rear as per the regulations.

M. Fabion, the Vulaines stationmaster, stated that, 'When the accident happened the station was completely deserted. It is not usually opened in the mornings until just after 6, but I heard the tremendous noise and ran and sounded the alarm to rouse the local inhabitants. The rescue operation was organised at once. I requisitioned cars, advised Dr Van de Bossche at Héricy, contacted the local police and the firemen at Fontainebleau, Héricy and Samoreau. There was no panic, as there often is in such cases.'

Driver Rodot was interviewed in hospital. He said, 'I was not to blame. The signal was open, I am absolutely certain. We had received a slow order, because there was a communication breakdown between the two block posts. The order was received at Champagne-sur-Seine, but the disc at Vulaines was clear, so we did not stop. Moreover, our speed did not exceed 60 kilometres an hour. I saw nothing in front of me. The collision occurred before I had time to apply the brakes. It happened on a curve in a slight cutting, and there was a thick fog.'

Close to Driver Rodot in hospital was Fireman Jules Forêt, who suffered a broken wrist and multiple bruises. Speaking with difficulty, he said, 'I think that it was a miracle that I escaped death. The steam which was pouring out almost suffocated me. I was trapped alongside the firebox and the burning coal was literally roasting me. I was flooded with water. I was trapped three hours before I could be released. I didn't see anything either. I felt a terrible shock. There was a sharp pain in my wrist, and I felt that I was suffocating. That's all.'

The Minister of Public Works issued a statement. He announced that three coaches of train No. 112 had become detached at Vulaines-sur-Seine, 75 kilometres from Paris, due to the breakage of a coupling. The passengers from the detached coaches were transferred to the front of the train which was waiting at Héricy. The three carriages stood protected by the *signal avancé* at Vulaines, awaiting the arrival of an engine to remove them. In addition, it was arranged to caution all trains approaching Vulaines. However, some minutes later train No. 106 collided with the coaches, killing five people and injuring thirteen sufficiently severely for them to be taken to hospital at Fontainebleu. The initial investigation had been unable to determine whether the disc approaching Vulaines had been at danger or clear, but in any case the driver had been running too fast in view of the caution order.

The day after the accident Chief Conductor Edouard Varly and Conductor Duclos declared to a newspaper reporter that they had not needed to protect their train in the rear as it had been covered by the Héricy signals. The reporter then asked the opinion of a signalling engineer. The latter stated that first of all he would explain that the initial cause of the accident had been a failure during the night of the automatic signalling between Héricy and Champagne-sur-Seine. Trains were then run between the two places at time intervals of at least ten minutes and were cautioned. Preceding train No. 112 was a goods train. The express then followed but must stop at the *carré* at Héricy as the shunting of the goods train ahead had been delayed by the bad weather. When the *carré* cleared and the train restarted it became divided between the Marseilles and Brigues portions. The front portion of the train had been stopped by the automatic application of the brakes but subsequently went forward without the detached coaches. In the position in which these were left the rear conductor should normally have gone back to protect them as they were outside what would in Great Britain has been referred to as the Héricy 'station limits' but the situation was complicated here by the closeness of Vulaines and Héricy to one another. This meant that the Héricy *disque de protection* was 2 km, 600 metres outside of Héricy station, and 1,100 metres outside of Vulaines station. Even if Driver Rodot had received a clear signal, as he alleged, he had not been proceeding at caution. On the other hand, the head conductor should have ensured that the stationary coaches were protected by detonators unless he could be certain that the *disque de protection* was at danger. [443] Driver Rodot also pointed out that had the signal been at danger the "crocodile" automatic train protection ramp should have sounded the whistle of his engine and registered an entry on the speed recorder, which it had not.

By Thursday evening the number of dead had risen to nine with the death of several of

[443] *Le Petit Journal*, Paris, France, 24th September 1926

the injured. [444]

General Girod, the deputy for the Doubs, gave notice of his intention to interpellate the government on the accident and the circumstances surrounding it as soon as the Chamber of Deputies reassembled. He intended to ask the Minister of Public Works how it was that with the present state of science applied to railways a stationary train could not be protected and to ask what measures the government and the railway companies proposed to do about the matter. [445]

By the end of October M. Milon, the Melun juge d'instruction (examining magistrate) had charged not only Driver Rodot but Conductors Varly and Duclos with causing the accident. [446]

Eventually only Chief Conductor Varly and Driver Rodot stood trial. On 26th January 1927, the chief conductor was sentenced to a month in prison and a 50 francs fine and the driver to fifteen days in prison and an identical fine. Both prison sentences were suspended. [447]

Two eastbound Canadian National Railway freight trains were involved in a collision a mile west of Farlane, Ontario, Canada on Thursday 4th November 1926. At 3.46 a.m. that morning train No. 210 was overtaken by train No. 465. Train No. 210 was a mixed fruit and cattle train. Marshalled next to the caboose at its rear was a cattle car in which were riding eight stockmen. Following the collision, the caboose and the stockman's car caught fire and it was known early on that four of the stockmen, who had doubtless been asleep at the time of the collision, had burnt to death, while by the next day the bodies of five victims had been recovered, burned beyond recognition. They were taken to Sioux Lookout, where an inquest was to be held on Saturday. Conductor Carroll, presumably in charge of train No. 210, and R.E. Hand, described as a policeman, were slightly hurt but the other trainmen escaped injury. The initial report on the collision stated that doctors had been despatched to the scene from Sioux Landing and Reddit and the injured would be taken to Winnipeg, while in addition to the inquest the railway would launch their own investigation into the cause of the accident. The line was soon cleared, and traffic was not disrupted. Those killed were named as William Moffat, of Centrali, Ontario and L. Senovich, P. Andrews, Joe McCoun and John Kiss, addresses unknown. Stockman Andrew Hicks and two other stockmen whose names were not stated escaped. [448] [449]

Sunday 30th October 1927 saw an accident in Italy involving a special train carrying Fascists to Bari to celebrate the fifth anniversary of Benito Mussolini's March on Rome (*Marcia su Roma*) between 27th and 30th October 1922, the event which brought the dictator to power. The Fascist's train was involved in a collision with an ordinary train near Triggiano, between Bari and Locorotondo.

---

444 *La Charente*, Angoulême, France, 25th September 1926
445 *The Times*, London, England, 25th September 1926
446 *La Peuple*, Paris, France, 29th October 1926
447 *L' Œuvre*, Paris, France, 27th January 1927
448 *The Lethbridge Herald*, Lethbridge, Alberta, Canada, 4th and 5th November 1926
449 *The Brandon Daily Sun*, Brandon, Manitoba, Canada, 4th November 1926

A large number of provincial Fascists had crowded both the ordinary and special trains to reach Bari, some even travelling on freight wagons. Train no. 2 arrived late in Triggiano at 07:14 that morning, only to find that there were about five hundred Black Shirts ready to climb aboard. The train was already too long for the platform, extending beyond it on the Bari side. Noting the lack of seats for all the waiting passengers, the political secretary of the Triggiano Fascists pressed the stationmaster to strengthen the train, and so two F wagons parked in the adjacent goods yard were added to it. The shunting operation was carried out, partly by hand, with difficulty. The length of the train was now such as to also extend beyond the station on the Capurso side. From Capurso, at 07:41, the special LB train arrived, also late. According to the post-accident investigation, its driver found the signal protecting Triggiano station clear, so he did not comprehend that the station was not ready to receive his train. After passing a sharp curve in the cutting which led to the station and realising the imminent danger of colliding with the rear of the still stationary train no. 2, he applied the air brake and reversed his locomotive. However, a violent rear-end collision was inevitable, given the insufficient braking distance available (about 150 metres) and the slight falling gradient. The engine was derailed, while some the vehicles of both trains were almost destroyed, and others were damaged; fortunately, they were not yet all filled with passengers. Worthy of note was the heroic act of two Triggiano citizens, Michele Campobasso, and Vito Giannelli, who immediately set to work to rescued the injured from the wreckage despite the impending danger: the boiler of the derailed engine, no longer under the control of its driver, could have exploded at any moment. The actions of the driver were also important; having received some first aid he ran to the locomotive to release the steam, thus removing the danger. The final toll of the disaster was nine dead (all from Noci) and about two hundred injured, making it the most serious train accident in Apulian history. The documents of the time identify the main person responsible for the tragedy as the Triggiano stationmaster, who had failed to replace the protecting signal. He had also have contravened the company directive which imposed, for that day, a limit on the number of vehicles on each train by authorising the addition of the two F wagons to train no. 2. At the end of the investigation, he was arrested and jailed. [450]

When news of the accident reached Bari all the celebrations were cancelled. The Podesta (senior magistrate), having informed the crowd of the tragedy all the Black Shirts, knelt in silence for five minutes. The festivities in other towns in the province were also cancelled. [451]

On the same day, a lorry carrying Black Shirts to a procession in southern Italy overturned near Potenza, killing four. [452]

Sportsonderzug (special sports train) 52 841 from München Hbf, Germany, on Sunday 15th July 1928, was heavily laden, and so was given a second locomotive. Shortly after starting from track 6 it came to a stand on the Donnersberger Bridge, it was initially assumed because one of the passengers had negligently pulled the communication cord, but it later transpired that the stop had its origin in the driver forgetting his timetable. In

[450] Webpage – Tutto Treno blog: FSE Triggiano Ricorda il Disastro Ferroviaria del 1927 con una Rarga (in Italian)
[451] *The Times*, London, England, 31st October 1927
[452] Ibid, 1st November 1927

order to avoid more delay the driver had instructed his fireman to carry out the brake test for him while he went back to get a copy of the timetable. The fireman had done so but being inexperienced in conducting the test he had forgotten to afterwards move the brake handle to the release position, hence the undesired brake application. A fairly minor matter in itself but it led to a major disaster due to a misunderstanding between two signalmen. The man in München No. 1 box thought that it was time for the special train to have cleared the next box at Donnersberger and he had an ordinary train ready to depart on track 8 and express D 49 timetabled to leave at 21:55. He therefore telephoned his colleague at Donnersberger and there was a misunderstanding, whereby although his instrument's red disc showed that the special train was still in the section the Donnersberger signalman cleared it back and accepted the ordinary train. As the special train began to move off again it was rammed by the regular one. Some coaches were telescoped, an oil lighting reservoir was damaged, and a fire broke out. Ten people died and 25 were injured.

Later there was considerable criticism of the rescue operation, as two hours after the accident some of the injured were still trapped in the wreckage. Medical evidence, however, refuted the suggestion that any of the passengers had been burnt alive. In a sense the fire prevented a more serious accident, as the driver of train D 49, seeing the flames on the line ahead, was able to stop short of the wreckage. [453]

Seven railwaymen subsequently stood trial for causing the accident but perhaps surprisingly were acquitted. It was held by the court that the installation at München Hbf was inherently unsafe as it lacked a centralised regulated train notification procedure and sufficient operating regulations.

An accident occurred in Paris, France, at 19:03 on Friday 3rd October 1930, at Pont Cardinet station, just outside of the Saint Lazare terminus of the Chemins de fer de l'État. A passenger train from Dieppe had deposited its passengers at the terminus and the empty stock was bound for the depot at Clichy. It was being propelled from Saint Lazare and was stopped at a signal 50 metres short of Pont Cardinet station when it was run into by passenger train No. 1163, which had left Saint Lazare at 18:59 for Mantes, running non-stop to Argenteuil. The latter train was made up of mostly steel rolling stock and capable of fast running; it was in charge of Conductor Maugan with Driver Coullaut and Fireman Columbu. The leading van of the Mantes train reared up in the air and the following second-class coach was telescoped into it. Many of the passengers got out of the train after the collision on the left-hand side and into the path of an incoming empty electric passenger train from Clichy, bound for Saint Lazare where it was to form a train at Houilles. Some of them were unable to get out of its path and were run down. Two passengers had been killed in the initial collision and two more were killed by the electric train.

The collision happened right in front of poste (signal box) No. 7 whose signalman immediately raised the alarm. A rescue operation was immediately begun, and firemen recovered the bodies of the four dead. Thirty injured were transported to Beaujon hospital. They included Driver Coullaut, who was badly hurt. Thirteen others with slight injuries were treated at the scene and then allowed to go home. On news of the accident

---

[453] *Von Siegelsdorf nach Aitrang,* by H. J. Ritzau, pgs. 42-3 (in German)

being announced at Saint Lazare a woman suffered a stroke and collapsed.

The Mantes train was particularly crowded as the accident happened just after the closure of the shops and offices. Panic had set in after the collision and possibly a hundred passengers threw open the doors of the carriages and got down onto the track despite warnings from the signalmen, other railwaymen, and their calmer fellow passengers. Other passengers were preparing to join those who had left the train while some of those on the track were running back towards the terminus. At that moment, the electric train had arrived, bound for one of the left-hand platforms of Saint Lazare and travelling at some speed. When he had seen what was happening the motorman of the electric train had braked hard but by then it was too late. Warned of the oncoming train some of the passengers on the track saved themselves by getting in close to the side of the stopped train. Some were struck by the train but escaped with slight injuries.

It soon appeared that the collision had been caused by Driver Coullaut overrunning signals. The signal d'avertissement, just outside of Saint Lazare station, was shown to have been green, indicating that the next signal, 150 metres short of Pont Cardinet station, was at danger, showing two lights, one red and one green. A third signal, at the station entrance, was also at danger and the empty train which was run into was stopped in front of it. It was not clear why the driver, who was left in a state of coma by the accident, had run past the signal. At the initial inquiry, the fireman stated that he had been busy with the engine fire, while the conductor, who had been riding in the front carriage and was somehow thrown from it without being much hurt, had noticed nothing wrong.

M. Dauty, director of the State railway, stated that the wrecked coach of the Mantes train had been a wooden one, and promised that all the rolling stock of the suburban trains would be of steel construction by the end of the year. The signals had been working normally and he could not explain how the accident had happened. [454] [455]

---

[454] *Le Matin*, Paris, France, 4th October 1930

[455] *L'Ouest Éclair*, Caen, France, 5th October 1930

CHAPTER 5

SIDELONG, HEAD-ON, DOUBLE AND FOULING COLLISIONS

Trains can come into collision with one another in a variety of ways. Rear collisions, that is ones involving one train running into the back of another, trains meeting one another head-on on a single track and trains colliding on rail/rail level crossings are detailed in other chapters. This chapter will deal with other types of collision such as those involving trains running into one another at junctions, meeting head-on on multiple tracks and colliding with vehicles left in sidings but which end up foul of running lines.

An addition to the entries in the 1881-1900 volume of this series, information on which has only just come to light, is a collision on Wednesday 11th November 1896 at a place named as Sokolk in the government of Grodno, in Russian Poland (now Sokólka in Poland). The St Petersburg – Warsaw express, travelling at high speed, came into collision with as goods train and the engine and all the carriages of the express were wrecked. Six people were killed and 13 were injured, 'some of whom will die.' [456]

On the evening of Tuesday 1st February 1916 Chemins de fer du Nord express train No. 502, which had left Boulogne at 13:27 and was due in Paris at 18:50, was running three-quarters of an hour late after having been delayed taking on extra coaches at Amiens, and did not arrive at Saint-Denis, France, until 19:25. Some railwaymen shunting the yard there had left a waggon foul of the main line just as the express was approaching, running at a considerable speed to try and make up time. The express engine collided with the stray waggon, was derailed, and seven following third-class carriages piled up behind it. A heavy restaurant car reportedly saved the rear vehicles of the express from being involved in the pile-up. In a repetition of the St Bedes Junction accident in England the previous December, escaping gas from damaged cylinders set the wreckage alight, though the fire was apparently fairly quickly brought under control by

[456] *St Johns Evening Telegram*, St Johns, Newfoundland, 19th November 1896

local firemen. It was initially reported that there were five dead and 15 injured. The bodies were taken to a local mortuary and the victims, some of whom were clearly in a serious condition, were conveyed to hospital.

A passenger from Boulogne told his story to a reporter. 'Our train, as you know, was divided at Amiens and some third-class carriages were added. This took too long, and we left there thirty or thirty-five minutes late. Suddenly, as everyone was getting their luggage ready because we were shortly to arrive in Paris, I felt the train stopping suddenly. I was thrown backwards. I realized that the driver must have reversed his engine. Bur he was not able to stop in time. A second later a terrific shock went through the train. There were six of us in our compartment, we were thrown one upon the other, and I was stunned for a moment. A little later, I found myself down on the track. I don't know how I got out of the train. Around me people were moaning, others were running to and fro. The carriages at the front of the train were on fire. The passengers from the rear of the train ran forward to try and help those in the burning coaches. The rescue work was quickly organised. We were taken to a nearby factory and given refreshments.'

At 22:00, the rescue work was still in progress and the number of known dead had risen to ten. Driver Quirinez and Fireman Gustave Logerot escaped unhurt. The newspaper report mentioned the great bravery of a Zouave (a soldier of Algerian origin) named Trancart who despite the flames managed to rescue five passengers from the wreckage. [457] [458]

On 3rd February the Nord Company announced that 13 passengers and the conductor of the train had lost their lives and there were about 20 persons injured.

The morning after the accident the ponderous *Pacific* class engine of the express and its tender were described as lying across the track in a V shape, with the wreckage of the third-class carriages which had been added to the train at Amiens piled up behind them. They had been telescoped and the fire had ravaged them. The number of known dead had increased from the ten previously reported, with the finding of the burnt bodies of two more men and two more women in the wreckage. One of the bodies was unrecognisable.

Interviewed after the accident, the crew of the express engine stated that all the signals had been clear for them when they had suddenly collided with a waggon while running at speed. For a couple of seconds nothing had happened after this, then their engine began to overturn to the right, at which point they had jumped off and saved themselves. The leading carriages of the train had caught fire at once and they could hear the cries of the trapped passengers.

Exactly how the accident happened was not clear, but it seemed likely that a new and inexperienced railwayman had omitted to couple up the last of a rake of waggons which was being shunted across the main line, so that this waggon was left behind foul of the line on which the express was approaching, unnoticed by anyone. [459]

A report from Petrograd stated that two express trains had collided near Vitebsk, Russia, on the Windau-Rybinsk railway, on about Saturday 5th February 1916. Sixteen

---

[457] *Le Figaro*, Paris, France, 2nd February 1916

[458] *The Times*, London, England, 3rd February 1916

[459] *Le Figaro*, Paris, France, 3rd February 1916

people died and 12 were injured. [460]

Less than a month after the Saint Denis disaster, on Tuesday 29th February 1916, a similar accident happened on the Chemins de fer de Paris à Lyon et à la Méditerranée (P.L.M.) at Maisons-Alfort, France. Omnibus train No. 28, which had left Montgeron at 18:20 and was due in Paris at 19:14, was crossing the Iles bridge, 500 metres from Maisons-Alfort, when its driver felt a shock, the fireman was thrown against the side of the tender and four carriages were thrown against the piers of the bridge, broken up and ended up lying across the track.

After the accident rescue work was begun by station staff who were soon joined by firemen. The rescue work was made more difficult by a strong smell of gas escaping from damaged cylinders, which made it impossible to use the lanterns brought by the rescuers. There was in the end no fire, but to prevent one breaking out the rescue operations had to be carried out in darkness. Electric torches were by this date coming into use, but none were apparently available to the rescuers.

Most of the victims of the accident had been riding in the first three carriages, which were smashed to pieces. After a while the escaping gas had dissipated, and the danger of a fire having passed lanterns could be used and the work of rescue was easier. The injured were rushed to medical establishments in Maisons-Alfort and the bodies of those killed were taken to the station outbuildings.

The accident was initially blamed on a problem with the track related to recent work carried out on it but it was soon realised that the damage to the track was a result of the derailment and not the cause. It then became evident that just as at Saint Denis the train had collided with a stray waggon, in this case one belonging to a freight train from Brunoy. Shunter Fernand Mougeot, aged 35, was held responsible for the waggon having ended up where it did and he was arrested.

Apart from the seven passengers who had been killed outright about 30 were injured. At the time of the initial newspaper report on the accident several of them were said to be critically ill and one to be dying. [461]

On the 3rd of March it was reported that M. Bonneleau, described as a P.L.M. employee, but possibly riding on the train as a passenger, had died in the Saint Antoine hospital, making the eighth fatality in the accident. [462]

A collision occurred at Pedaso station, Italy, on Saturday 4th March 1916. Troop train No. 7536 arrived at the station just before 22:30, bound for Foggia with soldiers returning from leave. It consisted of an engine, old third-class carriages and goods waggons. Approaching from the opposite direction was Ancona-Castellammare Adriactico accelerated train No. 1643, consisting of an engine and first-, second- and third-class carriages. The accelerated train was due to arrive at 22:21 and was slightly late, but the stationmaster decided to order the troop train to shunt to make room for train No. 6506 on the main track. The troop train was duly taken forward towards the home signal on the

[460] *L'Homme Libre*, Paris, France, 8th February 1916
[461] *Le Petit Parisien*, Paris, France, 1st March 1916
[462] *Le Matin*, Paris, France, 3rd March 1916

Ancona side of the station so as to clear the points before setting back into the siding. The shunt move was made under signal protection but the accelerated train arrived at high speed and failing to stop at the signal collided head-on with the shunting train. The old vehicles of the troop train suffered the most. The rescue work was carried out in semi-darkness, partly by uninjured passengers from the Ancona train. Later relief trains arrived from Ancona and Casrellammare.

Driver Incecco, Firemen Pietrangeli and Burattini, Conductor Gueraldo, an unnamed maintenance foreman and six soldiers lost their lives and Drivers Donaggio and Trasatti, Fireman Marcucci and 27 soldiers were injured, six of them seriously. It was not stated which railwaymen belonged to which train.

The stationmaster had contravened the regulations by ordering the shunting of the troop train in the path of an approaching train and of course the driver of the Ancona train was at fault for overrunning signals. [463]

A hospital train carrying sick and wounded soldiers from the Eastern Front was in collision with a collision at speed with a train of tank cars loaded with petrol between Medzilaborce and Łupków, Hungary (the former place now in Slovakia and the latter in Poland) on Tuesday 26th September 1916. A fire broke out following the collision and burnt for two days. Officially 140 people died but unofficial estimates put the number of dead at twice that number. [464]

Forty Carranzista soldiers (that is those belonging to the forces of Mexican revolutionary leader Venustiano Carranza) were reportedly killed and about 80 injured when their train was in collision with a pilot train and telescoped on Sunday 5th November 1916. The accident occurred at Chirimoya, near Chelaya, Mexico. Two days earlier a pilot train attempting to reopen traffic to Tampico had been blown up (presumably having struck a mine) and a following troop train had been shot at by bandits and several soldiers killed. The injured were taken to San Luis Petosi. [465]

Mourners for the late Emperor Franz Joseph I of Austria were among those involved in a collision on the Magyar Államvasutak at Herceghalom, Hungary, on Friday 1st December 1916, involving an express coming from Wien, Austria, carrying a large number of people returning from the funeral, and a train from Budapest. To cope with the heavy traffic the express, Schnellzug No. 3, was run in two sections and supplemented by two special trains. In accordance with special instructions the first part of the express was held back at Komorn station so that the second section could precede it. At Herceghalom station, 46 kilometres from Budapest, what was now the first portion of Schnellzug No. 3 had to meet Budapest – Raab – Graz passenger train No. 1308, which like the express was delayed. Through an error this train was diverted from track 2 to track 3, across the path of the express. Schnellzug No. 3 passed the Vorsignal (distant signal) at 76 kilometres an hour and despite the caution indication of the latter its driver failed to

[463] Wikipedia – Incidente ferroviario di Pedaso (in Italian)
[464] Wikipedia – List of Rail Accidents (1910-1919)
[465] *The Ogden Examiner*, Ogden, Utah, United States, 10th November 1916

reduce speed. He did react to the home signal, but at that point only 500 metres remained to the fouling point and it was not enough. Seven coaches of the passenger train were derailed and the second and third carriages of the express were telescoped into one another. Sixty-eight passengers, mostly soldiers, lost their lives and 200 were injured.[466] [467] It was suggested that the inferior wartime oil used in the Vorsignal was the reason for the driver of Schnellzug No. 3 failing to see it[468] though this was really no excuse as had he failed to see it he should have treated it as a caution signal in accordance with the usual regulations.

Kiltimagh station, Ireland, on the Sligo branch of the Great Southern and Western Railway, was the location of a serious accident on Tuesday 19th December 1916. At 3.59 p.m. on the afternoon of the day, Signal Porter Patrick Horan at Kiltimagh accepted an up ballast train from Claremorris and received the 'Train Entering Section' signal for it at the same time. It was 10 miles from Claremorris to Kiltimagh and he therefore waited 20 minutes before asking 'Line Clear' for the train from the next station at Swinford and obtaining the electric train staff for the Kiltimagh – Swinford single-line section. A couple of minutes later the Claremorris signalman advised him that the ballast train had in fact not started as it had failed to surmount the bank out of Claremorris. He therefore cancelled the ballast train to Swinford, replaced his up signals and accepted a down special empty waggon train from Swinford (8 miles on the other side of Kiltimagh) instead. It was too foggy for him to see his up distant signal. The next thing that he knew, the ballast train had run past his up home signal and was passing his cabin, its driver whistling for the guard's brakes and its fireman working on the hand brake. Meanwhile the down empty waggon train was moving slowly up to the down home signal. Then he heard the noise of the two trains colliding. Twelve men had been riding in the van behind the engine of the ballast train; five of them were killed and the other seven injured when the van was telescoped by the engine tender.

Driver Michael Hanrahan, in charge of the ballast train, told the Board of Trade inquiry into the accident that the weight of his train had caused it to stall on the climb out of Claremorris, although he had often driven trains equally heavy before (the train consisted of 21 loaded waggons and two vans, weighing 300 tons). He had had to set back into Claremorris and have a second attempt, which had succeeded. Approaching Kiltimagh it had been very foggy. He had only been able to see the Kiltimagh up distant at short range but it had been fully off and he had assumed that he had a clear road on to Swinford. He had not seen the home signal at danger until he was about 10 yards from it. He had then applied the brakes, but the rails had been slippery from frost and fog, and his train had still been travelling at 30 miles an hour when it struck the other train at 4.46 p.m..

Fireman Michael Fleming of the ballast train needless to say also maintained that the Kiltimagh up distant had been fully off. He admitted that he had made no mention of the distant having been off after the accident.

Conductor Richard Roche of the ballast train, on the other hand, stated that his

---

[466] *The Age*, Melbourne, Victoria, Australia, 4th December 1916
[467] *Eisenbahnunfälle Bind 1* by Ludwig Stockert, pgs. 46-47 (in German)
[468] Wikipedia – List of Rail Accidents (1910-1919)

impression had been firm and strong that the distant signal had been at danger. He did not think that the train had been going any faster than usual approaching Kiltimagh.

Neither Brakesman Michael Kiely, riding in the rear of the front van of the ballast train with the conductor, nor Brakesman Richard Hennessy, riding in the rear van, had seen the distant signal.

The inspecting officer, Lt-Col E. Druitt, commented that the responsibility for the accident depended upon whether or not the up distant signal had been at danger. There could be no doubt, from the evidence of the signal porter and of Stationmaster James Forde, who had been in the signal box, that the levers in the signal box had been in their proper position, and the distant signal could not have been off unless it had stuck off. Also against the possibility of a clear distant signal was the evidence of Conductor Roche that the distant had been at danger, supported by that of Porter M. Maloney, who had seen it at danger on his way back to the station at 4.25 p.m.

The evidence, the inspecting officer stated, was in his opinion strongly in favour of the distant signal having been at danger, and it was clear that Driver Hanrahan had either missed seeing it or had assumed that he would have a clear run through the station, and had thus approached the home signal at too high a speed, so that he had overshot it and collided with the empty waggon train 478 yards beyond it. The driver had a good record and conditions had been adverse to his making a good stop; the line fell steeply at this point, partly at 1 in 65, the rails had doubtless been very slippery, he had a heavy train and the home signal had been obscured by the fog, but he had not worked his train with sufficient care given the conditions.

In a final remark the inspecting officer noted that it had not been thought necessary to light the signal lamps at the time of the collision, although it was half an hour after sunset. Driver Hanrahan had not complained about the signals being hard to see because they were unlit though. [469]

The *Times* newspaper in London reported a collision near Hamburg, Germany, on the Bergedorf – Geesthacht on the evening of Saturday 30th December 1916. Seven people were reportedly killed and 100 were injured in a collision between a passenger train carrying 400 persons, mostly workmen, and an empty train. No account of this collision seems to have found its way into any of the various German books on railway accidents. [470]

An unsafe method of working, which had been in use for many years, was the basic cause of an accident at Queensferry Junction, Scotland, on the Edinburgh – Glasgow main line of the North British Railway, in the early evening of Wednesday 3rd January 1917.

Queensferry Junction was eight miles of Edinburgh, and lay between Gogar and Ratho stations. It was the junction for a single-line branch to Dalmeny and South Queensferry, which at the Queensferry Junction end became a double-track loop with connections to

---

469 Board of Trade report

470 *The Times*, London, England, 2nd January 1917

the up (to Edinburgh) and down (to Glasgow) main lines. On the Edinburgh (east) side of the junction there was a trailing crossing between the up and down main lines. Some up passenger trains from Dalmeny terminated at Ratho Lower station near the junction on the branch and the engine of the train then had to run round its coaches at Queensferry Junction before standing on the down loop line there until it was time to work the next train from Ratho Lower to Dalmeny. The proper procedure would have been for the branch train to have run into the up loop, the engine to have been detached, run out onto the up main line, back over the mains crossover, run through the down loop and onto the rear end of its train, then to have propelled the whole train via the same route to the down loop. In order to avoid this slightly cumbersome method of working, which involved two runaround movements, the alternative method of working which had been used for many years was for the up branch train to be brought directly into the down loop, so that its engine could be detached and run around its train via the main line in one move. The problem was that there were no fixed signals for a wrong direction move off the down loop onto the main line, so hand signals had to be relied upon. On the day of the accident the engine of the 4.19 p.m. train from Dalmenty to Ratho Lower moved forward off the down loop prematurely and reaching the down main line was struck head-on by the 4.18 p.m. Edinburgh to Glasgow express, heavily laden with business people, women and children, and soldiers and sailors, going home after the Hogmanay holiday. The collision was a violent one; the leading composite carriage was telescoped by the tender of the express engine and practically destroyed, the second carriage was badly damaged and the remainder slightly so. All the vehicles except the rear North Eastern Railway third-class coach were electrically lit and there was no fire, but 12 passengers lost their lives and 44 passengers and four railwaymen were injured. [471]

Two passengers described their experiences. ‘We were about a quarter of an hour out of Edinburgh, and we were just getting steam up, travelling roughly at probably at 30 miles an hour. It all happened in a moment-just a sharp, violent shock, as if the brakes had been put on in a violent hurry. Bags and bundles tumbled down from the racks onto the heads of the passengers, and we were all thrown into each other’s arms. The lights went out, the windows broke, and the doors were jammed. There was much alarm, and a good deal of screaming by women and children, who were standing in the corridors and between the seats of the crowded compartments. The soldiers and sailors behaved splendidly. They squeezed through the broken windows and ran up the line to give all the help they could.

‘There was a call for doctors from amongst the passengers, and I saw two immediately report themselves. They had in fact begun work before the call came. Then ambulances began to arrive from Edinburgh. I saw four bodies. One of them seemed to be that of a soldier. The light engine and the engine of our own train, which bore the name of *Dunedin*, were in a twisted tangle on the line. We walked along the permanent way to Dalmeny Station, and sat there for an hour waiting for a train. What happened was that the back part of the express came to carry us back to Edinburgh, and I got a seat in the compartment where I had been before. I heard the guard of our train, whose head was bandaged, ask the driver of the stationary engine what his engine was doing there. The reply was that he knew nothing about it until he and his stoker were thrown on their backs.’

[471] Board of Trade report

Another passenger told a reporter, 'The train was cram full. I was in the first compartment. It was supposed to leave Edinburgh empty, reserved for passengers from Haymarket, a suburban station about a mile from the starting point, but we managed to get into the guard's van in that coach. We had been travelling about a quarter of an hour, when the collision occurred. Our chief concern was lest our carriage should turn over. Luckily that peril was just avoided. When we came to a bumpy standstill we got out. At first an extraordinary silence prevailed. Then one began to hear the terrified screams of women and children. The permanent way was all cut up, and the wires were twisted. The fine thing was the work of the soldiers among the passengers. They worked splendidly under the command of one or two officers who happened to be in the train. So perfectly did they set about their task, indeed, that the civilians were content to stand aside, feeling that if they interfered they would hinder rather than help. The soldiers got hatchets, and for the best part of an hour they were at work rescuing passengers from the first two coaches, who were crushed together between the telescoped seats, with the tender of the engine on the roof of their coaches. The soldiers took cushions out of the carriages, and on them carried the injured persons to the ambulance cars.' [472]

Queensferry Junction signalman John Philp told the Board of Trade inquiry into the accident that he had accepted the 4.19 p.m. Dalmeny train from Kirkliston at 4.28 under the 'Warning Arrangement' and he had hand signalled it into the down loop at 4.35. He had taken the Kirkliston-Queensferry Junction tablet from, he thought, the fireman, and had intended to hold the train on the down loop until the Edinburgh express had passed, when he would be able to let the engine of the branch train out onto the down main line to run around its train. He had lowered the down home and starting signals for the express at 4.31 and the down distant signal at 4.32 once the signalman in Ratho box had lowered his distant. Almost as soon as the branch train had come to a stand its engine had been detached and without any tip from the box it had begun to come forward off the down loop onto the down main line. Seeing what was happening he had thrown back his down main signals and rushing to the window had shown the driver of the engine a red light and whistled, though with the strong wind blowing he doubted if the driver would have heard his whistle. Gogar box, 2½ miles away, had sent the express on at 4.33 and the collision had occurred at 4.36.

Signalman Philp said that the same movement had been made twice a day for the last 22 years and he had never previously known an engine come forward off the down loop without being authorised to do so by the signalman. He was absolutely positive he had given no signal to the driver of the engine on the day of the accident. As far as he could remember he had been instructed to run branch trains around in this manner by Signalman Melton when he had taken over the box. He was sure that the Ratho stationmaster, his immediate superior, was aware of the method of working. He knew that such a method of working was wrong in principle, but as it had been going on so long it had never struck him that fixed signals were necessary. Engines were signalled off the down loop at night by the display of a white light or if the engine was close enough to the box by word of mouth.

Driver John Ramsay told the inquiry that he had been on the engine of the branch train, with Fireman James Cairns. He said that the service between Dalmeny and Ratho Lower had only begun on Monday 1st and previous to that he had never worked a train

---

[472] *The Times*, London, England, 4th January 1917

between Dalmeny and Queensferry Junction. On Monday 1st his fireman had given up the tablet at the box and they had moved forward onto the main line without further instructions. On Tuesday 2nd, the day before the accident, after Fireman Cairns had handed the signalman the tablet they had been shown a red light and kept waiting for ten minutes before being allowed to go forward. On the day of the accident they had, as before, been shown a white light to enter the down loop. After his fireman had given up the tablet they had received no further instructions from the signalman, nor been shown a red light, so he had assumed that he was justified in going forward. Fireman Cairns had seen the express coming just as their engine got to the main line.

Signalman Philp recalled said that he had given Driver Ramsay a white light to come forward on the Monday, possibly the driver had not seen it. He had not thought it necessary to show the driver a red light on the day of the accident, assuming that he would not move without further instructions. [473]

Stationmaster George Orr at Ratho said that he had found Signalman Philp sober, cool and collected after the accident and added that he was a good signalman. He (the stationmaster) admitted that the use of hand signals 'was not good working nowadays.'

District Inspector David Urquhart said that the method of working at Queensferry Junction had been in use for 25 years and he had concluded that it was right, considering it a shunting operation within station limits, carried out in full view of the signalman and under his instructions. [474]

Colonel J.W. Pringle investigated the accident on behalf of the Board of Trade. He stated that Driver Moffat of the express had been too badly injured to give evidence but according to Fireman Hyslop his driver had shouted and applied the brakes immediately before the collision. Given that the signals had all been clear, the curvature of the line and the sudden appearance of the engine it was clear that the crew of the express were not at fault. The immediate cause of the accident was clearly a misunderstanding between Signalman Philp and Driver Ramsay and the inspecting officer considered them both to be at fault. Given that there were no trap points to protect the main line and that he had already accepted the express the signalman clearly should have displayed a red light to stop the engine, and his not having done so after he had done so the previous night was illogical. As for the driver, ordinary caution should have demanded that he get a distinct instruction to move forward onto the main line after the tablet had been given up. The origin of the accident was, however, the long-standing and irregular method of working, which the operating superintendent, when he became aware of it after the accident, stated was indefensible and ordered to cease.

Colonel Pringle recommended that the layout at Queensferry Junction should be remodelled to avoid the need for engines running around their trains to foul the main lines, or alternatively fixed signals and perhaps trap points should be provided. He was also critical of senior signalmen being appointed as district inspectors in their own areas, so that they were not likely to condemn dangerous practices which they had long adopted when they were signalmen. Finally he drew attention to the general danger arising from light engines running around their trains. [475]

[473] Board of Trade report
[474] *The Times*, London, England, 5th January 1917
[475] Board of Trade report

A rather unusual accident occurred at 11.20 on Wednesday 17th January 1917 at Massy-Palaiseau station, France, on the Grande Ceinture line on the outskirts of Paris. Train No. 11255, coming from Le Havre, and carrying slightly over 1,600 English *permissionaires* (soldiers on leave) (105 officers and 1,500 men) had stopped at the station. Unfortunately as it approached the station and while it was ascending a slight incline the train became divided, between the 23rd and 24th carriages at a point known as La Butte de Versailles. Going up the incline the rear carriages had lagged behind the main train but then the gradient changed to a falling one and the separated carriages gathered speed and collided violently with the main portion of the train. Two carriages, one of them belonging to the Belgian State Railway, were wrecked. Fourteen soldiers were killed and about 40 injured.

After the accident the station staff and the uninjured soldiers took care of the victims and the injured were taken first to the station waiting room and thence to Versailles Hospital, some by road ambulance and others on an ambulance train which was specially stopped to pick them up. The bodies of the dead, two non-commissioned officers, two corporals and ten privates, were also taken to Versailles Hospital. By 18:00 the line had been cleared and the re-formed train continued its journey to its destination. [476]

Although a passenger train was involved in the accident it was apparently not fitted with automatic brakes. Probably due to the exigencies of wartime it was made up, wholly or partly, of non-passenger rolling stock, or alternatively the brake fittings of vehicles, such as the Belgian State Railway one mentioned in the newspaper report, which were incompatible with those of French vehicles.

At 02:00 on the morning of Saturday 27th January 1917 an express from Aurillac to Paris on the Chemin de fer de Paris à Orléans had passed through Saint-Amand-Montrond and was approaching Châteauneuf-sur-Cher, France. Its passengers, who were largely soldiers on leave returning to the front, were either sound asleep or dozing. As the express reached Châteauneuf there was suddenly a terrific shock as it came into collision with a shunting goods train. The locomotive of the express climbed onto the waggons of the goods train and the three leading coaches of the express were smashed against the tender of the engine.

Eleven passengers – seven soldiers and four women civilians – died in the accident and 64 passengers, 48 of them soldiers, were injured. M. Drifort, the local juge d'instruction, opened an inquiry to determine the cause of the accident. Driver Joly of the express, insisted that the signals had been clear for him; other witnesses stated that the signals had been at danger to protect the shunting goods train and that the express had overrun them. [477]

Two passenger trains collided at Finkenrath, Germany, on Saturday 3rd February 1917. Seven people died and 25 were injured. [478]

---

[476] *Le Temps*, Paris, France, 19th January 1917
[477] *Le Petit Parisien*, Paris, France, 28th January 1917
[478] *Eisenbahn-Katastrophen in Deutschland* by Hans Joachim Ritzau, pg. 79 (in German)

A train carrying sick and wounded Russian prisoners who had been repatriated via Sweden was involved in an accident at Holmsveden, Sweden, on Monday 26th February 1917. Train 3048 was en route from Trelleborg and Haperanda. It was not meant to stop at Holmsveden, but to run through to Bollnäs, where the passengers would be given supper, and was travelling at about 40 kilometres an hour when owing to a wrongly set pair of facing points it was diverted at about 8 p.m. onto a siding, collided with the buffers, and continuing, though a coal bunker struck a pump house. The four leading carriages, containing in total 60 or 65 invalided passengers, were telescoped into one another, the first two completely and the other two partly, and were shattered. It was the practice for mentally ill patients to be placed in the leading carriages of the train, and there were 23 of these patients in the most badly damaged front carriage. By 2 a.m. the next morning the bodies of five dead had been removed from the wreckage and it was clear that at least the same number remained buried. Ten were said to be seriously injured and a number slightly hurt. The driver escaped unhurt and the fireman was only slightly hurt, as was a Swedish nurse accompanying the train, named Elisabeth Trägårdh. [479]

An Exchange Telegram from Stockholm stated that, 'The work of rescue was made difficult by the condition of the passengers, most of whom were blind; had lost arms or legs or were otherwise rendered incapable of helping themselves. Their hardships were increased by the fact that the wreck occurred in darkness at a distance from any important town.' [480]

After the accident relief trains were sent from Bollnäs and Gävle. Those injured in the accident were taken to Söderhamn and Gävle. Later in the evening a train carrying food was sent from Bollnäs and the passengers had their delayed supper. There were 228 persons on the train, including two officers and five Swedish nurses. By the following morning the number of known dead had risen to ten, with 14 seriously injured and 40 less badly hurt.

The accident was blamed on negligence by a porter, J.G. Lundkvist. who without checking the lie of the points and despite not being able to clear the home signal had given the train driver a clear signal with his handlamp. When Driver L. Brodén and Fireman J. Westberg saw that the points were wrongly set, the driver had immediately applied the vacuum brakes, but the siding was only 100 metres long and disaster was unavoidable. The engine ended up in a field and was not much damaged. Behind the four telescoped carriages six or seven others were more or less damaged; the rear half of the train escaped injury.

After the points to the siding had been used the regulations stated that they should be locked and the key taken to the station buildings and placed in a control apparatus, thereby unlocking the semaphore signal protecting the entrance to the station from the south. The siding involved in the accident was little used. When the Russian prisoner of war train was expected and the porter found that the semaphore signal would not work, instead of going down to the siding, when he would have immediately discovered that the points were wrongly set, he had contented himself with showing a hand signal to the

479 *Dagens Nyheter*, Stockholm, Sweden, 27th February 1917

480 *The Brandon Daily Sun*, Brandon, Manitoba, Canada, 27th February 1917

driver, and had thereby overridden the safety equipment [481] (qv Hall Road, England, 1905).

An immediate investigation into the accident was opened which laid the blame squarely on Porter Lundkvist, which seems perhaps a little unfair. It emerged that the acting stationmaster was Station Foreman August Rosén. The latter had himself checked the entrance and exit points to the station but not those to the siding. He had then been asked by the porter if all was clear for the prisoner of war train to pass through and had assured the porter that it was. Porter Lundkvist explained that at 6 p.m. some waggons had been taken from the siding by train No. 1651. Normally after the work at the siding was finished he would have locked the points himself but he had had to hurry away to another job and thought that a colleague had locked the points. When the prisoner of war train was due he had checked with the station foreman just in case. Then when the signal would not clear he had thought that it was because there was a fault with it; it had never occurred to him that the points might still be open to the siding. [482]

One of the injured passengers died later, bringing the final death toll to 11.

Bayerisch Staatsbahnen mixed train No. 926 had been set back across the opposite running line into a siding to attach a horse-box at Nannhofen, Germany, on Wednesday 18th April 1917 (date also given as 19th April) and was returning to its own line when it was run into by an Augsburg – München express (train No. 53). Four vehicles of the mixed train were destroyed and another was overturned while in the express the engine, tender and a mail van were overturned and a brake van was telescoped into the following passenger carriage. Twenty-four people were killed outright, nine were fatally injured and another 34 were hurt, some seriously. Driver Obermeier of the express stated after the accident that he had seen two white lights, and predictably was backed up by his fireman. It was clear, however, that he had been mistaken and that the Einfahrsignal (home signal) had been at danger. He had some excuse for his mistake as blizzard-like weather conditions had prevailed at the time. The fahrdienstleiter (signalman) who had carried out the shunting manoeuvre with the express close at hand in conditions of poor visibility must also be held partly to blame for the accident, though he had not actually infringed any rules. [483]

Driver Obermeier stood trial for causing the accident 3½ years later but was acquitted, it being pointed out that the contemporary of a white all-clear signal made things very difficult for drivers, added to which there were many cases of signal lights being extinguished, spectacle glasses broken, signal arms not working properly and so on. [484]

On 5th August 1917 two passenger trains met in a head-on collision in Milwaukee, Wisconsin, United States, on the Chicago Milwaukee and St Paul Railway. At the location of the accident there were usually four tracks in use, the two northerly, Nos. 1 and 2, being used for westbound and eastbound passenger trains, and the two southerly

---

[481] *Dagens Nyheter*, Stockholm, Sweden, 27th February 1917
[482] Ibid, 28th February 1917
[483] *Eisenbahnünfalle Bind I* by Ludwig Stockert, pg. 52 (in German)
[484] *Historiche Eisenbahn Katastrophen* by Bernhard Püschel, pgs. 109-10 (in German)

tracks for freight traffic. Because of work on the elevation of the tracks the automatic signalling was out of use, and since 11th June track No. 1 had also been taken out of use, east and westbound passenger trains both being worked over track No. 2. Trains over the single line were controlled by a towerman in Clinton Street tower, 3,250 feet east of Milwaukee station, assisted by two switchtenders.

On the day of the accident westbound passenger train No. 3 from Chicago to Milwaukee passed Washington Street tower, about 5,300 feet east of Milwaukee station, at 11.19 p.m. and collided with train No. 58 900 feet east of Clinton Street tower at 11.22 p.m., having almost come to a stand. Train No. 58 had left Milwaukee at 11.15 p.m. for Chicago and was running at about 18 miles an hour. The front ends of both locomotives were considerably damaged and John Baumgartner, the engineer of train No. 3 and five trespassers riding on the front end of the leading car of train No. 58 were killed. [485] Four of the trespassers were named as Onero Cantori, Henry Hanson and William McAulay, of Chicago; and Frank Miles, of Mamaroneck, New York, who was dressed in a soldier's uniform. [486] The latter had, according to the train crew, joined the train at Sparta, Wisconsin, and was on the way to his mother's funeral in New York. [487]

On duty that night at Clinton Street tower was Towerman Demonski. He told Interstate Commerce Commission investigators that he had telephoned Milwaukee station and been told that train No. 58 would leave on time. He had then rung the dispatcher about incoming trains Nos. 3 and 57, to be told that train No. 3 was a little late and train No. 57 was 40 minutes late. Learning this, he had telephoned the switchtender at Fowler Street and instructed him to let train No. 58 proceed and train No. 3 would be held for it. Next he had spoken to Switchtender Urich, who was sat outside the Clinton Street tower. "No. 58 is right on time, and No. 3 is a little late. No. 57 is about 40 minutes late, so you go down there and hold No. 3 for No. 58."

Urich had replied, "All right."

When the switchtender did not set off after a couple of minutes the instructions had been repeated to him. He had then left, but without his red lantern, so the towerman had called him back and said, "Now you understand you are to hold No. 3 for No. 58."

Urich had replied, "Yes, I understand, I am to hold No. 3 for No. 58."

The switchtender had then set off for the crossover two blocks away. Towerman Demonski was positive that he had told the switchman to hold train No. 3 and said that he had believed that the switchtender had thoroughly understood his instructions.

Switchtender Urich stated that he had been doing his job for two months. According to him the towerman had told him to, "Let No. 3 come and hold No. 57." He had left the tower at 11.11 p.m. and gone to the crossover with a red lantern to stop train No. 57. When train No. 3 had appeared he had thrown the switch and signalled for it to proceed. He said that when he had first set off and covered a block he had had his doubts as to his instructions and had returned to the tower but he was positive that he had been told to hold No. 57 and let No. 3 by. Soon after train No. 3 had passed him he had seen the headlights of train No. 58 but imagined it was the silk train on track No. 4, then the

[485] Interstate Commerce Commission report.
[486] *The Winnipeg Evening Tribune*, Winnipeg, Manitoba, 6th August 1917
[487] *Daily Ardmorite*, Ardmore, Oklahoma, United States, 6th August 1917

thought had come to him that it was working steam too fast for a freight train and he had shouted to the engineer of train No. 3.

Engineer Foley stated that he had passed Clinton Street tower at about 12 miles an hour and had seen train No. 3 but thought that it was stopped waiting for him to pass until he got closer. He had then shut off steam and applied his brakes.

Fireman Prohl said that his train had stopped at Allis station and the switchtender had then signalled for them to proceed. They had been three car lengths past the crossover when he had seen train No. 58 coming. His engineer had applied the brakes and he had just had time to jump.

The Interstate Commerce Commission concluded that although there was a conflict of evidence between Towerman Demonski and Switchtender Urich the towerman's account was more likely the correct one in view of the fact that the switchtender had admitted returning to the tower to check his instructions. Switchtender Urich was therefore in their view responsible for the accident. [488]

On Monday 13th August 1917 an up troop train and a down goods train collided head-on at Yemaras station, India, on the Great Indian Peninsula Railway. Five British soldiers of the Black Watch and Seaforth Highlanders were killed and five British officers, two Indian officers, 30 British soldiers and a Sepoy were injured. A fireman and three brakesmen on the troop train were also hurt but the railwaymen on the goods train escaped. There was considerable damage to rolling stock. The accident was caused by the driver of the troop train overrunning the outer and home signals. He was subsequently prosecuted and sentenced to six months' rigorous imprisonment plus a fine of 300 rupees or in default a further three months' rigorous imprisonment. [489] [490]

A mistake by a pointsman resulted in an express train being diverted into a siding at an unstated location midway between Petrograd (previously and currently known as Saint Petersburg, renamed Petrograd 1914, Leningrad 1924, reverting to Saint Petersburg 1991). and Moscow, Russia, on or about 13th August 1917. There it collided with a goods train and four crowded carriages were shattered, 60 people being killed and 150 injured. [491]

A shunting engine collided with a passenger train at Großneukirch, Germany (now Polska Cerekiew in Poland) on the Cosel – Bauerwitz branch, on Monday 17th September 1917. Six people died and 97 were injured. [492]

On about 25th October 1917 there was another collision in Russia at an unstated location between Moscow and Petrograd. This accident involved a Sevastopol express and a military train carrying demobilised territorial troops. Sixteen persons died and 80

---

488 Interstate Commerce Commission report.

489 Railways in India: Administration report for the year 1917-18

490 *The Times*, London, England, c15th August 1917

491 *The Daily Herald*, Adelaide, Australia, 16th August 1917

492 *Eisenbahn-Katastrophen in Deutschland* by Hans Joachim Ritzau, pg. 129 (in German)

were injured. [493]

A third disaster occurred in Russia on 2nd November 1917 when an express and a military train collided head-on on the Vladikavkaz railway in the Caucasus. This time there were reportedly 25 deaths and 70 seriously injured. Most of the victims were soldiers. [494] [495]

A Kowel – Demblin express on the Heeresbahn Nord was not normally booked to stop at the crossing place at Sporniak, Galicia (now in Poland), but on Friday 21st December 1917 it was arranged for it to cross an opposing goods train there. The Einfahrsignal (home signal) was at "Halt" against the express but it was not observed by the driver and the express continued through the crossing place at undiminished speed to collide with the penultimate waggon of the goods train, which at that moment was entering the loop from the opposite end. The engine, leading van and three carriages of the express and the last three vehicles of the goods train were wrecked and overturned. Seventeen people died, 56 were badly and 27 slightly hurt. [496]

Kaiserliche-königliche-österreichische Staasbahnen express trains Nos. 9 and 10 had to meet one another at Trzciana station, Galicia, on Monday 14th January 1918. There were four lines through the station, but two of them, including one of the main lines, were occupied by goods trains. A belated alteration in the arrangement for crossing the two trains by the fahrdienstleiter on the arrival of train No. 10 led to the accident. He left the Einfahrsignal (home signal) for train No. 10 at clear but belatedly replaced the corresponding signal for train No. 9 from "Frei in die Ablenkung" (clear for diverging route) to "Halt." The Vorsignal (distant signal) for this train was already at "Caution." Train No. 9 overshot the Einfahrsignal by 384 metres. Its driver then set back to bring his train to the rear of the signal that he had just overrun, train No. 10 ran through the station without being checked and collided with train No. 9. Fourteen passengers and two railwaymen lost their lives and 76 passengers and 12 railwaymen were more or less hurt. [497]

An accident reminiscent of that at Queensferry Junction, England, in 1917 inasmuch as it had its origin in a practice which, although it had not been officially authorised, had been in existence for some years, occurred on the Bengal-Nagpur Railway, between Benapur and Khargpur stations, India, on Thursday 17th January 1918. There was a steep incline just beyond Khargpur station, and goods trains were banked up the hill with an engine attached to their rear. For reasons which were not explained the practice had grown up of attaching a brake van to the rear of the banking engine. On the day of the accident the brake van was attached as usual and the train was pushed up the incline.

---

[493] The Times, London, England, c27th October 1917

[494] Ibid, c5th November 1917

[495] *The Railway Gazette*, Vol 27, pg. 522

[496] *Eisenbahnünfalle Bind I*, by Ludwig Stockert, pg. 48 (in German)

[497] Ibid, pgs. 17-18

When the end of the climb was reached the driver of the banking engine shut off steam and stopped pushing. The report on the accident does not say so but it was presumably nighttime as unbeknown to those on the banking engine the brake van had somehow become detached. The momentum of the van carried it forward for some distance and then it came gradually to a stand and was left behind in the section. A passenger train subsequently ran into the brake van and 16 people were killed and 33 injured. The official report on the accident noted that, 'the staff at fault have been dealt with departmentally.' [498]

At 5:10 a.m. on Friday 19th January 1918 leave train No. 3009 bound for Riga collided with passenger train No. 102 bound for Insterburg (now Chernyakhovsk in Russia) between Argeningken (now Nowokolchosnoje) and Pamletten, Germany, in what was possibly a head-on, single-line collision. At least 25 people died and 50 were injured. The line was blocked for 24 hours. [499]

A leave train coming from Paris and a goods train were in collision at Raphèle station, France, on the Chemins de fer de Paris à Lyon et à la Méditerranée on Thursday 16th May 1918. Seven carriages and six goods waggons were derailed; assistance was urgently requested from Arles, Lyons and Miramas and a 50 ton breakdown crane was sent to the scene. The casualties were initially reported as 8 dead and 29 injured, but it was feared that other bodies might be found in the wreckage. [500] [501]

Romania suffered a very serious accident on 29th June 1918 when an express train collided with a goods train between Mircesti and Roman. Forty-five people died and many more were injured. [502]

Two trains carrying soldiers on leave collided near Uggowitz, Austria, on the Villach-Pemtaffl line, on Wednesday 14th August 1918. The crossing of trains Nos. 917 and 918 had been altered from Tarvis to Uggowitz. All appeared ready for the arrival of train No. 917 at Uggowitz, where train No. 918 was standing waiting for it. The Einfahrsignal was cleared for the arriving train but the facing points were out of correspondence with the signal, and the arriving train was diverted into a head-on collision with the waiting one. The second and third carriages of train No. 917, both of older, lighter construction, were telescoped into one another. According to different accounts 11 or 20 soldiers, including some officers, were killed, and 6 seriously and 'many' slightly or 80 in total were injured. [503] [504]

---

498 Railways in India: Administration report for the year 1917-18
499 *Historiche Eisenbahn Katastrophen* by Bernhard Püschel, pg. 115 (in German)
500 *Le Populaire*, Paris, France, 18th May 1918
501 *Le Petit Parisien*, Paris, France, 17th May 1918
502 *The Times*, London, England, 1st July 1918
503 *Railway Gazette*, Vol 29, pg. 246.
504 *Eisenbahnünfalle Bind 1*, by Ludwig Stockert, pgs. 18-19 (in German)

A collision near Birdsell, Nebraska, United States, on 10th September 1918 involved a westbound passenger train and a work train. It happened on the Chicago Burlington and Quincy Railroad line between Seneca and Alliance, which was single track and worked by timetable and train orders, supplemented by a manual block system. Passenger train No. 43 was in charge of Conductor Content with Engineer Charles Fitzpatrick, and consisted of locomotive No. 2905, a milk car, a mail car, two baggage cars, a smoker, two chair cars, a dining car and three Pullman sleeper cars. The train left Hoffland, 5 miles east of the point of collision, at 2.48 p.m., 15 minutes late, and had reduced speed from 45 to 34 miles an hour when it collided with work train No. 714 at 2.55 p.m. The passenger train consisted of a mixture of steel and wooden vehicles, with all-wooden smoking and baggage cars marshalled between much heavier cars with steel underframes, and the smoking and baggage cars were not equipped with anti-telescoping devices. As a result the smoker was telescoped by the baggage car for 29 feet and ten of the 11 fatalities in the accident occurred within it. Two other passengers and two railroad employees were injured.

The work train consisted of engine No. 714, 12 ballast cars and a caboose and was hauling soil between a cutting just east of Birdsell and Alliance. The caboose and engine were attached to the east end of the train. The train was occupying the line under the authority of order No. 28, which read:

Engines 714 and 3142 work extra from 7.30 a.m. until 7.30 p.m. between Alliance and Hoffland protecting against each other and protecting against second-class trains. Extra 366 west meet extra 714 at Birdsell. All extras west except extra 366 west wait at Hoffland until 10.30 a.m.

A second order, No. 64, was issued in the afternoon, which read:

All extras west except 5290 west wait at Hoffland until 5.01 p.m. Extra 109 wait at Hoffland until 6.01 p.m.

These orders did not give extra 714 any rights against passenger train No. 43, and it should have been in clear of the main track at 2.33, the time train No. 43 was due to leave Hoffland, as it was expected to clear the time of all first-class trains. Instead it was still on the main line loading dirt, 1,000 feet east of the east switch at Birdsell, a closed train order station, at 2.55 p.m. Hence the collision.

The Interstate Commerce Commission (ICC) investigation into the accident showed that it had been the result of multiple failures by different parties. To begin with the conductor of the work train had failed, as he apparently frequently neglected to do, to check his time card at Alliance when he had checked the train register there, which showed the times at which trains had passed. He had thus overlooked the fact that train No. 43 had not gone by. The engineer of the work train, on the other hand, had not forgotten the passenger train, but here another factor came into play. By chance Alliance was the boundary line between Central and Mountain time, with Central Time, east of Alliance, being an hour later than Mountain Time. His watch had two hour hands so that it could show both times but on the day of the accident he had become confused and had forgotten that he was working east of Alliance. Going by Mountain Time rather than Central Time he had believed that it was only 1.55 p.m. when the accident occurred and that he thus still had over half an hour to be clear of train No. 43 by 2.33.

One of the brakemen on the work train was well aware of train No. 43 and that the work train was occupying the main track on its time. He did not know on what authority it was doing so but said nothing. His lack of regard for the safety of his train was, the investigators stated, inexcusable.

Fireman Peters and the other brakeman of the work train were completely inexperienced. Neither of them had been supplied with a time card or knew where to get one and they were thus unaware that their train was occupying the main track in the time of the passenger train. Although the fireman had been working for three weeks he had not been supplied with a watch either. The ICC concluded that this indicated a laxity in the supervision and training of employees which should be corrected.

Conductor Content of train No. 43 said that he considered that as his train had left Hoffland under a so-called permissive card, the wording of which told the conductor and engineer that the block was not clear and to expect to find a train anywhere in the section between Hoffland and Alliance, that his train should have proceeded with caution, especially approaching stations and when the view was obstructed. However, his engineer had not reduced speed at any point until he had braked in emergency immediately before the collision.

Brakeman Ehrhardt of train No. 43 did not think that its speed had exceeded 45 miles an hour and agreed that it had not been reduced until just before the collision. His understanding was that engineers did not run cautiously when under a permissive card but relied upon the other train having flag protection.

Flagman Dodd of train No. 43 put its speed at 30 miles an hour. He considered that trains running under a permissive card should approach places where the view was obstructed able to stop within the range of vision but said that his engineer had not done so.

Engineer Fitzgerald of the passenger train stated that he had left Hoffland some minutes late and had been running at 45 miles an hour near the scene of the accident, but because of the gradient speed had dropped to 35 miles an hour when he saw the steam shovel of the work train. He thought that he could have stopped within his range of vision had his fireman been looking out but at the moment of the collision the latter was pulling down coal. He did not think that having a permissive card meant that he should reduce speed where the view was obscured, but that it was to tell them that there was a train in the block and that he should look out for flag or other protection. At one time they did slow down at points where the view was obscured but this had meant that they had lost time.

The commission concluded that the accident was due to the work train occupying the main track in the time of train No. 43 without flag protection, for which its conductor was held responsible. At the same time it noted that the manual block system as operated on the division gave little or no protection against collisions between trains running in the same direction, as despite having received a permissive card Engineer Fitzgerald had been running his train at top speed and had not instructed his fireman to keep a sharp lookout, but had relied entirely on there being flag protection, and in doing so had apparently been following the usual custom. There was obviously an urgent need for this practice to change. [505] [506]

---

[505] Interstate Commerce Commission report.

[506] *The Omaha Daily Bee*, Omaha, Nebraska, 11th September 1918

With the First World War drawing to a conclusion a collision at Sames-Guiche station, France, on the Chemins de fer du Midi was only very briefly reported in the French press. At 20:00 on 1st October 1918 a train carrying refugees collided with a passenger train in the station on the Toulouse-Bayonne line. According to one report eight people died and an unstated number were injured. [507] Other reports stated that one train was carrying repatriated soldiers rather than refugees and that there were only three dead and seven injured. [508] [509]

A passenger train and a troop train collided at Junkenrath, Germany, between Köln and Trier, on Wednesday 9th October 1918 due to a mistake by a fahrdienstleiter (signalman). There were 16 deaths and 28 cases of injury. [510] [511]

At 5.15 a.m. on the morning of Saturday 12th October 1918 there was a collision at Celrá, Spain, between a special passenger train carrying immigrants from Port Bou, close to the French border, and Barcelona. The immigrants were probably workers who had been employed in French munitions factories during the war. When the special train reached Celrá the signals were against it as goods train No. 276 was busy shunting there, but a violent thunderstorm was raging, with heavy rain, and it seems likely that the poor visibility caused Driver Llámase José Delader of the passenger train to overrun the signals. The collision occurred 200 metres beyond the protecting signal. The engine and waggons of the goods train were derailed and damaged while the leading van and three coaches of the passenger train were reportedly completely destroyed, despite the fact that both drivers saw what was going to happen and did their best to prevent it. Seven passengers died and 19 were seriously and 42 slightly hurt. Driver Delader escaped with slight injuries while Driver Pedro Soler of the goods train was badly hurt. A special train, accompanied by Senior Doctor Res, took most of the injured to hospital in Gerona and another special train arrived in Barcelona at 2 p.m. with 453 uninjured passengers and five who were only slightly hurt. The line was blocked for a considerable time. [512]

On Friday 18th October 1918 a collision occurred on the Preußisch Staatsbahnen at Uerdingen, Germany. A hospital train which was entering the station came into sidelong collision with a goods train which was leaving it due to a set of points being wrongly set. There was much material damage and seven soldiers and a conductor died and seven soldiers were seriously and 21 slightly injured. [513] (Another source gives 11 fatalities.)

---

[507] *Le Figaro*, Paris, France, 2nd October 1918
[508] *Le Matin*, Paris, France, 2nd October 1918
[509] *Le Temps*, Paris, France, 3rd October 1918
[510] *Eisenbahn-Katastrophen in Deutschland* by Hans Joachim Ritzau, pg. 79 (in German)
[511] Katastrophen der deutschen Bahnen Teil II: Chronik 1845-1992 by H.J. Ritzau, pg. 44 (in German)
[512] *La Vanguardia*, Barcelona, Spain, 13th October 1918
[513] *Eisenbahnünfalle Bind 1*, by Ludwig Stockert, pg. 49 (in German)

A troop train collided with a shunting movement at Malmedy, Germany (became part of Belgium in early March 1925) on Wednesday 13th November 1918. Eight died and 29 were hurt. [514]

Two trains were involved in what was possibly a rear collision on the Gossensaβ side of Sterzing station, Italy, on Monday 18th November 1918 (date also given as 11th November). Two carriages of one train were wrecked and 21 Italian soldiers were killed and 47 others injured according to one report. [515] Another report stated that there were 18 dead and 30 injured. The soldiers were very likely repatriated prisoners of war. The accident occurred at about 23:00. [516]

On 10th December 1918 two trains collided at Herzele, near Audenarde, in Belgium. Eighteen people died and many were injured. Among the victims were five British and one French soldier. [517] [518]

On Saturday 13th December 1918 a double-headed military train of 40 vehicles with 600 passengers became divided at Syrau, Germany, between Plauen and Hof. The detached portion initially remained stationary but then began to run forward and collided with the front portion of the train, which had come to a halt. Seven vehicles in the middle of the train were wrecked. Eight people died and at least 16 were hurt. After the accident the undamaged forward portion of the train continued to Plauen, the rear vehicles were drawn back to Mehltheuer and the line was clear at 10 p.m. [519] [520]

A train carrying demobilised American troops collided with two German locomotives in Montieramey station, near Troyes, France, on Sunday night, 2nd February 1919. Seven died (some sources say eight) and 30 were injured. [521] The train was en route from Chaumont to Brest. It was alleged that the driver of the colliding train had disregarded the signals protecting the two engines. [522]

Laughlin Junction, Pennsylvania, United States, on the Baltimore and Ohio Railroad just outside of Pittsburgh, was the scene of a sidelong collision on Saturday 22nd February 1919. Here the double track of the River Subdivision was joined by the double-track branch from New Castle Junction by a "Y" junction, the western leg of the "Y" joining the Pittsburgh line 800 feet west of Laughlin Junction tower. The eastbound track of the

---

514 *Eisenbahn-Katastrophen in Deutschland* by Hans Joachim Ritzau, pg. 79 (in German)
515 *Tiroler Volksbote*, Innsbruck, Austria, 20th November 1918
516 Webpage – Pietri Grande Guerra – BOMBARDAMENTI-EXPLOSIONI e INCIDENTI ferroviari: le vittime
517 *The Times*, London, England, 14th December 1918
518 *La Presse*, Paris, France, 13th December 1918
519 *Eisenbahn-Katastrophen in Deutschland* by Hans Joachim Ritzau, pg. 119 (in German)
520 *Eisenbahnünfalle in Sachsen* by Claus Schlegel, pg. 121 (in German)
521 *L'Humanité*, Paris, France, 3rd February 1919
522 *La Dépêche de Brest*, Brest, France, 3rd February 1919

New Castle branch crossed the westbound track of the River subdivision (Pittsburgh) line by a movable crossing.

The events which led to the accident began at about 3.45 p.m. when engine No. 5032 was derailed on the crossing, badly damaging it. It was 7.05 p.m. before the engine could be re-railed, after which two movable points were installed and spiked normal, that is for westward movements over the River subdivision. At 8.15 Track Supervisor Donahue reported the line safe to run over at 10 miles an hour and a slow order was issued at 8.19 p.m. As a little work needed to be done on the damaged signalling the work was then turned over to Signal Supervisor W.D. Carroll. Westbound passenger train then passed through safely at 8.34 and engine No. 5026 at 8.41. Once this engine had gone, Signal Maintainer Christian ordered the spikes to be removed so that the work could be completed. This was done and shortly eastbound passenger train No. 166, known as the Versailles local, from Pittsburgh to Versailles, Pennsylvania, which had left Pittsburgh at 8.51, approached the crossing, as did engine No. 2034 on the westbound track, en route to work a train from Pittsburgh station and running backwards at about 5 miles an hour. The points of the movable crossing having been unspiked, the engine was diverted onto the eastbound line and into the side of the passenger train. Neither train was derailed but the engine severely damaged the sides of several coaches of the express and eight passengers and a railroad employee were killed and 81 passengers and six employees were injured. [523] The passengers sitting on the left-hand side of the coaches of the train were the ones affected while those sat opposite escaped. The accident occurred at a point where the railroad was elevated above the streets, which made the rescue work more difficult. [524] [525]

Engineer Joseph Donahue of engine No. 2034 testified at the Interstate Commerce Commission (ICC) inquiry into the accident that he had been given a slow order at Wheeling Junction which stated that he was not to exceed 10 miles an hour over the west leg at Laughlin Junction. At that point the order had been repeated to him by someone he thought was Train Master Grow; his fireman had also repeated the message to him. He had then reduced speed to 5 miles an hour and had seen semi-automatic signal 36 protecting the crossing go from stop to caution and then at once to clear. He had seen no stop signals when his fireman suddenly called out to him, he had applied the brakes, and the collision had immediately occurred.

Fireman Johnson confirmed repeating the slow order to his engineer at Laughlin Junction and seeing signal No. 36 clear. When within a few feet of the crossing he had seen someone giving stop signals and heard them call out, "That will do." He had shouted to Engineer Donahue who had at once applied the air brakes. He had seen no lights at the crossing other than that of the signal.

Engineer Berg of train No. 166 said that he had reduced speed upon seeing the Laughlin Junction distant signal at caution. Approaching the crossing at 8 or 10 miles an hour both signal arms there had been at danger, but they had then changed first to caution and then to clear. He had worked steam and got the speed up to 20 or 25 miles an hour when suddenly there was a crash and his fireman had shouted to him to stop. He had applied the brakes in emergency as there was a second crash.

---

[523] Interstate Commerce Commission report

[524] *Warren Morning Chronicle*, Warren, Pennsylvania, United States, 23rd and 24th February 1919

[525] *The Titusville Herald*, Titusville, Pennsylvania, United States, 24th February 1919

Fireman Kalfas of train No. 166 said he was sitting on his seat when he saw the tender of the engine strike the cars of his train and had shouted to his engineer, who had applied the brakes.

Train Master Grow stated that he had been advised at 8.17 p.m. that the westbound track was ready to return to service with a 10 mile an hour speed restriction but that trains could not run to or from the "Y" on the westbound track. He had notified the dispatcher to issue slow orders and had also called to the crew of each train as it passed. He had duly shouted to the crew of engine No. 2034, which was coming at 8 or 10 miles an hour, and it had reduced speed further. He had then gone into the tower and the first he knew of the accident was when the operator had remarked, "What's happened now?" He knew that the interlocking was out of order but not that the spikes would need to be withdrawn to complete the repairs nor that signal No. 36 could be cleared with the movable points out of place.

Signal Supervisor Carroll stated that extensive damage had been caused to the points by the earlier derailment and no repairs could be made until the engine had been rerailed. A slow order had been issued just after 8 p.m. and he had asked the operator to stop all westbound trains at the signal before clearing it in order to protect the men working on the track. He did not recall issuing any instructions about removing the spikes from the movable points to complete the repairs and had not intended to make any heavy repairs until all the important passenger trains had passed. No definite instructions as to when the spikes should be removed had, though, been issued. His account of what had happened when engine No. 2034 had approached was confused but it appeared that he had not realised that there was anything wrong until his men had shouted for the engine to stop, when he had waved a lantern at it two or three times until the lantern had gone out.

Signal Maintainer Christian said that Supervisor Carroll had told him to apply the connecting rods and when ready to connect one, no trains being about, he had ordered Repairman Foose to draw the spikes, having, he thought, opened the circuit to hold signal No. 36 in the stop position. According to him it was Supervisor Carroll who had first seen that there was an engine coming when he (Christian) had called out to stop it. He saw Repairman Spangler flagging the engine.

Repairman H.T. Foose said that Supervisor Carroll had been present when the spikes were drawn 8 or 10 minutes before the accident. He saw Repairman Spangler flagging the engine, but not Supervisor Carroll. There had been no time to restore the points to normal.

Repairman P.M. Spangler thought that the spikes had been drawn only a couple of minutes before the engine appeared.

Operator Jenks at Laughlin Junction tower stated that he had been aware that the moveable points were out of order, though he had not been actually advised that they were. He denied having been instructed to slow trains using the signal and had stopped one engine previously only because it was already in the section when the slow order had been issued. He was questioned about Rule No. 625, which stated:

When signals or switches are undergoing repairs, signals will not be displayed for any movements which may be affected by such repairs until it has been ascertained from the repairmen that the switches are properly set for such movements.

His answer was that he did not know what the repairmen were doing and as he had not been informed that the track was out of order he considered that he had complied with the rule.

The view of the ICC investigators was that Signal Supervisor Carroll was directly responsible for the accident. He had been in full charge of the work, knew what was to be done, and should have known beyond question that when the spikes were to be drawn, everyone was properly and definitely instructed as to what to do and that signal No. 36 had first been disconnected so that it could not be cleared from the tower. Responsible to a lesser degree was Signal Maintainer Christiansan who should not have ordered the spikes withdrawn until he knew that proper protection was afforded. [526]

Passenger train No. 417 from Paris to Beauvais was leaving Persan-Beaumont station, France, on the Chemins de fer du Nord on Sunday 25th May 1919 when it was run into from the side by a shunting engine, which smashed two compartments of a third-class carriage. At the time of an early report on the accident the bodies of four victims had been removed from the train, a fifth person had died in Beaumont hospital and ten other passengers were injured, some seriously. The accident occurred at 11.18.

The passengers who were killed were named as Mme Yvonne de Bréville, of Ansnières; M. Antoine Niolanne, his wife Mme Francine Niolanne and their five-year-old son Alfred, of Paris; and René Parigny, of Méru. One of the injured passengers may also have died later.

According to one newspaper Fireman Blanchard of engine No. 651 admitted responsibility for the accident. As the passenger train was setting off he had moved his engine forward to take water. For some reason that he could not explain he had opened the regulator too far and his engine on track No. 1 had shot forward and gone 50 metres past the two water pipes and into the path of the passenger train on track No. 3. [527] Another newspaper stated that there were two versions of how the accident had happened. One was that Driver Carron had left his engine and told his fireman to take it into a siding. The fireman had begun to move the engine and when he realised that it was on a collision course with the passenger train he had lost his head and jumped off. The other version, which resembled that given in the first newspaper, was that Fireman Blanchard had got down to oil the engine and had been just climbing aboard again when his driver had said, “Attends-moi là, je vais faire de l’eau.” (Hang on, I must get some water.) Then the driver had for some reason moved his engine too far forward.

The line was blocked after the accident, causing much disruption to traffic. Things were back to normal at 7 p.m. [528]

A mail train from Delhi collided with a goods train at Kinana station, near Rohtak, in the Punjab, India, on 29th June 1919. The accident, on the North Western State Railway, resulted in the death of 41 passengers and probably two railwaymen as well. The brief official report on the accident added that the driver of the mail train had been prosecuted

[526] Interstate Commerce Commission report
[527] *Le Matin*, Paris, France, 26th May 1919
[528] *Le Petit Journal*, Paris, France, 26th May 1919

in connection with the accident but had been acquitted. [529] [530]

Chemins de fer de Paris à Lyon et à la Méditerranée train No. 103 bis from Paris arrived at Tarascon station, France at 07:10 on Friday 26th September 1919, 2 hours late, and stopped on track No. 1. Sixteen minutes later train No. 1514 bis from Bordeaux, which was itself 36 minutes late, arrived. The signals were against it, but its driver disregarded them and made no effort to stop his train until he saw the Paris train in his path. He then applied the brakes in emergency but was unable to avoid a violent collision, possibly a sidelong collision, which split the Paris train into two. A second-class carriage was broken up, a first-class overturned and a third-class derailed. All three vehicles were crowded with passengers.

Those who lost their lives were: Derridji ben Ahmed, of the 3rd Tirailleurs at Bône; Mme Durand, of Marseilles; Mme Gautier, of Comps; Mlle Catherine Frizoni, of Nice; M. Jules Male, of Prunet; Sergeant Emmanuel Monis of the 21st Infantry at Montpelier and an unidentified woman aged about 30 with a ticket to Nice. Twenty-six other passengers were injured.

Driver Courment of the Bordeaux train was detained pending an inquiry into the accident. [531]

At 5.07 a.m. on Friday 24th October 1919, passenger train P 1002 collided with the engine of goods train G 8901 which was standing foul of the main line at Kranowitz, Germany, between Troppau and Ratibor, and fire broke following the collision. A barrel of 96% alcohol being smuggled in the luggage van of the passenger train exploded, and passengers who were carrying more alcohol concealed in their clothing were transformed into human torches. The majority of the passengers were said to have been smugglers, who were carrying spirits in specially shaped tins or in imitation loaves of bread. A reporter wrote, 'in their frantic attempts to rid themselves of these contrivances they poured the spirit out onto the floor' and 'many victims, enveloped in flames, escaped into the fields and there died a terrible death, their bodies being so burned as to be unrecognizable.' [532] At least 40 of them were believed to have died. Sixty passengers were seriously injured and were taken to hospital, where several more of them later succumbed from their burns. Other passengers, fearful of being apprehended for smuggling, made themselves scarce after the accident. The official number of dead was 34; the true figure probably lay between that and 60, with perhaps 130 injured. [533] It is not clear why the smuggling of spirits was so widespread in Germany at this time, as there was no prohibition in that country.

At 8.30 a.m. on Monday 10th November 1919, goods train G 7551 collided with a

[529] *Railways in India*: Administration Report for the Year 1919-1920 Volume 1
[530] *The Times*, London, England, 12th July 1919
[531] *Le Matin*, Paris, France, 27th September 1919
[532] *The Times*, London, England, 27th October 1919
[533] Katastrophen der deutschen Bahnen Teil II: Chronik 1845-1992 by H.J. Ritzau, pgs. 45-6 (in German)

shunting movement at Neumühl-Kutzdorf, Germany, on the Stettin – Küstrin line. The train was carrying a large number of unauthorised passengers, 45 of whom lost their lives, while 18 were injured. [534]

A collision between two trains in Turkey at a junction on Wednesday 10th December 1919 reportedly resulted in 35 people killed or injured. The location was not stated but trains of the Chemin de fer de Smyrne – Cassaba et Prolongements and Chemins de fer Ottoman d'Anatolie were apparently involved. [535]

Doubtless because of heavy Christmas passenger traffic, on the afternoon of 22nd December 1919 train No. 312 of the Chemins de fer du Nord was closely followed by train No. 213 bis from Lille to Paris. The latter train was stopped by signals just outside of Douai station, France, when it was run into in what was presumably a sidelong collision by train No. 4942 from Dunkerque which arrived on an adjoining line. The accident occurred at about 15:00. Four carriages of train No. 213 bis were completely demolished. The station staff immediately began a rescue operation. Eleven passengers were killed outright and 22 injured, most of them seriously. The victims were taken to all the hospitals in the town, most of them reportedly without their papers (identification papers?). The work of clearing the line was finished by 19:00.

The undamaged part of train No. 312 bis continued its journey to Paris, arriving at the Nord station at 19.55, and here a newspaper reporter managed to collar a passenger who had been riding in the third compartment from the end of the train for an interview. He said, 'I was quietly sitting in my compartment waiting for the train to start when suddenly at about 16.50. I was violently thrown against the door and slightly injured by bits of broken glass; I tried to get out, but it was impossible to open the door. I and the other passengers in my compartment had to get out through the glass portion of the door. On the track a sad sight met our eyes. The two compartments behind the one that I was in were completely broken up and scattered about the track. The injured tried in vain to escape from their metal prisons and the railwaymen broke up the roof of the carriages with pickaxes so as to be able to pull the injured out. The train was still hissing and trembling with anxiety I climbed into my compartment.'

Driver Jean-Baptiste Mérian of the colliding train was detained for examination by the investigating magistrate. [536]

A couple of days later it was reported that both Driver Mérian and Signalman Bajeux had been arrested and confined in Cuimoy prison. This led to protests by the Syndicat National de Chemins de Fer, who alleged that the basic cause of the accident was defects in the signalling system. [537]

Briefly reported in the London *Times* was a collision between a Königsberg train and a military train at Czersk on about Wednesday 28th January 1920. As the exact date of the

---

534 Ibid, pg. 46
535 *Railway Gazette,* Vol 32, pg. 34
536 *Le Petit Journal*, Paris, France, 23rd December 1919
537 Ibid, 27th December 1919

collision is unclear and Czersk returned to Poland from Germany (Prussia) on 29th January 1920 it is impossible to be certain if the accident happened in what was then Germany or Poland, but no record of the accident appears in any of the extensive German literature on railway accidents. The newspaper report stated that there were between three and nine dead and about 15 badly hurt. [538]

On Monday 17th May 1920 No. 8 up passenger train was due to cross No. 1 down Gujarat Mail train at Jakhwada station, India, on the broad-gauge tracks of the Bombay Baroda and Central India Railway. The route was set up for the mail train to hold the main track at the station and the up train should have taken the loop line, but it was wrongly diverted instead into a dead end siding, where it collided with a waggon. The engine, two bogie third-class coaches and a third-class mail composite coach were derailed and telescoped and 26 passengers were killed and 41 injured. Damage to the rolling stock and permanent way amounted to 23,500 rupees.

The railwaymen involved were prosecuted. A pointsman was sentenced to a year's imprisonment and others were fined. [539]

Considerable mystery surrounds an accident mentioned in *Railway Disasters of the World* which supposedly occurred at a location named Neon Darostan in Spain on 20th May 1920, when a head-on collision took place between a passenger and a goods train, completely wrecking the front portions of each train and derailing the remainder of the trains. This was reportedly a major disaster, with 40 people killed. To begin with, a comprehensive list of Spanish railway accidents (*Accidentes Ferroviarios en España*) which includes many much less serious accidents, makes no mention of this disaster at all. Again, it has been impossible to identify any location named Neon Daroston. The *London Times*, on the other hand, carried a brief Reuter report of an accident near Yaroslav, Poland, on about 20th May 1920 and it seems likely that this is the accident referred to in the *Railway Gazette* (and thence in *Railway Disasters of the World*) and that the location is not Neon Darostan but Near Jaroslav. Jaroslav (now Jaroslaw) is in the region of Poland formerly known as Galicia and it is conceivable that this Galicia in Eastern Europe was confused with the Galicia region of Spain, hence the error as to the country in which the accident occurred. According to the Reuter report only seven people died and 37 were injured.[540]

A collision occurred on Sunday morning, 27th June 1920 at Bœmish Brod station, Czechoslovakia, between a goods train from Prague and a special train of Sokols (a Czech gymnastic society) coming from Presbourg. Five persons were killed and 50 injured, 30 of them seriously. A train crew were arrested and blamed the accident on excessive fatigue. [541]

---

538 *The Times*, London, England, 30th January 1920

539 *Railways in India*: Administration Report for the Year 1920-1921

540 Ibid, 24th May 1920

541 *Journal de Genève*, Geneva, Switzerland, 29th June 1920

Atoka, Oklahoma, United States, on the Missouri Kansas Texas Railroad, was the scene of a collision on Sunday 4th July 1920. Northbound extra train 503 had arrived at 7.45 p.m. with cars of the E.B. Reed Carnival Company which was to perform at Atoka the next day, which as Independence Day itself fell on a Sunday was doubtless taken as a holiday by many people. The placing of the cars in the yard for unloading was a rather time-consuming operation and almost two hours later 13 vehicles were still standing on the southbound main track awaiting disposal in the yard. From the south end these were a sleeping car, two coaches, another sleeping car, a baggage car, a combination coach, a baggage car and six flat cars. The first six of these vehicles were owned by the carnival company and all were of wooden construction. At 9.30 p.m. they were run into by extra train No. 845, running northbound on the southbound track. Fire broke out immediately after the collision and five vehicles were damaged beyond repair as a consequence of the collision and fire. Seven passengers died and 20 were injured; the conductor and fireman of train No. 845 were also injured. [542]

A report of the aftermath of the collision stated that, 'Immediately after the crash one of the cars caught fire and amid the shrieks of the dying and the dull thud of falling timbers came the crackle of the flames as they crept upon those pinned under the wreckage, and unable to extract themselves. The train was one of bedlam for a few minutes. Hundreds of persons, including those members of the carnival crew who were not on the doomed cars, rushed to the scene and began to work frantically to rescue the victims.

'The freight train stopped a short distance above the wreckage and the crew returned to lend its aid. With no adequate water supply, and with the smoke and fire from the burning cars making the rescue work almost impossible, the workers turned their attention to the cars not yet on fire in an effort to save the injured. As soon as an injured form was brought out of the debris, the doctors gave temporary aid, dressed wounds, and gave it over to others for care. Some of the injured were able to crawl from under the wreckage while others were pinned and had to be pried from under the cars.' [543]

Extra No. 845 consisted of a locomotive, 67 cars and a caboose in charge of Conductor Little with Engineer Patrick and Fireman Adams. It had left Ray, Texas, at 6.35 p.m. and at Calera, 37 miles south of Atoka, it had been crossed over to the southbound track out of the way of northbound passenger train No. 2. It then continued northwards over the southbound track on the basis of an order issued by the dispatched for it to use the southbound track to the end of the double-track section at Stringtown, north of Atoka, with rights over all opposing trains.

Engineer Patrick told the Interstate Commerce Commission investigators that when he arrived at Atoka he had applied the brakes and reduced the speed of his train from 25 to 8 miles an hour, as he wished to stop and take water at the water crane there. He had then released the brakes to let his train drift down to the crane and had not seen the standing cars until he was about three car lengths away from them. He had then applied the brakes and thought that his train's speed had dropped from 8 to 4 miles an hour at the moment of collision and that he could have pulled up altogether but for the falling gradient and the momentum of his heavy train. He admitted that he had not been operating his train within yards limits so as to be able to stop within his range of vision as laid down in the rules.

[542] Interstate Commerce Commission report
[543] *Daily Ardmorite*, Ardmore, Oklahoma, United States, 5th July 1920

Conductor Little stated that he had been riding on the engine and had been the first to spot the standing cars when they were seven or eight car lengths away. He had called out to the engineer. Both he and the fireman had jumped from the engine and fallen. He agreed with his engineer's estimates as to the speed of the train and that his engineer had not had his train under proper control.

The rear brakeman was not even aware that there had been a collision until he went to the front of the train after it had stopped.

Yardmaster Gardner at Atoka was apparently unaware of the approach of extra train No. 845 until about 9.10; he was told by the conductor of train 503 that he would have to hurry up and clear the main tracks or the other train would be into them. He had then noticed, shortly before the accident, that the southbound automatic block signal was showing danger, which as no southbound train had passed, could only mean that a northbound train was coming on the southbound track. At this point he had told a switchman to run back and flag the oncoming train, but the collision had occurred very soon afterwards. The Atoka operator apparently knew of extra train 845, but not that it was running on the southbound track.

Dispatcher Carter testified that he had issued an order for extra train 845 to run from Ray to Atoka and later one for the train to use the southbound track between Calera and Stringtown. It was customary for the dispatcher to notify Atoka when first- and second-class trains were running against the current of traffic and frequently third- and fourth-class trains and extras were also thus notified. Dispatcher Carter could not remember if he had so notified Yardmaster Gardner as to extra train 845 and the yardmaster said that he had not.

The Interstate Commerce Commission blamed Engineer Patrick and Conductor Little for the collision as they had failed to run through Atoka yard under complete control as required by Rules 98 (c) and 98 (e).

Rule 98 (c) stated:

Third and fourth class and extra trains are required to approach and pass all water tanks, coal chutes, and yard limits under complete control.

Rule 98 (e) stated:

Yard limit boards define yard limits. Yard engines may work within yard limits against third or fourth class or extra trains, but will give way as soon as possible upon their approach.

The commission also noted that the order authorising extra train 845 to use the southbound track was issued at Durant, 5 miles north of Cadera so that the train had been operated over that distance against the current of traffic without any order to do so.

It was noted that had the cars affected been of modern steel construction they would not have been crushed by the colliding train and the wreckage would not have caught fire, either from oil stoves or lights. The deaths, injuries and suffering resulting from the fire would thus have been averted. [544]

The coroner's inquest on the accident recorded a verdict of negligence against

[544] Interstate Commerce Commission report

Engineer Patrick and Dispatcher Carter. [545]

At 19:15 on the evening of Saturday 9th October 1920 goods train I-D 264 from Mantes, heavily laden with coal and minerals, had become divided by the breakage of a coupling behind the seventh waggon, apparently unknown to any of the train crew. The breakage was noticed at Sartrouville station, and a warning was sent to the station at Houilles, France. It was then decided to shunt the goods train at Houilles so that it could be reconnected. Unfortunately when the front portion of the train was stopped there the rear portion collided with it with sufficient force to derail four waggons and throw them onto the opposite running line just in front of passenger train M-P 216, which had left Paris St Lazaire at 19:06 for Mantes. This train, which was well filled, did not stop at Houilles and was travelling at 50 kilometres an hour. Its driver, seeing the danger, applied his brakes, but the distance was too short to avoid a collision. There was a terrific impact which left the wreckage piled high in the air. The engine of the passenger train overturned, and the first eight coaches of the passenger train were twisted and wrecked. Early reports put the number of dead at 39 and by a later report this figure had increased to 45. In their initial derailment the waggons of the goods train also demolished signal box No. 7, either seriously injuring or killing (according to different accounts) the signalman. [546] [547]

The Minister of Public Works, M. Le Troquer, went to the scene of the accident, which he declared had been the result of an unfortunate chain of circumstances. It seems doubtful if an inspecting officer of the British Ministry of Transport would see it that way. Given that there was a falling gradient on the approach to Houilles station it was surely an unwise decision to stop the forward portion of the divided goods train, when a collision with the detached rear portion was a predictable result, and British signalling regulations contained definite instructions to signalmen that there were, if at all possible, to not stop the front portions of divided trains on falling gradients, but merely to indicate the situation to their drivers.

A collision occurred at Erie, Pennsylvania, United States, on the New York Central Railroad, on 20th October 1920. There were four running lines at this point, running east and west and numbered from north to south 3, 1, 2 and 4. Tracks 3 and 1 were used by westbound trains and 2 and 4 by eastbound trains. West of Erie station crossovers connected tracks 3 and 1, 1 and 2 and 2 and 4 and another crossover opposite the station connected track 4 with the station yard. A switch tender controlled the crossovers. The collision had occurred at the west end of the crossover between tracks 1 and 2.[548]

Westbound passenger train No. 23 from Buffalo, New York, to Cleveland, Ohio, in charge of Conductor Lilley and Engineer Brown, arrived at Erie on track 1 at 10.34 p.m. and departed at 10.38 p.m., an hour and 12 minutes late. A minute later it was struck by eastbound passenger train No. 60, a Chicago, Illinois, to Boston, Massachusetts, express.

---

[545] *The Daily Silver Belt*, Miami, Arizona, United States, 14th July 1920
[546] *Le Radical*, Paris, France, 11th October 1920
[547] *The Times*, London, England, 11th October 1920
[548] Interstate Commerce Commission report

[549] The latter train was in charge of Conductor Chapman and Engineer Hotchkiss and Fireman Batt. It approached Erie station on track 2 at an estimated speed of 15 miles an hour when it was diverted through the middle crossover and into the side of train No. 23. In the collision engine No. 4817 of train No. 60 grazed the side of the seventh coach of train No. 23, tore off the side of the Pullman sleeping car *Peosta,* then violently struck the end of the ninth car almost head-on, driving it and the tenth car backwards. Seven passengers were killed outright, one was fatally injured and 18 others were less badly hurt. Three railroad employees and a Pullman porter also died.

Engineer Hotchkiss of train No. 60 told the Interstate Commerce Commission investigators that he had received clear indications at signals 89.2 and 88.2 as he neared Erie. It was only about 10.40 and his train was not due to leave Erie until 10.42, so he was only drifting as he approached the station. He saw some men at the switches and was about to whistle to warn them when they stood clear anyway. Then without warning his engine was diverted through the crossover. He had at once applied the brakes but did not think that there had been time for them to take effect before his engine had struck the other train.

Fireman Batt corroborated his engineer's statements as to the speed of their train and the clear signal indications.

Engineer Gearhart, deadheading ('riding on the cushions') train No. 60; Division Engineer Upson, standing on the platform of Erie station; and Crossing Watchman Pfeiffer at Sassafras Street Crossing all confirmed that the signals had been clear.

None of the crew of train No. 23 knew of anything wrong before the collision.

The investigation determined that engines Nos. 4578 and 5941 coupled had arrived at Erie at about 10.25 p.m. on track 3 in charge of Engineer Newman. Engineer Wagner had then taken engine No. 5941 through all three crossovers to track 4, Switch Tender Bell opening but not closing the switches. Engineer Wagner had closed most of the switches, but not two, one of them being the west switch of the middle crossover diverting trains from eastbound track No. 2 to westbound track No. 1. At this time Signal Maintainer Larson and Assistant Signal Maintainer Carlson were working at the east switch of the west crossover and Signal Maintainer Truax was working at the crucial west switch of the middle crossover, installing new switch boxes. Switch Tender Bell told the inquiry that having ridden across on engine No. 5941 he had been about to get down and close the remaining two switches when Signal Maintainer Larson had shouted to him that he would do so. He had seen Signal Maintainer Larson reset the west switch and assuming that he would replace the other had retired to his shanty.

Signal Maintainer Larson testified that with his assistants he had been changing the switch boxes of the middle crossover. In order to avoid delays to trains he had bridged the wires of the signal protection relay, with the result that the automatic signals would show clear regardless of the position of the crossovers. When engine No. 5941 had crossed over he had told Switch Tender Bell that he would close the switch at the east end of the west crossover, the one at which he was working, which he had done. He did not remember saying he would close the other switch.

Assistant Signal Maintainer Carlson said that he had not paid any attention to the

[549] *Olean Evening Herald*, Olean, New York, United States, 20th October 1920

conversation between Larson and Bell and was not aware that the middle crossover was open or that the controls could be disconnected, so that the signal and crossover were out of correspondence. He had only joined the railroad on 7th July.

Signal Maintainer Truax said that he knew that the relays had been bridged but not that when this was done the signal could show clear with the crossover open, also that he had not known that it was open. Later he stated that he had forgotten that it was open.

The Interstate Commerce Commission found Switch Tender Bell and Signal Maintainer Larson principally to blame. It was clear that there had been a misunderstanding between Bell and Larson as to which switches Larson was to operate but in any case it was Bell's duty to open and close all the switches and not delegate part of his job to someone else. Signal Maintainer Larson should not have taken it upon himself to bridge the circuits so as to make the signal show false clear indications but should have spiked the points before disconnecting the wires to install new switch boxes. If the crossovers needed to be used the spikes could have been temporarily withdrawn.

Signal Maintainer Truax, the commission found, had also contributed to the accident by failing to check that the switch, which was only three feet from where he was working, was properly aligned for the approaching train. [550]

Two passenger trains reportedly collided 'due to a mistake by an unskilled railwayman' on or about 27th October 1920 at Lufany in Transylvania, Romania. Fifty persons were reportedly killed and over 200 injured. [551]

Severe weather conditions in the form of a blizzard were the main factor in a collision on the Chemins de fer du Nord at Armentières, France, on Sunday 12th December 1920. The blizzard was raging hard when passenger train No. 2402 from Saint-Omer to Lille via Berguette left the station. It had covered 400 metres and was gathering speed when it was run into from the side by coal train No. 7049 from Fives-Lille. The passenger train should have left Armentières at 20:43 but was somewhat late and did not depart until 21:00. The coal train was a heavy one and was moving at some speed at the moment of the collision. The engine of the coal train struck the tender of the engine of the passenger train and the two following third-class carriages were telescoped.

The rescue work began at once, with the first rescuers being uninjured passengers from the rear of the train and station staff, who were soon joined by firemen and local doctors who hurried to the scene. The injured were taken to the station and later by special trains to hospitals in Lille. The bodies of ten dead were then removed from the wreckage and taken temporarily to a room in the station. Five of them could not immediately be identified. Various railway officials, the prefect of the Nord department and the public prosecutor left for Armentières to investigate the accident.

The work of clearing the line went on through the day and the following night; meanwhile single-line working was put into operation. Those who lost their lives included Conductor Michaux of the passenger train, about 23 passengers, and Driver

---

550 Interstate Commerce Commission report

551 *Railway Gazette,* Vol. 33, pg. 836

Félix Germain and Fireman Alfred Bury of the passenger train. [552]

An investigation opened by M. Baret, the principal inspector of the Nord company, appeared to establish that the responsibility for the accident lay with Driver Wattrelez of the coal train. The latter declared that in the poor visibility he had thought his signal had been clear but it was held that he should have realised the signal was against him and stopped his train. Signalman Dumoulin, who controlled the junction, stated that his signals had been working perfectly half an hour before the collision and they were tested after the accident and found to be working properly. In any case, the clearing of the signal for the passenger train would have necessarily locked the carré applying to the coal train at danger.

Driver Wattrelez admitted that the advance signal had been at danger but claimed that his fireman, who had been looking out, had indicated to him that the carré was clear, so he had released the brakes. He conceded that the signal had not been properly off and that he should therefore, according to the regulations, have treated it as a danger signal. However, the detonators which should have automatically been placed on the rails when the signal was at danger, had not exploded.

Driver Wattrelez and his fireman were charged with manslaughter, but bailed.

The same newspaper report added that a tenth passenger, M. Jules Rémy, a miner from Somain, had died in hospital from his injuries. [553]

Brief reports appeared in the London *Times* and a number of Australian newspapers of an accident near Bommidi, India, on the broad-gauge South Indian Railway, on 14th December 1920. Thirty Indian passengers were killed and 35 injured, according to the newspaper reports, when a mail train and a goods train collided. [554]

The subsequent report of the government investigation of the disaster gave fuller details of what had happened. It was stated that a goods train was being moved from the platform line to a second line at the station to allow the Mettupalaiyam Mail to run through the station, but the latter train had collided with the shunting goods. Six waggons and the brake van of the goods train were damaged and the engine and tender of the mail train were overturned and partly buried beneath the wreckage of the goods train. The first two carriages of the mail train were capsized and badly damaged, the third carriage was forced into the ashpan of the overturned engine and caught fire and the fourth was derailed.

The bodies of 40 passengers were removed from the wreckage, nine of which were burnt beyond recognition, and 41 people were injured, eight of whom died on the way to hospital and four more of whom succumbed later, bringing the final number of dead to 52.

The driver of the mail train had overrun the signals while the station staff had failed to observe the rules for shunting operations. The staff responsible were being prosecuted at the time of the report. [555]

---

[552] *Le Petit Parisien*, Paris, France, 13th December 1920
[553] Ibid, 14th December 1920
[554] *The Times*, London, England, 16th December 1920
[555] *Railways in India*: Administration Report for the Year 1920-1921

A serious accident occurred in Kreuzwald station, France, on the Chemins de fer de l'Est, at about 05:00 on the morning of 15th December 1920. Train No. 552 was entering the station from Hargarten-aux-Mines. It failed to obey the signals, which stood at danger, and collided with goods train No. 7510, which was shunting there. The leading carriage of the passenger train was completely demolished in the collision and several other carriages badly damaged. The station officials immediately began a rescue operation. An initial report spoke of five dead and 60 injured. [556]

Another report published the same day stated that there were 17 dead and 25 injured, 12 of them seriously, and added that most of the victims had been miners from the Hargarten and Burlay region. [557]

A third report stated that nine dead bodies had been recovered at the time that the reporter left the scene but that it was feared that many others still lay in the wreckage. The injured were taken to a miners' hospital in Kreuzwald. The Metz public prosecutor had opened an inquiry but had not yet attributed blame for the accident. The driver of the passenger train stated that he had been unable to make out the signals in the fog. [558]

Scanty reports appeared in a handful of American newspapers regarding a collision between a passenger train and a freight train at Huachana, near La Oroya, Peru, on Sunday 30th January 1921. Many of the coaches of the passenger train were said to have been completely demolished and seven persons killed and over 40 injured. [559] [560]

Late in the afternoon of Tuesday 1st February 1921 a goods train bound for Limoges, France, became derailed near Bardys on the double-track main line of the Chemin de fer de Paris à Orléans. The accident was only a minor one but it meant that single-line working had to be introduced between La Jonchère and Bardys. Thus when omnibus train No. 1919 from Châteauroux to Limoges arrived at La Jonchère it was crossed over onto the other running line for the journey on to Bardys, where it was due at 19:15. Meanwhile in the opposite direction freight train No. 4116 from Limoges, due at Bardys at 19:04, was being shunted out of the way of the omnibus train. Detonators had been put down to protect the shunting train and the signals were standing at danger. However, the driver of the omnibus train disregarded both the exploding detonators and the signals and continued towards Bardys station. When he finally saw the shunting goods train it was too late.

The shock of the collision was severe and five coaches of the omnibus train telescoped one another. An early report spoke of three dead and 47 injured. Local railwaymen began the rescue work and news of the accident was sent to Limoges, from where a relief train was despatched. At the time of the report a number of injured had already been admitted to Limoges General hospital. The conductor of the omnibus had

---

[556] *Le Radical*, Paris, France, 16th December 1920

[557] *La Lanterne*, Paris, France, 16th December 1920

[558] *Le Gaulois*, Paris, France, 16th December 1920

[559] *The Titusville Herald*, Titusville, Pennsylvania, United States, 31st January 1921

[560] *The Oil City Derrick*, Oil City, Pennsylvania, United States, 31st January 1921

both legs severed.[561]

By the morning of 3rd February nine dead had been identified. They were MM. Auguste Malabre, a farmer from Folles; Bonnefon, a Limoges contractor; and Texier, a Bessines teacher; Mme Marie Chaumont, of Limoges; and MM. Valier, Tharaud and Roussy, PTT (Postes Télégraphes et Téléphones) employees from Lomoges, and journaliers (labourers employed on a daily basis) Lyraud and Grandjean from Bardys. There were 22 injured still in hospital, some of them in a very serious condition, while there were also a number less seriously hurt. The injured included Conductors François Geneyrand and Marin Boussard and Driver Jean Combrouze of the goods train and Conductors Jucherand and Pierre Fournier, Driver Armand Gourdy and Fireman Coinaud of the passenger train. (Which conductors were on which train is not entirely clear.)

The initial investigation into the accident indicated that 34-year-old Driver Gourdy had disregarded all the signals and had approached the station at 70 kilometres an hour (72 to be precise according to the speed recorder). Fireman Coinaud had apparently noticed his driver running at an excessive speed for an omnibus train at every station and he would certainly have run past Bardys station had his train not collided with the freight train instead. He had only applied the brakes when his train was 10 metres from the obstruction. Many witnesses had reportedly been interrogated by M. Demartial, the juge d'instruction (examining magistrate).[562]

Two of the seriously injured victims died in the days after the accident, increasing the number of dead to 11.

Questioned in hospital, Driver Gourdy denied that he had been told to slow down by his fireman and had ignored him. Following his interrogation the driver was arrested.[563]

At the end of March questions were asked in Parliament about the frequent accidents on French railways and M. Le Troquer, the Minister of Public Works, referred in detail to the Bardys disaster, but declared that he would not say anything which might prejudice justice.[564]

Driver Gourdy stood trial at the end of May. He was found guilty of having caused the accident and sentenced to two years' imprisonment and fined 300 francs.[565] No explanation seems to have come out for his rash driving; there was no suggestion, for example, that he had been intoxicated.

On the night of Saturday 5th to Sunday 6th February 1921 severe winter weather conditions were affecting the working of the Südbahn lines in Austria, with unusually heavy snow disrupting operations. Thus an international express passenger train from Italy via Klagenfurt, Leoben and Bruck due in Wien Südbahnhof at 9.40 p.m. on Saturday night was over three hours late. The snowfall had seriously interfered with communications, resulting in the train having to stop at many stations which it would have normally run through. At 12.30 a.m. on Sunday 6th the express was running into

---

[561] *Le Matin*, Paris, France, 3rd February 1921
[562] Ibid, 4th February 1921
[563] *La Charente*, Angoulême, France, 6th February 1921
[564] Ibid, 31st March 1921
[565] Ibid, 29th May 1921

Felixdorf, where it should have crossed express goods train No. 811 from Wien, the line being single track.

The goods train entered the station from the opposite direction and the driver of its engine, No. 629.12, overran signals and ran through the siding and out onto the main line again, into the path of the oncoming express. The latter was running at a considerable speed and there was a head-on collision. The leading carriages of the express were telescoped into one another, the rear vehicles were derailed and overturned. The two locomotives were badly damaged and some of the waggons of the goods train were wrecked. There was a scene of utter chaos in the station after the accident, intensified by the fact that the few station lights were unable to penetrate the darkness and driving snow.

An official report on the accident stated that five of the express passengers had been killed and 16 had been seriously and many slightly injured. Two of those injured later succumbed. [566] Among the injured was Mr William Upson, described as having been a member of the American Mission to Austria, so presumably a missionary. [567]

Very brief details reached the American press of a major disaster on the Ferrocarriles Nacionales de México (Mexican National Railway) on or about Sunday 3rd April 1921. A passenger train and a freight train collided at Villa Garta, Mexico, between Parradon and Monterey, and 33 people were killed. The only person to lose his life who could be identified at the time of the report was Victor Villareal, candidate for mayor at Piedras Negras, a Mexican settlement on the opposite side of the Rio Grande river to the Texas town of Eagle Pass. [568]

On the afternoon of Monday 2nd May 1921 an accident occurred near Welch, West Virginia, United States, on the Norfolk and Western Railway. (The Interstate Commerce Commission report states that the accident occurred on 3rd May but as it happened in the afternoon and a report about it appears in a morning paper of that date this is clearly wrong.)

The single-track Tug Fork Branch, the location of the accident, was 19 miles long and several branches ran off of it. Passenger trains served these branches and as they lacked turning facilities trains making round trips were required to run in reverse for considerable distances. The branch was worked by operating and timetable rules almost as a yard and was under the control of a yardmaster at Wilcoe.

The train involved in the accident was passenger train No. 160, consisting of engine No. 954, a mail and baggage car, an express car and three coaches, the vehicles all being of the open platform type and of wooden construction. In charge of the train was Conductor Compton, with Engineer Richardson and Fireman Atwell. The train had passed Havaco at 1.55 p.m. and its coaches were being propelled towards Welch when it was suddenly diverted into a siding at an open switch and struck the endmost of a row of loaded 100 ton steel coal cars, which was standing only about 14 feet from the east end of the crossover. The two rear (that is, leading in the direction of travel) coaches were

---

566 *Die Lokomotive,* No. 5, May 1921, pg. 76.

567 *The Times*, London, England, 8th February 1921

568 *The Boston Globe*, Boston, Massachusetts, United States, 4th April 1921

practically demolished and the rearmost coach telescoped the next one over two-thirds of its length. The remainder of the train was undamaged and the coal car only slightly so. Six passengers died and 41 passengers and one railwayman were injured.

The passengers who lost their lives were Mrs Fred Pearson, of Wilcoe; George Robinson, draftsman of the United States Coal and Coke Company at Gary, West Virginia; Ricardo Quitantidi, address unknown; Lawrence Harris, of Havaco; Arthur Hill, of Superior, West Virginia and an unidentified Italian woman from Havaco. [569]

Brakeman R.H. Powell told the Interstate Commerce Commission (ICC) investigation that he had been riding on the platform of the coach at the leading end of the train to control the backward movement. A signal whistle was provided here, together with a valve by which he could apply the brakes. Approaching the scene of the accident he had sounded the whistle to warn some section men (platelayers) working on the track and 60 feet from the connection to the siding he had seen that the switch was open. His train was moving at 15 or 18 miles an hour. He said that he had at once opened the valve of the backup hose to apply the brakes, opened the door of the coach to shout a warning to its occupants and jumped from the platform. Had the section men not been obscuring his view, he added, he would have been able to see the switch target at the points earlier.

Engineer Richardson said that he had felt the air brakes apply as his train neared the siding, though not very hard, and he had thought that there was probably a cow on the track. Shutting off steam he had looked back to see the rear coach entering the siding, his fireman had shouted and he had applied the brakes in emergency. Before the brakes had had time to take effect the collision had occurred.

Fireman Atwell thought that the brakes had come on as the train was going through the points and Conductor Compton said that it had not seemed like an emergency application. All agreed that the brakes had been working properly.

Section Foreman Fine told the inquiry that he had been transporting some ties (sleepers) from a point east of the point of the accident to a point west thereof, using a motor and pole car, and because of the coal cars on the siding he had had to use the main line as far as the west crossover switch. He had operated this switch to put the motor and pole cars into the siding, which extended almost a mile on either side of the crossover. He had not closed the switch but had given the switch key to Sectionman Hale to unlock the east end of the crossover. Their motor had duly entered the siding and had proceeded westwards along it to a point opposite the west crossover switch. There he had stopped the motor and instructed the sectionman to close the west switch, but did not actually see him do so or observe the position of the points or switch target.

Sectionman Hale insisted that he had closed the points but said that he had not locked them as it had been intended to use the crossover again shortly.

The ICC remarked that examination of the switch after the accident had shown it to be in good condition, unlocked and set for the siding. There had been no sign of tampering.

The commission concluded that the responsibility for the accident was shared by Section Foreman Fine and Brakeman Powell. Company Rule 617, relating to the duties of section foremen, stated: ‘They must not permit their switch keys to pass out of their possession, and must personally attend and supervise the opening and closing of

---

[569] *Raleigh Register*, Beckley, West Virginia, 5th May 1921

switches.' The ICC believed that either the section foreman had forgotten that the switch was open, or he had closed it himself as soon as the motor had passed through it and then forgotten that he had done so and given the key to Sectionman Hale, and that the latter had operated the switch again and failed to notice that it had been left open for the siding.

As for Brakeman Powell, the commissioners noted that the switch was visible from an approaching eastbound train for 250 feet. They did not believe that it had been entirely obscured by the sectionmen as the brakeman had alleged and considered that had he been keeping a proper lookout he could have applied the brake in time to stop the train and avert the accident. [570]

On the morning of Wednesday 11th May 1921 a collision occurred at Mataundh station, India, on the Jhansi-Manikpur section of the Great Indian Peninsula Railway, involving up passenger train No. 654 and down passenger train No. 653. The leading brake van and the postal third-class composite carriage of train No. 654 were described as having been completely smashed and some of the coaches of train No. 653 were also wrecked. About 10 minutes after the collision the coaches of train No. 653 caught fire and the leading two vans and following four carriages were burned. It was estimated that 11 passengers and four railway servants were killed, and some 50 people were injured, most slightly. The chief responsibility for the disaster was placed on the driver of train No. 654 for overrunning signals. The guard of that train was also held to blame for not having kept a proper lookout and stopped the train by applying the automatic brake. The contravention of certain rules by the station staff at Mataundh was also stated to have contributed considerably to the accident. All of those found to blame were punished departmentally as it was held that none of them could be prosecuted criminally except the driver of train No. 654, who perished in the disaster. The damage to the permanent way and rolling stock was estimated at Rs. 79,900. [571]

On the evening of Saturday 11th June 1921 a serious collision occurred on the outskirts of Madrid, Spain. The express for Andalusia which had left Madrid at 20:20 arrived at Villaverde at 20:35 and at the Los Rosales junction collided with the Toledo mail train, which was already crossing the junction. The driver of the express had braked hard before the collision but a luggage van, mail van, one first-class and two second-class carriages of the Toledo train were struck and damaged to a greater or lesser extent. There were only injuries in the front vehicles of the mail train but a number of the 18 passengers in the first-class carriage were killed. The driver of the express was trapped in the wreckage of his engine and burnt to death. The Villaverde station staff and local residents immediately went to the assistance of the victims. The Minister of Public Works was advised of the accident only five minutes after it had occurred and immediately went to the scene and organised the rescue operation.

Several army students who were on the Toledo train realised that Colonel Losala of the Academy of Infantry was among the injured. He was released, having suffered two broken legs. Six bodies were recovered from the wreckage, they were: Alfred Blumenthal, an American, aged 24; Gregorio Blasco, a company wages clerk {?} and his

---

570 Interstate Commerce Commission report

571 *Indian Railways in 1921-22*: Administration Report, Volume 1

11-year-old daughter Pilar; Francisco Pernal Sánchez, a postal official from Malaga; don Sergio de Novales, a former deputy, and don Nicasio Fernandez, a Torrejón doctor. In addition, a 12-year-old boy who had not been identified was also killed. Some of the injured were reportedly dead upon arrival at the hospital. Conductor Domingo Pespal chanced to be in the first-class carriage when the accident happened but was able to jump out of a door and escaped unhurt.

It was officially announced the following morning that there were in fact only nine dead and 23 injured, five seriously. The work of clearing the wreckage continued. [572]

It soon emerged that the accident had apparently been the fault of the driver of the Toledo mail train, who had overrun signals and intruded five metres onto the path of the Andalusia-bound express. The engine of that train was an American model, weighing 140 tons, and which had only been in service 15 months. The express had been travelling at 65 kilometres an hour and the accident might have been much worse but for the courage of Driver Manuel Montero, who had stuck to his post and by braking had considerably reduced the speed of his train. The Toledo train was seven minutes late and priority had been given to the Andalusia train over the junction. The driver and fireman of the Toledo train had been arrested. The work of clearing the line and especially of the massive engine of the Andalusia express took several days, meanwhile trains were diverted. By the time of the second report on the accident the number of dead had increased to 12. Examination of the signalling had shown that it was working perfectly. The postal van of the Toledo train was a vehicle which was 25 years old, and the Post Office had been trying for ten years to get the railway company to replace it with a more modern one. Reportedly the Post Office had been in the process of suspending the service to protect the lives of their employees until the old vans were replaced.

At his interrogation, the driver of the Toledo train stated that he had missed a signal approaching Villaverde and thus had not been able to pull up before fouling the junction. He had then attempted to reverse his engine, without success.

The fireman of the express said that when Driver Montero had realised that a collision was inevitable he had exclaimed that he would try and save his train even at the cost of his own life and had applied the brake so violently that the shock had derailed the two leading coaches of the Andalusia train. [573]

A week after the accident the railway union announced that they were convinced that the driver of the Toledo mail train was in fact completely free of blame, and that if he continued to be detained they would take action, even possibly going on strike. [574]

The final outcome of the case against the unnamed Toledo driver has not been determined.

On about Thursday 21st July 1921 a special train carrying a military band and soldiers from Saltillo to San Luis Potosi, Mexico, to participate in the festivities there, collided with another train in Carneros station. Three of the musicians were killed and about 30 injured. A wrecking crew were immediately sent out and the dead and injured were

---

[572] *La Vanguardia*, Barcelona, Spain, 12th June 1921
[573] Ibid, 14th June 1921
[574] Ibid, 20th June 1921

returned to Saltillo. Shortly afterwards a military train under the command of General Pineda which was also en route to San Luis Potosi 'to salute President Obregon and other distinguished visitors' was involved in another accident when it collided with an engine near the location of the first accident. This time the results were worse; nine soldiers were reportedly killed, and a large number injured. The previous Sunday (17th) three soldiers forming the armed escort to a train had lost their lives in the derailment of a train due to a broken rail near Talla station, between Monterey and Torreon. [575]

Italy suffered a major disaster on the evening of Saturday 27th August 1921. During the summer months many inhabitants of Roma travelled to Ladispoli, a popular bathing place midway between Roma and Civita Vecchia, and special trains were run to cater for them. It was one of these, accelerated train No. 4681, which was involved in a collision with a freight train at Magliana station on the Roma Termini-Fiumicino line.

At 18:00 on the day of the accident the passenger train, loaded with holidaymakers and bathers returning from Ladispoli to Roma was entering the station on track No. 1 when it was struck by freight train No. 6715 at the connection between tracks Nos. 1, 2 and 3. The engine of the freight train rammed the centre of the passenger train by advancing from track No. 3 towards the connection before the passenger train had cleared it. Despite a last-minute attempt to stop it the freight train engine pushed the carriages towards the outer embankment and overturned eight of them, causing the deaths of 37 people and the injury of over a hundred. The line was blocked and two relief trains were soon prepared from the Ponte Galera side, while a third train was sent from Civitavecchia. Signor Micheli, the Minister of Public Works, was active in organising the rescue operation.

A further accident was just avoided as the driver of a passenger train leaving Roma Trastevere at 18:05 was warned of the obstacle just in time and made an emergency application which caused panic and injuries among its passengers but avoided a further collision. Another fortunate circumstance was the fact that it had been a wet day which had prevented many others from making the trip to Ladispoli. [576]

Passenger train No. 4681 from Ladispoli to Roma Termini consisting of a steam locomotive, a luggage van and 18 passenger carriages, mostly third class. The freight train, from Civitavecchia to Roma, was composed of a steam locomotive and eight waggons.

The freight engine had been uncoupled from its train and there was an urgency to get it onto track No. 1 before the arrival of a *direttisimo* train bound for Pisa.

The judicial investigation of the accident was undertaken by Prosecutor Chimienti, who wished to interrogate various witnesses, including the Magliana stationmaster, the driver of the freight train and the pointsman in order to ascertain who had authorised the shunt move and why it had been carried out prematurely. It was not possible, however, in the days following the accident to trace either Stationmaster Neroni or Driver Martini, who had both disappeared. It was ascertained, though, that the locomotive had been driven by the student fireman. The latter was seriously injured and in the absence of the

[575] *The San Antonio Light*, San Antonio, Texas, United States, 24th July 1921
[576] *The Times*, London, England, 29th August 1921

driver he was placed in custody. The director general of the railways opened his own internal investigation into the accident.

On 30th August the report of the judicial investigation was filed. It blamed Stationmaster Neroni for having authorised the shunt move despite the regulations prohibiting such an operation during the arrival of a train; Driver Martini for having abandoned his engine and Trainee Fireman Filocomeni for having moved the locomotive too soon. The fireman was only 17 years old.

Despite the earlier figure of 37 the confirmed number of fatalities was possibly only 23 while the injured taken to various hospitals in Roma numbered 85; to these must be added about 40 slightly injured who were treated in the first aid post of Termini station and released. [577]

Train P 888 disregarded the Einfahrsignal (home signal) at signal at Iserlohn, Germany, on Sunday 4th September 1921 and came into sidelong collision with train P 1317. Five people died and 19 were injured. [578]

The circumstances surrounding a collision between two Oregon Washington Railroad and Navigation trains near Celilo, Washington, United States, on 1st December 1921 were quite complex. The events leading to the accident began with the washing away of a bridge at Eagle Creek on the Oregon Washington line, necessitating the diversion of their trains between Portland, Oregon, and Fallbridge, Washington, over the track of the Spokane Portland and Seattle Railroad, which ran on the opposite side of the Columbia River. At Fallbridge there was a connection from the Spokane line via a bridge over the Columbia River to the Oregon Trunk Railway, with a spur from the latter line to the eastbound Oregon Washington track known as the Oregon Trunk Connection. The tracks climbed steeply from the Oregon-Washington line over the Oregon Trunk Connection so that westbound trains bound for the Oregon Trunk Railway needed the assistance of a helper engine.

The trains involved in the accident were westbound passenger train No. 17, running at the time of the accident as extra No. 3211, from Omaha, Nebraska, to Portland, Oregon, and eastbound train No. 12 from Portland, Oregon, to Spokane, Washington. Westbound train extra No. 3211 consisted of engine No. 3211, a mail car, a dynamo and baggage car, two coaches, a chair car, a tourist car, a dining car, three Pullman sleeping cars and an observation car, in charge of Conductor McCrary with Engineer Gardner and Fireman W. E. Thompson. Eastbound train No. 12 consisted of engine No. 3202, a combination mail and baggage car, a dynamo and baggage car, a smoking car, a coach, a tourist car, two Pullman sleeping cars and an observation car, in charge of Conductor Allision with Engineer Allen and Fireman George Bristow.

Just after midnight on the morning of the accident westbound extra No. 3211 received Order No. 284 on form 19 at Biggs, 6½ miles east of the site of the accident. The order read:

---

[577] Wikipedia: Incidente Ferroviario della Magliana

[578] *Eisenbahn-Katastrophen in Deutschland* by Hans Joachim Ritzau, pg. 79 (in German)

Eng 3211 run extra Biggs to Oregon Trunk connection about one mile west of Ceilio. Run on Eastward track and has right over all opposing trains. This order annulled at two one 201 a.m.

The train left Biggs at 12.15 a.m. and collided head-on with train No. 12 at about 12.30 a.m. Engine No. 3211 of the westbound train was considerably damaged in the collision, the steel mail car was telescoped over most of its length by the tender, the third car came to rest, badly damaged, across the westbound track and the front end of the fourth car was telescoped about 12 feet. In the eastbound train, No. 12, the engine was quite badly damaged, its tender telescoped the first car about 18 feet, and the second, fourth and fifth cars were quite badly damaged, the front and of the fifth car, a wooden tourist car, being telescoped about 18 feet. [579]

According to the Interstate Commerce Commission three passengers, US Marine Mail Guard A.H. McBride, Pullman Porter C.J. Yarbrough, Fireman George Bristow of engine No. 3211 and a trespasser were killed and 65 passengers, six railroad employees, a mail clerk and a Pullman porter were injured. A newspaper named the passengers who had lost their lives as Mrs J.W. Walling, of Amity, Oregon; Jack Cole, aged 5, of McMinnville, Oregon, her grandson; and L.J. Kirk, of St Paul, Oregon. The newspaper report added that workmen clearing the wreckage had found the bodies of four transients who had been beating their way westwards at the front of train extra No. 3211, bringing the number of dead to 10. It was feared that the transients would probably never be identified. [580] It was believed that the men had been clearing away snow in the Biggs area before setting off westwards. On the arrival of the relief train from The Dalles an hour and a half after the collision with doctors and nurses, cots and stretchers had to be carried several hundred yards over deep snow and ice to reach the victims. [581]

Pilot Conductor Clark told the Interstate Commerce Commission (ICC) inquiry that he had been in charge of the helper engine at the connection and had been asked by Night Chief Dispatcher Walsh to report the arrival and departure of passenger trains from that point, so as train No. 12 had come over the bridge from Fallbridge he had advised the dispatcher that it was approaching. When train No. 12 had arrived at the Oregon Trunk end of the bridge the helper engine had been moved out of the way and the train had been backed over the connecting spur to the westbound track. Its fireman had then asked him if they were to run as train No. 12 from there. He said that he did not know, but that train No. 4 had left with indicators showing it to be train second No. 8, and that when he had advised the dispatcher that train No. 4 was coming over the bridge the dispatcher had told him that that train would get orders at Biggs, so he supposed that train No. 12 would be treated the same way. The pilot conductor said that he had then got off the engine of train No. 12 and as it was passing he had shouted to the conductor that he would report the train by at 12.22.

Engineer Allen said that he had intended to stop at the telephone booth and wait for the conductor to ring the dispatcher but he had understood from the pilot conductor that the latter had already spoken to the dispatcher and that his train was clear to proceed. When he had seen train extra No. 3211 approaching as his train was rounding a curve he

[579] Interstate Commerce Commission report

[580] *Hood River Glacier*, Hood River, Oregon, United States, 8th December 1921

[581] *The Morning Oregonian*, Portland, Oregon, United States, 2nd December 1921

had assumed that it was on the proper, westbound, track.

Conductor Allison likewise said that he had taken it that the pilot conductor had already spoken to the dispatcher and it was in order for them to go on.

Pilot Conductor Clark said that after train No. 12 had departed he had telephoned the dispatcher to advise him that it had left at 12.22, whereupon the dispatcher had told him that train extra No. 3211 was coming westward on the eastbound track from Biggs.

The Oregon Washington dispatcher's office was normally at Portland but a failure of telephone communications had resulted in it being moved temporarily to The Dalles, 12 miles west of the accident site. Night Chief Dispatcher Walsh stated that he had authorised train extra No. 3211 to use the eastbound track from Biggs to avoid delay as there was no direct connection from the westbound track to the Oregon Trunk connection, so had train extra No. 3211 used its own westbound track it would have had to proceed some distance further towards The Dalles to reach a mains crossover, then back along the eastbound track to the Oregon Trunk connection. It was customary for westbound trains to use the eastbound track from Biggs when they were being diverted onto the Spokane line and for the conductors of eastbound trains to be told to go to Biggs, an office which was open continuously, for orders. He stated that he had instructed Dispatcher Stiles to tell Pilot Conductor Clark to hold train No. 12 there as train extra No. 3211 was coming over the westbound track from Biggs. Dispatcher Stiles denied this.

Dispatcher Walsh admitted that order No. 284 had not been properly worded; that it should have been a form 31 order (requiring the train to stop and the order to be signed by the engineer and conductor) rather than a form 19 order (which could be hooped up to the engineer and conductor as the train was under way); that the rules only permitted trains to be operated against the current of traffic when the line was blocked, which had not been the case; and that it was a violation of rules to operate trains on verbal instructions. With regard to the latter transgression he said that he had considered it a safe practice in this situation as no train could depart from the Oregon Trunk Connection without calling the dispatcher.

In conclusion the ICC found that the accident had been caused by the failure of Conductor Allison and Engineer Allen to contact the dispatcher before leaving the Oregon Trunk Connection, and Chief Night Dispatcher Walsh authorising the movement of extra train No. 3211 over the eastbound track without properly protecting the movement. Responsibility must be shared by Pilot Conductor Clark who had misled the crew of train No. 12 into thinking that the dispatcher had cleared their train to proceed.

According to Rule No. 83, train No. 12 should not have occupied the eastbound track at Oregon Trunk Connection without first ascertaining that all trains of the same or superior class had arrived or left. This was normally done by means of a train register, or in the absence of a register by the issuing of a special train order (Rule No. 83C). In the absence of either train register or operator one of the train crew should have assumed the role of operator and telephoned the dispatcher for orders.

Under Rule No. 83A as Oregon Trunk Connection was its initial station train No. 12 should have obtained a clearance card (Form 2643).

Chief Night Dispatcher Walsh had committed a gross violation of Rule 23 of the Special Instructions for Train Dispatchers by moving train extra No. 3211 against the current of traffic. For safe working, when trains were worked against the current of traffic signatures must first be obtained from all engines or trains occupying either track within the limits of the order and the order placed at junction points with connecting lines. When train order No. 284 was issued there had been no operator to issue such order at Oregon Trunk Connection to either train No. 12 or the helper engine. The only precaution taken by Dispatcher Walsh had been the placing of order No. 284 addressed to eastbound trains at DK, the dispatchers office at The Dalles, where there was no train order signal and which was not a clearance station for passenger trains, the latter being WH office, more than a mile distant. Dispatcher Walsh claimed that he had been unable to contact WH over the despatching circuit and had intended to do so via the city telephone. It was doubtful if this order was put out at DK at all before issuing it at Biggs to train extra No. 3211, the inferior train, as required by the rules.

Regarding Pilot Conductor Clark, the ICC considered that although his intentions had undoubtedly been for the best his conduct had been such as to mislead the train crew and his responsibility for the accident was great.

The commission noted that, 'The investigation of this accident discloses a disregard for rules and many loose operating practices for which the officials of the Oregon Washington Railroad & Navigation Company are responsible.'

Beyond the actions of the persons directly involved in the collision the commission found much to criticise. It remarked that the detouring of trains was irregular but by no means uncommon. Good railroad practice required to begin with the provision of a competent operator at Oregon Trunk Connection, to receive orders and give instructions to train crews. Despite there being no operator there, no special instructions had been issued requiring train crews to contact the dispatcher. Expecting train crews to perform the exacting duties of operator was dangerous and could not be too strongly condemned, nevertheless the testimony of Superintendent Buckley showed that he was happy for train crews to act as operators.

Continuing, the ICC stated that the practice of running trains against the current of traffic was dangerous and should not be resorted to except when absolutely necessary, as recognised by the promulgation of Rule No. 23 for dispatchers. It seemed, however, from the testimony of Dispatcher Walsh, that the practice was commonly resorted to merely to save time, something which higher officials must have known about and condoned.

It appeared to be not uncommon for trains to leave the Oregon Trunk Connection and proceed over the 8 miles to Biggs without either a train order or a clearance card, simply on the authority of a verbal statement from the dispatcher such as, "Go over to Biggs and I'll fix you up there." No blanks or other stationary were in any case kept at the telephone booth at the junction with the Oregon Trunk Connection for receiving orders or clearance cards.

Conditions at the dispatcher's office at The Dalles were not such as to ensure safe and efficient operation. Unable to use the telephone despatching circuit, the dispatchers were having to revert to the telegraph. On the night of the accident Dispatcher Dolan, a new man, was technically on duty but orders were being issued and signed by Dispatcher Stiles, another new man. Neither man had been familiar with the prescribed form in which train order No. 284 should have been issued and there was no rule book to hand so they were being helped by Chief Dispatcher Walsh, but even he did not issue the order correctly and in issuing it had violated the hours of service law. Dispatcher Stiles was ignorant of Rule No. 23; according to Pilot Conductor Clark the dispatcher had told him to advise the conductor of train No. 4 that they could go to Biggs for orders, and so on. Train despatching at The Dalles was being done in an irregular and haphazard manner.

In the closing paragraph of its report the ICC wrote, 'The operating officials of the Oregon Washington Railroad & Navigation Company can not be too severely criticized for their failure to provide a safe method of working trains between Oregon Trunk Connection and Biggs. This is not an isolated case involving the improper handling of a single train, but represents a practice which has grown up as a result of detouring trains with considerable frequency. These officials can not escape their responsibility in connection with this accident on the ground that they were ignorant of these conditions and practices; it was their duty to know of them. The promulgation of rules and instructions is but the smallest portion of their duties; only when they know that such rules and regulations are being fully complied with can they be said to have discharged their duty.' [582]

Six people died and 18 were injured in a collision at Rilly-Semuy station, near Vouziers, France, on the Chemins de fer de l'Est, on 2nd May 1922.

On the day of the accident omnibus train No. 620, which left Hirson at 14:05 and was due at Bar-le-Duc at 20:29, was still at Rilly at 17:30, although it should have left there at 17:06. At Rilly the train had to pick up three more vehicles, which involved it setting back from main track No. 2 across main track No. 1 to collect the vehicles from the siding (track 3). As this manoeuvre was in progress the shunting train was run into by freight train 6/D H1, which usually ran through the station at 30 or 40 kilometres an hour on track No. 1.

Killed outright in the accident were M. Siroux Jules, aged 63, a mail carrier, of Ste Menehould; Mme Petit, aged 53, of Termes (Ardennes); M. Riegler, a watchmaker, of Montcornet, Aisne, a 10-year-old Belgian child named Vandenberghe and a fifth person whose identity was not immediately known. One of the injured passengers, M. Van Fillout, another Belgian, who had suffered a fractured skull, died on 4th May and at that time six others were still in a serious condition, including a woman who had to have her left leg amputated. [583]

An investigation into the accident blamed it on a signalman and on M. Gustave-Jean Lucier, the Rilly-Semuy stationmaster, aged 41, whose wife chanced to be among those injured. It was stated that the signals had not been maintained at danger against the freight train before the shunting move was begun. [584] Both the stationmaster and M. Augustin-Jean-Baptiste Champeau, aged 37, described as a *facteur mixte*, who was presumably the signalman, were charged with involuntary homicide and involuntary wounding, arrested and remanded at Vouziers.[585]

On 22nd June both the stationmaster and the signalman were found guilty and they

[582] Interstate Commerce Commission report
[583] *L'Homme Libre*, Paris, France, 4th May 1922
[584] *La Charente*, Angoulême, France, 6th May 1922
[585] *Excelsior*, Paris, France, 4th May 1922

were sentenced respectively to five and three months' imprisonment. [586]

A misunderstanding on the part of a train conductor and his failure to discuss the matter with other members of his train crew led to a collision near Grand Gorge, New York, United States, on the Ulster and Delaware Railroad on Friday 26th May 1922. Work train extra No. 21 consisted of engine No. 21, a flat car, a steam ditcher, two flat cars, and a caboose, in charge of Conductor Guy Mattice with Engineer A.J. Pelham. At Stamford, eight miles west of Grand Gorge, at 7.53 a.m., the train was given order No. 6, which read:

Order No. 1 is annulled. Engine 21 will work extra 750 am until 1130 am between Stamford and Roxbury protecting against Extra 22 east after 1020 am.

Roxbury was six miles east of Grand Gorge. The train reached Grand Gorge at 9.57 a.m., where it received train order No. 13, which read:

Eng 21 will work extra 1130 am until 3 pm between South Gilbaca and Arkville protecting against extra 22 east after 1130 am and against extra 39 west after 1240 pm.

Arkville was 17 miles east of Grand Gorge. The train left Grand Gorge at 10.58 a.m., moved eastwards two miles, stopped to pick up some scrap and then, at 11.22 a.m., began to set back towards Grand Gorge, when it was run into by train No. 22 at about 11.27 a.m. All the vehicles of the work train except the steam ditcher were demolished and the engine and leading two cars of train No. 22 were derailed. Six railwaymen died in the collision and five were injured.

Eastbound extra freight train No. 22 consisted of engine No. 22, ten cars and a caboose in charge of Conductor Joseph Redland and Engineer Charles Neeba. [587] At Kortright, 26 miles west of Grand Gorge, this train received train order No. 7, which read:

Order No. 1 is annulled. Eng No. 22 will run extra Kortright Station to East Meredith and East Meredith to Arkville. Work train extra 21 protects against extra 22 east between Stamford and Arkville after 10.20 am.

The Interstate Commerce Commission (ICC) inquiry determined that after the scrap had been picked up east of Grand Gorge, Supervisor William Lafferty (one of those killed) had asked Conductor Mattice if, before the train proceeded eastward, it could return towards Grand Gorge to drop off the section crew at the spot where they had earlier been picked up. Conductor Mattice had agreed to do so and Brakeman Holden was instructed to go to the engine and instruct Engineer Pelham to set back. Flagman Krom had suggested that he go back and protect but Conductor Mattice said not to do so as the train could back quicker than the flagman could do on foot. Conductor Mattice said that he had been thinking in terms of train order No. 13, and that he did not need to protect until 11.30. Engineer Pelham knew that his train needed protecting and assumed that his conductor had arranged it. Fireman Sember said he had not known of train order No. 13. Conductor Mattice said that he did not recall receiving any similar orders previously. The chief dispatcher said, however, that such orders extending the hours of work trains were common.

---

586 *Le Temps*, Paris, France, 23rd June 1922

587 *Catskill Mountain News*, Margaretville, New York, United States, 2nd June 1922

The ICC concluded that the accident had resulted from the failure to provide flag protection for the work train for which Conductor Mattice was responsible. It stressed that the accident showed the necessity for a thorough understanding between all train crew members of the contents of all train orders. [588]

At 02:30 on the morning of Tuesday 11th July 1922 the Asturias mail, train No. 423 from Madrid to Gijón, collided head-on with *rápido* (express) train 406 from La Coruña, Vigo and Gijón in Paredes de Nava station, Spain, on the line of the Compañía del Norte. The trains met on the points at the León end of the station. An early report spoke of the despatch of relief trains and the return of one of them with some of the injured, including Fireman Estaniciao Diaz Tazcón of the express. The engines of the mail train, No. 3012, and the express, No. 3139, were said to be wedged into one another. The express came off relatively lightly in the collision. Its leading van was demolished and sleeping car No. 844, marshalled second in the train, was slightly damaged, but the remaining five first-class coaches and van escaped unscathed. Much less fortunate was the mail train. This train consisted, behind the engine of a van, of a carriage containing a mixture of bunks and couchettes, ten first-, second- and third-class coaches with seating accommodation and a mail van. In the collision the leading van was telescoped by the tender of the engine and two of the passenger coaches also telescoped one another. Thirty passengers, all in the mail train, died, together with its Driver Vicente Avella and Fireman José Maria Fernández and the conductor of the express. Conductor José Fernández (who apparently had an almost identical name to the fireman, though this might be an error in the newspaper report) of the mail train and Fireman Jesús Villalba of the express escaped with slight injuries. Over 50 passengers were hurt to a greater or lesser degree.

Early reports blamed the accident on the driver of the mail train, who had failed to stop in Parentes station. That the accident had not been even more serious was said to be thanks to Driver José Villa Mazariego of the express, who had almost pulled up his train before the collision.

Conductor Fernández of the mail train, who miraculously survived the accident, gave a detailed account of it. He said that he had been well aware that according to the timetable his train was due to cross the express at Parentes. When his train began to pass through the station without stopping he had looked out of the window and seen the express at the points. He had then dived for the brake and applied it hard, becoming aware that his driver had belatedly also applied his brakes. Realising that a collision was then inevitable, he had closed his eyes and clung to the brake. Somehow he had escaped with some painful bruises. [589]

As is sadly often the case, some decided to profit from the disaster. In this case it was not a matter of parties robbing the bodies of the victims but people profiting in legal, if unethical, ways. For example it was reported that the doctors (literal translation of *médicos* but undertakers meant?) who carried out the embalming of the bodies of the dead first made enquiries as to the financial situation of the victims' families in order to determine how much money they could demand for their services. According to the author's less than perfect understanding of the Spanish text the Norte company also

[588] Interstate Commerce Commission report

[589] *La Vanguardia*, Barcelona, Spain, 12th July 1922

demanded payment in advance for the transportation of the bodies of the victims of the accident caused by its employees. [590]

On Friday 21st July 1922 the Baghdad – Basra express came into collision with a stationary train at a point eight miles south of Baghdad, Iraq. Five coaches of the express were reportedly wrecked and an Englishman, a Mr Thompson, described as an executive engineer at Hillah on the Euphrates, was one of those killed. He had perhaps been involved in looking for oil, as the first well was discovered in 1923. Others who lost their lives were an Arab officer and five other Arabs; whether the latter were soldiers or civilians the report did not say. Fourteen people were hurt. [591]

The day after the Iraq accident Logan, Missouri, United States, was the scene of a head-on collision between two passenger trains of the St Louis-San Francisco Railway at a crossing place on the single track. The line was operated by timetable, train order and an automatic block system. The trains involved were eastbound No. 2, the *Texas Special*, and westbound No. 9, the *Meteor*. Train No. 2 consisted of engine No. 1051, a combination coach and baggage car, a coach, and five Pullman sleeping cars, all of all-steel construction, in charge of Conductor Laughlin with Engineer Donohoe and Fireman C.S. Nealy (or Eley). Train No. 9 consisted of engine No. 1041, a baggage car, a combination mail and baggage car, two coaches, a dining car and three Pullman sleeping cars, in charge of Conductor Cantillion with Engineer C.H. Ring and Fireman A.W. Gelker.

At Aurora, 5½ miles short of Logan, train No. 2 received order No. 34, Form 19, to meet train No. 9 at Logan, train No. 9 to take the siding. Albion had been left at 3.32 a.m., nine minutes late, it stopped just short of the east switch at Logan, and while standing there was struck by train No. 9.

Train No. 9 left Billings, 5 miles from Logan, at 3.37 a.m., an hour and eight minutes late. At Billings it had received a copy of train order No. 36, Form 31 (this train was inferior to train No. 2 and so whereas train No. 2 received a Form 19 order, delivered via a hoop as the train was passing, train No. 9 received that same order as a Form 31 order, for which it had to stop and the conductor and engineer sign the order). Train order No. 36 read:

No 2. Engine No. 1051, meet No. 9, engine No. 1041, at Logan instead of at Aurora and meet No. 5, engine No. 1047 at Nichols. No. 2 take Siding on Passenger Siding Nichols. No. 9 take siding at Logan.

Approaching Logan train No. 9 passed signal No. 2065 at caution, passed signal No. 2619 at danger and instead of heading into the siding at the east switch continued on the main line and collided with train No. 2 at about 3.45 a.m. at a speed estimated at between 20 and 45 miles an hour.

Both engines and the first two cars of each train were damaged in the head-on collision, and the combination coach and baggage car of train No. 2 telescoped the adjoining coach for a distance of about 10 feet. Five passengers and Engineer Ring were

---

590 Ibid, 14th July 1922
591 The Times, London, England, 24th July 1922

killed and 107 passengers, three railwaymen and two Pullman porters were injured. The passengers who lost their lives were all from the same family, being Mr and Mrs Andrew Hammer, of Stoutland, Missouri, and three of their five children, Vera, 10; Clara, 8 and Dan, 2. Their son, Bunner, aged 12, and daughter Marjorie, aged six months, were among the injured. [592] After the accident striking shopmen from Monett rushed to help in the rescue work and a relief train was sent from Springfield. [593]

At the Interstate Commerce Commission (ICC) inquiry Conductor Cantillion stated that when Operator Allhands had given him train order No. 36 in the office at Billings he had read it out to Engineer Ring and the engineer had repeated it back to him, Porter Stevenson also being in the office at the time. The operator and the porter heard the engineer say that they were to head in at Logan. Engineer Ring then read the order to Fireman Gelker, and repeated that they were to go into the siding at Logan.

Fireman Gelker stated that speed had been slightly reduced at signal No. 2605. Then he had been busy with the fire and had not realised that they were so close to Logan until his driver applied the brakes in emergency. Seeing the other train only a few cars away and realising that there was no chance of stopping he (the fireman) had then jumped. Fireman Gelker and Conductor Cantillion put the speed of the train approaching Logan at 50 miles an hour. Porter Stevenson put the speed at 20 to 30 miles an hour as they reached Logan.

Conductor Laughlin instructed Head Brakeman Kines to go and open the east switch for train No. 9 to avoid delay but before the latter had had time to do so train No. 9 had arrived. Engineer Donahue saw the headlight of train No. 9 close at hand in time for him and his fireman to jump from their engine. Engineer Donahue was of the opinion that Engineer Ring had whistled three blasts for him to back his train up, but there had been no time to do so.

Head Brakeman Kines testified that he had got to within 10 or 15 feet of the east switch when train No. 9 had passed him, travelling he thought at 40 or 45 miles an hour.

The ICC concluded that the accident had been due to the failure of Engineer Ring to obey signal indications and head in at the east switch at Logan. It appeared that he had understood the order on leaving Billings. The only explanation which could be put forward was that train No. 9 was normally the superior train and hence took the main line. Had he reduced speed to 25 miles an hour on passing the caution signal he could have stopped at signal 2619, 291 feet east of the point of the accident, and prevented it. The accident thus, again, pointed to the need for an automatic train-control device to stop trains which were not obeying signals. [594]

France suffered another accident on Saturday 21st October 1922 at Rouffach station on the Chemins de fer d'Alsace et de Lorraine line between Colmar and Mulhouse. Strasbourg – Lyon – Bordeaux express No. 4, which had left Strasbourg at 22:30, carrying ordinary passengers and soldiers on leave, was travelling at 80 kilometres an hour when it was diverted into a siding 600 metres short of Rouffach station and collided

[592] *The Monett Weekly Times*, Monett, Missouri, United States, 28th July 1922

[593] *The Sedalia Democrat*, Sedalia, Missouri, United States, 28th July 1922

[594] Interstate Commerce Commission report

with a freight train. The express engine was badly damaged and the ends of first- and third-class carriages at the front of the train were crushed. A first report spoke of 11 dead, seven of them soldiers, and 12 injured, two very seriously. The injured were transported to hospitals at Rouffach and Colmar. The accident happened at 23:55.

Those killed were named as Sergeants François Gonnet and Lucien Chauvelot of the 152nd infantry regiment; Corporal Lareguestre and Private Hervy Louis of the same regiment; Corporal Charles Godiguon, male nurse at the Colmar military infirmary; M. Julien Chovo, workshop manager at the Postes Télégraphes et Téléphones at Belfort; Marcel Guyonnet, of Limoges; Kamel-Agreiss, State litigation lawyer and a member of the Egyptian mission to Cairo; Félix Vial, of Dijon; Mme Alberte François Gautheron, of Culoz and an unidentified woman.

An investigation into the accident quickly placed the blame on Signalman Siffert and Head Porter Kleix and they were arrested. [595] A later report on the 22nd stated that a M. Claudel, of Dijon, had succumbed to his injuries, increasing the number of dead to 12. [596]

On 24th November at Colmar the two railwaymen were found guilty of homicide through negligence and each was sentenced to 13 months in prison (suspended) and fined 100 francs. [597]

A passenger train returning from the coronation jubilee collided with another train near Costanza, Romania, on or about Monday 23rd October 1922, according to a Central News Agency despatch from Bucharest. Thirty people died. The crew of the colliding train were said to have been intoxicated. [598]

Another disaster occurred in France on Sunday 29th October 1922 on the main line of the Chemins de fer de l'État at Châtelaudren on the main line from Paris to Brest. Fast train No. 501 had departed from Paris Montparnasse at 20:05 on Saturday evening and was due in Brest at 07:15 on Sunday morning. It was considerably delayed on leaving Saint Brieuc. Running at speed past La Méugon and Plouvara-Plerneuf, it was only a few minutes from Guingamp. Suddenly, as his train approached the small station at Châtelaudren, the last before Guingamp, the driver of the express saw in front of him the red tail lamp of freight train No. 4555, which was being shunted off the main line but had not yet got in clear. He shut off steam and applied the brakes but the distance was too short and his engine struck the rear of the freight train with great force, destroying six of its waggons. The leading van of the express and the following second- and third-class carriages were telescoped into one another and overturned.

A reporter described the scene after the accident: 'It was, in the dark, a scene of indescribable chaos. There was smoke, terrible noises, cries, shrieks of pain from amongst the wreckage. Distraught passengers, roused from their sleep by the tremendous shock, were running along the track, shouting; others, who by a miracle had escaped death, were wringing their hands, calling for loved ones who were perhaps buried in the

---

595 *Le Radical*, Paris, France, 22nd October 1922
596 *Le Populaire*, Paris, France, 22nd July 1922
597 *L' Œvure*, Paris, France, 25th November 1922
598 *Lake County Times*, Hammond, Indiana, United States, 24th October 1922

wreckage; the enormous heap of partitions, boards and other debris. The terrible scene was illuminated by the light from the engine firebox.'

The one member of the station staff who was on duty witnessed the collision then sounded the tocsin to arouse the local inhabitants to the disaster which had happened in their midst, and this alerted M. Le Tellier, the mayor of Châtelaudren. When the latter arrived on the scene he found the uninjured passengers at work extricating their less fortunate fellow travellers, sometimes, according to the report, at the risk of their own lives, a party of Marines especially distinguishing themselves. Fireman and members of the Red Cross and the Société de Secours were also soon in attendance. As soon as news of the accident reached Saint Brieuc at 06:16 Operations Inspector Le Hardy, who chanced to be there, commandeered a light engine and set off for Châtelaudren. He was soon followed by M. Jamet, the Saint Brieuc stationmaster, who had made up a rescue train which carried station staff and Drs Héry and Prigent.

The investigation into the accident began at once, the parquets (prosecutors) of both Saint Brieuc and Guingamp going to the scene. It was determined that because of a violent snowstorm that had been raging at the time of the accident the *avertisseur* (distant signal) on the Saint Brieuc side of the station had malfunctioned when it was operated from Châtelaudren station. Driver Jaurand, when he had arrived at the *avertisseur* running at 100 kilometres an hour, had thus found it showing *ouvert* (open, that is, clear) and it was only when he had seen the *carré* showing *fermé* (closed, that is, danger) that he had had any warning, only 150 metres from the shunting train. The porter at Châtelaudren who was working the signals declared that he had been sure that *avertisseur* was working properly, though he had not heard the indicator bell which should have sounded in connection with it to prove that it was functioning. [599]

A newspaper reporter visited and interviewed M. Le Coquen, the railwayman on duty at Châtelaudren on the night of the accident, who denied that he was to blame for it.

M Le Coquen: "I must tell you that I was the only person one duty at the station at the time of the accident. Usually there is no problem, but yesterday, in addition to the express, which was due to pass 50 minutes late, I was advised of a freight train, the one which was involved in the accident. It was a short time before the arrival of the omnibus train due from Guingamp at 05:38. I worked out that I would not have time to shunt the goods train before the arrival of either the omnibus train or the express. I immediately telephoned to refuse the goods train but was told that it was too late, it was already on the way with 30 waggons."

"Was it not possible to stop the (freight) train?"

"No, at that hour there was no one on duty except at Guinchamp and Châtelaudren."

"Then you operated the signals {to protect the shunting goods train}?"

"Certainly."

"Did you hear the indicator bell ring?"

"I could not formally swear to it, but I am absolutely certain that I heard it. Had it not done so I would have instinctively worked the lever again."

The reporter added that M. Le Coquen had never been known to make a mistake

[599] *Le Ouest Éclair*, Caen, France, 30th October 1922

before and that the mayor of Châtelaudren refused to believe that he could be at fault. [600]

By the time of this report there were known to be 14 dead and about 40 injured being cared for in hospital at Saint Brieuc and Guingamp. Another of the injured died the next day.

M. Yves Le Troquer, the Minister of Public Works, went to the scene of the disaster a day or so after it had happened and rode on an engine to judge for himself the visibility of the signals, though the weather conditions then were doubtless much better than on the morning of the collision. Attention had now shifted to the actions of the Saint Brieuc assistant stationmaster, and his dispatch of the freight train in front of the express, despite the severe snowstorm and overruling the protests of his subordinate at Châtelaudren who was forced to foul the running line by a shunting operation in front of the oncoming express train, in contravention of a regulation drawn up by M. Troquer himself after the 1911 Courville disaster, which prohibited this being done during fog or falling snow. [601]

Eventually, at the end of July 1923, three different railwaymen stood trial and were found guilty at Guingamp of culpable homicide in connection with the accident. Châtelaudren porter Le Coquen received the most severe punishment and was sentenced to six months' imprisonment and fined 100 francs, Assistant Stationmaster Moulet at Saint Brieuc was given three months' imprisonment and Driver Jourand one month's imprisonment, presumably because although he had been deceived by the false clear signal it was judged that he had not been driving sufficiently cautiously given the weather conditions. [602]

What was perhaps an especially nasty accident happened on Wednesday 13th December 1922 at Humble, Texas, United States, on the Houston East and West Texas section of the Southern Pacific Railroad. The trains involved in the accident were passenger train No. 28, known as *The Rabbit*, and a light engine, No. 349. The passenger train consisted of engine No. 255, a combination baggage and mail car, two coaches and a Pullman sleeping car, in charge of Conductor Kempsey (or Campsey) and Engineer Holland (or Hottle). The train left Houston at 9.15 p.m. and collided with the engine, which had been moved foul of the main line, at about 9.50 p.m. In the collision the boiler check valve of the engine was broken off and searing steam at boiler pressure entered the front coach of the train. All the casualties were in this coach except the fireman of the train, who was injured jumping from his engine. Nineteen passengers, a newsagent, the conductor and the porter of the train were scalded to death and ten more passengers were injured. Apart from the conductor, fireman, and perhaps the porter and newsagent all the victims were black, as it was the 'Jim Crow' coach which was struck. [603]

The engine had been left on track No. 1 of the yard at Humble overnight, and it was the job of Engine Watchman Smith to resupply the engine with fuel oil and then take charge of it for the remainder of the night. Engine Watchman Smith told the Interstate Commerce Commission (ICC) inquiry that he had syphoned all of the oil from one tank car, disconnected the syphon and gone to his shanty to prepare his supper. His statements

---

[600] *Le Petit Journal*, Paris, France, 31st October 1922

[601] *L'Homme Libre*, Paris, France, 1st November 1922

[602] *Le Gaulois*, Paris, France, 1st August 1923

[603] *San Antonio Express*, San Antonio, Texas, United States, 14th December 1922

as to what had happened then were conflicting and did not agree with his initial statement shortly after the accident to the deputy sheriff. He implied that the engine had run away and that its throttle was defective to the extent that it would cause the engine to move of its own accord even with its brakes applied.

Engineer Holland told the inquiry that approaching the station he had reduced the speed of his train to 25 miles an hour. There had been a ground fog but about a third of a mile from the station he had glimpsed the train order signal at stop and had whistled for it to be cleared. Keeping his eyes on the signal he had only seen the engine fouling the track when his train was 100 feet from it, when he had applied the brakes in emergency and opened the sanders. The train order signal, 32 feet high, had been easier to see through the fog than the locomotive. He put the speed of his train at the moment of impact at 20 miles an hour.

Engineer Coker, who had stabled the engine, and Engineer Henning, who had taken train No. 215 100 miles on the day of the accident, stated that there had been nothing wrong with its throttle which would have caused it to move on its own.

The commission concluded that the responsibility for the accident lay with Engine Watcher Smith, who had probably moved the engine so that it could be refuelled from the other tank car while he was having his supper, and in doing so had moved it foul of the main track in front of the passenger train, the approach of which he had not observed. [604]

The next of what had been a series of bad accidents in France occurred on the evening of Sunday 18th February 1923 at Port-à-Binson on the Paris – Strasbourg main line of the Chemins de fer de l'Est. Involved in the accident was train No. 35, the 17:00 express from Paris Est to Strasbourg, which ran non-stop to Châlons-en-Champagne and had a maximum speed of 110 kilometres an hour. The train had suffered some delay on its journey and was attempting to regain time when it arrived at Port-à-Binson at 18:35. It was 500 metres from the station when its driver, named Besse, became aware of the waggons of a goods train obstructing the line. He immediately applied the brakes but the express was reportedly travelling at or near its maximum speed and a collision was inevitable. A waggon of the goods train, the leading van of the express and the tender of its locomotive were pushed almost 200 metres down the line towards Epernay and the front of the engine was smashed in. The remains of the van, which was almost completely broken up, were forced beneath the following third-class carriage. Conductor Tissier, who was riding in the leading van, by some means escaped with only slight injuries, and freeing himself from the wreckage he at once set off along the track to protect the opposite running line with detonators. The front van penetrated a compartment of the third-class coach filled with soldiers on leave and the whole of this coach was wrecked, while the following coach was said to have been left in 'a sorry state.' The carriages at the rear of the train were undamaged and their passengers escaped injury.

Rescue work was begun at once by railwaymen and uninjured passengers using lanterns and help soon came from the local inhabitants, together with Doctors Guénard and Janiver and Assistant Surgeon René Netter and Chief Surgeon Jean Pellot from Epernay hospital. Alerted by the Dormans police, local owners of cars and lorries

---

[604] Interstate Commerce Commission report

immediately put them at the disposal of the railway. Fifteen cars with medicines and nurses went to the scene and the injured were taken to the Auban-Möet hospital in Epernay and the bodies of the dead to the mortuary in the town.

During the night the Epernay *parquet* (prosecutor) went to Port-à-Binson and questioned the stationmaster and the signalman. The magistrates then decided to remove them to Epernay and after a fresh interrogation by the state prosecutor they were arrested. Stationmaster Millière, who had personally supervised the shunting of the goods train, stated that he had believed that the shunt move had been completed and the main line was clear, and had thus permitted the express to approach. The stationmaster and Signalman Guérin admitted that they had failed to observe the regulations. The details of their interrogation were given in full in the newspapers, thus:

'The *juge d'instruction* (examining magistrate) asked the signalman why he had reported the line clear.

"I asked the stationmaster, who was near me, 'Do you think that the {goods} train is stationary?'"

"Oh, yes, probably."

"So I cleared the signal {for the express}. The stationmaster being next to me, how could I not trust him?"

"But you should have waited for the signal from the pointsman near the shunting train?"

"Yes, I admit that I made a serious mistake. Bur I repeat that the stationmaster was next to me, so how could it not be all right?"

For his part, M. Millière conceded that he had been careless.

"I myself cleared the last signal, called the chequerboard, freeing the line. I thought that the goods train was stopped and cleared the signal automatically."

"You had sent the pointsman to the goods train, that is to the points, to give you the agreed signal?"

"That's right, but the man did not take with him the special key that he needed to open the box to give me the signal."

"That is not enough reason for you to avoid responsibility. The regulations demand that before clearing the line you await the agreed signal, that is to say the bell, from the pointsman. So if the latter could not send the signal, because of negligence on his part, you should have waited for him to come back and simply told you verbally. What do you say?"

"It's true. I was wrong."'

The two railwaymen were charged with homicide through negligence due to failure to obey the regulations, [605] but it has not be possible to find any report of a trial.

By the following day the casualties were stated to be 13 dead and 35 injured. [606] Another victim apparently died later. [607]

---

[605] *Le Petit Journal*, Paris, France, 20th February 1923
[606] *Le Figaro*, Paris, France, 20th February 1923
[607] *L' Œuvre*, Paris, France, 9th May 1923

On Saturday 17th March 1923 a troop train collided with a light engine at Friemersheim, Germany. At least 40 died and many were injured. [608]

A brief account reached London of a collision between two Régie trains (that is, ones operated by the occupying Franco-Belgian administration) between Wanne and Gelsenkirchen, Germany, on Thursday 28th June 1923. Five people reportedly lost their lives. [609]

On Sunday 1st July 1923 a passenger train and a goods train collided near Băile Felix, a spa resort in the Transylvania area of Romania. Seven people died and 20 others were injured. [610]

The day after the Băile Felix accident, Monday 2nd July 1923, a much more serious one occurred at Vintyleanca station, between Ploesti and Buzău, Romania, about 55 kilometres from Bucharest in the Wallachia region. Owing to the reported negligence of a shunter a Bucharest – Jassy (now Iaşi) mail train was diverted into a siding where it collided with a stationary goods train. Three passenger coaches were telescoped and a third-class carriage was said to have been reduced to matchwood. Sixty-three people died and over 100 were injured. Special transport took the injured to hospitals in Ploesti and Bucharest. [611]

Two hundred people reportedly died in a collision between two express trains near Plevna, Bulgaria, on the line between Sofia and Varna, on Monday 23rd July 1923. A report from Sofia on 26th July stated that 160 bodies had been recovered and that rescuers were still searching the wreckage for more bodies. Another 300 persons were said to have been injured. Faulty signalling was blamed for the collision. 'An engine and eleven coaches were derailed and flattened by the contact. From ten coaches, no one escaped alive, and dozens of victims were unrecognizable, mutilated under the grinding framework of the wooden cars.' [612]

Six people lost their lives and ten were seriously injured in a collision near Frydek, Czechoslovakia, in Moravia province, on about Wednesday 17th October, 1923. A large number of other passengers were slightly injured and it was feared that more bodies might be discovered when the wreckage was cleared away. [613]

---

608 *Eisenbahn-Katastrophen in Deutschland* by Hans Joachim Ritzau, pg. 84 (in German)
609 *The Times*, London, England, 30th June 1923
610 Ibid, 2nd July 1923
611 Ibid, 4th July 1923
612 *The Tyrone Daily Herald*, Tyrone, Pennsylvania, United States, 26th July 1923
613 Ibid, 17th October 1923

On Thursday 15th November 1923 the driver of an engine bound for Heilbronn failed to stop at a "Halt" indicating Ausfahrsignal (starting signal) at Stuttgart-Unterturkheim, Germany. He then continued for 1,100 metres along the opposite running line, without apparently realising it, and eventually came into head-on collision with passenger train P 1431, whereby 12 people were killed and 15 injured. For his gross negligence he was sentenced to a year and three months' imprisonment and, not surprisingly, lost his job as a driver. [614]

At 02:00 on the morning of Monday 24th March 1924 train No. 62, the Ostend – Basle – Geneva international express which had left the Belgian port at 15:45 on Sunday afternoon and Metz at 01:05 met with disaster at Benestroff in Alsace, France, on the Chemins de fer d'Alsace et de Lorraine. The signals were clear for the train to run through the station but instead it was diverted into a siding and into collision with a goods train. The express had been running at a good speed. When its driver felt the violent shock as his engine ran onto the open points he at once shut off steam and applied the brakes but the 100 metres that he had to pull up in was of course nowhere near enough and a powerful impact ensued. The engine demolished the waggons at the back of the goods train, its tender overturned and the four leading vehicles of the train telescoped one another, trapping their passengers in the wreckage.

As soon as the initial shock had passed the station staff, aided by local inhabitants, began the rescue work with the means at their disposal. The first to be released were 13 victims who were in considerable pain, and one of whom was critically injured. They were taken to hospitals at Metz and Strasbourg. The severely mutilated bodies of eight victims were also discovered. Two of them were Conductor Camille Ruckly of the goods train and Driver Nicholas Dutroux (travelling as a passenger on the express?); two others were soldiers named Camouf and Corrès. The remaining four victims could not be immediately identified. Driver Maugeot of the express suffered a broken leg and Fireman Lamberger was more seriously injured and was still trapped three hours after the accident. In addition three postal workers named Schaeffer, Andt and Robertsau were trapped in the Ostend – Basle mail van. The injured included an Italian workman and a Czechoslovak but the London *Times* assured its readers that although the train connected with the cross-Channel ferry from Dover there had been no British passengers among the victims. Relief trains were run from Metz and Sarreguemines and a special train from Strasbourg carried officials who were to undertake an investigation into the accident. [615]

The investigation quickly reached a conclusion as to the cause of the accident and Signalman Jean Farny, at *poste* (signal box) WB, was arrested. It was alleged that he had let the goods train into the siding and had then fallen asleep before he had reset the road, not waking up before the accident. [616]

By the following day, seven of the dead had been identified. Apart from Conductor Ruckly and Driver Dutreux they were soldiers Jean Baptiste Camons and Gustave Modeste Carrez; M. Henri Baer, a railway locksmith; M. Jean Neiss, a farmer and Mme

[614] Katastrophen der deutschen Bahnen Teil II: Chronik 1845-1992 by H.J. Ritzau, pg. 56 (in German)
[615] *Le Populaire*, Paris, France, 25th March 1924
[616] *Excelsior*, Paris, France, 25th March 1924

Noémie Lambert, a fairground merchant. The other victim remained unknown. [617] In addition Fireman Lamberger died of his injuries less than 24 hours after the accident. [618]

At the end of April Signalman Farny was released on bail as it transpired that he was not quite as much to blame as had initially been supposed. Due to a defect in the signalling the disc had shown clear for track No. 1 despite the road being set for track No. 4. [619] This apparent mitigating factor notwithstanding, at the beginning of July he was found to have been guilty of causing the accident and sentenced to six months' imprisonment and fined 300 francs. [620]

Early on Wednesday 23rd April 1924 a major disaster occurred in Switzerland at Bellinzona on the Gotthard main line of the Swiss Federal Railways. One of the trains involved was the second portion of northbound Milan – Chiasso night express train No. 51b and the other was a southbound express, train No. 70, consisting of portions from Basle and Zurich which had been combined at Arth-Goldau. Train No. 51b consisted of electric locomotives Nos. 12329 and 12342, and a number of vehicles, in particular an old first- and second-class coach equipped with gas lighting, placed second in the train behind the heating coach. In charge of its locomotives were Drivers Buffi and Burg, with Secondmen Pietro Boni and Auguste Schwarz. Train No. 70 was also double headed, with electric locomotives Nos. 12502 and 12322 and a number of coaches. Its locomotives were in charge of Drivers Mauricio Cavivioli and Jacob Briner, [621] with Secondmen Vittorio Brunetti and Gozzi.

The two trains arrived at Bellinzona on their respective tracks at the same time and should have just passed one another but instead train No. 70 was diverted to the right across the northbound track, the facing points having been left set for a southbound train to access the San Paulo goods station on the west side of the main line. The engines of the two trains collided on the crossing. All four engines were badly damaged and jammed together but otherwise only the heating coach and luggage van of the southbound train were much damaged. The northbound train suffered much worse. Its luggage van was at its rear and the German coach marshalled behind the heating van immediately caught fire due to damage to its gas cylinders, was soon blazing furiously and the fire spread to the following Italian coach. The northbound train was a very international one; in addition to 15 Swiss passengers there were 52 Italians, 45 Germans, 4 Americans, 2 Czechs, 2 English, 2 French and 2 Norwegians. At least 14 and possibly 17 of the passengers and 7 railwaymen died. The passengers who lost their lives included Herr Karl Helfferich, wartime German Finance Minister and a leading member of the *Deutschnationale Volkspartei* (the German National People's Party, a conservative party of the Weimar Republic which preceded the Nazi Party); Signor Planzi of Milan and Professor Weithaus and Dr Wertheim, both of Berlin. [622] The railwaymen who lost their lives were Drivers Cavivioli and Briner; Secondmen Boni and Schwarz, MM. Degottardi and Snozzi, who were operating the heating coaches and M. Scexo, travelling as a passenger in the heating

[617] *Le Grand écho du Nord de la France*, Lille, France, 26th March 1924
[618] *L'Homme Libre*, Paris, France, 25th March 1924
[619] *Le Nouvelliste d'Alsace*, Colmar, France, 28th April 1924
[620] *Le Matin*, Paris, France, 9th July 1924
[621] *Le Confédéré*, Martigny, Switzerland, 25th April 1924
[622] *The Times*, London, England, 24th, 25th and 26th April 1924

coach of the northbound train.

The inquiry into the accident determined that southbound goods train No. 8572 had been running 55 minutes late and the official at Ambri-Piotta, near the south end of the Gotthard Tunnel, had therefore given priority to train No. 70, which normally followed the goods train to Bellinzona. He notified several stations as far south as Biasca, but not Bellinzona, a little further south again. The official at Biasca did not, as he should have done, advise Bellinzona either.

The Bellinzona signalman had admitted southbound goods train No. 668 to the San Paolo yard at 02.06 and not having been advised of the altered order of the southbound trains, he left the points set from the southbound line to the goods yard for goods train No. 8572.

There was a fixed signal north of the connection to San Paolo goods yard which stood at danger. In clear weather it was visible for a considerable distance, but it was not provided with a distant signal. Further south the home signal for Bellinzona station stood at clear. According to the evidence of Secondman Brunetti he saw the danger signal as the southbound train approached it travelling at about 70 kilometres an hour. When Driver Cavivioli failed to apply the brakes the secondman attempted to warn him but the driver replied that the signal only applied to trains entering the goods yard. No, it applies to all trains, the secondman had protested, and the driver had then had misgivings and began to brake. By then though it had been too late and their train was diverted across the northbound line. As the headlights of the northbound train had appeared close at hand the secondman had jumped and escaped. Apart from the fact that the driver had probably 'read through' to the signal at Bellinzona station it was found that for the year prior to the accident he had only been engaged in shunting operations; he had been chosen to operate the express due to a lack of drivers to operate the heavy Easter traffic. He was, therefore, probably not as familiar as he might have been with the layout at Bellinzona, where a so-called temporary signalling installation had been brought into use on the opening of the San Paolo goods yard, but had not been updated for some time.

The fact that the signals could have been cleared for a northbound train despite the road being set for a train to cross its path, protected by only a stop signal, was not found fault with in the official report, which insisted that drivers must obey signals, but was much criticised elsewhere. The official report did note that the signalling at Bellinzona was in the process of being updated. It was also pointed out in the Swiss press that the accident should have been prevented by a system of automatic train control.

One positive result of the accident was that the Swiss Federal Railways forbade in the future the use of gas-lit rolling stock on any of their trains. [623]

The small railway network of Portugal did not see many serious accidents, but a fairly bad one happened on Tuesday 19th August 1924, on the outskirts of Lisbon, involving a Lisbon – Estoril express and a goods train. [624] Five persons died and many were injured. The passengers who lost their lives were: Joaquim Franco de Matos director Havas

[623] *Railway Accidents of Great Britain and Europe* by Ascanio Schneider and Armin Masé, pgs. 83-90

[624] *The Times*, London, England, 20th August 1924

agency in Lisbon, aged 85; João Anastacio Gomes, aged 80, consul general for Costa Rica in Portugal; Paul Marthe, aged about 70, identifiable only by a business card and Franciso Alberto de Almeida, aged 16, a bank employee from Cascais. Fireman Luciano Ramos of the express was the fifth fatality. His driver, Sancho dos Santos, aged 60, was in hospital in a very serious condition.

On the day of the accident there was single-line working between Belem and Pedrouços, the next station down the line, and this was an important factor in relation to the collision.

A newspaper report the day after the accident attempted to determine how it had happened and who was responsible for the accident. It wrote:

'The temporary stationmaster at Belem was not responsible.

Why?

Because he did nothing to contribute to the accident and did everything that he could to prevent it.

Who was to blame then?

Just Driver Santos of the express and the Belem head porter (*praticante* in Portuguese)

Why?

Because Driver Santos at the start of a long journey, instead of concentrating on driving his engine, was talking to his fireman, while the porter, without checking with the stationmaster, as he should have done, telegraphed back to Alcantara that the line was clear, when in fact it was occupied by the goods train.

The temporary stationmaster was named as Edgar José de Silva, aged 26, the chief porter as José Gomes Serra, aged 21, and the pointsman as Antonio dos Santos, aged 46. The last two were arrested after the accident, as presumably would have been the driver had he not been in a critical condition in the infirmary.

A reporter spoke with Stationmaster de Silva to get his account of what happened. The latter said, "I will tell you the truth, and only the truth. I don't know why anyone should think that I was in any way responsible as it has been shown that I am not in the least to blame." He spoke calmly and sincerely as he recalled the accident.

"At 5 in the afternoon I received a message that the "Mouse" as we call it, that distributes the posts for the electrication of the line, was at Pedrouços waiting to come down."

"And train No. 302, the goods train?"

"Was at Belem, shunting, as it had to pick up two waggons. Having received the message, and with still 20 minutes left before the express left Lisbon, I did what was natural. As the goods train was not ready to go I accepted the "Mouse," but the latter took longer than expected to reach us. In view of this I went to the office to tell the head porter to warn Alcantara that the line was blocked and to hold the express there, but to my amazement the head porter told me 'I have already accepted it.'"

"What did you think?"

"That there was likely to be an accident and as there was no time to get the goods train out of the way I did all that I could to prevent it."

"What did you do?"

"I left the signal showing red against the oncoming express and ran back 500 metres with a flag to signal it to stop. This is what must be done when there has been a serious mistake with the signalling. Guard Junqueira saw my flag but I found to my horror that the engine driver was disregarding the disc signal at danger"

"How far out is the disc signal?"

"One and a half kilometres."

"I began to wave my flag furiously at the engine and shouted to the driver to stop. Several soldiers joined me in shouting and Guard Junqueira did the same. But it was no use. The train still came on, without slowing up.

"And the driver?"

"He was not looking out. When the engine passed me he was washing his hands. He did not see the disc signal or the flag I was waving. He didn't even hear us screaming at him to stop. And the train continued at speed."

"And the station staff?"

"They did everything that they could. The pointsman blew a horn and waved a danger flag. The driver of goods train No. 302 shouted, like the others, and blew the whistle of his engine. Driver Santos, distracted as he was, saw and heard nothing. The rest you know..."

The stationmaster added, "What a disaster! What a disaster! But it was not my fault at all. I assure you."' [625]

It seems a little strange that the newspaper felt so free to discuss who was to blame for the accident when the matter was very much *sub judice*. There was however a note at the end of the article stating that although two railwaymen had been arrested the authorities had not yet determined who was at fault.

At 03:40 on Saturday 12th July 1924 Chemins de fer de l'Est express train No. 38 bis, which had left Strasbourg for Paris at 22:10 the evening before, travelling at 70 kilometres an hour, collided with a coal train which was being shunted at Vitry-la-Ville station, France. Four of the coal waggons were telescoped into one another and thrown off the track and the two leading vehicles of the express, a postal van and a first- and second-class composite, were 'reduced to matchwood' while the third vehicle, a third-class coach, was badly damaged. Five passengers died and about 15 were injured. Most of the passengers on the train were soldiers on leave from the troops occupying the Rhine. The victims were all French apart from Mr Gabrial Torlotting, of New York, who suffered a broken ankle. [626]

M. Cardin, operating manager of the company, explained the circumstances of the accident. The coal train, from the Ruhr, had to be shunted out of the way of the express Vitry-la-Ville, but the operation had been complicated by the fact that a waggon which had developed a 'hot-box' (overheated axle) also had to be detached. During the shunting

---

625 *Diário de Lisboa*, Lisbon, Portugal, 20th August 1924

626 *The Times*, London, England, 15th July 1924

operations the train was wrongly diverted onto the main line, where the station signal was already clear for the express. Upon seeing the *signal avancé* at caution the driver of the express had reduced speed, but then seeing the station signal at clear he had opened his regulator again, and the express had been travelling at 55 kilometres an hour when it had struck the waggons. [627] Maurice Starck, the night officer at Vitry-la-Ville, was arrested and taken to Châlons prison. [628]

On Friday 6th March 1925 two goods trains collided at Mogosoaia, Romania, on the outskirts of Bucharest, owing, it was said, 'to faulty shunting.' Both engines and ten waggons were completely smashed. A number of soldiers were riding on one of the trains and the casualties were therefore considerably greater than they would otherwise have been. Eight people died and 28 were hurt. [629]

Probably late on Monday 30th March 1925, though reported in various newspapers as having happened on Tuesday 31st, a long-distance train from Tashkent and Turkistan collided head-on with a suburban train in Sokolniki Park, near Kazansky station, on the outskirts of Moscow, USSR. Several coaches were smashed, including the sleeper from Tashkent, and the bodies of the victims were buried in the wreckage. Mounted militia were used to keep back crowds of curious spectators while railwaymen fought to rescue those trapped in the wreckage. Scores of ambulances went to the scene and the injured were taken to nearby Sokolniki hospital. Many of those who died were killed while they were asleep. Fifty people reportedly lost their lives and a large number were injured. [630] [631]

On about Tuesday 22nd September 1925 two armoured trains, running without lights, collided at an unstated location in Estonia. Five people died and 12 were injured. [632]

At about 1.13 a.m. on the morning of Friday 16th October 1925 engine No. 118 of No. 37 up parcels was moving, tender first, in the up direction along the down loop line at Halsa Junction, India, on the Eastern Bengal Railway, while train No. 8, the down Dacca Mail, was approaching. The engine was not stopped short of the fouling point with the down main line and a collision ensued. The tender of the engine of the parcels train was completely wrecked and the engine and three leading coaches of the mail train were badly damaged and derailed. Ten people were killed outright and 37 passengers were injured, two of whom subsequently died in hospital. The damage to rolling stock was estimated at 108,038 rupees.

The Halsa Junction assistant stationmaster and a person referred to in the report on the accident as a gunner, who was presumably a railwayman rather than a member of the

---

627 *L'Œuvre*, Paris, France, 13th July 1924
628 *Excelsior*, Paris, France, 13th July 1924
629 *The Times*, London, England, 7th March 1925
630 *The Evening Gazette*, Xenia, Ohio, United States, 31st March 1925
631 *Indiana Evening Gazette*, Indiana, Pennsylvania, 31st March 1925
632 *The Times*, London, England, 24th September 1925

military, were tried and convicted for having caused the accident under section 101 of the Indian Railways Act, 1890, the stationmaster being sentenced to six months and the gunner to a year's rigorous imprisonment. On appeal the Sessions Judge acquitted the stationmaster but confirmed the sentence on the gunner. [633]

On Christmas Day, Friday 25th December 1925, a train carrying passengers home for Christmas collided with a freight train at Arad, Romania, not far from Bucharest. According to initial reports, 10 died and 20 were seriously injured. [634]

A serious buffer stop occurred on about Thursday 4th February 1926 at Moscow Oktyabrsky station, USSR (now Moscow Leningradsky). Two carriages of an express train from Leningrad (now Saint Petersburg) were smashed, five people were killed and 25 injured. [635]

Due to a mistake by a fahrdienstleiter (signalman) passenger train P 9266 collided with stationary passenger train P 9265 at Wülknitz, Germany, on the Riesa – Elsterwerda line, on Monday 22nd March 1926. Six people died and 11 were injured. [636]

Although seven people died in the accident a collision at Edgeware Junction, Alabama, United States, on Tuesday 30th March 1926 was not reported on by the Interstate Commerce Commission and was only noted in a couple of newspapers. Edgeware Junction was the point where a short branch to Edgeware Mine diverged from the branch to Bayview, not far from Birmingham. Bayview was a company town, built by the Tennessee Coal and Iron Company, by this date part of United States Steel, and it would seem likely that trains run over the branch, now part of the Birmingham Southern Railroad, were reserved for company employees, hence the lack of interest of the Interstate Commerce Commission in the matter. The trains involved in the evening collision were an accommodation train, carrying 200 negro workmen, and a freight train. It was raining hard at the time of the accident, which was blamed on a confusion of signals. [637] Twenty-one other men, including, a newspaper report noted, two white men, members of the train crew and the only whites on board, were injured. Four of them were said to be in a serious condition but it was believed that they were not fatally hurt. [638]

What was described as the first serious accident on the Tangier – Fez line in Morocco occurred on Wednesday 21st July 1926. Mixed train No. 105 which had departed from Casablanca at 9 p.m. the previous evening had passed Meknès and reached Seba Ayoun when due to a set of points having being left open it was diverted into a siding and

[633] Report by the Railway Board on Indian Railways for 1925-26
[634] *Evening State Journal*, Lincoln, Nebraska, United States, 25th December 1925
[635] *The Times*, London, England, 6th February 1926
[636] Katastrophen der deutschen Bahnen Teil II: Chronik 1845-1992 by H.J. Ritzau, pg. 61 (in German)
[637] *Dothan Weekly Eagle*, Dothan, Alabama, United States, 31st March 1926
[638] *Bluefield Daily Telegraph*, Bluefield, West Virginia, United States, 31st March 1926

collided with a waggon loaded with cereals. The vehicles of the train reportedly shot past the engine and seven passengers were killed and 20 injured, six seriously. The casualties included a French gunner and a mail carrier. [639] There was also considerable material damage. [640]

The Timisoar (now Timişoara) express, and a local train were in collision at Contesci, Romania, near Bucharest, on Monday 13th September 1926. Eight persons were killed and 30 injured. [641]

There was another collision in Romania on Saturday 2nd October 1926, again involving an express and a slow passenger train, at Cotzoreni, near Craiova. Five persons died and 12 were injured in this accident. [642]

Eight people died and over 50 were injured when a passenger train collided with a freight train near Vila Anastácio, Brazil, on Wednesday 13th October 1926. Most of the victims were workmen. A mistake in signalling was said to have caused the accident. [643]

A collision occurred near Machungo, Manchuria, on Wednesday 8th December 1926 on the South Manchuria Railway. Twenty-five persons lost their lives and 54 were seriously injured. [644] Confusingly a number of newspapers reported another collision in Manchuria which supposedly had occurred three days later. The *Railway Gazette* seems to have regarded this as another accident, giving its location as near Tiehlong. However, the casualties in this reported second accident were supposedly also 25 killed and 54 injured. [645] To the author this seems to be too much of a coincidence, and it would seem almost certain that there was only one accident on the earlier date (reported in an American newspaper the same day). It has not proved possible to identify either Machungo or Tiehlong on a modern map.

A major disaster at Rockmart, Georgia, United States, on Thursday 23rd December 1926, involved two named trains of the Southern Railway, the *Royal Palm*, southbound from Chicago, Illinois, to Miami, Florida, and the *Ponce de Leon*, northbound from Miami to Chicago. The southbound train was waiting at Rockmart for its northbound counterpart but the latter train, instead of taking the siding, continued along the main track and a head-on collision resulted. Eleven passengers, a newsagent and 7 railroad employees were killed or fatally injured and 113 passengers, 6 Pullman staff and 4 railroad employees were hurt. Most of the fatalities occurred in the Pullman dining car.

---

[639] *Le Matin*, Paris, France, 22nd July 1926
[640] *La Charente*, Angoulême, France, 24th July 1926
[641] *The Times*, London, England, 15th September 1926
[642] Ibid, 4th October 1926
[643] *The Abilene Daily Reporter*, Abilene, Texas, United States, 13th October 1926
[644] *The Anniston Star*, Anniston, Alabama, United States, 8th December 1926
[645] *The Railway Gazette*, Vol 45, pgs. 709 and 772

Mr Gillespie, a former Detroit police commissioner who was a passenger on the northbound train and who escaped injury, described the crash to a newspaper reporter when he arrived in Atlanta after the accident. "The crash came at 6.40 p.m. entirely without warning and while out train was making about 50 miles per hour. I was in the second car behind the diner and every person in our car was thrown to the floor. The impact was terrific, and the train jerked, groaned, and creaked. There was a great crunching sound and windows caved in almost the length of the car. If the brakes on our train were applied at all, it was at the very moment of the crash. As soon as I realized what had happened I got out and began investigating the damage and how it all happened. The fireman of a freight train which was on a siding told me that the *Ponce de Leon* had sped towards his train and the oncoming *Royal Palm,* which was around a slight bend, and that as the *Ponce de Leon* approached the freight her engineer saluted and the freight engineer, realizing that for some reason the northbound train was unaware of the existence of the other southbound immediately signalled to him to make an emergency stop. It was too late, however, and the crash resulted a few seconds later."

Mr Gillespie then climbed through ankle-deep mud along an embankment to the wrecked cars ahead. "The day coach was virtually driven through the dining car, and telescoped almost the entire length, with its top extending several feet above the top of the diner. The lights in the diner were out and because of the extreme darkness and pouring rain it was impossible to see inside, but it took just one glance to prove that death had stalked heavily. I gathered two or three uninjured and commandeered an automobile which had stopped in passing and we went into Rockmart for aid."

Mr Gillespie obtained ladders and lanterns in Rockmart, requested relief from Rome, Atlanta and elsewhere, and followed by doctors and volunteer nurses returned to the accident scene, where three bonfires were lit to provide illumination. "The uninjured and the people of Rockmart co-operated in the work of rescue and went at it coolly an in {a} determined manner. As fast as bodies were removed they were laid aside in a nearby field and we did not take time to identify those we knew were dead. Lives had to be saved, and it was the injured that we worked after."

When light was available scenes of horror could be seen. "From two windows of the wrecked diner, I saw terribly torn parts of two negro waiters' bodies hanging, their white coats smeared with blood. From another window we dragged the decapitated body of an apparently middle-aged white woman. Two tots who expected to greet Santa Claus tomorrow no doubt, were so badly crushed that identification will be almost impossible. Through one window of the diner we saw a man still seated at his chair at the table. Apparently he was only dazed. We got him out somehow and found his body had been severed just below the waist."

In another incident of the disaster the same newspaper reported that a prisoner being taken for trial for forgery escaped from the train but returned to rescue the policeman who had been escorting him. 'H.F. Bullis escaped from the smoking compartment of a car in the wrecked *Ponce de Leon* last night at Rockmart, Ga, last night and then went back inside for Edward Wiseman, 27 year old Lexington, Ky, policeman who was returning home to Lexington to answer charges of forgery. Neither thought the occurrence worthy of note.

"Sure, said Wiseman, guys often come back to help the officers out of a jam like that."

Both men, injured, lay alongside of each other in an Atlanta hospital while a policeman loaned from the city sat placidly in between.

"Remember me calling you, Ed?" asked Bullis.

"Gee, I thought you were killed at first, as soon as I got through thinking I was killed."

"Sure, I remember," the policeman replied. Adding as an aside that as soon as hospital physicians would release them they would continue their trip.' [646]

The Interstate Commerce Commission (ICC) report stated that the *Royal Palm,* train No. 101, had consisted of engine No. 1456, a club car, five Pullman sleeping cars, a dining car and two sleeping cars, all of steel construction. It had been in charge of Conductor Randall with Engineer Corrie and Fireman Baker. The *Ponce de Leon,* train No. 2, had consisted of engine No. 1219, a combination car, a coach, a dining car, and seven Pullman sleeping cars, again all of steel construction. It was in charge of Conductor Arwood with Engineer S. J. Keith and Fireman H. Moss.

At Dalton, Georgia, the crew of train No. 101 were given train order No. 92, issued on Form 19 as they were the superior train. The order read:

No. 101, one naught one, engine 1456, meet No. 32 thirty two, engine 1326, at Shannon, No. 6 six, Engines 1260 and 1205 coupled at Atlanta Jct., First No. 2 two, Engine 1219, at Rockmart and Second No. 2 two, Engine 1265, at Braswell. No. 32 thirty two, No. 6 six and First and Second No. 2 two take siding.

Train No. 2 received an identical copy of train order No. 92 before it left Atlanta, Georgia, issued on Form 31 as it was the inferior train.

Engineer Corrie stated at the ICC inquiry that at Rockmart the operator had handed him a clearance card indicating that the block was clear on the arrival of train No. 2. Fireman Baker looked ahead and told him that that train was not yet in sight, so he had moved ahead a little, taken water, then moved slowly down the main track towards the end of the passing track at 4 to 6 miles an hour. His headlight had been burning brightly. The headlight of train No. 2 then appeared and as it neared the switch at the end of the passing track it occurred to him that it was coming in too fast to stop. Applying his air brakes he had shouted to his fireman to jump and had done so himself. He put the speed of train No. 2 at perhaps 60 miles an hour and saw no sign of braking. He heard the engineer of train No. 2 whistle twice, the second time the usual whistle that another section of the train was following.

The evidence of Fireman Baker, Conductor Randall, and Flagman Stansell more or less agreed with that of Engineer Corrie.

Baggagemaster Copeland of train No. 101 stated that after the accident he had helped to remove Road Foreman of Engines J.M. Pearce from the locomotive of train No. 2. The latter asked the baggagemaster what had happened and was told that train No. 2 should have taken the siding for train No. 101. The road foreman had replied that Engineer Keith, Fireman Moss, and everyone concerned had told him that he was to hold the main track and had added that on approaching Rockmart he had sounded that station whistle and had again asked the fireman, "We hold the main track," and the fireman had said, "Yes." The baggagemaster did not know what was meant by "everyone concerned." Road

---

[646] *New Britain Herald*, New Britain, Connecticut, United States, 24th December 1926

Foreman Pearce later died of his injuries in hospital and Fireman Moss was killed outright.

Engineer Keith of train No. 2 said his fireman had read the orders at Atlanta. At McPherson, 11½ miles south of Rockmart, he had got down to oil his engine while they were waiting to cross an opposing train. When he had got back up on his engine he had found the road foreman there and the latter had said that he, Pearce, would work the engine forward as far as Rome. Engineer Keith said that he had read the orders to his superior and then left the engine and climbed into the combination car, where Conductor Arwood had checked that he had advised the road foreman of the orders. The road foreman had appeared normal and had operated the engine normally between McPherson and Rockmart. It was not until the train had been passing the south switch at Rockmart that Engineer Keith had realised that something was wrong and jumped up to open the conductor's valve but there had been no time to do so. The train had still been travelling at 45 to 50 miles an hour and he had not noticed the brakes being applied. He was sure that the road foreman had understood the orders, or he would not have left him in charge of the engine. It was the practice when road foremen took over the engine for the engineer to ride in the train but not, at least as far as Road Foreman Pearce was concerned, to actually hand him the train orders. Engineer Keith said he did not consider this a violation of Rule 220, which read:

When a conductor or engineman or both, is relieved before the completion of a trip, all train orders and instructions must be delivered to the relieving conductor or engineman. Such orders or instructions must be compared by the conductor or engineman before proceeding.

Engineer Keith submitted that the road foreman was an official and not an engineer and so the rule did not apply, that if the road foreman had not understood the orders it was incumbent upon him to ask to see them and added that in this case he had started away at once.

Conductor Arwood confirmed that he had checked with Engineer Keith that he had told the road foreman the orders. He had been busy collecting tickets as the train had reached Rockmart. He thought that the brakes had been applied in emergency a few moments before the collision. The meeting point signal at Rockmart he had left to the baggagemaster to give while he (the conductor) was occupied collecting tickets. He put the speed of his train at the time of the accident first as 25 or 30 miles an hour, then change this estimate to 35 or 40 miles an hour.

Flagman McElhaney, who had been assisting the conductor collect tickets, put the speed at 35 to 50 miles an hour and Train Porter Robinson thought it was 50 miles an hour. Both these men heard the road foreman sound the station whistle. The flagman thought that the air brakes had been applied, then released, the train porter did not think that they had been applied at all.

Engineer Turner of southbound freight train extra 5243 stated that his train had been in the passing siding when train No. 2 was approaching and had been instructed to back his train up far enough to let train No. 2 in at the south switch. He said that he heard the road foreman on train No. 2 give a long whistle, then the whistle that a second section was following. By then he had realised that train No. 2 was coming in too fast so instead of acknowledging the 'second section following' whistle he had given a long whistle of his own as a danger signal. He put the speed of train No. 2 at 40 miles an hour or more and saw no sign of its brakes having been applied.

Conductor Sorrells of train extra 5243 said that he realised when he saw train No. 2 that it was coming in too fast, but he could not give an estimate of its speed. He had helped to extricate the road foreman after the collision. Asked what had happened, he told the road foreman that he should have taken the siding; the road foreman did not say anything to him to indicate that he thought that he was to hold the main track.

Dr Shaw, of Rome hospital, where the road foreman was taken, said that the latter had remarked that there was no excuse for his mistake.

Road Foreman of Engines Pruitt stated that he would not relieve an engineer en route unless the latter was taken ill or for some similar reason. Occasionally he would relieve an engineer from the starting point; in such cases he would not permit the engineer to leave the locomotive and always took possession of any orders himself.

Signal Supervisor Hinds stated that automatic block signals and train-control apparatus were being installed on lines in the area, but delays had occurred and none of the signals north of Atlanta were yet installed, though almost all the engines were equipped with the train-control apparatus.

The ICC concluded that the accident had resulted from a failure of Road Foreman Pearce to fully understand the train orders or else his forgetting them. Which was the case was unclear but having exercised his right as the engineer's superior to take over the engine it was incumbent upon him to be absolutely clear as to the orders. As he had whistled approaching Rockmart it was clear that he had not been physically incapacitated and knew where he was. The commission noted that ordinarily train No. 2 was superior by direction and hence took the main track, but that that superiority had been reversed by the train order.

The ICC was not impressed by Engineer Keith's argument that Rule 220 did not apply as Road Foreman Pearce was an officer rather than an engineer and pointed out that the moment the road foreman took over the locomotive he became the *de facto* engineer. The strict observance of the rule was even more necessary when the engineer left the locomotive. Furthermore, in permitting the engineer to ride back in the train the road foreman had violated instructions issued by the superintendent eight months earlier.

Why Fireman Moss, who had been shown the orders, had not drawn the road foreman's attention to them when he failed to slow down could only be a matter for conjecture.

Conductor Arwood denied that his preoccupation with collecting tickets had prevented him from attending to his duty of seeing to the safe operation of the train. If this was so he should have made it his business to determine the exact location of the train, in which case he might have been able to apply the brakes in time to avert the accident. If the work of ticket collecting interfered too much with the duties of a conductor, baggageman, flagman or train porter then separate ticket collectors or train auditors should be provided.

The conductor and engineer did not appear to have a correct understanding of the rules.

It was not good practice, the ICC contended, for road foremen, who were not subject to the same periodical physical examinations as enginemen, to take over the driving of engines except in emergencies.

At the time of the accident automatic signals and train-control apparatus had in fact already been installed between mileposts 9 and 135, a section which included Rockmart station, and only awaited painting and testing. It was unfortunate that delays had occurred in bringing them into use, as they would have prevented the accident. [647]

An express from Irkutsk to Moscow collided with another train at Arsaki station, USSR, about 60 miles north of Moscow, on Saturday 8th January 1927. Sixteen people died and many were injured. Some reports spoke of the accident as a derailment. [648] [649]

The headlines in an English national newspaper on Tuesday 15th February 1927 read:

TRAIN COLLISION AT HULL

TEN DEAD AND 50 INJURED

CHILDREN'S COURAGE

At about 9.10 a.m. on Monday 14th February 1927, St Valentine's Day, an outgoing London and North Eastern Railway passenger train from Hull Paragon station to Scarborough, England, collided head-on with an incoming passenger train from Withernsea, almost beneath the Argyle Street overbridge, only 500 yards from the end of the Hull Paragon station platforms. Severe telescoping affected portions of the first five coaches of the Withernsea train and the leading coach of the Scarborough train was likewise telescoped by the tender of its engine. Eight passengers were killed outright and four more subsequently died of their injuries. Twenty-four passengers were seriously injured and 22 were slightly hurt. Both engine crews were also injured. Those killed included Miss Gladys Boyd, who had been sitting beside her fiancée, Harold Audsley, who was injured; Mr Francis Jackson, assistant librarian at Hull library and Philip Escreet, aged 17, who had run and caught the train which his brother and father missed, thereby possibly saving their lives.

Several passengers told their stories to newspaper reporters. 'Mr F.W.P. Garbutt, of Oban House, Withernsea, who was in a rear compartment of the Withernsea train, said the crash of the collision came quite without warning, and he was thrown out on to the permanent way. He afterwards saw the buffer of one of the engines, and a lamp had been hurled twice as far away. Mr Garbutt also states that after a very slow approach round the curve the train put on speed again as soon as it had passed under Argyle Street Bridge.

'Mr W. Long, a commercial traveller, of Withernsea, said he was reading his newspaper when there was a loud report. The carriage rocked violently, and then tilted on its side. He and the other passengers in his compartment scrambled out unhurt. Other passengers have described similar experiences. They were nearly all persons who had been in the saloon car or the hinder coaches of the Withernsea train. First-hand accounts from passengers in the first coach are meagre, for the reason that nearly all who were in

[647] Interstate Commerce Commission report

[648] *The Times*, London, England, 10th January 1927

[649] *The Telegraph*, Brisbane, Queensland, Australia, 11th January 1927

that coach were killed or are injured and now lie in hospital. Mr George Bowman, one of the club saloon passengers, who was taken to the Poor Law Institution with injuries, said, "Suddenly there was a crash, and we were all squeezed together." Mr George White, of Withernsea, a director of a Hull firm, who was in company with Mr Bowman, said that the collision caused the saloon to "close up like a concertina."' [650]

The Withernsea train consisted of Class F 4-4-0 engine No. 96, three third-class bogie coaches, a third-class saloon, a third-class brake composite, a third-class coach, a first- and third-class lavatory composite coach, and two third-class brake composites, in charge of Passenger Guard Billamy with Driver Robert Kirby Dixon and Fireman Thomas Scott. The Scarborough train consisted of Class M 4-4-0 engine No. 1628, a third-class brake composite, a first- and third-class lavatory composite, a third-class coach, a third-class brake composite and a third-class coach, in charge of Passenger Guard Carr with Driver Samuel Atkinson and Fireman Charles Wilkinson.

From the station box at Hull Paragon past Park Street signal box, 238 yards from the station box to a point near West Parade Junction, 727 yards from the station box, where the Leeds and Doncaster, Leeds, and Scarborough, and Hornsea and Withernsea lines diverged, there were no less than seven running lines, known from north to south as A, B, C, D, E, F and G. Lines A, B, D and F were used by down or ingoing trains and lines C, E and G by up or outgoing trains. Park Street box worked various connections between the different running lines. Paragon station and Park Street boxes both had electro-pneumatic miniature lever frames.

West Parade Junction box was double-manned. Signalman Taylor, who was in charge, was the first witness whose evidence appears in the Ministry of Transport report. He stated that the Withernsea train had been offered to West Parade by Botanic Gardens box at 9.05 a.m.; it was about ten minutes late. He had immediately offered it to Park Street on line B and had it accepted, then at the same time Park Street had offered him the Scarborough train on line C, he had accepted it and Park Street had quickly sent it on (given the 'Train entering Section') signal. Botanic Gardens had then sent on the Withernsea train, and he had sent it on in turn to Park Street. His colleague, Signalman Hayward, had cleared the signals for the Withernsea train and he, himself, had pulled off (cleared the signals) for the Scarborough train. The view of trains coming from Paragon station was much restricted and the first that he had known of a problem was hearing the noise of the collision. He had immediately thrown back the signals which had been cleared for a down Leeds and Doncaster express on line D and the 9.05 a.m. up express from Hull to London on line E.

Signalman Hayward said that he had only found out that the Withernsea and Scarborough trains had collided when the fireman of the latter had come to the box for assistance.

Signalman Bradley in Paragon Station box said that three signalmen worked his box between 8.30 a.m. and 1.30 p.m. He had cleared the signals for the Scarborough train to leave platform 1 at about 9.06 a.m. The only route from platform 1 to Park Street was via line C. The Withernsea train normally arrived at platform 3 before the Scarborough train had left. It arrived at Park Street on line B and to reach platform 3 crossed to that platform at Paragon station or from line B to D at Park Street, in either case it had to

---

[650] *The Times*, London, England, 15th February 1927

cross the path of the Scarborough train. He had never known of a case of an extraneous electric current reversing a set of points. Including light engine and empty train movements there might be 700 trains in and out of the station in 24 hours.

There were three signalmen on duty at Park Street box, Alfred Campling, John Clark and Edwin Gibson. None was senior to the others, but each took it in tuns to act as chargeman, who acted as traffic regulator as well as taking care of B, C, D and E roads. Signalman Campling was chargeman on the day of the accident and worked levers 0 to about 70. Signalman Clark worked levers about 70 to about 130 and Signalman Gibson the rest.

Signalman Campling stated that he had put an empty carriage train from line C to the carriage road, Paragon box had offered the Scarborough train and he had offered it forward to West Parade. Then he had told Signalman Gibson, "Right away Scarborough," and the latter had cleared the signals for it. West Parade had then offered the Withernsea train on line B and he had offered it forward and had it accepted by Paragon station on line D. He had then said to Signalman Clark, "B to D for Withernsea when Scarborough is out." Signalman Campling said that he had then turned his attention to the 9.05 a.m. departure for London Kings Cross, an incoming train from Brough and an engine movement.

His first notice of a problem was Gibson remarking, "Where's that chap gone?"

"What chap?"

"Well, I thought the Scarborough train looked a little bit too far over."

Signalman Gibson stated that he had pulled levers 169, 171 and 174 for the Scarborough train, then replacing the levers in turn, first No. 174, then No. 169, then No. 171, the latter when the engine and about three coaches had passed the signal. He then concerned himself with dealing with the London train and heard Signalman Clark remarking that he could not pull a lever. He had then noticed the apparently strange position of the Scarborough train and had shouted, "Where's that chap gone to?" He could not understand how the Scarborough train could have got onto the wrong road except through No. 95 facing points. He had never known a set of points to operate except through movement of the respective lever.

Signalman John Clark was a special class reliefman with 46 years' service. He said that when Signalman Campling had said, "B to D as soon as C is out," he had begun to set the road for the Withernsea train, which meant pulling levers Nos. 97, 96, 94, 83 and 82. He watched Signalman Gibson and as soon as the latter had replaced lever No. 171 he pulled levers Nos. 97 and 96 but could not get lever No. 94 over, nor fully reverse lever No. 82 to free lever No. 83. Gibson then remarked upon the strange position of the Scarborough train. He could not believe that he had pulled lever No. 95 instead of 96.

Driver Dixon on the engine of the 8.22 a.m. Withernsea train said that they had been delayed by adverse signals. His train had approached West Parade box travelling at about 15 miles an hour. Passing under Argyle Street bridge he had seen the signal off for him for line B and then, dropping his eyes, he had seen an engine in front of him and had immediately applied the continuous brake. The trains had then been only an engine length apart. He had never experienced any problems working into Hull and preferred it to Leeds, York, Doncaster, and Scarborough stations.

Fireman Scott put the speed of his train slightly different to his driver. He saw the other train at about the same time as his driver and said that it was not possible to see that the trains were on the same road until they had passed under the Argyle Street bridge.

Passenger Guard Billamy said that the brakes had been tested before leaving Withernsea and had worked correctly at four stations along the single-track branch.

Driver Atkinson of the Scarborough train put the speed of his train at about 10 miles an hour approaching Park Street and said that he had been able to see all the signals at clear as far as Victoria Crossing (the box beyond West Parade). The speed of his train had then increased to about 15 or 20 miles an hour. He did not notice his engine passing through the crossover and his first notice that something was wrong was when he had lost sight of the West Parade home signal. He had crossed and recrossed the footplate to check and when he was certain that he was on the wrong line had applied the brakes in emergency. His train was then going at 20 or 25 miles an hour and he had not yet seen the other train.

He said that freeing himself from the coal which had half buried him, he had gone to Park Street box and said to Signalman Campling, "What are you playing at this morning?"

The latter had turned to Signalman Clark. "Did you pull that lever?" There had been no reply and he had left the box.

Fireman Wilkinson agreed that there had been no movement to indicate that the train was going through the crossover. His driver had crossed the footplate, had shut off steam and then shouted, "Look out!" and he had seen the Withernsea train in front of them.

Passenger Guard Carr stated that like his colleague on the Withernsea train he had tested the continuous brake before his train started. He did not notice the train going through the crossover either and said that his only warning of trouble was the application of the brakes by his driver.

Electrical Chargeman Edward Varey stated that in eight years he had never known of a set of points being moved other than by operation of the lever, except during testing.

Colonel J.W. Pringle stated that it was clear that the accident had been due to the Scarborough train being wrongly diverted onto line B and remarked that he could recollect no previous accident in similar circumstances. There could be no doubt that the train had been diverted through No. 95 facing and No. 83 trailing points. Levers Nos. 95 and 171 were mechanically interlocked so that it was impossible to clear signal No. 171 except when points No. 95 were normal, or to reverse the points unless the signal was at danger.

The inspecting officer stated that if signal No. 171 been replaced prematurely when the engine and three coaches had passed the signal the leading bogie wheels of the engine would have been 219 feet 2¼ inches beyond the signal and 37 feet 9¾ inches from the locking bar. Assuming a speed of 13¾ miles an hour for the train at that point it would have taken, Colonel Pringle calculated, 1.9 seconds to cover that distance, while tests showed that the operation of No. 95 points lever took 1.6 seconds. The inspecting officer considered that Signalman Clark must therefore have inadvertently operated lever No. 95 in the brief time available. The alternative explanation was that a false current had caused the points to reverse, but tests showed this to be an extremely remote possibility.

Rule No. 61b stated:

When a signal other than a distant signal has been lowered for the passage of a train it must not (except in the case of accident or obstruction) be again placed at danger until the last vehicle of the train has passed it, or the train has been brought to a stand; nor, in the case of a junction, until the last vehicle of the train has passed it and is clear of the junction points.

Had Signalman Gibson strictly obeyed this rule its wheels would have been on the locking bar; the points could not have been reversed and the accident would not have happened.

Rule No. 60 stated:

When a train is approaching facing points the signalman must see that the lever which governs them is close home to the frame, and that the catch is firmly down in the notch and remains so until the whole of the train has passed.

According to this rule Signalman Clark should not have operated lever No. 96 (working the opposite end of the slip points to those controlled by lever No. 95) either until the Scarborough train was past the crossing, though the movement of points 96 would not have diverted the train.

The inspecting officer considered that the root cause of the accident was psychological – Signalmen Clark and Gibson had acted hurriedly in order to avoid stopping the already late-running Withernsea train at the Park Street home signal.

Turning to related issues, Colonel Pringle rejected criticisms that the traffic had grown so considerably at Hull Paragon since the Great (First World) War and the station layout was so complicated that it was inherently dangerous. Figures showed that while the station was considerably busier than it had been in 1906 (174 ingoing and outgoing trains daily) it was in fact slightly less busy than in 1914 (215 trains in 1914, 210 in 1927). Drivers considered the station simpler and easier to work into and out of than comparable stations in the region and Station Inspector Brant stated that no difficulty had been experienced in working or crossing trains and the platforms designated were arranged so as to reduce the number of trains crossing on another's path as much as possible. The inspecting officer was of the opinion that the layout was no more complicated and the traffic no heavier than at many other large termini. Traffic working at termini was necessarily complicated by the need for all trains to reverse, the movements of empty coaches and light engines, and the need for the engine and carriage sidings to be accessible from all platforms.

The old practice of locking the doors of the leading compartments of trains as a protection to passengers had fallen into disuse with the substitution of carriages with steel underframes for all-wooden ones. It was difficult to predict where in a train telescoping would occur – in the Withernsea train there had been telescoping in all of the first five vehicles – and shorter couplings and stronger underframes and carriage ends should reduce telescoping.

Local newspaper reports that the railway company had objected to the inquiry being held in public were untrue.

In conclusion Colonel Pringle found that Signalman Clark was directly responsible for the accident by inadvertently moving No. 95 points instead of No. 96. The company should caution signalmen against disregarding the rules in order to avoid delay. District Superintendent Ferguson regarded all the men in Park Street box as first-class signalmen. Given the reduced visibility at the accident site Driver Dixon and Fireman Scott of the Withernsea train could not be held responsible for the collision. Driver Atkinson and Fireman Wilkinson of the Scarborough train had, perhaps, not been quite as alert as they might have been, or they might have realised sooner that they were on the wrong line, but such an extraordinary diversion could not have been anticipated and the clear signals as far as Victoria Crossing had misled them.

The inspecting officer recommended the installation of a track circuit between signal No. 171 and points No. 95 to prevent the points being moved as a train was approaching them. [651]

At the resumed inquest on the victims of the accident the jury returned a verdict of "Accidental death." Their verdict in full stated, "We conclude that Signalman Clarke, while endeavouring to set "B" road to "D," for the incoming Withernsea train, in accordance with instructions given him by Signalman Campling, accidentally pulled lever 95. We further consider that the pulling of lever 95 was made possible by Signalman Gibson restoring signal 171 to danger before the whole of the Scarborough train had passed the signal, as set out in Clause 61 B of the Railway Company's general rules.

"We also find that the railway company are aware this rule is broken frequently, and we are of opinion that sufficiently concise instructions are not given limiting discretion in the restoring of signals. We do not find that the mistake or action of any particular individual shows such negligence as to lead us to return a verdict of manslaughter, and therefore we return a verdict of accidental death." [652]

On Wednesday 9th March 1927, a passenger train from Buzea ran into a goods train in Crivina station, Romania, in a dense fog. Five people died and a number were injured. [653]

Train No. 4, the up Calcutta – Madras mail train, collided with a shunting engine and waggons in the goods yard of Bhadrak station, India, on the East Coast section of the Bengal-Nagpur Railway, at 22:30 on the night of Sunday 13th March 1927 due to a set of facing points having been left in the wrong position. Both engines were badly damaged, and the leading luggage van and two third-class coaches were telescoped. Fourteen passengers and a pointsman were killed and 66 passengers, both engine drivers and one fireman were seriously injured. 'The points locker responsible for locking the points was found principally to blame and was sent up for prosecution.' [654]

Five people died and eight were injured when a passenger train collided with a freight train near Baku, USSR, on Saturday 18th June 1927. [655]

---

[651] Ministry of Transport report
[652] *The Times*, London, England, 17th March 1927
[653] *The Times*, London, England, 11th March 1927
[654] Report by the Railway Board on Indian Railways for 1926-27 Vol. 1
[655] *Albuquerque Journal*, Albuquerque, New Mexico, United States, 19th June 1927

The 9th of July is Argentina's independence day, celebrating its separation from Spain in 1816. To celebrate the occasion in 1927 the Argentinian president, Marcelo de Alvear, sent out invitations to the heads of government in Brazil, Paraguay, Uruguay, and Chile to send parties of cadets from their military schools to take part in the celebrations, which included the unveiling of a monument to General Mitre. [656] Thus it was that a group of Chilean military cadets became the victims of a railway accident in Argentina.

The party of Chilean cadets from the Libertador Bernando O'Higgins Military School set off on 6th July from Chile and crossing the Andes on the Ferrocarril Trasandino arrived at Mendoza, in Argentina, where they received a warm welcome. [657] It was on the continuation of the journey to Buenos Aires that the tragedy occurred.

The cadets left Mendoza at 1 a.m. on Thursday 7th July 1927 on train No. 3092 of the Ferrocarril Buenos Aires al Pacifico. The train was double-headed and consisted behind the engines of a horse box, a baggage car, a first-class car, two dining cars and ten sleeping cars. It was in charge of Inspector Tomás Bunting, with Driver Avelino Bavio on the pilot engine and José Busso on the train engine, and Firemen Luis Bordí and Gumersindo Quintero. [658] All went well until the train arrived at Alpatacal station, where it had to cross ordinary passenger train No. 3 from Buenos Aires to Mendoza, the line being single. At the moment of the accident train No. 3 was either stationary or nearly so but train No. 3092 failed to stop at the protecting signal at red and entering the station at 65 kilometres an hour collided head-on with train No. 3. Driver Arturo Levet on train No. 3 shouted to his fireman to jump clear but himself stayed on his engine and died in the collision. [659] The collision occurred at 4.45 a.m. The two engines were destroyed, and the coaches overturned and immediately caught fire and a number of the victims were trapped in the wreckage. The violence of the collision was shown by the fact that the length of the two trains was reduced from 280 metres to 75 metres.

Zelmira Garrigós de Von der Heyde, a passenger on train No. 3, had been travelling from Buenos Aires Retiro to Mendoza with her brothers Carlos and Fernando. They were riding in a coach near the rear of the train and hence escaped death or injury. Señora Von der Heyde left an account of the accident, in which she wrote, 'On 7 July Carlos, Fernando and I took a train to Mendoza. We said goodbye to Alico {her husband or eldest son} at the station unaware of what awaited us. We slept peacefully until 4 a.m. when we were suddenly awakened by a terrible jolt and at the same moment we heard a fearful, deafening screaming. From the carriage window we could see the glow of a terrible fire. It was freezing cold and there were heartrending screams coming from people who were trapped in the wreckage and were pleading to be put out of their misery before they were burnt to death, while uninjured survivors stood by, unable to help them.' She wrote that she could not erase the scene from her memory and that she did witness at least one of the trapped passengers shot dead to save him from the encroaching flames. [660]

Other witnesses told similar stories and of officers shooting the six (some sources say

---

[656] *Railway Gazette*, Vol 47, pg. 249
[657] Wikipedia – Tragedia de Alpatacal
[658] Wikipedia – www.losandes.com la tragedia de alpatacal
[659] Webpage Diariouno.com: Levet, el maquinista que se inmoló en la tragedia de Alpatacal
[660] Ibid.

eight) horses trapping in the leading horse box to save them from being burnt to death. [661] When attempts were made to extricate Colonel José María Barceló Lira, the director of the military college, he cried, "Save my cadets first." He survived anyway but Brigadier Moena; First Sergeants Pino, Larenas, Collao and Montes; First Corporal Riveros; Dragoon Novoa; Soldiers González, Seguel and Rosas and Cadets Fonseca and Pérez lost their lives. A large number of railwaymen were also killed: Inspectors Domingo Doda and Bunting; Drivers Levet, Bavio, Busso (Guzzo?), Victor Lorocotondo, Saturnino Velasco, Damián Alustizo and José Álvarez; Fireman Bordin and Waiters Manuel Estévez, Pío Ferrari, Sabino Ferro, Miguel Ferra, Manuel Reys, and Camilo Bondin. Why there were so many drivers involved was not explained. Ten others were seriously and 21 slightly injured.

Despite the disaster the uninjured survivors continued their journey to Buenos Aires; five officers and 120 cadets with a banner and a band of whistles and drums, the remainder of the musical instruments having been lost in the collision and fire. The cadets were reportedly cheered and applauded incessantly on the Buenos Aires parade. The injured and the bodies of the dead were taken back to Mendoza. [662]

There was reportedly a fog at the time of the collision [663] but another explanation for the accident might of course be that Driver Bavio had dozed off and his fireman had likewise done so or had been preoccupied with his fire. The crew of the train engine might have been relying on the men on the pilot engine.

The Canadian Pacific Railway suffered a serious accident at Dockrill Siding, Ontario, Canada at 2.15 p.m. on Sunday 27th November 1927. Westbound passenger train No. 19 from Montreal, Quebec, to Chicago, Illinois, United States, which had left the Quebec capital at 9.45 a.m., collided head-on with a freight train on the main track at the crossing place. Seven or possibly eight people died in the accident, all passengers in the colonist car of the express except for Engineer William Burnett of that train.

The eastbound freight train, bound for Montreal, consisted of an engine, 44 cars of grain and flour and a caboose, in charge of Conductor Hogg, with Engineer Lean Bradford and Fireman Rhynard Post. It reached Dockrill at 2.07 p.m. to take the siding for the express, which was due at 2.13 p.m. [664] The accident originated in an error of judgement on the part of Engineer Bradford. From the west the siding was approached by an incline, followed by a fall to the west siding points. Anxious not to stall with a heavy train, ascending the incline the engineer put on a lot of steam but then found that he was unable to bring up his train on the falling grade which led to the siding itself until he had overrun the west switch by five cars. He attempted to set back but was unable to push his train up the grade clear of the west switch. Instead he uncoupled the front five cars and proceeded with them along the main track at full speed to the east switch in the hope of being able to back these five cars down the siding, recouple them to the cars stopped on the single line west of the west switch and draw the whole train into the siding. There was, however, nowhere near enough time to accomplish this manoeuvre and he had not

---

[661] *Gazeta de Noticas*, Rio de Janeiro, Brazil, 20th June 1927
[662] Wikipedia – Tragedia de Alpatacal
[663] *The Times*, London, England, 8th July 1927

[664] *Medicine Hat News*, Medicine Hat, Alberta, Canada, 1st December 1927

even reached the east switch when the express came around a corner and the two trains collided at a speed of possibly 100 miles an hour.

The two engines, their tenders, the steel baggage car at the head of the express, a wooden mail car, a wooden colonist car and four freight cars piled up in what was described as an inextricable, indescribable heap, while an hour and a half after the collision fire broke out in the wreckage. [665]

The Canadian Pacific stated the day after the accident that there were only five dead, but other reports spoke of seven dead while when the wreckage was being cleared on 2nd December an additional body, of a child less than a year old, was discovered. [666] Most the passengers killed and injured were Dutch, German, and Polish immigrants bound for Western Ontario, many of whom did not speak English. Sixteen seriously injured passengers were taken to the Western hospital in Toronto and a number of others were less badly hurt.

By the next day four of the passengers who had lost their lives were identified as Franz Rischur, of Mannheim, Germany, bound for Toronto; Wilhelm Bauer, bound for Newcastle, Ontario; A.R. Antopp, of Meaford, Ontario; and Jacobus Keukelaer, of Breskins, the Netherlands, bound for Stratford, Ontario. [667] The identification of Herr Bauer turned out the next day to have been incorrect; he was reported to be safe. The same day it was stated that an anonymous donor had sent $50 to Arthur Bonisteel, a Canadian Pacific trainman, in appreciation of his efforts in leading the rescue work. [668]

Early on the afternoon of Wednesday 11th April 1928 the Chemins de fer du Nord suffered a major disaster close to its Paris terminus. Under the Doudeauville overbridge suburban train No. 552 from Villiers-de-Bel and Pierrefitte, due at Paris Nord at 13:12, collided head-on with omnibus train No. 13 from Paris to Amiens, due to depart at 13:15. The tank engine of the local train was running bunker first and the 130 tonne Pacific engine of the Amiens train buried itself deeply in the coal bunker of the lighter engine. The two engine crews jumped when they saw the inevitability of a collision and all four escaped with their lives. In the Amiens train the two second-class carriages marshalled behind the leading two luggage vans telescoped one another over half of their length, the steel underframe of the first carriage cutting through the wooden compartments of the coach behind. At the front of the suburban train was a four-wheeled luggage van. This was forced against the engine and smashed to pieces; the conductor riding in it was killed instantly. Following the van were two wooden four-wheeled first-class coaches; these were telescoped into one another and were the location of most of the deaths and injuries on the Villiers train. [669]

After the accident rescue operations were immediately begun by station staff and firemen from the principal Paris fire stations. By 15:00 five bodies had been recovered in pouring rain and transported to the mortuary at Lariboisière hospital.

The collision occurred right in front of No. 4 signal box. The Villiers train was much

---

[665] *Lethbridge Herald*, Lethbridge, Alberta, Canada, 10th December 1927
[666] *Winnipeg Free Press*, Winnipeg, Manitoba, Canada, 3rd December 1927
[667] *Manitoba Free Press*, Winnipeg, Manitoba, Canada, 29th November 1927
[668] Ibid 1st December 1927
[669] Ibid, 12th April 1928

used by people coming into Paris after luncheon and was crowded. A panic broke out among the surviving passengers after the accident and they scrambled out of the train and scattered over the network of tracks, being collected together with difficulty by railwaymen and officials, many of them in a dazed condition.

By 16:00 it was reported that in addition to the five bodies recovered from the train five victims had died en route to Lariboisière hospital and four of the injured had died there, a total of 14 dead. The total number of injured was still unknown but there were known to be 17 badly hurt, several of whom were unconscious. An oxy-hydrogen torch had to be employed to release the remains of Conductor Edmond Blondeau from the twisted metal of the tank engine.

The Paris correspondent of the London Times noted, 'Two cases of great fortitude are reported. A postal official named Degremont, of Amiens, had both legs cut off in the wreckage, but his first thought was for the mails under his charge. He threw the registered bag out of the window, and before he died gave the list of its contents into safe keeping. Father Robert, a Capucin Friar, was pinned in the wreckage. He begged his rescuers to attend to other passengers as he, he said, was safe. He died before he could be freed.' [670]

An investigation was opened by M. Delalé, the juge d'instruction, and it quickly became apparent that the accident had been due to a mistake by the driver of the omnibus train, who had brought his train out of the terminus onto the incoming Soissons line, having started his train against a *carré* at danger. He pleaded that the signals had not been against him and that with the multiplicity of tracks he had not realised that he was on the wrong line. Following the interrogation the driver, Joseph Herbot, aged 45, was arrested on a charge of involuntary homicide. [671]

The work of clearing the line was begun as soon as all the dead and injured had been removed from the trains and by midnight the wrecked coaches had been removed. Men working by the light of acetylene flares then began work on the repair of the tracks and the removal of the damaged engines. Close by in a siding were a line of waggons filled with wreckage. Normal working was resumed at 05:00 on the following morning. As might be expected on the evening of the accident the services into and out of the terminus were greatly disrupted and a large number of suburban trains had to be cancelled. At 18:00, towards the end of the evening rush hour, the crowd of delayed passengers was so great that it was difficult to get into or out of the station. Emphasis was placed on getting the more important long-distance trains away, but they were all very late in arriving and departing, some of the afternoon boat trains from the Channel ports not arriving until 22:00. [672]

The day after the accident reporters interviewed Fireman Petit-Mercier of the Amiens train, who was badly hurt in the collision and was lying in bed, his head swathed in bandages. He told a reporter in a low voice, "I was busy with my fire when Driver Herbot suddenly shut off steam. Up until that moment I did not know that anything was wrong. I was thrown against the firebox door, and I only half know what happened after that."

"This is the first time that you have worked train No. 13 with your driver?"

---

[670] *The Times*, London, England, 13th April 1928

[671] *Le Grand écho du Nord de la France*, Lille, France, 12th April 1928

[672] *The Times*, London, England, 13th April 1928

"Yes."

The reporter pointed out that even if Driver Herbot had not worked this particular train before he should have been familiar with the station layout from working other trains.

"You did not realise yourself that you were coming out on the wrong road? The public imagine of course that you could have seen the mistake and warned your driver."

"No, for the first kilometres I was fully occupied with working on the fire."

Apart from the fireman, the reporter continued, it was the conductor who as responsible for checking the points and signals and communicating, if necessary, by whistle with his driver.

"The conductor," the fireman continued, "only goes up into the lookout when called by the driver. Otherwise on starting from the terminus he is fully occupied with sorting the waybills and luggage."

"But how do you explain you driver starting against a stop signal, when he is aware of the great responsibility placed on him?"

The fireman shook his head. He didn't know. He saw nothing. All that he knows is that his mate is a calm, balanced, methodical man who had served six years as a fireman and nine as a driver. His error could not be explained.

A senior official of the Ministry of Public Works was asked for a technical explanation of how the accident had happened. He stated, "The station is so arranged that by means of two crossovers any arriving train can reach any platform, any departure line. The day before yesterday {that is, the day of the accident} at 13:14, a special race train for Enghien left track 11 and via a slip road reached track 10 from whence it proceeded straight ahead on the Pontoise departure line. A minute later train No. 13, which runs express to Chantilly, and which should take the Chantilly departure track, was due to depart, but to reach the Chantilly line it had to cross the Pontoise line, on which the Enghien train was leaving, and to protect this train at the connection with the Pontoise line at the Saint-Ange bridge the signals stood at danger against train No. 13, while to prevent a sidelong collision between the two trains should the signal be overrun the points were set for an incoming track. The signalman began to prepare the road for train No. 13 to depart, for which he had to operate three sets of points, but meanwhile the driver of that train started away against the *carré* set at danger and thus brought his train onto the incoming line on which train No. 552 was arriving, and the collision followed. Driver Hebert alleges that he did not hear the detonators which exploded when his engine passed the signal at danger."

"What do you say about the driver running past the signal?"

"It's unbelievable. A railwayman would think it crazy."

After the accident, the drivers divided into two groups. Those who believed, against all the evidence, that the signals had been clear, and those who thought that they had been misread. It transpired that Driver Herbot had in fact for some reason run past three danger signals. [673]

---

[673] *Le Journal*, Paris, France, 13th April 1928

The number of dead had now risen to 15, with 33 injured.

The Times correspondent in Paris commented at some length on the old wooden coaches involved in the accident and work on their replacement by steel stock. 'The advantages of steel stock in cases of accident are, of course, realized by the authorities – it was the telescoping of the wooden coaches which caused the heavy loss of life yesterday – but there are difficulties which prevent the immediate general adoption of steel stock. The most serious difficulty is increased weight. A steel coach will weigh perhaps 45 to 50 tons, where a wooden shelled coach weighs 30. Consequently, the train which a given locomotive can haul must be reduced in length, the number of passengers who can be carried is reduced, and the cost of running the train is therefore greater. There are also limits to the size of an express locomotive. Weight can be increased but dimensions, except in length, have already reached their limit. A larger boiler will need a larger fire-box; as the size of the fire-box can be increased only in length, it soon tends to exceed the limit which the fireman can cover. There is no choice but to run shorter trains. On suburban lines, owing to the fact that engines are habitually worked to full capacity, the problem is still more difficult.

'Since the War steel stock has been under steady construction. Most of the best-known express trains to and from Paris have now steel coaches. The form adopted is usually a strong steel shelf with wooden internal divisions, for the true all-steel coach has been found to be disagreeably noisy. For some time fast trains have been composed of one type of coach, so that the danger to a wooden coach from a steel neighbour may be avoided. The *Wagon-Lits* Company is substituting steel for wooden coaches as fast as possible. Many of the well-known Blue trains, for instance, are now made up of extremely strong steel coaches with steel shells, only the internal fittings being of wood. These coaches are very heavy, and their roofs are of lap-riveted steel plate. Tests have shown that they will resist telescoping under the heaviest pressure.

'The Nord Company has been experimenting with steel stock since 1923. The first of a new batch of 40 steel coaches were put into service this year. These coaches are exceedingly strong; their side panels are welded steel stampings; their roofs and frames constitute, as it were, shock-resisting keels, and the ends are fitted with double steel bulkheads. A steel coach from the Nord works was shown at the Paris Exhibition of 1925. It was put into service after the Exhibition, and almost at once was involved in a serious accident. The coach was overturned, and others crashed into it, but it suffered no internal damage, and its occupants escaped with cuts and bruises.

'In addition to equipping its principal express trains with steel coaches the Nord Company is engaged in experiments with lighter steel stock for suburban trains. Wooden coaches are also being fitted with end plates and trusses of steel. The French railways, taken together, had ordered about 400 steel passenger coaches if various types by the end of last year. A proportion of the new steel third-class coaches are designed for rapid conversion into hospital cars in case of war. Their compartments are quickly removable, the sockets and brackets for beds and stretchers are already in place, and the complete change can be made in 24 hours.'

It has not proved possible to trace the eventual fate of Driver Herbot – whether he stood trial for causing the accident or whether the case was eventually dropped.

An accident in some ways similar to that near Paris Nord occurred in England on Tuesday 27th June 1928 at Darlington Bank Top station on the London and North Eastern Railway. Again the collision happened at a large station, though this time not a terminus, and again a driver overran a stop signal and fouling a running line used by trains moving in the opposite direction, came into head-on collision with one coming the opposite way.

The first reports of the accident reached London at 1.20 a.m. the next morning and the late editions of the London newspapers carried news of a serious accident at Darlington, in which six people were known to have died, the victims being passengers on a return excursion train from Scarborough to Newcastle. As it soon transpired, the accident was considerably worse than as first reported; 25 passengers were killed outright or fatally injured and another 45 were badly hurt. The driver and fireman of the excursion train and the driver of the parcels train with which it collided were badly enough hurt or shocked to have to leave duty. As in the Paris accident the heavy casualties were mainly due to telescoping, in this case of the second and third coaches of the excursion train. Fourteen of the dead, all but one of them women, came from the small mining town of Hetton-le-Hole; they were mainly members of the Mothers' Union. The accident has been likened to a pit disaster in reverse, with husbands mourning wives instead of wives mourning husbands.

The excursion train was not booked to stop at Darlington, and it was intended that it should not pass alongside the station platform but take the down through (avoiding) line to the east of the station platforms to reach the north end of the station. It was while it was in the process of taking the down through line at the south end of the station that it came into collision with an engine shunting some parcels vans which had fouled the main line.

A newspaper correspondent described the scene when he arrived there and the rescue operation. 'The wreckage had nearly all be cleared away from the rail tracks when I arrived, but sufficient of it was accumulated by the side of the line to show the terrible nature of the destruction. What was left was splinters of woodwork, loose stuffing of seats – hardly any piece recognizable as the side or door or window of a carriage. The line where the two trains had met was strewn for many yards with broken glass. The platelaying gangs were still hard at work relaying metals, and there appeared to be six or eight hours' work yet to be done before the tracks entering Darlington Bank Top station could be restored to normal. A telephone pole on the line on the line had been demolished and had to be re-set and restrung with wires.

'The coach immediately behind the passenger engine suffered little damage, but No. 2 coach was smashed to fragments, and it was to this and the third coach that the attention of rescuers was first directed. Those who were earliest on the spot say that a silence of perceptible duration followed the crash. Then the air was filled all at once with shrieks and moans and cries for help, as the injured began to come to their senses and to realize what had happened. In the confusion and darkness the noise of escaping steam added to the threats of danger, and the fear of fire was present to the minds of many. Though there was no real delay in the arrival of skilled assistance – railwaymen, police, doctors, ambulance people and nurses – the work of breaking into the wreckage was hampered by the necessity of assembling suitable tools and gear, and the darkness added to the horrors of the time.

'One of the first thoughts of the rescuers was to give water to the sufferers, failing stimulants, which were not at hand. Order began to be established on the arrival of two motor ambulances, and axes and saws were got from the station, but it was soon found that many of the victims could not be reached unless the roof of the second (telescoped) carriage could be lifted, and it was some time before a breakdown crane could be brought into position to do this. It is certain that hours elapsed before some of the trapped victims who are still living could be released.

'The majority of the injured on being released from the wreckage – some through the efforts of bystanders who managed to get crowbars and pickaxes at the station – were conveyed to the Greenbank Hospital, but a few were taken to the Feetham Hospital. The driver of the excursion train, Dominic McNulty, of Blaydon-on-Tyne, was found staggering about the station, but refused medical attention. His statement about the accident is:

'"The distant signal was set for us, and I went ahead. Just before reaching the station I saw something in front of us, and put on the brakes, but we could not prevent a collision, and I was thrown right off the engine. When I came to I saw the terrible sights behind, and I just joined in and helped to get some of them clear."

'McNulty was persuaded afterwards to go to Greenbank Hospital.'

Fireman Bryant of the excursion train related, "I was firing the engine at the time of the catastrophe. The signal gave us the road. Something, however, seemed to tell me that all was not right. Both the driver and myself looked and saw an engine directly ahead of us. I could not make out whether it was stationary or coming in our direction. The driver pulled up and did everything in his power to avoid the accident. Then came a terrific crash. The engine was separated from its tender and fell over on its side. I could see no more for steam and was so dazed I did not know what was happening."

A passenger, Mr William Davidson, of Swan Hunter and Co's shipyard, Wallsend, stated, "I got out and ran to the front. The people in the front coach were jumping out, and appeared to be all right, but terrible cries and groans were coming from the second coach, which was telescoped, and, though even from that coach some of the passengers were scrambling out, apparently without serious hurt, from the fifth compartment rearwards there was disaster. I jumped on the footboard, but it was impossible to do anything. I shall never forget the scene. Inside were rows upon rows of seats, one on the top of the other, with human beings sandwiched in between them. I am used to gruesome sights {he was ambulance supervisor at the shipyard} but never have I seen anything of this kind." [674]

The sequence of events which led up to the disaster began with the arrival at Darlington at 10.45 p.m. of an up train which as far as there, carried passengers, but which southwards therefrom ran solely as a parcels train to Northallerton and York. On arrival at Darlington, 11 minutes late, the passenger-cum-parcels train consisted of engine No. 2369, three carriage trucks, two vans, a third-class brake composite, three vans, a first and third-class composite coach, a third-class brake composite and a brake van. The first nine of these vehicles, with seven others to be picked up at Darlington, were to form the train south from Darlington.

At the Ministry of Transport inquiry, Station Foreman William Morley related that he

---

[674] *The Times*, London, England, 29th June 1928

had instructed Signalman John Robson in Darlington Platform East signal box, which controlled the centre of the station, to let the parcels train down beyond a scissors crossover to make room for another train. Signalman Robson therefore offered the train to Darlington South Junction box, who accepted it under caution and the train drew forward to the south end of the platform. Assistant Shunter Michael Moreland then uncoupled it between the third and fourth vehicles. The shunter then instructed the driver that he was to go forward with the front carriage trucks and then back into the up duplicate line, where there were three horse boxes and four vans to be picked up. On Signalman Thomas Walls in the South Junction box in due course, when he was able to, clearing his signal, the train thus pulled forward foul of the down main line, it having to move this far forward in order to clear the points leading from the up duplicate line. It was then backed into the up duplicate line and the shunter coupled up the seven vehicles. At 11.04 p.m. Signalman Robson offered the train again to the South Junction, who accepted it under the "Warning Arrangement," indicating that the main line at the South Junction was blocked because the down excursion was approaching. Signalman Robson cleared his signal No. 2 and then No. 8, the up calling-on signal, to show the driver that he had been accepted at caution. Thus far there was no problem, but Shunter Moreland then saw that his driver was taking the engine past a group of three signals controlled from the South Junction and two of which, Nos. 14 and 18, applied to the up duplicate line, both of which were at danger. The shunter then applied the Westinghouse brake from his van, imagined his driver was stopping, released the brake again and looked to see if one of the signals had been cleared. Finding that they were both still "on" and that his driver was still not pulling up he applied the brake a second time and the train had almost stopped, then there was a crash. The shunter said that he had not applied the brake fully the first time for fear of dividing the train; the second time he had applied it harder, but still not fully.

Darlington South Junction signal box was described as a very busy block post, manned by two signalmen and a booking boy. On the night of the accident these were Signalmen Thomas Walls and James Garrett plus an unnamed boy. Signalman Walls was in charge and working the northern half (levers 1-70) and Signalman Garrett was dealing with down trains and working levers 71-140. Signalman Walls explained that the parcels train always had shunting to do at Darlington which was sometimes done within station limits and other times at the south end, as on the night of the accident. He described the first shunt move and said that he had then expected the parcels train to come to a stand at the up duplicate line home signal, No. 18, until it was cleared.

Signalman Garrett said that the excursion train had been offered by Croft Junction at 11.01 p.m. and accepted a minute later when the "Blocking Back" signal from the Platform East box for the parcels train had been cleared back. He had then cleared his down through line signals and the train passed the box at 11.08, three minutes late. Just before it did so, he and Signalman Walls had heard a strange clicking noise and realising that something was wrong he had thrown back the down through signals and pulled the detonator placer. The clicking noise was caused by the parcels train running through trailing points Nos. 53 and then 68.

The excursion train consisted of engine No. 2164, two third-class coaches, a third-class brake composite, two third-class coaches, two third-class coaches, a first-class coach, a third-class brake composite, three third-class coaches and a third-class brake composite, in charge of Passenger Guard Frederick Hadwin with Passed Driver Dominic

McNulty and Fireman Vernon Bryant. The train had left Scarborough on time at 8.55 p.m., and the brakes had worked correctly on the journey when required, such as for permanent way checks between Scarborough and Seamer and a signal stop at Sunbeck Junction. The Darlington South Junction distant signal had been off, and the junction had been approached at about 45 miles an hour. Driver McNulty said that when the headlights of engine No. 2369 had appeared on his line he had applied the continuous brake; there had been no time to reverse his engine or apply sand. Fireman Bryant and Guard Hadwin generally confirmed their driver's evidence.

In charge of the engine of the parcels train was Driver Robert Bell, with Fireman J. McCormack. Driver Bell said that he had had considerable experience as a fireman of the line between Darlington and York, but his driving experience had been mainly limited to local trips in the Newcastle area. He had not signed the Road Book for the line between Darlington and York but would have been happy to do so as he knew the road and signals, and he had been confident (over-confident, perhaps?) to drive the parcels train without a conductor (pilot). It was, however, the first time that he had driven the parcels train involved and he had never done any shunting at Darlington. Describing the shunt move which had led to the accident he explained that having set back onto the up duplicate road and attached the seven vehicles the shunter had signalled for him to go ahead, and the calling-on signal had been lowered. He believed that the clearing of this signal authorised him to draw his train forward foul of the down through road far enough to clear the points, so that he could propel the ten vehicles attached to his train onto the remainder of his train standing on No. 1 platform line. He could not explain why he thought that the clearing of signal No. 8 gave him authority to pass signals Nos. 14 and 18 when he knew that the clearing of signal No. 11 applying to the No. 1 platform line did not give him leave to pass signals Nos. 17 and 19 at danger. This was just the impression that he had got. He had not noticed any brake application from the van until immediately before the collision and said that the speed of the train had only been 6 or 7 miles an hour, so he could have pulled up easily before fouling the down through line had he been warned by a brake application from the shunter earlier.

Fireman McCormack denied any knowledge of the signals in the Darlington yard.

Passenger Guard J.H. Sharp was in charge of the parcels train but said that on arrival at Darlington he had been busy loading parcels into the vans and had left the shunting to the shunter.

Stationmaster Beeby said that trains were usually shunted at the north end of the station and at the south end only if a second train needed to be accommodated in the platform. He considered that Shunter Moreland should have stopped the train at once if he saw the driver running past a signal. His view was shared by Station Foreman Morley and Guard Sharp.

The inspecting officer at the inquiry, who was again Colonel J.W. Pringle, stated that the accident was clearly due to Driver Bell misunderstanding the limited authority conferred by calling-on signal No. 8 and running past signals Nos. 14 and 18 for reasons which were difficult to understand. The only explanation was that his knowledge of signalling was insufficient to enable him to correctly interpret the signalling of a yard with which he was unfamiliar. He should have asked for a conductor or for the assistance of Shunter Moreland.

Shunter Moreland was very familiar with the yard and should have applied the brakes powerfully as soon as he saw his driver running past the signal. With the train only travelling at seven miles an hour he could have stopped it easily in the 117 yards between the signals and the fouling point with the down through line. His explanation that he feared dividing the train was not acceptable.

Driver Bell was principally responsible for the collision. Fireman McCormack was not responsible but for someone who had been a fireman for seven years his statement that he knew nothing about signals was hardly satisfactory. Shunter Moreland must share some responsibility for the accident by not taking the initiative and applying the brakes. The crew of the excursion train and the signalmen were entirely free from blame.

The inspecting officer drew the company's attention to the fact that despite his knowledge of the signalling at Darlington, Driver Bell had said that he would have been quite happy to sign the Road Book. Clearly something more than a mere signature was required to show that a driver was really familiar with a stretch of line.

Colonel Pringle was not happy with the Platform East calling-on signals Nos. 8 and 11 which he considered confusing to drivers, and he recommended the substitution of a three-position light signal.

The failures of Driver Bell and Shunter Moreland notwithstanding, automatic train control would have prevented the accident.

The first coach of the excursion train alone had been gas lit. The gas equipment had not been damaged and no fire had broken out.

Telescoping had caused most of the casualties. The best method of preventing telescoping was one for serious consideration by the companies, but in any case there was a large amount of older rolling stock in good condition and fit for service which it would be impossible to scrap for a number of years. This last remark seems somewhat strange coming from an inspecting officer, as if he was simply accepting that telescoping was bound to occur in accidents for some years to come and that was that. [675]

The inquest on the victims of the accident was concluded on 17th August, the jury taking only seven minutes to find a verdict of "Accidental Death," adding that they considered the accident was due to an error on the part of Driver Bell in misreading the signal. The jury foreman remarked that, "My colleagues wish me to say that they are very satisfied with the way Mr Bell gave his evidence. It was a very great trial to him." [676]

Ten persons were killed, and many hurt when a passenger train collided with a freight train at Grónjec, Poland, near Warsaw, on Wednesday 18th July 1928. Four passenger carriages were demolished in the accident. [677]

BP (Beschleunigter – accelerated) passenger train 911 collided with goods train G 7535 at Dinkelscherben, Germany, between Augsburg and Ulm, on Tuesday 31st July

[675] Ministry of Transport report

[676] *The Times*, London, England, 29th and 30th June, and 2nd, 3rd and 18th July 1928

[677] *Brainerd Daily Despatch*, Brainerd, Minnesota, United States, 18th July 1928

1928. Sixteen people died and 48 were injured. [678] Modernisation work was in progress at a signal box and the interlocking at the Ulm end of the station had been temporarily disconnected. A set of facing points had been operated by the signalman to divert the non-stop passenger train onto the through line to overtake the goods train but he had failed to check that they had gone to their correct position. [679] The train driver saw that facing points were in the wrong position but too late to avert the accident. [680]

A mistake by a signalman named Fibich was the reported cause of a disaster at Zaječí station, near Lundenberg, Moravia, Czechoslovakia, on Monday 10th September 1928. A Paris – Prague – Vienna – Budapest international train, travelling at a speed variously reported as 30, 40 or 60 miles an hour, was diverted onto the wrong line and collided head-on with a goods train. The leading three coaches of the passenger train were telescoped and almost all their occupants killed or injured. [681] Most reports stated that there were 20 dead and about 30 injured but a possibly later report from Berlin stated that 24 bodies had been recovered, eight more were still buried in the wreckage and five people were critically injured and not expected to survive. Fifty others were slightly hurt. The injured were transported to hospital at Brno, the capital of Moravia, which was not far away. [682] (The German name Brünn is used in most of the reports on the accident instead of Brno while a number of the reports confusingly give the location of the accident as Saitz, the German name for Zaječí.)

The driver of the express applied his brakes, jumped and escaped; the fate of his fireman was not recorded. Another lucky survivor was Constantin Argetoianu, a conservative Romanian politician who was later briefly Prime Minister, and who had been riding in one of the wrecked front coaches but had gone to the dining car a few minutes before the collision. [683] As might be expected a large number of the passengers on the train were foreigners, many of them Germans.

Distraught by what he had done, Signalman Fibich attempted to hang himself after the disaster but was prevented from doing so by his colleagues and was subsequently arrested by the police. [684]

The year 1928 was not a good one with regard for railway safety in Great Britain, with a number of bad accidents, such as that at Darlington Bank Top (see above). On Saturday 13th October there was another major disaster, this time on the former Midland, now London Midland and Scottish Railway main line from Birmingham and Gloucester to Bristol, at Charfield, England. The overnight mail and passenger train from Leeds to Bristol collided with a goods train which was being shunted out of its way and a goods train which happened to be passing in the opposite direction also became involved in the

[678] *Eisenbahn-Katastrophen in Deutschland* by Hans Joachim Ritzau, pg. 41 (in German)
[679] Katastrophen der deutschen Bahnen Teil II: Chronik 1845-1992 by H.J. Ritzau, pg. 71 (in German)
[680] *Von Siegersdorf nach Aitrang* by H. J. Ritzau, pgs. 43-4 (in German)
[681] *The Times*, London, England, 11th and 12th September 1928
[682] *The Gettysburg Times*, Gettysburg, Pennsylvania, United States, 11th September 1928
[683] *The Daily Republican*, Rushville, Indiana, United States, 11th September 1928
[684] *Dunkirk Evening Observer*, Dunkirk, New York, United States, 11th September 1928

accident. The wreckage piled up against an overbridge and almost immediately caught fire from escaping gas. Sixteen passengers lost their lives and 24 passengers, 13 postal sorters and four railwaymen were injured.

A newspaper correspondent described the accident and what followed. 'The passenger coaches immediately behind the express engine were telescoped over the tender and piled up under the bridge, which the front part of the engine had just cleared. The first coach, torn from its under-carriage, was driven with great force over the top of the tender, while its roof was crushed and partly removed by the stone arch of the bridge. The second coach, unable thus to enter the bridge opening, reared itself up at an acute angle, and came to rest with the front overhanging the bridge parapet, having scattered some of its contents. The third coach in like manner jammed itself up against the mass of wreckage and was in effect telescoped. The fourth and last passenger coach was less seriously damaged, though some people travelling in it were injured. Behind came three mail vans, in which sorters were at work.

'In the awful confusion and darkness which followed, those passengers who were not seriously hurt tried to release some of the less fortunate people whose legs were trapped between seats driven together by the collision or were pinned down by heavy pieces of debris. There was no light, and many of the passengers, who had been sleeping, were violently awakened by the shock, only to find themselves in black darkness, and in some cases in a suffocating atmosphere caused by gas from the damaged containers. One man with whom I talked had the impression that his head and shoulders were driven through the roof of the carriage. He pulled himself out, slid to the ground, and then went to help get others out. A woman travelling with two children found that the little girl's leg was jammed between two seats, and after tugging until she released the crushed limb, she smashed a window and crawled with the child out on to the side of the coach.

'The sleeping people were scantily dressed. Most of them were without boots, and there was no possibility of getting luggage out. Those who got out began straight away to help their fellows, and very quickly there was aid from the startled village nearby. But rescue work had but a brief opportunity. Within a very short space there were two explosions, due no doubt to gas being lighted by fire scattered from the two wrecked engines, and at once the mass of wreckage under the bridge took fire and was soon blazing furiously. It threw up a large column of flame on either side of the bridge, melting telegraph wires high above and setting fire to the posts bearing them. There were still people calling for help in those smashed coaches when the fire started, but after that there was never a hope of reaching them, for the arch of the bridge was practically filled with wreckage and no one could come alongside it to prise up the steel and timber which held the unfortunate people prisoners.

'Quickly the fire spread to the rest of the train, and several injured persons were got out only just in time. One woman, crying out for help, told how the fire was coming nearer and nearer, and how she was protecting her face with her travelling rug. Almost all the injured still detained in the Bristol hospitals are singed or scorched. One terribly injured woman at the Bristol General Hospital may count as the least of her injuries the fact that half the hair of her head is burned off. Very few minutes elapsed from the time of the collision to the time when the fire made further rescue work impossible.

'A Penzance man who was going home with his wife and widowed sister from Sheffield, gave me an example of how fiercely advancing fire cut off suddenly the work of rescue. Having wormed his way out of the wrecked compartment he dragged out his wife and took her to the side of the track. On his return, two girls in a neighbouring compartment called to him for help, and after releasing them he went back to get his sister out, only to find that fire had taken hold of the coach, and that he could not approach it.

'Fire brigades from all the towns in the district were summoned. From Bristol, Gloucester, and Stroud they came clattering in the fog along the narrow Gloucestershire lanes. The quickest of them took half an hour to arrive, and the pile of burning wreckage was a mass of white-hot metal long before then. The only water supply was that of the railway tanks at Charfield Station, 100 yards away. All the brigades used that, and though it was a vastly better supply than they would have found in more remote spots, it was not sufficient to keep them all working at full pressure.

'Except for the additional horror of destroying by fire the bodies of the killed, this mattered little. There was never any hope of saving the rolling stock, and the bridge had already been damaged by the collision and by the fire under it before the brigades arrived. So intense was the heat that the bricks of the arch had been calcined when I examined them yesterday, and the tar surface of the roadway had been melted and had run in streams to the edge of the road. The bridge looked as though very little vibration would suffice to make it collapse, and the police have not allowed traffic to use it since. While the wreckage blazed, driving everybody far back from the scene, those who had escaped or had been rescued were being attended to in the village.

'All the injured were taken first to the little hotel close to the station, and doctors from the surrounding countryside gave first-aid, putting broken legs into splints and preparing the sufferers for their journey of from 16 to 20 miles to hospital. The less seriously injured were soon picked up by private motorists and taken to Berkeley or Bristol. Later some were taken to hospital in ambulances, but for the worst cases an ambulance train was run up from Bristol into Charfield Station and the sufferers were taken in to the Bristol Station, only a short distance from the General Hospital.

'Among the wreckage it was impossible to do anything for five hours. The crash occurred at about 5.30. It was about 10.30 when the breakdown gangs, which had been assembled north and south of the scene with great promptitude, were able to make a start at clearing away the heap beneath the bridge. Nor, owing to the peculiar situation of the debris, could the usual methods be adopted. The wreckage was jammed tightly under the bridge. The derricks generally used to lift telescoped coaches off the lines and deposit them at the sides could not be brought into service because they would have had to lift the bridge too in order to get the twisted frames away. Chains were attached therefore to portions of the wreckage, and engines on either side of the bridge hauled pieces out until there was room for the derricks to get to work.

'There was still fire in the heap, and the firemen remained at their posts, playing on the burning material, both as it was dragged from under the bridge and later when the huge cranes were pushed up and began the removal of the bigger pieces. In these circumstances it was about six hours after the accident that the first of the remains were recovered, and in most cases they were mere fragments and quite unrecognizable. The body of the guard in the front part of the mail train was recovered late yesterday afternoon, but it could never have been identified except by the brass buttons of the uniform and by inference from the position in which it was found. {This must in fact have been a railwayman travelling as passenger, as the Ministry of Transport report does not mention any railwaymen on duty losing their lives in the collision.} The little Railway Hotel, which gave help to the injured early in the morning, has been made the mortuary since, and all that has been found of the dead awaits there the identification which must depend more on deduction than recognition.

'The Penzance clerk who saved his wife but lost his sister came back here, after having his dislocated shoulder attended to in Bristol, to see if he could find the sister or news of her. There was nothing in the improvised mortuary which he could recognize, nor as the day passed were any of the fragments that came in from time to time identifiable, so that he went away again, daring to hope in a dazed fashion that somehow she had escaped. Her remains were found last night. So far as can be ascertained the remains represent 14 bodies. The work of identification can be completed only by a process of elimination, and will no doubt proceed slowly.' [685]

The 10 p.m. mail train from Leeds to Bristol had consisted of 4-4-0 engine No. 714, a parcels van, a van, a first- and third-class composite, two third-class coaches, a first- and third-class composite, a Post Office sorting van, a Post Office tender, a Post Office van, and a brake van.

Charfield was a standard ex-Midland Railway signal cabin controlling the station layout, which included a down refuge siding. The line was worked by Midland Railway rotary interlocking block instruments and an outer home signal was provided, enabling the signalman to accept another down train when the line was clear to his inner home signal, while some track circuits were provided, including one which extended on either side of the outer home signal (No. 75) and a second in advance of the inner home signal and extending beyond the points leading to the down refuge siding (No. 661). The position of the arm of the down distant signal was repeated in the signal cabin and the signal lamp was repeated, so that a buzzer sounded continuously if the lamp went out. The down outer and inner home signals were mechanically interlocked with the points leading to the down refuge siding, the inner home signal was electrically locked by occupation of track No. 661 and the down starting signal could not be lowered without a release from the next cabin at Wickwar. The down distant signal could not be cleared until the outer and inner home signals and the starting signal had been cleared. Berkeley Road South Junction, the next cabin on the Gloucester side of Charfield, was out of switch at the time of the accident and the next cabin open was Berkeley Road Junction, about 5¼ miles distant.

The main line to Bristol was quite busy early on this Saturday morning, with faster trains interspersed with slower ones which had to be shunted out of their way. At the Ministry of Transport inquiry into the accident it was stated that at 4.34 a.m. the 9.15 p.m. Great Western Railway (GW) down fitted freight Oxley Sidings to Bristol, consisting of 2-6-0 engine No. 6381, 49 loaded waggons and a van, had arrived at Berkeley Road Junction, and Signalman R.H. Smith had shunted it onto the branch line there to give precedence to the 12.45 a.m. London Midland and Scottish (LMS) parcels train from Leicester to Bristol. He had duly accepted this train from Coaley Junction at 4.41 a.m., it passed him at 4.48 a.m. and was cleared back from Charfield at 4.53 a.m. He had then brought the GW train off the branch. This train was accepted by Charfield at

[685] *The Times*, London, England, 15th October 1928

4.53 and cleared back from there at 5.14.

Signalman Henry Button at Charfield testified that at 4.42 a.m. he had cleared back to Berkeley Road another goods train, the 10.35 p.m. LMS through freight from Washwood Heath to Bristol. He had decided that he must shunt this train out of the way of the Leicester parcels train, and it had been in the refuge siding clear of the down main line at 4.47. Meanwhile at 4.42 he had accepted the parcels train from Berkeley Road Junction, as he had been justified in doing as soon as the Washwood Heath train had passed his inner home signal. In consultation with the Control office at Fishponds he had then decided to let the Washwood Heath train follow the parcels and to shunt the GW Oxley train at Charfield for the Leeds – Bristol mail. The parcels train duly passed him at 4.53, he accepted the GW train from Signalman Smith at Berkeley Road Junction at the same time and let the Washwood Heath train out of the refuge siding to await the clearance of the parcels train from Wickwar. It was at this point that things began to go wrong. The Washwood Heath train was not booked to stop for water at Charfield and its driver had not told Signalman Button that he needed any, but it had stopped to take water anyway. As a result, although it had been accepted by Wickwar at 4.55 it did not start away until 5 a.m. Signalman Button said that had he known that the Washwood Heath train wanted water he would have kept it in the refuge siding and let the Oxley goods precede it to Wickwar, followed by the mail train. The Washwood Heath train having passed his starting signal, Signalman Button replaced it and cleared his inner and outer home signals for the Oxley goods. When the Oxley goods operated the treadle 15 yards in advance of the inner home signal he had cleared the train back to Berkeley Road Junction and accepted the Leeds mail at 5.13 a.m. (Signalman Smith recorded the time as 5.14). The train passed Berkeley Road Junction at 5.16 a.m., travelling, according to Signalman Smith, at 60 miles an hour. The "Obstruction Danger" signal was received from Charfield at 5.20.

Signalman Frederick Brooking at Wickwar stated that the Washwood Heath train had reached him at 5.18 but he had decided that he must shunt it for the mail train and so he did not clear it back to Charfield until 5.32. The train had passed over his release treadle at 5.18 and he could have accepted a following goods train under Regulation 5 (the "Warning Arrangement") at that time but not a passenger train until the line was clear, as no outer home signal was provided at Wickwar. Neither the Oxley train nor the mail train was offered to him anyway.

Back at Charfield Signalman Button was positive that the repeaters for his down signals had indicated that they had returned to "Danger" after the GW goods train had arrived and he had replaced them. Knowing that the mail train was going to be stopped waiting for the GW goods to finish setting back into the refuge, he had had his eye on the indicator for track circuit No. 75 at his outer home signal. As he had watched the indicator it went from "Clear" to "Occupied," and then back to "Clear," which could only mean that the mail train had run past the signal. Meanwhile the 4.45 a.m. LMS empty waggon train from Westerleigh to Gloucester, consisting of 0-6-0 engine No. 3593, 45 waggons and a van, had been accepted from Wickwar on the up line at 5.13 and was just passing through the station. The double collision occurred close to his cabin, and he sent the "Obstruction Danger" signal both ways at 5.20. Half an hour after the collision the driver of the mail train had come to his cabin and told him that his down distant signal had been "Off," to which he had replied, "Impossible," and the driver had gone away.

The Oxley goods train was in charge of Guard William Fortune, with Driver Gilbert and Fireman H. Sutton. Driver Gilbert said his train had been running late. He had experienced no fog until Naas Crossing (about three miles on the Berkeley Road side of Gloucester). He related being shunted at Berkeley Road Junction and said he had seen the signals there and at Berkeley Road South Junction. He had seen the Charfield down distant signal, and it had been showing a full red light. He had closed the regulator. The outer and inner home signals had both been green and the Charfield signalman had told him that his train was to be shunted. He had been duly setting back at 5 miles an hour and his engine had been about under the overbridge when they were run into.

Fireman Sutton said that the fog had been 'pretty thick' at the Charfield down distant signal so that he had only seen it from a distance of about 20 yards, but he had had no doubt that it had been at "Danger." The two home signals, which had been off, had also been somewhat difficult to see.

Guard Fortune said that his train had started slightly late from Oxley, had been delayed south of Wednesbury, at Standish Junction, Coaley Junction and Berkeley Road Junction. His work had prevented him looking out for any of the Charfield signals. When his train had stopped to shunt he had got out onto the platform to follow it back into the platform.

The Westerleigh empty waggon train was in charge of Goods Guard John Callaway, with Driver George Honeyfield and Fireman Arnold Clarke. Driver Honeyfield was of the opinion that the fog had been sufficiently bad for fogsignalmen to be on duty but said that he had been able to make out the up signals at Charfield from close range. The mail train had passed him as his train was passing under the overbridge at Charfield station and he had felt a slight pull on his train but had not realised it was anything serious. The up advanced starting signal had gone back in his face, and he had stopped his train, when he had found that only eight waggons were still attached to his engine, the last two of them derailed and damaged.

Fireman Clarke agreed with his driver's account.

In contrast to his driver Guard Callaway did not consider that the fog was thick enough to warrant fogsignalmen.

In charge of the Leeds mail train was Passenger Guard William Millier, with Driver Ernest Aldington and Fireman Frank Want.

Driver Aldington testified that his brakes had been in order, he was well acquainted with the road and drove the mail train regularly. The signals at Berkeley Road Junction and Berkeley Road South Junction had all been clear. Approaching Charfield he had moved over to the fireman's (left-hand) side of the footplate to get a better view of the distant signal. His train had been travelling at about 45 miles an hour and when about 75 yards from the signal he had seen a green light, while his fireman had remarked, "He's got it off, mate." He admitted having failed to see either the outer or inner home signals, the atmosphere being thicker there. Reassured by the clear distant signal he had nevertheless kept his regulator open, and the crash had followed, his engine turning over on its right side. He confirmed his speaking to Signalman Button afterwards.

Fireman Want stated that he did not know the road between Gloucester and Bristol as well as the section on the Birmingham side of Gloucester. When his driver asked him to look out for the Charfield distant signal he had done so, though he not been sure where it was, and it had been foggy. However, he had seen, he alleged, the signal from a distance of 60 yards and had said, "Right away," to his mate and gone back to his firing. He had put one shovelful of coal on the fire, closed the fire-box door, put down the shovel and the collision had occurred.

Guard Millier said that the mail train had left Gloucester four minutes late. It had been travelling at between 45 and 50 miles an hour at the time of the collision, which had thrown him down. He could not speak as to the signals.

Immediately after the accident Signalman Button had advised Fishponds Control that there had been a serious accident and that ambulance appliances and breakdown trains should be sent from Bristol and Gloucester. Meanwhile Guard Callaway had called up Stationmaster T.E. Brown. The latter had a very busy time. He had to check that the line was protected in both directions; advise the Heads of Department of the accident by telegram; set up a chain of men with buckets to fight the fire; find axes and bars for helpers attempting to release victims; telephone the police station at Wootton-under-Edge asking them to quickly send doctors, nurses and ambulances; telephone the fire brigades at Dursley, Stroud and Bristol, and advise Bristol Hospital to prepare for casualties. After all that he probably needed a strong cup of tea!

Allowing for errors and inaccuracies the signal cabin train register entries suggested that the mail train had probably been somewhat exceeding the temporary speed restriction of 45 miles an hour, due to work on the track, in force between Berkeley Road Junction and Mangotsfield. District Engineer Grimoldby said that the line was in fact amply strong enough for a considerably higher speed.

The position of the Charfield down distant signal when the mail train approached it was of course the crucial question. After the accident Signalman Button found its repeater in his cabin showing "Off." He attributed this to wreckage from the collision resting on the signal wire. District Controller John Robinson found the signal arm hanging off to the extent of 20° from the horizontal, not enough to give a full clear signal. He could only get the arm to go to "Danger" by disconnecting the chain from the balance weight with a hammer and chisel. He was satisfied that wreckage on the wire had pulled the arm off. Signalling Inspector William Whitworth agreed and said that wreckage had also cut the wire to the up advanced starting signal, which was why it had flown to danger in front of Driver Honeyfield.

Chief Lineman Reynolds, District Assistant Ault and District Inspector Hewitt of the Electrical Department confirmed that tests had shown nothing wrong with the down distant signal repeater in Charfield cabin.

The question of the thickness of the fog and the necessity for fogsignalmen was examined at length in the Ministry of Transport report. The evidence of the various drivers, firemen and guards as to its intensity varied considerably. Signalman Button was definite that he had been able to see the back light of his down inner home signal 214 yards away which was his fog object (the object which if it could not be seen from the signal box indicated that the fog had become dense enough to require calling out the fogsignalmen). Stationmaster Brown agreed that the fog had not been thick enough to need fogsignalmen. Several local residents described the fog as only having been a typical autumn ground mist. Police Constable Strong, who had been a passenger on the train, thought on the other hand that the fog had been thick.

Also examined in depth in the report was the question of the gas lighting of coaches. Most of the vehicles of the mail train had had wooden bodies with steel underframes and most (eight) were lit by oil gas. Mr H.C. Warboys of the Bristol Area Carriage and Wagon Department, described the damage to the rolling stock of the mail train. The gas cylinders of four coaches had suffered damage and large volumes of gas must have escaped immediately after the collision.

Carriage and Wagon Superintendent Lemon stated that electricity had been the standard illuminant for all main line stock since 1914 (after the 1913 Ais Gill collision described in the last volume of this series) with the exception of kitchen cars, electricity not being practicable for cooking. The War had slowed the building of new rolling stock as then the company were engaged in constructing ambulance trains, but building had resumed since and about 50% of the Company stock was electrically lit. About 1,000 gas lit vehicles per annum were being broken up.

As regards all-steel vehicles, Mr Lemon stated that the Company was experimenting with them and 435 in all. Echoing the comments of the *Times* correspondent in Paris after the Paris Nord accident (see above) he said that there were several objections to all-steel vehicles:

They weighed 10 % more than wooden vehicles with steel underframes of the same capacity.

All-steel stock cost 10% to 20% more to construct.

Extricating passengers in the case of accidents would be more difficult.

All-steel stock tended to be hot to ride in in summer and cold in winter, plus it was subject to more vibration at high speeds.

He did not think that the anti-telescoping collision proof cars used in the United States, with heavy ends designed to collapse under stress, would be very suitable for use in Great Britain, and added that in a sudden stoppage from 60 or 80 miles an hour passengers would be likely thrown about a coach and killed or seriously injured. Money would be better spent on making collisions impossible.

As to the incidence of fires, the inspecting officer stated that in the 30 years since September 1898 of accidents investigated there had been 22 fires in 21 accidents. Of these 14 had been directly due to gas and in 15 reports the inspecting officers had called attention to the dangers of gas lighting or directly recommended the substitution of electricity.

Beginning the conclusions to his report Colonel Pringle stated that more than 12 years had elapsed since an accident with so many deplorable features had occurred on a British railway.

Colonel Pringle concluded that the atmospheric conditions at Charfield station had not been such as to require the calling out of fogsignalmen. Signalman Button had therefore been justified in accepting the mail train under full "Line Clear." The evidence suggested that the fog had been densest in the vicinity of down outer home signal, out of sight of the signalman, and none of the enginemen had told him that fogsignalmen were required.

The lever of the down distant signal must have been continuously at danger since the passing of the parcels train at 4.53. From 5.13, when the points had been operated to admit the Oxley goods train to the refuge siding, the mechanical and electrical interlocking would have maintained the outer and inner home signals at "Danger." The signal repeaters, which tests showed were working correctly, showed that the actual positions of the outer and inner home signals had corresponded with the danger positions of their levers, while the evidence of Driver Gilbert and Fireman Sutton proved that the distant signal had then been at "Danger" in correspondence with its lever.

After the collision Signalman Button had been able to see the arm of the down inner home at "Danger" and the repeater showed that the outer home was also at "Danger." The repeater for the down distant signal had then shown it to be "Off" but this had been shown to be due to wreckage which had fallen onto the signal wires.

It had to be considered if anything could have caused the distant signal to show a false clear indication between 5.10 a.m. and 5.19 a.m. There was no evidence of the signal wires having become jammed, except of course after the collision. It was unlikely that a person standing on or stumbling over the signal wire would have caused it to go to "Clear" and any such effect would have been momentary. There was no evidence of any heavy weight having fallen from the Oxley goods train. The likelihood of deliberate interference with the signal seemed remote.

Colonel Pringle decided that he had, therefore, to reject the evidence of Driver Aldington and Fireman Want that the down distant signal had been "Clear." No fogsignalmen had been out south of Gloucester and with the fog getting thicker and signals harder to see Driver Aldington should have reduced speed in accordance Rule No. 143. Both driver and fireman must have missed seeing the distant signal, or Driver Aldington had been relying on Fireman Want, who had admitted that he did not know the line very well. Beyond the distant signal the fog had got thicker but Driver Aldington had not reduced speed, his fireman had gone back to firing, and the outer and inner home signals had been missed. The regulator had still been open and the brakes unapplied when the collision had occurred. Driver Aldington had never previously been stopped at Charfield and had probably been trying to make up time. He must take the principal responsibility for the collision, with a minor degree falling on Fireman Want, who did appear to have been a reliable observer. No blame was attached to Signalman Button or anyone else (it was reported that Signalman Button's hair had gone completely white from the shock of the collision).

The sudden stoppage of a train travelling at a considerable speed had caused the disastrous telescoping of its first five vehicles, made worse by their composite (wood and steel) construction and the lack of shock absorbing buffers. As in previous reports the inspecting officer was anxious to add that medical opinion was that all of those who had died had done so from multiple trauma and none had burnt to death.

When the momentum of a train travelling at high speed was suddenly checked it did not seem that serious ill effects could be prevented, even in well-fitted stock with shock-absorbing devices. Colonel Pringle stated that he was in agreement with Mr Lemon that the wholesale replacement of current rolling stock with all-steel vehicles should be postponed until questions such as maintenance costs and durability had been satisfactorily answered. Meanwhile the risk of telescoping could be reduced by measures such as the use of steel vans at the front and rear of trains.

The accident would have been prevented by Automatic Train Control as used on some sections of the Great Western Railway.

The ignition of the large amount of gas escaping from damaged cylinders by smouldering ash, live embers or sparks generated during the collision was clearly the cause of the fire.

As to recommendations, Colonel Pringle stressed the need for the eventual adoption of a system of Automatic Train Control to prevent accidents resulting from trainmen failing to obey signals.

Meanwhile the liability to telescoping could be reduced by including all-steel vehicles in the front and rear of all express and important passenger trains, the fitting of shock-absorbing buffers or collision fenders to all rolling stock used on such services and the marshalling of older vehicles and those not fitted with anti-telescoping devices in the centre of trains.

Electricity should be substituted for gas for illumination on all but unimportant branch lines. The carriage of gas in cylinders on trains was always a source of danger, as proved by Continental experience as well as domestic (he was aware of the 1924 Bellinzona tragedy – see above). Progress in substituting electric for gas lighting had not been as rapid as desirable – a table at the end of the report showed the proportion of electric and gas lit rolling stock owned by different railway companies, ranging from 100% electrically lit coaches on the Metropolitan Railway to 100% gas lit on the Midland and Great Northern Joint Railway. Charfield was, however, to be the last British disaster in which a gas-fed fire played a part. [686]

The inquest on the majority of the victims was held at Staple Hill, in the suburbs of Bristol. At its conclusion on 30th November the foreman returned a verdict that Mr James Charles Gaston, one of the victims, had died from shock and multiple injuries in a collision between an LMS mail train and a GWR goods train. Continuing, he announced, "We are satisfied that Signalman Button was not at fault and that his signals and apparatus were in good working order. We are satisfied that in spite of conflicting evidence about the fog, it was not necessary to call out the fog signalmen. We are unanimously of the opinion that the collision was caused by the negligence of Driver Aldington in passing the signals when at danger." The coroner said that he must interpret the verdict as one of manslaughter against Driver Aldington. [687]

In mid-November Driver Aldington appeared before the magistrates at Wootton-on-Edge but after hearing all the evidence the magistrates dismissed the case against him, a decision which was greeted by a burst of applause from the back of the court. [688]

At the time that the first inquest had been held two bodies remained unclaimed, supposedly those of a boy of about 12 and a girl aged about 9 recalled by a porter collecting tickets at Gloucester, who said that they had been apparently travelling without their parents. A separate inquest upon them was held at Wootton-on-Edge on 3rd November. It heard a sad story from one of the surviving passengers. An engineer from Bradford had been en route to Devon for a holiday with his fiancée, Miss Florence Cross, 38, of Belper, Derbyshire, a music and elocution teacher. 'Mr Helman Brooke, a mechanical engineer, of Lidget Hill, Bradford, who was travelling with her, is still in hospital at Bristol and was brought to Wootton-under-Edge in a motor ambulance. He

686 Ministry of Transport report

687 *The Times*, London, England, 17th, 18th, 19th, 20th and 31st October and 1st November 1928

688 Ibid, 21st and 30th November and 1st December

looked very ill and was deeply affected while giving his evidence. "Miss Cross and I were going to Teignmouth for our holiday," he said. I travelled from Bradford and picked her up at her parents' home in Belper. We were in the second compartment of the first passenger coach with only a luggage van between ourselves and the engine of the express. I was lying on the seat facing the engine and Miss Cross was opposite to me. I remember passing through Gloucester. I had just placed a blanket over Miss Cross, and was lying down again, when I heard her scream. I remember nothing further until I found myself lying on the top of the coach. I had been flung clean out of the compartment into the open air. I crawled over the wreckage, calling Miss Cross by name, but I heard nothing of her, and I never saw her again. Eventually I crawled on to the roof of another coach which was alongside and was rescued from there by means of a ladder." It was stated that some keys had been found which might have belonged to the dead woman, and also a piece of crocodile leather similar to that of which her handbag was made.'

After the taking of some evidence the inquest was adjourned for a month in the hope that the victims might be identified. They could not be and on 10th December the jury simply returned a verdict that they had died of multiple injuries sustained in the collision. [689]

On Friday 26th October 1928, the westbound Simplon – Orient express was stopped in Resca station, Romania at about 2.30 a.m. for a fast Bucharest train to pass. Owing to a mistake by a pointsman the Bucharest train was accidentally diverted onto the line occupied by the Simplon train. In the resulting collision the Paris sleeper and brake van of the Simplon train and a third-class coach and brake van of the Bucharest train were wrecked, severe telescoping being again a feature of the accident. Fifteen Romanians in the third-class carriage and ten in the sleeping car were among the 34 who reportedly lost their lives. The victims included Signor Rocca, an Italian engineer, his wife, and daughter. Mr Rocca's father, General Rocca, had been killed a short time earlier in a road accident. [690] Fifty others were said to have been injured. Another victim whom it was initially believed had lost his life was an American businessman, a Mr Hirschler, but he later returned to Budapest with only a slight head injury, having been trapped for six hours in the wreckage of the sleeper. [691] There were allegations that the station staff had done little to help after the accident and passengers had to telegraph neighbouring stations for help. Resca was in any case somewhat remote from any town, the nearest being Caracal and Slatina, several kilometres away. Meanwhile the passengers were assisted by artillery cadets. The injured were taken to hospitals in Slatina and Crajova (now Craiova). The pointsman and station foreman made themselves scarce after the accident, which probably explains why there was no one left at the station to help. They later re-appeared; the pointsman was dismissed, and the traffic agent (station foreman?) was suspended for two months. The driver of the Bucharest train was suspended for the same period and barred from promotion for a year. [692]

---

[689] Ibid 5th November and 12th December 1928

[690] *Berkeley Daily Gazette*, Berkeley, California, United States, 26th October 1928

[691] *The Times*, London, England, 27th and 29th October 1928

[692] *Railway Gazette*, Vol 49 pg. 567

The express train which had left Paris at 00:15 for Bruxelles came into sidelong collision with a goods train at the entrance to the station at Hal, Belgium, at 06:35 on Wednesday 17th April 1929. The express was coming into the station at full speed as the goods train was leaving it and crossing the main line from Paris en route to Tournai. The express engine cut the rear of the goods train in two, jumped the rails and continued for some hundred metres, then turned completely around and fell onto the barriers of a level crossing and then into the adjoining street. The leading luggage van of the express was crushed by the following Belgian mail van and the French mail van, the third vehicle of the train, climbed upon the first two vehicles, demolishing them further. Crowds of local people who heard the noise of the accident rushed to the scene.

Railwaymen began the rescue work, Red Cross ambulances arrived at 07:15 and a relief train was sent from Bruxelles. By 09:00 seven bodies had been recovered and there were known to be a dozen seriously injured. The bodies of four Belgian mail clerks had been found, together with those of a Belgian conductor and of a brakesman from the goods train who had been literally cut in two. The fate of two other Belgian mail clerks was unknown, it was believed that their French colleagues in the adjoining mail van had all escaped. Some of the passengers further back the express were apparently unaware of the accident; one of three Englishmen on the train was still asleep when found by rescuers. [693]

By noon it was believed that the number of dead had increased to 12, but the number was later corrected to ten, comprising the two railwaymen and eight of the 13 Belgian mail clerks who had been in their van. Meanwhile the public prosecutor was conducting an initial examination of Driver Alfred Daniel of the express. There had been a thick fog at the time of the collision, and it seemed likely that he had overrun the signals.

A large amount of mail was being carried on the express. The French mail was almost undamaged, the Belgian mail on the other hand was badly damaged, the van having caught fire. The mail was forwarded to Bruxelles, partly by train and partly by road vehicle. Most of the French mail clerks had been asleep in their van when the accident happened. [694]

Thirty-five people were injured, 19 fatally, when an engine collided with a passenger train standing in Plastow station, near Cracow (Kraków) Poland, on about Wednesday 3rd July 1929. [695]

A passenger train collided with some waggons of a goods train near Briansk, USSR on Monday 21st October 1929. Three carriages overturned and were smashed. The casualties were not known at the time of the only report into the accident. [696]

At 00:07 on Friday 10th January 1930 a collision occurred at Clutterbuckganj station, India, on the East Indian Railway. No. 557 up passenger (mixed?) train was standing in

693 *The Times*, London, England, 18th April 1929
694 *Le Peuple*, Paris, France, 18th April 1929
695 *Linton Daily Gazette*, Linton, Indiana, United States, 6th July 1929
696 *The Times*, London, England, 23rd October 1929

the station. The signals were against train No. 72, the down Peshawar – Lucknow express, but the driver ran past them and the facing points being wrongly set, his train was diverted into a head-on collision with the up train. Eleven people were killed outright and 20 were injured, one fatally. The damage to rolling stock amounted to 83,300 rupees. The driver and pointsman were tried under Section 101 of the Indian Railways Act, convicted, and sentenced to six months' rigorous imprisonment. The poor pointsman was also tried and convicted under the Indian Penal Code and sentenced to eight months' imprisonment, the sentences to run concurrently. [697]

A passenger train collided with a freight train at Tcherni station, USSR, on Monday 19th May 1930. (The location of the accident is also given as Chernaya). [698] The accident, on the Moscow – Kazan line, killed 28 people and injured a number of others, at least 29 of them seriously. An investigation into the accident was begun and the railwaymen believed responsible were arrested. [699]

Sasso station, just outside of Bologna, Italy, was the scene of a major disaster early on the morning of Thursday 3rd July 1930. Diretto train No. 39 from Bologna to Roma (Rome) started at 06:20 and ten minutes later, at 06:30, travelling at 50 kilometres an hour, it arrived at Sasso station and collided head-on with a freight train standing at the platform. The approach to the station from the Bologna side was on a sharp curve and with the signals standing clear for him the driver of the passenger train had little warning of the obstruction, but he applied the brakes at the last moment. The passenger train consisted of an electric locomotive, a luggage van, six first, second and third-class coaches and a sleeping car. The freight train, marked D D Q, consisted of an electric locomotive, about 20 waggons loaded with food and earthenware, and a van. It had arrived at the station five minutes earlier.

As a result of the violent collision the two engines ended up wedged into one another. The waggons of the freight train and their contents were damaged, but not badly. In the passenger train, on the other hand, the leading van was forced up into the air and crushed the following third-class carriage, in which most of the casualties occurred. This coach had been crowded with passengers, while some who could not find a seat had gone to ride in the luggage van instead. Some lives were no doubt saved by the actions of Conductor Manaresi, who had been looking out of the window and seeing that an accident was about to occur had shouted to the passengers to save themselves by getting down on the floor or jumping from the train. He, himself, escaped with slight injuries.

Men who had just arrived to begin their labours at a nearby construction site witnessed the accident and rushed to the scene carrying a long ladder, with which it was possible to climb the wreckage to reach the injured. A newspaper reported that one of the passengers had been trapped with his left leg caught between two pieces of metal and was groaning in agony. The first rescuers helped to free him, cutting through the steel with an oxy-acetylene torch. Fortunately, there were several doctors travelling on the train, including two Hungarians named Ferenc Ivanic and Kopita Imre, who were able to assist

---

[697] Report by the Railway Board on Indian Railways for 1929-30 Volume 1
[698] *Railway Gazette*, Vol 52, pg. 833
[699] *The Bakersfield Californian*, Bakersfield, California, United States, 21st May 1930

in the rescue work. Fifteen passengers were killed outright and 41 injured were taken to various Bologna hospitals, 12 of them being seriously hurt. Teams of railwaymen worked all morning and most of the afternoon clearing the wreckage and normal working was resumed at 17:00.

As might be expected trains into and out of Bologna over the line through Sasso were much delayed. Diretto train No. 38 from Firenze to Bologna started on time but only got as far as Pistoia when news of the accident was received, and it had to return to Firenze to take an alternative route via Faenza. Likewise direttisimo No. 48 from Venezia (Venice) in the opposite direction, due in Firenze at 09:45, had to be diverted from Bologna via Faenza.

A newspaper reporter in Firenze (Florence) interviewed some of the surviving passengers. The first of these to arrive in Firenze was a Signore Ermici, a merchant from Livorno, who had come from Sasso by car. Visibly shaken up by his experience, he spoke to reporters briefly. He said that he had been riding in the last carriage of the train. As they reached Sasso station the brakes had come on very hard and there was a violent shock. At the same moment, the wall of his carriage broke off and came in on him, leaving him trapped between his seat and the detached wall. It was not long before rescuers freed him. He had only suffered a few scratches and was able to continue his journey by car.

Train No. 39, reformed using three coaches of a local train and with a different engine, arrived at Firenze Centrale just before 11:00. It was carrying survivors of the accident and a large crowd of people, including railway officials and reporters, were waiting for it. One of the survivors who was himself a railwayman from Firenze was Forunato Orapani, who had worked as a brakeman on a night train to Bologna and was returning 'on the cushions' to his home depot, accompanied by another Firenze railwayman, Salvatore Avuri. The two had taken seats in the leading third-class carriage but shortly after leaving Bologna Avuri left his mate to go to the luggage van, saying he hoped to lie down there and get a little sleep. As the train neared Sasso station Brakeman Orapani saw people making desperate signs for the train to stop and looking out of the window he saw why; his train was coming in on an occupied track. Realising that there was no way that the driver of his train could pull up in time and that he was in great danger if he stayed where he was, he opened the carriage door and holding his hands to his face jumped into the void. He thought that the train had then been travelling at not less than 60 kilometres an hour, but he escaped almost unhurt. He did not know what had happened to the other occupants of his carriage. It seems likely that his mate was one of those who lost their lives, as a list of the fatalities includes a Sabatino Avuri.

Soldier Brunello Innocenti was another survivor. He told a reporter, "I left Milano {Milan} at midnight and was riding in the centre of the train. We were due at Pistoia at 8.10 and were running at 75 kilometres an hour, our usual speed. We were nearing Sasso when there was a deafening screeching. It was the brakes being applied. The others in my compartment were wondering what was happening when a tremendous blow threw them one on top of the other. Immediately afterwards there were cries of pain and fear. I cannot exactly describe the scene before my eyes when I managed to extricate myself from the coach after much effort; there were two or three wrecked coaches, a heap of twisted steel and a many victims moaning and groaning. None of the passengers in my coach was hurt." [700]

A postal worker who was a passenger on the train suffered serious injuries in various parts of his body. Rescuers were naturally keen to take him away for treatment, but he insisted on remaining with the packages which had been in his custody until a colleague could relieve him. Only later, when post office officials arrived at the scene, did he agree to go to hospital. In a similar vein another seriously injured passenger refused to leave the scene while his relatives were still in the wreckage. There was also at least one lucky escape, Marcello Buzzoni got into the first carriage at Milano but changed to the third carriage at Bologna.

With the death of several of the injured, including one who survived until 9th July, the number of dead finally reached 18.

The accident was apparently the result of a technical failure, described as 'a polarity inversion at the station indicator panel,' which prevented the facing points from being reversed but did not prevent the protecting signal from being cleared. [701]

A passenger train and a goods train collided between Petrova and Viseul Bistra, Romania, on Wednesday 16th July 1930. According to *The Times* in London this was only a minor accident, resulting in 22 *injured.* [702] The *Railway Gazette*, on the other hand, stated that there were 22 people *killed* in the accident. [703] Frustratingly, it has not proved possible to trace any other report on the collision to determine where the truth lies.

About nine people died (some sources said eight, others ten) and 15 were injured in a head-on collision between two express trains at Secelanu station, Romania, on the Bucharest – Konstanza line, on Thursday 14th August 1930. [704] [705]

At 13:30 on Monday 18th August 1930 a train carrying 58 miners left Creutzwald-La-Croix station, Lorraine, France, for the La Houve colliery, where they were to descend to the seam at 14:00. As they neared the mine their train came into collision at a junction with a coal train. A loaded coal waggon crushed to pieces one of the carriages carrying the miners. Personnel from the mine were the first on the scene to begin the rescue work.

A retired policeman who was working as a mine controller was killed instantly and three miners were so badly injured that they died very shortly afterwards. Fifteen other miners were taken to hospitals at Creutzwald, Freyming and Merlebach in a very serious condition and 30 others were badly hurt. Families of the miners rushed to the scene of the accident and also besieged the hospitals seeking news of their loved ones. By the end of the next day two more miners had died of their injuries and six others, who legs had been trapped between the seats, had had to undergo amputations.

---

[700] *La Stampa*, Turin, Italy, 4th July 1930
[701] www.sassomarconifoto.it/index.php/storia/momenti-di-storia/disastro-ferroviario-a-sasso/
[702] *The Times*, London, England, 18th July 1930
[703] *Railway Gazette*, Vol 53, pg. 138
[704] *North Adams Transcript*, North Adams, Massachusetts, 14th August 1930
[705] *The Times*, London, England, 15th August 1930

A pointsman had authorised the coal train to approach from No. 2 office and then, in response to a telephone request, had allowed the miners train to approach from No. 1 office. Both trains were being propelled rather than pulled so their drivers had a poor view of the line ahead. [706] The pointsman was arrested after the accident but denied responsibility for it. [707]

An accident reminiscent of a couple in the United States in the nineteenth century inasmuch as it involved a circus train and the escape of wild animals, occurred at Guadalupe, Mexico, on Wednesday 26th November 1930. A press agency report stated, 'Lions, bears, tigers and elephants, wild in a new-found freedom from circus cage and corral, roamed the Guanajuato mountains today as charro and hacienda owners banded together to track them down and either kill them or return them to captivity.' Meanwhile at Irapuato 18 members of Beas Circus, one of the largest in Mexico, were lying in hospital beds, suffering from severe burns after a fire, which had consumed their train early the day before. At least 12 of their colleagues had died and some of the injured were not expected to survive.

The fire which had caused the deaths and injuries arose from a collision which had also released the menagerie. The circus train was being switched when it was shunted into some tank waggons containing gasoline (petrol). The contents of the damaged tank waggons sprayed onto the wooden coaches of the circus train and the whole train was soon ablaze.

'Circus performers and employees, clad in nightclothes, escaped from the coaches as best they could, only to meet a new peril, the maddened animals which were getting loose from their cages as the fire burned the cars around them. Finally guns were secured and some of the more ferocious beasts were killed. The others fled.' The escaped elephants stampeded through the town, causing a swathe of destruction.

It was reported that a passenger train from Mexico City to Guadalajara was stopped on a nearby track and some of the lines entered its coaches. Frightened passengers smashed the windows of their berths to escape. The train eventually arrived in Guadalajara ten hours late. Passengers on the train spoke of having seen the charred remains of 12 bodies removed from the wreckage. Others were believed dead. The bodies of three of the performers were unrecognisable. [708]

---

[706] *Le Nouvelliste d'Alsace*, Colmar, France, 23rd August 1930

[707] *La Charente*, Angoulême, France, 21st August 1930

[708] *The Lincoln Star*, Lincoln, Nebraska, United States, 27th November 1930

CHAPTER 6

SINGLE LINE COLLISIONS

The operation of trains in opposite directions over a single line of rails carries the obvious danger of head-on collisions between them. By 1916 there were many hundreds of miles of single-track railways throughout the world, especially in places where the density of traffic did not justify the provision of a second line. Most of them were worked with the aid of the electric telegraph, with decisions as to the meetings between trains being either made locally or, as in the United States, by a central dispatcher. In Great Britain, some lines were worked by telegraph, supplemented by a staff and paper tickets, while an improved method of single-line control had been brought into use in 1880 after the invention by Edward Tyer of the electric train tablet system, but this system was not used extensively outside of the British Empire. The invention of a failsafe track circuit in 1872 by the American William Robinson [709] led to the introduction in the United States of automatic block signals, by which train movements on a single line could be safeguarded by signals warning of the approach of an opposing train, but the cost of installing such signals was high and as late as 1906 the Interstate Commerce Commission was stating that out of 194,727 miles of railroad there were still only 41,916 miles protected by them. [710] With the protection of trains running over a single line left solely to the exchange of telegraph or telephone messages, rather than by the more secure electric train tablet or its later equivalents, and not supplemented by some form of automatic signalling, collisions due to human error in the transmission or interpretation of telegraph messages became all too common.

A new accident towards the end of 1910 has just come to light, too late to be included in the 1901-15 volume. It occurred in China on the Peking (now Beijing) – Hankow line at a location given as near Pin Lok, the old name for Changsha, the capital of Honan province. The accident was said to have happened on the nineteenth day of the tenth Chinese month, which should equate to Sunday 20th November 1910 in the Gregorian

[709] Wikipedia – Track Circuit
[710] Wikipedia – Automatic Block Signalling

calendar. A passenger train was nearing Pin Lok when a goods train, which had apparently set off prematurely, collided with it. The vehicles of the two trains were said to have become 'inextricably mixed up together and the rear vehicle of one train becoming detached ran back down an incline for ten li' (rather more than three miles). The driver and fireman of the passenger train were killed outright, as were two soldiers riding in the front of the train. The passengers in the third-class carriages, which were presumably marshalled at the front of the train, were the ones who were injured, 138 of them being hurt. Of these 13 died almost as soon as they were released and 14 more succumbed in a carriage in which they had been placed for medical treatment, bringing the death toll to 31.

It was stated that this was the most serious railway accident to have happened in China up to that date. The driver of the goods train, who had probably jumped from his engine before the collision, along with his fireman, was held responsible for the accident and was taken into custody, pending the results of an inquiry. [711]

Five people, all railwaymen, died in a collision on the Chicago Burlington and Quincy Railroad near Mascot, Nebraska, United States, on Monday 24th January 1916. Westbound passenger train No. 7 consisted of engine No. 2529, a combination baggage and mail car and two coaches, all of wooden construction, in charge of Conductor Thomas Bell with Engineer James Duffy and Fireman Lee Fogg. It had left Lincoln, Nebraska, at 7 a.m. for Oxford and collided with train No. 2043 at 12.53 p.m.

Eastbound local freight train No. 2043 consisted of engine No. 2043, four cars and a caboose, in charge of Conductor Brooks with Engineer John Eckman and Fireman Walter Moll and had left McCook, Nebraska, also at 7 a.m., for Hastings. Both engineers, both firemen and Head Brakeman Carl Neubauer of the freight train died, and six passengers, Mail Clerk O.K. Porter, Conductor Bell, Express Messenger William Lawson, and Brakeman Frank Quinn were injured. Engineer Duffy lived for some time after the accident and asked that his wife, who was seriously Ill in an Omaha hospital, not be told of his having been in an accident [712]

Conductor Bell of train No. 7 stated at the Interstate Commerce Commission (ICC) that he had been told that Atlanta, the last station before Mascot, would be unmanned until 1 p.m., but in fact when they had reached there at 12.44 p.m. the operator had been on duty. The signal board had been clear, and the operator had told him that everything was all right. After the accident he (the conductor) had spoken to the two rear brakemen of train No. 2043, who had admitted that the accident was entirely their fault, and that the whole crew of the freight had forgotten the passenger train.

Conductor Brooks of train No. 2043 said that his engineer had whistled upon nearing Mascot station, had received a clear signal from the station and one of the brakemen had given the engineer a proceed signal from the caboose. After the accident, his watch had shown 12.42 or 12.43 and somehow it had lost ten minutes since leaving McCook. This had nothing to do with the accident, however, which had been due to him and the remainder of his crew overlooking the passenger train, which they met every day, usually at Oxford. His statement was corroborated by Rear Brakemen Neubauer (a brother of the

---

711 *The Straits Times*, Singapore, 13th December 1910

712 *The Nebraska State Journal*, Lincoln, Nebraska, United States, 25th January 1916

head brakeman?) and Sentence.

Dispatcher Gosch said that he had expected the freight to take the siding at Mascot for the passenger train and had therefore given the operator at Atlanta the block for train No. 7 before Mascot had reported the freight train as passing there. When Mascot had reported the freight train as passing he had immediately realised the possibility of a collision and had tried to contact Atlanta, but had been unable to get a reply, evidently because the operator was out dealing with train No. 7. When he did get through to Atlanta it was too late to stop the train. There was no rule requiring a block for opposing trains. Train No. 7 was superior by class, and it was up to the crew of the freight train to clear its time at every station by 5 minutes.

The ICC concluded that the accident was due to train No. 2043 occupying the main track in the time of train No. 7, which was superior, for which Conductor Brooks and Engineer Eckman were directly responsible. Rear Brakemen Neubauer and Sentence were both qualified conductors, and 'it seems unexplainable that they should have been so neglectful of the safety of their train as to have failed to consult their watches and the time-card in order to know definitely that their train had a right to the track it was occupying.' [713]

By the beginning of the twentieth century the majority of single lines in the British Isles were worked by the electric train staff system or one of its variants. These method of working was extremely safe and in theory should have made accidents impossible, but no system is absolutely foolproof and three accidents, all by chance in Wales or the Welsh Borders, occurred between 1916 and 1921, on lines worked by the electric tablet system. The first accident happened near Torpantau, Wales, on the Brecon and Merthyr Railway, on Thursday 3rd February 1916.

Torpantau station stood at the summit of the Brecon and Merthyr's main line and was a crossing place, the crossing places on either side being Pentir Rhiw to the north and Pontsticill Junction to the south. The line climbed steeply from Pentir Rhiw to Torpantau and then fell equally steeply towards Pontsticill, with gradients of 1 in 55 and then 1 in 47. The trains involved in the accident were the 6.30 p.m. goods train from Talyllyn Junction and a special goods train from Pontsticill.

The Talyllyn train consisted of tank engine No. 7, 17 loaded and five empty waggons and a brake van and was in charge of Guard William Griffiths with Driver Thomas Lloyd and Fireman James Morgan. The train was banked in the rear from Talybont to Torpantau by engine No. 19 with Driver William Vaughan and Fireman Trevor Price. The special train from Pontsticill consisted of engine No. 11, six loaded and four empty waggons and a brake van and was in charge of Guard E.A. Ball with Driver W. Morgan and Fireman H.G. Jones.

On duty in Torpantau box on the evening on which the accident took place was Signalman Edgar Thomas. The Talyllyn train left Pentir Rhiw at 8.24 p.m. and reached him at 8.50 p.m. He had kept his distant and starting signals at danger and lowered his down home signal to admit the train to the station. The up special train he had accepted from Pontsticill at 8.42 and he received the "Train Entering Section" signal for it at 8.47.

[713] Interstate Commerce Commission report

Soon after the down goods train arrived Guard Griffiths asked him if the banker was going through to Pontsticill. "No, she's going back," Signalman Thomas had told the guard. Shortly afterwards, Signalman Thomas told the Board of Trade inquiry into the accident, he had heard the train engine of the goods train give a crow (one long and one short whistle). Going out of his box he heard the train engine give another crow, the banking engine responded with one short whistle and the train started. Signalman Thomas said that he had immediately showed a red hand lamp to the train engine and then, when the train had kept going, to the rear engine, and he had shouted to the guard, whom he could not see, that the Merthyr engine was on the way up. He had then gone back into his box and had reversed the facing points at the end of the loop to prevent the train running through them.

As he came out of his box again the banker had stopped outside and he had exclaimed to Driver Vaughan, "Where's that man going to?"

"He's right away," the driver had replied.

"He hasn't got the tablet and I've a goods coming up," the signalman had answered.

Returning to his box once more Signalman Thomas had telephoned his colleague at Pontsticill, Signalman J.T. King, and asked him if the up goods had gone. "Yes, they've left about eight minutes," was the reply. Signalman Thomas then told Signalman King that the down goods had left Torpantau without the tablet and against the starter at danger.

Guard Griffiths of the down goods said that when his train come to a stand at Torpantau he had got out of his van and begun to walk towards the signal box, pinning down waggon brakes. Having put down six, he found out from the signalman that the banker was not going forward. He had gone back to the rear of his train, uncoupled the banker and shouted to Driver Vaughan to give a crow to the front driver to indicate that the brakes were down, and the train was ready to go. The crow was sounded, the train engine had whistled, and the train had started. He had got into his van, taken off his wet mackintosh and sat down. It had been a very rough night, blowing, raining and very dark. He had not heard the signalman shouting, did not tell the front driver that the banker was coming off at Torpantau or give him a hand signal to start, nor did he see the Torpantau starting signal.

Driver Vaughan of the banker said that they had come to a stand opposite the stationmaster's garden gate about 8.45. The guard had got down, gone forward pinning down brakes, then returned and told him that the banker was not going through, uncoupled his engine and told him to whistle the front man up. Neither he nor his fireman could see the position of the starter from where they were. He had whistled and a minute or so later the front engine had whistled back, and the train had started. Following the train down to the signal box he had handed the signalman the tablet and the signalman had asked him where the train was going, as it did not have the tablet and there was a train coming up from Pontsticill. Driver Vaughan said he had then rushed back to his engine and opened the brake whistle but had failed to attract the attention of the departing train, though he had whistled until it was out of sight. After the signalman had spoken to him he saw that the starting signal was at danger. Later he had gone down the line with the Torpantau stationmaster until they met Guard Griffiths coming back, who told them that it was no use going any further as his driver and fireman were pinned under the wreckage. They had then returned to Torpantau and telephoned for help.

The trains were travelling at 12 or 14 miles an hour when they met on a curve. The trailing wheels of the engine of the down train were the only ones derailed but the leading waggon of this train mounted and pushed the coal bunker of the engine against the firebox, killing Driver Lloyd and Fireman J. Morgan. Fireman Jones of the special up train saw the light of the other train at the last moment, shouted to his driver, "For God's sake, jump!" and did so himself. Driver W. Morgan did not see the other train at all and had no time to jump but survived, though injured.

Lt-Col E. Druitt, the Board of Trade inspecting officer, stated that there was no reason to doubt Signalman Thomas's statement that the up starter had been at danger when the down train had passed it. He surmised that Driver Lloyd of the up train had been misled by the crow given by the banking engine into thinking that the banker was going forward with his train and that Driver Vaughan on the banker had the tablet, though this did not excuse his passing the starting signal at danger. It was unfortunate that Signalman Thomas had reversed the loop points to prevent their being burst as the train doing so might have alerted Driver Lloyd that something was wrong.

Guard Griffiths, the inspecting officer commented, should not have told Driver Vaughan to give a crow whistle when the banker was not going on but should have gone forward and told Driver Lloyd that the banker was not following and given him a proper start signal in accordance with Regulation 171(g) (with a green signal from his hand lamp). The habitual non-observance of Regulation 6(b) (showing the tablet to the front driver of a train which was assisted in the rear before giving it to the driver of the banker) by Signalman Thomas also contributed to the accident in the opinion of the inspecting officer. Also had Guard Griffiths gone forward to give his driver the proper departure signal he would have noticed that the starting signal had not been cleared.

Lt-Col Druitt noted that Guard Griffiths was a man of long experience, but Signalman Thomas had very little service, having started as a porter at Talybont-on-Usk on 16th March 1914 and been at Torpantau only since 2nd December 1916. He said that he had spent considerable time in the signal box at Talybont-on-Usk and three days with the Torpantau signalman before taking over. The inspecting officer said that he was aware of the difficulty of providing signalmen at the present time (that is, because of the First World War) bur he thought the training that Signalman Thomas had received was hardly sufficient for a lad under 19 years of age, even though the traffic was very small, with only four trains in each direction between 8 p.m. and 8 a.m., only one of which needed a banker.

In conclusion, Lt-Col Druitt stated that, 'The Company should take steps to ensure that their signalmen are fully instructed in all the regulations for tablet working, and that these are strictly carried out in future at all their signal boxes on the single portions of the railway.' [714]

In Austria on Thursday 11th May 1916, the crossing of Kaiserliche-königliche österreichische Statsbahnen goods trains Nos. 502 and 521 was altered to Lawoczne. This meant that goods train No. 521 also had to cross passenger train No. 1712 at Lawoczne rather than at the next station at Slawsko. The Lawoczne fahrdienstleiter (signalman) was

[714] Board of Trade report

busy making out this crossing order when train No. 521 departed without clearance. Meanwhile the departure signal for the passenger train had already been received from Slawsko. The trains met on the single line two kilometres from Lawoczne, the passenger train travelling at 40 kilometres an hour and the goods train at 17. The passenger engine went down a four-metre-high embankment and was badly damaged, the goods engine was less so, but two waggons were destroyed and five damaged. Eight passengers died, 13 passengers and four railwaymen were seriously injured, and 25 passengers and two railwaymen were slightly hurt. Sharing responsibility for the accident with the crew of the goods train was the Lawoczne pointsman, who had prematurely set the road for its departure. [715]

Accidents caused by deliberate interference with the instruments for single-line working, such as that on the Oudh and Rohikhand Railway in India in 1908 (see previous volume), were very rare, but two more occurred in 1916 and 1918, the first in Japan and the second in England. The Japanese accident was easily the more serious of the two.

On 29th November 1916 20 people died and 180 were injured when two passenger trains collided head-on between Shimoda and Furumagi stations, in Aomori Prefecture, Japan, on the Tohoku main line. The instrument at Furumagi station was correctly locked because a token had already been removed for a down train at Shimoda station. Due to a misunderstanding an unqualified employee at Furumagi station wrongly assumed that the instrument had failed and against the rules managed to 'burgle' it and was thus able to extract a second token which he gave to the driver of an up train. [716] One train was reportedly carrying soldiers to Aomori, a seaport on the north coast. [717]

January 1917 saw another collision on an American interurban line. On 27th of that month a passenger car and a freight car of the Cleveland Southwestern and Columbus Railway met one another near Pumping Station, Ohio, United States. Five passengers and one of the motormen died in the accident and 44 passengers and one employee, probably the motorman of the northbound car, were injured. The line was worked by timetable and train orders, the orders being sent by telephone. Motormen and conductors were obliged to call at telephone booths at designated points, where they would receive orders from the dispatcher. In the absence of orders extra trains were required to clear the times of superior trains by five minutes, and regular trains must not pass a meeting point designated in the timetable without an order. There was no superiority by direction.

The southbound car, No. 140, was a combination passenger and smoking car of wooden construction with a steel underframe and was in charge of Conductor Pease with Motorman Kemmerer. It had left Cleveland, Ohio, at 12.15 p.m. for Wooster, Ohio, and left Berea at 1.29 p.m., six minutes late, as train No. 35, with an order to meet northbound train No. 406, but no order regarding northbound train extra No. 407.

The northbound train, extra No. 407, was a baggage car used for freight traffic, and

---

[715] *Eisenbahnunfälle Bild I,* by Ludwig Stockert pgs. 7-8

[716] Webpage – Japanese Railway and Transport Review No. 33: Japanese Railway Safety and the Technology of the Day.

[717] *The Evening Tribune*, Albert Lea, Minnesota, United States, 2nd December 1916

was in charge of Conductor Fred Crittle with Motorman John Lawrence. It had left Wooster at 1 p.m. with an order to run extra to West Park (north of Berea) but no orders regarding train No. 35. There was also a third person riding in the car, Motorman Lance, who was deadheading (riding on the train as a passenger to take up duty).

Northbound extra No. 407 made its last stop at Brunswick sub-station and should have waited there, as train No. 35 was due there at 1.35. Its crew failed to consult the timetable and proceeded northward, colliding with train No. 35 two miles further on near Pumping Station at 1.52. The southbound train was travelling at between 18 and 25 miles an hour at the moment that they met, the northbound train at from 25 to 30 miles an hour. [718] In the collision the northbound car telescoped the southbound one, which was crowded with passengers, for a distance of 15 feet. Fire then broke out in the wreckage of the two cars, caused by an overturned stove at the front of the freight car. Motorman Kemmerer was crushed to death where he sat and his body burnt, and 'An unidentified Italian, wedged between two sears, slowly burnt to death. His screams were heard a quarter of a mile away as he writhed and struggled to pull himself from the creeping flame. [719]

Conductor Pease told the Interstate Commerce Commission inquiry that he had suddenly felt the current reversed, and within five seconds the cars had collided.

Motorman Lawrence of the northbound car testified that he had reminded his conductor of the meet with train No. 35 before they reached Brunswick. At Brunswick Motorman Lance had either put off or received a letter. His car had passed Norton switch at about 1.45 p.m. and shortly afterwards he had seen that his watch read 1.49. Motorman Lance had been stood on his right and Conductor Crittle on his left. Suddenly the conductor had cried, "My God – thirty-five!" and he had replied, "Here he comes!" He had immediately applied the air brakes, grabbed the whistle, and stepped back. The brakes had acted well but the distance had been too short. He and the conductor had overlooked train No. 35.

Conductor Crittle stated that after leaving Wooster they had received orders at Seville and Medina but nothing regarding train No. 35. Leaving Brunswick, he had ridden in the front vestibule of the car and had suddenly thought of train No. 35. He had, it seemed, no sooner mentioned train No. 35 to his motorman when it came into sight and he heard his motorman apply the brakes and whistle, the collision happening almost at once. No one was to blame but Motorman Lawrence and himself, though the presence of Motorman Lance might have distracted them.

The commission came to the obvious conclusion that Conductor Crittle and Motorman Lawrence were responsible for the accident. Both were considered competent employees, though the conductor had failed three times and the motorman had failed once to register trains in the previous three years.

Only two days later another motorman died in a further single-line collision on the Cleveland Southwestern and Columbus near Winters, Ohio. [720]

Twenty people died and 56 were injured in an accident between Ghunsore and Pindrai

---

[718] Interstate Commerce Commission report

[719] *The Evening Independent*, Massillon, Ohio, United States, 29th January 1917

[720] Interstate Commerce Commission report

stations, India, on the 2 ft 6 in gauge section of the Bengal-Nagpur Railway, on 23rd July 1917. Rather unusual circumstances led to the accident. A passenger train proceeding between the two stations was held up at a point where the line had been breached, probably by floodwater as this was the middle of the monsoon season. The train staff sent a message to the nearest station for assistance, but before this arrived they managed to temporarily repair the breach themselves and the train proceeded on its journey. Travelling too fast, after five miles it then came into collision with a train which had been despatched to rescue it, no protection having been provided for the failed train as per the rules.

The guard of the passenger train was found to be principally responsible for the disaster. He was prosecuted and sentenced to three months' rigorous imprisonment. [721]

The Shore Line Electric Railway of the United States suffered a major disaster on 13th August 1917 near North Branford, Connecticut. Nineteen persons died in the accident and 35 more were injured.

The Shore Line was another interurban system. The accident happened on its Saybrook division, a single track which ran between New Haven and Chester, Connecticut. Regular trains were run over the line at hourly intervals between 5.25 a.m. and 11.55 p.m., but no timetables were provided for either employees or the public. The trains were given numbers and their times were entered on a handwritten schedule, which gave the times at different stopping places, with meeting points shown in red ink. The schedules were posted at the termini but not provided to the conductors and motormen who were expected to memorise their contents and be thus guided as to meeting points. No written orders were provided. Telephones were provided at crossing points. Trains were expected to wait at crossing places until the designated time when, if the opposing train had not appeared, the conductor or motorman must ring the dispatcher who would either advise them to wait or to proceed to the next crossing place. A written record of such orders was kept only by the dispatcher. Extra trains were not given orders but ran on schedules like the regular trains, these extra schedules being placed on the bulletin board or attached to the regular schedule as a supplement. One of the trains involved in the accident was an extra train known as the "half hour car" which made one regular trip five days a week between Saybrook and New Haven, leaving Saybrook Car House at 3.43 p.m.

On the day of the accident the "half hour" train left Saybrook on time, consisting of car No. 16 in charge of Conductor Tryon with Motorman Negus. It was scheduled to meet eastbound train No. 22, consisting of car No. 2, in charge of Conductor Arthur Jones with Motorman Harry Morris, at North Branford turnout, at 4.55 p.m. Instead it ran past the turnout, stopped at North Branford station, 1,600 feet further on, picked up a passenger and set off again, colliding with train No. 22 about 500 feet further on at 4.57 p.m. [722]

'Both cars, of heavy construction, were running at high speed, it is said, and the force of the impact was such as to lock them together – a mass of twisted iron and steel and splintered wood. Many of those killed died instantly, and others within a short time of

[721] Railways in India: Administration Report for the Year 1917-18
[722] Interstate Commerce Commission report

being taken from the wreckage. Most of the dead are women, with a few men and some children among them. Few of them have been identified so far, the injuries being such as to make identification difficult in a large number of cases. As fast as the bodies were taken out of the wreck they were sent to the Branford morgues. The injured were rushed to New Haven hospitals. Most of the persons killed lived, it is believed, in Guilford, Madison, Saybrook and nearby towns on Long Island sound.' [723]

Motorman Negus was unable to give Interstate Commerce Commission investigators any reasonable explanation for his failure to stop at the North Branford turnout. He said that the last thing that he remembered before coming to his senses immediately before the collision was passing a section gang a mile before the turnout, though he admitted that if he had merely allowed his car to coast downhill through the turnout uncontrolled it would probably have gained such a speed as to be derailed on the turnout points and in any case would lose the trolley pole if going more than 4 or 5 miles an hour. On the same trip at a previous meeting point, he said, he had also dozed off, waking just in time to avoid a collision with an opposing car. He did not remember stopping or restarting from North Branford station. He did not remember running back through the car just before the collision; afterwards, he had come through sitting on the ground only slightly hurt. He admitted having tied down the controller handle, and having done so before, though it was against the rules. He did it, he said, to rest his arm. (The controller handle had an automatic cut-off feature such that if the pressure on it was released the power was cut off, though the brakes were not applied). He said that other motormen did the same thing and other rules were violated by employees on a daily basis. He had never made a protest against working long hours but had felt tired for several days before the accident.

Conductor Tryon testified that after leaving West Point Road he had sat down to write some reports and remembered nothing more until after the accident, when he had asked the motorman what had happened and then gone back to telephone the dispatcher. He had not reminded the motorman of the North Branford meet, it not being customary to do so. In explanation of his loss of consciousness he said that he was subject to fainting spells and severe headaches and had lost consciousness while on duty before. He had never reported these attacks, but his superiors were aware of them, as they had witnessed one. He had worked 13 hours a day for the week preceding the accident, which was more than he liked, and many of the men had complained of the long hours. After the accident he had telephoned the dispatcher, taken the names of witnesses and flagged an approaching car but then he had apparently lost consciousness again. Why, he could not explain. It was not the long hours as such which were the problem, he explained, but the arrangement of the runs, which made it hard for him to get enough rest. For example, the previous few days his duty had begun with a train from Guilford at 5.25 a.m., which meant leaving his home at Ivoryton the night before, arriving at Guilford at 11.15 p.m. and sleeping the night in the car. Every day, he said, motormen started their cars after passengers had got on and off without receiving a bell signal from the conductor, though this was against the rules.

Track Labourer Adam Bania stated that he had arrived at North Branford on the 4.10 train from Guilford to clean the switches and light the lamp at the turnout and having done so was waiting for the eastbound car to take him back to Guilford. He was sure he had heard the westbound car whistle for two level crossings near North Branford station

---

[723] *The Manchester Journal*, Manchester, Vermont, United States, 16th August 1917

and saw Motorman Negus look back as a passenger boarded the car. Someone then said, "All right," though he did not hear a bell, and the car started, the collision shortly following. Afterwards he had spoken to the motorman and Negus had said, "It was our fault," and asked why he had not waited for train No. 22 the motorman had explained that he had been thinking that he was driving a regular car, which did not normally cross anything at North Branford, rather than an extra, which did. Track Labourer Bania said he had not spoken to the conductor after the collision.

A passenger, Frank Harrison, said that he had boarded the car at Guilford Green. Conductor Tryon had collected his fare, after which he had not seen the conductor again. When the car ran past the North Branford turnout he had been surprised, knowing that cars usually crossed there. He thought Motorman Negus whistled for the North Branford level crossings, he was certainly awake, looking out of the window, and sitting up straight with his hand on the controller. Someone shouted. "All right." at North Branford station but he did not hear a bell. When he saw the approaching car, he ran for the door and the motorman had been in front of him and had jumped out first. Mr Harrison said that he had been looking out for the opposing car as he had been expecting danger, they not having met it as usual. Motorman Negus had reduced speed for the points at the turnout and had appeared normal in every respect.

The passenger who had boarded the car at North Branford was named George Cook. He said he caught the car daily at 4.55. Before boarding he had been talking to the trackman and they had been surprised to see the westbound car appear before the eastbound and concluded that the latter must have been held up at New Haven. He had barely taken his seat in the car when the motorman rushed past him. He had jumped up to see what the trouble was, saw the eastbound car and was heading for the vestibule door when the cars met, and he was thrown to the floor. He did not see the conductor before the accident.

Dispatcher Mooney stated that the company had discontinued printed timetables in December, 1913, he did not know why. Since then the men were wholly dependent on their memories in making meets, and there was no custom requiring the conductor and motorman notifying one another when passing meeting points, though he considered that it was desirable. He was not aware of any previous case of a car overshooting a meeting point.

Superintendent Rudd said that he was aware that rules were violated; there were always violations of rules. The printed timetables had been discontinued, against his advice, by the general manager as being unnecessary. Previously releasing the controller handle had caused the brakes to apply automatically but motormen had been using this to make ordinary stops, shaking up the passengers needlessly, so the brake feature had been abandoned. Inspector Hoyt had spoken to him about motormen tying down the controllers and he had spoken to the men concerned, but he did not consider it a serious offence, and none had been suspended for doing so. A notice forbidding the practice had also been put up.

Inspector Hoyt testified that he understood that the printed timetables had been dispensed with to save money. He believed that every employee ought to have a copy of the schedule and had recommended that this be done. He also thought that some sort of block system should be employed. In the year and a half that he had been inspector he had reported two cases of cars running past meeting points. One of these had been the employee's third offence and he had been discharged. He had spoken to the superintendent about controller handles being tied down and a notice had been issued but he was not aware of anyone being disciplined.

Superintendent of Transportation Cain stated that he had only been employed by the Company on 1st August and Mr Stratton, the general manager, only since 1st July, so they had had insufficient time to do more than observe the general situation and make tentative plans. He had observed many infractions of the rules and a reorganisation was already being prepared. He proposed to unify the various lines to improve employee supervision, introduce a standard operating code and educate employees as to the importance of their work. He was heartily in favour of the block system and planned to introduce written orders and printed timetables, also a general order prohibiting dispatchers from discussing other matters whilst issuing orders.

The Interstate Commerce Commission concluded that the direct cause of the accident was the failure of Conductor Tryon and Motorman Negus to wait at North Braford for train No. 22. It was not believed that Motorman Negus had fallen asleep, his statement to Track Labourer Bania that he had forgotten he was not driving a regular car was likely nearer the truth. That Conductor Tryon had fallen asleep was, however, likely, passengers having testified before both the coroner and the Public Utilities Commission that they had seen him asleep. Contributing to the accident in a large manner were the slipshod and unsafe operating methods in vogue on the road, together with the absence of any mechanical safeguards. The crude pen-written schedule drawn up for employees covered a period of 19 hours, with 22 trains each way on weekdays and 20 on Sundays. Between New Haven and Chester some trains had five meets and others six, indicated only by being marked in red ink. Nor were written orders issued when meeting points were changed.

In conclusion the commission wrote, 'Such unsafe methods of train operation are inexcusable and display a lamentable lack of forethought for the safety of passengers and employees. The evidence also shows that there was a habitual violation of rules by employees, and that no proper effort was made by those in authority to preserve discipline or secure the enforcement of rules. The statement by former superintendent Rudd indicate that he had no conception of his responsibility as an operating officer, nor of what was required of him in the direction of securing safe operation.'

The commission's report ended with a review of the hours worked by Motorman Negus and Conductor Tyron. For example, on the day of the accident the motorman had come off duty at 1.40 a.m., slept in the car barn at Saybrook and come on duty again at 6.10 a.m. after a break of only 4½ hours, in violation of the Hours of Service Law. He had never complained about his long hours, saying that he was willing to take what work he could so as to pay off debts that he had incurred. His living conditions were bad, with no boarding place; he got his meals wherever he was at the time and slept in the car barn except when he was invited to stay at a fellow employee's house. Conductor Tyron had worked from 6.30 a.m. the previous day until 12.05 a.m. on the day of the accident, with a little over four hours off duty during this time. Sleeping in the car barn at Guilford, he had come on duty again at 5.25 a.m., had been relieved at 2.13 p.m. but had resumed duty again at 3.40 p.m., after a break of only an hour and 27 minutes, and but for the accident would have worked until 7.37 p.m., again in violation of the Hours of Service Law. [724]

On Monday 17th September 1917, the meeting point of two German passenger trains was altered from Wilsenroth to Willmenrod. The driver of train No. 3982 was given a written order to this effect, despite which he left Willmenrod before the arrival of the opposing train and the two collided on the single line. Seven people died and 18 were injured. [725]

As in the Rockledge accident rather less than two years earlier it was mainly negro passengers in the 'Jim Crow' coach near the front of the train who died in a collision near Kellyville, Oklahoma, United States, on 28th September 1917, on the St Louis-San Francisco Railway. The Interstate Commerce Commission report gives a (presumably final) death toll of 32 passengers without subdividing them by race but an earlier newspaper report, when 27 bodies had been recovered, stated that 17 of them were negroes, three of them Indians (also segregated to the Jim Crow coach?) and seven white. One railroad employee also lost his life in the accident and 57 passengers, and five employees were injured.

The trains involved in the Kellyville disaster were westbound passenger train No. 407 and eastbound freight train extra No. 1322. Train No. 407 consisted of engine No. 1035, a baggage car, mail car, two coaches, a chair car, three sleeping cars and an official car, in charge of Conductor Temple with Engineer John Ruhl and Fireman Carterton and was en route from St Louis, Missouri to Lawton, Oklahoma. All the vehicles of the train were of wooden construction except the all-steel baggage car. Eastbound freight train extra No. 1322 consisted of engine No. 1322, 29 empty passenger cars (the train was the empty coaching stock of a troop train), two loaded box cars and a caboose in charge of Conductor Huff with Engineer O'Hara and Fireman Tolbert.

At Sapulpa, Oklahoma, train No. 407 received two train orders; No. 56, which read:

Extra 1322 East has right over No. 407, Depew to Kellyville,

and No. 61, which read:

No. 406 gets this and meet No. 407 at Kellyville.

Train No. 407 left Sapulpa at 2.12 p.m., arrived at Kellyville at 2.27 p.m., left there at 2.30 p.m. and collided with train extra No. 1322 about two and a half miles further on at 2.42 p.m. Train extra No. 1322 had left Oklahoma City at 10.30 a.m., received a copy of train order No. 56 at Depew, left there at 1.47 p.m., and passed Bristow, 13.4 miles west of Kellyville, at 2. 12 p.m. In the collision the front end and the tender of locomotive No. 1035 were damaged, the leading baggage car of train No. 407 was only damaged slightly, but the following wooden mail car overrode and telescoped the coach following it, the two forward compartments of which were set aside for negro passengers, and it was here where all those killed and most of those injured were found. The other cars of the passenger train were only slightly damaged. In the freight train locomotive No. 1322 was badly damaged, its tender telescoped the leading baggage car, the third and fourth cars

[724] Interstate Commerce Commission report
[725] *Eisenbahnunfälle Bild I* by Ludwig Stockert, pg. 16

were partly telescoped and the fourth, fifth and sixth cars derailed and destroyed.

Engineer Ruhl told the Interstate Commerce Commission (ICC) investigators that he had received train orders Nos 56 and 61 at Sapulpa, had read them, handed them to his fireman, the latter had read them and handed them back to him and he had re-read them out loud. When they reached Kellyville, he saw two trains in the siding there; one was No. 406 and the other he took to be extra 1322. He saw the figures 13 on the sand dome, thought that he saw a 2 and was convinced it was train extra No. 1322. He had intended to check the number on the caboose but at that point the conductor had whistled, and he had omitted to do so. Unbeknown to him the second train in the siding was not No. 1322 at all but another extra, No. 1343, which the crew of train No. 407 had no orders respecting. Engineer Ruhl stated that after leaving Kellyville he had left the cab of his engine to close the drain cocks on the air pump when he had seen the other train approaching. He had rushed back to the cab, shut off steam, applied the brakes in emergency, put down sand, whistled, reversed his engine and jumped.

Fireman Carterton agreed that his engineer had shown him the orders. He had seen train No. 407 and a second train at Kellyville but did not notice the number of that one, nor did his engineman say anything about it. His engineer saw train No. 1322 first. He thought that his train had been travelling at 35 miles an hour when the other train was first seen but speed had been reduced slightly before the collision.

Conductor Temple of train No. 407 said that he had received the orders at Sapulpa, read them aloud and given copies to his brakeman. At Kellyville he saw train No. 407 and a freight train. Looking through a glass door for the number of its engine he had seen only one figure, which he took to be a 2. He had assumed that his engineer had seen the number and gave him the signal to pull down to the lower end of the yards and then a stop signal to put off a drunken man. He had not discussed the train orders with the brakeman as the man had them to read for himself. He saw the crews of the two trains waiting at Kellyville but did not talk with them or ask what trains they were in charge of. He admitted that he was required to check that the two trains at Kellyville were the ones they were ordered to meet there but said that he had relied upon his engineer.

Brakeman Leitwein of train No. 407 said that he had seen the two trains at Kellyville and had seen that one was train No. 406 but did not see the number of the other train. The conductor had handed him two orders at Sapulpa, and he had intended to read them when he had time.

Engineer O'Hara of train extra No. 1322 said that his train had been travelling at about 25 miles an hour when his fireman, who had been on the inside of the curve, had called out and applied the brakes in emergency. The speed of his train had dropped to 8 or 10 miles an hour before the collision.

Fireman Tolbert of train extra No. 1322 said that he had seen train No. 407 when they were about 400 yards from bridge 447.8 and had shouted to his engineer.

Conductor Huff of the freight train spoke of receiving the order giving his train right over train No. 407 to Kellyville at Depew. At the time of the collision he had been riding in the caboose and did not see the other train before the accident.

Engineer Trone of train No. 1343 stated that the engines of trains Nos. 407 and 1343 had stopped about two car lengths apart and there had been nothing to prevent Engineer Ruhl from seeing the number of his engine. Likewise, Conductor Pugh of that train testified that indicators on his caboose were clearly showing the train's number.

Dispatcher Coombs said that he was not sure that he had known that train No. 1343 would be at Kellyville when train No. 407 received its orders at Sapulpa, and it had not occurred to him that the crew of train No. 407, having orders to meet two trains at Kellyville, would be confused if more than two trains were there. Had a middle order been issued to the operator the collision would probably have been prevented but the company rules did not provide for the issuing of such orders.

The ICC found that Conductor Temple and Engineer Ruhl were mainly responsible for the collision. It was their absolute duty to identify train extra No. 1322 before leaving Kellyville and such carelessness and dereliction of duty on their part in not having done so was inexcusable. Brakeman Leitwein was also at fault for failing to read train order No. 56 in time, had he done so the collision might have been prevented.

The ICC noted that no block system was in use on the Oklahoma division and the train order system, which in any case was inherently weak, was not used in the most efficient manner, as the operating rules of the St Louis-San Francisco Company did not provide for the issuing of a meet order to the operator at the point where the trains were ordered to cross (a middle order) as envisaged by Rule No. 208 of the Standard Code of rules of the American Railway Association. Until either a block system was installed, or middle orders introduced accidents such as that at Kellyville were likely to occur. [726]

An Oklahoma newspaper reported that Engineer Ruhl was devastated by his mistake. 'Suffering mental agony, John F. Ruhl, engineer of the passenger train that collided with the empty troop train, is confined to his home in Sapulpa under the care of physicians.

'"It was terrible, the moans of the crushed and dying and the sight of my train torn to pieces, Ruhl said, "I know I'll see it forever."

'"We had orders to meet regular passenger train No. 8 as well as a special empty troop train, No. 1322. As my train entered the Kellyville yards, the passenger train was standing on the siding and behind it was the troop train, as we supposed. I looked for the engine number of the trooper, and as these numbers are not so conspicuous as they were prior to the time they were taken from the side of the tender and supplanted by the word 'Frisco,' in passing I was not sure beyond the first two numerals, '13,' that it was the train that we had been ordered to meet. I thought to satisfy myself as we passed the rear of the special, but before I could pass I was slowed down by the conductor who stopped the train to allow a passenger to avoid being carried beyond his station. This took my attention from the caboose attached to the troop train and I failed notice it when we passed by."

Ruhl could not account for the time that elapsed between the first moment after the crash until he was found wandering miles from the scene of the wreck. "All I remember was the flying glass, there seemed to be tons of it, the cries and groans, and then I saw my fireman stumble and fall, and everything went black. It was awful."' [727]

By the end of 1917 the Baltimore and Ohio Railroad line through North Vernon,

---

[726] Interstate Commerce Commission report

[727] *Daily Ardmoreite, Ardmore*, Oklahoma, United States, 1st October 1917

Indiana, United States, was already equipped with track circuiting and automatic block signals, the latter consisting of three-position upper-quadrant semaphores. The line was single track and the semaphores at the ends of the crossing places were "stop and stay" signals. There were also intermediate signals which trains could pass in the stop position under flag protection. During normal operation when a westbound train occupied a track circuit at Oakdale station, three miles east of North Vernon, "stop and stay" signal E71-28 at the east end of the crossing place at North Vernon was set to the stop position. At the same time signal E72-05 near the end of the station platform was set to caution and an annunciator bell sounded in the North Vernon office.

On the day of the accident eastbound passenger train No. 2 consisted of engine No. 2128, a baggage and mail car, a passenger and baggage car, a coach, a sleeper, a dining car and a parlor car, in charge of Conductor Shaner with Engineer Day and Fireman Matthews. It arrived at North Vernon at 7.4 p.m. and their received train order No. 142 on Form 19, which read:

Second 2 No. twenty three 23 eng. 1428 has right over No. two 2 No. sixty eight 68 and No. forty six 46 engines 2128 1420 and 1448 Milan to North Vernon.

Five other orders were received at the same time, one directing the train to run 30 minutes late North Vernon to Milan, the others relating to cars on different sidings. As the station at North Vernon was some distance east of the office two copies of the orders were hooped to Conductor Shaner as the train passed the office. At the station he gave one lot of copies to the engineer but failed to read the orders himself. The train left North Vernon 51 minutes late, at 7.12 p.m., passed signal E71-28 at the end of the passing track showing stop and at a speed of 40 miles an hour collided with train westbound second No. 23 a mile further on.

Westbound train second No. 23 consisted of engine No. 1428, a baggage car, two coaches, a sleeping car, and a dining car in charge of Conductor McEvily with Engineer French and an unnamed fireman. It passed Nebraska, 10 miles from North Vernon, at 6.55 p.m., and was running at 20 or 25 miles an hour when it collided with train No. 2 at 7.15 p.m. Engineer Day and Fireman Matthews, Engineer French and his fireman, three other railwaymen and one passenger were killed and 40 passengers, four railwaymen and nine 'other persons' were injured.

Operator Watkins at North Vernon told the Interstate Commerce Commission (ICC) inquiry that when he had hooped the orders to Conductor Shaner signal E72-05 had still been clear. Train No. 2 had stopped in the rear of it. At 7.09 or 7.10 the annunciator bell had rung, and he saw that signal E72-05 had gone to caution. He had especially noticed this because the dispatcher had asked him if train No. 23 was there yet. He had seen train No. 2 leave at 7.12 and had assumed that it was moving down to the end of the crossing place.

Conductor Shaner was too badly injured to attend the hearing but had apparently said after the accident that he had only begun to read the orders when the collision had occurred.

Sandwich Vendors Herbert and V.C. Kutchback, who did business at North Vernon, had seen that signal E72-05 was at caution, and the former said that he had remarked to Head Brakeman Kempton, "There must be something in the block." The brakeman had looked towards the front of the train and replied that it must be the accommodation.

Engineer Smith, deadheading, said that he had also noticed signal E72-05 at caution.

Post-accident tests of the signalling showed it to be working normally. In view of this and the statements of the operator, vendors and deadheading engineer the ICC concluded that it was clear that the accident had been due to the failure of the crew of train No. 2 to obey order No. 142 combined with their running past signal E71-28 at stop. Conductor Shaner's admission shortly after the accident that he had not read the orders before leaving North Vernon was an admission of negligence and culpability. Head Brakeman Kempton had also been negligent in knowing that signal E72-05 was at caution but failing to stop his train passing signal E71-28 at stop.

Continuing, the ICC wrote, 'This accident was due entirely to the failure of employees properly to perform their duty. It seems almost incredible that an experienced conductor would leave a station without reading his orders as Conductor Shaner did in this instance, or that Engineman Day would overlook both order and signal. The block signal system provided for the protection of traffic on this line is complete, adequate, and well-maintained, and had its indications been obeyed this accident would not have occurred, notwithstanding the failure of Conductor Shaner and Engineman Day to read their orders before leaving North Vernon.'

The ICC reiterated the need for automatic train control devices to prevent accidents of this nature.

Finally it noted that the absence of more fatalities and serious injuries among the passengers was doubtless due to the steel and steel-underframe cars of the trains. [728]

A newspaper report stated that Fireman Matthews had had a premonition of his own death and had hunted out his life assurance policies and made sure that his wife knew where they were in case they were needed. Mrs Matthews had collapsed upon being informed of her husband's death.

At the beginning of 1918 China was in its Warlord Era, with different parts of the country being ruled by different military cliques, generally divided into the Northern and Southern factions, who were in conflict with one another. On about Saturday 5th January 1918 the Northern troops were in retreat, and at Changsa station their leader commanded the stationmaster to immediately despatch a train to the north, carrying their soldiers. Knowing that a southbound train was proceeding towards his station over the single line the stationmaster naturally refused but the troops piled onto the train anyway and forced its driver to set off at speed, doubtless at gunpoint. The predictable collision with the southbound train occurred and 300 soldiers were reportedly killed, and many others injured.

A reporter commented that, 'The marvel is that similar accidents do not occur more frequently. During the past few years there have been instances of troops overriding engine drivers to do their will. Not long ago the writer was on a train crowded with nearly 2,000 Chinese between Peking and Fengtai when, as a precaution, the foreigner in charge stopped the train, although the signals were open. As we lay at a siding a train full of Chang Haun's soldiers dashed round the bend half a mile in front. If we had gone on

[728] Interstate Commerce Commission report

there would have been a terrible accident.' [729] The accident has obvious similarities with the 1917 Modane catastrophe in France, where the military allegedly forced a driver to set off down a long steep incline with an overloaded train so that it ran away.

Early on the morning of Friday 18th January 1918 two goods trains collided at Parkhall, England, near Oswestry on the Cambrian Railways. As no passenger trains were involved the casualty list was not high, but one railwayman was killed and four were injured, three seriously. The accident was unique on the railways of Great Britain as regards its cause, which, as in Japan and India previously, was tampering with the instruments, permitting the extraction of two tablets for the same stretch of single track at the same time.

The single line between Oswestry and Ellesmere, specifically Oswestry North and Ellesmere Junction signal boxes, was worked by Tyer's No. 6 tablet instruments, with an intermediate crossing place and tablet station at Whittington, slightly under 2 miles from Oswestry. The small number of night-time trains had resulted in the installation, as an economy measure, of separate McKenzie and Holland long section instruments in Oswestry North and Ellesmere Junction boxes, enabling Whittington signal box to be switched out at night and thus saving the wages of a signalman. The short and long section instruments were of course interlocked with one another, so that once a tablet had been removed from either instrument at one end of the section a tablet could not be removed from either instrument at the other end. The tablets were also different in size and shape.

On the evening of the day before the accident Whittington signal box was switched out as usual at 10.26 p.m. and the long section instruments were brought into use. At 10.32 p.m. Signalman Edwards at Oswestry North box obtained a tablet for an up goods train, which duly cleared Ellesmere Junction at 11 p.m. Then at 12.30 a.m. on the day of the accident Signalman Evans at Ellesmere obtained a tablet for a down goods train, which reached Oswestry at 1.08 a.m. Up to this point there was no dispute as to what happened in the evidence given by the two signalmen at the subsequent Board of Trade inquiry.

According to Signalman Evans at Ellesmere at 1.29 a.m. he had obtained the release from Signalman Edwards at Oswestry of a tablet (No. 9) for another down goods train from Wrexham, gave the tablet to that train and it proceeded towards Oswestry. Signalman Edwards at Oswestry, however, denied having released a tablet for any second down train or that he was aware of any tablet being taken out at Ellesmere. He stated that at 1.49 a.m. he had wished to send an up goods train Ellesmere and had sent the 4-1 bell code for it to Ellesmere. Signalman Evans at Ellesmere admitted receiving the "Call Attention" signal from Oswestry and stated that after a moment's delay he had telephoned Oswestry and asked, "Has the goods arrived?"

Oswestry had replied, "Yes, it has." According to Signalman Edwards his words had been, "Yes, long ago," but Signalman Evans denied hearing the words "long ago."

Signalman Evans asserted that the Oswestry signalman (Edwards) then cleared the tablet instrument for the up goods train, whereupon he had released a tablet for the down train. Signalman Edwards on the other hand denied having cleared back the up goods

[729] The Singapore Free Press and Mercantile Advertiser, Singapore, 11th January 1918

train, asserting that his instrument had displayed "Clear" rather than "Tablet out from," and that he had merely asked for and been given a release from Ellesmere and had duly obtained a tablet (No. 21) which he handed to the fireman of the up train. The "Train entering Section" signal for the up train was sent and acknowledged at 1.51 p.m. The entries in the train registers of the two boxes supported the statements of the respective signalmen.

At this time, the down train was quite close to Oswestry and the two met at Parkhall three minutes later. The up train was travelling down a falling gradient of 1 in 108 at about 20 mile an hour and when Driver Williams saw the down train he had only time to close the regulator, apply the brake and jump. Fireman T.H. Dyke of the down train was killed, and Driver Williams and Fireman Lewis of the up train and Driver Sockett of the down train were all seriously injured. The two tablets found and examined after the collision and the statements of the two drivers, Fireman Lewis and the two guards, Davies of the up and Jones of the down train, all indicated that each train had the correct tablet.

At about 2.10 a.m. Guard Davies got back to Oswestry North box and told Signalman Edwards what had happened, whereupon the latter telephoned Ellesmere. Signalman Evans at Ellesmere had exclaimed, according to Edwards, "Good God! Hadn't you cleared it?"

Edwards had replied, "Certainly not; I know nothing about the goods." Signalman Evans, on the other hand, stated that at 2.02 a.m., that is before Guard Davies could have reached Oswestry North box, Edwards had rung him and asked, "Where's the Wrexham goods?"

He had replied, "It's there {at Oswestry.}"

Edwards had said, "No, it's not." Edwards had advised him of the collision a few minutes later.

After being told of the collision by the guard Signalman Edwards had walked (or more likely run) to the scene of the collision, arriving there about 2.20. He found Driver Williams and asked him if he had his tablet. "Yes, and so has the other driver," Williams had replied, in a tone which can only be imagined. Edwards only stayed two minutes before returning to his box, being met on the way by Inspector Pugh. The Board of Trade inspecting officer who investigated the case, Colonel J.W. Pringle, commented that Signalman Edward's action in leaving his box was certainly undisciplined but not necessarily proof of a guilty conscience.

The problem, of course, was, how had two tablets been extracted for the same section of line at the same time? Something which was supposed to be impossible. This could clearly only have happened either through some serious fault with the instruments or through deliberate interference with them by one or other or both of the signalmen. After the collision, the Cambrian Railways Company had the instruments examined by an independent expert, Mr. J. Sayers, Telegraph Superintendent of the Midland Railway. Mr Sayers was unable to find any fault with the instruments such as could permit two tablets being released at the same time and considered that they had been adequately maintained. He was not, however, very impressed with the design of the McKenzie and Holland long section instruments.

The inspecting officer commented that this was one of the most difficult investigations with which he had had to deal. Even given the faulty design of the instruments commented upon by Mr Sayers and possible contact between the telegraph and tablet wires at Ellesmere due to a heavy snowstorm, he could see no combination of circumstances which would reconcile the statements of the two signalmen or clear them from suspicion of having failed to observe the indications of their instruments and a failure to work according to the regulations.

The telephone enquiry made by Signalman Evans at 1.49 a.m.:

"Has the goods arrived?" instead of specifically, "Has the Wrexham goods arrived?" the inspecting officer characterised as slipshod and indicating the need for greater accuracy in making telephone enquiries. The reply that it had, even if not including, "Long ago," should also have given Evans concern given the weight of the heavy (46 vehicles) train, the short time which had elapsed since it had left Ellesmere and the length of the section. It was also strange that Signalman Edwards should have taken the question to apply to a train which he had cleared out 40 minutes earlier. Colonel Pringle concluded that one or both of the signalmen had something to conceal which was not to his or their mutual advantage (basically, of course, he meant that they were lying, though that word was not used in his report) and that they had not been working the traffic in accordance with the regulations. It was possible that the poor design of the instruments and contact between the telegraph and tablet wires due to the snow had interfered with the indications and locking of the instruments. The men should have called out the linesman or else resorted to working with a pilotman but instead might have resorted to illegal means of releasing tablets, relying upon the entries in their train register books and telephone messages. The signalmen's reluctance to contact the linesman or introduce pilot working was no doubt due to the fact that it was the middle of a dark and cold winter's night. Colonel Pringle regretted that he was unable to clear either signalman from responsibility for the accident.

The inspecting officer recommended that the phonopore (telephone) wires should be moved from the tablet wires to the telegraph wires, that the long section instruments should be redesigned to correct the defects pointed out by Mr Sayers, and that signalmen should be reminded of the importance of accuracy in telephone conversations. [730]

Finland, which had been a part of the Russian Empire for over a century, gained its independence at the beginning of 1918. Almost immediately it was embroiled in a civil war between the conservatives (White Guards) and the socialists (Red Guards) supported respectively by the Germans and the Bolsheviks. [731] On about Sunday 3rd February 1918 a major railway disaster occurred, presumably a single-line collision, near Kemi in northern Finland. It involved a train carrying Red Guards and what was described as a government train sent to intercept the Red Guards. The trains were travelling at full speed when they met, and all their coaches were smashed. 'A great number of persons were killed or injured.' [732]

[730] Board of Trade report.

[731] Wikipedia - Finnish Civil War

[732] *Laredo Times*, Laredo, Texas, United States, 10th February 1918

On 16th February 1918 two trains collided between Tamarite de Litera and Binéfar stations in Huesca province, Spain. The trains involved were Barcelona-Zaragoza main train No. 222 and livestock train No. 295 from Galicia, bound for Barcelona. The livestock train had irregularly left Binéfar station where it should have awaited the arrival of the mail train. It was supposed that the driver of the livestock train had failed to observe the stop signal at the station due to thick fog. [733] The collision occurred at 3.30 a.m. at kilometre 142, on the line of the Caminos de Hierro del Norte de España. The livestock train had been eagerly awaited in Barcelona, where there was a meat shortage.

An initial report from Barcelona spoke of nine dead and 31 injured, 12 seriously. The only named victim in the first reports was Hernández Salazar, one of the drivers. Both drivers and firemen survived. Those on the mail train could not avert the collision but they caught sight of the opposing train in time to jump and were injured. The engine crew of the livestock train presumably also jumped. Relief trains were dispatched from Barcelona and Lérida, the injured being taken to the latter place. [734]

On 17th February it was reported that there 16 dead and over 30 injured and the next day the number of reported fatalities had risen to 21. Most of the cattle on the livestock train were also killed (which as they were en route to the slaughterhouse was a little academic for them). [735]

On 18th February it was reported that the judicial authorities were still investigating the accident and the station employees at Binéfar, together with the driver and fireman of the livestock train, had been arrested.[736]

The First World War, by then in its closing stages, played an indirect part in a major catastrophe near Nashville, Tennessee, United States, on 9th July 1918, for many of the victims were negroes on their way to their jobs at a munitions works.

The Nashville Chattanooga and St Louis Railway line out of Nashville was double track for 2½ miles out of Nashville but then became single at Shops. The collision occurred about two miles beyond Shops and halfway between there and Harding at a location known as Dutchman's Bend. Involved in the accident were passenger trains Nos. 1 and 4. Southbound train No. 1 from Memphis to Nashville consisted of locomotive No. 281, a baggage car, five wooden coaches and two Pullman sleeping cars and was in charge of Conductor Tucker with Engineer William Lloyd and Fireman Tom Kelly. Northbound train No. 4 consisted of locomotive No. 282, a combination mail and baggage car, a baggage car and six coaches, all of wooden construction, in charge of Conductor James Eubank with Engineer Dave Kennedy and Fireman Luther Meadows. Southbound trains were superior to northbound ones. Train No. 4 was timetabled to leave Nashville at 7 a.m. and No. 1 to arrive there at 7.10 a.m.

Before leaving Nashville train No. 4 had received train order No. 29, which read:

No. 4, engine 282, hold main track, meet No. 7, Eng. 215, at Harding. No. 1 has engine 281.

---

733 *La Vanguardia*, Barcelona, Spain, 17th February 1918

734 *La Correspondencia de España*, Madrid, Spain, 16th February 1918

735 *El Telegrama del Rif*, Melilla, Spain, 18th February 1918

736 Ibid, 19th February 1918

Tran No. 4 left Nashville seven minutes late at 7.07 a.m., passed Shops at 7.15 a.m. and collided with train No. 1 two miles on while travelling at a speed of 50 miles an hour. In the collision engine No. 281 of train No. 1 was derailed on the west side of the track, the baggage car of that train was demolished, its first coach ended up across the track and was penetrated by the combination baggage car of train No. 4 and the second coach went down the embankment. The front of the third coach was jammed into the second and the remaining cars remained on the track and were little harmed. The engine of train No. 4 was derailed to the east side of the track and badly damaged, the combination car was pushed into the first coach of train No. 1 as described above, the baggage car and leading coach were telescoped into one another, and the next coach was partly telescoped into that in front. The rear coaches of the train were not derailed and were little damaged. According to the Interstate Commerce Commission report 87 passengers and 14 employees were killed and exactly the same number of passengers and employees injured, [737] though these figures appear suspect; firstly inasmuch as they imply that there were no less than 28 employees on the two trains, which seems an awful lot, and secondly in that exactly the same number of passengers and employees were apparently both killed and injured, which seems rather a coincidence. Contemporary newspapers do speak, however, of 100 dead in all, including all four of the engine crews.

Relief trains were rushed to the accident site and hundreds of men were soon engaged in the rescue work. Many automobiles also went to the scene.

'A large number of the dead and injured in the Nashville Chattanooga and St Louis railroad wreck are believed to have been labourers sent from Memphis to work in the Dupont powder plant. These workmen occupied coaches ahead of the Pullman sleepers, and H.M. Baugh, secretary to Superintendent E.M. Wrenn of Nashville, said over a long distance telephone that 90 per cent of the deaths and injuries occurred among the labourers and in the day coaches.

'"Fate took a hand in the forwarding from Memphis of the largest number of men we have sent to the Dupont company," said M. M. Glidwell, in charge of the powder company's Memphis employment bureau. Fatalities among their number are said to be heavy.

'"We were without sufficient funds to pay for the men's transportation. Late in the afternoon they sent a delegation to me insisting that they got away at once. My hands were tied, but I told them to be at the depot for the departure of the L & N at 8.45. Thirty minutes before train time I received $1,000 by telegraph, rushed to the depot and within 20 minutes I had the men in their cars. I am sorry now that they left, for I expect many are injured and probably some are dead.

'"The white men in the list were expert carpenters, plumbers, machinists and other sorts of high-class laborers. Most of them were from Memphis and towns near this city. The negro men were from north Mississippi and Shelby county. They were farmers who had laid by their crops and wanted to go to Dupont's for two month's work."

'The cars the men travelled in were placed in the train next to the baggage cars. They were wooden cars, and reports indicate that the dead and injured were taken out of them.

'The wrecked Nashville train was attached to the Louisville and Nashville Cincinnati

[737] Interstate Commerce Commission report

train that left Memphis Union at 8.45 in the evening. At McKenzie, the Nashville cars were handed to the Nashville Chattanooga and St Louis railroad and hauled to Nashville as No. 1.' [738]

The negro labourers had been riding in two Jim Crow cars at the front of train No. 1.

Dispatcher Phillips told the Interstate Commerce Commission (ICC) that he had added the fact that train No. 1 had engine No. 281 to the order as it was customary to advise the crew of train No. 4 in this way to aid them in identifying that train and advise them that it had not reached the start of the double track at Shops. The trains were expected to pass one another on the double track; if they did not, train No. 4 must wait for it at the end of the double track at Shops unless directed by train order to proceed, as train No. 1 was superior by direction. No train register was provided at Shops.

Conductor Eubank of train No. 4 told of giving train order No. 29 to Engineer Kennedy, who had read it back to him. He said that he had remarked to the engineer that No. 1 was somewhat late that morning, but he did not believe that the mail would delay them so that the dispatcher would have to change the meeting point to Vaughns Gap, He supposed from his remarks that the engineer would understand that train No. 4 was likely to have to wait at Shops for train No. 1. The conductor said that he had read the order to the train porter shortly after they had started, and to the flagman while they were on the double track, telling him to look out for No. 1. The conductor said that he knew that train No. 1 had not left Shops when they left Nashville. Having spoken to the porter and flagman he had then busied himself collecting tickets. He was busy doing so when he heard a train go by, which he had assumed was No. 1, though he could not check from the packed car. Collecting all the tickets on the crowded train would take him all his time to the next stop and he relied on the other crews members to identify train No. 1 and stop if need be at Shops. His train had been travelling at 45 miles an hour when he had felt the brakes being applied, closely followed by the collision. Engineer Kennedy had seemed normal and in apparent good spirits.

Flagman St Clair of train No. 4 remembered Conductor Eubank giving him the train order but not his mentioning train No. 1. He saw a switch engine with 8 or 10 cars on the double track. He was very inexperienced, this being only his first trip as a flagman on a passenger train, after two trips as such on a freight train.

Conductor Tucker of train No. 1 said that it had been running 30 minutes late. It had been running at between 50 and 60 miles an hour when the collision occurred, with no preceding brake application. No order had been given to them regarding train No. 4, as they were the superior train.

Conductor Riggle, who worked train No. 4 alternately with Conductor Eubank, testified that he always kept a special lookout for train No. 1, and made it a special point not to pass Shops until he knew personally that it had arrived. He might leave it to the other trainmen to identify it but would not let his train pass Shops until they had assured him that they had done so, if he had not seen it himself.

Engineer Winford, who alternated with Engineer Kennedy, said it was sometimes necessary to wait at Shops for train No. 1. The signals at Shops were usually clear for passage onto the single line but he understood that this gave him no rights over a

[738] *The Bolivar County Democrat*, Rosedale, Mississippi, United States, 13th July 1918

superior train.

Operator J.S. Johnson stated that he came on duty at Shops tower at 7.06 a.m. or 7.07 a.m. and train No. 4 had passed at 7.15 a.m., travelling at about 25 miles an hour. At that moment he did not know if train No. 1 had arrived. When he reported the passing of train No. 4 to the dispatcher he was told to try and stop it. He blew the emergency air whistle when No. 4 was about a train's length past his tower, but no notice was taken of it. Nothing was said about train No. 1 when he took over that morning and the fact that the train order signal was in the proceed position did not give train No. 4 rights over train No. 1. Operator Johnson was not criticised by the Interstate Commerce Commission, and this accident shows the great difference in train working between the United States and say, Great Britain. To British eyes the idea that a signalman could just let a train proceed onto a stretch of single track without being aware if an opposing train had arrived, and even have his signal at the entrance to the single line standing at clear when an opposing train was approaching, seems incredible. Following American practice the signal at Shops was provided solely to show the engineer if there were any orders for his train, and not to indicate line occupancy, and the role of the operator was limited to delivering orders and manipulating the double to single line switches (points).

Travelling Engineer Fahey testified that he considered Engineer Kennedy to have been a very careful man, well versed in the rules and a person who would strictly adhere to them. The travelling engineer said that he had observed other enginemen passing Shops solely upon the word of the operator that a superior train had arrived and as a result the superintendent had issued bulletin 1268 on 28th May 1918. This read:

Understand some engineers on outbound trains are passing Shops without any definite information as to whether superior trains have arrived, other than to ask operator at Shops tower. This must be discontinued. Superior trains must either be registered before the northbound train departs or be identified by some member of the superior train, meet the superior train between Nashville and the Shops or have an order at the Shops stating that the superior has arrived.

NOTE – See that train dispatchers understand this and have the orders ready at the Shops so that they can be handed on to outbound trains.

The Interstate Commerce Commission (ICC) concluded that the accident had been caused by Engineer Kennedy and Conductor Eubank occupying the main track on the time of a superior train and that they were responsible for it. They had violated Rule No. 83, which read in part:

A train must not leave its initial station or any division or a junction, or pass from double to single track, until it has been ascertained whether all trains due which are superior or of the same class have arrived or left.

The evidence showed that the conductor and engineer had worked the same train for several years and were aware when they left Nashville that train No. 1 had not arrived. As Engineer Kennedy was killed it was impossible to account for his mistake. Conductor Eubank had relied completely upon other members of the train crew identifying train No. 1 and had assumed that a train passed in the yards was that train. That he was busy collecting tickets could not be considered a valid excuse given that his first duty was to provide for the safety of his train and ensure that it did not pass onto the single line until No. 1 had arrived. In assuming that a train on the other track was No. 1 without positively identifying it or being notified of its identity by another crew member he had been extremely negligent, particularly in a yard where switching movements were constantly being made.

The ICC noted that although the operating rules of the company required employees not to enter the single track until it was known to be clear of superior trains no means had been provided for them to do so. The May bulletin contemplated outgoing trains being given orders at Shops that all superior trains had arrived, but this was not being done. The custom of merely notifying train crews of the numbers of the locomotives of opposing trains was utterly inadequate as a safeguard given that such numbers must be read under all conditions of weather and traffic.

According to the records during June 23 trains operated into and out of Nashville, plus a number of yard movements between Nashville and Shops. Under these circumstances it was absolutely essential for the crews of outgoing trains to have definite information regarding opposing trains, either by the use of a train register at Shops or by the issuing of orders at Shops with definite notice of the arrival of trains which had not already reached Nashville.

The accident would, without doubt, have been prevented had there been a manual block system on the single track north of Shops. According to the timetable there were 27 train order offices in the 172 miles between Nashville and Hickman, Kentucky, of which 14 were open continuously. Four scheduled passenger trains and 12 scheduled freight trains operated over the line daily. 'With this volume of traffic, and in view of the universally recognized features of safety afforded by the block system, there can be no valid excuse for the failure or neglect on the part of the railroad company to utilize existing facilities for the purpose of operating a block system on that line.'

The ICC remarked that all the vehicles of the two trains except two sleeping cars had been of wooden construction and six cars had been entirely destroyed. 'This accident presents a more appalling record of deaths and injuries than any other accident investigated by the Commission since the accident investigation work was begun in 1912. Had steel cars been used in these trains, the toll of human lives taken in this accident would undoubtedly have been very much less.' [739]

As pointed out by Robert B. Shaw in his book [740] barring sudden illness or suicidal intent the failure of Engineer Kennedy to wait for train No. 1 can only be explained in one of two ways, either he misread train No. 29 as reading, *Meet No. 1 at Harding*, which seems unlikely in view of his alleged conversation with Conductor Eubank, or else he mistook the switch engine and cars for the passenger train. There was only the conductor's word that he had discussed the order with his engineer, so it was possible that he had not. With the death of Engineer Kennedy, the reason for his error must remain a mystery.

The Nashville disaster is often said to have been the worst in United States history in terms of fatalities, though this seems doubtful in view of the Eden, Colorado, bridge accident of 1904.

---

[739] Interstate Commerce Commission report

[740] A History of Railroad Accidents, Safety Precautions and Operating Practices by Robert B. Shaw, pg. 128

July 1918 also brought another interurban accident in the United States, this time near Chelsea, Michigan, on the Detroit Jackson and Chicago Railway. On 20th July 1918, a passenger train and a freight train collided head-on, killing 15 passengers, and injuring 60.

The Detroit Jackson and Chicago line ran from east to west between Ann Arbor and Jackson, Michigan. It was single track and worked by timetable and train order, the orders being issued by telephone. There were no block signals. On the day of the accident eastbound train No. 176 was run in two sections. The first section, consisting of car 7791, left Jackson at 7.09 p.m., carrying green flags to indicate that there was a second following. It was in charge of Conductor Kidwell with Motorman Rose. The second section, consisting of car 7776, in charge of Conductor English and Motorman Taylor left Jackson nine minutes later at 7.16 p.m.

Arriving at Francisco, 15½ miles from Jackson, at about 7.35 p.m., it was getting dark, and Motorman Rose removed the green flags and substituted green lights because of the near end of daylight. About 7 miles further east, at Chelsea, first 176 met westbound extra No. 1936, which was loading and unloading freight. First No. 176 whistled to indicate that a second section was following, and the signals were acknowledged by Motorman Charles Fiske of extra No. 1936.

Leaving Jackson, train second No. 176 was in possession of an order which read:

Run Second Section Train 176 from Jackson to Ann Arbor. Meet Train 173, Car 7521 at Francisco, Meet Train 175, Car 7796 at Warsaw.

When this train reached Francisco, it was given another order, which read:

Train No. 176, Car 7776, at Francisco, Meet Train No. 173, Car 7521 at Hoppe instead of Francisco.

Eastbound train second No. 176 duly met train No. 173 at Hoppe as per the order and proceeding from there collided with train extra 1936 at a stop named Heffner, a mile west of Chelsea. Train second No. 176 was on a limited stop service, calling only at principal stops, and was travelling at 50 or 55 miles an hour, reduced to 25 or 30 miles an hour at the moment of impact. The force of the collision drove train extra No. 1936 back about 50 feet and crushed the front of it but car 1936 was the heavier and more solidly constructed of the two and car 7776 was telescoped about 20 feet of its length. Neither car was derailed.

Extra No. 1936 was in charge of Conductor Buck. It had left Ann Arbor at 6.18 p.m. with an order which read:

Run extra Ann Arbor to Jackson. Meet extra east, Car 1947, at Grass Lake.

Conductor Buck told the Interstate Commerce Commission investigators that he had been in the car loading and unloading freight at Chelsea when first No. 176 had passed. He had not seen it but heard it whistle to indicate that another section was following and had answered the whistle. He thought he heard it pulling out at 7.45, on time, but did not see it go. When, a bit later, he got out of his car he saw a car leaving eastwards; he could not see any green flags on it and was sure that it was the indicated second section leaving. He called the dispatcher to say his train was ready to go, was told that his orders were good and left at 7.55 p.m. He had not spoken to his motorman before leaving. A mile from Chelsea they were running at 40 miles an hour when the motorman had suddenly reversed his car and speed had fallen to about 10 miles an hour when the trains met.

Like his conductor Motorman Fiske said that he did not see the first section of train No. 176 but heard it whistle to indicate that there as another section following. He had kept on loading freight, and it had appeared to him that the second section was close behind the first (presumably meaning that he saw or heard the first section leave the station and assumed it was the second section). They finished working with the freight at 7.53 or 7.54, his conductor spoke to the dispatcher, and they set off. He had not checked with the conductor if the second section had gone, being confident that it had. About a mile from Chelsea he saw the car of train No. 176 and just had time to apply the brakes and jump. He did not think they had travelled faster than 25 miles an hour. It had still been daylight at the time of the collision, with the low sun shining in his face.

Motorman Rose said that it had still light when he had changed the flags to lights but he thought it close enough to sunset to make the change. His whistle regarding a second section had been properly responded to. They had only stopped at Chelsea 2 minutes, and he did not see how their train could have been mistaken for the second section.

Conductor Kidwell of train first No. 176 said that he did not think that it had been too early for the flags to be change to lights at Francisco as it would be sundown by the time they reached Chelsea and the lights could be more easily seen than flags. They had not remained at Chelsea more than two minutes and his motorman's whistle had been answered promptly by the other train.

Motorman Taylor of second No. 176 stated that after leaving Hoppe he had been running at a top speed of 55 miles an hour when, rounding a curve, the other car had appeared from behind a row of trees. He had immediately cut the power and reversed his car, bringing its speed down to 25 or 30 miles before the collision.

Conductor English of second No. 176 on the other hand did not consider that the speed of his train had been much reduced before the cars met. In his opinion it was too early to change the flags to lights. The sun was still up, shining over the top of the trees, making it hard to see a train which had the sun behind it. The weather had been very bright and clear.

Train Dispatcher Peppier stated that when Conductor Buck had rung him from Chelsea he had told the conductor that the order he had received at Ann Arbor was still good. Nothing was said about train second No. 176, and he did not know whether it had passed. It was the strict responsibility of Conductor Buck and Motorman Fiske to be absolutely sure that second No. 176 had gone before setting off.

The Interstate Commerce Commission (ICC) found Conductor Buck and Motorman Fiske responsible for the accident, it having been their obvious duty to be sure that second No. 176 had passed, as per Rule 230:

Trainmen must know when meeting or passing trains that the trains met or passed are the trains that should be met or passed at that point.

Contributing to the accident had been the failure of Conductor Kidwell and Motorman Rose to obey Rule 87, relating to the display of signals, which read:

Night signals are to be displayed from sunset to sunrise. When weather or other conditions obscure day signals, night signals must be used in addition.

Night signals had been substituted for day signals on train second No. 176 before sundown and when the weather was bright and clear which had probably misled Conductor Buck. Rule 105 of the company called for flags to be used during the day only and lights at night whereas on steam railroads it was found to be safer to display flags all the time plus lights at night. [741]

The high casualty list was mainly due to the fact that car 7776 was a light wooden car, not provided with anti-telescoping devices, which could not resist the heavier car 1936, which had a steel underframe. [742]

A passenger train and a troop train collided at Hoenningen, Germany (between Dümpelfeld and Liers), on Friday 16th August 1918. More than 31 people died and 73 were injured. [743] The troop train from Jünkenrath was carrying soldiers who were returning to their units after training at Eisenborn, and in addition to its passenger coaches it included two waggons of ammunition. The passenger train was coming from Remagen. Fire broke out following the collision and boxes of cartridges and hand grenades later exploded, making the rescue work hazardous. Two fire engines went to the scene and a bucket brigade was formed to carry water from the Ahr River to fight the fire. There were no doctors, only some medical orderlies and two Augustinian sisters (nuns) who had been on the passenger train, but local villagers gave much assistance. At 10 p.m., four hours after the collision, rescue trains reached the scene from Koblenz and Jünkenrath and the injured were taken to hospitals in Adenau, Bad Neuenahr and Gerolstein. By the evening there were 10 dead, 45 injured and 17 missing. That night another 15 charred corpses were pulled from the wreckage. The exact number who lost their lives could not be determined; it was at any rate more than 31. The line was double track, but temporary single-line working was in operation at the time. [744]

The St Louis-San Francisco Railway between Newberg and Springfield, Missouri was single track and worked by timetable and train orders, but additional protection was provided in 1910 by the provision of automatic signals, capable of displaying "Stop", "Caution" and "Proceed." These were placed between one and two miles apart in each direction and an overlap was provided to provide for the contingency of a train entering a single-line section at the same moment from each end. The more sophisticated signalling failed, however, to prevent a disaster near Marshfield, Missouri, United States on the evening of Tuesday 17th September 1918.

The trains involved in the accident were eastbound passenger train extra No. 1260, a troop train running from Waco, Texas to St Louis, Missouri, and westbound freight train second No. 39. Train No. 1260 consisted of locomotive No. 1260, six Pullman sleeping cars, a baggage car, another six Pullman sleeping cars and a caboose and was in charge of Conductor Wrinkle with Engineer Douglas, Fireman Lawrence Armentrout and an un-named student fireman. The freight train consisted of locomotive No. 56, four loaded and 60 empty cars, in charge of Conductor West with Engineer Beiseigle and Firemen

---

[741] Interstate Commerce Commission report

[742] *The Kokomo Daily Tribune*, Kokomo, Indiana, United States, 31st December 1917

[743] *Eisenbahnunfalle in Europa* by Erich Preuß, pg. 69 (in German)

[744] www.aw-wiki.de – Das eisenbahnunglück am 16 August 1918 in Liers

Wilfong and Mack.

Train extra No. 1260 left Springfield, Missouri, at 6.55 p.m. with train order No. 115, giving it a right over all first-class trains between Springfield and Newburg and restricting its speed to 30 miles an hour. At Strafford, 11½ miles east of Springfield, it received an order to run 15 minutes late and passing Marshfield it picked up order No. 127, which read:

No. 9, engine 1069, meet passenger extra 1260 east at Conway.

Beyond Marshfield extra No. 1260 passed automatic signal No. 2126 in the stop position and rather less than 1½ miles east of Marshfield and 13½ miles short of Conway it collided head-on with train second No. 39. The latter train had left Newberg at 10.45 a.m., and at Lebanon, 62¾ miles west of Newberg, it received train order No. 113, which read:

Second 39, engine 56, has right over second No. 32 to Marshfield and hold main line.

It had left Lebanon at 6 p.m., passed Conway at 7.14 p.m., passed automatic signal 2103, three miles east of Marshfield, at caution, and was running at 12 or 15 miles an hour when it met extra 1260.

The front ends of both locomotives were damaged in the collision, the tender of locomotive No. 1260 was derailed and stood on end and the leading wooden sleeping car was crushed between the tender and the following steel under-framed sleeping car. It was in the wooden sleeping car that most of the deaths and injuries occurred, though Fireman Armentrout, the student fireman and Head Brakeman G.W. Richardson also lost their lives and Engineer Douglas was seriously injured. In all 12 passengers, all soldiers, were killed and five railroad employees (including Engineer Douglas) and 35 passengers were injured. Behind the two leading sleeping cars of the troop only the baggage car was badly damaged, while in the freight train three of the front cars were derailed and practically demolished.

Conductor Wrinkle of train extra 1260 told the Interstate Commerce Commission (ICC) investigation that they had passed through Marshfield travelling at about 20 miles an hour. Rear Brakeman Carner said that the train order signal had been at danger. Conductor Wrinkle said that he had not noticed any of the automatic signals around Marshfield and had had no warning of the accident until the brakes were applied in emergency. Soon after the accident he saw Conductor Beiseigle of train No. 39 who told him that he had no orders regarding train extra 1260.

Rear Brakeman Carner testified that he had been riding on the rear platform of the caboose and had seen the train order signal at danger when the conductor asked him about it. He saw the automatic signal before Marshfield at caution but did not notice signal No. 2126.

Engineer Douglas stated that the automatic signal preceding Marshfield had been at caution and the train order signal there at danger. Having collected his train order at Marshfield he had looked ahead to automatic signal No. 2126, and it had been clear. He had not noticed it again. There had been no difficulty in seeing the signal. Its blade had been clear, he did not remember seeing the signal light. He was sure he had not confused the signal with that applying in the opposite direction. Between picking up the train order and passing signal 2126 he had read the order, given it to the head brakeman, looked at the lubricator, read his schedule and running orders, turned on the headlight and hooked up the engine.

Conductor West of second No. 39 stated that his train had waited at Phillipsburg for extra No. 1066 east, which had rights over all trains, and had left there at 7 p.m. At Lebanon he had received a copy of train order No. 113. He had been riding in the caboose when his train passed Conway at 7.13. The train order signal there had been at clear; the station was lit up, but he had seen no one about. Signal No. 2103 had been at caution, going to red after part of his train had passed it.

Engineer Beiseigle said that he had received train order No. 113 at Lebanon, his train had passed extra No. 1066 at Phillipsburg, just over 12 miles west of Lebanon, and signal No. 2103 had been at caution. His fireman had remarked that train No. 32 was heading in at Marshfield and he had thought the same and had not shut off steam, though the speed of his train had dropped anyway going up the hill. When he had seen the reflection of the headlight of extra No. 1260 he had shut off steam, applied the air brakes in emergency and he and the two firemen had jumped. He had received no orders regarding extra No. 1260 and knew nothing about the train.

Fireman Wilfong of second No. 39 said that he saw signal No. 2103 at caution. His engineer had then partly shut off steam and made a slight brake application. When he saw the smoke of train extra No. 1260 he thought that it was train No. 32 stopped at Marshfield.

Fireman Mack of second No. 39 agreed that signal No. 2103 had been at caution, that his engineer had partly shut off steam and applied the brakes gently and that the smoke had seemed to be from train No. 32 at Marshfield.

Rear Brakeman Sater of second No. 39 said that he had seen the train order boards at Phillipsburg, Conway, and Niangua all in the clear position. He thought that the speed of his train had been reduced to 5 miles an hour before the collision. He knew of no train order regarding extra No. 1260.

Dispatcher Chronister at Springfield testified that he had been advised of extra No. 1260 just after 4 p.m. and that it would arrive at Springfield about 6.20 p.m. He had then begun to make arrangements for it to cross opposing trains, preparing train order No. 115 and telephoning it to different stations from about 6 p.m. The order was long and had to be transmitted slowly. It was delivered to extra No. 1260 at Springfield and made complete at 6.42 p.m. He alleged that the order had been transmitted to Conway, repeated, and made complete at 6.08 p.m., he did not know who to other than it was the second trick operator. He had not intended extra No. 1260 to pass Conway before second No. 39 arrived there but had not asked the Conway operator to place the train order board against extra No. 1260 nor issued any order to hold the train there, and he had not taken any steps to hold second No. 39 at Conway other than by order No. 115. He admitted that if train second No. 39 had been delayed between Lebanon and Conway after the schedule time for extra No. 1260 to pass through Conway and communication with Conway had failed there would have been no way of preventing a collision between the two trains east of Conway. He had only worked as a dispatcher for the Frisco since 23rd August and was not familiar with Frisco dispatching rules, though he was for the general rules for dispatching.

Operator Foster at Conway stated that he had been on duty from 3 p.m. to 11 p.m. but been absent from the office between 5.40 or 5.45 and 6.15 or 6.20, having been excused by the dispatcher to go and have his supper. When he had left the office, he had locked it. There had been no outstanding orders and the train order board had been clear. He had been in the office and had reported the passing of train No. 39 at 7.14 and had asked about train No. 9, to be told that it was late. Told of the dispatcher's statement that he had accepted order No. 115 at 6.8 he positively denied it and stated that he had never had that order and signed an affidavit to that effect.

Mr H.G. Herston, an insurance agent, stated that he had walked back to Conway station with Operator Foster, arriving there probably at about 6.30 p.m.

Operator Windle at Springfield testified that the dispatcher had given him order No. 115 and he had heard the dispatcher trying to contact Conway. He thought that the dispatcher had eventually managed to contact all the offices but could not be sure.

Operator Henron at Dixon stated that he had heard the dispatcher trying to contact Conway but could not say if he had got through.

Operator Robertson at Lebanon said that he had heard several of the operators repeat order No. 115 but he could not say if the Conway operator was one of them. He had heard the Conway operator speaking to the dispatcher after the collision and denying that he had taken the order.

Operator Hathaway at Richland stated that he had not received a copy of train order No. 115.

Foreman Charles Williams of the wrecking crew recalled hearing some soldiers who had been riding in the second coach of extra 1260 saying that the signal beyond Marshfield had been red when their train had passed it. He did not get their names.

A man named F.E. Richardson, in charge of a pump house at Marshfield, spoke of seeing signal 2126 at red after the engine and one car of extra No. 1260 had passed it.

A man named R.D. Dailey made an affidavit that he had seen both the blade of signal No. 2126 horizontal and its light showing red when train extra No. 1260 had passed it.

The ICC concluded that the accident had been caused by Dispatcher Chronister failing to transmit order No. 115 to Conway for delivery to train second No. 39 and to Engineer Douglas passing signal No. 2126 at danger. The statement of Mr Herston that he had walked back to the office with Operator Foster at about 6.30 p.m. indicated that the operator, as he had maintained, had not been in the office at the time he was alleged to have accepted order No. 115.

A very careful examination of signals Nos. 2126 and 2103 had shown them to be working normally. With second No. 39 east of signal No. 2103 signal No. 2126 should have been at danger and in view of the fact that it had been working properly and the statements of Messrs Richardson and Dailey it was believed that it had been at danger when Engineer Douglas passed it, despite his statement to the contrary.

The accident was another of the type which would have been prevented by a system of automatic train control. [745]

A coroner's jury at the inquest on the victims of the disaster held at Marshfield, who could obviously not have been privy to the findings of the ICC inquiry, which had not at

[745] Interstate Commerce Commission report

that time been conducted, managed to exonerate the crews of both trains from blame for the accident. [746]

A catastrophic single-line collision occurred in India on 19th June 1919, again on a line worked by the supposedly almost infallible electric token system. Down goods train No. 127 on the East Indian Railway left Firozabad with the wrong token and shortly afterwards collided near there with up passenger train No. 7 bound for Calcutta, which was in possession of the correct token for the Firozabad – Makhanpur section. The exact mechanism of the accident was not outlined but presumably in some way the token for the rear section was taken forward into the next section, as was to happen again near Abermule, in Wales, a year and a half later.

In the initial collision four enginemen and 15 passengers were reportedly killed but the collision broke the gas reservoirs of the three leading carriages of the passenger train, the escaping gas became ignited and set fire to the wooden coaches and fanned by a strong wind the fire eventually destroyed six of the ten carriages of the train. As a result, many more passengers died, the exact number according to the official report being about 100 but the number was possibly as high as 300. [747] Five railwaymen in all lost their lives and three railwaymen and 44 passengers were injured.

The official report on the accident stated mysteriously that, 'To ensure greater safety in working, certain suggestions made by the committee which assembled to investigate the cause of the collision, have been adopted and others are receiving attention.' What the suggestions were was not detailed.

As for the railway servants involved, 'The staff at fault were prosecuted and sentenced to various terms of imprisonment.' [748]

The next accident to be dealt with is unusual in that although it happened in the United States it involved two Canadian Pacific Railway trains, a westbound passenger train and an eastbound freight train. The passenger train was No. 39, a regular working but which on the day of the accident was much busier than usual, as the steamer Empress of France had docked at St John, New Brunswick, and a large volume of passengers had disembarked and were on their way westwards to Montreal, Quebec. These steerage passengers were largely British emigrants to Canada, but there were also some returning Canadian soldiers, doubtless mostly longing to be home after the ordeal of the war. To cater for the heavy traffic train No. 39 was being run in multiple sections.

The stretch of line on which the accident took place was the Moosehead subdivision of the Brownville division of the company, which ran westward from Brownville Junction, Maine, to Megantic, Quebec. The line was worked by timetable and train orders, with orders being transmitted by telephone. The accident occurred on 20th December 1919 at a point 2¼ miles west of Onawa, Maine, United States, and 20 miles west of Brownsville Junction.

Westbound passenger train third No. 39 consisted of engine No. 783, a box baggage

---

[746] *The Commercial Tribune*, Cincinnati, Ohio, United States, 20th September 1918

[747] *Railway Gazette*, Vol 32, pg. 62

[748] Railways in India Administration Report for the Year 1919-20

car, a coach, two colonist cars, two tourist cars, a colonist car, a café car, a tourist car and two cabooses, in charge of Conductor Dillon with Engineer Wilson and Fireman Hatcher. Before leaving Brownville Junction it was given three train orders which read as follows:

Train order No. 20:

First No. 39, engine 818, late; Second No. 39, engine No. 852, on block; Third No. 39, engine unknown, run four 4 hours and twenty 20 minutes late, Barnard to Boundary.

Train order No. 28:

Order No. 20 is annulled. First No. 39, engine 818, late; Second No. 39, engine No. 852, on block: Third No. 39, engine unknown, run five 5 hours late, Barnard to Boundary.

Train order No. 38:

Engine 783 display signals and run as third No. 39, Brownville to Megantic.

The train left Brownville Junction at 6.25 a.m., 5 hours, and 5 minutes late, passed Barnard at 6.51 a.m., passed Onawa at 7.9 a.m. and was travelling at 25 or 30 miles an hour when it collided with eastbound freight train first No. 78.

Freight train first No. 78 consisted of engine No. 2516, 30 loaded cars, two empty cars and a caboose and was in charge of Conductor Manuel with Engineer Bagley and Fireman Henniger. It left Megantic at 6 p.m., 30 minutes late, set out six loaded cars at Holeb, and at Moosehead, a non-telegraph station 42 miles from Holeb, it met first and second No. 39 and a westbound freight train, extra No. 3470, from whose crew it received a copy of train order No. 28. Greenville, 10¾ miles east of Moosehead, was reached at 6.30 a.m., and here they were given another copy of train order No. 28 and train order No. 20. The train left Greenville at 6.40 a.m., with the intention of passing train third No. 39 at Morkill, just under 7 miles further east. They reached there at 6.57 a.m., going into the siding, and the Morkill operator handed up copies of order No. 47 to the engine and caboose. This order read:

Train order No. 47:

Train No. third 39, engine 784, late. Fourth No. 39 run engine unknown, run eight 8 hours late, Barnard to Megantic.

After stopping there only two minutes train No. first 78 left Morkill at 6.69, ran past Bodfish, a non-telegraph siding 5¼ miles further east, and was running at 20 or 25 miles an hour when it collided with train third No. 39 a couple of miles further on.

The engines of the two trains were locked together in the collision, the baggage car of the passenger train was destroyed and the coach behind it telescoped the following colonist car for two thirds of its length. The second colonist car was slightly damaged, the remainder of the train very little. Nine cars of the freight train were derailed. Some newspaper reports mentioned a fire, but none is mentioned in the official report on the accident. Both engineers, both firemen and 19 passengers were killed were killed more or less outright, and nine railroad employees and 59 passengers were injured. [749] Two of the injured passengers later died in hospital at Montreal, bringing the number of dead

[749] Interstate Commerce Commission report

up to 25. [750]

A Canadian newspaper reporter interviewed one of the surviving passengers. '"We were all very much relieved when we got off the ship at St John, as we thought we had concluded the worst part of the trip from Barrow-in-Furness to our destination at Moose Jaw," said Miss Dora Hayhurst, while relating her experiences. Miss Hayhurst is a typical rosy-cheeked young English girl, sixteen years of age, and told a very clear story of the accident. Continuing, she said, "People who have been crossing the Atlantic for many years told us that it was the toughest passage that they had ever had, and I can assure you that I could not possibly imagine anything worse, therefore you can imagine we were very pleased to disembark at St John, as we looked forward to the remainder of our journey to Moose Jaw without any apprehension. We little realized that the most harrowing part of the whole journey was yet to come. We had just breakfasted and were sitting looking out of the car window when, without the slightest warning, there was an awful crash, and the air was literally filled with flying timbers and debris of all kinds. I was thrown to the floor between seats, and they provided a sort of arch over me, which prevented heavy parts of the falling car roof from reaching me; on the other side of the car seats did not hold fast throughout the whole length of the car, and some therefore were injured. Immediately after the crash there was a momentary impression caused by the action of the car, which closed up like a concertina, and then, when this pressure was relieved the debris and broken parts of the car came down in a shower."' [751]

There were also specially tragic aspects of the crash. 'The saddest case in the whole of the terrible tragedy is that of little Ivy Bingham, aged seven, who is the only survivor of a once happy family of four in Isabella, Man. The child's mother and her two young sisters aged three years and two months respectively were instantly killed, while Ivy came through without a scratch. The little girl went on to Montreal on a special train, which took the uninjured passengers. Some of her fellow passengers undertook to deliver the child to her friends at Isabella.

'Another very pathetic case is that of the Petley family. Mrs Petley was on her way from the old country to join her husband in Winnipeg. She was accompanied by her little son, John, aged five years, and her daughter, Eileen, aged four. Both the children were killed, and the mother had one of her legs fractured.

'Included in the passengers was one family party of nine people, including three generations. Every one of the party received some cuts or bruises but were only slightly injured. They were Mrs E. Bromley, her daughter and son-in-law; Mr and Mrs William Falkard; Mrs Falkard's two sisters, Mrs E.M. Falkard and Doris Day Hurst, and four children. The whole party were bound for Grindrod, B.C. Mr Falkard is a veteran of the South African war and received a Mons star in the recent European campaign. His escape, and that of his relatives, was remarkable.

'Einar Nillson, who was killed, and Nils Nillson, severely injured, were brothers. They were Swedes, and on their way to Rainton, Sask. The surviving brother speaks no English.'

'A special train, consisting of two Pullman cars, left Brownville Junction Sunday for Montreal with the injured passengers. At Montreal ambulances await the train's arrival

---

[750] *Manitoba Free Press*, Winnipeg, Manitoba, Canada, 23rd December 1919

[751] *The Brandon Daily Sun*, Brandon, Manitoba, Canada, 23rd December 1919

and the injured will be taken to the Royal Victoria Hospital. The bodies of the Petley children also went on this train.

'Those who survived the terrible calamity speak in the highest terms of praise of the treatment afforded them by the C.P.R. officials and residents of Brownville.' [752]

As it had occurred in the United States the accident was investigated by the Interstate Commerce Commission (ICC) even though it involved two Canadian trains.

Conductor Dillon of train third No. 39 told the investigators that he had had a copy of order No. 28 according to which his train was due to leave Onawa at 6.59 a.m. It actually did so at 7.09 a.m., the collision occurring between 7.13 and 7.14, when his train was moving at 25 or 30 miles an hour. He had felt no brake application.

Flagman Williams of train third No. 39 agreed that it had passed Onawa at 7.09 and that there had been no brake application. He had been riding in the cupola of one of the cabooses but could not see the opposing train because of the curvature of the line.

Head Brakeman Cook of train third No. 39 put its speed at 21 or 22 miles an hour. After the collision he had spoken to Conductor Manuel of train first No. 78, who had said that he had overlooked their train.

Head Brakeman Austin of train first No. 78 testified that when they had left Greenville they had expected to meet train third No. 39 at Morkill and he had operated the points to let their train into the siding there. The Morkill operator had come out of his office with a form 19 train order and handed him a copy but said nothing about either to him or the engineer. Head Brakeman Austin said he had given the order to Engineer Bagley, who had remarked, "We have eight hours on third No. 39," and had directed him to operate the points to let their train back onto the main line, which he had done. His first warning of the accident had been Engineer Bagley exclaiming, "Look out boys!" and applying the air brakes in emergency.

Flagman Gardiner of train first No. 78 said that as they had left Greenville Conductor Manuel had calculated that they could reach Morkill for third No. 39. At Morkill the operator had come out and handed an order to the engineer of the train and a second copy to Engineer Chase, who had been deadheading and was standing on the steps of the forward caboose. Conductor Manuel told him that the order gave them more time on train third No. 39 and to go up onto the top of the train, turn up the retainers and give the engine crew a proceed signal, which he had done. Returning to the caboose when the train was about a mile beyond Morkill he had picked up the train order and misread it as giving them eight hours on train third No. 39. The brakes had come on a few seconds before the collision when his train had been travelling at 25 or 30 miles an hour.

Engineer Chase testified that at Morkill the operator had given him the train order, apparently thinking that he was the conductor, but had made no comment regarding it. Going into the caboose he had passed the order to the conductor, who had read it and given it to the flagman. Engineer Chase said that he had not himself read the order but assumed that it was a meet order with third No. 39 at some point east of Morkill. He had then laid down to try and get some sleep. He admitted that he should not have collected the order but have left it to the conductor.

---

[752] Ibid, 22nd December 1919

Conductor Manuel said that on receiving the order at Greenville that train third No. 39 was five hours late he calculated that they could make Morkill with train first No. 78 and had been expecting that at Morkill they would be advanced further, so that the dispatcher could get their train to Brownville Junction before the train crew had exceeded their legal maximum of 16 hours on duty. At Morkill he saw the operator give the order to Engineer Chase and heard the former say something about eight hours, from which he had assumed that third No. 39 was running eight hours late. This idea had been reinforced by the operator coming out so promptly to deliver the order. Engineer Chase had then given him the order and he had glanced at it and read it as saying what he had expected it to say, that third No. 39 was eight hours late. He had told the flagman this and instructed him to turn up some of the retainers. There had been nothing confusing about providing it was read carefully.

Operator Kingdon at Morkill stated that he had told the dispatcher when he had heard train first No. 78 approaching and the dispatcher had then issued order No. 47. He had handed one copy of the order to the brakeman riding on the pilot of the engine and the second to a man riding on the front steps of the caboose, to whom he had said, "Eight hours on fourth No. 39." Another man had been riding on the rear steps of the caboose, whom he thought might be the conductor, had said, "Eight hours." The train pulled through the siding and left at 6.59. It was a couple of minutes before he could report the departure of the train to the dispatcher, the circuit being engaged. When he told the dispatcher that train first No. 78 had left the dispatcher had replied, "They have gone, have they?" Operator Kingdon said he had thought it strange that the freight train had first taken the siding to meet the passenger train and had then proceeded on receiving an order which only gave them time against a following section, but he did not know how late third No. 39 was and assumed that the train crew knew what they were doing and had given the matter no further thought. It did not occur to him to ask the train crew where they were going for third No. 39 when he saw the train departing. He said that if the train was stationary it was the practice to give the order to the engineer and conductor but if the train was moving it was handed to anyone able to take it.

Operator Valley, on duty at Morkill from 8 a.m. to 4 p.m., stated that he had arrived at Morkill from Greenville on train first No. 78 and knew that train third No. 39 was running five hours late and that the crew of train first No. 78 were planning to meet it at Morkill. He had seen Operator Kingdon give the train order to Engineer Chase, heard the operator say, "Eight hours," and knowing nothing of a fourth section assumed that it must be an order giving first No. 78 eight hours on third No. 39. Conductor Manuel had asked him to close the points at the east end of the siding after the train had gone and he had done so. Walking back to the office he thought it very strange that train first No. 78 could have eight hours on third No. 39 at Morkill when it had only had five hours on it at Greenville, seven miles back, and he had told Operator Kingdon that he did not understand it. The latter had made no reply but after the collision had said that he had not understood what Operator Valley was talking about.

Dispatcher Shaw at Brownsville Junction spoke of issuing train orders Nos. 28 and 20 to train third No. 39, expecting train first No. 78 to wait at Morkill for it, and of issuing order No. 47 to Morkill. He thought that the Morkill operator had first advised him that train had left at 7.01 a.m., as the operator had testified. He had then assumed that train first No. 78 intended to go to Bodfish for third No. 39. He had been unusually busy at that moment, dealing with nine trains and having to issue several orders, so it had not struck him that first No. 78 would have had to have left Morkill at 6.45 in order to reach Bodfish in time to clear third No. 39, due at that point at 7.05 a.m. Had he realised that first No. 78 did not have time to reach Bodfish he could probably have got third No. 39 stopped at Onawa. He had only realised that first No. 78 was running on short time when the telephone circuit failed at 7.15 due to the collision. It had not struck him as unusual that first No. 78 had first taken the siding at Morkill and then proceeded as this often happened due to crews changing their minds as to how far they could go.

Superintendent Boyle was of the opinion that Dispatcher Shaw had not been too busy to check the progress of train first No. 78 and anticipate where it should meet third No. 39 and said that it was the practice of dispatchers to check such movements.

The Interstate Commerce Commission (ICC) blamed the accident on the failure of Conductor Manuel and Engineer Bagley to correctly read the train order. Flagman Gardiner must also share some responsibility as he had also misread the order. The commission also considered that Dispatcher Shaw had been extremely irresponsible in his duties. Under Canadian Pacific rules train first No. 78 needed to reach Bodfish five minutes before the scheduled arrival time of train third No. 39 at 7.05 a.m., that is at 7 a.m., whereas it had only left Morkill, 5¼ miles away, at 6.59. The dispatcher had been advised that train second No. 78 had left Morkill not later than 7.02 a.m. and third No. 39 did not pass Onawa until 7.09 a.m. and so could easily have been stopped there.

Operator Kingdon had had his doubts as to whether train first No. 78 should have gone forward but the remarks made by Dispatcher Shaw when he had issued order No. 47 (about letting first No. 78 go) and the dispatcher's apparent lack of concern when he had reported that first No. 78 had left, had made Operator Kingdon believe that he was not cognizant of all the facts of the situation.

Had Conductor Manuel taken the order directly from the operator it was possible that he might have read it more closely and might have heard the full remark made by the operator to Engineer Chase mentioning a fourth section of train third No. 39. Despite the interest of Engineer Chase in talking the order he had not bothered to read it or learn its contents. 'The practice of employees not on duty, or others, meddling in train movements in which they are no way concerned, except in cases of emergency, cannot be too strongly condemned.'

The ICC noted that the coach and colonist cars behind the baggage car of train third No. 39 had been of wooden construction with some steel parts. Had they been of all-steel construction the number of casualties would have been materially reduced.

During the first 15 days of December an average of 41 trains ran over the Moosehead division. Traffic was less heavy in the summer but as a whole conditions warranted the installation of a block system, which would probably have prevented the accident. [753]

Five train passengers and the same number of railroad employees lost their lives in a collision on the Rutland Railroad, near Riverside, Vermont, United States, on Sunday 14th March 1920. The accident occurred on the Bellows Falls sub-division of the Vermont division of the company, a single-track main line worked by timetable and train orders transmitted by telegraph.

[753] Interstate Commerce Commission report

One of the trains involved in the collision was southbound freight train extra No. 28, which from Summit, about 25 miles short of the location of the accident and where a helper engine had been detached, consisted of engine No. 28, 46 cars and a caboose in charge of Conductor Phillips with Engineer George Cady and Fireman Ralph Bean. The train left Rutland, Vermont, at 12.55 p.m. and reached Summit at 3 p.m. Two train orders were received at Summit, one of which, No. 410, stated that an opposing passenger train, No. 165, would run one hour and 15 minutes late, Bellows Falls to Rutland. The train left Summit at 3.22 p.m. and at 4.07 p.m. arrived at Gassetts, 16¼ miles further south. Here it received another train order, No. 414, which read:

No. One sixty five 165, wait at Bartonsville until four fifty 4.50 p.m. for extra No. 28 south.

A copy of the order was given to the fireman as the engine passed and a second copy, together with a clearance card, was delivered to the conductor by means of the train order hoop. It was 8½ miles from Gassetts to Bartonsville and another 9¼ miles from there to Bellows Falls. The freight train left Gassetts at 4.11 or 4.12, ran through Bartonsville without stopping and collided with train No. 165 at a point 3,000 feet north of the north switch at Riverside and less than two miles short of Bellows Falls at 4.39 p.m. while travelling at an estimated speed of 45 miles an hour.

Northbound passenger train No. 165 was a Boston to Montreal express. It consisted of engine No. 52, a milk car, a combination car, a coach, a dining car, and a Pullman parlor car, and was in charge of Conductor Brislin with Engineer Cornelius Sullivan and Fireman William Farrell. It received train orders Nos. 410 and 414 at Bellows Falls and left there at 4.33 p.m.

Engine No. 52 of the passenger train was thrown to the west side of the track in the collision, turned around and badly damaged, the milk car was destroyed, and the front end of the combination car was badly damaged. Engine No. 28 of the freight train was thrown down the embankment on the east side of the line, eight cars of the freight train were practically destroyed and four more derailed. [754] Engineers Cady and Sullivan, Fireman Farrell, Brakeman Francis Rafferty of the passenger train and Head Brakeman A.W. Granger of the freight train. Fireman Bean jumped, landed in the Connecticut River with water up to his waist, and escaped injury. The passengers who lost their lives were J.S. Lent and F.S. Lent of Rockingham, Vermont (father and son); Enoch Bullock, of Ludlow, Vermont; George Simonds, of Portland, Maine, a travelling salesman and Dr W.J. Hunter, an itinerant physician. [755]

At the Interstate Commerce Commission investigation (ICC) Conductor Brislin testified that he had been in the dining car of the express when the brakes had come on and the trains had collided. Afterwards he had spoken to Conductor Phillips of the freight train and the latter had said, "I couldn't quite figure out where you were to wait for us."

Fireman Bean stated that he had taken the train order from Operator Spafford at Gassetts, without a clearance card, and was positive that Spafford had said, "Wait at Bellows Falls until 4.50. I haven't got time to make out a clearance. That is all there is, go ahead." The fireman said he had handed the order to the engineer, who had read it out to him and the head brakeman as, "No. 165 will wait at Bellows Falls until 4.50 for extra

---

[754] Ibid.

[755] *The Berkshire Evening Eagle*, Pittsfield, Massachusetts, 15th March 1920

28 south." The three of them had discussed the order, and whether to go to Midway 4¾ miles south of Bartonsville, or Riverside, 7¼ miles south of there, for the express, and Engineer Cody had decided that they had time enough to get to Riverside. According to the fireman the train had drifted all the way and its speed had not exceeded 30 miles an hour.

Engineer Cody had suddenly remarked, "Look out," and whistled for brakes. Fireman Bean said that the engineer usually handed the orders to the head brakeman and himself to read but he had not done so on this occasion. Fireman Bean said that he realised that according to Rule 211 he was required to read the order. Shown a copy of it at the investigation, he said that he thought he would have read it correctly, despite it being illegible.

Conductor Phillips of freight train No. 28 testified that the air brakes of the train had been tested before they left Rutland and had been satisfactory; they had not been re-tested after the helper was detached at Summit. The order given to him at Gassetts had been very poorly written, but he thought that it read Bellows Falls; he had discussed it with the flagman, who had agreed. He was aware of special bulletin order No. 363, restricting the speed of freight trains to 25 miles an hour, but had not thought of it at the time. The conductor said he had calculated that his train had time to reach Riverside, though he thought they might have to protect the head end of their train when going into the siding. At Bartonsville he had again read the order and still read it as Bellows Falls, as had the flagman, who pointed out that there were three men on the engine who must also have done so. The conductor said he had not worked with Engineer Cady before and had not been relying on him for the safety of the train. He did not think his train had exceeded 25 or 30 miles an hour at any point.

Flagman Brooks of extra No. 28 said that leaving Gassetts he had read the order as reading Bellows Falls. At Bartonsville the conductor had told him to read it again, which he had done, though he had paid more attention to the train number than to the station named. He had also remarked to the conductor that there were three experienced men on the engine who must have read the order. The flagman added that he and the conductor had read the order hurriedly; had they read it more carefully they would probably have read it correctly. He did not think that the train had been going too fast; he had not noticed it riding in a four-wheeled caboose.

Assistant Superintendent Kramer stated that a few hours after the collision he had asked Conductor Phillips why he had run past Bartonsville and the conductor said that he had thought of opening the emergency valve to apply the brakes and stop the train but had been afraid of pulling a drawbar, also that he and the flagman had agreed that the place on the order must be Bellows Falls as those on the engine were experienced men.

Operator Spafford said that he had completed the order as the freight train was approaching and to avoid delay had not made out a clearance card for the engine crew. According to him his words to the fireman as he had delivered the order had been, "You have got no clearance card, that is the only order you have got, it is all right to go ahead." He denied mentioning the contents of the order and admitted having violated Rule 211 in allowing the train to pass the train order board without a clearance card. He stated that when the dispatcher had first given him the order he had read it back as No. 314, but the dispatcher had corrected him to 414. The number on the order itself he had not bothered to alter. He was not aware that the first 4 meant that the order meant the Bellows Falls sub-division, the other figures being the number of the order. He said that he had issued orders to Engineer Cady for the last 25 or 30 years and the engineer must be familiar with his handwriting.

The ICC concluded that Conductor Phillips and Engineer Cady were chiefly responsible for the accident for having misread the train order, while Fireman Bean and Head Brakeman Granger were at fault for not having read it themselves as Rule 211 (a) dictated. The commission considered that although the order was illegible, if had been read carefully it should have been decipherable. From the copy reproduced in the accident report this seems debatable to the author, as the order is extremely illegible, but then perhaps it would be less so to someone familiar with the operator's handwriting. In any case, as the commission pointed out, had Engineer Cady been in any doubt he should have stopped and queried the order. It was possible, though, that he had been influenced by Operator Spafford's remark as to Bellows Falls, if, as Fireman Bean claimed, that was what was said. No such excuse could be made for the conductor and flagman, and if Conductor Phillips had been in doubt as to the meaning of the order, as his post-accident comments to Assistant Superintendent Kramer implied, then he had been grossly negligent in not stopping his train at Chester (the telegraph office between Gassetts and Bartonsville), to clarify the matter. Both the conductor and flagman were at fault for twice failing to read the order with proper care.

Engineer Cady had, the ICC continued, violated Rule 221 by leaving Gassetts without a clearance card when the train order board had been displayed, and had also disobeyed bulletin No. 363, limiting the speed of his train to 25 miles an hour. The collision had occurred at 4.39 p.m. and train No. 28 had passed Chester, approximately 11 miles back, at 4.25 p.m., giving an average speed between the two places of 47 miles an hour. Rule No. 507 had also been broken as the air brakes had not been tested at either Summit or Gassetts, although the train line had been broken at both places (for the helper engine to be detached at Summit and for the engine to take water at Gassetts).

Operator Spafford, while having written an order which was understandable if it was read with care, was nevertheless open to censure for the manner in which he had handled it. 'His failure to correct the number of the order after his attention had been called to the mistake by the dispatcher was negligent in the extreme, while the manner in which he wrote the order, and his action in verbally authorizing the extra to pass the train order board without a clearance card indicated a lack of realization of the importance of his duties he was supposed to perform. Carelessness of this nature is one of the causes frequently leading to accidents of this character and should not be tolerated.' (A clearance card was given to a train crew by the operator together with the train orders. The card listed the train orders that the crew should have received at that point and gave them clearance to enter the single track provided that they had all the orders listed on the card, if any.)

In the final paragraph of the report the ICC noted that the accident again drew attention to the weakness of the train order system and the need to adopt some sort of block system. There were a total of five first-class and two second-class trains each way daily over the Bellows Falls sub-division, and 18 stations, all of which were day telegraph offices except the four between Chester and Bellows Falls. The installation of a block system would not be difficult, was recommended, and would undoubtedly have prevented the accident. [756]

What was known as the West Side Branch of the Junction City sub-division of the Portland division of the Southern Pacific Railroad suffered a collision near Bertha, Oregon, United States, on 9th May 1920. The line involved was a single-track electrified one running for 55¾ miles between Portland and Whiteston, Oregon, over which trains were worked by timetable and train orders, orders being despatched by telephone.

One of the trains involved in the accident was westbound electric passenger train No. 107, which consisted of a motor combination coach, a trailer coach, and a motor coach, all of steel construction, in charge of Conductor Johnson and Motorman Bland. This train had left Portland at 10 a.m. in possession of a copy of train order No. 204, providing for his train to take the siding and meet train No. 124 at Bertha. Train No. 107 was due at Bertha at 10.22 a.m., and it was approaching there at about 10.23, and travelling at 20 or 25 miles an hour, when it collided with train No. 124.

Eastbound passenger train No. 124 had run as train No. 123 from Portland to Reedville, a non-telephone station 15½ miles from Portland, continued as an extra to Main Street, another non-telephone station 4 miles further on, and then returned to Reedville, before leaving there for Portland as train No. 124. This train consisted of a motor combination coach and a motor coach, both of all steel construction, and was in charge of Conductor Austin Pharis and Motorman Silas Willett. Conductor Pharis had been given a copy of order No. 204 before leaving Portland and also a copy of order No. 218 stating that train No. 124 had car No. 502 rather than 515. This train left Beaverton, 5½ miles from Bertha, made several intermediate stops, called at Bertha, left there, and was travelling at about 20 miles an hour when it collided with train No. 107 half a mile further on.

Car 503, the leading vehicle of train No. 107, was raised above the floor of car 502, the leading one of train No. 124, and that car was badly telescoped. Motorman Frederick Peebler, who had been deadheading in train No. 107, was killed, as were Motorman Willett, Newton Hoover, another off duty railroad employee, and 5 passengers on train No. 124, while 92 passengers and 10 railroad employees, 5 of them off duty, were injured. One of the injured passengers later died. [757] The passengers who lost their lives, all Oregonians, were Mrs Charles Crooks and Ina Hatch of Hillsdale; Mrs C. R. Arundell and her son Robert Arundell, aged 4, of Dosch station; and Fleurot Dosch Josselyn, aged 7, also of Dosch station. The passenger who died later was Miss Camille Dosch, aunt of Flleurot Dosch Josselyn.

A local newspaper described the accident and wrote, 'Out of the crash and terror of collision, when two trains plunge into head-on wreckage, as they did in yesterday's disastrous smash at Bertha station, emerge brief stories of men and women under severe trial – stories that are creditable to the race in their narrative of unselfish service or loyalty, or there glimpses of humor amid stark scenes of death.

'Spared from the fate that crushed life from the body of his fellow engineer, on the inbound train, the surviving engineman did not by word or inference seek to cast the

---

[756] Interstate Commerce Commission report

[757] Ibid.

blame of the wreck upon the shoulders of the dead craftsman. "I only know that I was not to blame," he said, stoutly declining to offer a single conjecture regarding the probable cause of the crash.

'From the mass of twisted steel and iron and ragged plush, where the two front coaches met, searchers dragged the lax body of a victim and laid him down in the grass as one of the dead. Ten minutes later, before the stretcher bearers reached him, he rose and declared his freedom from injury. Clenched in one hand was a fragment of glass. He carried it away as a souvenir.

'There thronged the tracks and right of way, soon after the collision, a vast crowd of curious, who gazed at the flaccid arm and hand of the dead engineer, pinned between the wreckage of the two coaches. Well-dressed women picked up bits of twisted iron and splinters of glass as mementoes of the catastrophe.' [758]

At the Interstate Commerce Commission (ICC) inquiry Motorman Bland testified that it had been Motorman Peebler, from his position on the left-hand side of the car, who had first seen train No. 124 and had called out, whereupon he (Motorman Bland) had applied the brakes in emergency.

Conductor Johnson of train No. 107 said that the collision had occurred only a few seconds after the brakes were applied. Both motorman and conductor said that Bertha was the usual crossing place of the trains and that there had been nothing unusual about train order No. 204.

Conductor Pharis stated that he had received train order No. 204 before leaving Portland, had understood it and it was similar to previous ones he had received. He said that when Motorman Willett had whistled for Bertha he had pulled the cord once to indicate to the motorman that they were approaching a meeting point, and the motorman had acknowledged his signal. They had stopped at the station, some passengers had been picked up, the air brakes had been tested and he had given the proceed signal on the air whistle. Although he had been looking out to identify train No. 107, Conductor Pharis said that he had not realised that his train was running past its meeting point until it was at the east switch and moving at about 35 miles an hour. He knew that train No. 107 had not arrived and had been trying to reach the emergency cord to stop the train, but he was not sure if he had or no. The conductor said that Motorman Willett had appeared well, though he had complained of having felt ill the night before and of not having had a good night's rest.

Brakeman W.A. Fisch of train No. 124 claimed that he had not seen the order, though the conductor claimed to have shown it to him. The brakeman agreed that it was customary to meet train No. 107 there. He thought that the train had reached 40 miles an hour and had noticed no brake application before the crash.

A bridge carpenter, an assistant bridge foreman and a regular passenger who had been riding on train No. 124 all put the speed of the train before the collision at 40 or 45 miles an hour, though Motorman Senders, another passenger, only put it at 30 miles an hour. Motorman Senders had talked to Motorman Willett earlier that day and had thought him to be in a normal condition, Willett had said nothing about feeling unwell.

The ICC stated that evidence as to whether the brakes had been applied on train No.

[758] *Morning Oregonian*, Portland, Oregon, United States, 10th and 11th May 1920

124 before the collision was contradictory.

Dispatcher McLardy stated that he had cleared train order No. 204 at Portland. It had been issued by the previous dispatcher and he considered it correct and proper. It referred to both the outward and return trips of trains Nos. 123 and 124 because Reedville and Main Street were both non-telephone stations.

The ICC came to the conclusion that the failure of Conductor Pharis and Motorman Willett to obey train order No. 204 had caused the accident.

Why Motorman Willett had failed to wait, even after just exchanging signals regarding the meeting point with the conductor, must remain mere conjecture. Should he have collapsed there was also a 'dead man's control,' but that had not helped. Conductor Pharis had clearly not been paying proper attention to the operation of his train. 'Knowing that after leaving the station at Bertha his train would have to stop within a distance of less than one-half mile for the purpose of meeting train No. 107, it is inconceivable how Conductor Pharis could have returned to the inside of his train and have become so engrossed with other matters as not to know that the meet order was being disregarded.'

The ICC report concluded with the usual comments regarding the weakness of the train order system and a recommendation to install an adequate block signal system to ensure the adequate protection of trains. [759]

The inquest jury on the victims of the accident returned a verdict blaming Motorman Willett, Conductor Pharis and Brakeman Fisch for the accident. Their verdict in full read:

"We, the undersigned jurors, find these deaths were caused by a head-on collision of Southern Pacific trains No. 124 and No. 107, and that the wreck was caused by the failure of the train crew of No. 124 to execute orders.

"It is our opinion that Engineer Willett of train No. 124 was rendered mentally or physical incapacitated by some unknown cause prior to reaching the clearance of the siding and remained so up to the time of the collision." [760]

The last significant single-line collision of 1920 was another on an American interurban line. The interurbans seemed to have more than their fair share of single-line collisions during their rather brief existence, doubtless due to their primitive operating methods. On 6th September 1920 two Denver and Interurban Railroad trains collided near Globeville, Colorado, United States. There was a short stretch of double track at Globeville station, beyond which the line became single. It was on this stretch of single track between Globeville and Denver & Interurban Junction that the accident occurred. The line was worked by timetable and train orders. Northbound trains were superior to southbound ones.

Southbound passenger train No. 308 on the day of the accident consisted of car No. 153, in charge of Conductor C.W. Grenamyre with Motorman Stevens. It was due in Globeville at 11.30 a.m. and left Boulder Junction at 10.41 a.m., passed Westminster, the last telegraph station open before Globeville and two miles north therefrom, at 11.24

---

[759] Interstate Commerce Commission report

[760] *Morning Oregonian*, Portland, Oregon, United States, 12th May 1920

a.m., six minutes late, and the collision occurred at 11.34, while it was reportedly travelling at 35 miles an hour.

Northbound passenger train extra No. 158, consisted of motor No. 158 and a train and was in charge of Conductor J.W. Schulze and Motorman Lawrence Cripps. This was an excursion train between Denver and Eldorado Springs and had already made one round trip between these places. On its second trip it arrived at Globeville at 11.27 a.m. and there its conductor and motorman were given a clearance card and train orders Nos. 8 and 52. Train order No. 8 was a slow order applying to a stretch of track further along the line. Train order No. 52 read:

Motors 158 and 152 run as 2 passr extras on second trip with right over extra 703 south Louisville Junction to Marshall. Passr extra M. 158 north and No. 309 may run on a 5 min. block. Passr extra 158 and 152 north via Louisville Junction.

Extra 158 passed Globeville at about 11.30 a.m., passed from the double to single track and was running at about 25 or 30 miles an hour when it collided with train No. 308 half a mile north of Globeville. In the collision motor No. 158 telescoped motor No. 153 of train No. 308 about 15 feet and all the fatalities were in train No. 308. Conductor Grenamyre, a section foreman and 11 passengers were killed, and five railroad employees and 209 passengers were injured. [761]

'Most of the casualties occurred on the inbound car but few on either car escaped injury. Every available automobile of the Denver police and fire department was rushed to the scene, but they were too few to transport the injured to hospitals, so private cars were pressed into service. A corps of doctors and nurses also was rushed to the scene.

'At least two persons were killed when they jumped in an attempt to save themselves, according to J.W. Zarina, of Louisville, Colo., who was on the inbound car. One of them was Conductor G.W. Grenimyer, 61, who was in charge of the car. He was the oldest conductor on the road in point of service, according to company officials. At the time of the accident his wife was in Denver hospital in a dying condition.

'Zarina said he was talking to the conductor and to Frank Dolby, also of Louisville, when they sighted the outbound car rounding the curve.

'"Grenimyer turned to me as white as a sheet, he said, and cred, 'My God, what is that?' A moment later he shouted, 'We'll all have to jump!'"

'The conductor, according to Zarina, was the first to jump and was closely followed by Dolby. Both of them were instantly killed. Zarina jumped later and escaped with injuries.

'Four of the dead and many of those injured were from Louisville and were on a car coming to Denver to attend a Labour day celebration here.

'As soon as word of the wreck was received in Denver a squad of the soldiers here in connection with the Tramway strike were rushed to the scene. They found great difficulty in holding back the crowd of the curious who threatened to hamper the work of the rescue.' [762]

Motorman Stevens of train No. 308 told the Interstate Commerce Commission (ICC)

---

[761] Interstate Commerce Commission report

[762] *Cheyenne State Leader*, Cheyenne, Wyoming, United States, 7th September 1920

inquiry that when he saw the opposing train it had been 460 or 500 feet away. He had opened his sanders, applied the brakes and given one long whistle.

Conductor Schultze of train extra 158 told the inquiry that when they had arrived at Globeville on the second trip he had gone into the office and collected the two train orders and a clearance card. He found that the operator had already registered for them and giving the orders to Motorman Cripps he had read them, and they had then left the office and boarded the train, not having checked the train register. (In North America, the train register was not a record book kept by a signalman to record train movements, as in Great Britain, but rather a record book provided at specified points on a single line where conductors recorded their trains. The register was checked by the crews of opposing trains to ascertain if trains superior by class or direction which had to meet had arrived there). Conductor Schultze said that he had not forgotten train No. 308 but had had it in his mind that it was due at Globeville at 12.30 p.m. rather than 11.30 a.m. On the train he had given the orders to Flagman Murphy, reminded the latter to change from the low-tension to the high-tension at the end of the double track, collected some tickets and then gone forward to ask the motorman about where they were meeting opposing trains, not having his own timetable. As he reached the motorman the latter saw the opposing train and told him to jump.

Motorman Cripps stated that while he had been reading the orders in the office at Globeville he had heard the noise of a passing car. A man had then come into the office (the conductor of the following train, No. 309) and Motorman Cripps had assumed that he must be the conductor of train No. 308 and that that train had arrived. They had set off from Globeville about 11.29 or 11.30. When he saw train No. 308 approaching he had given a blast on the whistle, applied the air brakes, opened the sanders and jumped.

Flagman Murphy said that he had put the orders given to him by the conductor into his pocket and occupied himself with changing the trolley, no stop being made at the end of the double track. He had thus not had time to read the orders before the collision.

Operator Obland at Globeville said that when train No. 158 had started he had assumed that it was just moving forward to the end of the double track. Train No. 309 had moved forward to where train No. 158 had been standing, he had inquired of its crew if train No. 308 had arrived, and with that had realised that it had not, and that train No. 158 was not waiting. He had then run out of his office and towards train No. 158 but had been unable to attract the attention of its crew.

The ICC noted that not only had Conductor Schulze failed to check the train register as required by the rules, but he had failed to fill out and deliver to his motorman a copy of form 1514-Revised, which was a written check on the register, and the operator had registered the names of the crew of the train and its arrival time, saying that it was his custom to do this for all trains. The conductor's register at Globeville had spaces for the names of the conductor, motorman and flagmen, the train's arrival and departure times, train numbers, signals displayed and the time that crew members had checked their watches. Conductor Schulze said that he had checked the register and signed his name on the first trip of the day, but had left the other entries for the operator to fill in. He had not given the motorman the register check form (1514-Revised) on either trip.

The conductor said he had not had a timetable and that when he had asked for one for himself and for Flagman Murphy at the car barns he had been told that they were kept at

Globeville and at Globeville that there were none left. He had made no more effort to get one. The day of the accident was his first day working on the line apart from two days working as assistant conductor and only Flagman Murphy's third. Motorman Cripps had been off sick with ptomaine poisoning the day before the accident and only back at work because he had been told that there was no one to take his place.

The ICC concluded that Conductor Schulze and Motorman Cripps were responsible for the accident, having failed to wait at Globeville for train No. 158. Although he had no experience on the Denver and Interurban Conductor Schulze had 17 years' service on the Colorado and Southern Railroad and should therefore have realised the importance of checking the train register, especially as his train was an extra and it was going from double to single track, also of giving his motorman a copy of form 1514-Revised, as required by Rule No. 83A.

As Motorman Cripps knew that his conductor did not have a timetable he should have realised that the latter could not properly check the register and he should not have left Globeville without a copy of form 1514-Revised. He was thus equally at fault with the conductor with regard to the accident.

Operator Obland, while not directly responsible for the accident, was open to severe criticism for registering the train crews. Had he not done so Conductor Schulze would probably have properly registered and filled out form 1514-Revised.

The ICC commented, 'The manner in which the train register at Globeville is handled cannot be too strongly condemned. The evidence indicates that the operator registers for train crews, and that it is not customary for train crews to check the register, in fact, it was stated during the investigation that the crews of regular trains identify a train as it passes them en route and use this identification as a check on the register. The conditions disclosed by this investigation strongly indicate that trains are commonly operated on this road without attention being paid to the rules governing the train register. Attention is also directed to the lack of timetables for the use of employees in train service. Flagman Murphy had worked on the Denver & Interurban Railroad on three different days within a period of two weeks, while Conductor Schulze was working his first day in a month and his third day in over 12 years, neither of these employees had timetables, although the evidence indicates that Conductor Schulze made a reasonable effort to obtain them on the day of the accident.'

As usual the commission drew attention to the weakness of the timetable and train order system and the fact that an adequate block system could have prevented the accident. [763]

Having suffered a single-line collision due to interference with the tablet instruments at Park Hall in 1918, the Cambrian Railways went on to experience a much worse accident less than three years later, on Wednesday 26th January 1921, this time from a different cause. The scene of the accident was a point between Abermule and Newtown stations, Wales, on the main line of the company. The trains involved were the 10.25 a.m. up express from Aberystwyth to Shrewsbury and the 10.05 a.m. down slow train from Shrewsbury to Aberystwyth. The express consisted of 4-4-0 engine No. 95, a London and

[763] Interstate Commerce Commission report

North Western (LNW) van, a Cambrian first and third-class composite coach, a Cambrian tea car, a Great Western (GW) first and third-class composite coach, an LNW brake van, an LNW first and third-class composite coach and an LNW brake composite, in charge of Guard Edward Shaw with Driver John Pritchard Jones and Fireman John Owen. The local train consisted of 4-4-0 engine No. 82, a Cambrian first and third-class composite coach, an LNW third-class coach, a Cambrian first and third-class composite coach, an LNW first and third-class composite coach and two LNW brake vans in charge of Guard Edwin Chetwood with Driver George Jones and Fireman Albert Evans.

Abermule station, the scene of the mistake which led to the accident, was a crossing place on the single line, 3¾ miles from Montgomery station to the north-east and almost 4 miles from Newtown to the south-west. It was the junction for the short branch to Kerry. The station staff on duty at Abermule at the time of the accident were Relief Stationmaster Frank Lewis, Signalman William Jones, Porter Ernest Rogers and Booking Clerk Francis Thompson. A proper signal box was provided adjacent to a level crossing at the Montgomery end of the station, but the Tyers No. 6 tablet instruments were located in an instrument room next to the booking office in the main station buildings. Even though the station stood on the main line the traffic through it was hardly intensive, with an average of only 23 trains in each direction on weekdays and none between midnight and 4 a.m. or 12 noon and 2 p.m.

At 11.50 a.m. on 26th January the relief stationmaster was away having an early dinner, the porter and clerk were also having their dinner in the booking office and the signalman was in the adjoining telegraph room. Signalman Jones accepted the slow train from Montgomery at 11.52, released the Montgomery-Abermule tablet and telephoning Moat Lane Junction, 8½ further west, learnt that the express had passed there, so he assumed the two trains would cross, as usual, at Abermule. Telling Porter Rogers that the local train had been off Montgomery three minutes and the express had passed Moat Lane the signalman then went to close the crossing gates and then to his box to clear the signals for the local train. Stationmaster Lewis had by now returned but the signalman did not tell him of the train movements.

When Newtown belled the express at 11.56 Porter Rogers went into the instrument room, accepted it and released the Abermule-Newtown tablet. He then went to the south-west end of the station, where the points at that end of the station were worked by a ground frame, being too far away to be worked directly from the box. The down local train arrived at 12.02 and as the stationmaster was not present Clerk Thompson crossed from the up platform and took the Montgomery-Abermule tablet from the engine crew. He was on his way to the instrument room to replace it in the instrument when he met the stationmaster returning from the goods yard. Handing the tablet to the latter, he said, “Change this tablet, Frank, while I go and collect the tickets.” He then returned to the down platform to collect a ticket from the single passenger who had alighted.

Relief Stationmaster Lewis had been pounced on, he said, when he had returned from his dinner, by Permanent Way Inspector Thomas with an enquiry about a waggon for stakes, a matter of urgency as the inspector wished to catch the down train. The two men were in the goods yard when the local train pulled in. Returning to the station the stationmaster said he had met the clerk, who had handed him the tablet holder with the words, “Take the tablet for the down train, he is going on.” Clerk Thompson had a stutter which made it hard to make out what he was saying and was perhaps mumbling because he was embarrassed at having been caught handing the tablet, being aware that this was against the rules and having, it was stated at the inquest, been told off by Mr Parry, the regular stationmaster, for doing so.

"Where's the express?" the stationmaster asked.

"About Moat Lane," the clerk had thoughtlessly and misleadingly replied. Assuming that the express must be running late and that the clerk had already changed the tablet the stationmaster, anxious to avoid delay, went to the engine of the local train and gave them back the same tablet that they had handed to the clerk shortly before.

Porter Rogers had gone to the west end ground frame to set the road for the up express but seeing the stationmaster giving the local train the "Right away" assumed that something must have delayed the express at Newtown leading to a change of plan. He released the slot on the down starting signal for the local train. He, of course, was the only person who knew that a tablet had been released for the express.

Driver Jones of the local train was oiling his engine when Fireman Evans took the tablet from the stationmaster. The stationmaster had not bothered to look at the tablet and the driver and fireman set off without doing so either, thus signing their death warrants. The train started away at about 12.03. Meanwhile Station Foreman Brock at Newtown had given the "Train entering Section" signal for the express at 11.59 but with no one in the Abermule instrument room it went unacknowledged. It was only when Clerk Thompson went to the instrument room to give the "Train entering Section" signal to Newtown and the "Train out of Section" signal to Montgomery for the local train that he realised that there had been a mistake, and the stationmaster realised that he had given the crew of the local train the wrong tablet. Stationmaster Lewis rang Newtown to ask if the express was still there, to be told that it had left at 11.59. A vain attempt was made to warn the crew of the local train by lowering and raising the up distant signal, but the train had probably passed that signal by then anyway.

The express had passed Cil Gwrgan Crossing when Driver Pritchard Jones saw first the steam of the local train and then the front of its engine. He applied the brakes, opened his whistle, and jumped. Seeing his driver braking and whistling Fireman Owen had looked up, seen the local train close at hand and jumped off as well. Driver George Jones and Fireman Evans on the local train were apparently completely unaware of the express; they were both killed. The crew of the express had both examined their tablet before they left Newtown. After the accident Driver Pritchard Jones was more seriously injured than his fireman but was very concerned lest they had had the right tablet. Hunting through the wreckage Fireman Owen was able to find both tablets and show them to his driver to reassure him that they were in the clear. Traffic Controller Morgan, who had been a passenger on the local train, then took custody of the tablets and later gave them to Chief Traffic Inspector George, who returned on foot with them to Abermule. [764]

The length of the trains was reduced from 252 yards to 180 yards in the collision, the two locomotives, three coaches of the express and one of the local train forming a tangled mass of wreckage. Nine of the vehicles were gas lit but fortunately there was no fire. Fourteen passengers, including Lord Herbert Vane-Tempest, 'The King's host in Wales,' who was a Cambrian Railways director; [765]Driver George Jones and Fireman Evans of

[764] Ministry of Transport report

[765] *The Cambrian Railways Volume II* by Rex Christiansen and R.W. Miller, pg. 101

the local train and Guard Shaw of the express lost their lives. Thirty-three passengers, Driver Pritchard Jones, and Fireman Owen of the express and Guard Chetwood of the local train were more or less injured.

A reporter wrote, 'From Newtown and the countryside people were attracted in large numbers by the news of the disaster, and from an elevated position on the far side of the line they looked down upon the efforts of breakdown gangs from Oswestry and Machynlleth, working under the direction of Mr MacDonald, the traffic manager of the Cambrian system. Doctors were busy in rendering first aid to the injured, and motor ambulances were quickly brought up to convey them to the Newton Infirmary. At 5 o'clock it was stated that 14 persons had been killed, and at that time the task of clearing the Paddington coach had not been completed. The work was one of great difficulty, and not a little danger, as there was a risk of the carriage collapsing. Further compartments were seen to be jammed together, and it was hard to make headway. One of the first to go to the accident was a local hotel-keeper. He told me that it was an appalling sight. The Paddington carriage had been telescoped, and the seats all crushed up together as though it were a concertina. One poor man was pinned to the side of the window, and late in the afternoon he was still living, having been liberated and taken to the hospital. "When I saw him," said my informant, "he was hanging out of the side of the coach, but his legs were pinned underneath, and there were three bodies – those of two women and a man – on the top. One was that of a nurse. The guard of the London train was pulled from right underneath the first compartment, his head having been smashed in. Then, I could see three men's heads and one woman sandwiched within a few inches of one another. There was little fire, the water from the engines having evidently put out the boiler fires. The wreckage showed no sign of fire. One man who was in the first carriage of the local train told me that he received a shower and found that it was the water coming from the engine, which was on the roof of the compartment. He and a companion escaped through the window.'

A passenger was also interviewed. 'Mr A. Bell, of Greenwich, one of the passengers by the express, stated that he was in a compartment next to the engine of the express when the accident happened. Just as he was opening the door to go out into the corridor there was what seemed to him a terrible explosion, and he was knocked down. The roof of the compartment collapsed, and the sides fell in. How he got out he does not know, but he found himself in a dazed condition in a hedge with injuries to his right knee and to his left leg and with his clothing torn to shreds. He was suffering considerably from shock.' [766]

Colonel J.W. Pringle enquired into the accident for the Ministry of Transport and in his report wrote, 'This deplorable accident is unique in the annals of British history. Such a tale of failure in broad daylight, and of misunderstanding on the part of so many men, would have been incredible, prior to the accident. Indiscipline and slipshod methods of custody and of transference of tablets, on the part of the traffic staff, combined with failure on the part of enginemen to examine tablets, could alone have made this disaster possible.'

The inspecting officer's conclusion as to the cause of the disaster was the obvious one that it had been due to the crew of the local train departing from Abermule with the

[766] *The Times*, London, England, 27th January 1921

wrong tablet.

Turning to the responsibility for the accident, Colonel Pringle noted that it rested equally with the traffic and locomotive staff. The ultimate responsibility for leaving a tablet station with the correct tablet rested with the engine driver. It was common practice for the fireman to actually exchange the tablets, but this did not excuse the driver for the duty of examining the tablet and checking that it was the correct one. Driver George Jones and Fireman Evans had clearly failed to do so and had paid with their lives for their neglect.

As to the traffic staff only the stationmaster, John Parry, or his relief, and the two signalmen were authorised by the regulations to handle the tablets and reminders had been sent drawing the attention of staff to these regulations. On the day of the disaster Stationmaster Parry had been on holiday and Relief Stationmaster Lewis was taking his place. Stationmaster Parry stated that although Porter Rogers was not strictly an 'authorised person' he had been permitted to handle the tablets under the direct supervision of himself or the signalman. Telegraph Clerk Thompson (who was only 15 years old, Rogers was only 17) had no authority to do so. Relief Stationmaster Lewis, on the other hand, stated that when he had taken over the running of the station it had been common practice for Thompson to handle the tablets and he (Lewis) had merely warned the boy to be careful. Rogers freely admitted that he had handled the tablets and used the instruments in the absence of both stationmaster and signalman and Thompson adhered to his statement, in the presence of Stationmaster Parry, that the latter had shown him how to work the instruments. The inspecting officer concluded that it was clear that a 'slipshod, happy-go-lucky practice' had grown up at Abermule whereby anyone was allowed to operate the tablet instruments, and which had 'rendered possible the extraordinary series of assumptions and misunderstandings which alone made this disaster possible.' He considered Stationmaster Parry mainly responsible for this state of affairs, but that Relief Stationmaster Lewis, Signalman Jones and Traffic Inspector Lloyd were also partly responsible.

Stationmaster Lewis had made his first mistake in not going to dinner earlier, so as to be at the station to personally supervise the crossing of the express and the local train, as per the specific instructions of the company. His next mistake had been to accompany the permanent way inspector to the goods yard without first ascertaining the whereabouts of the express and arranging to be called when it was belled from Newtown. His most serious mistake was to have failed to examine the tablet given to him by Thompson. Lewis had given his evidence in a straightforward manner and fully admitted his responsibility.

Signalman Jones, on the other hand, was a less credible witness and seemed to consider himself free from blame. He was aware, however, that the express had left Moat Lane, should have collected the tablet from the down train in the absence of the stationmaster and had done nothing to find out what had happened to the express which he knew should have crossed the local train at Abermule. The inspecting officer considered him almost as much to blame as the relief stationmaster.

Rogers and Thompson, Colonel Pringle remarked, were 'both well-intentioned lads, but not of an age to understand the danger that may arise from dual responsibility, ill-considered action and careless speech.' Rogers was seriously to blame for not informing the stationmaster or signalman that he had accepted the tablet. Thompson was to blame for leaving the booking office unattended, for taking the tablet from the crew of the local train and for misleading Relief Stationmaster Lewis as to the whereabouts of the express.

In summary, the inspecting officer wrote, 'The case as a whole is an illustration of how fatal understandings may arise from allowing a number of people to participate in the use of single line instruments, and in the duty of receiving and handling tablets, and as such it forms an object lesson to all traffic managers, stationmasters and signalmen concerned with single-line working, of the importance of the strictest discipline in carrying out the regulations for the custody and transference of tablets.'

The report went on to consider methods of improving methods of single-line working. The painting of tablets or staffs in different colours the inspecting officer argued would not be helpful as they would get dirty, the colours would fade and so on while on long stretches of single-track engine crews would find it hard to memorise the colours of the different sections. He was equally unenthusiastic about the use of tablet and staff instruments for alternate sections. Reliance, he thought, must be placed on enginemen carefully examining tablets.

Had the tablet instruments been placed in the signal box at Abermule the collision might have been prevented, but on the other hand having the instruments in the signal boxes had not prevented the 1918 Park Hall collision. Some boxes would also need enlarging to accommodate the instruments (not that difficult a task, surely?).

The interlocking of the tablet instruments with the starting signal was another suggestion which he thought was more reasonable, at least in some locations, though it was not a requirement of the Ministry of Transport and he thought that the decision should be left to the railway companies themselves.

The consequences of the accident would have been much less serious had the fourth vehicle of the express not been completely telescoped by the third vehicle. Six principal causes of vehicles overriding one another could be noted: -

(a) Long buffering allowing freedom for vertical movement.

(b) The shape and curvature of the faces of the coach buffers

(c) The marshalling of light vehicles between heavy vehicles

(d) The tilting of a coach due to the derailment of wheels at one end of it

(e) Inequality in the heights of the buffers due to heavy loading, the age of springs, tyre wear.

(f) The effect of a sudden brake application.

It was ten years since the accidents at Willesden, Hawes Junction and Coke Ovens, Pontypridd had led inspecting officers to recommend companies to reduce the risk of telescoping by the use of uniform rolling stock, improvements in buffering, the deepening of headstocks and anti-collision devices. The railway companies had resisted, arguing that were telescoping prevented disastrous effects of a different character would result, but this, Colonel Pringle thought, was untrue. Some improvements had been made by some companies, but the inspector considered that the whole question of equipping rolling stock with devices to prevent overriding and telescoping should be again reviewed by a committee of experts currently engaged in the standardisation of carriage and waggon stock.

In summary the inspecting officer recommended that: -

(a) At stations where the single-line instruments could not be located in the signal box, and in other special circumstances, the starting signals should be interlocked with the single-line instruments.

(b) With an increase in the distance in which points could be worked from a signal box the west end points at Abergele could be so worked and the ground frame abolished.

(c) An expert committee should examine means of preventing telescoping. [767]

At the inquest on the victims of the disaster Mr Henry Warwick, the Superintendent of the Line, stated that all cases of irregularity in working the tablet instruments were dealt with severely. There had only been one case on record of a driver entering a section without a tablet; the stationmaster involved had been dismissed and the driver reduced to the lowest grade. This may have been true for the Cambrian, but such things did happen from time to time, sometimes it would seem rather frequently. Thus on the London and South Western Railway on 26th November 1897 a train left Sidmouth without the tablet; on 18th March 1898, at Tipton St Johns, the junction with the Sidmouth branch, a train started with the wrong tablet; on 7th January 1899, a train from Sidmouth left Tipton St Johns without changing the tablet and finally, less than a week later, on 13th January 1899, a train left Netley, near Southampton, without the tablet. [768] No accidents fortunately resulted from any of these incidents. Such a series of incidents within a short time was probably exceptional but doubtless such things happened elsewhere from time to time and if no accident resulted they were covered up by the staff concerned, or at any rate only mentioned in internal company correspondence. The tablet system was safe, but certainly not infallible.

The verdict of the inquest jury, announced on 3rd February, was that Acting Stationmaster Lewis and Signalman Jones had been guilty of gross neglect, and Porter Rogers and Clerk Thompson guilty of negligence and excess of duty, also that there had been a lack of supervision of tablet working by superior officers. They recommended that the tablet instruments should be moved to the signal boxes and all level crossings connected by telephones to the stations on either side. The coroner pointed out that the jury's verdict amounted to one of manslaughter against Lewis and Jones. The foreman replied that the jury had understood that but pressed by Mr J.H. Thomas, M.P., general secretary of the National Union of Railwaymen, to give the matter more thought, the jury retired again, and the second time returned a verdict that Lewis and Jones were guilty of great neglect calling for most severe censure. This censure was administered by the coroner, who added that no doubt the two had already been punished by the agony of mind which they had suffered. [769]

No. 3 up mail train and No. 364 down goods train collided with one another between Tawwi and Peinzalok stations, Burma, on the Burma Railways at about 22:35 on Monday

[767] Ministry of Transport report

[768] *South Western Circular* (published by the South West Circle railway society), Vol 18, No. 8, pg. 361

[769] *The Times*, London, England, 28th and 29th January and 3rd or 4th February 1921

25th July 1921. About 100 of the passengers were killed and 54 injured. The drivers of both trains and their first firemen were also killed. The second firemen of both trains and the brakeman of the mail train were among the injured. The damage to the trains and permanent way was estimated at 1,000,400 rupees. Many of the passengers on the mail train had been bound for Mandalay to attend the races there and the dead included four Burmese jockeys. It would appear that the part of the mail train carrying the first-class passengers must have escaped much damaged as the reports spoke of third-class passengers being killed and injured and the first-class passengers having returned to Rangoon. [770]

The accident was attributed to a breakdown in the block system which had allowed two tokens to be withdrawn for the same section at the same time. Not though, it would seem, in this case, due to a railwayman deliberately burgling the instrument but rather accidentally. The report on the accident stated that, 'Remedial measures for the better maintenance of wires near stations have been taken by the Railway Administrations concerned.' [771]

On more than one occasion down through the years the opening of a new railway line has been marred by a bad accident. This was the case in 1921 in Norway. On Saturday 17th September, the Dovrebanen opened between Dombås and Støren, completing a direct standard-gauge line between Kristiania (Oslo from 1925) and Trondheim. The next day a train carrying a number of prominent Norwegians who had taken part in the opening celebrations was involved in the first bad railway accident in the history of the country.

King Haakon VII had opened the new line. He chose to remain in Trondheim for a few days but many of the dignitaries who had accompanied him had business back in Kristiania which could not wait that long, and they had to return to the capital. To cater for them a special train was arranged which the railway authorities in Trondheim had originally planned to have departed at midnight Sunday/Monday, after the arrival of the connecting train from Støren with the day train from Kristiania via the narrow-gauge Rørosbanen, this being train No. 361, which was due in Trondheim at 11.58 p.m. They were overruled by their headquarters, who wished for as early as possible an arrival in the capital and decreed that the special train, designated Litra D, should start from Trondheim at 11.45 p.m., so as to arrive back in the capital at 1.30 p.m. on Monday. Train No. 361 was to be crossed just outside of Trondheim at Marienborg at 11.52 p.m.

There were railway workshops at Marienborg and there were in fact two tracks between Trondheim and there but the original narrow-gauge track, now converted to a mixed gauge with the addition of a third rail, was only used to serve the workshops, while the new standard-gauge track was worked as a single line. There was no station at Marienborg, just a crossing loop on the standard-gauge line, with semaphore innkjør (home) signals at each end and a hut for the signalman, from where the signals were operated. A little north of the station the old and new lines passed through separate tunnels at Nidareid.

On duty at Marienborg on the night of the accident was Telegraphist Peter Wiig. It was difficult for one man to deal with the crossing of two trains on his own as the points

770 *The Times*, London, England, 27th and 30th July 1921
771 Indian Railways in 1921-22 Administration Report Vol 1

at each end of the loop were worked by levers on the spot, so he had asked for assistance, but without success. Also the signals, he considered, were rubbish; drivers often complained that they showed half green and half red, especially that at the north end, though the one at the south end could not be relied on either. He was anxious that night not to delay the southbound special. Selsbakk, the next station on the south or Kristiania side, reported to him that train No. 361 had left there, and he advised it forward in advance to Skansen on the Trondheim side, Skansen in return reporting that train Litra D had left there. The points at the north end of the loop were set and locked for the main track and he cleared the south semaphore for train No. 361 and went to the south points, 50 metres from his cabin, and set them for the crossing loop. Afraid that the driver of the northbound train would not find the signal showing a proper clear and thus stop unnecessarily, delaying the special, he had then positioned himself at the south points displaying a green handlamp to the approaching train to assure the driver that it was clear for him to enter the loop. He was sure that the driver of train No. 361 was aware that he had to cross train litra D at Marienborg. Train 361 came in slowly; he restored the points for the southbound train and picked up his lamp. Turning round he saw the train proceeding through the north points and out onto the single line towards Skansen and Trondheim. Going pale, he made his way hurriedly back to his hut, shouting as he did to a pair of railwaymen who were stood nearby, "They should have crossed the extra here!" Back at the hut he had telephoned Skansen, 900 metres away. "Stop the extra train," he had shouted, "train No. 361 has passed here!"

"The extra has just passed here," came the reply. With that, there was nothing more to do.

Train Litra D consisted of Class 30b engines Nos. 364 and 365, a luggage van, six sleeping cars and a conductors van, with Driver Aksel Flataas and Fireman Sigurd Dahl on the pilot engine. There were 96 passengers on the train. The train set off somewhat late at 11.54 p.m. It crossed the Ravnklo bridge, passed through Skansen station, where the signal was clear, and entered Nidareid tunnel. The tunnel was full of smoke and as the leading engine left it the driver opened the window to try to see through the smoke.

Train No. 361 was in charge of Conductor Halvor Skott with Driver Martin Øien and Fireman Karl Stuevold. The three had taken over the train at Tynset and before they did so the driver and conductor went through their notices, of which, with the opening of the new line, there were a large number. Conductor Skott saw that there was one which referred to an extra train to be crossed at Marienborg but seeing that the notice was headed '*Natt til mandag 19*' read it as meaning that the crossing was on Monday night and remarked that that would be one for his colleague Harsjøen tomorrow. Terms such as *natt til Fredag* – night to Friday – meaning Thursday night and *natt til Mandag* – night to Monday – meaning Sunday night are commonly used in modern Norwegian but were rather new in 1921, hence the confusion. His driver apparently did not see the notice at all or else also misread it.

At Støren the crew and passengers of train No. 361 had to change onto a standard gauge train for the journey on to Trondheim. The new train, also No. 361, consisted of Class 21a engine No. 182, a luggage van, a brake van and two carriages. The stop at Støren should have only lasted two minutes according to the working timetable. The change-over took a good deal longer than two minutes and the train left Støren 20 minutes late, but Driver Øien was confident that the time could be regained on the run to

Trondheim. At Selsbekk train No. 448 from Trondheim was crossed and Conductor Flaate climbed onto the engine for the onward trip to Trondheim. As they proceeded the conductor asked the driver and fireman where they were going to cross the extra train.

"What extra train?"

Driver Øien had replied, "I don't have any orders about any extra trains."

"Oh, I just thought I had heard something about one," the conductor said, and the matter was dropped.

As train No. 361 slowly approached Marienborg the south semaphore was visible showing green from a good distance. "Clear," the fireman and conductor said to one another, and Driver Øien put on a little steam. Shortly another green light was seen, which all three men on the engine took to be the signal given by the telegraphist from his hut that they were clear to proceed to Skansen, not noticing that it was in the wrong place, on the right-hand side of the track. In the conductor's section of the luggage van was Conductor Skott, while Conductor Rugelskjøen was busy sorting luggage in the other part of the van. Conductor Skott lowered the window and saw the south semaphore was showing green. To keep out the smoke he then closed the window momentarily and when he attempted to lower it again to see the *klart gjennom* (clear through) hand signal at the telegraphist's hut the window jammed and would not open. By the time that he had got the window open his van had passed the hut. There was no signal to be seen but he knew that telegraphists often did not continue to exhibit the signal until the whole of the train had passed, as they were meant to.

As train No. 361 reached the north end of the crossing loop there was a jerk as it ran through the locked points in the wrong direction. Driver Øien shouted that there was something wrong with the points and they looked out but could see nothing. Nevertheless, he applied the brakes a little and glanced back at the north semaphore, which had wrongly shown "Clear" a couple of weeks earlier, but it was properly at red. The next moment he looked ahead and saw the bright white light in front of the leading engine of train Litra D, emerging from the new Nidareid tunnel, 200 metres away. Driver Øien jammed on the brakes and opened his whistle while the fireman and conductor jumped from the engine. The oil lamp of train No. 361 was considerably less powerful than the electric light of the special train and Driver Flataas only saw it at the last moment. He was just applying the brakes when the trains collided at 11.57 p.m.

Engine No. 182 was the most seriously damaged of the three and Driver Øien was badly hurt. He managed to get down from his engine but then lay on the ground, unable to move. Frøken Lutz, sat on her verandah with her sister, witnessed the collision and ran to the scene. "Get the fireman to see to the boiler," the driver groaned to her. The fireman found that he could not do anything due to damage caused by the collision, but the fire was low anyway and there was no danger of a boiler explosion. In the local train the luggage van was the only vehicle much damaged; Conductor Skott was quite badly hurt, Conductor Rugelskjøen less so. The extra train suffered worse; its leading luggage van was thrown against the side of the cutting and the two following coaches telescoped one another.

Six passengers died in the accident and ten passengers, and three railwaymen were injured. Those who lost their lives were: Captain Thoralf Bjørnstad, an official from the State Railways headquarters; Architect Erik Glosimodt, who had designed the stations at the south end of the Dovrebanen; Senior Engineer H.G. Hammer, district engineer for the Sørlandsbanen; Telegraph Director Thomas Heftye, a former defence minister who had held a number of other positions; Major N.C. Ræder, former chief of the General Staff communications department, and Colonel Nils Johannes Sejelsted, director of the Norges Geografiske Opmaaling (map agency).

Editor Nordahl-Olsen of *Tidens Tegn* newspaper was a passenger in the first sleeping car. He described his experiences: "...We were coming from the wonderful celebrations held by the Trondheim authorities and the atmosphere was high. As soon as I got to my cabin I took off my tailcoat and white tie and changed into a blue hunting jacket and black tie to feel more relaxed and comfortable. With that the whistle blew and the train set off amidst waves and farewell shouts from those on the platform. Journalist Wigum, who had stood alongside me at the window as we left the station, joined me in my cabin and we sat on the edge of the bunk and chatted. We had not been sat very long when out journey was fearfully interrupted. There was a tremendous crash, the lights immediately went out and we were thrown about. 'God help us!' I cried out to Wigum, 'We've gone into the river,' and we waited expecting water to come flooding over us at any moment. We were both completely calm. Glass and bits of wood flew about us, and I felt a severe blow to my face and blood running down to my chin. I found that I had suffered a rather deep cut in my lower lip. How I got it I don't know.

"When everything had calmed down we tried to orienteer ourselves. Wigum said that he was fine, but he had bled heavily from his head. Otherwise he was unhurt. I realised that I still had all of my limbs. When the disaster occurred Wigum was sat to my right, and we were sat facing the engine. A little light coming through the window then showed us that Wigum had been thrown up in the air and I was beneath him. We talked quite calmy to each other.

"The hole above Wegum was large enough for him to thrust a hand out of and wave for help but it was impossible for us to get out by ourselves. Then I began to worry more and more that the carriage could catch fire. One coach of the train was lit by acetylene gas, though the remainder of it had electric lighting. When the first rescuers arrived we thought that their lights were the beginning of a fire. Then we shuddered, it cannot be denied."

After the accident it took some time for the rescue operation to get under way. Some young men who were soon on the scene lacked tools, but a shopkeeper named Walseth borrowed a car and emptied his store of axes, saws, and files. Later a rescue train arrived from Trondheim. Lighting the scene was a problem at the start, with only paraffin lamps, but later carbide lamps were set up to illuminate the scene. It took some time to realise the extent of the disaster, and that what appeared to be one coach was in fact two telescoped together. It was also a laborious task to cut into the wreckage and it was close on 4 a.m. before the last trapped passenger could be released. The injured were taken to Trondheim and St Elisabeth's hospitals. King Haakon was told of the accident at 1 a.m. and went to the scene, remaining there until 4.30.

Immediately news of the disaster became public there were naturally demands to know how it had happened and it soon became apparent that it was the local train which had been at fault. Driver Øien was seen by a newspaper reporter before he was taken to hospital but all he would say was, "*De viste klartsignal*" (they gave me a clear signal). The newspapers soon began to cast the blame beyond the crew of train No. 361. It was questioned why the departure of train Litra D had been brought forward 15 minutes, why there had been only one person on duty at Marienborg, what if almost the whole of the Norwegian government had been on the train and had lost their lives in the accident and so on.

Initially six railwaymen were charged with negligence in connection with the accident but the charges against Conductors Flatte and Rugelskjøen were later dropped. The case against Conductor Skott, Driver Øien, Fireman Stuevold and Telegraphist Wiig was proceeded with in February 1922. It emerged that the expression '*natt til*' had hardly ever been used in train notices previously and never for trains departing before midnight. The case ended with the acquittal of all four railwaymen. All four remained in the railway service; Conductor Skott was later promoted to Head Conductor, Telegraphist Wiig later became a stationmaster, Fireman Stuevold became a driver and Driver Øien continued as such until his retirement.

As a result of the accident slight alterations in the regulations were made in order to try and avoid a recurrence. It was laid down that drivers and conductors must confer and discuss the journey before it began and that the tracks used by trains at each station were to be specified in the working timetable, rather than being decided by the stationmasters on the spot. [772]

The Philadelphia and Reading Railroad experienced a major disaster near Woodmont, Pennsylvania, United States, on the morning of 5th December 1921. The circumstances of the accident were complicated. Woodmont was a station on the single-track Newton branch, which left the main line between Philadelphia and Trenton at Cheltenham, running north-east for 16½ miles to Newton. The branch was worked by timetable and train orders, with eastbound trains being superior. There was also an automatic block system between Cheltenham and Bryn Athyn, using enclosed disc signals, and a manual block system between Bryn Athyn and Newton with an intermediate block station at Churchville. Just over half a mile on the west (Cheltenham) side of Bryn Athyn was Huntingdon Valley station, where there were situated automatic signals 713 for eastbound trains and 713-A for westbound trains. At Bryn Athyn station there was a semaphore block signal applying to eastbound trains, a flag being used to give orders to a westbound train. Just east of the station was a spur track, the entrance to which faced towards Newton, used to cross trains, and to the rear of the connection to the spur track was automatic signal 716, the first applying to westbound trains.

On the day of the accident westbound passenger and milk train No. 154 was in charge of Conductor Stout with Engineer F. Rock and consisted of engine No. 265 and five vehicles. It had left Newton for Philadelphia at 6.50 a.m. and stopped at signal 716 at 7.28 a.m. The signal was at danger and in any case the train did not have sufficient time to reach Huntingdon Valley to cross eastbound train No. 151, which was superior by direction. A flagman was sent ahead and after an appropriate interval the train pulled forward to Bryn Athyn station to allow passengers to get on and off and its crew were given Form 19 train order No. 9, which read:

Disregard signal 716 and run carefully. Complete 7.31 a.m.

[772] *På Sporet* (Journal of the Norsk Jernbaneklubb) No. 32 (December 1981)

Train No. 154 was too long for the spur track, so it was then backed to a point sufficiently far on the east (Newton) side of the connection to the spur so as to allow train No. 151 to back into it. This was duly done and train No. 154 then proceeded westwards without stopping again at Bryn Athyn.

Behind train No. 154 was westbound train No. 156, in charge of engine No. 278, four coaches and a combination coach and baggage car, all of wooden construction, in charge of Conductor Smith and Engineer J. Rock. It left Newtown for Philadelphia at 7.30 a.m. and arrived at Churchville at 7.41 a.m. The block signal here was at "Stop," and Conductor Smith was given Form 31 train order No. 11, reading:

No. 151 will meet No. 156 at Bryn Athyn. No. 151 take siding. Complete 7.44 a.m.

The conductor showed the engineer the order, the block signal was cleared, the train left at 7.45 a.m. and collided with eastbound No. 151 in a rock cutting near Woodmont and about 1¼ miles short of Bryn Athyn at about 7.55 a.m. The combined speed of the two trains when they met was between 55 and 65 miles an hour. In the collision both engines were forced up into the air and wrecked. The combination car of train No. 151 was partly telescoped by the tender of the engine and the second coach was partially telescoped by the combination car. In train No. 156 the leading coach was completely telescoped by its engine and the next coach was partially telescoped. Fire broke out immediately after the collision and completely destroyed the leading vehicles of each train and damaged others. It was believed that it was the fire rather than the initial collision which caused many of the fatalities. Twenty passengers, the firemen of each train and three railwaymen off duty were killed and 65 passengers and five railwaymen were injured.

Eastbound train No. 151 from Philadelphia to Newton consisted of engine No. 167, a combination passenger and baggage car of wooden construction and two passenger coaches of wooden construction with steel underframes. It was in charge of Conductor Evans and Engineer Yeakel. At Huntington Valley it found automatic signal 713 at danger. A flagman was sent ahead, and the train followed him to Bryn Athyn, station. There the block signal was at danger and train No. 154 was stopped facing them on the Newton side of the connection to the spur track. Conductor Evans told his engineer to pull forward and back into the spur track, but the latter refused to do so without a train order, so the conductor went to the office and received Form 31 train order No. 11, reading:

No. 151 will meet No. 156 at Bryn Athyn. No. 151 take siding. Complete 7.46 a.m.

Train No. 151 then moved forward and backed into the siding, train No. 154 proceeded westwards, and train No. 151 then proceeded eastwards at, according to the record, 7.46. A stop was made at Paper Mills station, and it collided with train No. 156 half a mile further on.

Conductor Evans told the Interstate Commerce Commission (ICC) inquiry into the accident that when he had received the order at Bryn Athyn he had taken it to apply to train No. 154. He gave the order to his engineer, but they did not discuss it and he did not show the order to the other train crew. As soon as train No. 154 had left his train proceeded, despite the block signal still being at "Stop," and without a clearance card or other permission from the operator. He considered that having the train order had been sufficient.

Engineer Yeakel said that when his conductor had handed him the order he had just glanced at it and like the conductor had assumed that it applied to train No. 154.

Operator Clayton at Bryn Athyn gave 'conflicting and vacillating statements.' He said that Conductor Evans had read the order in front of him, apparently correctly. Initially he stated that he had not reported train No. 154 back to Churchville and Churchville had not offered train No. 156. In a second statement he admitted clearing train No. 154 back to Churchville as soon as it had passed automatic signal 716 and had accepted train No. 156 under a clear block. He had not noticed train No. 151 leaving until it was too late to stop it. He admitted that had he waited for train No. 154 to reach his station before clearing it back the accident might have been averted. He could not explain altered entries in his block record book for trains 151 and 154.

Operator Tomlinson at Churchville also made conflicting statements. Firstly he said that he had allowed train No. 156 to proceed before Bryn Athyn had cleared train No. 154 on the understanding that that train was at Bryn Athyn station. Later he stated that Bryn Athyn had in fact cleared train No. 154 while the conductor of train No. 156 was in his office. His block record book only began on 1st December. When he was required to produce the old book it was found that pages after 30th November were missing, while more columns were filled in in the new book relative to the old one. Eventually he admitted that he had made a copy the original for the sake of the investigation and had destroyed several pages of the original book.

Conductor Stout of train No. 154 was clearly unclear whether the block for westbound trains ended at Bryn Athyn station or automatic signal 716. Engineer F. Rook thought that the train needed to clear Bryn Athyn station before it could be cleared back to Churchville and Engineer J. Rook of train No. 156 agreed.

Dispatcher Rich testified that there had been problems with the telegraph wires at another point and when the conductor of train No. 154 had reported that automatic signal 716 was at danger he had assumed that the problem must be affecting all the signals in the area and had issued train order No. 9 to avoid delay. He admitted that the signal might have been at danger because opposing train No. 151 was in the section (as was, of course, the case). Subsequent inspection by the signal maintainer showed that the signal was working properly.

The ICC concluded that the accident had resulted from the failure of Engineer Yeakel and Conductor Evans to obey train order No. 11, probably because they had been anticipating what it would say. Had Engineer Yeakel read the order aloud in the presence of the conductor the mistake in thinking that the order applied to train No. 154 would doubtless have been discovered. Had Conductor Evans had a proper understanding of the rules he would not have allowed train No. 151 to leave Bryn Athyn without a clearance card.

It was also clear that train No. 156 had received a clear block signal at Churchville while train No. 154 still occupied the Churchville – Bryn Athyn block section. Train order No. 11 had been completed to Churchville at 7.44 a.m. and train No. 156 had left there shortly after. The same order was made complete to train No. 151 at 7.46 after which baggage had to be dealt with, the train move forward and then set back into the spur track before train No. 154 could move forward past signal 716. The statements of Operators Clayton and Tomlinson were so conflicting and their block records of so little value that it was impossible to determine if the Bryn Athyn operator had cleared the block before train No. 156 left Churchville or if the Churchville operator had let the train go with a clear block signal knowing that the block as still occupied.

The ICC commented that, 'The investigation of this accident disclosed that in this instance all the safeguards provided for the operation of trains were rendered ineffective by errors on the part of employees involved, misunderstanding of operating rules and improper practices which had grown up in service.'

Apart from the specific failures which had led to the collision there was a general laxity in the observation of block signal rules.

Notwithstanding the requirement for a positive block to be maintained for opposing trains the timetable provided for regular meets between scheduled trains at Southampton, a non-block station between Bryn Athyn and Churchville, trains being allowed to enter the section on a clearance or caution card.

There was no clear and uniform understanding as to whether the siding at Bryn Athyn was within the station limits or in the Churchville – Bryn Athyn block section or whether train No. 151 needed a clearance or caution card to pass the block signal to go into the siding.

The Bryn Athyn train order signal was not clearly defined and there was conflicting evidence as to whether the block signal or a flag should be used for eastbound trains.

Train No. 154 was given a train order to disregard automatic signal 716 on the assumption that it had failed when it had not, creating a dangerous situation for no other reason than to avoid delay. This could lead to a disrespect of all automatic block signals and could not be too strongly condemned.

There was no system for the regular re-examination of employees. Some had not been examined since 1914 and the dispatcher had never been examined.

The ICC stated that, 'The investigation disclosed extremely lax supervision by responsible officers over the conduct of employees in the observance of ordinary safeguards in train movement. There was no system of instruction of employees to provide a correct and uniform understanding of the rules, nor any effort on the part of responsible officers to correct existing operating practices which were contrary to rules and a constant menace to the travelling public. The laxity with which the manual block system was operated on this line is a condition which could have been easily discovered had any attempt been made by officials to check it up.'

The ICC considered that the manual block system which was nominally in effect on the line was entirely inadequate in view of practices revealed by the investigation to furnish the required protection. It recommended that the carrier take prompt measures to ensure that employees properly understood and obeyed the operating rules provided to safeguard train operation and that a complete automatic train control system be installed over the whole line.

Most of the rolling stock of both trains had been of wooden construction with Pintsch gas lighting. It was possible, though there was no direct evidence, that the gas had contributed to the post-accident fire which caused most of the deaths. Had all-steel equipment been in use the fire would probably have been prevented. The railroad company had not bought any wooden cars for many years and had been buying 50 steel cars a year up until the war, since when the company pleaded that its financial position had prevented the purchase of new equipment. The commission pointed out that the accident was likely to cause a considerable financial burden on the company which could have been avoided with the use of steel cars. It recommended the substitution of all-steel rolling stock as soon as possible and that in the meantime wooden cars should not be marshalled at the head of trains or between steel cars. [773]

Despite its severity the accident seems to have received little attention in the press, perhaps because it only involved a pair of local trains.

An accident near Martin, Utah on Monday 28th August 1922 is described in the Tunnel Accidents chapter.

A major disaster occurred in Cuba on Sunday 6th May 1923 on the line of the Hershey Electric Railway, a standard-gauge single-track interurban line between Havana and Matanzas. A few Americans were passengers on one of the trains, including Messrs George A. Glenn and C. J. Schwingle, described as appraisers from Milwaukee. Mr Glenn described his experiences of the accident. “My partner and I were on our way to Havana from Matanzas. Our train should have taken the siding for the one from Havana to pass. We were travelling about 45 miles an hour when we began to round the curve. When the impact came, I turned a complete somersault, and my companion was knocked unconscious. I carried Schwingle to safety. We were in the rear of the last car, which was not crushed, although the front portion had been telescoped.”

Mr Glenn added that fire had broken out immediately after the collision and the wooden coaches of the trains had burnt very easily. The two motormen survived the accident; the one on the train from Havana was thrown through a window and escaped unhurt, while the motorman on the train from Matanzas was only slightly hurt. Less fortunate were the conductors, the man on the train from Matanzas was killed and his colleague on the train from Havana was possibly fatally injured. At least two women and a child were pinned in the wreckage and burned to death and in all 25 people reportedly lost their lives and over 50 were injured. The trains collided near Canasi. A special train took most of the injured to Matanzas. [774] Some accounts put the number of fatalities as between 12 and 15, while at least one report stated that there were 50 dead. [775]

As a result of the collision the power supply circuit breakers were tripped, cutting the current, but an inexperienced employee at the control centre, unaware that should this happen the current should be left off as the loss of power might have been due to an accident, apparently repeatedly restored the current, electrocuting some of the passengers. [776]

How the accident came about does not seem to have been reported.

---

[773] Interstate Commerce Commission report

[774] Syracuse Herald, Syracuse, New York, United States, 7th May 1923

[775] The Creston Advertiser-Gazette and Plain Dealer, Creston, Iowa, United States, 7th May 1923

[776] Webpage – Excelencias del motor.com: Tren de Hershey ¿con sabor a chocolate?

On Monday 13th August 1923 northbound passenger train No. 6 on the Atchison Topeka and Santa Fe Railway (ATSF) consisted of engine No. 1349, two baggage cars, a smoking car, a chair car, a coach, four Pullman sleeping cars and two business cars, in charge of Conductor Pegan with Engineer A. Hansen and Fireman J.T. Pearson. [777] At Rocky Ford, 17½ miles south of Fowler, Colorado, United States, the train received order No. 318, on Form 19, which in part provided for a meet with train No. 13 at Elder, just over 3 miles south of Fowler. Then at Manzanola, 5½ miles short of Elder, they were looped another Form 19 order, No. 324, which read:

No. 6 eng. 1349 meet second No. 13 eng. 351-327 coupled at Hamlet instead of Elder.

The train left Fowler, 4½ miles short of Hamlet and the last open telegraph office, at 4.30 a.m., and had only covered half a mile when running at between 20 and 30 miles an hour it collided head-on on the single line with southbound train second No. 13.

Because of washouts on their own line southbound Colorado and Southern Railroad passenger train No. 609 was being detoured over the ATSF between Pueblo and Trinidad, running as train second No. 13. It consisted of engines Nos. 351 and 327, a baggage car, a mail car, a baggage car, two coaches, three Pullman cars and two business cars, in charge of Conductor Staley with Engineer T. Schmanke and Fireman Heath on the pilot engine and Engineer C.H. Gray and Fireman G.L. Chowning on the train engine. This train received order No. 324 at Nepesta, just over 7 miles from Fowler, and passed there at 4.22 a.m. Hamlet was a blind siding, 2¾ miles from Nepesta. The train passed there without stopping and collided with train No. 6 while travelling at 35 or 40 miles an hour. The second car of train No. 13 and the second and third cars of train No. 6 were telescoped in the collision. Engineer Hansen, Fireman Pearson, and a baggage man on train No. 6 and Engineers Schmanke and Gray of train second No. 13 were killed and 41 passengers, four railwaymen, an express messenger, a mail clerk, and a news agent were injured.

Conductor Staley told the Interstate Commerce Commission (ICC) inquiry that he had not been familiar with the ATSF line. A Santa Fe crew had been on the pilot engine. He had received the train order at Nepesta, had noted the distance from there to Hamlet and calculating that it would take 4½ to 5 minutes to cover that distance he had been waiting for the whistle signal indicating that the train had reached Hamlet. Not having heard any whistle after six or seven minutes and his train not reducing speed he had become uneasy. He had indicated a meeting point to the engine crew, got no response, signalled to the engine crew again, then seen train No. 6 approaching. At this point he had opened the conductor's valve and the collision had occurred a few seconds later.

Fireman Heath stated that he had taken the order at Nepesta and handed it to Engineer Schmanke, who had remarked, "We meet them at Fowler." He (the fireman) had then read the order and also read it as 'Second No. 13 meet No. 6 at Fowler instead of Elder.' Nearing Fowler he had seen a light which he had at first supposed was an automobile on the highway before realising that it was the headlight of train No. 6. His engineer had applied the brakes at about the same moment, but he was not sure if this was because he had seen the opposing train or was merely preparatory to stopping at Fowler. He had shouted to the engineer and jumped.

Fireman Chowning on engine No. 327 said that he had taken the train order at Nepesta

[777] *The Davenport Democrat*, Davenport, Iowa, United States, 13th August 1923

and given it to Engineer Gray, who had read it while he had worked on the fire. His engineer had not given him the order to read nor told him what it said. Looking out he had seen a light, shouted, and jumped. He had not noticed any brake application or whistle.

Conductor Pegan told of receiving the order at Manzanola changing the meet to Hamlet, of his train gaining speed after leaving Fowler and the brakes then being applied in emergency.

Flagman Burch of train No. 6 spoke of meeting Fireman Heath after the collision, when the fireman had said to him, "We had orders to meet you here, I think," that is, at Fowler.

Operator Cole, at Nepesta, spoke of hooping the train orders to the two trainmen of train second No. 13.

Operator Burkepile, of Fowler, spoke of hooping a permissive card to the engineer, conductor, and flagman of train No. 6 and said that he had been picking up the hoops when he had heard the noise of the collision.

The ICC found Engineer Schmanke and Fireman Heath responsible for failing to obey the train order. The copy of the order held by Engineer Schmanke could not be identified afterwards owing to its condition but those held by the conductor, flagman and operator were legible and it was hard to account for his and Fireman Heath's mistake. The accident had again resulted from a failure of the human element and would have been prevented by an adequate train control system. [778]

At about 04:20 on Friday 12th December 1923 No. 12 down passenger train and No. 111 up goods train collided head-on on the single line near the up outer home signal of Tarki station, India, on the Rawalpindi – Jhelum line. In a familiar scenario the two front coaches of the passenger train were telescoped into one another and caught fire from escaping gas. The engine of the passenger train and the leading engine of the goods train were badly damaged. Seven passengers lost their lives and 32 were seriously and 15 slightly injured. The Tarki stationmaster was found to have been to blame for the accident and was prosecuted. [779]

Two interurban trains of the Union Traction Company of Indiana collided head-on near Ingalls, Indiana, United States, on Saturday 2nd February 1924. At least 16 passengers died, and 35 passengers and one railwayman were injured.

The Interstate Commerce Commission (ICC) inquiry into the disaster noted that the single-track line was operated by timetable and train orders. An automatic block system was also installed but had been out of use since 27th December of the previous year, thus for just over a month. There was no superiority of trains, but their meeting points were assigned in the working timetable. Should a train arrive at a meeting point and the opposing train not be there the crew contacted the dispatcher by telephone for orders. The train orders were made on a special form, 19 and 31 orders not being used. Two copies of

---

778 Interstate Commerce Commission report
779 Report by the Railway Board on Indian Railways for 1923-24

orders were made, the conductor giving one copy to the motorman and retaining the other. If there was a trailer the conductor of the leading car showed the orders to the conductor of the trailer.

The trains involved in the accident were eastbound No. 24 and westbound No. 21, due to cross according to the timetable at Mur at 3.50 p.m. Both trains had trailers, and both consisted of wooden cars with steel underframes. Conductors Sawyer and Campbell were in charge of train No. 24, with Motorman Flinn and Conductors Mitchell and Engle were in charge of train No. 21, with Motorman Anthony. Train No. 24 left Indianapolis at 3.17 p.m., 17 minutes late. At Lawrence it met train No. 917, and the conductor of that train gave the crew of train No. 24 order No. 45 directing them to meet a train at Goul. The dispatcher and the crew of train No. 45 stated that this order related to train No. 23, which was following train No. 21, but the crew of train No. 24 read it as relating to train No. 21. They thus departed from Lawrence, proceeded past their scheduled meeting point with train No. 21 at Mur without stopping, passed Dent, and collided with train No. 21 near Ingalls.

Train No. 21 was also late, being 18 minutes behind schedule when it left Anderson at 3.33 p.m. At Pendleton it received train order No. 46, providing for a meet with train No. 24 at Dent. Had the crew of train No. 24 contacted the dispatcher at Mur, as they should have done when train No. 21 was not there, they would have been given order No. 46 at that point. The motor car of train No. 24 telescoped that of train No. 21 and fire broke out in the wreckage soon after the collision, destroying both motor cars and the trailer of train No. 21. Many of the passengers died in the fire; the exact number who died was not known.

Trains Nos. 23 and 24 normally met at Belt, but as train No. 24 was late Dispatcher Boyle decided to advance train No. 23 to Goul with train order No. 45. The order had first to be delivered to the restricted train, in this case No. 24, so the dispatcher sent the order to Lawrence, where it was received by Conductor Danforth of train No. 917 for delivery to train No. 24. Conductor Danforth wrote down the order as for trains Nos. 23 and 24 to meet at Goul, registering as train No. 24, and Motorman Kapp of train No. 917, who was with him in the telephone booth, confirmed that it had been plainly written and repeated back to the dispatcher correctly. Conductor Danforth gave his own name to the dispatcher but wrote the names of Conductor Harris and Motorman Standish on the order, these being the employees he anticipated would be working train No. 24. The latter train shortly arrived, and Conductor Danforth said that he had given both copies of the order to Motorman Flinn, telling him, "Here is an order for you to meet train No. 23 at siding Goul." Motorman Kapp confirmed this, and both stated that Motorman Flinn had repeated the order correctly. Conductor Danforth acknowledged that he should have given one copy of the order to Conductor Sawyer.

Motorman Flinn alleged that the order as received by him had read:

No. 24, car No. 286, will meet No. 21, car No. 275, at Goul.

He agreed that the order had been well written. He said that he had read the order to Conductor Sawyer and then placed both copies in the train number sign box, where they were destroyed in the post-collision fire. Conductor Sawyer at first more or less agreed with this but later stated that he had just seen the orders handed to the motorman and the train had started. Conductor Campbell in the trailer did not see the orders at all, nor had he asked to see them. Motorman Flinn expressed the view that the order he had been given at Lawrence had been incorrect.

Dispatcher Boyle testified that after issuing train order No. 45 to the crew of train No. 917 to deliver to train No. 24, he had issued order No. 46 to train No. 21 at Pendleton to meet train No. 24 at Dent, and the following train No. 26 at Lawrence. In error he had written the point at which the latter order was issued to train No. 21 as Mur, the regular meeting points of trains Nos. 21 and 24, but he said that this oversight had had no bearing on the order, and he defended his issuing of order No. 45 to train No. 24 via the conductor of train No. 917 at Lawrence. It appeared that the issuing of orders to one train crew via another crew was common practice, and the practice was approved by Superintendent of Transport Keever and General Manager Nicholl. Mr Keever referred to Rule No. 218:

To issue train orders at stations where there are operators or agents, the dispatcher will instruct the operator or agent to display a stop signal for a train to stop for orders. A train must not pass a station where such signals are displayed without reporting to the agent or train dispatcher and until such signals have been taken down, removed, or cleared. When a train order is given through an agent, he must read it back and complete it in the same manner as a motorman or conductor. The conductor and motorman receiving such an order must read it aloud back to the agent, and receipt for same by writing their names upon it before detaching the copies intended for them.

The trainmen were apparently considered to be 'agents' as far as this rule was concerned.

It was stipulated in the timetable that all trains between Indianapolis and Anderson must register at Long and Lawrence, but the crew of train No. 24 had not registered at Lawrence, nor had they checked the train register. The practice of one crew registering for another had apparently existed for years. Conductor Danforth said that there was no rule authorising this but pointed out that nor was there any rule specifically prohibiting it. Dispatcher Boyle and Superintendent Keever agreed.

Little importance was, it seemed, placed on the name of the person signing a train order. Motorman Kapp stated that he would accept and order with some other crew's name on it and it was customary to do so. Dispatcher Boyle considered that any name might be put on the order and Superintendent Keever that the name on the order meant nothing.

Superintendent of Power Distribution Penisten stated that the automatic signals had been taken out of use to carry out alterations which it was not practicable to carry out while they were in use. They were about to be brought back into service at the time of the accident. Regarding failures of the 50 automatic signals Mr Penisten said that there had been 125 during October, November, and December 1923, compared with 166 for the corresponding period in 1922. He considered the figures comparatively good. There was no fixed programme of signal maintenance, the work being done as required.

The ICC investigation showed that it was commonplace for dispatchers to issue orders for trains to pass automatic signals at danger (see 1921 Woodmont accident above). A check of dispatcher's books showed for example that between 27th September and 5th October 1923 over the 30½ miles between Long and Belt 26 such orders had been issued to 40 trains, while between 30th November and 1st December of 288 orders issued 82 were orders to disregard signals.

General Manager Mitchell, asked about the average of approximately one signal failure per day, replied that he did not think that was bad for an electric line.

In conclusion the ICC stated that the immediate cause of the accident was an error in the transmission, receipt or reading of a train order. The underlying cause was lax operating practices. The orders in possession of the crew of train No. 24 had been destroyed in the fire. For Conductor Danforth to have mistakenly copied the order he would have had to wrongly write down both the train number and the car number. If the dispatcher had made the mistake this would not explain why Conductor Danforth and Motorman Kapp had read the order correctly, while there were no erasures or alterations in the train dispatchers book. The weight of evidence therefore suggested that Motorman Flinn had misread the order. Conductor Sawyer did not see the order but was equally responsible for the error. Had the order been directly received by the crew of train No. 24 the accident would have been averted.

According to the rules, as train No. 24 was more than five minutes late it should have reported to the dispatcher at least two sidings short of its meeting point and it was incumbent upon the dispatcher to issue such orders or instructions, but this was not done by the dispatcher at Lawrence to avoid further delay, instead he had intended to issue the order at Mur.

Had the automatic block system not been suspended, or had a manual block system been substituted, the accident would probably have been averted.

The commission listed the various lax operating practices which they had uncovered:

(1) Trainmen taking orders for other crews.

(2) Trainmen checking registers and registering for other trains.

(3) Trainmen signing the names of the crew to whom they expected the order to be delivered.

(4) Poor maintenance of the automatic block signals.

(5) Almost daily train orders for crews to disregard signal indications and proceed with caution.

(6) The operation of trains possessing caution orders at schedule speed.

(7) The rules for automatic block signals were not in the rule book.

In addition, the commission's inspectors had observed:

(1) Signal mechanisms showing want of proper care and attention.

(2) Trains operating at night at full speed over main track switches, the lights of which were extinguished.

(3) A facing point switch at a location where trains run at full speed on a falling gradient was not connected to the signal system.

(4) Operation of a train at night as a first section but with no electric lights indicating this because they were out of order.

(5) A train shown as having two different meeting points with an opposing train in the working timetable.

(6) The use of the dispatcher's wire for miscellaneous personal conversations

(7) An order marked as OK in the train dispatcher's book which had not actually been sent to or repeated back by the train crew.

(8) Numerous alterations and erasures relating to orders in the train order book.

In a scathing paragraph at the end of the report Director W.P. Borland wrote, 'While it is believed the immediate cause of this accident was the failure of a crew to read a train order correctly, underlying that is the extreme carelessness and neglect shown in general operation. It is scarcely conceivable that any official having due regard for the safety of those who travel on his railway, or those who are employed thereon, should permit the existence of the conditions which this investigation disclosed. And yet General Manager Nicholl, who is the responsible operating officer of this railway, stated that he contemplated no changes in the operating practices. Such an attitude on his part is not only inexcusable but is a striking indication, if one were needed, for the general disregard of first principles of safe operation manifested by officials and employees responsible for train operations. This investigation discloses that action should be taken forthwith to do away once and for all with the many dangerous, careless, and lax operating practices existing on this road. If this is not done, accidents will again occur with possibly greater loss of life than in the instant case as a direct result of the inherently dangerous practices apparently countenanced by the present operating officials.' [780]

On Friday 29th August 1924, a major catastrophe occurred in India, at a location given in the official report as mile 644/2, between Harappa Road and Mirdad Muafi stations and in the British press as Montgomery. The accident happened on the Lahore-Khanewal section of the North Western Railway and involved two passenger trains, Nos. 43 up and 24 down. The engines of both trains were badly damaged and four bogie coaches next to the engine of train No. 24 were telescoped while one bogie coach of train No. 43 was smashed and a second slightly damaged. No less than 107 passengers and two railwaymen were killed while 88 passengers and five railwaymen were injured. Sixty of the most seriously injured were brought to Lahore where the doctors had pooled their resources to deal with the situation. The first-class passengers, who presumably had been in the rear coaches, escaped death if not injury. [781]

The trains were meant to cross one another at Harappa Road but train No. 24 was running late, so Control decided to advance train No. 43 to Mirdad Muafi. The assistant stationmaster at Mirdad Muafi, on receiving notice of the alteration, cancelled train No. 24 to Harappa Road and accepted train No. 43. However, he had already made out an "Authority to Proceed" order for train No. 24 which had been left at the facing points and had cleared his signals for that train. He neglected to retrieve the order and replace the signals to danger before accepting the up train. The predictable result was that down train No. 24 picked up the order, ran through the station under clear signals and entered the single line section, its crew blissfully unaware that train No. 43 was on its way from Harappa Road.

The guilty railwayman was prosecuted, convicted, and sentenced to two years'

[780] Interstate Commerce Commission report

[781] *The Times*, London, England, 31st August and 1st September 1924

rigorous imprisonment for his negligence. [782]

The Illinois Traction System interurban line suffered an accident near Davis, Illinois, United States, on Friday 20th March 1925. A passenger train and a work train collided at about 1.05 p.m., killing six passengers and the motorman of the passenger train.

The single track involved was worked by timetable and train order with an automatic block system, but the signals were out of use on the day of the accident.

Southbound passenger train No. 7 consisted of wooden motor car No. 250 in charge of Conductor Roberts and Motorman William Beele. It had left Carlinville, 4½ miles short of the scene of the collision, at 12.55 p.m., 20 minutes late, and was practically at a standstill when struck by the other train 600 feet south of the south end of Davis siding.

Work extra No. 1513 was a line car in charge of Conductor Fore and Motorman Lindsey. It had been working at Benld, 10½ miles south of Davis, until 12.10 p.m., when it was given an order to run to Moody, the next station north of Davis. When the two trains collided the open platform at the front of the line car telescoped motor No. 250 for 25 feet.

Motorman Lindsey told the Interstate Commerce Commission (ICC) inquiry that they had started from Benld shortly after receiving the order to go to Moody. Approaching Loveless, 1¾ miles from Davis, he had asked Conductor Fore what the first train was that they were to meet, and the conductor had replied that it would be train No. 95, due at Davis at 1.49 p.m. He had not seen the passenger motor until two linesmen sitting on some reels loaded onto the front of the car had shouted a warning and jumped. Then he had seen the danger, placed the brakes in first the service and then the emergency positions and, he thought, reversed his motor.

Conductor Fore stated that at Clark, 3½ miles from Davis, he had looked at the timetable. Motorman Lindsey had suggested the first train they would meet would be train No. 95 and looking at the timetable he had agreed and remarked that they could easily get to Moody before that train. He thought that he and the motorman had seen train No. 7 about the same moment when the linemen jumped and he said that with the linemen sat where they were the motorman could not have seen the passenger train any earlier, though from his position he (the conductor) could have had he been looking out. Neither he nor the motorman had mentioned train No. 7, which they had both forgotten about.

Line Foreman Rek said that he and Lineman Larimore had been standing behind the reels loaded on the front of the car. They knew nothing of the train orders. He had seen the car of train No. 7 rounding a curve about ten car lengths away, shouted a warning and jumped.

Groundman Lewis, riding at the rear of the cabin of the line car, said his first warning of the accident was seeing the motorman starting towards the rear of the cab. He put the speed of the line car at 30 to 35 miles an hour.

The conductor of the passenger train was too badly hurt to make a statement.

A severe sleet storm on 18th December 1924 had caused much damaged to the infrastructure of the line, damaging telephone wires, trolley lines, automatic highway

[782] Report by the Railway Board on Indian Railways for 1924-25

crossing signals and so on and disabling automatic block signals over 121 miles. All but the 12 miles of signals between Gillespie and Carlinville (the stretch on which the collision had occurred) had been restored by the date of the accident.

The ICC reached the predictable conclusion that the accident was the fault of Conductor Fore and Motorman Lindsey, who had forgotten train No. 7, which would have passed Davis already at 12.44 had it not been late. Had the automatic signals been working or an automatic train stop, or train-control device been in use the accident would have been averted. [783]

On Sunday 22nd March 1925 there was an accident at Ricohoc, Louisiana, United States, on the line of Morgan's Louisiana & Texas Railway and Steamship Company, part of the Southern Pacific system. The line was worked by timetable, train orders and an automatic block system and the collision occurred 887 feet east of the east passing track switch, at about 3 a.m., in a dense fog.

The passenger trains involved in the collision were eastbound No. 12, the *New Orleans Limited*, and westbound No. 109, the *Sunset Express*. The trains were due to meet at Ricohoc at 2.56 a.m. and train order No. 3 required train No. 12 to wait there until 3.05 a.m. for train No. 109.

Train No. 12 consisted of engine No. 601, a combination baggage and mail car, a baggage car, two coaches, a chair car, a tourist car and three sleeping cars, in charge of Conductor Crane with Engineer Matthews and Fireman Thibodaux. It passed Bayou Sale, just under 4 miles west of Ricohoc, at 2.54 p.m., passed through Ricohoc before 3.05 a.m. and had almost stopped when it collided with train No. 109.

Train No. 109 consisted of engine No. 615 a combination baggage and mail car, a baggage car, two coaches, a chair car, a tourist car and four sleeping cars, in charge of Conductor Dunshie with Engineer Conery and Fireman Frank Newby. It had left Patterson, just over 5 miles east of Ricohoc, at 2.52 a.m.

Both engines were derailed and badly damaged in the collision. In train No. 109 the engine tender telescoped the combination baggage and mail car for about 20 feet. Worse, in train No. 12 the baggage car telescoped the coach behind it, the Jim Crow coach reserved for black passengers, over two-thirds of its length. Six passengers, five railwaymen and two other persons were killed and three passengers, six railwaymen and four other persons were injured. One of those listed in the Interstate Commerce Commission report as an 'other person' was newsboy Frank Stafford, making his first trip. [784]

Conductor Dunshie of train No. 109 stated that his train had had ample time to get to Ricohoc to meet train No. 12. His engineer had sounded the proper whistles for a meeting point and had begun to reduce speed when the collision occurred, at, the conductor thought, about 2.58 a.m., when his train was travelling at about 20 miles an hour.

Conductor Crane said that his watch had been five seconds slow when he had compared it to the clock at Lafayette. He had received four train orders there and

[783] Interstate Commerce Commission report
[784] *New Plymouth Sentinel*, New Plymouth, Idaho, United States, 26th March 1925

delivered them to Engineer Matthews but did not see him read them. As the train neared Ricohoc he had whistled to remind the engineer of the meet and the latter had replied. The train did not slow down though, and he had whistled again. Once more the engineer had replied but not with the correct code and becoming alarmed, he had gone to the window and seen that the train was going over the points at the far end of the loop, whereupon he had jumped for the brake valve. He put the time of the collision at 3 a.m. The block signal at the east end of the loop had been at danger; Engineer Matthews told him afterwards that he had not seen the signal.

Engineer Matthews of train No. 12 said that his watch had been 12 seconds slow at Lafayette. He had read the orders and understood them. Approaching Ricohoc he had acknowledged the signal from the conductor. The weather had been very foggy, but he had just glimpsed the distant and home signals west of the station. Knowing that the loop was a long one he had made a light brake application, had been distracted by the safety valve popping and had then lost his bearings in the fog, only realising that he was going to overshoot the loop when the stop signal at red had loomed up out of the fog and his fireman had shouted, "Look out, Doc!" At that point he had applied the brakes in emergency and the fireman had jumped, telling him to do likewise. He had not seen any immediate danger at first but then saw the headlight of train No. 109 at close quarters and jumped as well. He put the time of the collision at 3.03 a.m. He said that he had not forgotten the wait order but had lost his bearings in the fog.

Fireman Thibodaux said that his watch had been correct at Lafayette. He remembered his engineer answering the conductor's signal as they approached Ricohoc. At the west switch he had expected the engineer to shut off steam but when they passed the water tank with no reduction in speed he had told the engineer that it was only 2.56, train No. 109 was not there, and they had an order to wait until 3.05 for it. Passing the telephone booth he had said that they had better stop. The engineer apparently did not hear him so had shouted loudly, "Stop her," put out the fire and got down the gangway. The engineer had then applied the brake in emergency and he (the fireman), had jumped just before the collision. Engineer Matthews had seemed to be perfectly normal. It had been very foggy, and the signals could only be seen at close range. His watch had been broken when he jumped and stopped at 2.57, which he put as the time of the collision. He had taken no action to stop the train himself as his engineer's hand had been on the brake valve and he had expected him to brake in time. The engineer had not had his train under control as he approached the eastbound home signal or the station.

Brakeman Ferris of train No. 12 said that he had been aware of the meet order. When they had passed the west switch at 45 miles an hour with no reduction in speed he had assumed that train No. 109 must be there, then when he saw that it was not he had assumed that an order had been received at Bayou Sale changing the meet to Patterson.

The ICC placed the primary responsibility for the accident on Engineer Matthews for failing to approach the home signal with his train under control, failure to wait for train No. 109 as per the train order and passing the block signal at the east end of the loop in the "Stop" position. It was difficult to understand how such an experienced engineer could have become so confused as to his whereabouts, notwithstanding the fog. Given the fog he should have exercised extraordinary care and having answered the conductor's signal he should have brought his train under control.

Fireman Thibodaux was blamed for not having been more alert to the impending danger and taken more decisive action when his engineer approached Ricohoc without having his train under control and had failed to reduce speed. By the time the fireman had alerted the engineer it was too late.

An automatic train control system would have prevented the accident.[785]

A coroner's jury agreed with the verdict of the ICC, holding that the accident was 'due to the fact that orders given train No. 12 at Lafayette were not properly executed.' They stopped short of finding that Engineer Matthews should be tried for manslaughter.[786]

The Chicago Milwaukee and St Paul Railway experienced a head-on collision early on Sunday 15th November 1925 at O'Connor Siding, just over three miles east of Sacred Heart, Minnesota, United States. Engineers William Fillmore and Abe Young, Head Brakeman D. Harmer and mail clerk Charles Davis died in the collision and 15 passengers, five railwaymen and a mail clerk were injured.[787]

The single line on which the accident occurred was operated by timetable, train orders and a manual block system. O'Connor Siding was situated between Sacred Heart and Renville stations, which were just under seven miles apart. The sidetrack (loop) at O'Connor was over two miles long but part way along it a crossover connected it with the single-track main line, and only that part of the sidetrack east of this crossover formed the crossing place proper, the portion west of the crossover being used as a storage track.

One of the trains involved in the accident was westbound passenger train No. 17, the *Columbian*, from Minneapolis, Minnesota, to Tacoma, Washington. This train consisted of engine No. 6305, a mail car, a baggage car, a mail car, a dynamo baggage car, two coaches, a tourist car, a dining car and two sleeping cars, in charge of Conductor Harrington with Engineer Fillmore and Fireman Matthew Jarvis. The other train was eastbound freight train No. 264, consisting of trains Nos. 8100 and 8052, 95 cars and a caboose, in charge of Conductor Trueman with Engineers Ervin Bucklin and Young and Firemen Bernard Donovan and Skramstad.

At Olivia, 11 miles east of Renville, the crew of train No. 17 received train order No. 2, Form 19, directing them to wait at O'Connor Siding until 1.30 a.m. for train No. 264. The train left Olivia at 1.10 a.m., 18 minutes late, and at Renville its crew were given a clearance car, Form A stating that the signal there was at "Stop" for the train to meet train No. 264 as per the order and that the block was otherwise clear. Renville was passed at 1.26 a.m., the west passing switch at O'Connor Siding was passed after 1.30 and the train was travelling at between 15 and 20 miles an hour when it collided with train No. 264. Train No. 264 had received a copy of train order No. 2 at Minnesota Falls at 1.11 a.m. and passed Sacred Heart at 1.25 a.m., at which time, according to the rules, would have been clear of train No. 17. In the collision engine No. 6305 was badly damaged and its tender and the leading car of train No. 17 were destroyed. Engine No. 8052, the train engine of train No. 264, was overturned and the leading 10 cars (waggons) of the train were derailed, three being destroyed and the remainder more or less badly damaged.

---

785 Interstate Commerce Commission report

786 *The Daily Herald*, Mississippi Coast, Mississippi, United States, 24th March 1925

787 The Brainerd Daily Dispatch, Brainerd, Minnesota, United States, 16th November 1925

Fireman Jarvis of train No. 17 told the Interstate Commerce Commission (ICC) investigation that he remembered receiving the train order at Olivia and the clearance card at Renville. He had then seen the headlight of train No. 264 and had supposed that it was waiting at Sacred Heart. It was his first trip as fireman on this train and he had been keen to get a good fire before the crew change at Montevideo. He had been working on the fire when his engineer had made a service brake application, at once followed by an emergency application. Getting up on his seat, he had seen train No. 264 a short distance away and jumped. He had not said anything to Engineer Fillmore about train No. 264 not being at O'Connor Siding or about seeing the headlight. His engineer had appeared fit and had been awake and attending to his duties at all times. Engineer Fillmore was one of those killed.

Conductor Harrington of train No. 17 said that they had passed through Renville at about 10 miles an hour. It had then been 1.28 by his watch and they had passed the east passing track switch at O'Connor at 1.32. He had then been going back through the cars when the emergency brake application and collision had thrown him down. Getting to his feet he had noticed that it was 1.38 on his watch.

Head Brakeman Harry Cadwell thought it had been 1.29 when he received the clearance card at Renville and said that he had been confident that they could not reach O'Connor Siding before 1.30, the time up to which they were to remain there according to the order, so he had not bothered to consult his watch again. Baggagemaster Osterwind and Flagman Rohl had not seen the order.

Engineer Bucklin, of train No. 264, interviewed in hospital, testified that his train had left Montevideo at 12.15 a.m. and at the end of a short section of the double track at Minnesota Falls he had received train order No. 2. He had remarked to his fireman that he did not think that there was time for them to get to O'Connor but decided to go as far as Sacred Heart, 5½ miles from Minnesota Falls. The loop at Sacred Heart would only accommodate 78 cars, so a flagman would have to be sent out and train No. 17 "saw by." (Train No. 264 would enter the loop and run to its far end, leaving its rear end still foul of the main line. Train No, 17, cautioned by the flagman, would stop on the main track opposite the freight train and having cleared the single line, train No. 264 would pull forward and proceed, as would train No. 17 as soon as the rear of the freight train had cleared the single line.)

It was 1.25 when his train reached Sacred Heart and he then decided there was insufficient time for the flagman to go ahead on foot and warn train No. 17 so illogically he had elected to proceed to O'Connor Siding, although according to the rules his train should have been clear of the main line at 1.25 (five minutes before train No. 17 was authorised to leave O'Connor Siding). His argument was that from a quarter mile east of Sacred Heart the line was tangent (straight) for 6 miles, so he would be able to see what was ahead of him. The headlight of train No. 17 had duly appeared, and he said that he had watched it carefully; at one point it had disappeared, either because of a dip in the track or it had momentarily gone out. Suddenly he had realised that the headlight was very close and had applied his brakes in emergency. Asked why he had not stopped at Sacred Heart and sent the leading engine ahead with the flagman he replied, "Well, why didn't I stay at Minnesota Falls? That is where I should have stayed."

Fireman Donovan of engine No. engine No. 8100 recalled saying to Engineer Bucklin that they only had 16 minutes to get to O'Connor Siding and stated at the inquiry that had he been engineer he would have wanted 30 minutes to get from Minnesota Falls to Sacred Heart with the load that they had and two engines. He had not, however, discussed with his engineer where they were going for train No. 17, even after the headlights of the latter train had appeared, saying that he had supposed that his engineer knew what he was doing. In fact he had given no warning or suggestion that they stop before train No. 17 had come very close and he had jumped.

Fireman Skramstad, on engine 8052, said that at Minnesota Falls he had told Engineer Young, who was not very familiar with the route, that they would not have time to reach O'Connor Siding and that Engineer Bucklin probably intended to head in at Sacred Heart. They had passed there, and he had told his engineer that they were probably going for the west storage track at O'Connor, which was a mile or two further on (actually it was three). He said that he and his engineer, who was killed, had been relying on the crew of the pilot engine, who knew the road better. He had then told Engineer Young that they had run out of time and watching the headlight of train No. 17 had seen it getting brighter and brighter. Finally his engineer had applied the brakes in emergency, he had shouted to his mate that there was going to be a collision, jumped, and the trains had collided.

Conductor Trueman of train No. 264 stated that on receiving order No. 2 at Minnesota Falls he had known that they did not have enough time to get to O'Connor Siding but had thought that they had time to reach Sacred Heart. When he had seen that they were not stopping at Sacred Heart he had gone out and placed a fusee on each side of the caboose as a stop signal to the engine crew. He had then gone back inside and had been going to open the conductor's valve to apply the brakes, but Flagman Null had feared that doing so would derail the train. The conductor said that he had then made up his mind that the leading engine had been detached and gone forward to O'Connor Siding to warn the passenger train.

Operator Martinson at Renville said that train No. 17 had passed him going slowly, apparently trying to kill time.

Operator Greens at Minnesota Falls said that the headlight of train No. 264 had burning brightly.

The ICC concluded that the accident had resulted from train No. 264 attempting to reach O'Connor Siding and get clear of superior train No. 17 in insufficient time (a practice commonly known as 'running on short time') and that Conductor Trueman and Engineers Bucklin and Young were responsible. There was little that could be said to explain the action of Engineer Bucklin in trying to reach O'Connor Siding or of the failure of Conductor Trueman to stop the train. There had been no misunderstanding of the train order or of the applicable rules. Engineer Bucklin had decided to go to Sacred Heart although he knew that his train was too long to get in the siding. Having succeeded in getting there his decision to go on further had been completely unjustified. Likewise nothing had justified Conductor Trueman's assumption that the leading engine had been detached and sent forward and he was therefore equally responsible.

Although Engineer Young was apparently less familiar with the route he had had the train order, a timetable and a watch so he could have determined that there was insufficient time to get to O'Connor Siding.

The weight of the evidence was that train No. 17 did not pass O'Connor Siding before 1.30 a.m. and that the collision occurred at 1.35 a.m.

The difficulty of the enginemen in determining how far away an opposing train was on a long straight stretch of track reflected the situation which had occurred in a collision near Thorpe, Wisconsin, in December 1922 and tests showed that it was indeed impossible to tell how far away an opposing train was in such circumstances.

An adequate block system would probably have prevented the accident and an automatic train stop or train control device should have prevented it. [788]

The Detroit Monroe and Short Line Railway, part of the Detroit United Lines, suffered a bad accident on Thursday 2nd September 1926. The interurban line involved was single track, worked by timetable and train orders, with no block signal system and no superiority of trains.

The trains involved in the accident were northbound passenger train No. 223, consisting of motor car No. 7529, and southbound train No. 224, consisting of motor car No. 7096. In charge of train No. 223 was Conductor Charles Leatherman with Motorman Andrew Schlegel, which train No. 224 was in charge of Conductor Charles Adams and Motorman William Owen. [789]

Train No. 223 was about a mile short of Pine Siding, near Monroe, Michigan, United States, where it was timetabled to meet train No. 224, when the two trains collided. Both cars were telescoped in the collision and nine passengers, and a student conductor were killed and 32 passengers and four railwaymen, probably the two motormen and the two conductors, were injured.

Train No. 224 left Oakwood, Michigan, at 12.45 p.m. and at Rockwood its crew received two orders:

Meet No. 221, car No. 7520. At Newport

Meet No. 15, car No. 7312, at Pine.

The train left Rockwood at 1.11 p.m., on time, met the two trains mentioned in the orders and continued beyond Pine Siding without having met train No. 223.

At the Interstate Commerce Commission inquiry Motorman Owen of train No. 224 stated that he was employed as both motorman and conductor at different times. He recalled receiving the two orders at Rockwood and seeing the motorman of train No. 15 at Pine, but he could not remember the car number. He admitted entirely forgetting the regular meet with train No. 223. He said that his train had been travelling at about 45 miles an hour when he had seen the other train and by applying the brakes in emergency, opening the sanders, and reversing the motor he had brought the speed down to about 15 miles an hour when the trains struck one another. He added that on the day preceding the accident he had finished duty at 5.45 a.m. after working ten hours, had something to eat, visited a dentist, had something more to eat and then resumed duty at 1.54 p.m. until 11 p.m. Going to bed at 11.30 p.m. he had got up at 5 a.m. to resume duty at 5.50 a.m. Apart from his lack of rest had had been feeling bad from continuing toothache.

Conductor Adams of train No, 224 stated that he had only been working the train since 14th August. Like his motorman he had completely forgotten about train No. 223.

788 Interstate Commerce Commission report

789 *The Times Recorder*, Zanesville, Ohio, United States, 3rd September 1926

He had noticed nothing wrong with his motorman.

Motorman Schlegel and Conductor Leatherman of train No. 223 were too badly hurt to give much useful information.

The ICC reached the obvious conclusion that Motorman Owen and Conductor Adams were to blame for the accident. Both were experienced men. Motorman Owen had been on duty for 16 out of the previous 24 hours, in violation of the Federal hours of service law, and had been nervous and depressed from unremitting toothache as well as tired. Conductor Adams had no excuse for his oversight other than becoming confused by meeting another train at Pine.

No recommendations were made by the commission. [790]

The American interurban collisions continued with one on the Dayton and Troy Electric Railway near Dayton, Ohio, United States, on Saturday 25th June 1927. The line on which the accident occurred was single track, worked by timetable and train orders without a block signal system. Superiority of trains was not conferred by direction. Some of the sidings on the line were of considerable length and were operated as second main tracks, one of these being siding 100, which extended 3¼ miles north of Dayton. The collision occurred on the single track about a quarter mile north of the end of siding 100.

The trains involved in the accident were northbound No. 204, consisting of motor No. 250, in charge of Conductor Stevens and Motorman Covert Kessler; and southbound No. 75, consisting of motor No. 200 in charge of Conductor Hart and Motorman John Pemberton.

Train No. 204 left Dayton at 5 p.m. It failed to stop at Idlewild, at the north end of the double track, for its timetabled meet with train No. 75 and the two collided a little further north at about 5.12 p.m. The front of the northbound car was demolished, and the southbound car was telescoped about a third of its length. Five passengers and Motorman Pemberton were killed, and 35 passengers, Motorman Kessler, Conductors Hart and Stevens and an off duty railwayman were injured.

At the Interstate Commerce Commission inquiry Motorman Kessler stated that he had received a clearance card at Dayton. He had been operating the train for less than a week and previously had always passed train No. 75 before the end of the double track. Not having done so he had overlooked it. After the accident he reportedly stated, "I thought I had passed the southbound car on the siding and that the track was clear." [791] His train had been travelling at 35 to 40 miles an hour at the end of the double track but had reduced speed for a car on a level crossing just north of the Miami River bridge immediately before seeing the opposing train and had reduced the speed of his train to 10 miles an hour when they met.

Conductor Stevens said that he had thought of train No. 75 as they left Dayton but had then been busy collecting tickets and had forgotten about it. In contrast to the motorman he thought that the speed of the train had increased on the descending grade north of the Miami River bridge and that although the brakes had been applied at the last moment

[790] Interstate Commerce Commission report

[791] *The Star Journal*, Sandusky, Ohio, United States, 27th June 1927

speed had not been much reduced and had been still 35 miles an hour when the trains collided. He said that normally if he did not spot train No. 75 on the double track he would inquire as to it from the motorman, but he had been still engaged collecting fares when his train entered the single track.

Conductor Hart of train No. 75 stated that his train had been travelling at 35 miles an hour when his motorman had made a heavy brake application just north of Station Stop Two which he had at first assumed was to stop to pick up passengers. His train had been almost stopped at the time of the collision.

The ICC found Motorman Kessler and Conductor Stevens responsible for the accident and made the usual comments that it would have been prevented by an adequate block system or automatic train control or train stop device. [792]

Somewhat unusual circumstances lay behind a single-line collision in South Africa on the evening of Wednesday 27th July 1927. A new crossing loop had been opened at Glenroy and a southbound goods train from Roodekop had in its possession the Mapleton-Glenroy staff and was to cross the Natal mail train from Durban to Johannesburg at the latter point. Forgetting, presumably, about the new loop, the driver continued southwards until he collided with the passenger train several miles further south near Heidelberg. The two locomotives and the vehicles of the two trains formed a huge pile of wreckage which it was estimated would take 36 hours to clear. Intense cold and driving rain hampered the rescue operation and some of the injured died of exposure before they could be removed to hospital. Two Europeans – a fireman and a ticket examiner – and at least 22 'natives' died and 60 'natives' and three Europeans were injured. {Note – *Railway Disasters of the World* states that the new loop was at Mapleton, but this makes no sense as Glenroy is south of Mapleton and Heidelberg is south of Glenroy.} [793] [794]

What was at the time the British colony of Ceylon (now Sri Lanka) suffered a major disaster on the evening of Monday 12th March 1928 at Katukurunda. On the day of the accident an express left Marara at 4.55 p.m. for Maradana (Columbo). Designated train No. 250, this was made up of an engine, seven passenger coaches, three goods waggons and a fish van. Coming in the opposite direction was train No. 547, a Colombo – Aluthgama slow train consisting of an engine, eight coaches and a fish van. [795] Both trains had European drivers; Driver David Crowe was driving the engine of the express and Driver Bennett the engine of the slow train.

The trains normally crossed on another at Payagala but train No. 547 was 18 minutes late when it reached Kalutara South station just before 8 p.m. Because it was delayed its crossing with train No. 250 was moved to there and the latter train proceeded from Payagala. Meanwhile Assistant Stationmaster Jayawardene at Kalutara South indicated to Driver Bennett of the slow train that he must wait there for the express, but in some way

---

[792] Interstate Commerce Commission report
[793] *The Times*, London, England, 28th and 30th July 1927
[794] *Railway Gazette*, Vol 47, PG. 334
[795] Webpage – Travel Kalutra: Katukarunda Railway Accident – First head-on collision in Sri Lanka

the driver seems to misunderstood what he was to do, and instead set off towards Payagala. As the line was apparently worked by the electric train tablet and the express train was in possession of the tablet, Driver Bennett must also have left without one. The assistant stationmaster ran several hundred yards along the track after the departing train, waving his cap in the fading light to try and alert the guard and no doubt shouting at the top of his voice. It was all no use. The guard failed to spot him, and the train continued its journey. Reportedly the assistant stationmaster sat down on the track and prayed for divine intervention to prevent the disaster. A few minutes later he heard the crack of the collision; the two trains had met on a curve. The two engines were badly damaged and the express engine overturned, while several coaches were telescoped. The sound of the collision was apparently heard about eight miles away. Years later an old lady who had been a child living in a nearby village at the time of the accident related that, "We all ran out of the house leaving our dinner plates on the table as the entire village was shaken by a loud noise." [796]

The rescue operation was hampered by the darkness. The only light available was that from Cadjan (interwoven palm leaf) torches. Driver Crowe was killed outright, Driver Bennett was trapped in his cab and could not be freed in the darkness. Like several trapped passengers he had to wait until daybreak. A surgeon then decided that his left leg must be amputated on the spot. This was done and he was finally released and taken to Colombo but died in the hospital there a few minutes after admission. Villagers removed the bodies of 28 passengers before the authorities arrived and a police party assisted by several hundred volunteers freed 40 or 50 injured passengers. The sudden influx of so many casualties overwhelmed the local hospitals.

With Ceylon still being a colony the accident led to a question in the British House of Commons the next day, where Mr Fletcher, the Colonial Secretary, promised that a full inquiry would be held. This was undertaken by Lt Col F.H. Eustace, of the Indian Railways. Naturally, it found that the main responsibility lay with the negligence of the dead Driver Bennett, but that there had been contributory negligence by the guard of train No. 547, and it was held that Assistant Stationmaster Jayawardene had not been 'fully alive to his responsibilities.' [797]

For the last time in this book we look at a single-line collision on an American interurban. This location this time is given in the Interstate Commerce Commission report as near Delong Siding, and in a newspaper report as near Fithian. The collision happened on the main line of the Illinois Traction System between Danville and Champaign, Illinois, a single-track electric line worked by timetable and train order with no block signal system and no superiority of trains.

The trains involved in the accident on Thursday 31st May 1928 were eastbound limited train No. 74, consisting of car No. 271, in charge of Conductor Clarence Cummings and Motorman Frank Craig; and westbound train No. 77, consisting of car No. 279, with Conductor Lochbaum and Motorman C. M. Cummings. Both cars were of wooden construction with steel sheafing and anti-telescoping devices. Eastbound train No. 74 left Champaign, 18¼ miles west of Delong Siding, at 4.23 p.m. Delong was its

[796] Webpage – 150 Years of the Ceylon Railway, How it all began.
[797] *Railway Gazette*, Vol 48, pg. 384 and Vol 49, pgs. 173 and 229

timetabled meeting point with train No. 77 but it failed to stop at the siding, colliding with the opposing train a short distance further one. The underframe of car No. 271 overrode the floor of train No. 279 for 25 feet, demolishing it, while the front of car No. 271 of train No. 74 was badly damaged. The accident happened at about 5.01 p.m. [798] Motorman Craig and six passengers were killed and 19 passengers and both conductors were more or less injured. The passengers who lost their lives were Dr Charles Johnson, of Champaign; Dr George Johnson, his son, of Terre Haute, Indiana; Dennis Pyle of White Heath, Illinois; Ira Clodfelter, of Monticello, Illinois; Fred Gulick, of St Joseph, Illinois; and E.R. Townsend of Chicago. Illinois.

C.J. Smith, a Fithian farmer, was ploughing his field 200 feet away when he heard screams above the roar of the interurbans. "There was an awful smash, and then everything was quiet. I ran as fast as I could to the two coaches, and when I got there, people were just beginning to move. It looked like everybody that wasn't dead had been stunned."

Most of those who were killed had been in the smoking compartment of train No. 77. The first rescuers on the scene were motorists on state highway 10, which ran parallel to the interurban line. They turned their cars into ambulances and hearses to carry the victims to Danville, Champaign and Urbana and ambulances proper, with doctors and nurses, went sent from the three places, while three wrecking cars were sent from Danville. The two doctors had attended a meeting of the Wabash Valley Æsculapean Society in Danville. [799]

Conductor Cummings, of eastbound train No. 74, told the Interstate Commerce Commission (ICC) inquiry that approaching Delong Siding he had exchanged the usual whistle and bell signals. He had then gone to the rear platform as it was usual for the first train to arrive at a crossing place to run past the switch and back in. He had then realised that his train was not slowing up and he could see train No. 77 approaching. Seeing this he had rushed inside and seized the bell rope and the emergency brake valve. Motorman Craig had been in good mental and physical condition when he had last talked to him at Urbana. Ever since he had become a conductor in December 1923 it had been the practice for the first train to enter the siding, pulling by it without protection and backing in. He had never been censured for operating trains this way.

Somewhat unusually, three passengers on train No. 74 gave evidence at the inquiry. Mr B.W. Taintor spoke of there having been a person sat on a camp stool between the seats and the open door of the motorman's compartment. This person had been conversing with the motorman between Urbana and Ogden, 2¾ miles short of Delong Siding, though not apparently after that. As the train approached the west switch at Delong he had noticed the opposing train. When his train had failed to reduce speed at this point he had shouted a warning to the motorman and hurried back through the car, noticing that the air brakes were being applied as he did so.

Mr. O.A. Acuff stated he had seen train No. 77 before his train reached Delong Siding but had not realised that his own train was not going to stop until it was on its way through the siding when he had asked the motorman, was this not the crossing place? The motorman, apparently deep in thought, had not answered. Mr. Acuff said that he had then

[798] Interstate Commerce Commission report
[799] *The Decatur Daily Review*, Decatur, Illinois, United States, 1st June 1928

reached into the motorman's cab, tapped his arm and cried, "My God, man, stop this car – look ahead of you!" Finally the motorman had looked up, shut off the power, which had been full on, and applied the brakes.

Mr. C.H. Redenbaugh said that he rode the train practically every day and so was well aware that it crossed train No. 77 at Delong. Reaching the west switch at high speed he had looked for the other train in the siding, had seen that it was not there and then that it was approaching. Thinking that his own train was not going to be able to pull up in time he had rushed back through the car; he had not noticed the brakes being applied until the conductor's valve was opened.

Motorman C.M. Cummings of train No. 77 said that he had glimpsed train No. 74 when it was east of Rumpler Siding. It had then disappeared from view before reappearing around a curve. He had assumed that it was going to run past east switch and back in, but it seemed to be going very fast. Having previously braked lightly he had then braked harder. Still train No. 74 came on, he saw fire flying from under it, made an emergency brake application and reversed his car. The speed of his train had fallen to about 15 miles an hour when he had jumped. He had seen train No. 74 backing in on previous occasions and thought that the practice had official sanction.

The ICC concluded that the accident was due to train No. 74 failing to obey a meet order. The evidence indicated that Motorman Craig had been in some unknown way incapacitated or else so deeply engrossed in his thoughts as to not realise where he was.[800] Interestingly a newspaper report stated that Motorman Craig had been pulled alive and conscious from the wreckage by Mr. W.E. Richards, an Ogden businessman, and had died in the latter's car as he was being driven to hospital. Before he died he had made a statement. "I saw him coming and tried to put on the brakes, but the air must have given out. They didn't work." Superintendents A.W. Manley and Mike Connor stated, however, that there had been nothing wrong with the brakes, Connor stating that he did not believe that the brakes of either car had failed, that they had worked, the wheels had locked and then slid along the rails. [801] There is no suggestion in the ICC report of any brake failure. The commission did point out that had the conductor not been expecting the train to pull forward before setting back anyway he would have taken action sooner, though it considered that he had been slow in any case to do so.

The ICC suggested that the primary responsibility for the accident rested, however, not with the crew of train No. 74 but with the higher officials for failing to enforce the rules and considering that some rules served no purpose. The commissioners were not impressed with Superintendent of Transportation Morris, who questioning Conductor Lochbaum suggested that the latter considered that Rule No. 55, regarding protecting trains, could be disregarded because Rule No. 64 required all trains to approach crossing places under full control. The commissioners likened this situation to where a train is stopped in automatic block signal territory and the flagman relies on the engineer obeying signals while the engineer relies on the flagman having protected his train, each depending on the other being 100% perfect. [802]

---

[800] Interstate Commerce Commission report

[801] *The Decatur Daily Review*, Decatur, Illinois, United States, 1st June 1928

[802] Interstate Commerce Commission report

One of the problems with steam locomotives was that on long journeys they might need to stop several times to be replenished with coal and water. In some countries, such as Great Britain, this problem was solved as far as water was concerned by the provision of water troughs. In Sweden in the late 1920s engineer Fredrik Ljungström attempted to solve the problem in a different way, he designed a steam turbine locomotive which had, he hoped, a higher efficiency and a lower coal and water consumption. A steam turbine engine, No. Å 1474, was duly brought into use to work express trains 21 and 22 between Stockholm Central and Bollnäs and was found to be capable of making the whole 317 kilometre journey without requiring further coal or water, whereas a B class locomotive had to take on coal and water at Krylbo and water again at Ockelbo.

On the evening of Thursday 21st June 1928 train No. 24, the Jämtlands express, left Stockholm Central at 20:00 for Östersund, Storlien and Trondheim in Norway. The train had a conventional steam engine of class B, which was changed together with its crew at Krylbo. The train then set off again with Driver Per Westberg and Fireman Karl Simon Hellberg in charge of B class engine No. 1385. Ten minutes after the departure of train No. 24 train No. 22 arrived at Krylbo. This train had left Stockholm Central at 20:15 for Övre Norrland and was hauled by the steam turbine engine, No. Å 1474. After a brief stop train No. 22 also continued northwards from Krylbo, two stations (block sections?) behind train No. 24. All went well as far as Storvik, but after leaving there the feedwater pump began to malfunction, so that not enough water was being fed into the boiler. The driver and fireman made repeated attempts to cure the problem, but they could not, and on arrival at Ocklebo station it was clear that the engine could not take the train any further and another engine was required. The train clerk at Ocklebo contacted Bollnäs but apparently wrongly stated that it was train No. 24 which needed assistance. The clerk at Bollnäs immediately contacted the engine shed there and Driver Erik Lundqvist and Fireman Ivar Hammarlund were awoken from their sleep and instructed to proceed with a light engine to Ocklebo to assist the disabled train. [803] [804]

A little later the Ocklebo clerk sent another message to his counterpart at Bollnäs correcting his first one, it was train No. 22 which needed assistance, not 24. Half an hour later Driver Lundqvist had B class engine No. 1312 ready to go, and he read the order he had been given by the Bollnäs dispatcher by the light of the open firebox door. The order read:

Assistant engine Bollnäs – Ocklebo Leave Bollnäs 01:31 Meet train No. 24 Assist train No. 22 from Ocklebo to Bollnäs [805]

At this point Driver Lundqvist made the same mistake as the train clerk at Ocklebo and mixed up the two trains. Instead of waiting at Bollnäs for train No. 24 to arrive he left at once at about 01:20. The Bollnäs clerk, seeing the engine setting off, made several fruitless attempts to stop it by trying to telephone the crossing keeper at Länninge, five kilometres from Bollnäs. The express and the engine collided on a curve near the level crossing, the express travelling at 84 kilometres an hour and the engine at 40 at the moment of impact. Both engines were badly damaged in the collision and the leading vehicles of the express were telescoped and wrecked. The rear vehicles of the train were

---

[803] Wikipedia – Järnvägsolyckan i Lennige

[804] Webpage – 75 år siden tågsammanstötningen i Länninge

[805] Webpage - Järnvägshistorisk forum: Länningeolyckan

little damaged. A fire caused by gas escaping from broken cylinders was extinguished before it could cause too much harm. The exact number fatalities is unclear as different sources give different figures, but it appears that at least 16 passengers died, together with Firemen Hellberg and Hammarlund and Driver Vestberg of the express. About 50 passengers were injured. The dead included an M.P., Johan Johannson, a member of the Swedish Agrarian Party. [806]

Driver Lundqvist survived the accident and was subsequently imprisoned but was released after about a month. An investigation into the disaster failed to come to any clear conclusions as to its cause. [807]

Ten people died and ten more were injured in a collision between two express trains in Spain on Saturday 29th September 1928. The trains were train No. 95 from Madrid to Algeciras and train No. 96 from Malaga to Madrid. The accident was due to the failure of train No. 95 to wait at Linares-Baeza station for train No. 96. The trains collided about a kilometre from Las Madrigueras at kilometre 323. Engines Nos. 1738 and 1744 of trains Nos. 95 and 96 were embedded in one another and parts of both trains were badly damaged, but the strength of the rolling stock was believed to have prevented many casualties. The accident happened at 04:30 so there is a possibility that the driver of train No. 95 might have fallen asleep.

Newspaper reports appear a little confused as to the name of the crews (or possibly it is the writer's poor understanding of Spanish at fault!) but it appears that train No. 95 was in charge of Conductor Ramón Agulló with Driver Garcia and Fireman Antonio Samos; and train No. 96 was in charge of Conductor Pedro Casado with Driver Manuel Sánchez and Fireman Matias de Aza. All survived the accident but Brakeman José Monllor was amongst the fatalities. [808]

Another single-line collision with disastrous consequences occurred in Canada near Drocourt, Ontario, early on the morning of Wednesday 20th March 1929. Again the victims were mostly immigrants on their way to a new life in Canada and, as a newspaper remarked, 'many of them without relatives or friends in the Dominion.' The accident happened this time on the Canadian National Railway, and involved two transcontinental trains, the westbound and eastbound *Nationals*, trains Nos. 3 and 4, between Toronto, Ontario; Winnipeg, Manitoba; and Vancouver, British Columbia. The trains normally crossed one another at Waubamik, near Parry Sound, Ontario, but train No. 4 was running two hours late while train No. 3 was on time. The meet was therefore moved westwards from Waubamik to Drocourt. Train No. 3 was inferior by direction and should have taken the siding at Drocourt for train No. 4. Instead train No. 3 ran through Drocourt without stopping at 3.57 a.m. and collided with its opposite number in a rock cutting a mile further on. The engineer of the westbound train saw the danger at the last moment, applied his brakes in emergency and jumped, followed by his fireman. The crew of the eastbound train apparently remained oblivious of the impending collision and stayed at their posts.

---

806 *The Times*, London, England, 23rd June 1928
807 Webpage – Järnvägshistorisk forum: Länningeolyckan
808 *La Vanguardia*, Barcelona, Spain, 30th September and 2nd October 1928

Train No. 3 was the worse sufferer in the collision as the leading baggage car was telescoped into the following colonist car. To make matters worse fire then broke out in the wreckage, started this time not by gas but probably by the upsetting of the coal stove in the colonist car used to passengers to cook their food. [809] Train No. 4 was less affected as the leading express car, filled with fish, absorbed most of the impact. Fireman Horace Smith and Express Messenger Edwin Strutt were the only two to die on train No. 4 but on train No. 3 Conductor Benjamin Barstead, Brakemen Roy Ferguson, Leslie Savage and W.G. White, newsagent George Millar and about 17 passengers lost their lives. As in a number of other accidents it was impossible to tell exactly how many of the immigrants had died as there were few or no records of their names and their bodies were destroyed in the post-accident fire. The conductor's ticket stubs, which might have given information as to the number of passengers, were burned and lost. Engineer Paul Gouvreau of train No. 4 was among those who were badly hurt while Engineer Vernon Alexander of train No. 3 was barely hurt.

It was soon apparent that the collision had been caused by train No. 3 failing to wait for its counterpart at Drocourt. In a dying deposition made the day after the accident Brakeman Savage said that he had been the first person to realise that his train had run past its meeting point, and he had shouted to Conductor Barstead, "He has run past the switch." A quick look out of the window confirmed to the conductor that this was indeed so and he had made a grab for the brake valve to stop the train a moment before the trains met. [810] Conductor Barstead was of course equally responsible with the engineer for the safety of the train. Why he had failed to realise that they had passed Drocourt before his brakeman spoke could not be ascertained as he was killed outright. On the engine Fireman Ernest Riley thought that they had passed Drocourt but he had only been on the run a fortnight and was unsure without checking his time card. Finally Rear Brakeman Charles Gorrie at the back of the train also became aware that they were not stopping where they should and was likewise reaching for the brake valve at the moment of impact. Engineer Alexander was apparently on the verge of a nervous breakdown the next day, a friend of his commenting that, "He is crying all of the time. He says he doesn't know why he did not stop at Drocourt. He says he was talking to the fireman and was past Drocourt before he knew it." [811]

An inquest at Parry Sound on Fireman Smith of train No. 4, as representative of the victims of the accident, was quickly held. The verdict of the jury was that, "The said Horace Smith came to his death in the early morning of March 20 in a head-on collision between trains No. 3 and 4 on the Canadian National Railways. The said collision was due to the neglect of train No. 3 in running past the meeting place at which they had orders to stop." Following the inquest not only Engineer Alexander but Fireman Riley and Rear Brakeman Gorrie were arrested on a charge of manslaughter. [812]

On 14th May the trial began of Engineer Alexander alone, the grand jury not having reported on the indictments against the fireman and brakeman. The engineer was defended by Arthur Roebuck, a leading criminal lawyer whose fees were paid by the engineer's union, the Brotherhood of Locomotive Engineers. On the first day of the trial

---

809 *Manitoba Free Press*, Winnipeg, Manitoba, Canada, 21st March 1929

810 Ibid, 23rd March 1929

811 Ibid

812 *Medicine Hat News*, Medicine Hat, Alberta, Canada, 26th March 1929

Roebuck questioned General Superintendent W.T. Moodie as to the division of responsibility between the engineer and the conductor, attempting to shift blame for the disaster onto the dead conductor. He also got the superintendent to admit that the engineer had an excellent record, the meets were safer at manned stations such as Mowat, 8 miles further east (though this would have delayed train No. 3) and that the accident might not have occurred had the line been fitted with automatic block signals. [813]

On the second day of the trial Engineer Alexander gave evidence in his own defence. He said that he had seen the milepost at Drocourt, had blown his whistle and had seen the lights at the far end of the station but the conductor had not answered his whistle. Crown Attorney W.L. Haight, cross-examining, asked why he had not stopped. The engineer replied, "I was busy with the engine and worrying over its condition. I did not get the conductor's signal. It slipped my mind." Fireman Riley spoke of their having had problems with the engine starting to "foam," with water in the boiler bubbling or frothing and escaping through the throttle into the cylinders. The trouble had first begun on the steep grade out of Parry Sound and was aggravated as always when the engine was using a lot of steam. Fireman Riley agreed that there had been no response from the conductor to the engineer's whistle. He said that when he saw the corner of a building flash by he had thought it might be a station, had looked at his watch and had been checking his time card when he had seen the headlight of the other train.

The Crown Attorney said that it all boiled down to what Engineer Alexander had told Dr Limbert after the accident – that he had forgotten the meet. "Yes, it slipped my mind," the engineer replied. Mr Justice Raney asked the engineer why, when the boiler trouble had recurred, he had not stopped his train in the siding at Waubamic to fix it. The engineer replied that in ten minutes he could have improved matters but, "It is not done. I never knew of it being done in my whole railroad experience." [814]

Addressing the jury on the third and final day of the trial, Mr Justice Raney suggested that they ignore the charge of manslaughter and concentrate on the lesser one of criminal negligence or causing grievous bodily harm through neglect of duty. Stating that he had no intention of sending the engineer to prison were he found guilty, Mr Raney added, "I would not like to think that an engineer would be justified in running a lame engine which required so much attention as to imperil the lives of passengers. Alexander could have run it into a siding, spent the necessary time to correct the engine into a workable condition and then proceeded. He might then have reported the cause of the delay, and the officials could investigate who was responsible for the faulty engine going on the road. Somebody was responsible for this disaster. You are not asked to place the responsibility on Alexander entirely. You could not honourably do so. But you must determine if he was partially responsible regardless of the fact that someone else may also have been responsible."

The jury found Engineer Alexander "Not Guilty," adding a recommendation that block signals should be installed at all stations where there were no operators and that trains should not be allowed to cross at stations which had neither block signals nor operators. Following the acquittal the charges against the fireman and the rear brakeman

---

[813] *Manitoba Free Press*, Winnipeg, Manitoba, Canada, 15th May 1929
[814] Ibid, 16th May 1929

were dropped. [815]

Three passengers from Czechoslovakia were among the five travellers who died in an accident in Austria on Thursday 22nd August 1929. Thirty-four others were injured, seven seriously and two, it was initially feared, beyond recovery.

The trains collided at a point between Schwarzach-St Veit and Loifarn. They were Prague – Villach express D115 and local passenger train P716 from Spittal-Millstättersee to Schwarzach-St Veit. The trains normally crossed one another at Loifarn at 11.27 but the express was running 15 minutes late. The fahrdienstleiter (signalman) at Loifarn therefore suggested to his colleague at Schwarzach that the crossing be altered to there. It was here that the cause of the accident lay. There was a misunderstanding. Schwarzach replied that the crossing could not be altered but because of an electrical malfunction {?} the Loifarn fahrdienstleiter supposed that his proposal to alter the crossing had been accepted and allowed train P716 to depart. Meanwhile train D115 was on its way from Schwarzach and the two collided on a curve just under 2 kilometres from Schwarzach at 11:35. The local train, going down an incline, was travelling at 49 kilometres an hour and the express at 21 kilometres an hour when they met. The local train had just left a tunnel; had the collision occurred in the tunnel it would doubtless have been much worse. On the other hand, Wächterhaus Kt. 2 (crossing keeper's cabin?) was close to the accident site and had it been staffed the accident might have been prevented, or its effects minimised, but as an economy measure it was unmanned.

The tender of the local train was forced up into the air by the leading mail van, which was very badly damaged, but the four postal staff in it somehow survived. The baggage van second in the train was almost destroyed and the third vehicle in the train, a second-class carriage, was quite badly damaged and one person died in it. The express suffered worse. Again the leading postal van was wrecked and again the postal official in it somehow survived more or less unhurt. The front of the dining car behind was partly telescoped into the postal van, the third-class carriage which came next was not much damaged but the following second and third-class Czechoslovakian composite was almost demolished, and it was in this vehicle that the other four fatalities occurred. The remaining vehicles of the express were little damage.

After the accident medical staff and supplies were sent from Schwarzach hospital and relief trains from Schwarzach, Bischofshofen, Saalfelden, and Badgastein. By 16:30 passenger trains were able to work to the scene from either side with passengers walking from one to another via a hastily prepared footpath. The line was reopened to traffic at 10:30 the following morning. [816]

Engelbert Polansky, of Vienna, who had been a dining car cook for 19 years, described his experiences. "We left Schwarzach station. Dinner was already being prepared, but there were not yet any guests in the dining car, only the staff. Suddenly there was a jolt, and I was thrown to the ground. Getting up I found that all the plates and food had disappeared from the oven. I was thrown down again, then again. On the fourth blow I hit a wall. No one knew what had happened. None of us thought of a collision, though we knew we were on a single track. Each of us thought at first that it was a

[815] Ibid, 17th May 1929

[816] Wikipedia – Eisenbahnkatastrophe 1929 auf der Tauernbahn – Salzburgwiki

colleague giving us a push for fun. Only when we looked out did we realise that there had been a terrible accident. Only one of us had a slight injury to their left foot. There was terrible devastation in the dining car. It was only dented and damaged at the front end, where the toilet was, but everything had been prepared for lunch and bottles, sugar and everything lay smashed on the floor. All the cupboard doors and shutters were open, and their contents mixed up." The cook added that it was a stroke of luck that because the train had been diverted previously over a different route the dining car was marshalled with the toilet rather than the kitchen at the front end, otherwise the six dining car staff might have lost their lives. [817]

The passengers who died were Dr Bianca Bienenfeld, a Viennese woman physician; Dr Otto Wols, a lawyer; Herr Gustave; Richard Rabe and Frau Ida Wagner, all of Czechoslovakia. Erica Wilbrich, a New York authoress, was among the injured. Other foreigners were believed to be among the injured. [818]

Unusual circumstances led to an accident near Deblin, Poland, in the early hours of Wednesday 9th October 1929. The driver of a goods train failed to wait at a wayside station for an opposing passenger despite attempts to warn him by a pointsman throwing one or more stones at the engine and its guard also attempting to stop it, presumably by applying the brake in his van. The trains collided and at least five people were killed and 25 injured. It was reported that two lives had been saved by the use, for the first time in a railway accident, of military aeroplanes to take nine of the injured to hospital. These were probably fixed-wing aircraft as helicopters were still in the very earliest stages of development.

It was said that the driver of the goods train had been fast asleep and that the shock of the collision had caused him to bite off the tip of his tongue, but that he had since stated he had been suffering from the strain of a bereavement as his brother-in-law had thrown himself under a train the previous day. [819]

Bad weather contributed to an accident on the León – Gijón line in Spain on Wednesday 17th December 1930. Gijón – Madrid express train No. 502 normally crossed goods train No. 541 at Santibañez but the passenger train was running 40 minutes late, so the meet was changed to Cuadros, one station on the Gijón side of Santibañez. Visibility was very poor due to a blizzard and the express overran the station, ignoring signals set to stop it, and collided head-on with the goods train a short distance beyond. Twelve people died and another 18 were injured. Those who lost their lives included Conductor Juan San José, Driver Antonio Rufino Camuria and Fireman Amancio Quiñones of train No. 502 and Driver Andrés Pardo of train No. 541. [820]

---

[817] www.ansichtsvertenversand.com - Eisenbahnkatastrophe auf der Tauernbahn
[818] *The Times*, London, England, 23rd and 24th August 1929
[819] *The Times*, London, England, 11th October 1929
[820] *La Vanguardia*, Barcelona, Spain, 18th and 19th December 1930

CHAPTER 7

RUNAWAY TRAINS AND VEHICLES

A variety of situations can lead to a train or part of a train running away out of control. If there are railway staff in attendance the train or engine can only run away out of control if there is a failure of the brakes or if the brakes are incapable of holding it on a steep gradient. In the case of unattended vehicles, the action of gravity alone is sufficient to precipitate a runaway. In contrast to the situation on a road, a train or vehicles can sometimes run for many miles without mishap, an accident occurring only if and when the runaway meets with another train, comes to the end of the line or gets to a curve too sharp for the runaway to negotiate at the speed at which it is travelling.

An American interurban line suffered a bad runaway accident on Saturday 12th August 1916, the runaway ending in a collision. The company involved was the Southern Cambria Traction and the accident occurred between Brookvale and Echo, Pennsylvania, United States. The Southern Traction line ran between Johnstown and Ebensburg through the Allegheny Mountains and was thus plagued by severe gradients. On the day of the accident a car driven by Motorman Angus Varner got out of control, for reasons that were not clear, and it ran through Brookvale station at 30 miles an hour, the motorman being seen by lineside witnesses to be frantically waving his arms. The runaway car continued at speed down the steep incline till shortly before Echo it collided with another car which was standing on the line near the car sheds. The cars were demolished in the collision and 14 persons were killed outright and 11 more died shortly after being removed from the wreckage. Sixty-three others were injured, two of whom subsequently died, bringing the total number killed to 27. Motorman Varner was among those who lost their lives and with his death was lost the one person who might have been able to explain what had happened. Other victims included the entire Dishong family. Darrell Dishong and his children Alda, 12, and Chester, 5, died shortly after the accident and Mrs Dishong died later in hospital. [821] Most of the victims were in the outbound car (the one which the runaway crashed into) and were on their way to an annual reunion of the Conner-Dishong-Riblett families.

[821] *The Boston Sunday Globe*, Boston, Massachusetts, United States, 13th August 1916

The collision occurred at an isolated location in the mountains, and it was half an hour before help reached the scene. 'A score of physicians and nurses were sent from Johnstown. All motor ambulances, motor trucks and several pieces of motor fire apparatus were impressed into use. Thousands of persons stormed the local offices of the company and packed Main Street, waiting for the return of the ambulances.

'The scene when the injured arrive here was indescribable. Men and women battled to get a glimpse of the injured as they were rushed through the crowded streets and business in the heart of the town was suspended.' [822]

Another newspaper reported, 'The story of Conductor G.R. Morgret, in charge of the excursion car, was vivid. He said that he had no warning of the impending collision until the runaway car had reached him. His first thought after the collision was that the mountainside had toppled over the car. As he reached the ground he viewed the wrecked car and saw his motorman, Taylor Thomas, wedged between the wall of the two cars. He pulled at a board and the body fell to the ground. Morgret then aided in rescuing those prisoned in the wreck. He carried out several bodies.

Only two miles from Echo and on the day after the disaster a car was derailed at Riverview and narrowly escaped going down a high embankment but as it was, no one was injured. The company declared that this time there had been a deliberate attempt to wreck the car by placing a bolt or spike on the rail and they offered a reward of $100 for the apprehension of those responsible. [823]

No report of the outcome of the inquest on the victims of the disaster has come to light but given that the circumstances of the accident were a mystery the verdict was presumably one of 'Accidental death.'

The single track, metre, gauge, electrified Bigorre Tramway, in France, opened in 1914, had sharp curvature and very steep gradients, as severe as 100‰ (1 in 10) in the three kilometres between Artigues and Gripp. It is therefore not surprising that it suffered the occasional runaway, in particular two bad accidents in 1916 and 1921. On the evening of Thursday 31st August 1916 a train consisting of two cars and carrying 60 excursionists was making the descent from Artigues through Gripp to Sainte-Marie-de-Campon. On the steepest portion of the route, above Gripp, the brakes for some reason failed to function, the speed of the train rapidly increased, and on a curve the train left the rails and went off the track to the left to crash into a bank.

Six passengers lost their lives in the accident and 17 were injured, three very seriously. Those injured included a doctor back from the front, both of whose legs were crushed.

An inquiry was opened into the accident. It was stated that there had recently been several previous cases of the brakes failing to work. [824]

After staying neutral for the first two years of the First World War, Romania joined

[822] *The Washington Post*, Washington, D.C., United States, 13th August 1916
[823] *Indiana Evening Gazette*, Indiana, Pennsylvania, United States, 14th August 1916
[824] *Le XIXe siècle*, Paris, France, 2nd September 1916

the conflict in August 1916 on the side of the Allied powers and enjoyed initial success before being overrun by German forces under Field Marshall August von Mackensen. It was in these circumstances that train E-1, known as *The Courier*, left Galaţi for Iaşi (Jassy) on Friday 11th January 1917. The train was made up of 26 vehicles. Its departure was delayed for several hours because the station had been bombed by German aircraft, damaging the locomotive, which had to be replaced. Riding on the train were refugees from German occupied Brăila and from the Muntenia (Greater Wallachia) region who were fearful of German bombing attacks, Romanian soldiers on leave and students, Russian officers and men and members of the French military mission. As the train proceeded on its journey it rapidly became overcrowded, more vehicles were added to the formation and there were stops, sometimes lasting hours, while passengers found room. The travelling conditions were horrendous. Many if not all the vehicles of the train were only gas-lit boxcars with no windows and no heating. Due to the overcrowding some of the passengers resorted to riding on the roofs of the vehicles where they died of exposure or fell off during the journey. Yvonne Blondel, daughter of a former French ambassador to Bucharest, who was one of the passengers, wrote later, "To our horror, a man and a 10-year-old boy were taken down frozen. Other shadows that were staggering, hardened by cold, recounted that, at some curves, many people – men and women – had been thrown off the train."

Having got as far as Bârlad the train had to wait there overnight because the line ahead was blocked by snow, which it took some considerable time for a force of railwaymen and soldiers to clear. The journey continued the next morning after the line had finally been cleared. At Bârnova the brakes of the train were checked and were working normally. The last stop that the train made before it met with disaster was at Bălteni. Between there and Ciurea there was an incline of about 16 kilometres with a ruling gradient of 1 in 40 and it was on this bank that the double-headed train got out of control. According to a newspaper report the air brakes operated only on the first two vehicles of the train because a brake valve had been closed between the second and third vehicles.

Alerted by the alarm whistles from the engines of the train the station officials at Ciurea reportedly operated a set of points to divert the runaway from track 1 to track 2, because standing on track 1 were some waggons loaded with 'tar' (or oil?). It would seem more likely that if track 1 was occupied the route would already have been set for track 2 if that were clear. At any rate, the speed of the train was such that only the engines and one vehicle of the train managed to negotiate the crossover to track 2, the remainder of the train leaving the track and at least one of the vehicles striking some fuel tanks (probably the oil tanks on track 1) triggering an explosion and huge fire. The speedometer of one of the locomotives was later found jammed at 95 km/hour. A surviving passenger stated, "I sensed perfectly how the train jumped off the rails like a monstrous reptile of iron and steel, pulling all its travellers to mutilation or on the great journey to beyond...I had the feeling that I was thrown into the bottom of a pit, a rain of objects sliding around my body...How long did this torment last? A few minutes but to me it seemed interminable..." The accident occurred at about 1 a.m. on Sunday 13th.

Yvonne Blondel was saved by two soldiers from the French military mission who were able to pull her from the wreckage as her clothing caught alight. Other passengers were killed in the initial impact or died in the fire. Those riding on the roof were either thrown beneath the vehicles and crushed to death or if they were lucky were thrown into the snow. According to Ms Blondel looters appeared on the scene after the accident and robbed some of the passengers. Rescue operations were soon in hand with railwaymen, soldiers from a nearby ammunition depot, two companies of Romanian and two companies of Russian soldiers all taking part. The survivors were taken on to Iași station, only a few miles distant, where the injured were given first aid. [825]

In the hours after the accident Dimitrie Gercanu, the Minister of Public Works; the public prosecutor, the prefect of Iași and security agents who questioned survivors all went to the scene. There were accounts of the aftermath of the disaster. "Passing through Ciurea I looked at the disaster; waggons crushed, burned, and teams of workers drawing more dead from under the wreck. Behind the station the dead were lined up in several rows...several hundred. With eyes removed, heads broken, arms detached, hands, legs, burned bogies. Women, officers, soldiers..." Another account stated, "A whole string of waggons burnt, not even with their metal skeletons, soaked like wax by the fire that consumed it...around the station everything seemed ruins and grave..." Hundreds of bodies were found among the remains of the burnt out vehicles. Repeated calls to identify them were made and 374 identified victims were buried in mass graves behind the station. Officially it was stated that there were 374 dead among the 2,000 passengers on the train, with 756 more injured. [826] Among the injured was the French Attaché, Marquis de Beltoy. [827] The real total number of dead has been estimated as being between 800 and 1,000, making it one of the worst catastrophes in railway history.

Rumours circulated after the accident of fortunes destroyed in the fire which followed the accident or looted from the victims and the names of famous people were knowingly falsely added to the list of those killed. Thus, in by then German-occupied Muntenia a memoirist wrote joyfully of the deaths in the disaster of Romanian politicians Take Ionescu, Mikhail Cantacuzeno and Emil Costinescu, the first two of whom had not even been on the ill-fated train. Emil Costinescu had, but he survived the accident. With the country as a whole facing a crisis due to the war the accident was little noted in the Romanian press and it is unknown if any official investigation was carried out. [828] It seems most probable, given the gross overcrowding of the train, that one of the passengers in trying to find a spot to station himself or herself for the journey accidentally and without realising it closed the brake valve, as has happened elsewhere at different times.

A runaway accident in England on 15th September 1917 resulted in four deaths and 46 cases of injury but because it did not happen on a public railway there was no investigation by the Board of Trade, only a coroner's inquest. The accident occurred on the Catterick Camp railway of the War Department, a 4-mile-long steeply graded military line in Yorkshire, opened in 1915, and which made a trailing connection at Catterick Bridge station with the North Eastern Railway's Richmond branch. Trains were run to and from the camp line directly to and from Darlington.

At about 4 a.m. on the day of the accident between 400 and 500 soldiers were boarding a train at Catterick Camp station for the journey to Catterick Bridge and on to

[825] Wikipedia – Ciurea rail disaster
[826] *The Times*, London, England, 22nd January 1917
[827] *The Washington Post*, Washington, D.C., United States, 21st January 1916
[828] Wikipedia – Ciurea rail disaster

Darlington. Unfortunately, at this point the engine had not been coupled onto the train and before all the men had taken their seats the coaches began to move off, slowly at first but then with increasing speed. A few of the men, realising that something was wrong, jumped out before the coaches had got up speed, but most remained seated. The coaches then gathered speed and at one point were moving at almost 50 miles an hour. The rearmost coach left the track first at a level crossing, then at a sharp curve the leading coach detached itself from the others, ran on for another two miles to Catterick Bridge station, jumped the rails at the junction, demolished a set of buffer stops and came to a stand on the ballast. The second coach was also derailed and rolled over an embankment into a field. Four other carriages were damaged and the men in them injured. The soldiers in the coach which had reached Catterick Bridge station escaped with shock and bruises but a few minutes after the accident one of them, Private Hugh Cameron, who was possibly in a confused state, was run down and fatally injured by the engine of a shunting train. Two of the men in the coaches which were derailed on the branch, Lance-Corporal Pepper and Private Green, were killed outright. The fourth fatality was a Private Muirhead, who died of his injuries two days after the accident.

The soldiers who had escaped injury set to work to assist their less fortunate comrades and assistance soon also came from the camp. The injured were motored to the military hospital and a North Eastern Railway breakdown gang, doubtless sent from Darlington, was soon on the scene to clear up the mess. [829]

The inquest on the victims was quickly held. Lance-Corporal Wilson of the Royal Engineers testified that he had been in charge of the train on the day before the accident. There had been difficulty with its Westinghouse brake. He had left the train on the loop line overnight with its brakes applied. In addition, he had applied four hand brakes and chained the wheel in the brake van. Asked by the Coroner if he had locked the brake van he stated that he usually locked it but that night he had lost the key and had not finished until 10.30 p.m. The Coroner was not impressed and told him that it did not matter if it had been 12.30 a.m., he should have found the key.

Colonel Colville, who had had charge of the entraining troops, testified that his men had reached the station at 3.50 a.m. The man in charge of the train had directed them to the station and given them permission to board, saying that the engine would be up in five minutes. The colonel said he had ordered the men to get on and then, walking to the end of the train with the adjutant, he had noticed it beginning to move. Asked by the Coroner, he said that it would have impossible for anyone to have entered the brake van. He had then been informed by the signalman that the position was serious and that nothing could be done. There had been no horseplay among the men and all of them denied having touched any of the hand brakes.

Sergeant William Field of the Royal Engineers testified that the train had passed his signal box at 60 miles an hour (surely an exaggeration?). Had the train been going more slowly he could have stopped it by applying one of its hand brakes, but it was impossible to do so at the speed it was travelling. There had been no brakes applied when the train passed him.

Captain Eggleton, in charge of the camp railway, stated that it was customary to apply the hand brakes of coaches stabled at the station and he had corroboration that the brakes

---

[829] *The Times*, London, England, 17th September 1917

of the accident train had been applied the night before. He concluded that the hand brakes must have been tampered with, something which had happened once before when one of the soldiers had turned a brake wheel in ignorance of its function.

The Coroner said that unless one of the German prisoners from a nearby prisoner of war camp had interfered with the hand brakes – and he did not see how that was possible – he did not think that anyone could have deliberately interfered with the brakes. Possibly, one of the soldiers, who were no doubt high spirited, had turned the wheel accidentally. He did not consider that anyone could be held responsible.

A verdict of accidental death was returned. [830]

Eleven months almost to the day after the Ciurea catastrophe in Romania a very similar catastrophe, with a similarly high death toll, occurred near Saint-Michel-de-Maurienne, France, on the Chemins de fer de Paris à Lyon et à la Méditerranée (P.L.M.). The accident resulted indirectly from the defeat of the Italian army by Central Powers forces at the Battle of Caporetto, which had ended in late November. In order to reinforce the Italian front line a French and British Expeditionary Force was then sent to northern Italy. A month later the front had stabilised and in order to improve the morale of the French troops a decision was made to grant leave to those who had also previously fought on the Western Front. Thus, on the night of 12th December 1917 military train No. 612 was returning to France from Vicenza in Italy. Passing through the Mont Cenis tunnel the train reached Modane, where two additional coaches were added for the continuation of the journey to Chambéry, from where the soldiers would disperse to their homes in various parts of France, with 15 days' leave covering the Christmas holiday.

On leaving Modane the train consisted of 19 Italian vehicles, consisting of 15 bogie coaches, two four-wheeled coaches and a baggage car at each end. The vehicles were all wooden ones, with steel underframes. It was carrying 942 men of the 46 and 47 divisions but few officers, as they had mostly left it at Modane to catch a Modane – Paris express. Such a heavy train would normally have been provided with two locomotives, especially in view of the fact that there was a long descent over the 15.5 kilometres from Modane (1,057 metres above sea level) to Saint-Michel-de-Maurienne (711 metres above sea level) with a ruling gradient of 30‰ (1 in 33) [831] and that the train was only provided with Westinghouse air brakes on its leading three vehicles, though it was also supplied with seven brakemen to operate the hand brakes. The maximum permissible load for a train hauled by one engine over this portion of the line was 144 tons and this train weighed no less than 526 tons, making the likelihood of being able keep control of it on the long descent remote. A second locomotive had originally been assigned to work the train but had instead been requisitioned to work a munitions train. According to a popular version of the events the driver of the train, Louis Girard, had at first refused to take the train out because of its heavy weight and inadequate brake power, but had been forced to do so at gunpoint by a military officer. Recent research has, however, shown this story to be untrue, and there was in fact no member of the military present at the station when Driver Girard was given the departure signal by the ordinary P.L.M. official at the station

[830] *The Times*, London, England, 18th September 1917
[831] Webpage – L'accident de Saint-Michel-de-Maurienne le 12 Décembre 1917

and his conductor. [832]

The train stopped at Modane for an hour to let other trains pass and left there at 11.15 p.m. For the first 1.5 kilometres from Modane the line was level, but it then fell for over a kilometre at 20‰ (1 in 50) to just before Le Freney. With the hand brakes screwed down as hard as possible the speed down this first descent was limited to 9 km/h but after a short level stretch past Le Freney there was another two-kilometre drop at 29.8‰ (1 in 34) and the speed of the train began to increase, despite the air brakes of the 4-6-0 locomotive and leading three coaches being hard on. The brakemen, alarmed by the increasing speed of the train, screwed down the hand brakes as hard as they could but the rolling stock was badly maintained due to the war, the brake acted unevenly, and the speed of the train continued to increase. The extreme braking meanwhile made the brake blocks and the tyres of the wheels red-hot, and the wooden floorboards of the coaches began to catch fire. By the time that the train had reached La Praz, just under 6 kilometres from Modane, it was very obviously running away, travelling at about 100 km/h, and Signalman Matthieu there telephoned St-Michel-de-Maurienne station just in time to enable them to stop an overcrowded ambulance train setting off towards Modane. An even worse accident was thus averted, though it was a very close run thing, as the Saint-Michel-de-Maurienne stationmaster had already whistled the ambulance train away before the telephone call was received, but the departure signal was not noted by the crew of the banking engine at the rear of the train, so it did not start at once. [833]

For the last 7.5 kilometres from Plaine des Berchettes to Saint-Michel-de-Maurienne the line falls continuously at 29‰ and 30‰ (1 in 34 and 1 in 33). The speedometer needle of the locomotive was past the highest speed on the dial: 120 km/h. The train passed La Saussaz and was nearing Saint-Michel-de-Maurienne but at this point a bridge over the River Arc was followed by a relatively severe curve, with a high retaining wall on one side and a slope on the other. The leading bogie of the locomotive had become derailed about the start of the curve and the engine became detached from its train and continued its journey, despite being partially derailed. Behind it the leading baggage car of the train failed to take the curve and collided with the retaining wall at an estimated speed of about 150 km/h. The remaining 18 coaches of the train then piled up behind the baggage car in a gigantic heap of wreckage. As mentioned earlier some of the coaches had already caught fire and the blaze now enveloped the whole of the wreckage. To add to the disaster a number of the soldiers had with them grenades and shells which they had been bringing home from Italy as war souvenirs and many of these exploded in the heat, adding to the destruction. The detached engine ran through Saint-Michel-de-Maurienne and Driver Girard finally managed to bring it to a stand about six kilometres further down the line at Saint-Jean-de-Maurienne. He then returned to Saint-Michel-de-Maurienne and thence to the scene of the derailment, accompanied by railwaymen from the two stations and by some Scottish soldiers who had been waiting at Saint-Michel-de-Maurienne to proceed to Modane.

The rescue operation was hampered by the rocky terrain in the vicinity, the pile of wreckage and the heat. Local people who had turned up to help with the rescue work were also reportedly stopped from getting to the site by the military police. A total of 424 bodies removed from the wreckage could be identified, while another 135 could not be

[832] Webpage – La verité sur l'accident du train de 12 Décembre 1917 à St-Michel-de-Maurienne
[833] *Railway Accidents of Great Britain and Europe* by Ascanio Schneider and Armin Masé, pgs. 222-225

identified. Thirty-seven bodies were found along the right of way between La Praz and the site of the derailment, these being soldiers who had either jumped from the runaway train or had been thrown off it. Some of the men who had jumped survived, however. Finally, another hundred or so soldiers died of their injuries in the next 15 days either in hospitals or while being taken to them. The final death toll was thus probably about 675. In addition to the soldiers two of the brakemen lost their lives. The military hospital at Saint-Michel-de-Maurienne and the nearby Bozon-Verduraz pasta factory were converted into field hospitals for those injured in the accident and also did service as mortuaries. [834]

Wartime censorship and the fact that the accident involved a troop train meant that news of it was supressed to a degree, though not entirely, several French newspapers mentioning it briefly whilst a Swiss newspaper published on 14th December, wrote of possibly 800 or 900 dead. [835]

Driver Girard and five other P.L.M. employees subsequently had to appear at a court martial on a charge of having caused the accident. They were all acquitted, no doubt justifiably, as the accident was basically a result of the exigencies of wartime transportation.

The accident remains up to the present (June 2021) the worst in French railway history and one of the worst anywhere in the world in terms of casualties.

A Preußisch Staatsbahnen troop train was proceeding from Gǔsten towards Aschersleben, Germany, on Thursday 7th February 1918 when a drawbar broke, and the rear vehicles of the train began to run back down the hill to Gǔsten. Reaching the station, there they collided with a goods train standing at the Einfahrsignal (home signal). Four coaches and a goods waggon were derailed and much damaged, 15 soldiers and two guards were killed, and 36 soldiers were injured, 21 of them seriously. [836]

An Italian troop train from San Pietro di Carso was involved in a serious accident on 17th February 1919. 'Due to a shunting error' at Nabresina a raft of goods waggons was pushed out onto the main line at the summit of an incline of about 16 kilometres in length from Trieste, Nabresina station being almost 4,000 metres above sea level. The waggons then ran away down the bank and collided with the troop train. Sixteen soldiers were killed outright, 18 were seriously injured and 250 slightly hurt. [837] [838]

By Wednesday 5th March 1919 the First World War had been over almost four months but its ramifications were felt in an accident which occurred on the morning of that day between Famechon and Namps, France, on the Amiens – Rouen line of the Chemins de fer du Nord. Troop train H.C. 53 was bringing British soldiers to Dieppe for demobilisation and had just left Famechon tunnel when it was struck by 26 carriages which had become separated from a preceding troop and had run back down an incline

---

834 Wikipedia – St-Michel-de-Maurienne derailment

835 *Neuchatel Feuille Davis de Neuchatel*, Neuchatel, Switzerland, 14th December 1917

836 *Eisenbahnunfälle Bind 1* by Ludwig Stockert, pgs. 38-39 (in German)

837 *The Age*, Melbourne, Victoria, Australia, 25th February 1919

838 *Le Petit Parisien*, Paris, France, 19th February 1919

for some 15 kilometres. The runaway vehicles had attained a considerable speed and the shock was considerable. Twelve coaches were wrecked.

Assistance came to accident site from Romescamps, Amiens and Beauvais. The bodies of 11 soldiers and the French conductor of the train were soon removed from the wreckage and it was known that 75 others were injured, 25 of them seriously. By the following day, another seven bodies had been discovered and it was stated that a number of the injured were in a critical condition. The accident completely blocked the tunnel and trains had to be diverted via alternative secondary routes [839] until the line was cleared on the evening of the 6th and traffic was greatly disrupted.

The accident was attributed to the breaking of a coupling of a German coach, one of a number of those which had been handed over to the Allies as part of the war reparations, another way in which the war had indirectly led to the accident. It was stated that the second troop train had been following the first at an interval of half an hour, implying that time interval working was in operation, perhaps because of wartime damage to the signalling. [840]

"The brakes wouldn't hold, they wouldn't work." Those were the dying words of Engineer Frank Clifford, of the New York Central Railroad, after a collision at Dunkirk, New York, United States, on Tuesday 1st July 1919. Eight passengers, three railroad employees and a trespasser died in the accident and 105 passengers, two railroad employees, three Pullman porters and one other person were injured.

Two trains were involved in the accident, the first of which was westbound train No. 41, which consisted of locomotive No. 4811, a mail van, two baggage cars, three day coaches, and five sleeping cars, in charge of Conductor Talcott with Engineer Thayer. The rearmost sleeping car had a wooden superstructure with a steel underframe. This train had left Buffalo at 12.40 a.m., 38 minutes late, arrived at Dunkirk at 1.33 a.m. and left there after station work had been completed at 1.43 a.m. It had only proceeded two train lengths when it was brought to a stand by the flagman in the rear Pullman sleeping car, *Valdivia*, opening the emergency valve after having discovered a hot box. The train was then set back into the station again, reaching there at about 1.50 a.m. By 2.18 a.m. the car inspectors had almost finished changing the brass in the journal box, but the train was still standing in the station, when it was run into by train No. 7.

Westbound train No. 7, known as *The Westerner*, consisted of locomotive No. 4808, an express car, a coach, four Pullman sleeping cars, a combination car, a coach, and four Pullman sleeping cars, in charge of Conductor Loomis and the aforementioned Engineer Frank Clifford and Fireman Frederick Stoltz. The first car was of wooden construction, the remainder were all of steel construction. It had left Buffalo Clinton Street station at 1.24 a.m., four minutes late, and was still travelling at an estimated 40 miles an hour when it collided with train No. 41. The front train was driven forward about 175 feet. The *Valdivia* sleeping car was telescoped over two-thirds of its length, practically demolished and thrown into a street running alongside the railroad. He penultimate sleeping car of the leading train was also overturned but not greatly damaged. The force of the impact sheared the boiler of engine No. 4808 from its frame and led to a considerable escape of

[839] *Le Figaro*, Paris, France, 6th March 1919
[840] *The Times*, London, England, 7th and 8th March 1919

steam. In train No. 7 the leading express car was crushed between the engine tender and the steel day coach behind and demolished and 15 feet of the day coach were telescoped. The remainder of train No. 7 was little damaged. Fire broke out in the wreckage but was extinguished by the local fire brigade before it had time to do much damage. Engineer Clifford, Fireman Stoltz, and Express Baggageman Charles Peck of train No. 7 were the three railroad employees who lost their lives. [841]

J.H. Charles, of Toronto, Ont., was a passenger in the front train, being in the next-to-last Pullman car. He was among the most fortunate and although badly shaken up was not thrown out of his berth. He found his suitcase and grip near at hand plus his clothes and quickly dressing, crawled out of the car. 'Mr Charles is a survivor of the *Lusitania* which was sunk by a German submarine. He was thrown into the water and clung to a piece of wreckage for about an hour when he was rescued by a lifeboat and landed at Greenston. {Sic – this should be Queenstown, Ireland.}

'Mrs Mary Chappell., of Columbus, O., was sleeping in the next to last Pullman of the forward train when the wreck occurred. John Brooks, of this city {Dunkirk} assisted her out of one of the windows and she was taken to the Knights of Columbus where she dressed. At her request she was taken to the hospital in an automobile and although somewhat bruised spent two hours in doing all she could to help there, remaining until a few minutes before her train departed.

'O.L. Hartman, 470, West Delevan Avenue, Buffalo, on the same Pullman, described the scene at the time, "Men, women and children who had been thrown out of their berths were struggling to get up. Many of them were tangled up in the bedclothes. I crawled out of the car and did what I could to assist others out. I was there when the engineer was pulled out of the wreckage and heard someone say he had heard the engineer tell that the air would not work."' [842]

Giving evidence at the Interstate Commerce Commission (ICC) inquiry into the accident, Flagman Frawley of train No. 41 testified that he had noticed the hot box before the train reached Dunkirk and when it had stopped there he had given a stop signal to the other members of the train crew, but it had not been observed. When the train had restarted he had sounded the air whistle twice, but the train had still not been pulled up, so he had opened the emergency valve. He had then gone back about a train length, lit a fusee and signalled the train back to the station. After that he had gone back to X tower, rather more than half a mile east of the passenger station, to explain to the towerman what was happening. The towerman had told him that train No. 7 had left Buffalo 15 or 20 minutes previously and he had begun to walk back further. At Roberts Road Crossing, 1,400 feet from the rear of his train, he had seen train No. 7 approaching and began giving it stop signals. He heard the engineer of train No. 7 whistling repeatedly as the train was passing Middle Road Crossing, 2,600 feet east of Roberts Road Crossing. He had continued showing a red fusee and had thrown his lanterns at the engine of the train as it passed him. At this point the engine whistle was still being sounded repeatedly and he had heard the engineer calling for brakes. He thought that the train was travelling at 35 or 40 miles an hour and there was no fire flying from its brakes, except perhaps those of the engine which he had not noticed. He did not go back as far as the home signal to see

[841] Interstate Commerce Commission report

[842] *Dunkirk Evening Observer*, Dunkirk, New York, United States, 1st July 1919

what indication its lamp was showing but its arm was in the horizontal or danger position.

Conductor Talcott of train No. 41 stated that after the train had backed to Dunkirk station he had realised what the problem was. Realising that train No. 7 was approaching and not stopping, he had shouted to the men working around the journal to get out of the way and had given his own engineer a proceed signal in the hope that he might be able to pull away in time. He had seen the flagman signalling to train No. 7 about Roberts Road Crossing and heard the engineer sounding his whistle.

Brakeman Nichols of train No. 41 stated that he had also seen the flagman giving stop signals.

Engineer Thayer stated that after backing his train he had alighted from his locomotive and had been standing in the street when he saw train No. 7 approaching, apparently moving very rapidly.

Towerman Siebert, on duty that night in X tower, stated that after train No. 41 had passed at 1.33 a.m. he had replaced lever No. 2, restoring the home signal to danger. After the train had left the station and then set back to it Flagman Frawley had come back to his tower. When the annunciator had indicated the approach of train No. 7 he had told the flagman, who had then set off back, swinging his lanterns. The home and distant signals were at danger. Notwithstanding this, train No. 7 had run past his home signal and passed his tower at 2.18, its engineer giving repeated short whistle blasts. It was moving at 40 or 45 miles an hour, with fire flying from the driving and tender wheels of the engine but not from the remainder of the train.

Standing about 800 feet of the distant signal for X tower on another track was freight train extra No. 4734. Engineer Smith of that train testified that the distant signal for train No. 7 had been displaying yellow, or caution, when that train passed it. Its engineer had not been working steam and he had heard him whistling. After the accident he had drawn the rear portion of train No. 7 back from Dunkirk to Erie and had experienced no problems with its brakes.

Head Brakeman Sargent of freight train extra 4734 said that he had heard the engineer of train No. 7 whistling for brakes at the Middle Road Crossing.

Conductor Eggleston of the freight train agreed that the engineer of train No. 7 had not been working steam, that he had been whistling for brakes at the Middle Road Crossing and that the brakes of the rear portion of train No. 7 had worked normally after the accident.

Flagman Hopp of the freight train stated that the engineer of train No. 7 had shut off steam when passing the caboose of his train. Both he and Brakeman Madden of the same train had heard the engineer of train No. 7 whistle for brakes.

Railroad Patrolmen E.M. Retzer and Burke confirmed Engineer Clifford's dying statement that his brakes had failed to work.

Flagman Galbraith of train No. 7 stated that its brakes had been tested after leaving Buffalo Clinton Street and had worked properly at crossings in Buffalo Creek with the Delaware Lackawanna and Western and Erie Railroads, while there had been a slight reduction at the Silver Creek water pan (troughs). He had noticed the whistling at Middle Road Crossing but thought that it was because there was someone on the line.

Conductor Loomis of train No. 7 confirmed the brake applications at Buffalo Creek and Silver Creek. He stated that he had noticed no brake application nearing Dunkirk, nor had he noticed any whistling.

Riding in the car with Conductor Loomis were Second Conductor C.S. Tifft and Brakeman Colmorgan. Both stated that they noticed the brake applications at Buffalo Creek and the whistle being blown nearing Dunkirk. Second Conductor Tifft said that the brakeman had said something to him about the whistling, but he had not paid any attention to the remark.

Baggagemaster Wilson of train No. 7 confirmed the testing of the brakes after leaving Clinton Street and the brake applications at Buffalo Creek and Silver Creek. He said that he had not heard the whistling.

Clarence Aber, a passenger on train No. 7 who was a former engineman, made a written statement as to having noticed the brakes of train No. 7 being tested after leaving Clinton Street.

Road Foreman of Engines Aske testified that he had examined the brake shoes of locomotive No. 4808 immediately after the accident. They had clearly been applied hard enough and long enough to become hot while those from the engine truck were blue from heat.

Assistant Car Foreman Kelly and Car Inspectors Drury, Polanowsky, Damback and O'Neill testified to inspecting and testing the air brakes of train No. 7 before it left Buffalo, when they had all been in good condition.

Wreckmaster Schack stated that when he reached the scene of the accident at 5.30 a.m. he had found the angle cock of the air brakes closed at the rear of the tender, though that could have been a result of the collision. He had also found, crushed between the tender and the remains of the express car, the body of a trespasser.

The dead trespasser was identified as a Charles Schiller, a Dunkirk resident. An associate of his, a chauffeur named J.F. Helfeldt, stated that he had been at Dunkirk station the previous evening with Schiller, who had suggested that Helfeldt accompany him east to Buffalo, and thence take a fast train west to Chicago. Helfeldt said that he had refused, whereupon Schiller had started towards the front of train No. 44, due to leave Dunkirk at 7.52 p.m., though he did not see Schiller board that train and never saw him again before the accident.

The ICC concluded that the brakes of train No. 7 had clearly been working properly upon leaving Buffalo and as far as Silver Creek water pan, 4 miles east of Dunkirk. It was also apparent that Engineer Clifford had shut off steam and applied the brakes in good time upon approaching Dunkirk had the angle cock not been closed, and when he had realised that something was wrong he had whistled for brakes. How the angle cock came to have been closed was a matter of conjecture, but the commission speculated that Schiller had become wet through from water flying up as the train passed over the water pan at Silver Creek and at that point had decided to abandon the idea of going to Chicago and had closed the angle cock in the expectation of being able to stop the train at Dunkirk, which it was booked to run through without stopping, so that he could alight and go home.

The commission went on to consider why train No. 41 had developed a hot box only 40 miles out of Buffalo and noted that the *Valdivia* sleeping car had been thoroughly overhauled at Buffalo and received new journal brasses only six days before the accident but did not come to any conclusion as to why the hot box had developed so quickly. [843]

In 1919 the American Smithsonian Institution organised an expedition to Africa, mainly directed at exploring the crop-producing possibilities of the continent, with botanist Homer Shantz as the lead researcher. [844] Also on the expedition, financed by Universal Studios and scouting for film locations, was silent film actor William Stowell. [845] The expedition having started in Cape Town, South Africa, by 24th November 1919 its members found themselves en route by rail from Sakania to Elizabethville (now Lubumbashi) in the Belgian Congo (now the Democratic Republic of Congo). The party travelled in a special coach attached to the rear of a goods train. At a location described as 'in the Engo forest' the train stopped for fuel and while stationary was run into from behind by a 'man-controlled water truck' which had broken away. The rear coach was 'smashed to matchwood' and three Americans, Dr William R. Armstrong of the Smithsonian Institution, actor William Stowell, and a photographer named Pliny Horne, all of whom chanced to be from Los Angeles, California, were badly hurt. Dr Armstrong died three hours later and Stowell the same night. By chance, another photographer named Henry Kobler and a newspaper correspondent named Edward Thierry had left the coach and gone forward a short while before; they escaped with a shock. It was initially reported that botanist Homer Shantz had died in the accident. [846] As an afterthought a newspaper report on the accident added that four 'natives' had also been killed. [847]

Spain suffered a runaway train accident on 23rd January 1920 involving an overnight working of the Compañía de los Ferrocarriles Andaluces. Given the poor condition of the company's trains and infrastructure the accident apparently came as no surprise to the locals. A newspaper remarked upon the company's impossible timetable, its chronic delays, the fact that no civilised country would permit its animals to be carried in its passenger coaches and so on. This was certainly not a luxury express service, but rather a mixed train, consisting of an engine and tender, a leading brake van, 23 goods waggons, a first-class coach, a second-class coach, two third-class coaches and a rear brake van. It was stated that the train should have had 12 brakemen to control it, which seems a large number. How many brakemen there actually were is not clear.

The accident happened on the Bélmez – Córdoba line at about 2.30 a.m. Train No. 2 ascendente (trains travelling to Madrid or to the starting point of the line if not Madrid were designated as ascendente, equivalent to the English use of up, meaning trains going to London; and those from Madrid or the starting point as descendente, equivalent to down in England) had passed through La Belanzona station and was descending an incline when it ran away. It was initially stated to have reached a speed of 120 km/h, though a later report modified this to 100 km/h. The train travelled as far as Pradillos

---

843 Interstate Commerce Commission report

844 www.naturalhistory.si.edu/research/botany/about/hisdtorical-expeditions/smithsonian-african-expedition

845 Wikipedia – William Stowell

846 *The Albany Advertiser*, Albany Western Australia, Australia, 3rd December 1919

847 *Nebraska State Journal*, Lincoln, Nebraska, United States, 8th December 1919

(No. 1) Tunnel where its rear vehicles left the track on a curve; the four passenger carriages, the rear van and three goods waggons, two of which contained pigs, ended up piled in a heap of wreckage. The engine became detached from its train and ran three kilometres before also leaving the rails.

On news of the accident reaching there a relief train was sent from Cercedilla station with rescue personnel, doctors, and medical supplies, whilst at dawn motor cars were sent to the scene from Córdoba. Meanwhile the surviving uninjured passengers had gone to the aid of their less fortunate fellow travellers. The rescue operation was made difficult by the darkness, the confined space within the tunnel and the mass of wreckage but by 2 p.m. two seriously and three very seriously injured passengers had been extricated, along with the bodies of three railwaymen and four passengers.

Those who lost their lives were Fireman Juan Avalos, Brakemen Agusto Jurado and Antonio Ropero, and passengers Miguel Molino (described as a pomegranate dealer), Francisco Montes and two others who had not at the time of the report been identified. Very seriously injured were passengers Nicolás Pérez, Federico Garrido and Juan Sánchez and seriously hurt were passengers Luis Ramfrez and Francisco Granado. [848] It was feared that there were more bodies buried in the wreckage. Although his fireman was killed the driver survived. He alleged that the train had run away because the brakes had failed to act when he had applied that at the start of the descent. The driver and conductor were arrested after the accident but were soon released when it became apparent that the runaway was unlikely to have been their fault.

By 4 p.m. a total of nine bodies had been recovered but the work of clearing the wreckage was far from complete and there was still talk of the possibility of finding more victims. [849]

A third newspaper report on 27th January described the funerals of the victims and gave a more detailed account of how the accident had happened. It stated that the train had begun to get out of control after leaving La Mocha halt. With the excessive speed of the train the goods waggons had begun to pitch and roll and as the train entered the tunnel the second-class coach had fouled the tunnel wall. This had caused a sudden interruption to the momentum of the train which had resulted in a coupling breaking and the engine separating from the vehicles. The locomotive had continued its run for 3.3 km to an iron bridge near Mirabueno Siding where it had fouled the bridge girders and was also derailed. Meanwhile the rear portion of the train, including the passenger carriages, had piled up in the tunnel as described above. There was no mention in this paper of the discovery of the bodies of any further victims in the wreckage. [850]

Both the railway company and the government came in for criticism after the accident, though Señor López, the Minister of Public Works, visited the injured in hospital and gave each of them 300 pesetas, as well as making a donation to the hospital. [851]

---

[848] *El Guadalete*, Jerez de la Frontera, Spain, 24th January 1920

[849] *El Defensor de Granada*, Granada, Spain 24th January 1920

[850] Ibid, 27th January 1920

[851] Webpage: Sevilla.abc.es-Cien años del accident ferroviaro de Los Pradillos de Córdoba, una catástrofe que se pudo evitar

At least four men died in an accident to a logging train near Edgewick, Washington, United States, on Friday 30th April 1920. An engine of the North Bend Lumber Company with just one car attached got out of control when a line shaft working the braking power snapped. There were 45 men riding in the car, 25 of whom jumped off before the engine left the track. Those who lost their lives were the engineer Harry Fleming and timber workers Joseph Finnigan, William Hallwood and John Sundquist. Seriously and possibly fatally injured were John Dahlstrom and Ted Mueller, while 22 others were less badly hurt. The train became derailed on a curve three miles from North Bend and overturned at the bottom of the incline. [852] [853]

Early on Monday 21st June 1920 cattle train No. 295 from Coruña, bound for Barcelona and consisting of an engine, ten cattle trucks and a brake van, had passed through Sabadell station, Spain, and was about a kilometre down the line at kilometre 245 when it was run into from behind at 03:35 by 13 vehicles which had escaped from Tarrasa station. The van at the end of the cattle train was broken to pieces in the collision and the wreckage blocked both tracks of the Norte line and demolished one or more telegraph poles, interrupting communications. The runaway coaches were gas lit and the gas cylinders exploded after the collision, causing a fire which destroyed almost all of them, as given the isolated location and the fact that it was 3.35 in the morning when the collision occurred, it was necessarily some time before help arrived. The first rescuers, station personnel from Sabadell and members of the Red Cross, concentrated their efforts in any case on saving those who were still alive rather than fighting the fire. As soon as news of the accident reached Barcelona a relief train was sent from there.

Five people died in the accident. Three of them were quickly identified as Antolin Corcola, described in a newspaper report as a military conductor, probably a member of the Civil Guard; and Remigio Garcia and Josė Carpio. The other two were also described as mozos (young men) and all four of the latter victims were probably drovers in charge of the cattle. Three other drovers were injured, one seriously.

The first train to leave Barcelona, a mixed one for Zaragoza and Bilbao, departed at the usual time of 06:20, crowded with passengers, who had not been told of the accident down the line. Shortly before reaching Moncada the train was stopped for almost two hours and on its passengers asking the cause of the delay they were then told of the accident. Another two hours was then lost at the site of the accident while the passengers were exchanged with those from the mail train from Zaragosa and Bilbao, stopped on the other side of the obstruction. One of the passengers on the train from Barcelona was a young man, who learnt at Moncada of the death of his father in the accident and in his grief began to run down the line towards Sabadell. He was brought back by two members of the Civil Guard and eventually continued the journey on the train. The departure of a special train of pilgrims, due to leave Barcelona for Limpias at 7 a.m., was postponed for 24 hours because of the accident and the blockage of the line.

The Tarrasa stationmaster told newspaper reporters how the accident had come about. In accordance with the usual practice during the night the staff made up the trains that

[852] *Morning Oregonian*, Portland, Oregon, United States, 1st May 1920
[853] *Daily East Oregonian*, Pendleton, Oregon, United States, 1st May 1920

were to leave for Barcelona the first thing in the morning. On track one were placed the carriages for the 7 a.m. departure, on track two the stock for the 6 a.m. departure and on track three, as usual, goods waggons which were to be added to the passenger trains. The cattle train had passed, and the night foreman and his men had stopped work for a break and a meal. While they were eating their food, completely unaware of any problem, they were advised by telegram by Sabadell that a number of runaway coaches had passed through there going towards Barcelona, and which must have come from Tarrasa. The night foreman had then gone to have a look and had found that five large coaches, seven small coaches and a van, all the stock which had been prepared for the 6 a.m. train, had vanished from track two. It was impossible, he said, to understand how it could have happened. The marshalling of the stock had been completed some time earlier and no one had been in the area.

Just after the message had been received from Sabadell the pointsman at the Barcelona exit from the station had appeared and said that he had seen the empty coaches going by and had thought at first that it was a shunting manoeuvre. The 13 vehicles weighed about 300 tonnes, the line fell steeply towards Barcelona, and running freely the vehicles could have reached a speed of (according to the newspaper) 400 kilometres an hour (250 miles an hour). [854] This must be a considerable exaggeration on the part of the reporter – even were such a speed attainable the vehicles would surely have become derailed on a curve well before they caught up the cattle train unless the line was dead straight.

The Chicago Milwaukee and St Paul Railway was unusual for an American railroad of this date inasmuch as its main line over the mountainous section through Montana and Idaho was already electrified by 1920. It was on this stretch that serious runaway accident occurred on 30th September 1920, at Piedmont, Montana.

Over the 5½ miles eastwards from Vendome to Piedmont the line fell continuously, with gradients between 1.74 per cent (about 1 in 57) to 2 percent (1 in 50). Helpers (banking engines) were used in both directions. These helpers were usually marshalled about the middle of trains, their exact location depending on the comparative strength of the cars. The helper engines remained on the trains after the summit of the line had been reached so that the regenerative braking feature of the electric motors could be used to better control the speed of the descending trains, though as this feature was only operative at speeds between 17 and 22 miles an hour the air brakes had to be used to stop the trains. [855]

At about 11.50 p.m. on the day before the accident train extra 10203, in charge of Conductor Lyons with Engineer Hyrup, arrived at Vendome where it had to cross an eastbound gravel train. This having been done the conductor signalled the engineer to restart. Engineer Davies on the helper engine found that he was unable to restart his train with a current of 200 amperes and so increased it to 220 amperes, whereupon the rear portion of the train, that is the portion to the rear of the helper engine, parted between the seventh and eighth cars and began to run back towards Piedmont. About 30 cars then ran back down the incline and had almost reached Piedmont when they collided with the rear

[854] *La Vanguardia*, Barcelona, Spain, 22nd June 1920
[855] Interstate Commerce Commission report

of the gravel train that had been crossed at Vendome. Fortunately for Conductor Curwin of the gravel train, who was standing at the rear of his train, had time to get out of the way.

Eight people died in the accident; seven of them were trespassers but the eighth was an off-duty engineer, killed as he heroically attempted to prevent the accident. The Interstate Commerce Commission report makes no mention of Engineer A.E. Mutz, but his actions were well recorded in the newspapers. Engineer Mutz was off duty and with his wife had been at Piedmont station waiting to catch a passenger train east to Three Forks when he saw the runaway cars coming down the hill. He had immediately jumped into an electric engine and attempted to pull the gravel cars into a siding clear of the main line. He succeeded in getting a good part of the train into the siding before the runaways struck the remainder of it. Tragically, the collision threw an insulator from an overhead power line through the window of the engine that he was operating, killing him. [856]

Four of the trespassers who lost their lives were named as James Murray, of Butte; Art Digman, of Missoula, and Roy Somnos and R. Fassin of St Paul.

Head Brakeman Brown of train extra No. 10203 told the Interstate Commerce Commission (ICC) inquiry that following the direction of Engineer Hyrup of the helper engine he had got off the train as it pulled into the siding at Vendome to check for leaks in the brake pipe. He had found and repaired two leaks and was back on the helper engine when the back-up movement had commenced, and the train had parted. He had then gone back with Fireman Merrill and while the fireman had rolled the pulled-out coupler off the track he had signalled for the helper engine to set back and had started to put down hand brakes. Then the cars had begun to move back, he had signalled the helper to stop, and he and the fireman had pinned down five or six hand brakes before the cars got beyond their control. The head brakeman thought that it had been three or four minutes before the cars began to move, Fireman Merrill thought it had been only a minute or so and Conductor Lyons that it had been six or seven minutes. The head brakeman stated that the cars being back up had not touched those which ran away.

The only railroad employee at the rear of the train when the cars began to run away was Rear Brakeman Joseph Gormley, who was too badly hurt in the accident to be questioned for some time but who said he did not remember anyone from train No. 10215 shouting to him about any leaks, and he had not carried out any repairs.

Conductor Lyons stated that on the arrival of the remaining 41 cars of his train at Butte tests showed that only 30 of the brakes were in proper working order but no brake test having been carried out at Three Forks, it was impossible to speak of their condition at Vendome.

Car Inspector Bond at Three Forks stated that when Conductor Lyons had told him that his train was made up he had coupled up the air hose, signed a copy of form 975 and left it in the caboose. There was a space at the bottom of the form to indicate the number of brakes cut out but before he had filled this in he had gone to attend to a preference freight which had just arrived and by the time he had finished with that job train extra No. 10203 had departed.

Conductor Lyons stated that trainmen were not required to make a brake test at Three

[856] *Morning Oregonian*, Portland, Oregon, United States, 1st October 1920

Forks, and he had not seen one made. He admitted that he had not at any time looked at form 975 or made any other attempt to determine the condition of the air brakes. At Piedmont, the road motor had pulled the first 34 cars out onto the main line, then the helper motor had pulled ahead and coupled the rear 62 cars to the rear of the first 34. The air had been working through to the caboose, but no full brake test had been done as per the rules.

Engineer Hyrup agreed that no brake test was carried out at Three Forks and said that the car men who were supposed to make the test never told him how many brakes were cut out or otherwise inoperative. He thought that had the brakes at the rear of the train been in good condition they would have held the train at Vendome for at least 15 minutes.

Engineer Elliott and Head Brakeman Lee of eastbound train extra No. 10215 testified that when train extra No. 10203 had pulled into Vendome they had noticed a bad leak 10 or 12 cars from the caboose and the head brakeman said that he had called Rear Brakeman Gormley's attention to it.

Engineer Davies of helper motor No. 10227 stated that the gauge of his motor had showed only 40 pounds of brake pressure upon leaving Piedmont and he did not suppose that the brake pressure in the caboose, 62 cars back, had been more than 30 pounds.

Car Inspector Bond said that he had told Rear Brakeman Gormley that the air was coupled up and that if he was not around, to check that the air was through to the caboose before departing. He said that it was customary in the case of through trains to inspect them upon arrival and just to see that air was through to the caboose when they departed. In the case of trains such as extra No. 10203 which started from Three Forks he made a slightly fuller test but not as thorough as it ought to have been due to lack of time and assistance.

The ICC noted that while it was impossible to determine the state of the hand brakes on the runaway gravel cars an examination of 64 gravel cars at Three Forks, none of which were on the repair track or marked "bad order," the hand brakes of 23 were inoperative.

Timetable Rule No. 1 of the Special Rules for Operating Trains over the Mountain Grades stated that when there was no helper at the rear the last car must be one equipped with a good brake and must be manned at all times.

Rules Nos. 4 and 5 stated that before descending or ascending a grade a full air test must be carefully supervised by the conductor, the air cut out on any car with faulty brakes, an "Air cut out" card attached, and the engineer notified of all such cars.

The evidence indicated that Rules 1 and 5 were often disobeyed, but Conductor Kirwan stated that it was not customary to observe Rule No. 5, while Assistant Superintendent Hamilton and Travelling Engineers McAvoy and Maxwell stated that they did not consider it necessary to observe Rules 1 and 5 at Piedmont or Newcomb and no one had ever been disciplined for not observing them.

The commission concluded that the accident had been caused by inoperative or ineffective brake equipment on train extra No. 10203. Investigation had disclosed that no proper inspection of the brake equipment of the train had been made at its initial terminal or at any other point. A proper brake pressure had not been maintained and the efficiency of the brakes once the cars had separated had been greatly reduced if not entirely destroyed, as indicated by the fact that the mostly empty cars had started down the grade in a very short interval of time. Although the tests should have been performed by the car men at Three Forks this in no way relieved the train crew of starting before they were sure that 85 per cent of the brakes were in service.

Rule 932 stated:

See that trains are properly made up and inspected, that at least 85 per cent of all cars in trains are equipped with air brakes in service; that proper signals are provided, and that air-brake, signal-whistle and steam-heat tests are made before starting. Make same tests whenever any cars are taken from (except rear) or added to trains.

Rule 933 stated:

Do not start trains from inspecting stations until inspector's work is completed.

For his failure to observe these rules the ICC considered Conductor Lyons to be open to the most severe censure. He and Engineer Hyrup were also at fault for failing to comply with Rule No. 5, but here they were only following common practice, and the opinion of higher officials that the observance of rules 4 and 5 was not necessary at Piedmont. Regardless of this opinion, had rule No. 5 been observed the dangerous position of the brake system would have been discovered and the accident averted.

The ICC then carried out a trial by having inspectors ride on train extra No. 10207, due to leave Three Forks at 11 p.m. With the discovery of various faults, it took nearly five hours to complete the testing of the air brakes to obtain the minimum percentage of effective brakes required by law. After the train had finally departed it was found impracticable to carry out Rules 1 and 4 or 1 and 5 and flagging rule No. 99 at Donald, between Piedmont and Newcomb, due to the number of train crew needed. Assistant Superintendent Wiltrout stated that he considered rule No. 99 to be the most important but he could not say that it was always done.

Investigation also showed that at terminals where car men were on duty it was the custom for them to make the air brake test, the conductors depending on the information in form 975 to determine whether 85 per cent of the brakes were operative, despite the fact that the forms only showed brakes cut out, and frequently no brakes would be cut out, but the brakes still not be effective on 85 per cent of the train.

The ICC noted that Three Forks was situated between two mountain grades, the one on which the accident had occurred, 35 miles to the west and another 75 miles to the east. The commissioners were not impressed by the statement of Conductor Kirwan that the only time he had known of a train being delayed on account of an airbrake test was when some Government inspectors had been present.

'It also appears that at many points in this territory important rules are frequently violated, apparently with the knowledge and tacit consent of supervising officials. It was shown that in many cases violations of the rules were due to the fact that there was an insufficient number of employees in the crew to comply with all of the rules at the same time.

'It is a matter of surprise that such dangerous practices should have been allowed to continue until finally they resulted in the occurrence of a disastrous accident, causing the loss of many lives. The continued existence of such conditions is a serious menace to the lives of the travelling public, as well as of employees, and the responsible operating officials of this railway should take prompt and efficient measures to improve air-brake conditions on this line and to insure the strict enforcement of their rules, if the occurrence of future accidents of this character is to be avoided.' [857]

What at the time that it happened was the worst railway accident to have ever occurred in Western Australia involved a runaway train on a private railway running for about six miles from Wokalup to Mornington Mills and owned by the Millar's Timber and Trading Company. The terminus of the line was situated in the hills at a considerable height above Wokalup and the line therefore fell practically continuously from the terminus to the main line.

On Saturday night, 6th November 1920, a train of 26 bogie and three small waggons was made up at Mornington Mills and was fully loaded with wooden railway sleepers, at least 4,000 of them, and general freight for shipping from the port of Bunbury. The total weight of the train was about 650 tons. It was in charge of Guard John Paulson, with Driver Clarence Cooke and Fireman Thomas Wilton on the engine, which was named *Jubilee*. The engine was to make the descent running tender first. When the train set off at about 9 p.m. it was also carrying a number of mill hands, who were taking advantage of the chance of a free ride to Wokalup, or were bound for nearby Harvey, where a circus was showing that evening. There was no brake van attached to the train, so the men were riding, as they normally did, on one or other of the loaded trucks.

All went well on the beginning of the journey down to Wokalup. About 2½ miles from there was a curve to a right, followed by a bridge over several spans across a stream and then a curve to the left. Here also the line was falling very steeply, with gradients between 1 in 30 and 1 in 40, and the train having got out of control by this point its engine left the track a couple of hundred yards beyond the bridge and the following waggons piled themselves up upon one another, forming a heap 25 feet high. Driver Cooke was thrown from his engine and somehow survived, but Fireman Wilton was buried in the wreckage and died. The men riding on the trucks were hurled from them as they left the track and buried beneath the mass of sleepers. The one exception was a man named John Flynn, who had been riding on the last truck and had somehow escaped serious injury. After the accident he ran all the way to Wokalup, where he arrived exhausted and raised the alarm.

By the time that the rescuers reached the scene of the accident darkness had fallen. They spent some minutes calling out, hoping for a response from one or more of the victims buried in the wreckage, but got no reply. They then came across Driver Cooke, lying 50 yards from his engine with a broken leg but otherwise not too much hurt. He asked for and was given a cigarette and was made comfortable. The body of Fireman Wilton was next discovered. He was severely mangled and very obviously dead; his hand was still clutching the hand brake. Guard Paulson was still alive, as was a passenger, Lloyd Morey, though both were injured. They and Driver Cooke were placed on a repairer's trolley and taken to Wokalup and thence via a special train to Bunbury. At Bunbury, a motor ambulance met the train at the Stirling Street Crossing and took them to the hospital, where the guard died shortly after his admission.

[857] Interstate Commerce Commission report

News of the accident reached Mornington Mill by 11 p.m. and a special train took 150 men to the scene. By the light of hurricane lamps, they worked until 2 a.m. without success, when efforts were abandoned until daybreak. By early Sunday afternoon the bodies of six victims had been recovered; five were dead and the sixth, Edward Delaney, died of his injuries in Bunbury Hospital. The mill hands who lost their lives were: Dennis Leatch; Ellis Lake; James McNamara, a returned soldier who had only arrived from Perth on the Friday evening; William Malacari, Frank Winfield and the aforementioned Edward Delaney, another returned soldier. It was known that another man, going by the name of Wallace, but whose real name was Walter Maggs, had also been on the train, and so the search for him was continued, and at 3.45 p.m. that Sunday afternoon his body was finally located. It was so pinned down by sleepers that even then it was found very difficult to extricate it and finally Dr Kennedy had to amputate a leg from the corpse so that the remains could be removed. It was stated that Maggs' wife was coming from South Australia to visit him and was due to arrive in Perth on the Trans-Continental train that Sunday. But for chance there could have been two more victims, as two other mill workers had attempted to board the train as it was pulling out of the terminus. One had tripped and fallen as he was running for the moving train and the other had been prevented from boarding it by Guard Paulson. [858]

Driver Cooke had recovered enough from his injuries to relate his story to a newspaper reporter a few days after the accident. He stated that the train had had a load equivalent to 53 trucks (assuming a bogie truck was equivalent to two ordinary trucks then 26 x 2 = 52 +3 small trucks =55 so perhaps his calculation was a little out, but the newspaper report states 53). It had certainly been a big load, but not, he considered, excessive, and he had taken similar loads before. Before they had started Guard Paulson had come to him and told him that he had pinned six brakes down. He had told the guard that that should be all right, but it would be well for the guard to stand by to apply more brakes should he (Driver Cooke) whistle. A mile and three-quarters from the mill he had noticed that the load was beginning to push the train and had whistled for brakes. This was about half a mile short of a place they knew as 'The Knob' where the train was always slowed down, and the remainder of its brakes pinned down. By the time that they had reached 'The Knob' the train was going too fast to enable the guard to do any pinning down and he had exclaimed to his fireman and Traffic Inspector Christopher Wilton, "Good God! She's pushed us over the Knob!" The traffic inspector was the fireman's father. Soon the train was going twice as fast as it should have been, but he decided to stick with her in the hope that Guard Paulson could pin down some more brakes. At the point, the traffic inspector had told him that he intended to jump for it, and he had warned the inspector, "Stop on here, man; you'll break your neck jumping off!" In fact, the driver was wrong, the inspector had jumped anyway and had survived more or less unhurt. Driver Cooke said that he had put down sand and had reversed the engine, but the locomotive was jumping so much that he had put it back into forward gear. Until they reached the bridge he had been fairly confident that the engine would stay on the track, but he said he had been 'a little frightened' of crossing the bridge – doubtless an understatement. The speed of the train not having diminished by the time that they reached the bridge he realised that there was little hope of avoiding a derailment. He had told Fireman Wilton to get in the gateway but not to jump until the engine left the rails.

---

[858] *Blackwood Times*, Bunbury, Western Australia, Australia, 12th November 1920

As they came to the cutting beyond the bridge the engine tender heeled over, and he was thrown from the cab as if by a catapult. For a long time, he seemed to be flying through the air, then he had hit the ground and rolled over several times; he had a dim recollection of sleepers flying past him and heard steam and water being blown out of the engine. A little later Joe Flynn had come to him, and he had told Flynn to hurry to Wokalup and get assistance.

With regard to the brakes Driver Cooke said that it was usual to pin down the brakes of several trucks at the mill, then to slow up for all the remaining brakes to be pinned down before the train reached 'The Knob.' Just now and then steam had to be applied to take the train over 'The Knob.' Two or three of the trucks in the train had vacuum brake cylinders. He did not know where they were in the train, they never bothered with them as they would not be sufficient to hold the train. He had had 25 inches of vacuum down on the engine and the brake had been applied all the way from about a mile and three-quarters from the mill. [859]

The inquest upon the victims of the accident was not concluded until 23rd February 1921 when the jury returned a verdict that 'the accident was caused by an insufficient brake power being applied at the proper time, and in the absence of any regulations governing the running of the train, the jury did not feel justified in attaching blame to anyone. A rider was added that regulations should be formulated for the safe running of trains on this line.' [860]

On Thursday 16th December 1920, the 5.10 p.m. narrow-gauge mixed train from Hüpstedt to Silberhausen, Germany, was fully loaded. Behind the engine were a two-axled passenger coach, a brake van and 19 waggons loaded with potash and equipped with six hand brakes which should have been operated by five brakesmen. It was frosty, with an icy east wind, and snow lay in the cuttings from the previous day's fall. The line fell towards Silberhausen at 1 in 50 and 1 in 55. At the beginning of the fall the brakes were applied hard enough to almost bring the train to a standstill, but soon afterwards it began to get out of control. The train ran through Silberhausen station, continued to the end of a siding beyond and smashed through the buffers. The engine, passenger carriage, brake van and nine of the potash waggons fell 25 metres to the valley below, the remaining ten waggons piling up at the end of the siding. Fourteen or 17 passengers died in the accident according to different accounts, but the higher figure is probably correct and includes three victims who died on the way to or after reaching hospital.

The cause of the runaway was shrouded in controversy. The statements of waiting passengers who witnessed the accident from the station did not tally with those of the railwaymen involved. The speed at which the train ran into the station was disputed, as also was how many people jumped from the train before it crashed.

The railway company had only a handful of employees: a conductor, two drivers/firemen, a clerk at Hüpstedt who also acted as a shunter and a ganger who also acted as brakesman. If he needed assistance Ganger/Brakesman Hebestreit could try and get men from the nearby village, but with the potash mines busy, experienced brakesmen

[859] *South Western Times*, Bunbury, Western Australia, Australia, 13th November 1920

[860] *Murchison Times and Day Dawn Gazette*, Cue, Western Australia, Australia, 25th February 1921

were hard to find. The near stoppage of the train near the start of its journey was probably due to the mishandling of the brakes by the inexperienced assistant brakesman. He might then have released them a little, jumped off the train and gone home, possibly by agreement with the ganger, who thought that he would be able to control the train single-handedly by going from waggon to waggon pinning down brakes. Another possibility was that when the train began to get out of control the assistant brakesman had panicked and jumped off. Hebestreit stayed with the train all the way; he survived but was injured. [861]

At 18:00 on the evening of Monday 23rd May 1921 a special train was bringing home a party of Bagnères-de-Bigorre master bakers who had attended a dinner held by their association at Artigues. Both places are in the mountainous Hautes-Pyrénées department of France, and the line fell steeply towards Bagnères. Their special train on the Tramway de la Bigorre metre gauge electric line consisted of a tractor, carriage, and trailer. As noted above in the account of the August 1916 accident the gradient of the line increased between some waterfalls and the hamlet of Gripp, France, and it was on this stretch that control of the train was possibly lost so that it accelerated until it became derailed and collided with a wall bordering the line. [862] Nine passengers died and 11 or 13 were injured.

Alerted to the accident, doctors from Bagnères hastened to the scene to treat the injured, who were taken to hospital in Bagnères. At the time of the newspaper report an inquiry was in progress conducted by the parquet (examining magistrate) and company engineers. The Paris correspondent of the *Times* in London stated that the train had got out of control, [863] but a French newspaper report suggested that recent heavy rain had caused the track to subside.

However, two days later the newspaper confirmed that the inquiry into the accident had established that the train had indeed run away because it had been too heavily loaded, and that on a sharp curve the wheels had left the rails despite the brakes working properly. The line had severe curves, the sharpest having a radius of only 16 metres, and very severe gradients; probably significant was the fact that there was a gradient of 100‰ (1 in 10) for 3 miles between Artigues and Gripp. [864] The paper noted that six people had died in the previous derailment on the line on 31st August 1916 and pointed out that nothing had been done since to prevent a further accident, such as attempting to improve the most dangerous points of the route or limiting the loading of trains. [865]

Runaway train accidents did not always end in deaths and serious injuries of course. On the morning of Saturday 9th July 1921 a special train loaded with 500 tons of coal left Bathurst, New South Wales, Australia, for Blayney. It was in charge of Guard Cook with Driver T. Austin and an unnamed fireman. Beyond Georges Plains, a few miles west of Bathurst, it had to ascend the steep incline of Tumulla Hill, and to enable it to do so a banking engine was provided to assist it from the rear, manned by Driver W. Moffat and

---

[861] *Von Siegelsdorf nach Aitrang* by H. J. Ritzau, pgs. 233-7 (in German)
[862] *L' Œuvre*, Paris, France, 25th May 1921
[863] *The Times*, London, England, 25th May 1921
[864] Wikipedia – Tramway de la Bigorre
[865] *L' Œuvre*, Paris, France, 27th May 1921

Fireman J. Thompson. As the train reached the steeper part of the hill the load proved too much even for the two engines, and the decision was taken to detach the front engine and run the train back down the bank to Georges Plains, where presumably the plan was to divide it and work it up the hill in two goes. Unfortunately with the front engine detached and unable therefore to provide any braking the banking engine proved unable to hold the train on the falling gradient, and it began to run away. Meanwhile a mixed train from Bathurst had arrived at Georges Plains. Stationmaster L. Markham, warned by the whistling of the banking engine, ran towards the points to divert the runaway to the loop line, then, realising that there was no time to do so, he shouted to the ten passengers on the train to jump for their lives. They did so, as did Driver Liddy, Fireman Hanrahan and the guard of that train. Driver Moffat, on the runaway train, which was now travelling at 20 mile an hour, also jumped, and escaped with a severe shaking and abrasions. Fireman Thompson was either afraid to jump or decided it safer not to, instead he jammed himself in a corner of the cab, was thrown from the engine on impact and miraculously escaped unhurt. The runaway engine was severely damaged, and the guard's van was telescoped into its tender. The engine of the passenger train sustained much less damage, but its buffers and front were smashed. Altogether it was a lucky escape. Train services were not disrupted, being continued via the loop line while the mess was cleared up. [866]

Another runaway accident in Australia, near Macedon, Victoria, on Saturday 4th February 1922, did not have such a happy ending. The 8.15 p.m. through Bendigo goods train left Melbourne Spencer Street on Friday night in charge of Guard Thomas Evans. It consisted of 45 waggons weighing 624 tons and was double-headed by an A2 class and a DD class engine, with Driver T. Garland and Fireman Walter Lawler on the pilot engine and Driver Thomas Monaghan and Fireman C. Straughair on the train engine. The line climbed very steeply through Macedon, with a gradient of 1 in 50 from Gisborne on the Melbourne side to Woodend on the Bendigo side, and as in the previous accident the severe gradient and the heavy train proved too much even for two engines. The train passed through Macedon at 1.03 a.m. and about a quarter of an hour later had only covered another mile and a half. Half a mile from the summit the drivers stopped to get up more steam to surmount the bank. When the train restarted the strain on the couplings proved too much and 13 waggons and the guards van broke away and began to run back down the bank.

Stationmaster J. Pennington heard the runaways roar through his station at a time he put as 1.17 a.m. and what he described as an agonising cry. He went to ring the stations on either side and almost at once there was a thunderous crash from the Gisborne side of the station. Another goods train had left Gisborne seven minutes earlier, so he knew that something bad must have happened. He awakened Porter E. Jacobs and the two made their way along the line with handlamps. They had covered about a mile when they met a figure coming towards them which turned out to be a dazed Fireman Philip Chandler, on his way to get assistance. The fireman said that he had stumbled across the body of a man 200 yards from the collision point, evidently the body of the guard who had jumped from his van. The fireman and the porter then went back to the accident while the stationmaster made his way back to the station to block the line.

---

866 *The Wyalong Advocate*, Wyalong, New South Wales, Australia, 12th July 1921

Driver George Kellam and Fireman Chandler of the second goods train said afterwards that they had heard the runaway waggons coming but had not taken much notice of the sound at first, thinking that it must be a train on the up line, and that the red light that they saw was the semaphore at Macedon. The crash came before they even had time to shut off steam and they were very lucky to escape with nothing more than a severe shaking. At the rear of the train Guard William Holbery was thrown against the window of his van. Picking himself up he made his way back down the line like a good railwayman and put down detonators to protect the rear of his train. He then staggered on to Gisborne station, arriving there at about 2.45. Explaining what had happened to Porter G. Wilson he then collapsed, exhausted by his long walk in the dark, suffering from a head injury and having had his false teeth broken. Stationmaster D. King was roused by Porter Wilson and made his way to the scene with some of the station staff.

Living near the scene of the accident at milepost 42 was Mr E. Williams, a eucalyptus oil distiller, who spoke to a newspaper reporter. "The noise of the runaway trucks did not waken me because I am accustomed to the noise of the trains. Bur the crash certainly woke me up with a start. The house trembled to the very foundations, as if there had been a mild earthquake, and the sound echoed and re-echoed in the still night air along the cutting and through the ranges. I could not realise what had happened. My dogs were roused by the echoes, and as the noise subsided I heard a man call out, 'Have you drawn the fires?' This turned out to be the fireman speaking to the driver of the train. I arose at once and went on to the line, and the first thing I saw was the body of Evans lying between the two tracks. This must have been after the fireman had passed on his way to Macedon. Nothing could be done until assistance had been sent for."

Describing the damage the reporter wrote, 'The guard's van was smashed to splinters as if it had been a matchbox. The funnel was off the A2 class engine, which was hauling the second train, the front was smashed in, and piled on top of the engine were three trucks, one on top of the other. Iron and steel fittings were so entwined and so compact as to give the impression that they had been welded together. Solid wheels were cracked like a china plate, and the thick steel bottoms and sides of trucks were bent as if they were mere cardboard. Nuts and bolts were found on the roadside nearby. Among all this was coal, bricks, and fruit cases scattered indiscriminately. The remaining trucks were lying about in fantastic shapes on the line and down the 15ft. embankment. The engine had been derailed, and the cowcatcher and front equipment were protruding from the debris. Small portions of the guard's van, including Evans's seat were found on both sides of the embankment a considerable distance away, so that even if he had remained in the van in the hope of rescue farther down, the line he would have been crushed to death. From the New Chum Bridge to the 42-mile post it is estimated that the gradient drops about 300ft. in the three miles, so that with the heavy trucks let loose on this grade the speed when the collision occurred must have been, experienced railwaymen say, nearly 100 miles an hour, The second train was climbing the grade at 15 miles an hour.' The estimated speed of the runaway waggons, while not as high as that given in the report on the Sabadell accident in Spain the year before, is probably well above the true figure. The 41 waggons behind the engine of the second goods train were not much damaged though the first three, like the engine, were derailed.

Upon news of the accident reaching Melbourne a breakdown train was sent to the scene. Clearing the line necessarily took time and the passengers, parcels, and mails of the morning up and down mail trains had to be transferred from one train to the other, a process which took an hour. Mails and newspapers to places beyond the scene of the accident were considerably delayed. The up line was clear by midday and both lines were clear by 2.30 p.m. [867]

The events leading up to the accident were gone through at the inquest on Guard Evans, where various facts not brought out in the initial newspaper reports emerged. The A2 engine, it was stated, had a defective smokebox which had interfered with the firebox draught, and Driver Garland had to help Fireman Lawler with the fire. Even so, the train had stalled half a mile from the summit, the boiler pressure on the A2 engine having fallen to 140 lbs per square inch, when it should have been 180 to 185. The hand brakes of the two engines were applied when the train stopped, and the drivers had conferred; Driver Monahan was in favour of dividing the train, but Driver Garland wished to save time by getting up steam on his engine and then pressing on. His view prevailed. When the pressure on the A2 engine had gone up 160 lbs an attempt was made to move but failed. A few minutes later the engines were blowing off steam and another attempt was made. This time the train moved a few yards, then stopped, and Driver Garland thought that part of the train had broken away. Taking a lamp he went back and found that he had been right, and exclaimed, "The train has broken in half. I heard the trucks rolling down the hill. God help the guard!"

In addition to the engine being in a poor condition it was stated that the coal had been poor. Driver Garland said it seemed to be slack bore-hole coal, though there was plenty of good coal available at North Melbourne. The load of the train had also been slightly over the maximum authorised load of 622 tons.

Driver Garland agreed with the evidence of his colleague and produced a broken drawbar pin from the truck which had broken away. He said that had a flaw and the iron was crystallized. The marshalling of the train was also good for station work but not from the driver's point of view. The heaviest waggons should have been placed immediately behind the engine and the lightest ones at the rear. [868]

On the final day of the inquest Alfred Smith, Chief Mechanical Engineer of the Victorian Railways, stated that the drawbar pin of the truck which had broken away had been ⅛ inch too small. He had not been able to trace how this had happened. The coroner returned a verdict of misadventure, stating that the accident had been due to a number of causes which had fitted together like a mosaic. Guard Evans had been killed in a fall from his van, which had broken away from a goods train, owing to the breaking of a drawbar. The drawbar pin had been deficient in construction, and unsuitable for its purpose by reason of its size. There was no evidence to determine how it came to be inserted in the drawbar. [869]

Two trains crowded with pilgrims bound for the Catholic shrine at Lourdes featured in a major disaster on the Chemins de fer du Midi in the early hours of Tuesday 1st August 1922 on the line between Agen and Tarbes, France. The first of them, train No. 7509 from Moulins, was being followed a few minutes later by one from Montluçon. The two trains passed through Mirande and some kilometres further on reached Mielan station, the first

[867] *The Argus*, Melbourne, Victoria, Australia, 6th February 1922
[868] *The Age*, Melbourne, Australia, 8th March 1922
[869] *The Recorder*, Port Pirie, South Australia, Australia, 16th March 1922

train reportedly passing there at 04:16. From there the line climbed steeply over the Gers hills before falling again to Villecomtal-sur-Arros and Tarbes. The first train almost reached the summit of the incline but then stalled. The train having come to a stand, the conductor went to the rear to protect it, but it then began to run back, despite the desperate efforts of the driver to stop it by applying the brakes. It ran back slowly at first but then more and more quickly. Some of the passengers on the runaway train reportedly realised what was happening and attempted to jump from the moving train but did not have time to do so. The Montluçon train was said to have been 30 minutes behind (see below) and the collision occurred near level crossing No. 110 at Languian-Mazous at 04:30.

The accident scene was described as appalling. The conductor's van at the rear of the Moulins train was crashed and the two last carriages, one first-class and the other a packed third-class, were opened up over their entire length and smashed up against the engine of the Montluçon train. The front vehicles were likewise telescoped and 'smashed to splinters.'

Uninjured passengers at once set to work to help their less fortunate companions and on receipt of the news of the accident a rescue train was immediately despatched from Auch, and a fleet of military ambulances took the injured to hospital in Tarbes. The Gers prefect and the Mirande parquet (examining magistrate) went to the scene.

The following day it was reported that there were 40 dead and 50 more injured. [870] The dead included at least two nuns and two crippled children, clearly bound for Lourdes in the hope of a miracle cure. Those injured included the Abbė Delaselle, a Chevalier of the Legion of Honour and the organiser of the pilgrimage. Mgr. Penon, the Bishop of Moulins, who suffered slight head injuries, organised a rescue party of uninjured children and himself dragged dead and injured from the wreckage. [871]

The surviving pilgrims continued their journey to Lourdes in the afternoon. A newspaper reporter noted that they no doubt thanked la Vierge (the Virgin Mary) for their survival from the accident and added sarcastically, 'L'egoisme et le fanatisme s'associent.' (Egoism and fanaticism combined.)

A couple of days after the accident it was blamed on the use of rolling stock in a deplorable condition, drawn by a locomotive, the brakes of which were not working properly, plus the failure to employ a second engine on such a steep incline. The driver of the Moulins train had apparently complained about the air compressor at Agen and it had got worse after that point, so that the sanders would not work on rails made slippery by drizzle.

By 4th August, a preliminary report on the disaster had already appeared, but it was alleged by a (Communist) newspaper that this tended to mitigate the flagrant responsibility of the Midi Company. The report stated that the engine of the Moulins train was a new one, weighing over 100 tonnes and perfectly capable of pulling a 256 tonne train up a 25‰ (1 in 40) incline. The air pump had malfunctioned after Mirande and there had also been a slight problem with the "giffard" (the injector) which had needed a makeshift repair. The problem with the air pump had necessitated a stop at Lass, just short of Mielan, and twice after passing there. The air pressure had been changing constantly and was clearly insufficient as at the bottom of the incline at Mielan it had

---

870 *Le Radical*, Paris, France, 2nd August 1922
871 *The Times*, London, England, 2nd August 1922

only been 2 kg rather than the 4 or 5 kg required. The manometer had clearly indicated the problem and we do not understand (the report continued) how a travelling inspector and driver could choose to continue up a long and severe incline followed by an equally long descent with insufficient air pressure. The newspaper commented that it recognised an attempt by the government to place all the blame for the accident on the railwaymen directly involved and was not at all impressed by a closing comment from M. Maison, the government official who had produced the report, that the initial investigation did not throw any blame on the Midi Company for the disaster. [872]

The number of dead had by now been corrected to 29, with 50 injured.

Another newspaper commented that while the Orleans company had used two engines to haul each train of pilgrims as far as Agen, when the Midi company had taken them over there only one engine took them forward on the grounds of economy. The Moulins train had engine No. 4128, in charge of Driver Geneste and Fireman Greole, plus Travelling Inspector Lessouret. The Montluçon train was hauled by a sister engine, No. 4129, with Driver Bloy, Fireman Lasserre and Travelling Inspector Raymond. According to this newspaper the interval between the two trains was 20 minutes (other reports stated 15 minutes and 30 minutes). This report spoke of Travelling Inspector Lessouret attempting to scotch the wheels of the engine by placing large stones under them and three times being thrown down by the engine going over them. The speed of the runaway train at the moment of impact was stated to have been 35 kilometres an hour (a more realistic figure than the estimates in the reports on other runaway accidents, so probably correct). The gatekeeper at crossing No. 110 was able to telephone Mielan station with news of the accident immediately but because the telegraph was out of use at night a message could not be sent to Tarbes until 06:30 and a relief train from there did not start until 07:20. Driver Bloy, who like his fireman and the travelling inspector had stuck to his post, was pronounced a hero for closing up on the train in front when he realised that it was running away and thus preventing it reaching a much higher speed, derailing and falling into a ravine. The two train crews all appear to have survived.

This newspaper too blamed the Midi company for attempting to evade its own clear responsibility for the disaster. After the line had been cleared and was back in use M. Paul, the director of the Midi, attempted to demonstrate with an ordinary train that Driver Geneste should have had no problem ascending the incline with the pilgrim train and a single engine in poor condition.

It roused one to anger, the newspaper concluded, that two trains carrying excursionists or pilgrims, like those in the disaster, each of which was made up of a dozen coaches, filled with thirteen or fourteen hundred passengers, were not operated in the safest manner. [873]

Just before Christmas 1922 Spanish troops took part in manoeuvres in the Chita area, the troops from Valencia, Játiva and Alcoy then returning to their respective bases in military trains. The soldiers of the Vizcaya battalion and the Otumba regiment returned in the same train on Friday 22nd December 1922. The men of the Otumba regiment alighted at Játiva and the train then proceeded towards Alcoy with the men of the Vizcaya

---

[872] *L'Humanitè*, Paris, France, 4th August 1922

[873] *L' Œuvre*, Paris, France, 3rd August 1922

battalion. The line climbed steeply over this section and near Agres station the train first stalled on the gradient and then began to run backwards at increasing speed. At Ontinyent station a goods train was engaged in shunting operations and the runaway train collided with it with a tremendous crash. The vehicles at the rear of the military train were completely destroyed in the collision.

The rescue operation lasted until 21:00. By then 12 corpses had been removed from the wreckage and there were found to be over 100 injured, 20 of them seriously. The dead included Lieutenant Colonel Don José Cañamaque Calleja, the commander of the Vizcaya battalion. [874]

The Bom Jesus do Monte funicular railway, connecting the town of Braga, Portugal, with a shrine at the summit of the mountain, was the scene of a serious accident on Whit Sunday, 20th May 1923. A car which was descending the railway ran away and crashed, killing six passengers, and injuring 30, most of them seriously. Most or all of the victims were from the town of Oporto. [875]

In a typical accident of its kind 29 vehicles of a down special goods train which had been detached from it at Pilkhani station at 19:30 on Monday 8th October 1923 ran back down an incline and collided with No. 18 down passenger train between Pilkhani and Sarsawa, India, on the Ludhiana – Saharanpur section of the North Western Railway. The three front third-class bogies (coaches) of the passenger train and five waggons of the goods train were smashed and 23 passengers were killed and 21 injured. The guard of the goods train and the Pilkhani assistant stationmaster were found to be jointly responsible for the disaster, prosecuted and sentenced to rigorous imprisonment and a fine. [876] [877]

In contrast to the Indian disaster there was a very lucky escape from what could easily have been another major disaster on the St Louis-San Francisco Railway near Carl Junction, Missouri, United States, on Monday 29th October 1923. At about 6 p.m. in the evening an engine was busy switching seven cars at the plant of the General Explosives Company in Carl Junction when the cars were pushed too far and ran out onto the main line. Three of the cars, each loaded with 25,000 pounds of dynamite, then ran back down the main line and collided with an oncoming Columbus, Kansas, to Joplin, Missouri, passenger train. Engineer Frank Davis and Fireman M. Searcy had no time to jump bur escaped unhurt anyway. Their engine was overturned into a ditch and fire from it quickly set the freight cars alight. The passengers on the train had barely recovered from the shock of the collision when they learnt what their train had collided with and that the runaway cars had caught fire. Naturally, they made a rapid exit from the train, fearing an explosion. One woman injured her spine; nine other passengers suffered only scratches and minor bruises, most probably sustained in their hasty escape.

A wrecking train and the Joplin fire department attended the accident, but the heat was

---

[874] *La Vanguardia*, Barcelona, Spain, 23rd and 24th December 1922
[875] *The Times*, London, England, 22nd May 1923
[876] Report by the Railway Board on Indian Railways for 1923-24
[877] *The Times*, London, England, 10th October 1923

too intense to enable the firemen to get close enough to the blaze to fight it and they spent their time keeping onlookers away for fear of an explosion. For the same reason, the wrecking crew did not attempt to save the cars. Meanwhile, 'Residents of Carl Junction spent six hours of uneasiness as the cars loaded with dynamite burned. Throughout the city, lights were burning in every house and anxious faces were pressed against window, the watchers hoping that an explosion would not occur but expecting it every moment.' [878]

Conductor O.B. Underwood and Mail Clerk J.M. Elrod remained in the mail car sorting mail until ordered to leave for their own safety by Mr Owens, the superintendent of the explosives company. Later six men risked their lives, rather unnecessarily one would have thought, to retrieve a casket containing the body of a Miss Margaret Burke, which was being returned to Joplin for burial. [879]

According to a local newspaper report the dynamite cars did not actually run very far before the collision, but other newspapers reported the accident as involving runaway cars, so it has been included in this chapter.

A very unusual and possibly unique accident happened in Portugal on Thursday 14th August 1924. As a Lisbon – Madrid express was running between Maro Miranda and Torres Novas, hauled by engine 103, 18 years old but apparently still in good condition, the locomotive, weighing rather more than 100 tonnes, somehow became separated from its 30 tonne tender. At the moment that the separation first occurred the train was running on a level stretch of track and neither the driver nor the fireman realised that there was any problem, but when it then came to an incline the driver opened the regulator and at this point the locomotive suddenly shot away from its tender, throwing both driver and fireman back 'like mannequins' onto the latter, while the engine, freed of its load, shot ahead 'like a meteor.' The driver and fireman reportedly clung on to what they could on the tender until it and the train came to a stand after running about 800 metres.

The stationmaster at Torres Novas was stood awaiting the arrival of the express when suddenly the detached engine appeared, running at a considerable speed. When he had recovered from the shock he hurried to his office to send a message to the next station at Entrocamento, but the two stations were close to one another and by the time that his message was received there the runaway engine was passing through, 'going like an arrow.' Entrocamento was a junction, and the engine took the single-track Oporto line, on which was approaching the Oporto – Lisbon mail train. The rogue engine, travelling supposedly at 140 kilometres an hour, although there was not much water in the boiler, met engine No. 18 of the mail train on a curve. The second- and third-class carriages of the mail train were smashed, and fire broke out in the wreckage. Twelve people died and 16 were injured. A newspaper reporter found the fireman of the Oporto train in hospital suffering from a fractured skull and broken legs. He told the reporter the following day that he had neither slept nor eaten since the accident, he had a fever, and his face was still covered in blood and coal dust. The fate of the other members of the train crews was not reported. A following Lisbon – Paris express was stopped near Santarem and a further

---

[878] *Joplin Globe*, Joplin, Missouri, United States, 30th October 1923
[879] Ibid, 1st November 1923

accident prevented. [880]

The railway company promised to carry out a full investigation into what might justifiably be described as a freak accident. [881]

To the north of the busy English railway centre of Doncaster an avoiding line had been opened in 1910 between Bentley Junction to the north east and Hexthorpe Junction to the south west, to allow goods traffic to run between Hull, Immingham, and Grimsby to the east and Sheffield, Barnsley and beyond to the west without having to negotiate the bottleneck of Doncaster station. The line ran mostly on embankment and climbed at varying gradients, including a stretch of 1 in 92, to Hexthorpe Junction. As it was not normally used by passenger trains it was worked by permissive block.

At 2.38 a.m. on Thursday 16th April 1925 the London and North Eastern Railway 10.30 p.m. Barnetby – Ashton Moss down express goods train, consisting of 2-8-0 engine No. 6206, 38 waggons and a brake van in charge of Goods Guard Edward Quickfall, with Driver Coates, came to a stand at the Hexthorpe Junction down avoiding line inner home signal to await the passage of the Leeds – Sheffield newspaper train. It was a rough, dark night with a gale-force westerly wind. The paper train duly passed on the main line at 2.44 a.m. and Signalman Leedell at Hexthorpe Junction cleared the home signal for the Ashton Moss train to proceed at 2.45 a.m.. It took a minute or two for the goods train to start and the signalman was immediately suspicious at how easily it had started away. Driver Coates also quickly realised there was a problem and stopped after travelling only 50 yards. His train had become divided behind the sixth waggon and 32 waggons, and the van began to run back down the bank.

Following the Ashton Moss train through the section was an Immingham Dock – Denaby empty waggon train, consisting of sister engine No. 6203, 81 empty waggons and a van. Signalman Leedell vainly lowered and replaced the down outer home signal several times in an effort to warn Guard Quickfall, and then waved his handlamp to try and alert Driver Calthorpe of the Denaby train, the headlights of which were already in sight. The headlights shortly went out and looking at his block instrument the signalman saw that its needle had dropped to "Normal." The runaways had collided with the Denaby train, demolishing a telegraph pole, and severing the block communications at 2.52 a.m.

The Denaby train had been at held at Bentley Junction for the Ashton Moss train and had left there at 2.32 a.m. after being cautioned that the other train was in front. Driver Calthorpe told the Ministry of Transport inquiry into the accident that he had been moving at about 12 miles an hour when he had seen the brake lights in front of him and had realised from the speed at which his train was closing up on them that it must be a breakaway. He had brought his train to a stand and jumped, shouting to his mate, Fireman Lawton, to do the same. The latter did not have time to do so but neither was hurt. The brake van and five waggons of the runaway were destroyed in the collision and 14 waggons, and the front of engine No. 6203 were damaged. The collision happened near Sprotborough Junction signal box, England, which was out of switch at the time, about 1,120 yards from where the breakaway had taken place.

---

[880] *The Times*, London, England, 16th August 1924
[881] *Diario de Lisboa*, Lisbon, Portugal, 15th August 1924

The Ashton Moss train had attached 11 waggons at Keadby and at Stainforth Goods Guard John Brown, Driver C. Brown and Fireman Ronald Scorah had joined Guard Quickfall in his van, having worked another train to there, where they had been relieved. Guards Quickfall and Brown and Fireman Scorah were all killed in the collision; Driver Brown somehow survived to give evidence at the inquiry. He said that Guard Brown had given Driver Coates the right of way at Bentley Junction and the four of them had then sat on lockers around the stove chatting. When they arrived at Hexthorpe Junction Guard Quickfall had remarked that his driver had brought the train up very nicely without a jerk and he would not put the brake on, adding that they would be held there for the parcels train. That train had passed, he (Driver Brown) had seen the inner home signal come off and he had remarked to Guard Quickfall that they were on their way. Driver Brown said that he had then sat down, the conversation had continued, and his fireman had passed cigarettes around. The start had been a smooth one. After a while, Driver Brown continued, the thought had occurred to him that he had not heard the different noise made when the wheels of the van entered the cutting beyond Hexthorpe Junction. No one else had noticed or said anything and clearly none of them had realised that they were running back, rather than going forward, but curious as to where they were, he had reached for the door handle of the van at the moment that it struck the engine, and in some manner he had fallen through the floor of the van.

Concluding the inquiry, Lieut. Colonel A.H.L. Mount said that divisions on the bank were not common, only two others having been recorded since 1st January 1924.

The waggon, the drawbar of which had failed, was a 5½ ton four-wheeled open waggon, dating from 1921 and built at Newton Heath. At the time of the accident it had been loaded with potatoes. Its drawgear consisted of standard short Gedge drawbars. The bar that had failed had done so at a defective weld and it was remarkable that it had lasted as long as it had. Carriage and Waggon Superintendent Reid of the London Midland and Scottish Railway stated that there had apparently been a previous fracture and an extraordinary repair had been attempted which had not only been a failure but had made matters worse. The bad workmanship had not been detected because the weld had apparently been carried out at a private works. The inspecting officer considered that the attention of railway companies in general should be drawn to the difficulty of allocating responsibility for inferior repair work in such a situation.

Continuing, Lt-Col Mount remarked that the primary cause of the division of the Ashton Moss train was therefore the failure of a very defective drawbar, but that a share of responsibility for the collision which had followed must rest with Guard Quickfall and that the accident emphasised the need for constant and unremitting vigilance on the part of the guards of unfitted goods trains, especially at night and in poor weather conditions. Had the guard applied his brake when his train first came to a stand, or had he gone out onto the verandah of his van to exchange signals with his driver according to the regulations during the seven minutes that his train was waiting he might have realised that part of his train had broken away, and if he had not been able to stop the runaway then at least all four men would have had chance to jump off. Driver Coates considered the guard to be a careful man and it could only be supposed that the weather conditions, the hour of the night and the presence of the other men in the van had resulted in a temporary lapse of duty on the part of the guard. [882]

An accident occurred on the Austrian section of the single-track Außerfernbahn line near Ehrwald at about noon on Wednesday 7th April 1926. The 21 passengers on the electric train saw an object fly past their windows shortly after leaving Schanz. The train then ran away, reaching it was said a speed of 120 kilometres an hour before it left the track on a curve and fell down a five metre embankment. Eleven railwaymen and passengers were injured but no one was killed. The 'object' which the passengers had seen going by had in fact been the body of the train driver, who had leaned out of the window of his cab too far and had either struck his head against a lineside object or just overbalanced and fallen out. Supposedly he had become aware of a problem on the outside of the train and had leaned out to investigate (but see East Manors accident, next). [883] [884]

What was described as 'a strange railway accident' occurred late on the evening of Saturday 7th August 1926 on the London and North Eastern Railway at East Manors station, just north of Newcastle Central station, England. [885] An electric train from Whitley Bay (actually the 9.47 p.m. circular service, Newcastle to Newcastle via Monkseaton and Heaton) had come into sidelong collision with a goods train. The passenger train was probably running at at least 35 miles an hour. As a result of the collision the leading coach of the passenger train was jammed up against the parapet wall beneath Manors Junction signal box but there was no telescoping, despite an initial newspaper report stating that there had been. Fortunately, the train was lightly loaded and only 16 passengers were injured, the only three in the leading coach being taken to Newcastle Infirmary. The motorman, William Skinner, was initially missing. His body was eventually discovered the next day two miles back the line. In the goods train two waggons were demolished, and seven others damaged.

Guard George Paterson, who had been riding in the brake compartment at the rear of the train, told the Ministry of Transport inquiry into the accident that the run had been normal as far as Heaton, except that the motorman, as he had on previous occasions, had run very fast into stations. Motorman Skinner had got out onto the platform at Heaton. After the train left there Guard Paterson said that he had been occupied making notes for his next trip and had not noticed where they were until the train was halfway along the platform at Manors East station, where it should have stopped. He had begun to make his way to the rear motorman's compartment to apply the Westinghouse brake but before he could get there the collision had occurred.

The signalmen in the different boxes could give little useful information. The train had been running under clear signals as far as Manors Junction, just beyond Manors East station, where Signalman Francis Topping decided to give priority to a goods train on the up main line. He had, therefore, in accordance with instructions, accepted the passenger train from Argyle Street box but kept his signals at danger until the train had operated a

---

882 Ministry of Transport report

883 www.ovb-online.de – Katapult in ein neues Zeitalter

884 *The Times*, London, England, 9th April 1926

885 Ibid, 9th August 1926

treadle in the rear of his up Tynemouth home signal, when he had cleared a draw ahead signal to allow the train to pull up to the starting signal at danger. When he saw the passenger train he had realised that it was not stopping, and he saw it hit the goods train.

Inspector Gill said that he had arrived at the scene shortly after the accident and searched the cab, expecting to find the motorman's body there, but it was not. What he did find was two handkerchiefs tied around the dead man's handle to hold it down independent of the motorman doing so.

Inspector Murray, who arrived somewhat later, confirmed that the two handkerchiefs had been tied as his colleague had described. He carefully cut them loose and removed them for safe custody.

By 3 a.m. the body of Motorman Skinner had still not been found. Police Sergeant Sandels was then sent back along the line to search for it and found the motorman's body lying under a road overbridge. The motorman had severe head injuries and there were signs that his body had struck three different bridge piers.

Major G. L. Hall, the inspecting officer, stated that there was no doubt that Motorman Skinner had deliberately secured the dead man's handle with the handkerchiefs either immediately after leaving Newcastle or at some point during the journey. He would not have had time to do so at Heaton. There was little point in speculating on the motorman's reasons for leaning out of his cab, which might or might not, Major Hall stated, have been legitimate. There was in any case no excuse for disabling a device upon which the security and even the lives of the passengers might at any moment depend. It was a most serious offence and very fortunate that the consequences were not worse.

It would not be reasonable, the inspecting officer concluded, to expect any of the signalmen to have realised that anything was wrong with the train in time to prevent the accident. Had he noticed that the Manors Junction home signal was at danger Guard Paterson could probably have stopped the train, but he had had no reason to suspect any irregularity and could not easily have seen the signals from his van. In conclusion the inspecting officer decided that Motorman Skinner was solely responsible for the accident, and noted that it was very difficult to prevent deliberate tampering with the driver's safety device. Such cases were rare, though there had been one on the London Brighton and South Coast Railway in February 1916, when the motorman had lost his life, but there had been no accident to the train.

Automatic train control of the brake trip type such as used on the underground railways would probably not have prevented the accident.

Improved facilities for guards to view trains from their vans were desirable.[886]

Although the Board of Trade inspector was too coy to state what he thought was the reason for Motorman Skinner's conduct it seems clear that the motorman was in fact a voyeur who had been trying to watch a courting couple engaged in some sort of sexual activity in the leading compartment of the electric train.

On about 1st September 1926 a passenger train on the Peking – Sulyuan railway on its way from Kalgan to Hankow, China, was derailed with disastrous results, 76 passengers

---

[886] Ministry of Transport report

being killed and 140 injured. The train apparently ran away due to defective brakes and three coaches crowded with passengers were hurled into mountain ditches. A report from Shanghai stated that there had been high military officials on board the train but no foreigners. [887]

On Tuesday 14th September 1926, readers of the morning papers in Sydney, New South Wales, Australia, were shocked to learn that late the previous evening there had been a serious accident to the North West Mail train. Details were still scanty, but it appeared that some waggons of a goods train had run back down an incline and collided with the mail train at a point between Murulla and Blandford. Several persons were feared killed, a number of others were injured, and the line was completely blocked. It was too early to obtain in general the names of the dead and injured but the family of Mrs A. Coppleson, who had been riding with her two grandchildren, Bill, and Clifford Richards, in the sleeping car, was assured that all were safe. [888]

By the next day's newspapers it had become clear that the accident was much more serious than had been at first realised. The number of dead listed in a paper on 15th September totalled 26 (it later rose to 27), with 34 more injured passengers in Murrurundi, Scone and Quirindi hospitals, plus four others whose wounds were slight enough for them to be able to return home. Most were passengers, but the injured also included Driver William Charlton of the passenger train.

This newspaper gave a very detailed account of the disaster, thus: 'The goods train had been pulled on to the Murulla siding to allow the north-west mail to pass, when the coupling joining the six rear trucks to the forepart of the train suddenly snapped. It is understood that the guard of the goods train was standing on the side of the line, directing the shunting operations, when the trucks began to move down the line. Although this official made a frantic effort to clamber on to the runaway trucks he was unable to do so, and they disappeared at a terrific speed round a curve in the line.

'It was known that the mail train had passed through Blandford station, and those on the siding were sick with fear and apprehension, but powerless to avert the impending catastrophe. A few seconds later there was a deafening crash as the runaway trucks struck, with terrific force, the mail train, and telescoped several carriages. The red-hot coals from the wrecked engine fell on to and ignited several bales of wool with which the trucks were loaded. The flames most fortunately did not reach the smashed carriages, or a much more terrible disaster would have resulted.

'The front of the engine was smashed in, while the two rear trucks of the runaway were reduced to a mass of twisted and shapeless steel. Next to the engine was a first and second class compartment, the occupants of which, with few exceptions, escaped with minor injuries. Next to this carriage was a combined first and second class sleeper, with the roof caved in, while the undercarriage had telescoped the third coach. It was from this latter carriage, which was for second-class passengers, that nearly all the dead and injured were removed. The coach, which crumpled up like a concertina, was smashed to matchwood, and that any one passenger should have escaped uninjured was miraculous.

---

[887] *The Telegraph*, Brisbane, Queensland, Australia, 3rd September 1926
[888] *Sydney Morning Herald*, Sydney, New South Wales, Australia, 14th September 1926

'Following the collision the scene was one of havoc and confusion. The screams and groans of the injured, pinned beneath the wreckage, intermingled with the agonised cries of children, many of whom were returning to school in Sydney, were heartrending. The passengers who had themselves escaped injury worked with all their might, to free those who had been entrapped in the wreckage. The first work of rescue was, however, seriously hampered by the lack of axes, saws, and other suitable implements with which to cut through the wreckage, which held its victims pinned mercilessly beneath. The burning bales of wool, which formed part of the consignment of the goods train, added to the terrors of the survivors, who feared that the wreckage might catch fire.

'Mr. H.H. Wright, of Bickham, Blandford station, realising that something terrible had happened, roused half a dozen of his men, and armed with axes, the party was among the first to the rescue. The news of the disaster quickly spread, and within an hour 50 or more willing helpers were assisting in the task of hacking away the splintered timbers.

'A preliminary inquiry into the deaths of the persons killed was opened before Mr. G.B. White, District Coroner, at Murrurundi, this afternoon. After formal evidence had been called the further hearing was adjourned until September 29th.'

The paper continued with the stories of survivors and witnesses describing the horrors of the accident, of which one must suffice: 'Mr. Frank Scott, of 135 Phillip-street, city, was injured, and was one of the first to see the smash. He had been on a holiday visit to Delungra, where his mother lives, and was returning to Sydney by train. He told a graphic story of the rescue work, for he was engaged on it continuously for nearly seven hours. The carriage he was travelling in was the fifth from the engine and was not broken in the smash. Mr. Scott said he was asleep when the collision occurred. He first knew that something was amiss by a loud grinding noise and by bumps— "One, two, three," he said. He and his fellow-passengers were thrown about on their seats, and the luggage from the racks rained punishing blows on them. With another man he rushed to the carriage door and leapt to the ground. The two made their way along towards the front of the train, past the fourth carriage, which was not greatly damaged, to the next one, which they found to be a total wreck. It was jammed between the adjacent carriages, which, being stronger, had crushed it like a nut between the jaws of a cracker, till it was reduced to less than a third of its original size. In the midst of this splintered wreckage were the mangled bodies of dead and dying. Screams of pain were mingled with the high-pitched calls for help, and in an undertone there could be heard the spasmodic groans of the dying.

'Mr. Scott and his companion for a moment were rooted to the spot horror-stricken, but they could see no way to help the injured, for they seemed hemmed in and covered with the wreckage as securely as though they were in a prison. Just then the utter blackness of the night was relieved by a red glare, and about 50 yards ahead of them the two men saw flames leap into the sky. They raced along the side of the line and came upon the burning wool. Lying beside some scattered wool, which had fallen from a broken bale, they found a man bleeding from a gaping wound in the back of his head. He was conscious, and kept muttering, "Where am I?" It later transpired that this man had come down with the wool trucks and the guard's van from the other train. Mr. Scott bound his head with a handkerchief and took him out of reach of danger from the burning wool.

'Mr. Scott next found the engine driver and fireman—the former wounded in the head, and the other only slightly injured. He and his companion later returned to the badly-broken carriage, and by this time between twenty and thirty people were round it, bent on rendering help. The spectacle and the agonised shouts and cries were enough to appal the steadiest mind and numb the body with horror, said Mr. Scott, but the heartrending appeals for help galvanised the rescuers into activity that was almost frantic. The rescuers were hampered most of all by the darkness, for the blazing wool was too far from the broken carriage to illuminate the scene.

'The first thing the rescuers did was to make two fires from splintered wreckage—one on either side of the broken carriage. This light helped them considerably, but it was nearly an hour after the smash before the rescue work could be undertaken effectively. By that time three axes and a saw were available, but great care had to be exercised in their use. "We would be going to get a man free," Mr. Scott said, "and when we chopped at a piece of wood which was holding one man down we'd hear a groan come from another man or woman underneath whom we did not know was there. Often we had to give up using the axe and wait till we could get the saw, because the chopping hurt someone underneath. Some were caused so much pain when we tried to free them that they begged us to leave them alone."

'A great many of the victims were pinned by their legs, with their bodies hanging over the side of the carriage, and the rescuers took turns in holding up those unfortunate people so as to take some of the weight off their injured limbs. This was heart-breaking work, because the person holding an injured man in this way could do nothing to relieve the suffering, and through weary minutes and sometimes hours had to wait, witnessing intense agony, until help arrived.

'In one part of the carriage were a man, his wife, and son. They were packed into a small space and imprisoned with heavy wreckage. The father was dead and the mother only half conscious, but the boy was conscious, and crying pitifully for water. Mr. Scott secured some water and tried to pass it down to him through the numberless obstructions. In the darkness he could not perceive the boy's hand, and as he tried to push the pannikin of water to him his hand knocked against pieces of wreckage and the water was upset. He made several attempts to give the boy water, but each attempt failed. Almost frantic with his realisation of his failure, Mr. Scott had at last to turn a deaf ear to the boy's agonised appeals and go on with the work of clearing the wreckage.

'Gradually and causing as little pain to the victims as they could, the rescuers cleared the wreckage, and one after another was lifted out. To facilitate the work, the people clearing the wreckage stayed by the carriage and passed the dead and injured to others standing by who carried them to rugs and cushions. Every minute brought fresh helpers. The injured were treated as soon as possible after they were freed from the wreckage. The motor train from Murrurundi took the three undamaged carriages back full of wounded, and later an engine and carriages came from the same place to take away the dead.'

The Railway Commissioners issued an official statement on the disaster:

'When the up goods train from Werris Creek was being drawn into the crossing loop at Murulla for the purpose of allowing No. 8 north-west mail (2.15 p.m. Moree to Sydney) to pass, a draw-hook on a truck attached to the goods train broke. Subsequently, while endeavouring to remove this rear portion with improvised coupling, the rear portion of the goods train ran back about a mile and a half into the section, Murulla-Blandford, and collided with the mail train, damaging three cars, and smashing one second-class car on the mail. Portion of the goods train which collided with the mail afterwards took fire, and a number of vehicles were destroyed.

'The Commissioners express sincere regret for the unfortunate accident, and everything possible is being done for the injured.' [889]

The inquest proper on the victims of the accident began at Murrurundi on 29th September. The first witness was Traffic Inspector Edward Heery, who told the inquest that he had arrived at Murulla at 1.15 a.m. on the motor train from Muswellbrook. He had checked the train register book and interviewed Signalman Walter White, who had told him that No. 62 up goods had arrived at Murulla at 10.40 p.m. and had stopped on the main line because it had become divided. Train No. 95 had arrived at 9.58 p.m. and had not departed until 10.52, when the signalman was finally sure that train No. 62 was clear of the points of the crossing loop at the Blandford end. When train No. 95 cleared the next station at Blandford it was followed from Murulla by the Brisbane express. Signalman White had then accepted train No. 8 from Blandford at 11.32 p.m. Meanwhile the crew of train No. 62 were trying to make good the break between the two portions of their train and the draw bar having broken, they resorted to using a tail rope. Shortly after the mail train had been accepted from Blandford the driver of the goods train set back the front portion of his train onto the detached rear portion but in doing so pushed the rear of his train out of the loop onto the main line. Signalman White instructed the driver to pull forward again and as the driver had done so the tail rope had snapped, and the detached waggons had begun to run back down the bank towards Blandford.

Traffic Inspector Heery testified that Guard Davies had told him that when the train had first become divided both portions had come to a stand because of the action of the Westinghouse brakes. While the train was at a stand the air brakes had gradually leaked off and he had not been able to reapply the air to the rear portion because he could not get the two portions close enough. He had applied the hand brakes to the rear portion instead but had released them after the tail rope was attached. When the waggons first began to run away he had thought that it was just his driver setting back. By the time that he realised what was happening he had only been able to pin two brakes down before the runaways were going too fast.

Driver Turner told the traffic inspector that as far as he had been aware air was being supplied to the rear of his train. The load of his train was 734 tons against a maximum of 780.

The traffic inspector stated that it was the guard's duty to secure the train with hand brakes and sprags after it had become divided. [890]

Traffic Inspector Heery continued his evidence at the next sitting of the inquest. He said that there was no evidence of any failure of the Westinghouse brakes. If the truck which broke had been in such a condition that it could not be recoupled it should have been remarshalled next to the brake van. It would not take more than 20 minutes to remarshal the trucks. The guard could have coupled the three-link coupling hanging on the drawbar, of which the hook was broken, to the automatic coupling with a pin which was provided for the purpose, and which he exhibited in court. The regulation stated that the Westinghouse brake must not be relied upon in cases such as had existed at Murulla, but sprags and handbrakes were to be used.

---

[889] Ibid, 15th September 1926

[890] Ibid, 30th September 1926

Continuing, witness said that he had never known remarshalling of a train to be done at Murulla, but he knew that trains had been shunted from one line to another there. Witness then described how he would marshal a train at Murulla, with a train at Blandford and another at Wingen.

The inspector said that a wire or hay rope was used to tie the end of the tail rope to the buffer. Upon witness being asked by Mr Sproule to inspect a broken drawhook, he said that it appeared to have a flaw, but he would not say it was sufficient to cause the break. An inspector was allowed at least 16 minutes in which to examine the train after the engine was attached. A tail rope used on a grade similar to that at Murulla was allowed to hold up to 110 tons with a single strand. If it were wound round three times it should hold 330 tons. The weight of the break-away portion was 261 tons, and if the tail rope had been properly fastened there would have been no break.

Inspector Heery said if the hand brakes had been put on three trucks and on the brake van, it should have brought the runaway portion to a standstill. Had the hand brakes of the two bogie F. trucks, which are very severe, been applied, they would have pulled the train up.

Witness told Mr. Bretnall that he had had 20 years' experience, 17 years in the traffic branch through all grades, from porter to night officer, and up to Inspectorship. Failures of the Westinghouse brakes were very rare; in fact, he had not known of one. It was the duty of the staff to make good the air brakes after the draw gear broke. He had seen a pinch bar used as a substitute for a pin to secure couplings. A guard had in the van both Westinghouse and hand brakes. Witness informed the Coroner that between Murrurundi and Murulla there were no special danger provisions regarding traffic. It was not regarded as a dangerous part of the line.

Arthur Aubrey Jones, of Concord, Sydney, mechanical draftsman, in the employ of the railways, said he had been engaged for 20 years in the chief mechanical engineering office on the designs of locomotives and rolling stock. The breaking strain of a standard draw, gear similar to the hook in question, was 45 tons. He had been present at the tests of draw gear at the hydraulic testing machine. The maximum pull on the goods truck draw gear, under the most adverse conditions, would be 900 tons, the maximum load on the level.

Witness, upon examining the broken draw hook in Court, said it was a piece of mild steel, showing a crystalline fracture and disclosed signs of a flaw, which seemed to have occurred at its manufacture, and had apparently existed for a long period. The flaw would not be ascertainable on examination. It would not seriously affect the breaking strain of the drawgear. Witness said that there would be a full factor of safety on the load it would have to carry. In regard to the tail rope produced in Court, witness said the breaking strain of similar ropes varied from 11 to 14 tons. Manufacturers of steel ropes guaranteed their breaking strain. That rope would safely carry on a grade of 1 in 81, a load of 114 tons. The speed of the runaway portion of a train of 261 tons on a grade of 1 in 81, for a distance of 915 yards, without brakes, he had computed at 26 miles an hour. Assuming that the end truck had a brake applied, and the next truck partly applied, the computed speed would be approximately 22 miles an hour. Witness was present at a test of an LHG type van at Clyde on Monday, with a load of 220 tons on a grade of 1 in 80. The train was completely stopped by application of the hand brake in the van after travelling a distance of 219 feet. The test was made of a train actually in traffic, and without previous notice to the staff. [891]

Draftsman Jones continued his evidence at the resumption of the inquest, when he testified that if the rope were tied in the manner shown in the diagram in the regulations he was of opinion that its safe load would exceed 114 tons, and possibly double that weight. If the link on the tail rope were attached in a certain manner to the automatic coupler and tied securely, witness believed that the safe load would be 280 tons. It should be strong enough to haul a load on a grade of 1 in 81.

In reply to the Coroner, witness said that the break in the draw hook would be caused by undue stress set up in the train, which found the weakest point throughout the coupling system. It would possibly be caused by a sudden jolt. Witness said that draw hooks were condemned when their size was 2¼ in. The hook in court was 2½ in. The steel of the broken hook in court showed signs of wear. This would not be unusual after a life of five years. It was apparently manufactured at Eveleigh workshops. A hook measuring 2½ in would still have a fair margin of safely. Even if the flaw in the draw hook were visible, witness said that he believed he would pass it for service. The single wire rope in court would carry a load of 328 tons on the level and on a grade of one In 100 a load of 132 tons.

Asked if it were possible to design or construct rolling stock which would withstand the impact of collision, he replied that it might be possible to design a structure which would do so, but that would probably not result in lessening the casualties. It would not be applicable to a railway vehicle because of its enormous weight, difficulties in transport, and because of the heavier rails and stronger bridges necessary to carry it.

Samuel Buchanan Irwin, chief clerk to the carriage and waggon superintendent, said the truck on which the drawhook had broken had been constructed by the Clyde Engineering Company, and had made Its first run in 1911. It was of 16 tons capacity and was fitted with a Westinghouse brake. It was in the Clyde repairing shed in June 1923, and also in February last. The original drawbar had been replaced, but he could find no record of the date.

Driver Charlton, Fireman Horace Findlater and Guard James Marx of the passenger train related their experiences of the accident. Driver Charlton stated that just prior to reaching the distant signal at Murulla the fireman had put out his hand and called out: "Stop."

"I could see immediately that something extraordinary had occurred, and I immediately applied the emergency air brake. Just I did that I leaned out of the cab and saw what I took to be a brake van showing two red lights coming towards us. It was very close. Our speed was 30 miles an hour. I do not remember anything more, as I became unconscious. When I recovered consciousness, I was still on the engine, but after being helped to the ground I could see my engine on the road, and the trucks on fire." [892]

The inquest continued for several more days, with evidence of tests being made with a train similarly made up to determine how easy it was for a runaway to occur. [893] Then on the penultimate day of the hearing evidence was heard from Signalman White and Fireman Jeffery Lloyd of the goods train.

---

[891] Ibid, 1st October 1926

[892] Ibid, 2nd October 1926

[893] Ibid, 6th, 7th and 8th October 1926

Signalman White testified that train No. 62 had first come to a stand on the main line while there was a stock train in the loop. It should have run right through but stopped because it had become divided and had then had to wait on the main line while it crossed two trains. At about 11.23 Driver Turner had shouted to him that they were now ready to go but he had told the driver that he must now wait to cross train No. 7. A few minutes later he had realised that the rear part of the train was running away and along with the guard had run after them to apply the brakes, but they could not stop them.

A long statement made by Fireman Lloyd to the police was read out. In court the fireman denied that the train had moved forward at all after the tail rope had been attached and suggested that the rear waggons had moved off on their own.

Driver Ernest Turner and Guard David Davies gave evidence on the last day of the inquest.

Driver Turner had made a statement made to the police, in which he said that he had been previously unaware that the guard had not coupled up the air hoses, and that the air was not in the rear of the train. He had also made a long statement from witness to his departmental officer, in which he stated that the guard remarked that they would do better without the air in the rear part of the train, as the rope might stretch and part the air hose. He added that after the breakaway he ran down after the trucks and waved his hand lamp, in an endeavour to attract the attention of the North-west mail, which he knew was in the section. The night officer ran past him after the runaway trucks.

The driver said that the first indication of a break in the train was shown by his air gauge, which rapidly dropped immediately after the tablets were exchanged, and the train was automatically stopped. After asking the guard to secure the rear portion of his train, he had asked the guard, "Have you got a pin?"

The guard had replied, "We are not issued with a pin."

It was pointed out that he had made no mention of that in either of his statements. He demonstrated how he and the guard passed the hook and tail rope three times through the top link of the chain. Driver Turner said that he had not coupled the air hose, that he had forgotten to do so, and if he had thought of it, he would have coupled it. He was sure he put the tail rope three times round the link and pulled it as tight as he could. In answer to a further question he said that it was not his duty to couple up the air hoses. He left it for the guard to do, but it was his duty to see that his train had a full pressure of air before the hand brakes and sprags were taken off.

Guard Davies said the break in the coupling was not his fault. He had no automatic pin, and he had to use the tail rope supplied. It was put on securely and broke through no fault of his. No effort on his part could have prevented the accident. The police had interviewed him while he was in bed, and got a statement, which he had signed. In this statement, witness described the steps taken by himself and the signalman to stop the runaway trucks. He said he did not test his air brake in the van, because he know there was no air.

Mr. Sproule (for the Australian Railways Union) said that in view of the warning given to witness by the Coroner, that he need not answer questions which might incriminate him, he would advise Davies not to give any further evidence.

The inquest ended with Driver Turner and Guard Davies being committed for trial on a charge of manslaughter.

The Coroner (Mr G.B. White) added the following rider: "I find that there was no automatic coupling pin in the brake van of the goods train. I also find that if such automatic coupling pin had been in the brake van and had been used as officially suggested, such accident and loss of life would have been avoided. I further find that on the evidence of the Chief Mechanical Engineer for Railways and other departmental experts an omission of great importance occurs in the list of the guard's equipment for goods trains, as no mention is made of the automatic coupling pin as part of such equipment. The grave consequence resulting therefrom is that departmentally no person is regarded as culpably responsible for the absence of such pin, and essential principles of safe working have been overlooked." [894]

Driver Turner and Guard Davies duly stood trial in early December at the Central Criminal Court at Darlinghurst, before Mr Justice Ferguson, charged with the manslaughter of John Giles, one of the passengers on the mail train. The prosecution was not exactly conducted with great zeal. At the beginning of the trial the Crown Prosecutor (Mr. W.T. Coyle. K.C.) said that the facts were painful to anyone. The two accused were men of most excellent character, against whom nothing could be said. They had been in the railway service for some considerable time and had risen to positions of trust. The men had made a great mistake – one that was not likely to occur again. Errors of judgment did not make criminals of men. but gross negligence made them liable. If the accused failed to link up the air brakes when the appliances were there, the men were guilty of gross negligence. [895]

The trial was concluded on Wednesday 8th December. In summing up the case to the jury Mr Justice Ferguson said that the case was one in which the jury should clear its mind of two possible sources of complication. In the first place the Crown Prosecutor had properly mentioned that this was not an inquiry into railway management as there was no material before them to show how a railway business should be conducted. The only matter for determination of the jury was whether Turner and Davies were guilty of manslaughter. The person whose death was investigated by the trial was John Giles. It was a deplorable fact that many lives were lost in the disaster, but it was not for the jury to be influenced by that. The charge of manslaughter could only be established if the Crown proved gross negligence which was criminal in character and deserving of punishment. The jury was not to be swayed by the number of passengers who were killed. Its duty would be the same if there had been only one death and that of an obscure person.

It was not suggested, the judge continued, that the accused were responsible for the original breakaway of the trucks. That was possibly the result of a defect in the drawbar.

It was admittedly impossible for them to make a connection as provided by the regulations, but they did the best they could. The night was dark and misty when they made the connection which was stronger than the regulations provided. They did their best with the wire rope, but they did not couple up the air hose and this was the negligence that the Crown charged against them. It was not disputed that at great risk to their own lives the guard and signalman did everything that was possible to stop the trucks, but unfortunately they did not succeed.

---

[894] Ibid 11th October 1926

[895] *Daily Telegraph*, Sydney, New South Wales, Australia, 7th December 1926

The judge's summing up was clearly favourable to the driver and guard and it took the jury only 15 minutes to find them "Not Guilty."

Mr Justice Ferguson, addressing the two railwaymen, said it was not often that he commented on the verdict of a jury but in this case he must say that he entirely agreed with it.

The two men were surrounded by friends on leaving the court and congratulated on the outcome of the trial. [896]

Experiments were being conducted on Tuesday, 16th November 1926 near Tiflis, Georgia, USSR, to test the efficiency of the new Kazantzev brake, which the Russians hoped they could adopt in place of American brakes. Iggor Kazantzev, the inventor of the brake system, was present while the trials were being conducted. The brakes unfortunately failed the trial and a serious accident resulted. They could not hold the train and 36 cars fitted with the new brakes ran away and toppled over an embankment while going at full speed. Some of the coaches were reduced to splinters, seven people were killed outright, and a score seriously injured. [897]

A train carrying phosphates from the Kourigha mines to Casablanca, Morocco, ran away and left the rails on a curve on Thursday 10th February 1927. Three people died and three more were injured. [898]

On Monday 25th May 1927 train No. 2, the Almedricos – Águilas mail train, had just left Pulpi station, Spain, and had covered 200 metres when it was struck from behind by a freight train which had run away, the driver of which was whistling continuously to indicate that he had lost control of his train. The driver of the mail train, becoming aware of the danger, attempted to accelerate his train out of the way but could not do so. Likewise a pointsman, seeing the runaway train approaching, attempted to divert it to another line but there was no time to do so. The passengers in the mail train heard the frantic whistling and some of them looked out of the windows to see the freight train coming up behind them. In a few moments, the collision had occurred, the engine of the freight train smashing through the last three coaches of the mail train. The accident happened at 15:00.

Freight train No A-244 consisted of an engine and a dozen waggons, loaded with 600 tonnes of iron ore. It had come from Serón station on the Lorca – Baza line. [899]

Seven passengers were killed outright, three more died of their injuries in the following days and 20 others were injured, 15 of them seriously. In addition seven passengers were reportedly missing. A Civil Guard and a Post Office employee were amongst those killed but the driver and fireman of the freight train escaped with injuries.

---

[896] *Sydney Morning Herald*, Sydney, New South Wales, Australia, 9th December 1926

[897] *Cincinnati Commercial Tribune*, Cincinnati, Ohio, United States, 17th November 1926

[898] *The Times*, London, England, 12th February 1927

[899] Federación Castellano Manchega de Amigos del Ferrocarril – Accidentes Ferroviarios en España

The brakes of the freight train had supposedly failed. The driver of the freight train apparently realised that his brakes were not working when his train was three kilometres from Pulpi. [900]

The Minister of Public Works ordered a strict inquiry into the cause of the accident. [901]

At about 22:15 on Monday 21st May 1928, No. 345 up fast goods train became divided between the 39th and 40th waggons near the summit of a bank about a mile from Thawatti station, on the Rangoon – Mandalay main line of the Burma Railways in Burma. The rearmost 35 waggons ran back to Thawatti station where they collided with No. 1 up mail train which had come to a stand on the 2 line. The engine and tender of the mail train were derailed, forced back 100 yards, and partly overturned. The leading luggage van was thrown into the air and landed on the vehicles of No. 340 down goods train, which was standing on the platform line, while the leading third-class bogie carriage was derailed. Two passengers in the mail train were seriously and three slightly injured. Damage to rolling stock and track amounted to 13,500 rupees. [902]

While shunting operations were in progress at Sessano station, Italy, at about 11:00 on Tuesday 18th September 1928, a waggon loaded with bags of cement in some manner became detached and ran away down a steep incline towards the station at Carpinone. A warning was immediately sent to the stationmaster there, but a train had already left in the direction of Sessano, consisting of a locomotive propelling a van carrying workmen who were to be employed along the line. The collision severely damaged the van and nine of the labourers were killed, though Foreman Michele D'Oliva and labourer Antonio Forte were only slightly hurt. [903]

An accident to a ballast train above Danea station, India, on the Gomoh – Barkakana section of the East Indian Railway, at about 16:15 on Wednesday 6th March 1929, was blamed on the train being overloaded. The driver lost control of it on an incline, and it ran into a catch siding. Two ballast coolies, a fireman, a keyman and a child lost their lives, six coolies and an assistant permanent way inspector were seriously injured, and 35 coolies were slightly injured. [904]

An accident occurred on the London Midland and Scottish Railway Bedford – Bletchley branch near Ridgmont, England, on Monday 21st October 1929. The 6.45 p.m. up brick train from Forders Sidings to Bletchley was in charge of Guard J.H. Coley with Driver J.S. Cutler and Fireman F.A. Hill. Riding in the van with the guard was Shunter F.J. Burridge, on his way back to Bletchley at the end of his day's work. The train consisted of G 1 class 0-6-0 engine No. 9059, 50 loaded waggons equivalent to 51 and a

---

[900] *La Nanguardia*, Barcelona, Spain, 26th and 27th May 1927
[901] *The Times*, London, England, 27th May 1927
[902] Report by the Railway Board on Indian Railways for 1928-29
[903] *Il Molise Fascista*, Italy, 20th September 1920
[904] Report by the Railway Board on Indian Railways for 1928-29

brake van, close to the maximum authorised load for the engine of 55 waggons, including the van.

Approaching Ridgmont station the up line climbed for about a mile and a quarter at 1 in 264, then at two and a quarter miles at 1 in 129. Driver Cutler told the Ministry of Transport inquiry that he had had great trouble in getting his train up the bank, and it had finally stalled near the Ridgmont up distant signal. The engine had been steaming properly and there had been no slipping. He had eased back a couple of waggon lengths to get a fresh start, started forward again and then saw from the receding lights of the brake van that part of the train had broken away. He had then taken what was left of his train on to Ridgmont signal box where he had told the signalman what had happened, put the 16 waggons in the siding and proceeded to Bletchley for the breakdown train. He had worked brick trains for some years and had not had one stall before, nor had he had to divide a train to take it up the bank in two portions for a long time. He did not know why the engine could not handle the train.

Shunter Burridge said that he had made the brick train up to the fullest possible load as per his orders. As soon as the train came to a stand on the bank Guard Coley had put on the break and it was never taken off. As soon as the waggons began to run back Shunter Burridge had jumped out of the van with a brake stick and pinned down as many brakes as he could, he thought between 10 and 20, before the end of the train had passed him, travelling by then he thought at 30 miles an hour.

Guard Coley said that he had remained in the van, attempting to screw down his brake harder until he saw the headlights of a following train and jumped.

Relief Signalman A.E. Hull was on duty in Ridgmont box. He stated that he had received the "Train Entering Section" signal for the brick train at 6.59 p.m. and the engine with 16 waggons had reached him at 7.38 when Driver Cutler had told him what had happened. He had at once sent the emergency telephone call to Millbrook; Signalman Gurney there had answered at once and he had explained the situation, then sent the "Vehicles Running Away on Wrong Line" bell signal.

At Millbrook Signalman Gurney had been out closing the hand-operated level crossing gates for a down passenger train but had just returned to his box when the emergency telephone call came from Ridgmont. He went to the lever frame but had no time to reverse the crossover to switch the brick train onto the down line before the runaway had reached him. The signals were against the passenger train, which had already been cautioned by Porter F. Thorn at Lidlington station, where there was no signal box. The porter had tried to telephone Millbrook but could get no answer, doubtless because Signalman Gurney had been out tending the gates. Signalman Gurney had been 29 years at Millbrook; he said that trains needed to be divided because the engine was overpowered about once every three months.

Following the brick train was the 2.30 p.m. up goods train from Cambridge to Bletchley, consisting of *Experiment* class 4-6-0 engine No. 5468, 35 waggons and a van, in charge of Guard A. Goodman with Driver A.H. Hankins and Fireman J.W. Scott. Driver Hankins testified that the Forders Sidings signals had all been clear for him. Wootton Pillinge box had been out of switch, but he could see the Millbrook distant signal at "Caution" and had been preparing to stop at the Millbrook home signal when the three red lights of the brake van of the brick train had come into view 300 yards beyond

Wootton Pillinge crossing. When he had realised that the van was on his line he had reversed his engine. He estimated the speed of his train at 25 miles an hour at the moment of impact. The engine of the Cambridge train overturned in the accident and the brake van and 12 waggons of the brick train, and three waggons of the goods train piled up in a heap. Fireman Scott was killed, and Driver Hankins and Guard Goodman were injured, as was Guard Coley.

Bletchley Carriage and Waggon Foreman Scoble told the inquiry that he had gone to Ridgmont after the accident and had found the coupling to be entirely missing from the rear waggon of the front portion of the brick train, an ex-Caledonian Railway coal waggon No. 55933. Walking down the line he had found the top and centre links near the Ridgmont up distant signal; each had contained a flaw which could not be detected by ordinary examination. The waggon had been built at St Rollox in 1892 and the coupling was of a now obsolescent type.

In his conclusions to the report the inspecting officer, Lt-Col E. P. Anderson, stated that the direct cause of the accident was the breakage of a coupling with undetectable internal flaws. The evidence made it clear that the coupling must have broken due to a jerk on restarting after the train had stalled on the bank. The train had not been overloaded, and he could only attribute its stalling to the fact that it had not been handled very skilfully by Driver Cutler, who if he had started away faster on the easier gradients near Forders Sidings would probably have reached the summit without difficulty. He considered that the driver had also shown a lack of judgement in not at once dividing his train instead of attempting to restart it on the bank.

The load of the train was within the capacity of the engine as fixed by practical tests and the designed power of the class. Loads of 50 waggons or over had also been worked without difficulty by engines of the same class on eight occasions since 1st August. The brake power of the engine and van was also sufficient, but after the breakaway the brake power of the 20 ton van was not, and Shunter Burridge had therefore acted correctly in trying to pin down waggon brakes, while Guard Coley deserved special credit for remaining at his post in such trying circumstances. No blame attached to any of the signalmen. Signalman Gurney at Millbrook would have incurred grave risk had he tried to divert the runaway waggons onto the down line before he was sure that the passenger train had stopped.

The line had been built before the present requirements for the installation of runaway catch points on gradients steeper than 1 in 260 a full train's length in the rear of the first stop signal of each block post on each section of ascending line. On the section in question the first catch points were 6¼ miles in the rear of the point where the waggons broke away, in the second block section in the rear when the intermediate boxes were switched out, as at the time of the accident. Lt-Col Anderson therefore recommended the installation of additional catch points at Millbrook and Lidington, in view of the operation of heavy trains up the long bank. [905]

Just a month after the Ridgmont accident another serious runaway occurred in England, though from a different cause. To reach Bath the Somerset and Dorset Joint Railway had to cross the Mendip Hills. The main climb from Evercreech Junction ended

[905] Ministry of Transport report

at Masbury Summit, from which the line fell for about 14 miles to Midford station, but there was then a climb of about two miles, mostly at 1 in 100, through Combe Down Tunnel, before the line fell steeply, partly at 1 in 50, for another two miles to Bath Junction. Combe Down Tunnel was 1 mile 67 yards long. There were two tracks from Evercreech Junction to Midford, from where the line became single to Bath Junction through Combe Down Tunnel and the shorter Devonshire Tunnel.

On Friday 20th November 1929 Driver Harry Jennings worked a goods train hauled by engine No. 89 from Bath to Evercreech Junction, accompanied by Fireman M.C. Pearce and Guard Wagner. To complete their day's work they then had to work another train back to Bath. This train was fully loaded, and 13 minutes were lost on the climb to Masbury Summit. Driver Jennings had remarked to his guard at Evercreech that his engine was not steaming well and at Radstock, closer to Bath, where empty waggons were dropped off and others picked up, Driver Jennings had told the guard that they were doing as well as could be expected. At Wellow, between Radstock and Midford, the guard suggested that the signalman there be asked to hold the train there until it could get a clear run through Midford, and a message was sent via a porter to the Wellow signalman, but the train left there before the matter could be arranged, and at Midford it had to cross a down passenger train from Bath. In the end, in fact, it had to cross not only the 5 p.m. down passenger train from Bath but also the following 5.15 p.m. down goods train, then to cross over onto the down line to allow the 1.20 p.m. up passenger train from Bournemouth to overtake it. Having arrived at Midford at 5.02 p.m. it finally left there at 6.13 p.m. Driver Jennings went to Midford box to receive the Midford – Bath Single Line Junction tablet from Signalman Larcombe and told the latter that he had a heavy train and an engine that was steaming badly, so that he foresaw difficulty in getting through Combe Down Tunnel.

The train set off at about 15 miles an hour, but speed had fallen to about four miles an hour by the tunnel portal. Fireman Pearce told the Ministry of Transport inquiry that the engine had not been slipping at all and the sanding gear had been in good order. Shortly after entering the tunnel the atmosphere had become very hot and smoky, causing him to cough violently. He had wrapped his coat around his head and sat down on the tender. His driver had been standing up on his side of the engine and had seemed unaffected. The fireman remembered nothing more until he regained consciousness after the accident.

Guard Wagner thought that the atmosphere in the tunnel had been denser than usual, and the train had been going slower than usual. He said that more than once he had turned his handlamp to the wall to make sure that the train had not stopped. As the train left the tunnel he had been surprised that the couplings had not tightened up due to the tender brake being applied and he realised that the train was gathering speed more than usual. He applied the brake gently, then harder. There were no whistles from the engine. The train was going faster and faster and as it was passing Bath Junction it became clear that there was nothing more that he could do, and a crash was inevitable. He had jumped.

Bath was a terminus, so the runaway train could go no further. Travelling at probably 50 or 60 miles an hour the train entered the goods yard, became derailed on the second set of facing points, and the engine continued until it collided with and partly demolished a shed. The engine overturned and the following waggons piled up in a large heap. As the train entered the yard Inspector John Norman shouted a warning to two shunters and a number taker who were able to run to safety; the inspector himself was killed in the accident, as were John Loader, a stationmasters' clerk at Gloucester who had just disembarked from a train and was making his way home across via the goods yard, and Driver Jennings. Fireman Pearce was trapped in the overturned engine along with the body of his driver and escaping steam and hot water; the engine fire and leaking gas from a broken pipe made his rescue very difficult and hazardous. A number of railwaymen received awards and commendations for their efforts in the rescue operation. [906]

An eye-witness to the accident told a reporter, 'Sparks were flying from the wheels. Suddenly there was a terrific crash. I realized that there had been an accident and ran up on to the line. I saw a great engine lying on its side across some of the metals, with the front of it projecting into the end of a shed. Volumes of steam and smoke were coming from the engine. I ran to the front engine of the train, and with a police constable and a railway official I helped to pick up a body from the permanent way. The man was dead. We then went to the cabin of the engine, where we helped to remove the driver and his mate. The driver's legs were hanging over the side of the footway, and his body lay on the footplate. His clothes were on fire. On a further search we found the body of a railway official at the back of the shed.' [907]

Concluding the report into the accident, the inspecting officer, Colonel A.C. Trench, remarked that there could be no question of any defects of the points or track or of the faulty operation of the points or signals having caused the accident, which was clearly due to the excessive speed of the train. The question therefore was what had caused the excessive speed of the train. Careless or reckless driving on the part of Driver Jennings could, Colonel Trench stated, be dismissed at once as a cause as the driver was intimately acquainted with the line, the engine, and the working of similar trains, plus had a very good record.

The possibility of overloading of the train or brake failure could also be discounted. The weight of the train was 169 "quarters" as against the maximum permitted of 172 "quarters," four "quarters" being the weight of a loaded 8-ton mineral waggon. Equally there was no suggestion of a brake failure, the engine brakes having been adjusted on the morning of the accident. Also the hand brake had not been used, the engine had not been reversed and the engine had not whistled.

The above possibilities having been ruled out, the only reasonable one remaining was that neither of the men on the footplate had been in a condition to control the train on the descent from Combe Down Tunnel. This possibility was supported by the statement of Fireman Pearce that the tunnel had been very hot and smoky, and he had had to wrap his coat around his face, the fact that after the accident the driver and fireman had been found lying on the footplate and that none of nine or ten witnesses who had seen the runaway had seen anyone on the footplate.

It must be concluded that the driver and fireman had been overcome in the tunnel by smoke and fumes and the train had emerged from the tunnel uncontrolled, reaching a speed of 50 miles an hour or more at the foot of the gradient (at Bath Junction) despite the brake application by the guard. This conclusion was confirmed by medical evidence at the inquest which showed that Driver Jennings had a blood carbon monoxide concentration of 75%, so that he must have been unconscious or almost so.

---

[906] Ibid

[907] *The Times*, London, England, 21st November 1929

Turning to recommendations, the inspecting officer noted that there had been no similar accident previously in the 50 years that the tunnel had been open. All witnesses agreed, however, that conditions in the tunnel could sometimes be very bad, this being when, as was true to a marked degree on the day of the accident, there was a combination of adverse factors, such as smoke from previous trains, an engine labouring heavily and possibly slipping and weather with high humidity and no wind.

Improved ventilation was the obvious solution, but the provision of either air shafts or fans providing a forced draught would be very costly.

It was only northbound trains that had problems and drivers indicated that problems only occurred if they were stopped at Midford and could not get a run at the gradient.

The simplest remedy would be to reduce the maximum load of goods trains or else employ a banking engine from Midford. Detaining northbound goods trains at Wellow until they could get a clear run through to Bath was desirable but might not always be possible given the limitations of single-line working. The Company were considering the possibility of reducing the maximum permissible load of trains.

Due to the limitations of the turntables available the larger engines had to work northbound trains tender first. It was unclear if this worsened the problem of engine smoke within the tunnel. [908]

A bad accident occurred in Brazil on Sunday 9th March 1930. The mountain town of Teresópolis, just outside of Rio de Janeiro, was about 3,000 feet above sea level and its cool climate and attractive location made it popular as a summer resort. It was served by a funicular railway, the journey between it and the capital taking about two hours. At 5 p.m. on the day of the accident a descending train consisting of engine No. 22 and two cars left Teresópolis, heavily loaded with passengers. It apparently suffered a brake failure, and to the horror of its passengers it went faster and faster until it left the track on a curve between Alto and Plumby stations [909] and fell a reported 500 feet to the valley below. According to some initial reports in American newspapers 30 people were killed and another 15 seriously injured, [910] but the actual number of dead was only eight. The rescue of the injured was made very difficult by the inaccessibility of the accident site. [911] The passengers on the train included members of the Fluminense football team, which had played a friendly match at Teresópolis, and one of the teams' players, Jorges Tavares Py, was among those who lost their lives. Other players were injured, and the team was seriously weakened, but it played the remainder of its game that season. Among the others who lost their lives were two children, the driver of the train, Manoel Virgínio da Salva and Fireman Mario dos Santos. [912] [913]

---

[908] Ministry of Transport report

[909] *Albuquerque Journal*, Albuquerque, New Mexico, United States, 11th March 1930

[910] *Huntingdon Daily News*, Huntingdon, Pennsylvania, United States, 10th March 1930

[911] *Atlantic News-Telegraph*, Atlantic, Iowa, United States, 10th March 1930

[912] www.flunomeno.com – O terrivel acidente de trem solfrido pelo Fluminense em 1930

[913] www.tremdaserradoriodejaneiro.blogspot.com – Acidentes na Estrada De Teresópolis

CHAPTER 8

BRIDGE FAILURES

Some of the most serious railway accidents of the nineteenth century had resulted from the failure of bridges. Bridges may fail for a variety of reasons. The inherent design of the structure may be faulty, the materials used in its construction may be poor and inferior, the materials may deteriorate over time, or its foundations may be washed away by a stream or river which has become exceptionally swollen following abnormally heavy rainfall. With improvements in bridge engineering, failures due to faulty design had greatly diminished by the start of the twentieth century, but accidents, including major disasters, continued to occur from such causes. Later in the twentieth century a new danger, mainly though not entirely affecting road underbridges, would be so-called 'bridge bashing' by vehicles which were too high to pass beneath the structure.

Penistone viaduct, England, on the Lancashire and Yorkshire Railway branch from Huddersfield close to its junction with the Great Central Railway Sheffield – Manchester main line, was the scene of a spectacular accident on 2nd February 1916. The previous month a crack had been discovered in the parapet wall at the southern end of the structure. The foundations were examined but no obvious cause for the crack was discovered, and temporary repairs were made.

At 4.15 p.m. on the day of the accident Driver Lockwood and Fireman Butler were busy doing shunting at Penistone, and during the course of the operation they had brought their engine, 2-4-2T engine No. 661, out onto the viaduct. Suddenly they became aware that the track in front of the engine was sinking, and fortunately they both had time to jump from the locomotive and run back the line towards the station and off the viaduct. A few seconds later a portion of the viaduct collapsed, leaving the engine briefly suspended across the gap on the now unsupported rails before it too fell 80 feet to the valley floor below. A gap 30 yards wide was left by the collapse. At the moment of the collapse the engine had been pushing waggons back towards the station and although one was derailed none followed the engine off the viaduct.

The driver and fireman had had a lucky escape and 20 minutes later a passenger train carrying pupils from Penistone Grammar School had been due to pass over the viaduct.

Following the collapse a temporary road motor service was introduced between Penistone and Darley Dale stations and later a temporary halt, named Barnsley Road Halt, was brought into use at the Huddersfield end of the viaduct. Through traffic, such as expresses from London to Bradford, had to be diverted over other routes. It was decided that recovery of the engine was too difficult, and it was cut up on the spot. Its chimney was reputedly made use of as a flowerpot at Brockholes station. The repaired viaduct was brought back into use on 14th August. [914]

It was assumed that the collapse was due to the effect of recent heavy rains on the foundations of the failed pier. [915]

A southbound passenger train on the South Dakota Central Railroad met with an accident at about 7.30 a.m. on the morning of Saturday 4th March 1916 near Thomas, South Dakota, United States. As the train was approaching a bridge its engineer, Robert Maxwell, realised that the structure was on fire, but it was too late to pull up. The bridge collapsed under the train, the engine rolling over to one side while the leading mail and baggage car and the following passenger coaches plunged straight down. Almost immediately the wreckage caught fire in a number of places, this fire having apparently not originated from the burning bridge but from overturned stoves.

Only one person, Mail Clerk George Loveday, was killed outright but Mr F.E. Dooling, a travelling salesman of Denver, Colorado, was trapped in the wreckage by heavy timbers and burnt to death despite heroic attempts to save him. Fourteen passengers, Conductor W.H. Davison, Engineer Maxwell, Fireman W.J. Dunn, and Expressman P.T. Burgess were injured. The conductor was especially praised for his work in getting passengers out of the wreckage.

A relief train was sent to the scene from Watertown and the injured taken to the Luther hospital there. One of the passengers on the train, Mrs Ethel Blackburn, herself a former nurse, was however very critical of the railroad physician who had accompanied the relief train, whom she said had done nothing either to dress the wounds of the injured passengers or to examine them to determine if they were fit for the journey to Watertown. Her comments were supported by a couple of other passengers.

The bridge had reportedly been discovered to be on fire by the crew of another train a day or so earlier. They had stopped their train to deal with it and believed that they had put the fire out. [916]

Despite its apparent magnitude, for some reason a bridge collapse on the Chicago Rock Island and Pacific Railroad early on the morning of Friday 2nd June 1916 does not seemed to have formed the subject of an Interstate Commerce Commission inquiry. The accident happened at about 2.50 a.m. at a bridge over Flood Creek, near Packard, Iowa, United States. (The stream was also referred to as Cold Water Creek.) Northbound passenger train No. 19 was crossing the bridge when the entire structure collapsed, high water having reportedly undermined its supports. The engine plunged into the creek,

---

914 Webpage – Huddersfield Exposed: Penistone Viaduct
915 *The Engineer*, London, 11th February 1916
916 *The Saturday News*, Watertown, South Dakota, 9th March 1916

taking with it the baggage car, smoker, and day coach. At the time of the initial report the bodies of two women had been recovered from the day coach and another seven persons were said to be missing, feared drowned.

When the train overturned in the creek many of the passengers were able to break windows and climb out onto the sides of the cars, where they waited until were eventually taken off by rescuers in boats.

'The storm which wrecked the train was of the proportions of a cloudburst and nearly paralysed wire communication. It was a long time after daybreak before the first rescue train dispatched from Waterloo reached the scene.'

The same cloudburst caused numerous washouts and great disruption to traffic on the Chicago Burlington and Quincy and the Chicago Milwaukee and St Paul Railroads. [917]

By the following day it was stated that the bodies of six victims had been recovered, six more were visible in the wreck and it was feared that the final number of dead might reach 25. [918]

By Monday 5th the number of confirmed dead had risen to 13 and a newspaper report that day gave a very detailed account of the accident. The town of Packard was described as no more than a cracker box station, an elevator, two store buildings – one of them empty, and a church. A mile and a half east of Packard the railroad crossed the appropriately named Flood Creek by a two-span girder bridge. In summer, the stream usually only carried a small amount of water. At 7 p.m. the night before it began to rain and before the storm had ended government water gauges had recorded 4½ inches of precipitation. Ten feet of water roared down the valley of Flood Creek and at the bridge, the embankment from behind the west abutment of the structure was washed away until the brick and mortar pier stood marooned.

'When the train reached the bridge the unsupported track and ties of what had been the west approach embankment sank under the weight of the locomotive, tender and smoking car. As the day coach struck the depression the momentum of the train was thrown almost horizontally the bridge piers and the structure toppled forward. The smoker, brought up suddenly, just as it left the bridge, was thrown directly across the track. The front trucks of the first two Pullman sleepers following the day coach, balanced crazily in the air and stopped, but the day coach itself, loaded to capacity, toppled over the side of the second girder and fell twenty feet to the bed of the stream.

'After the accident a nearby farmhouse was turned into a temporary hospital and morgue until relief could be dispatched from a point further north, where every available section man was at work repairing a washout. Wire communication was woefully uncertain for the storm had played havoc with the poles and cables along the way, but at last doctors from the nearby towns arrived, relief trains were dispatched, and the work of recovering the dead, caring for the injured and declaring the right of way was begun.'

The first assistance came from the small towns of Greene to the north and Clarksville to the south, later also from the larger communities of Waterloo, Mason City and Cedar Rapids. Thirty of the seriously injured victims were taken to the tiny hospital at Greene, most having to be then transferred to hotels or private homes due to lack of space.

---

[917] *Mitchell Daily Republican*, Mitchell, South Dakota, 2nd June 1916
[918] Ibid, 3rd June 1916

Notable for his heroism in the wreck was "Big Jim" Weatherall, the negro train porter from Cedar Rapids, who was instrumental in rescuing 12 passengers from the submerged day coach, and others from the water. 'They think a lot of "Big Jim" up here, and the story is told and retold of how he stood on the side of the car, broke the windows and lifted out the helpless women and children. And later, with W.T. Shalt, the Chicago news butcher, he obtained a boat and saved many passengers from the water lower down.'

The intrepid reporter visited the wreck, where he was prevented from coming very close because he had a camera and the railroad did not want people taking photographs of it, and thence to the hospital at Greene, where he interviewed a number of the survivors. 'Up in the rooms of the hospital I talked with Mrs William Alford, of Louisville, Ky, who had a narrow escape from drowning in the day coach but escaped with slight injuries. "I was dozing when the accident occurred," said Mrs Alford. "The first I knew of what had happened was when the water began to rush into the car. It reached my ankles, my knees, and finally it was up to my neck. I sank twice in attempting to reach a window and when I did reach one I was too weak to break the glass. But I flattened my hand against the pane and waited. Someone must have seen it though for just as I became exhausted the window was broken in and someone pulled me out."

'Although her mother and her two daughters lay dead in the morgue nearby, Mrs John Kober, of Waterloo, sat dry-eyed in the station and told of her experience. "We were on our way to spend the summer in Minnesota," she said, "and as I had been ill my mother insisted that I take a sleeper. I was awakened by the shock and dressed when I learned there had been an accident. The porter assisted me to cross on the fallen girders of the bridge to the Greene side and there I identified the body of my mother."

'And so it ran – sorrow, sorrow, sorrow on every hand. I stood in a telephone exchange and heard John Kober of Waterloo, his voice broken and his hand trembling, make arrangements for the transport of his two girls back to Waterloo, steadying himself now and then to be able to tell his listener to "be brave."

'I saw friends of Mrs T. E. Robinson turn from laughter to tears as they identified the body of the woman they had come to believe safe.

'I heard the telephone girls with tears in their own voices, plead with unthinking patrons to be patient until the death messages had been taken care of.

'I saw the long weeping line that sat in the morgue – a tiny establishment in the rear of a furniture store – and I saw them start up fearfully as the grey hearse that plied back and forth between the town and the wreck drove up.

'I saw the body of S.C. Lapp of North English, taken from the water some distance down from the wreck. He had not been listed among the missing, but business cards in his clothing quickly established his identity, and the dripping body was placed on a cot and hurried away.

'I saw Engineer Wescott at the wreck, just as he left his engine at Manly, waiting, waiting, for the body of his wife. He had been there all through the night.'

The reporter believed that he had come across some evidence of mismanagement by railroad officials inasmuch as other trains had been diverted to alternative routes or held back before the accident, but it emerged that this was because the track had earlier been washed away north of Greene, blocking the line. The washout was then repaired and as train No. 19 carried mostly local passengers it was allowed to proceed towards Greene first. A work train in front of it had also crossed the bridge safely only 20 minutes before the accident.

The identification of the bodies of some of the victims was not easy. No less than five Waterloo residents independently identified a body as that of Mrs Ella Sneicher, of Waterloo. It turned out that they were all wrong, for Mrs Sneicher answered a telephone call to her residence, indignantly denying that she was dead! [919]

Conductor Tiffany was apparently in charge of the train. The names of the engineer and fireman do not seem to appear in any of the newspaper reports, unless Engineer Wescott was driving the engine involved.

The final number of dead was 17. [920]

The First World War involved the colony of German East Africa, which by September 1916 was largely in British hands, though the German forces commanded by Lieutenant Colonel Paul von Lettow-Vorbeck were never defeated. The area captured by the British included the Tanganjikabahn (Central Railway) of the Ostafrikanische Eisenbahngesellschaft from the port of Dar Es Salaam to Dodoma and Kigoma in what is now Tanzania. The retreating German forces destroyed engines and rolling stock and also bridges but the damage was repaired by railway engineers, at first just sufficiently to enable the line to be used by motor lorries provided with flanged wheels, capable of pulling 15 tons, then further to enable the use of heavier steam trains.

On 5th May 1917, however, a re-built bridge collapsed beneath a train at Mkata. Its failure was not due to enemy action but because the original structure had been built with foundations which were insufficient to cope with a river swollen by heavy rain. Sixteen gunners from the 24th (Hazara) Mountain Battery (Frontier Force) and four Askari from the King's African Rifles were drowned when the cattle trucks in which they were riding plunged into the river. Many other men were badly injured by being flung against weapons and stores in the trucks. The report on the accident does not mention the engine crew so probably the front of the train had cleared the bridge before it collapsed. The damaged bridge and line were repaired by the 26th Railway Company. [921]

Heavy rain had caused flooding in the state of Pennsylvania, United States, in mid-February 1918. Engineer T.B. Mullen, in charge of a locomotive pulling a passenger train on the Tionesta Valley Railway, aware of possible danger, was driving cautiously on 20th February as he approached a trestle at Gilson, three miles from Sheffield. His slow running saved the passengers of his train but not himself as the trestle collapsed beneath his engine, which fell into the flooded stream and was entirely submerged. The entire trestle collapsed, and the engineer was pinned beneath his engine and drowned. Fireman W.S. Buzzard escaped from the engine with difficulty and other members pushed a board within his grasp to enable him to struggle out of the water. The fireman was injured, but

---

919 *The Evening Gazette*, Cedar Rapids, Iowa, United States, 5th June 1916

920 *Railway Age Gazette*, Vol 61, Chicago, Illinois, United States, pg. 152

921 Webpage – The Indian Railway Corps, East African Expeditionary Force, 1914-1919, by Harry Fecitt

not seriously. None of the passenger coaches was derailed. [922]

A passenger train had just left Sirari station, India, on the South Behar Railway, on Saturday 7th September 1918, and was proceeding slowly between the down home and outer signals when the abutment of a bridge that it was crossing collapsed, flood waters having washed away the foundations of the abutment and wing walls. Two passengers died, three were injured and the rolling stock and track were badly damaged. The line had been patrolled the same morning and it was decided that all reasonable precautions had been taken and no one was responsible for the accident. [923]

The Interstate Commerce Commission does not seem to have been very concerned about investigating bridge failures for in the same way that it did not concern itself with the 1916 Packard, Iowa, accident, it did not investigate an accident on the Chicago and North Western Railway at a bridge over Big Cottonwood Creek, near Whitney, Nebraska, United States, on Wednesday 15th June 1921. As is often the case early reports of the casualties were exaggerated, with reports of 12 or even 20 dead, but the final toll of five dead, 13 seriously injured and about 20 slightly hurt was bad enough.

A westbound train had passed over the bridge without incident at 8.30 p.m. but a sudden cloudburst after that time caused the creek to swell to a torrent which it was believed caused the failure of a wagonway bridge upstream of the railway, and the remains of this bridge had in turn struck and weakened the supports of the railway structure. Engineer C. A. Bandholts of eastbound passenger train No. 606 from Lander, Wyoming to Omaha, Nebraska, had no warning of a problem when his train reached the bridge in the darkness at 11 p.m. His locomotive cleared the bridge and remained on the rails but its tender broke loose and rolled over and over down the embankment. The leading mail car was badly crushed, and the baggage car which came next was 'smashed to splinters' by the smoker car third in the train. These three vehicles and the steel chair car which came next all fell 40 ft into the stream, while a Pullman car and the private car of Divisional Superintendent M.E. Pringle remained on the track on the approach to the bridge.

Those who lost their lives in the accident were Baggagemaster Robert Scott, mail clerk B.F. Stiles and passengers Frank Bosner, of Lander, Wyoming, C.M. Buck, of Grand Island, Nebraska, and F.M. Stewart, of Gordon Nebraska, a travelling salesman. The passengers who lost their lives had all been in the smoker. Engineer Bandholts, Fireman B.R. Lyons, Conductor Jim Finnegan, Brakeman C.N. Barker, Pullman Conductor H.H. Fickbohm, Flagman D.G. Roberts, and Express Messenger Walter Goodall were all more or less hurt. [924]

A surviving passenger, C.E. Warnecke, who walked and ran seven miles for assistance after the wreck, embraced his wife and then related his experiences to a reporter upon his arrival at Omaha. "I was in the coach, the fourth car. I understand the car had been condemned at Chadron a week ago. The crash happened in an instant. There

---

[922] *Harrisburg Telegraph*, Harrisburg, Pennsylvania, United States, 21st February 1918

[923] Railways in India Administration Report for the Year 1918-19

[924] *The Alliance Herald*, Alliance, Nebraska, United States, 17th June 1921

was a heavy grating sound – the coach sagged and crashed through the broken bridge. The lights went out. Water poured through the broken windows. There was a mad scramble of passengers for safety. Really the women were the bravest. The screams of the men could be heard above the women's faint cries for help. I was dashed forward over several seats and found myself near an aged woman entangled beneath broken seats. Water was pouring over her."

Here Warnecke described the aged woman's rescue – how he and another passenger dragged her through an aperture in the flooded coach. "I learned afterwards she was Mrs Gavison, of Ainsworth, Neb. Moyer and I carried her ashore, stumbling over the wreckage, in the surging waters. We learned she was not injured severely, then made a dash to give the alarm. My knee hurt and several times I thought I'd drop from exhaustion. Through the black of the night over a prairie trail Moyer and I kept on a run for some light of a ranch house. At last we found a place and phoned to Crawford." [925]

Had they gone east instead of west Whitney station was only two miles distant and in fact Superintendent Pringle and Flagman Roberts did make their way back to Whitney, probably arriving there before Warnecke and Moyer had found their ranch house. The Whitney agent contacted Chadron and a number of Whitney residents set out for the wreck with heavy ropes and lanterns.

A relief train took the most seriously injured passengers to Chadron at 6 a.m., and thence to hospital in Hot Springs, South Dakota. It was feared at the time that several more might die, but all seem to have pulled through. By then a wrecking train had also reached the scene from Alliance, Nebraska, and began the clearing-up operation. [926]

An inquest jury at the end of the month found that the bridge had been in good condition shortly before the accident. [927]

Very heavy rainfall over three days during the early part of the monsoon season caused what was described as an unprecedented flood and the collapse of bridge No. 36 west of Amroha station on the Oudh and Rohilkhand Railway on Saturday 25th June 1921. The engine and six leading carriages of No. 7 up passenger train from Moradabad and Gaziabad fell with the bridge into the Sot Nuddi River. The engine crew and an estimated 39 passengers lost their lives, and another 30 passengers were injured, one of whom subsequently died. The collapsed bridge resulted in a breach 150 feet wide. The damage to the bridge, rolling stock and track was estimated at 99,318 rupees. A goods train had passed over the bridge safely an hour earlier. [928] As with the Sirari accident the committee of enquiry set up to investigate the accident were satisfied that the line had been properly patrolled on the night of the accident, that those who had carried out the patrol had been fully aware of their duties, and that the bridge had been thoroughly inspected by an officer a month earlier. [929]

---

[925] *The Omaha Daily Bee*, Omaha, Nebraska, United States, 18th June 1921
[926] *The Alliance Herald*, Alliance, Nebraska, United States, 17th June 1921
[927] *The Shelby Sun*, Shelby, Nebraska, United States, 30th June 1921
[928] *The Times*, London, England, 29th June 1921
[929] Administration Report of Indian Railways in 1921-22

On Sunday 16th Jul 1922, an accident occurred to a fast train between Bucharest, Romania, and Vienna, Austria, near Sinaia, Romania. A bridge collapsed and the train, or more likely part of it, fell into a stream. Two people died and 24 were seriously injured. Traffic on the main line was interrupted for many hours. [930] In what appears to have been a strange coincidence the same bridge again collapsed beneath a Vienna express on Tuesday 19th September of the same year. This time the engine and a waggon, probably the leading luggage van, fell into the water. No casualties were mentioned in the report on the second accident. [931] Possibly damage to the bridge caused by the first accident was not properly repaired.

Following the defeat of Turkey in the First World War the Allies promised Greece parts of the former Ottoman Empire in Anatolia and to take control of this territory Greek forces landed at Smyrna (now Izmir) in May 1919. Greece and Turkey then fought a war which ended with the defeat of Greece and their evacuation of Smyrna in September 1922. Many atrocities were committed by both the retreating Greeks and the advancing Turks and a few days after the Greek withdrawal, the Greek and Armenian quarters of Smyrna were destroyed in a great fire. [932] In the following couple of months, at unknown dates, two serious train accidents in the Smyrna area resulted in the death of 20 Turkish soldiers and civilians. One of the accidents involved the collapse of a bridge, the other was a collision between two trains. [933] No more exact details seem to be obtainable.

A slow mixed train which had left Baghdad on Thursday night ran into what was described as a 'cyclonic rainstorm' 50 miles north of Basra, Iraq, early on the morning of Saturday 14th April 1923. A flooded river due to the heavy rain had weakened a bridge, which collapsed after half the train had crossed it. Twenty coaches and waggons fell into the gap and were 'telescoped and mixed in inextricable confusion.'

Three Indian soldiers died and ten were injured, as were six British, mostly severely enough to require stretchers. A relief train with doctors was sent from Basra. A bypass line was constructed, with presumably also a temporary bridge, and the line was reopened the next day. [934]

The United States suffered a major bridge disaster on the evening of Thursday 27th September 1923. Eastbound Chicago Burlington and Quincy Railroad passenger train No. 30 consisted of Colorado and Southern Railroad engine No. 350, a baggage car, a combination baggage and mail car, a coach, a chair car and three Pullman sleeping cars, in charge of Conductor Goff and Engineer Spangler. The Casper, Wyoming to Denver, Colorado train was wrecked at bridge No. 98, over Cole Creek, a mile and a half west of Lockett station, Wyoming, at about 9.05 p.m.. It was travelling at between 30 and 40 miles an hour when it approached the bridge around a curve. The bridge had either already collapsed or gave way beneath the train and the engine and first four vehicles

---

930 *The Times*, London, England, 17th July 1922
931 Ibid, 20th September 1922
932 Wikipedia – Occupation of Smyrna
933 *The Times*, London, England, 23rd December 1922
934 Ibid, 16th April 1923

went into the creek, as did the front of the fifth vehicle while the sixth was derailed.[935] According to the Interstate Commerce Commission (ICC) report 17 passengers, a brakeman, a mail clerk, and a Pullman porter lost their lives while seven passengers, the conductor, the engineer, and the fireman were 'missing', giving 30 fatalities in all, though some newspaper reports spoke of the recovery by 5th October, of 37 bodies,[936] implying that the real death toll must have been higher than that reported by the ICC.

The newspapers contained stories from the some of the survivors. 'Charles McLocken of Denver {who} escaped from one of the day coaches, said he didn't believe that more than fifteen in the two passenger coaches were saved. "I believe there were at least 100 in the two coaches, he said. "I was sitting about the middle of the front day coach," McLocken said. "The train appeared to be travelling at normal speed. Suddenly I felt a jar as apparently the baggage car in front of our coach went into the water. Almost instantly the front end of our car started to tip downward. I jumped thru a window and cleared the car just as it became submerged. With a few others I made my way to shore. We helped pull two or three injured from the water and then sent couriers to nearby phones to summon aid."

'A flagman, Harry Cruze, was a member of the train crew who escaped. He was standing on the rear platform of the last Pullman when the train plunged into the torrent. "I grabbed a rope and threw a line to passengers clinging to portions of the wreck above the water," Cruze said. About a dozen were pulled to safety over this life line."[937]

'First news of the wreck was reported by Daniel J. McQuaid, of Denver. He was in the smoking compartment of the rear coach when the train fell into the stream. He managed to escape from the car and swim ashore. He walked four miles, finding a house from which he could telephone news of the disaster.'[938] One report stated that McQuaid had reported the disaster from the Midwest camp of the Big Muddy oil fields.[939]

Bridge No. 98 was described as a pile driven trestle 111 feet long with a maximum height of 20 feet, consisting of nine bents (cross frames) 14 feet apart. Each bent consisted of five piles driven in as far as they would go. Stringers (cross beams) rested on caps above the piles and the ties (sleepers) rested on the sleepers. Upstream of the bridge a rip-rap 6 feet deep, 3 feet wide at the top and 7 or 8 feet thick at its base was provided as a protection against scouring. The bridge dated from 1914. It was noted that its location at the end of a curve meant that it was not visible from an engine headlight for more than 500 or 600 feet and any unsafe condition would undoubtedly not become obvious until the train was much closer.

Cole Creek was described as a small stream, about 14 miles long, which joined the North Platte River a quarter mile south of the railroad. The stream drained an area of 130 square miles, 80 of which consisted of adobe soil, which drained water very readily, and the remainder of very sandy soil which had formed a number of sand bars. The average annual precipitation was 13 inches, but over 5 inches of rain had fallen in September and almost 3 inches on the 27th and 28th. Inspection of the area several days after the accident

---

[935] Interstate Commerce Commission report
[936] *Winifred Times*, Winifred, Montana, United States, 5th October 1923
[937] *Evening State Journal*, Lincoln, Nebraska, United States, 15th September 1923
[938] *Centralia Daily Sentinel*, Centralia, Illinois, United States, 28th September 1923
[939] *The Kevin Review*, Kevin, Montana, United States, 4th October 1923

revealed three points where sand bars had blocked the stream, impounding large volumes of water, and a local inhabitant spoke of seeing an area of 340 acres flooded to a depth of 12 or 14 feet at 4 p.m. on the day of the accident; all the water had disappeared by 9 a.m. the next day.

Roadmaster Waygood told the ICC inquiry that on account of heavy rain he had boarded a westbound freight train to examine the line. At 3.35 p.m. there had been very heavy rain at North Glenrock, the first station east of Lockett and 8½ miles from Cole Creek and he had instructed the section foreman to inspect the track and notified the dispatcher at Casper. Sand had been washed onto the track east of Lockett in a cutting and he had also instructed the Lockett section foreman to patrol the track while the storm lasted. The freight train had passed Cole Creek at about 6.15 p.m., at that time there had only been 8 or 10 inches of water in it and arriving at Casper at 7.50 he had told the dispatcher that he did not consider the situation serious or slow orders necessary. Reaching Cole Creek soon after the disaster he had found the water in it to be 10 or 12 feet deep and all the bridge had gone. He said that in nine years he had never seen more than 2½ or 3 feet of water in the creek, and that amount only once.

Section Foreman Kellner testified that inspected the bridge at 8 p.m., when the stream had been 14 or 18 feet wide and about 18 inches deep, its normal condition. There was no drift or debris lodged against the bridge. No dangerous condition appeared to exist, and he decided that there was no need to do any further patrolling that night.

Master Carpenter Meyer stated that the bridge had been inspected on 16th September, when it had been in first-class condition.

Divisional Superintendent Grisinger considered that every precaution had been taken to guard against the action of unfavourable elements, and that from the reports that he had received he had not considered the conditions warranted holding train No. 30 at Casper or sending a special patrol in front of it.

Brakeman Crews (the ICC spelling of Cruze, it is not known which is correct) stated that the train brakes had been tested before leaving Casper. There had been no stops between there and Cole Creek, and the brakes were not applied until they came on in emergency immediately before the train went off the bridge.

The ICC concluded that the accident had been caused either by the bridge having been washed away before the train arrived or by its giving way beneath it, it was impossible to say which. The bridge had been in good condition at the time of its last inspection. There was no evidence of heavy rain in the area immediately before the accident and it was probable that a large amount of water, impounded by a sand bar upstream, had suddenly been released by the sand bar giving way, and the sudden torrent had either destroyed or fatally weakened the bridge. The findings of detached parts of the boiler at the scene had given rise to questions regarding a possible boiler explosion but there was no evidence of such an event. [940]

Four Baltimore and Ohio Railroad railwaymen died in an accident near Locust Grove, Ohio, United States, on Saturday 29th March 1924. Half a mile east of Locust Grove a stream known as Hog Run was crossed by bridge No. 545, a pile driven trestle of similar

[940] Interstate Commerce Commission report

dimensions to the one at Cole Creek, it being 111 feet long and 15 feet high.

Eastbound freight train No. 184 consisted of engine No. 2355, 21 cars and two cabooses, in charge of Conductor Farabee with Engineer Louis Kastla and Fireman H.F. Gartner. The train was crossing the bridge at a speed of between 8 and 15 miles an hour at 8.30 a.m. when the structure collapsed beneath it and the engine and leading seven waggons fell into the creek. The driver and fireman, together with Brakemen J.A. Birwell and Robert Powell, lost their lives. [941]

There had been heavy rain the night before; over 2 inches of rain had fallen in 16 hours at Granville, Ohio, about 8 miles away. The crew of the train were aware of possible problems and the engineer had told the conductor that he would take no chances, and if there were any places along the track that looked questionable he would stop the train and examine the ground before proceeding.

Conductor Farabee told the Interstate Commerce Commission (ICC) inquiry into the accident that they had had orders to reduce speed to 10 miles an hour five miles back from the bridge and to look out for high water. He put the speed of the train at 12 or 15 miles an hour at the time of the accident.

Supervisor Mahurd, who was riding in the caboose with the conductor, put the speed of the train at 8 or 10 miles an hour. He had gone forward to the front after the accident and said that there had been no sign of a derailment having preceded the bridge collapse. He had never experienced any problem with high waters at that point in the previous 18 years.

Section Foreman Bushey said that he had examined the bridge at 4 p.m. the previous day when there had been nothing wrong.

Master Carpenter Zinsmeister stated that the bridge had been in good condition. Part of it had been renewed in October 1920 and it had been inspected 40 days before the accident. He said that driftwood had blocked the eastern end of the bridge and a large tree had lodged against the trestle, diverting the floodwater, which had then undermined the west side of the structure.

A local resident, Mr O.W. Davis, witnessed the accident. According to him the train had been moving at 6 miles an hour as it approached but had increased its speed to 15 miles an hour as it reached the bridge. After the accident he had gone straight to the scene to find the engine submerged, and the roadbed washed out at the west end of the bridge.

A freight train had passed over the bridge just after midnight.

The ICC concluded simply that high water had weakened the bridge which had collapsed under the train, with no further comments or recommendations. [942]

What was then the colony of French West Africa suffered a major disaster on Tuesday 19th August 1924. Heavy rain had caused breaches in the Kayes – Thiès line at several points and at kilometre 673, the Paporah Bridge, which consisted of three 15-metre piers (spans?), gave way under a train. Twenty-eight 'natives' and the head of the Remiglio

[941] *New Castle News*, New Castle, Pennsylvania, United States, 29th March 1924
[942] Interstate Commerce Commission report

district lost their lives. [943] [944]

On the night of Saturday 21st March 1925 the Cape – Natal express met with an accident at a bridge over the Tugela River near Hartshill, South Africa. The scene of the accident was about 10 miles south of the town of Ladysmith and close to Colenso. There had been heavy rain in Natal for some days and in the usual scenario floodwaters had fatally weakened the bridge which gave way under the weight of the train. The engine and leading third-class coach plunged into the river, the breaking of a coupling at the rear of the leading coach preventing the remainder of the train from following, though the next coach was left partly suspended over the gap. The driver, fireman and 'several natives' lost their lives. [945]

Eastbound passenger train No. 18 of the Chicago Milwaukee and St Paul Railway from Seattle, Washington to Chicago, Illinois, otherwise known as the *Columbian*, was in an accident on Wednesday 22nd April 1925 at bridge No. 88, situated about a mile and a half west of Rahway, Montana, United States. In contrast to the previous few accidents described this one was not caused by floodwater damaging bridge foundations, the stream which the bridge crossed was in fact dry at the time of the accident, a so-called dry run.

The bridge was a 5 bent (cross frame) wooden trestle, 78 feet long, resting on untreated cedar piling. To protect against fire the bridge was sheathed with galvanised iron. A water barrel and bucket were provided at each end of the structure. The bridge dated from 1908; the piling had been renewed in 1915 and the decking in 1917. It was last inspected only four days before the accident when it had been in good condition.

The train involved consisted of engine No. 6345, a mail car, an express car, a baggage car, two coaches, a tourist car, a dining car, and a sleeping car, in charge of Conductor McGee, with Engineer Alex Bugby and Fireman Hans Struve. It was travelling at about 25 miles an hour when it reached the bridge at 10.12 a.m. As the engine came onto the bridge the structure gave way and the engine fell about ten feet to the bed of the stream below. The leading mail car was slightly damaged by its colliding with the tender of the engine, but the severing of the brake pipe brought the remainder of the train to a stand before it reached the bridge. The engineer and fireman survived the fall of their locomotive but were pinned beneath it and scalded to death by escaping steam before they could be released.

Conductor McGee told the Interstate Commerce Commission (ICC) inquiry that he had been at the front of the first-class coach when his train had stopped suddenly. Going forward and seeing what had happened he had returned to the baggage car and had raised the alarm using the emergency telephone apparatus there.

Head Brakeman Dutcher said that he had gone forward after the accident and had found that about ten feet of the west end of the bridge had burnt out, and that the metal sheathing was cold, indicating that the fire which had affected the two west spans of the structure must have gone out some considerable time previously.

---

[943] *La Croix*, Paris, France, 23rd August 1924
[944] *Le Petit Journal*, Paris, France, 22nd August 1924
[945] *Beaudesert Times*, Queensland, Australia, 27th March 1925

Roadmaster Mattheisen chanced to be a passenger on the train. Inspecting the bridge he found that portions of it were still burning. He said that he had passed over it three times the previous afternoon but had seen nothing wrong. Section foremen were, he continued, expected to inspect their lengths once every 24 hours, usually first thing in the morning. There were no other track controls except in unusual conditions.

Section Foreman Knutson testified that three cars of cinders had been unloaded in the vicinity of the bridge the previous afternoon, but he had seen no live coals among the cinders. He normally patrolled the section every morning but had not on the day of the accident because he had been occupied in laying out work for new men in his gang, plus he had been assured that everything was normal as an adjoining section gang had passed over the section the previous evening and had not reported any problems.

Section Foreman Schosneck of the adjoining section confirmed that he had been over the length the evening before and found nothing wrong.

The ICC concluded that the bridge had caught fire either from the cinders, from a burning pile of ties (sleepers) or from a spark from a passing train. The cinders had been in the cars several days before unloading and were thoroughly rain soaked. Grass around the tie pile was burnt but there was an area of unburnt grass between the edge of the burnt area and the bridge. Engine No. 6431 had been the last to cross the bridge prior to train No. 18, 11 hours previously. Like all the other engines used on the division it was fitted with a spark arrester which was less than a year old and in good condition. The metal sheathing on the bridge also made it unlikely that a spark from a passing train could have set it alight.

The ICC noted that while the cause of the fire was unclear it would have been discovered in time had the track been patrolled in accordance with what was said to be the normal practice instead of the section foreman postponing the patrol for less important duties. The railway company was also to blame as there was no rule or instruction which actually mandated such daily patrolling. [946]

A newspaper report noted that had westbound train No. 15 not been late it would have crossed train No. 18 at Ahles instead of at Rahway and would have been the train which met with the accident. [947]

A local train, consisting of an engine, a third-class coach, a first- and second-class composite and a luggage van, left Catanzaro, Italy, at 20:00 on Friday 9th October 1925 and continued close to the coast towards Badolato. A severe storm was raging, and heavy rain had also fallen during the previous few days, transforming streams into torrents. A bridge over the Ponzio River, a few kilometres from Badolato, about 30 metres long and 11 metres high, had been swept away a short time before the train reached it. The train driver had no warning of the collapse, and the engine, tender and third-class carriage were precipitated into the torrent and the composite carriage ended up suspended over the gap. Fifteen people reportedly died, mostly soldiers and local tradesmen, and presumably also the driver and fireman. The few passengers who escaped injury made their way to a neighbouring station on their hands and knees and rescue operations had to be delayed

[946] Interstate Commerce Commission report
[947] *Baker Sentinel*, Baker, Montana, United States, 24th April 1925

because several other bridges had been washed away. Two days after the accident some bodies had been recovered but others had not, and it was believed that they had been swept down to the sea. The engine, tender and leading coach were half-buried in the sand. [948]

Initial reports on railway accidents were often wildly exaggerated. Thus, the first reports of an accident between Sarajevo and Mostar, Yugoslavia, on about Thursday 26th October 1927 were that a passenger train had fallen over a precipice with the loss of 260 lives. [949] A follow-up report stated that the accident had happened to a freight train, not a passenger train. A bridge had collapsed and the train, or part of it, had fallen 150 feet into a ravine, killing two guards and severely injuring a fireman. [950] [951]

Westbound Chicago Rock Island and Pacific Railroad passenger train No. 5, the *Eastern Express*, met with disaster shortly before daybreak on Thursday 18th July 1929 about two miles west of Stratton, Colorado, United States. The train consisted of engine Nos. 932 and 907, a mail car, a baggage car, a combination baggage and chair car, a chair car, seven Pullman sleeping cars and an observation sleeping car, and was in charge of Conductor Bascom with Engineers Ryan and Peterson and Firemen Claycomb and McNeill. It had left Stratton at 3.48 a.m. and was travelling at between 45 and 55 miles an hour when it reached bridge 474.1 over Spring Creek at about 3.55 a.m. The engine and first five cars had crossed the bridge when it gave way and the tender of engine No. 907, and the first nine cars were all derailed. The sixth and eighth cars ended up partly in the creek and the seventh car completely so; it was in this vehicle that the fatalities occurred. [952] Eight passengers and two Pullman porters died, and 80 passengers were injured.

Bridge 474.1 was described as a deck plate girder bridge 85 feet in length with U-type concrete abutments at each end. South of the railroad a highway bridge spanned the same creek.

Survivors of the accident gave their stories to newspaper reporters. Pullman Conductor E. W. Bremken was asleep in an upper berth. He said, "It was terrible – the worst wreck I've ever been in. I don't know exactly what happened next. When I woke up I found myself across the aisle on top of the porter. The bridge must have given way just like paper. Both locomotives got across, maybe a car or two, and then everything seemed to go blotto. Nine cars were sprinkled crazily about with one of them lengthwise in the creek where it was covered with water. Everybody in that car, eight passengers and one of the best porters I ever knew must have been drowned. Rescue workers worked for hours with acetylene torches before they could even open up the roof of the car."

Mr O.W. Waddle and his wife were on their way from Kansas City for a Colorado vacation. Mrs Waddle related, "We waded out through that raging torrent and then stood shivering in the dark on the bank listening to the screams of the injured without being able to do anything to help them. It was terrible – just awful. For hours we waited there.

948 *The Times*, London, England, 12th October 1925
949 *The Morning Republican*, Findlay, Ohio, United States, 27th October 1927
950 *Corsicana Daily Sun*, Corsicana, Texas, United States, 27th October 1927
951 *The Times*, London, England, 28th October 1927
952 Interstate Commerce Commission report

Everyone seemed stunned. There was no wrecking equipment, nothing but splintered cars, mud, and water. Even when dawn came, it was impossible to get to the water-covered car because of the rushing torrent swishing over it." Mr Waddle added, "It was a real wreck, but we're not going to let it ruin our summer."

Miss Gwendolyn Glasgow, a 'pretty 20-year-old titian haired girl of Courtland, Kansas,' said, "When I looked up after the crash, all I remember seeing were two legs sticking straight up out of the mud. Someone pulled on the legs and out came a small boy with a very dirty face, but with no other visible injuries. Although we could hear screams above the roar of the stream, and although we knew people were dying in the darkness below us, several people just howled with laughter when they saw that small boy's muddy face. I thought it was funny, too, and then I fainted." [953]

Engineer Ryan on the leading (pilot) engine told the Interstate Commerce Commission (ICC) inquiry that the headlights of his engine had been burning brightly and he had noticed nothing unusual as he crossed the bridge. As his engine left the bridge the brakes came on and he saw the air pressure fall to 20, then to zero. Looking back, he saw that the tender of the second (train) engine was derailed. There had been water running under the bridge, something which he had not seen previously in 25 years of operating over the line; there had been a large body of water on the north side of the track. He had not encountered any rain en route and there was no indication of a storm at the bridge.

Fireman Claycomb could not add much to the inquiry as he had been busy firing at the time.

Engineer Peterson said that he had observed a large body of water north of the track as they approached the bridge. His first warning of danger had been the derailing of the tender of his engine.

Fireman McNeill on the second engine said that he had just finished firing when they got to the bridge, and he had noticed water on the south side of the track. There had been a severe crash under the tender at about the same moment as the brakes had come on.

Conductor Bascom said his first warning of trouble had been the application of the brakes. Half an hour after the accident, when it became light, he could see that the car in the water was still submerged, it was about an hour and a half before the water level dropped sufficiently for any part of the car to become visible. He said that he had seen water under the bridge before, but no more than about 2 feet deep.

Engineer Foley of eastbound train No. 6 testified that he had passed over the bridge at about 3.07 a.m.; he was sure there had been no water under the bridge at that time and no evidence of a storm. He had seen very little water in the creek in 11 years. His fireman, named Skeels, agreed.

Dispatcher Nolan stated that there had been no slow orders in the vicinity of Stratton, and he had received no reports of rain at any point.

Section Foreman Mulnix stated that he had never seen an unusual flow of water under the bridge in 16 years. He patrolled the track whenever there was heavy rain, but none had been reported. He considered that an unusual amount of water had backed up and

---

[953] *Albuquerque Journal*, Albuquerque, New Mexico, United States, 19th July 1929

overflowed the adjoining highway bridge, and that when that bridge and embankment had failed a rush of water had struck the railroad embankment immediately east of bridge 474.1, causing it to fail in turn along with the east abutment of the bridge.

In its conclusion the ICC noted that the cause of the derailment and bridge failure was obvious, but there was controversy over the role of the adjacent highway bridge. Railroad officials blamed the presence of the highway bridge, 170 feet upstream, for the washing away of the railroad embankment. The State of Colorado Public Utilities Commission, on the other hand, was of the opinion that the railroad embankment would have been washed away even if the highway bridge had not been there. An engineer physicist reached a similar conclusion and the ICC concluded that the basic cause of the scouring out of the approach embankment to the bridge was the high level of water in the creek, which the drainage prism had been inadequate to deal with, resulting in the scouting out of the embankment and bridge abutment. The presence of the highway bridge was not significant. [954]

A passenger train from Tunis, Tunisia, to Constantine, Algeria, consisting of two engines, postal and luggage vans and two large carriages, met with an accident between Petit and Nador stations, Algeria, 14 kilometres east of Guelma, on Sunday 6th January 1930. It was derailed due to the collapse of a bridge pier.

M. Jouaville, operations manager of the Chemins de fer Algériens de l'État, chanced to be a passenger on the train, and immediately made his way to Petit station, from where news of the accident was sent to Guelma and a rescue operation was begun. A company of *tirailleurs* (light infantry) was directed to the scene. At the same time, a relief train was hurriedly made up and proceeded to the scene with doctors and medical equipment, returning to Guelma later with the injured.

A reporter described the scene of the accident as horrific. The engines had gone down into the ravine one on either side of the track. The leading vans of the train had been telescoped by the following vehicles, a third-class coach and a first- and second-class composite. The third-class coach suffered the most and a dozen corpses were removed from it, some so badly mutilated as to be difficult to identify. There were 17 dead, seven of whom were still trapped in the wreckage at the time of the report. The dead who could be identified were Mme Fernand Guillemin, the wife of a Constantine electrician; M. Maurice Adda, of Tunis, his wife and their five-year-old daughter Nellie; Mme Touiotou, of Tebessa and her five-year-old child; M. Louciel, of Tunis; Mme Ronuau, of Souk-Ahras and an unnamed Israelite. Sixteen others were injured, one it was believed fatally and two very seriously. [955]

A postal courier whose van was overturned escaped with bruises and continued with his mail to Constantine for distribution. The work of clearing the line went continuously once all the casualties had been removed, supervised by the police and with the aid of the military. The line was expected to be blocked for some days, but arrangements were made to trans-ship passengers and goods from trains on either side of the obstruction. [956]

---

954 Interstate Commerce Commission report
955 *Le Petit Parisien*, Paris, France, 9th January 1930
956 *Le Journal*, Paris, France, 8th January 1930

CHAPTER 9

OPENING BRIDGE ACCIDENTS

Accidents involving opening bridges appear to have occurred in only a handful of countries, with the majority occurring in the United States. It seems that most countries managed to a large degree to avoid building such bridges, which from a railway point of view are a constant source of expense and worry. Shaw states that the early railroads on the eastern seaboard of the United States were forced by opponents of the new form of transport to build such bridges over even insignificant waterways where they were never really needed. [957]

The engineer, fireman and two yard brakemen drowned on Saturday 26th October 1918 when a Norfolk and Western Railway engine was backed off a cantilever bridge spanning the eastern branch of the Elizabeth River, Virginia, United States, into water more than 30 feet deep. The freight train of 44 heavily loaded cars, inbound from the West, was drawn by two locomotives. In setting back to reach a siding entering the yard, one of the locomotives plunged into open draw. The engineer had apparently overlooked a warning light found burning afterwards 1,000 feet from the structure. [958]

A serious drawbridge accident at Clayton, Louisiana, United States, on the Missouri Pacific Railroad, on the morning of Saturday 3rd February 1920 was clearly caused by the negligence of the train engineer.

Bridge No. 48 over the Little Tensas River was about 900 feet north of Clayton station. It was a deck girder structure about 450 feet in length, with two fixed spans and a central draw span, 150 feet long, pivoted in the centre and operated manually from the central pier. Protection for the open span was provided by two bullseye lamps on the centre pier which displayed green lights when the draw was closed and red lights when it was open for shipping. Red flags were also flown on each side of the span when it was open.

---

[957] A History of Railroad Accidents, Safety Precautions and Operating Practices by Robert B. Shaw, pg. 187

[958] *Alexandria Gazette*, Alexandria, Virginia, United States, 28th October 1918

Northbound passenger train No. 806 from Natchez, Mississippi to Monroe, Louisiana,[959] consisted of engine No. 8657, a baggage car, a combination mail car and coach (this was the Jim Crow car, reserved for the coloured passengers) and two coaches. It was in charge of Conductor Duffy Brown with Engineer James Taylor and Fireman Sam Castel. The train left Vidalia, Louisiana, at 6.05 a.m., 20 minutes late, stopped at Clayton station, left there at 7.04 a.m., 16 minutes late, and travelling at a speed of between 6 and 15 miles an hour ran off the open draw about a minute later. The locomotive, baggage car and combination mail car and coach fell into the river and were badly damaged, the leading wheels of the next coach were derailed. Five black passengers, Fireman Castel and Express Messenger Sam MacCallum were killed and six passengers, the engineer and another railwayman were injured.

Engineer Taylor told the Interstate Commerce Commission (ICC) inquiry that it had been a wet and slightly foggy morning, making it difficult to see clearly. He said that as he had left Clayton station he had become aware that the engine lubricator was not feeding properly. He had picked up an oil can, found that there was little oil it, had spotted a second oil can and picked it up to give to his fireman to fill the lubricator. He had then looked out of the window but did not notice that the draw was open until his engine was 100 feet from the bridge; he had then made every effort to stop the train, which he said was travelling at 15 miles an hour. He denied having seen either a red flag, red lights on the centre pier or the boat for which the draw had been opened and said that he considered that the bridge should have been protected by torpedoes. Only once or twice previously had he seen the bridge open.

Conductor Brown said that the brakes had been tested at Vidalia. At Clayton he had alighted to deal with the passengers and because of the people on the platform and a truck filled with baggage he had been unable to see the signals protecting the bridge. His first warning of the accident had been the brakes applying in emergency. As far as he was aware the engineer usually complied with the rule that drawbridges must be approached under full control prepared to stop until the bridgetender signalled that the draw was closed.

Rear Brakeman Hobson said like the conductor that the passengers and truck had prevented him from seeing lights at the draw bridge.

Train Auditor Box, who had been riding in the combination car, put the speed of the train at the drawbridge at only 6 miles an hour. He had not seen the protecting lights or had any warning of the accident.

Roadmaster Hinkley, who had also been a passenger on the train, testified that he had seen the steamboat for which the draw had been operated a minute or so after the collision when it had been abreast of the track, with its pilot house 10 or 12 feet above the tracks and its smokestack higher still.

Bridge Tender Loomis stated that at 6.45 a.m. on the day of the accident the steamboat *F.D. Norwood* had whistled for the bridge, whereupon he had placed red flags about 250 feet on either side of the span and opened the draw at 7 a.m. The lamps on the centre pier had been burning brightly. He said that he had seen train No. 806 stop at Clayton station, restart, proceed towards the bridge, run over the red flag, and run off the draw into the river. He and some men at the south end of the bridge had shouted and

---

[959] *The Monroe News-Star*, Monroe, Louisiana, United States, 3rd February 1920

made every effort to stop the train without effect and the engine had still been steaming at the last moment. According to his records he had opened the draw 13 times in November 1919, 25 times in December 1919, 18 times in January and three times in first three days of February.

The ICC reached the conclusion that the accident was due to the failure of Engineer Taylor to keep a proper lookout. He had evidently allowed his attention to be diverted from 'his paramount duty of seeing and knowing that the bridge was safe for the passing of his train.'

Rule 98 of the railway company read:

Trains must approach the end of double track, junctions, railroad crossings at grade, and drawbridges, prepared to stop, unless the switches and signals are right and the tracks clear. Before crossing the track of a railroad, or entering thereon, and before crossing any drawbridge, except where an interlocking system or other system of signals covered by special rules are in use, every engine must be brought to a stop at a distance not less than 200 nor more than 800 feet from such crossing or drawbridge and must not proceed until the crossing whistle (as provided in Rule 14-b) has been given and track plainly seen to be clear. Where required by law, trains must stop.

In addition timetable Rule Q-22 read in part:

All trains must approach draw-bridge and railroad crossings at grade under full control. At interlocked draw-bridges and crossings, interlocking rules will govern. At draw-bridges and at railroad crossings not interlocked, trains unless otherwise directed will make full stop 200 feet from clearance point and must not proceed until signalled to do so by Watchman or Flagman.

The engineer had 'failed absolutely to comply with the latter rule.' [960]

On Monday 19th February 1923, an opening bridge accident occurred at Milford Haven, in Wales. As the accident did not occur on a public railway it was not investigated by the Ministry of Transport, even though it resulted in two fatalities, but it was reported upon in the national press.

A light engine which had been shunting in Messrs T.W. Ward's shipbreaking yard was returning to the depot at the end of the day's work. To do so it had to cross a 30-foot-long swing bridge linking the harbour with Pill Creek but for some reason the bridge had been left open. It was not stated in the report what sort of signalling was provided to protect the bridge but as the line was basically only a siding very likely there was none, and in the darkness the driver could not see that the bridge was open. Four men were riding on the engine at the time, and they were plunged into the cold water with it. Two of the men, William Williams, and William George, were imprisoned in the engine and lost their lives. The alarm was raised immediately after the accident and men searched all night for the two, eventually recovering the body of William George.

Ivor Morgan, who had been the driver of the engine, and a workman named Smith, were fortunately good swimmers and were able to swim to safety before collapsing from exhaustion. 'Morgan, who was suffering badly from shock, but had escaped without

[960] Interstate Commerce Commission report

bodily injury, stated that he was engaged in the cab of the engine when it plunged headlong for 40ft into the sea. The engine had turned somersault after somersault, but Morgan and his colleagues were locked in it. Immediately the engine struck the surface of the water he slipped out of the trap, and being a powerful swimmer, made headway for some destroyers which were standing some distance away.' [961]

A rather unusual drawbridge accident happened at the Hilton Bridge over the Cape Fear River near Wilmington, North Carolina, United States, on the Seaboard Air Line Railroad, shortly after 3 p.m. on Thursday 5th April 1923. A shunting move was being made onto the bridge at the same moment that it was being raised for the passage of a steamer. Engineer E.C. Tenant declared afterwards that he had been upon the rising bridge before he could stop; presumably, he had failed to see the signals displayed indicating that the bridge was open. The engine was hurled upwards, a box car was buckled, and other cars were derailed. Engineer Tenant escaped unhurt, but negro Fireman Wesley Young and white Bridge Tender Major Watkins were badly, though not fatally, injured.

It was hoped that the bridge could be reopened by 4 a.m. the next morning. Meanwhile incoming trains were stopped north of the bridge and their passengers and crews taken across the river by boat and brought into Wilmington, and the trains worked back with northbound passengers from Wilmington, while 'Passengers for Florence and southern points on the Atlantic Coast Line were sent north via Wilson, N.C., and on to their destinations. Trains due to leave here {Wilmington} this afternoon were held in Union Station. The accident does not affect northbound traffic and the road to Goldsboro is open.' [962]

There was a lucky escape from what could easily have been a major disaster on the Rockaway Beach branch of the Long Island Railroad on Saturday 21st July 1928. A drawbridge across an arm of Jamaica Bay had been opened for the tug *Dynamic* when the 7.45 a.m. ten-car electric train from Flatbush to Rockaway Beach [963] approached the spot between Broad Channel and Hammel stations. According to railroad officials Motorman Carl Muller overran a protecting signal and the leading car, the smoker, plunged off the bridge into the water while the second car was left suspended off the bridge. The smoker was partly submerged and sank into the mud, but its occupants (numbering 15 or 24 according to different reports) were able to swim to the tug and all escaped, as did Conductor Harry Fisher and the motorman, who broke a window to get out. Later firemen cut off the roof of the smoker to be sure that no one was left trapped inside. About a dozen passengers were injured, but none seriously, all were X-rayed to determine their injuries, still a fairly new procedure at this date. Fortunately, the tide had been out when the accident happened, or it might have been a lot worse. [964] [965] The accident occurred at 8.20 a.m.

A major alert had been declared when news of the accident was received and eight

---

[961] *The Times*, London, England, 21st February 1923

[962] *The Bee*, Danville, Virginia, United States, 6th April 1923

[963] *Anaconda Standard*, Anaconda, Montana, United States, 22nd July 1928

[964] *Blytheville Courier News*, Blytheville, Arkansas, United States, 21st July 1928

[965] *Altoona Mirror*, Altoona, Pennsylvania, United States, 21st July 1928

ambulance from four Queens hospitals were sent to the scene together with police and fire reserves from every station in the borough. A score of lifesavers, who had been posing for photographs on Far Rockaway beach, also hurried to the scene. The injured were taken to Far Rockaway Hospital while District Attorney Newton of Queens ordered an immediate investigation.[966]

Robert Saulner, of Brooklyn, a passenger in the first car, described the accident. "We were rolling along about as usual, when suddenly I heard a ripping. Then there was a terrific jar as if the brakes had been jammed on suddenly. The next thing I knew the floor of the car had slanted downward and I looked out of the window and saw we were down in the bay. I made a dive for the platform. My idea was to get out into the bay where I could swim for it. I didn't know how deep the water was, and I didn't want to be trapped. I got out but I was so excited that I don't know how. I remember seeing the water rushing into the lower part of the car. It seems to me, everybody started smashing windows. It was a good thing there were no women in the car. Everybody seemed to go about the business of getting out, and all seemed able to take care of themselves. I think we were all pretty lucky."[967]

[966] *New Britain Herald*, New Britain, Connecticut, United States, 21st July 1928
[967] *The Charleston Daily Mail*, Charleston, West Virginia, United States, 21st July 1928

CHAPTER 10

TRAIN ACCIDENTS INVOLVING BRIDGES

Bridges could feature in train accidents in one of two ways. Either the bridge itself could collapse for a number of reasons such as faulty design, faulty construction or because its foundations had been damaged; or a train could fall from a bridge, the structure itself remaining intact. Bridge failures have been dealt with in the previous chapter; this chapter will concern itself solely with accidents where the bridge itself was not at fault. It may seem perverse to list such accidents separately instead of in the chapters dealing with primary causes of the accidents, the procedure adopted elsewhere in this book, but the fortuitous presence of a bridge has such a potential for transforming what might have been only a minor accident into a major disaster that it has been decided to treat such accidents separately, in the same way as accidents occurring in tunnels have been given a chapter of their own.

Fifty plantation workers were returning from Mokae after the day's work on a train of the Kaeleku Sugar Company when it became derailed on a 25 ft high trestle across the Haneoo Gulch, two and a half miles from Hāna, on Maui island, Hawaii, on Thursday 22nd June 1916. The engine and leading three cars of the train fell from the trestle to the lava bed of the gulch below, the cars falling on top of the engine, killing Fireman Tokushige and seriously injuring Engineer Nishiyama, both Japanese. Four labourers, Murai, a Japanese cane-cutter; Ah-Sing, a Chinese cane-cutter, and two Filipinos, names unknown, were killed or fatally injured, and 27 other labourers were less seriously hurt. [968] [969]

A Union Pacific Railroad passenger train was involved in an accident about midway between Beloit and Asherville, Kansas, United States, on the morning of Tuesday 15th January 1918. Train No. 945 consisted of 4-4-0 engine No. 945, a combination mail and baggage car, a baggage car and two coaches, all of wooden construction, in charge of Conductor Tozier with Engineer Reed and Fireman Blevans. It had left Beloit at 8.25

[968] *Honolulu Star Bulletin*, Honolulu, Hawaii, 23rd June 1916
[969] *Honolulu Hawaiian Gazette*, Honolulu, Hawaii, 27th June 1916

a.m. for Solomon and Salina and was derailed about five minutes later at bridge 54.92 over Plumm Creek. The locomotive kept the rails, the combination car was derailed and tilted to the left, the baggage car overturned, and the two coaches fell from the bridge into the creek, landed on their roofs and were demolished. [970] Three passengers were killed outright, 22 were injured, one fatally, and three railwaymen were hurt. The passengers who lost their lives were Alfred Cochrane, a businessman of Abilene, Kansas; Ross Carlton, a Beloit businessman; Mrs Della Charbonneau, of Concordia, Kansas, travelling between her nephew in Beloit and her son in Glasco, Kansas; and Mr N.H. Gilfillon, of Beloit, described as a travelling man, who died of his injuries the same evening. [971]

Bridge 54.92 was a wooden trestle 173 feet long and between 8 and 27 feet high, supported by 12 piling bents and two bulkhead bents.

Engineer Reed told the Interstate Commerce Commission (ICC) inquiry into the accident that the bridge had been under repair since the previous November, but he had not received any slow orders relating to it. He put the speed of his train at the time of derailment at between 20 and 25 miles an hour and said that he had not noticed any roughness of the track.

Fireman Blevans agreed that there had been no unusual roughness and said that his first notice of the derailment had been the application of the air brakes.

The ICC concluded that the train had derailed due to the failure of the ends of two rails of different shaped heads which had been spliced together at a joint, the severe thrusts produced by westbound trains having resulted in the fracture of first the outer splice bar at the joint and then, practically simultaneously, of the ends of the two rails. Contributing to the failures had been the fact that one of the rails was not spiked to the tie and that no slow order had been issued restricting the speed of trains. [972]

Indirectly, at any rate, the First World War played a role in the next accident to be described in this chapter. It was 10th May 1918, by which date the United States had joined the conflict, and a number of military training camps had been established around the country. At 8.56 a.m. on the day in question a special troop train left Camp Jackson, South Carolina, for Fort Sevier, in the same state, and it was to this train that the accident occurred, almost immediately after it had set off.

The train consisted of Atlantic Coast Railroad locomotive No. 973, three box cars, two wooden coaches, a steel underframe coach, three wooden coaches, a steel coach, three wooden coaches and a steel underframe coach, in charge of Conductor Felder with Engineer Sheppard of the Atlantic Coast line. It had only proceeded a train length and a half at low speed when it became derailed and the fifth vehicle in the train, Baltimore and Ohio Railroad wooden coach No. 3039, fell 20 feet from a trestle, landed on its roof and was completely demolished. Nine soldiers riding in this coach died and 25 were injured.

The wooden trestle concerned crossed Wild Cat Creek, at the boundary of the camp, still on United States Government property and was 600 feet long with a maximum height of 36 feet. It was only a year old, dating from the opening of the camp, and was of modern design and substantial construction.

Conductor Felder told the Interstate Commerce Commission (ICC) inquiry that he had

---

[970] Interstate Commerce Commission report

[971] *Abilene Weekly Reflector*, Abilene, Kansas, United States, 17th January 1918

[972] Interstate Commerce Commission report

been riding on the engine when he had seen a stop signal from Switchman Faisy which he had repeated to the engineer and the train had pulled up in half a car's length. Going back he had found a mismatched joint in the track, with only one bolt in the angle bars.

Engineer Sheppard stated that his train had been travelling at about 8 miles an hour and had stopped in 80 or 90 feet. Looking back he had seen the coach fall from the trestle. Like the conductor he had inspected the track afterwards, had found the defective joint and concluded that it had caused the derailment.

Switchman Faisy said that he had been riding on the train but had alighted short of the trestle. He had seen that two wheels of the Baltimore and Ohio coach were off the rails and had signalled to the engineer to stop.

Assistant Trainmaster Jones of the Atlantic Coast Railroad stated that an hour after the accident he had found a rail joint with one angle joint but a spike on each side of both rails, which were apparently secure. The first wheel marks of the derailment began four feet south of this joint.

Trainmaster Gaffney of the Southern Railway, on the other hand, stated that the joint was loose, allowing the rails to swing from side to side, causing a mismatched joint, and in his opinion a wheel flange had mounted the rail at that joint and caused the derailment.

First Lieutenant DeBaun, Company D, 306th Engineers, stated that he was in charge of track maintenance at the camp and had examined it three or four weeks previously. The track had been all right except that the joint bars had not been fully bolted because the bolt holes in the rails and the angle bars had not been spaced alike, a problem which he could not correct, not having a track drill.

Captain Suplee, Construction Quartermaster at the camp, said that he had inspected the track after the derailment but could find nothing that could have caused the derailment, which he did not consider had begun at the joint with only one track bolt.

First Lieutenant Rooks and Private Bost, of Company C, 321st Infantry, stated that they had been riding in the Baltimore and Ohio coach and it had not become derailed until it began to overturn, while Private Bost saw the sixth coach leave the rails before this happened.

The ICC concluded that the derailment had been occasioned by the improperly maintained joint located about 160 feet north of the trestle. The faulty condition of the joint had likely thrown the two rails out of alignment and the evidence indicated that this had derailed the forward trucks of the sixth car, which had then cause the fifth car to overturn and fall from the trestle. The maintenance of the track within the camp was under the supervision of the military authorities and the responsibility for the accident therefore lay with them. The age and condition of the Baltimore and Ohio car, dating from 1884, had doubtless greatly increased the number of casualties, and the indiscriminate mixing of wooden, steel underframe and steel cars in the train was an arrangement which greatly increased the danger to passengers in the wooden cars. The heavier steel cars should have been marshalled ahead of the lighter ones. [973]

The accident which occurred on the Illinois Central Railroad near Aplington, Iowa, United States, on Wednesday 29th May 1918 was very similar to a number described in the Bridge Failures chapter inasmuch as it involved the collapse of a bridge as the result

[973] Ibid.

of damage caused by a stream swollen to an abnormal degree by exceptional rainfall. The train involved was, however, derailed before it reached the bridge and so the accident has been included in this chapter.

Westbound passenger train No. 11 consisted of engine No. 1091, a mail car, two baggage cars, two coaches and three sleeping cars, in charge of Conductor Quinlan with Engineer William Alderman and Fireman August Golinvaux. [974] It was en route from Chicago, Illinois, to Omaha, Nebraska, had left Parkersburg, Iowa, at 1.54 a.m. and was derailed 6½ miles from there at bridge W-307-7, a mile and a half west of Aplington, at about 2.10 a.m. The bridge was a concrete arch structure with a span of 20 feet across a branch of Beaver Creek. Engineer Alderman, Fireman Golinvaux, Baggageman Frank Bigley, Express Messenger G.H. Dake and Mail Clerks James Rae and Albert Windmeier were killed, and 22 passengers and five mail clerks were injured.

Conductor Quinlan told the Interstate Commerce Commission investigators that his train had left Waterloo, Iowa, ten minutes late. He had received no slow orders and no notice of any severe weather west of Waterloo, though he had heard someone mention heavy rain at Iowa Falls, 20 miles west of Waterloo. It began raining at New Hartford, 13½ miles from Waterloo, but not excessively before his train became derailed while travelling at between 30 and 40 miles an hour.

Flagman Hockeme agreed that a car repairer had mentioned rain west of Waterloo and that there had been no indication of any unsafe condition prior to the derailment.

The last train to pass the derailment site before the accident was eastbound No. 12. Engineer Token of that train said there had been intermittent rain from Webster City, 49 miles west of Aplington, but nothing abnormal, nor had he noticed anything wrong when he had crossed the bridge at 11.10 p.m.

Track Supervisor Toohey said that he had ridden over the bridge twice two days before the accident and noticed no problems. He considered that the west approach embankment to the bridge had been washed away before train No. 11 had reached the spot.

Section Foreman Van Meter stated that there had been heavy rain at Aplington on the night of the accident and concerned about the state of a bridge two miles east of Aplington he had proceeded there at midnight with three of the sectionmen in his gang and had remained at that bridge until 2.30 a.m. He had never considered bridge W-307-7 west of Aplington to be at any risk.

Bridge Supervisor Jordan stated that he had made a quarterly inspection of bridge W-307-7 in March when it had been in good condition. Examining the accident site at 9 a.m. he had found the west approach of the bridge washed away from the foundation and west wall of the arch and that the south end of the foundation and wing wall of the arch had settled considerably. He was of the opinion that a washout had caused the derailment and the collapse of the bridge arch had followed the derailment.

District Engineer Laden stated that he had examined the bridge at 3 a.m. and had concluded that an extraordinary amount of water being forced against the south-east wing wall of the bridge had caused a very strong current to strike the west wall and the west approach embankment to crumble and wash out.

Roadmaster Rogers agreed that it had been raining excessively but he had not thought

[974] *Waterloo Evening Courier and Reporter*, Waterloo, Iowa, United States, 30th May 1918

the conditions such as to make it necessary to issue a warning about them or the possibility of washouts, especially as all the bridges in the vicinity had been of substantial construction. Arriving at the scene of the accident at 3 p.m. he had found a large amount of water at the site, and he basically agreed with the district engineer that a large volume of water had undermined the wing walls and then washed away all of the embankment before train No. 11 arrived. The falling of the derailed train upon the bridge arch had then caused it to collapse.

Several local residents spoke of exceptionally heavy rainstorms which had lasted until after midnight and that all the creeks had been full of water.

The ICC concluded that the accident had been caused by a washout at bridge W-307-7. It drew attention to Rule 16 of the Illinois Central rules regarding the maintenance of way and structures, which in part read:

In case of threatening or prevailing storms, track must be patrolled and all bridges, culverts, or particular localities in track liable to be affected by such storms must be closely watched. Force will be freely used to patrol the track under such conditions.

The commission found that in view of the severe weather conditions Section Foreman Van Water had been at fault for not patrolling the whole of his section of track that night and that he had exercised poor judgement in not utilising all of his seven men to patrol the track on both sides of Aplington. Had he done so the accident would undoubtedly have been prevented and the accident drew attention to the 'imperative necessity of strict compliance with all rules and regulations promulgated to provide proper safety in situations of this character.' [975]

An accident occurred at Oswiecim, Galicia (better known by its German name, Auschwitz, and now in Poland) on the Kaiserlich-königliche-österreichische-staatsbahnen at 3 a.m. on Saturday 3rd August 1918. Goods train No. 76, which was running very late, came to a stand on a bridge over the River Vistula at the Einfahrsignal guarding Oswiecim station. As the signal cleared and the goods train began to move forward, mixed train No. 62 came up behind it, travelling at speed. The driver of the mixed train suddenly saw the tail lamp of the goods train in front and made a violent emergency brake application, thereby avoiding a heavy collision with the train in front. A bad accident might thus have been entirely prevented, but the mixed train was composed of 13 air-braked passenger and service vehicles, followed by 48 goods waggons equipped only with hand brakes. With the sudden stoppage of the front vehicles of the train the unbraked rear waggons were forced into them and seven passenger coaches were thereby derailed, five of them falling from the bridge into the river. Eleven people died, 8 were seriously and 22 slightly injured. The principal blame for the accident lay with a signalman who had permitted the mixed train to proceed towards Oswiecim before the goods train had been reported back to him as having arrived. [976]

Early on Saturday 10th August 1918 a major disaster occurred on the Paris Bastille – Verneuil l'Etang line of the Chemins de fer de l'Est near Solers, France. Two trains were heading towards Verneuil, some minutes apart, when the first came to a stand out of course due to the breakage of a brake pipe. The conductor went back and put down

---

975 Interstate Commerce Commission report

976 *Eisenbahnunfälle Bind 1* by Ludwig Stockert, pg. 27 (in German)

detonators to protect the rear of his train but despite this the following train failed to stop and rammed the rear of the leading one. To make matters worse a train coming from the opposite direction shortly arrived and collided with two vehicles which had been thrown onto the opposite track by the initial collision, and a portion of one or other of the trains then broke through the parapet of the bridge and fell 12 metres to the Yerres River below. Initial reports spoke of 20 dead and 30 injured, while 80 horses being carried on one of the trains were also killed. [977] The number of dead later rose to 24 and of injured to 61.

Not mentioned in contemporary newspaper accounts was the fact that all three trains involved were troop trains. The first two were carrying artillerymen from the 101st heavy artillery regiment from Lizy-sur-Ourcq to Révigny, the third was carrying Moroccan infantrymen from Maron to Rethondes in preparation for an Allied attack scheduled for August 18th. [978] Given that the accident happened at 03:30 it seems probable that the driver of the second train had fallen asleep and hence did not respond to the explosion of the detonators; very likely he had been working excessive hours with the pressure of increased wartime traffic.

Goods train Gz 10889 collided head-on with passenger train P 5750, which was waiting to cross it, at Tollschütz, Germany, on the narrow-gauge line between Döbeln and Mügeln, on Tuesday 18th November 1919, probably because ice formation in the brake pipe had made the goods train's brakes inoperative. As the line was worked without signals the goods train should, according to the regulations, have come to a stand short of the beginning of the crossing loop. When the engine crew of the passenger train saw that the goods train was not stopping they jumped from their locomotive and the presumably unbraked passenger train then began to run back at increasing speed down an incline until it left the rails on a sharp curve and all but one of its vehicles fell from a bridge. Five people died and 24 were injured, most of them seriously. [979]

On 27th March 1920 eight waggons ran away down a steep incline from Pontebba station, Italy, near the old Austrian frontier, and collided with an oncoming express passenger train. The collision chanced to occur on the high Ponte di Muro viaduct at Dogna and three of the carriages fell from the structure into the Udine River. Thirteen passengers died, mostly by drowning, and a large number were injured. Almost all of those who died were Egyptian students on their way to Berlin University in Germany. [980] [981] [982]

Westbound passenger train No. 1 of the Minneapolis and St Louis Railroad was involved in an accident near Livermore, Iowa, United States, on Saturday 3rd July 1920. The train consisted of engine No. 157, a combination mail and baggage car, a baggage car, a smoking car and two coaches in charge of Conductor Conklin with Engineer Mike Murphy and Fireman Wilson. It was en route from Des Moines, Iowa to St Paul,

[977] *La Liberté*, Paris, France, 11th August 1918
[978] Webpage – La République: Une plaque en souvenir du drame ferroviaire de 1918
[979] *Eisenbahn-Katastrophen in Deutschland* by Hans Joachim Ritzau, pgs. 119-20 (in German)
[980] Webpage – Incidenti Ferroviari in Italia – Epoch – Treniebinari.it
[981] *Farmer and Settler*, Sydney, New South Wales, Australia, 31st March 1920
[982] *Brisbane Courier*, Brisbane, Queensland, Australia, 30th March 1920

Minnesota, had left Fort Dodge, Iowa at 11.10 a.m., 15 minutes late, and became derailed at bridge No. 131, 7 miles from Livermore, at 12.5 p.m., while travelling at about 30 miles an hour. The entire train derailed to the left and the first four coaches fell into the Bloody Run Creek, the bridge being almost entirely destroyed. The smoking car was practically destroyed and the front part of the next part of the next coach was badly damaged; almost all those killed and injured were in these two vehicles. Eight passengers died and 39 passengers and five railwaymen were injured. [983]

Engineer Murphy spoke of the accident to a reporter from his home in Fort Dodge. "I was in the cab with my head out of the window, and Wilson, the fireman, was on the deck, when I heard a 'z-r-r-r-r' as though a rod from the engine was dragging on the ties. The engine was only about 20 feet onto the 120 ft trestle when I heard the noise. I reached for the emergency brakes, but they began to act automatically with the breaking of the train. We were going about 30 miles an hour, down grade, The train began to come to a stop, but I knew right away that we were off the rails, and I thought the engine was going to go right off the side of the trestle and pull the train in after it. The planking on the edge of the ties is the only thing that kept the engine from tumbling over the edge in the middle of the bridge. We didn't go off, however, until we were clear across. Just as we hit solid ground on the soft earth and the engine started to sink in on the right side. I knew there was nothing more I could do so I hollered, 'Come on, Wilson, let's jump.' There was only one way we could jump, and that was out of the door the same way the train was falling. I jumped, with Wilson right after me. The embankment was about 30 ft high, and I ran or slid down the side as fast as I could. I remember seeing coal and water pass me as I slid and parts of the engine of the engine just missed me as they rolled along past me. I got to the foot of the bank and got through the fence and looked back. I saw that the engine was buried in mud at the side of the bank, and wasn't to fall any further, so I ran back to the baggage car, as soon as I saw that Wilson was clear of the wreckage. In the baggage car there were two baggagemen and the conductor. I let them both out and let them get to a place to lie down. They were both hurt and dazed. Then I let out the mailman and his assistant, who were caught in the wreckage of the mail car. It wasn't two minutes after I left the engine that I got to the smoking car. It was a good thing Wilson, and I weren't hurt, because we were the only two on the train that were able to help the injured. Those that weren't hurt were dazed, stunned or hysterical. The smoking car was smashed flat except at the front end. I got down and looked in and saw all the dead and injured piled up toward the front of the car. At least three were already dead. All the occupants of the car lay there, literally chewed up by the seats of the car which had crunched them as the car collapsed, breaking their arms and legs, and killing them outright. Both dead and injured were lying in a pile, heads down. I started to move some of the dead from on top of the injured, and to get the injured outside on the grass. There was broken glass and jagged timber all around, and everybody in the car was either dead or injured. If somebody hadn't given them help right away, a lot more of the injured would have died before relief came. The first man I rescued was Father Harry, of Marcus. I didn't know him at the time, though I remember I thought it was a member of the clergy because of the black vestment about his throat. Then I carried out several others. It was hard work to untangle the pile. Many times I took hold of an arm or a leg only to feel it give like a rope – fractured in several places. Before long – probably fifteen minutes – some farmers came, and I told them to go to the next coach and get the women out. They

[983] Interstate Commerce Commission report

were fine, they pitched in and helped and worked hard till the doctors came. The first doctors came from Livermore, four miles away. Then they became to come from Humboldt and Fort Dodge. I worked till 4.30, then I came home in a Fort Dodge automobile which was out there. I've been in one passenger wreck and several small freight wrecks before, but I never saw a wreck like this one in all the twenty-five years I've ridden the rails. I guess it's the worst wreck around here for years." [984]

Bridge No. 131 was a wooden pile bridge, 260 feet long, 27 feet high with 21 bents.

Engineer Murphy told the Interstate Commerce Commission (ICC) inquiry that he had noticed no unusual conditions before he reached the bridge. At the bridge he had heard a noise and 20 feet beyond its east end he had felt a sudden jerk and applied his brakes; he thought his engine had become derailed at that point. His engine was found to be in good condition immediately before the accident.

Examination of the track after the accident showed a considerable variation in the elevation of the rails.

Section Foreman Fine stated that he had done no work on the track in the previous six months except to raise low joints. He had inspected the track the previous day and considered the line and surface fair and the track safe for the maximum speed of 40 miles an hour permitted by the timetable rules. The track supervisor had gone over the line on 30th June in a gasoline motor car and had found nothing unusual.

Chief Engineer Kenley considered that the front tender truck of the engine had derailed first, and that the tender had rolled due to track irregularities, and that at the second joint south of the bridge, which was high, the weight had lifted off the right wheels of the front truck, allowing them to climb the rail.

The ICC concluded that the accident had been due to irregularity in the surface of the track and variation in the elevation of the rails. It noted that the railway company had experienced 24 tender derailments in the previous year, 14 of them in the previous four months. In addition it had investigated a previous derailment just west of Livermore in August 1917 which like the present one had been found to be due to the uneven and insecure condition of the track.

Contradicting the assertion of the section foreman, the commission considered that the track had not been in a safe condition for the safe operation of trains at 40 miles an hour and that a speed restriction should be imposed until the track had been put into a proper condition.

Had the bridge been provided with inside guide rails the derailed truck might have been held more closely in alignment and the accident results much less serious. [985]

The serious accidents involving bridges in the United States continued with one on Sunday 12th March 1922 on the Atlanta Birmingham and Atlantic Railway near Ben Hill, Georgia. Involved in this accident was eastbound train No. 2, which was only a short one consisting of engine No. 124, a baggage car and two coaches, all of wooden construction. It was in charge of Conductor H.S. Dixon with Engineer P.R. Bosworth and Fireman Powell, and had left Atlanta, Georgia, at 7.45 a.m. for Fitzgerald, Georgia, about 80 miles to the south. The train was travelling at about 30 miles an hour when its rear car became

---

984 *The Humboldt Republican*, Humboldt, Iowa, United States, 9th July 1920

985 Interstate Commerce Commission report

derailed about two miles east of Ben Hill. Some 460 feet east of the initial point of derailment was a single span plate girder bridge over Camp Creek, 80 feet long and about 40 feet high. As the derailed car reached the bridge it slewed to the right, overturned, and fell into the creek, turning over several times according to surviving passengers and landing upside down in the shallow water. Seven people died, three of them off-duty railwaymen, and 16 passengers were injured. [986] Those who lost their lives were W.M. Brooks, Western Telegraph Company linesman of Manchester, Georgia; R.C. Driver, of Roanoke, Alabama; Mrs Dr. I.H. Etheridge, of Atlanta, Georgia; W.E. McIntosh, of Roanoke, Alabama; Atlanta Birmingham and Atlanta cashier G.T. Elmore, of Atlanta, Georgia; Alabama; Atlanta Birmingham and Atlanta section foreman W.E. Johnson and Atlanta Birmingham and Atlanta operator R.W. Lanier, of Senoia, Georgia. [987]

Engineer Bosworth told the Interstate Commerce Commission (ICC) inquiry that the train had been drifting downhill to the bridge, at which point he and his fireman had felt a severe jerk. Looking back he had seen that the rear car was derailed and had made an emergency brake application.

Conductor Dixon stated that he had just left the rear car when the severe jerk had occurred. He saw the rear car rocking and went for the emergency valve, but the brakes applied automatically as the rear car broke away before he had time to reach it. Neither he nor Flagman Greatorex had noticed anything unusual prior to reaching the bridge.

It quickly became apparent that the derailment had been due to a broken wheel, a fragment of which had still been warm when examined by the conductor, fireman and Chief Engineer Beal two hours after the accident.

The ICC concluded that the derailment had been caused by a broken wheel, apparently precipitated by the air brakes on the car sticking. The wheel was quite new, having been cast on 18th October 1920. The ICC report included a lengthy report on the failure made by an engineer physicist. [988]

Train No. 2 of the Minneapolis St Paul and Sault Ste Marie Railroad was involved in an accident at Chippewa Falls, Wisconsin, United States, on Saturday 20th December 1924. This train consisted of engine No. 2719, three mail and express cars, a baggage car, two coaches, a sleeping car and a café observation car and was in charge of Conductor Ray with Engineer Caldwell and Fireman Kakaska. The train left Minneapolis on time at 7.45 a.m., passed Irvine, a mile, and a half from the scene of the accident, at 12.32 p.m., an hour late, and became derailed two minutes later while travelling at about 25 miles an hour. The rear truck of the rear car left the rails at a switch about three-quarters of a mile beyond Irvine, ran for about 885 feet to the centre of a bridge over the Chippewa River, then broke free from the remainder of the train and plunged over the side of the bridge to the river 40 feet below. The car ended up upside down in 6 or 7 feet of water. All of the passengers killed and most of those injured were in the observation end of the car which hit the water first. According to the Interstate Commerce Commission (ICC) report eight passengers died and five passengers and two railwaymen, a cook, and a porter, were injured, though at least one newspaper report speaks of ten fatalities. [989] [990]

---

[986] Ibid.
[987] *The Montgomery Monitor*, Mount Vernon, Georgia, United States, 16th March 1922
[988] Interstate Commerce Commission report
[989] Interstate Commerce Commission report

The river bridge was 771 feet long, with concrete abutments and five steel spans on concrete piers.

Engineer Caldwell told the ICC inquiry that he had noticed nothing wrong with the switch as his train approached it and the first warning of the accident that he had had was when the brakes applied in emergency and Fireman Kakaska had looked back and told him that a car was in the river.

Conductor Ray said that he had likewise been unaware of any problem until the brakes had been applied. There had, he believed, been 13 passengers in the parlor car when it went into the river. Afterwards he had gone back to look at the switch but by that time it had been repaired.

Flagman Richmond stated that he had been in the sleeping car. He had felt a lurch and heard the sound of escaping steam whereupon he had gone to a passageway to open the emergency valve, but he thought that by then the car had already gone into the river. Instructions had been issued that the flagman was not to ride in the rear observation car after passengers had complained of the smell of the flagman's lanterns.

Section Foreman Franz testified that he had installed a new switch point with a new bolt. He had been eating lunch when the accident occurred. Going to the switch he had found the switch rod out of place, the head rod bolt used to fasten the head rod to the clip was missing and the points were open about an inch. He had not been able to find the missing bolt and had replaced it with another bolt as a temporary repair.

The switch had been operated at 11.20 and 11.55 when it had apparently been working normally.

Roadmaster Dixon said that he had likewise been unable to find the missing bolt. He thought that it might have been broken either by pressure on the heel of the switch point or by a wheel flange passing between the partly open switch point and the stock rail.

The ICC concluded that the derailment had been caused by the switch point being disconnected from the head rod, due to the absence or failure of the head rod bolt, adding that it could not be definitely ascertained why said bolt had been absent or had failed.

Early on the morning of Wednesday 25th March 1925 a Hendaye – Paris express of the Chemin de fer de Paris à Orléans left the rails four kilometres south of Poitiers, France. The 12 coach train, which had left Hendaye at 16:47 the previous afternoon, had just passed through Saint-Benoit station, travelling at its normal speed of 77 kilometres an hour, when Driver Marcel Puet felt what he described as a slight shock caused by the breaking of the left-hand rail and had applied the air brakes in emergency. The engine came to a stand on the track some way down the line with the rear bogie of its tender derailed. The leading van overturned on a bridge across the River Clain, little damaged, but the next, first-class, carriage, broke through the parapet of the bridge, fell ten metres, and ended up submerged in the river. The following first- and second-class composite carriage fell on top of the first-class one and a third-class carriage, the fourth vehicle of the train, rolled down the embankment. The next three carriages were derailed but remained on top of the embankment and the rear five carriages stayed on the line, their occupants unaware that there had been an accident. The accident happened at 02:04.

Driver Puet kept his *sang-froid* after the accident, the newspaper reporter noted, and

---

[990] *The Sunday Messenger*, Athens, Ohio, United States, 21st December 1924

sent his fireman, Pierre Helanday, back to protect the train in the rear, and himself went forward, presumably on the engine, to the *semaphore* (signal box?) at Salles, two kilometres south of Poitiers.

A rescue train was despatched from Poitiers at 06:14.

An initial report stated that four bodies had been recovered from the leading passenger carriage and there were known to be 30 injured. One of the passengers to lose their lives was Doctor Adolphe Pédebidou, aged 71, a highly respected senator in the French parliament. The others were M. Jacques Loew, aged 30, a jewellery commissioner, of Paris; M. Serge Singher, a 24 year old law student, of Le Mans; and Raymond Gall, from her photograph a pretty-looking five-year-old girl, of Paris.

M. Victor Gall, the father of the little girl, spoke to a reporter. "No, I was not asleep. I was very conscious of the moment that the carriage shook. I seized my wife's arm and pointed to our little Raymond. I cried to my wife to watch out for her. I grabbed my baby…and there was this terrible fall. I felt we were falling into an abyss. I don't really know what happened. I found myself in the water. I heard a cry from my wife, Barbofant, and not seeing a drop I could pull her out of the carriage. Having put her down in a safe place, I went back to look for my child. I found the little one, I brought the baby out and gave it to its mother. I went back,. I searched… I dived… Ah! The misery!"

He had found his daughter a little later, cold. [991]

At least one report mentioned a fifth victim, whose identity could not immediately be established. He was described as having been wearing a sweater and a Basque beret. [992]

The London *Times* correspondent in Paris noted that among the injured were Mr Clifford Collinson, of Westbourne Square, Paddington, London, who had a contusion to his right foot, and Mrs Bellamy, of London, who had cuts and bruises. [993]

Newspaper editorials in the days after the accident blamed the broken rail on the use by the company of rails which were too light and had been laid 25 or 30 years previously, together with the use sleepers of inferior pine instead of oak or beech, and inadequate track fastenings. It was pointed out that the 'double mushroom' rails used by the company weighed only 38 or 42 kilograms/metre as opposed to the Vignoles rails weighing between 46 and 55 kilograms/metre used by other railway companies. Sand had also been substituted for rock ballast in many places to save money and the running of expresses with heavy engines over poorly maintained track had exacerbated the problem. [994] The track at Saint-Benoit had in fact been under repair at the time of the accident but no speed restriction seems to have been imposed.

The little Central American country of Costa Rica suffered by far the worst railway accident in its history on Sunday 14th March 1926. In order to raise money for a home for the elderly Professor Francisco Gomez Alizago had arranged with the Northern Railway company to run a special excursion train from Alajuela and Heredia to Cartago, where the excursionists could visit the statue of La Negrita at the Basilica de Nuesta Señora de Los Ángeles, which supposedly had great healing powers. Tickets for the trip were sold at the very low price of one and a half colón per person, each bearing the inscription 'For

---

[991] *Le Matin*, Paris, London, 26th March 1925
[992] *Le Figaro*, Paris, France, 26th March 1925
[993] *The Times*, London, England, 26th March 1925
[994] *L' Œuvre*, Paris, France, 2nd and 4th April 1925

the benefit of the elderly of Cartago.' The number of tickets sold was not restricted, with the result that probably a thousand were sold. On the morning of the accident whole families of mostly peasants, dressed in their Sunday best, walked to Alajuela station, intending to visit the old metropolis, the charity fair and then the basilica.

Engine No. 9 arrived at Alajuela station at 7 a.m. with three carriages, and these were soon full, with passengers crammed on the platforms at their ends as well as inside. The train started at 7.30 a.m., in charge of Conductor Gonzalo Facio, with Engineer Josefino Victor Manuel Calvo. A stop was made at San Joaquin de Flores, where more passengers forced their way onto the already full train, and then at Heredia. Three more carriages were waiting there, already filled with passengers, and Conductor Facio decided that they should be placed at the front of the train, ahead of the Alajuela ones. Five minutes later the train was on its way again, leaving a number of passengers who had tickets behind because there was no room for them. Likewise the train ran non-stop through Santo Domingo station, leaving more disappointed waiting passengers there. At 8.20 a.m. the train approached a bridge across the Virillá River and Engineer Calvo gave his locomotive more steam to ascend a slight rise. Then as the train was on the bridge tragedy struck. The weight of the passengers in the rear coach, which probably was not evenly distributed about it, caused it to lean to the left, what was probably a poorly secured rail gave way under the strain and the rear coach went over the side of the bridge and fell into the river, 190 ft below, taking the two coaches ahead with it. The engine and three leading coaches cleared the bridge and stopped 100 metres along the line. The driver and conductor testified later that they had considered going back to help the victims but fearful that they might be attacked by angry survivors had thought it better to continue to San José for assistance.

Many passengers were crushed in the coaches which fell into the ravine and the river was stained with the blood of the victims. The exact number who lost their lives is unknown, some newspapers stating that 300 had died. The most likely figure is probably 248 dead, with 90 or so more injured. Professor Alizago was amongst those who lost their lives.

A large crowd of people soon gathered at the scene, some trying to help the injured, though it was said that others merely impeded the rescue work, and as has happened elsewhere there were stories of a few wretches stripping the victims of their belongings.

The government declared three days of national mourning.

The railway company denied all responsibility for the disaster and survivors and relatives spent ten years fighting for compensation. [995] [996]

At midnight on the night of Thursday 10th to Friday 11th June 1926 the head office of the railway at Sydney, New South Wales, Australia, received an urgent telephone call from the stationmaster at Aberdeen to say that the express which had left Sydney at 3.30 p.m. on Thursday for Brisbane, Queensland, had become derailed while crossing a wooden viaduct between Kayuga and Aberdeen. Both engines and four carriages were derailed, the line was totally blocked and likely to remain so for some time. Several passengers had been injured.

The train consisted of two engines, a brake van, a first-class sitting carriage, five

---

[995] Webpage – Tragedia del Rio Villa

[996] *The Bryan Daily Eagle*, Bryan, Texas, United States, 15th Mar 1926

sleeping cars and a second-class sitting carriage with 143 passengers on board. The order of the train behind the engines was reversed at Newcastle. The pilot (leading) engine, AD class No. 1617, crossed the viaduct and overturned on the approach embankment beyond, the train engine, a NN class 4-4-0 No. 3504 and four carriages fell with the viaduct into the Hunter River. Initially it was reported that five passengers had died and 80 were injured, a little later it was found that in fact only four passengers had been killed outright. The passengers who died were Barbara Dalzell, aged 17, of Rockhampton, Queensland, returning from a holiday in Scotland; Ernest Baden Powell Hewitt, aged 25, a wool presser, of Greenwich, on his way to a shearing job at Charleville; Slyman Mansour, of Waterloo, Sydney, going to Brisbane with several members of his family, and Henry White of Melbourne, on his way to Brisbane with his wife.

Nurse Bennett, a passenger on the train, described her experiences. "We were laughing and chatting when we felt the train leave the rails. Instantly we were all thrown into confusion. The passengers were heaped upon one another, and we heard the splintering of woodwork, and a moment later the cries of the injured. We hastily extricated ourselves and crawled through the wreckage. The scene I looked upon was terrible. One of the carriages was smashed to matchwood, and from various parts of the train men and women were crawling and tumbling out. The most painful sight was that of the injured. Some limped, and others with expressions of despair on their faces lay on the ground. It was remarkable how in such a remote place how quickly assistance was forthcoming. Motor cars came from all directions. Fortunately, I escaped serious injury and was able to do a little."

Mr Basil Levy, a wholesale jeweller, of Pitt Street, Sydney, another passenger on the train, also told his story to reporters. "Fortunately, the train did not take fire, but the rescue work was very difficult. For the first few minutes we were spellbound. I jumped out of the carriage as soon as possible, the only injury I suffered being a cut finger. The cries of the injured were heartrending. Many people were calling for help, and their agonised cries were appalling. The uninjured passengers worked with a will and then arrived six doctors and a number of nurses. The doctors and nurses worked magnificently, and soon the injured were being motored to Scone."

The accident happened at about 10 p.m. As soon as news of it was received at Murrurundi a breakdown train was organised there.

After the accident while the line was blocked trains from Sydney were worked to and from Aberdeen, and those on the northern (Brisbane) side were worked as far as Scone. [997]

Despite the early statements that the accident had been caused by the collapse of the viaduct it soon became evident that this was not the case, Mr Flannery (the Minister of Railways) stating publicly the day after the accident that the viaduct had not been at fault. It became clear that the initial derailment had occurred 100 yards short of the viaduct, the collapse of which had been caused by it being struck by the engines and vehicles of the derailed train. The cause of the derailment, however, was not immediately clear. [998]

The inquest proper upon the victims of the accident opened on Monday 28th June at Muswellbrook. Brief testimony was given by surviving passengers and lineside residents who had witnessed the accident, then a journalist, Frederick Gardiner, spoke of the

---

[997] *The Sydney Morning Herald*, Sydney, New South Wales, Australia, 11th June 1926

[998] Ibid, 12th June 1926

viaduct being riddled with white ants (termites) and of a bridge inspector having told him that the bridge was rotten before ordering him away upon finding that he was from the Press.

Edward Butcher, recently appointed northern divisional engineer, spoke of having examined the track on the morning after the accident and having found differences in the elevation of the two rails on the Aberdeen side of the viaduct. He gave it as his opinion that the track irregularities combined with a train travelling at high speed could have caused an oscillation sufficient to cause the derailment. He did not consider that the absence of a check rail or the fact that it was a wooden rather than a steel or concrete viaduct had anything to do with the accident.

Richard Sexton, chief railways engineer for Queensland, agreed with Mr Butcher that the eastern rail was seven-eighths of an inch lower than the western rail on the Aberdeen side of the point of derailment. This difference in two rails was fairly constant for 8 ft, and regularly decreased until at the point of derailment both rails were practically level. This "slack" in the eastern rail extended back towards Aberdeen, decreasing to nothing 108 ft back from the point of derailment. From the point of derailment 20 ft north appeared the first marks of the derailed wheels. He concluded that one pair of wheels only were derailed. The marks continued down to where the main road had been disturbed to lay the deviation. He found only one broken sleeper before reaching this viaduct. This was undoubtedly broken by a derailed wheel. He would say the depression might be known as a hole in the road. The existence of this swing in the alignment, with a canted road and low rail being on the outside of the swing, produced a condition which, with high speed, might set up an element of danger. He would consider a speed of 40 to 45 miles an hour as likely to set up a dangerous condition on such a road. A vehicle with higher centre of gravity would roll more outwards than one in which the centre of gravity was lower. He was advised that the grade down to the end of the viaduct was 1 in 80, and the ganger had told him that he had fully metalled the road five days before the accident. He saw no sign of any spreading of the gauge. He made no examination of the bridge.

Examining a sleeper put in as an exhibit, witness said it was in a good serviceable condition, and the sample of pile showed signs of white ants in the centre. The weight of the engine was 20 tons on the driving axle, with a tender of 120 tons. The road would carry that class of engine, but it would roll. The rolling of an engine, together with the defects in the road, would find the weak spots very quickly. If this engine travelled at 40 or 45 miles an hour it might set up dangerous conditions, and cause derailment.

In reply to the Coroner witness said he would personally not be satisfied with the road in that condition. He considered that the driver, if he noticed the line in that condition, should report it to the ganger. [999]

The resumed inquest the following day heard evidence from various examiners that the engines and vehicles of the train had all been in good condition before the start of the journey. Sleeping car conductors James Bennetts and William Gorman then testified, followed by Guard Peter Donohue, who had been in charge of the train, who said that his van had overturned, and he had been unable to get out until assisted by Driver Monkley.

Night Officer Charles McDonald at Aberdeen spoke of exchanging the tablet with the driver of the second engine just before the accident, then Thomas Fensom, a farmer of Delungra, near the site of the accident, spoke of finding white ants and dry rot in the

[999] Ibid, 29th June 1926

viaduct, adding that he had, at his own expense, photographed the damage in the public interest.

However, William Evans, sub-inspector of bridges, stated that his opinion was that the cause of the accident was derailment before reaching, or on, the viaduct. The wheels had been locked by the emergency brake, which had caused them to skid on the transoms. This had broken the transoms on the viaduct, causing the wheels to drop down between the girders. The wheels had been hitting the headstocks above the piles and breaking them, thus taking away support to the girders.

When asked by Mr. Sayegh to inspect the pile exhibited in court, and describe its condition, witness said the timber was sound for railway purposes. It had an old pipe caused by white ants.

Mr. Sayegh: Do you say that timber eaten by white ants is equally as sound as one without any traces of them?

Witness said it would be a strong as one showing a pipe caused by natural causes. The pile would probably be 40 years old. He had seen evidence of dry rot in the viaduct, but the decay was not sufficient to warrant renewal of any timber. The viaduct did collapse, but it was caused by the derailed vehicle. If a steel viaduct had been there it would not have averted an accident. Witness informed the Coroner that the pile exhibited was 85 per cent. solid.[1000]

The inquest continued the next day when Richard Ernest Sexton, Chief Engineer of the Queensland railways, said that, as instructed by the Coroner, he had carried out a number of tests of the track. Witness said that he would say that the permanent way was not in a satisfactory state as to fettling. This should have come under the notice of the fettling gang. The road was "run" every day, and the "slacks" in surface and in alignment, and particularly the swing near the point of derailment, would have been apparent to any fettler "running" the road either by tricycle or trolly, or on foot. He would not say in what condition the track was at the time of the accident, as his inspection was made 17 days after the accident. He would describe the formation as a plastic formation, largely due to frequent rains. In dry weather that material would be dry, but in wet weather in a plastic condition. He would expect the swing or lateral movement to develop gradually. It would yield under pressure of a passing train, and the greater the speed and weight, the greater would be the depression. The ganger should have seen the depressions, as he himself saw them. That section of line would not be described as what railway men called a "muck length" There was no indication of clay having been spewed. The trenches disclosed sufficient metal ballast for maintenance purposes.

Evidence as to the state of the viaduct was tendered by Oswald Gordon Morton, Muswellbrook shire engineer, who said he had made four inspections of the wooden viaduct since the accident. He found two piles defective and affected by the action of white ants. All the timber appeared to be iron-bark, but much of it was very old. Out of 230 pieces of timber but 34 pieces remained in position.

In reply to Mr. Rogers (Crown Law Office), he would say that the attack of white ants had reduced the efficiency of the piles affected by 20 per cent, although the structure itself would not be affected by more than 10 per cent. In the viaduct there would be a high degree of safety under normal conditions by the load being on the rails. With the derailment of the train the condition would be completely altered.

[1000] Ibid, 30th June 1926

In reply to Mr. Cargill (for the Railway Commissioners), witness said that he would not cut out defective timber for the faults disclosed. They were defective in themselves, but not when considered with the viaduct as a whole.

The engineer of the Upper Hunter Shire at Scone, Bertram George Millard, also described his examination of the viaduct. He said that he saw evidence of white ants having been in the timber at one time. He estimated the decreased efficiency of the affected piles at 10 per cent. "It is quite wrong to say," he added, "that the viaduct collapsed. The structure was subjected to side strain and appeared to be wrenched over. The weight and force of the derailed vehicles smashed the timbers and dragged them over. It would carry the traffic normally with vehicles on the rails."

William Poole, member of Messrs. Julius, Poole, and Gibson, and of the Institute of Civil Engineering, said his examination of the timber in the viaduct showed that it was of the best selected throughout, and that the workmanship was of the highest type. The only timbers showing material deterioration were the upper ones of the middle girders. In only one case did a girder show surface deterioration. The transoms were sound and originally had a high factor of safety. The piles were in sound condition, and that exhibited in Court was the worst pile in the structure. His test showed it to be nearly equal to the best ironbark. The worst part of the pile was 20 per cent sound and had an efficiency of 80 per cent. The load to cause the pile to fall would be about 410 tons. It would carry four times the weight of an NN engine travelling at high speed.

Michael Taper, bridge carpenter, said he examined the viaduct about the middle of May and marked timbers requiring attention. There was nothing of an urgent nature. None of these timbers had been dealt with before the accident. Some new timber was put in three or four weeks ago, this having been ordered after his previous examination. He had seen traces of white ants, but none of the timber was badly affected.

James Swan, bridge ganger in the railway service said that he carried out repairs about three weeks before the accident. He saw no evidence of white ants when carrying out that work.

At the conclusion of the inquest for the day the Coroner (Mr. Swaddling) ordered the removal from the Court House of the section of hardwood cut from one of the piles of the Aberdeen viaduct. His object was to have this timber, which was an exhibit in the court, split up for closer examination as to the existence of living white ants. Inspector Evans, with an axe, split off large pieces from the old pile. Not a single living ant was found, and an examination of the pieces of old ant-eaten timber indicated that white ants had apparently not been present during recent years. Some of the decay was, in the opinion of experts present, caused by dry rot. [1001]

The next day of the inquest Thomas Waters, sub-inspector of the permanent way, stationed at Werris Creek, said that his district embraced the Aberdeen viaduct. He last examined the viaduct on May 24th, and he had found it in very fair running order. There were a few slight depressions in the line which could be more readily detected with a heavy locomotive.

In answer to Mr. W.A. Halse Rogers, witness said that the NN type engines gave more trouble than other types by "spreading" the rails. Witness did not consider that two engines of different classes coupled together gave more trouble than one alone. As far as his experience went, a derailment would not be brought about by irregularities on the line.

[1001] Ibid, 1st July 1926

John Angus Rich, an extra ganger, said that he had examined the line two days prior to the derailment, and between Aberdeen and the bridge the track was in good order. He did not consider minor depressions south of the bridge to be serious.

To the Coroner, witness said that if the depressions had been immediately adjacent to the bridge he would have considered they required immediate attention, since the bridge end of the depression would be rigid and give a jerking effect.

Ganger Ernest James Hodson said that he had been working on the length of line in question for seven years. He considered that the southern end of the viaduct required less attention than the northern end.

Albert Cecil Fewtrell, supervising engineer in the Existing Lines Branch, said that he had been divisional engineer of the northern section from January 1st, 1920, until March 31st last.

To the Coroner, witness said that the permanent way where derailment had occurred had given him no trouble whatever during his term. Witness examined the permanent way after derailment and found that the wheels had ridden over the rails and rail joints, and 24 feet south from the scene the east rail was depressed by one and seven-eighths inches extending back for 70 or 80 feet. The derailment had been caused by successive oscillations of the vehicles, irregularities of the road, and the high speed, which would be between 30 and 60 miles an hour. In his opinion, the leading bogie of the tender of the NN engine left the rails, first cutting through the transoms, causing the rails to spread, and tilting over to the right. These factors caused other wheels to leave the rails. Regarding the high speed of the train, witness said that a sudden application of the emergency brakes would cause the wheels to crash through transoms eight inches thick. As used on steel bridges, they would have probably withstood the force of the wheels.

William John Ellis, retired railway employee, said that he examined the derailed engine on June 23rd. The engine was "unbalanced" through various defects in the engine not caused by the accident, and which had existed prior to the accident. Witness described the damages suffered by the engine in the accident. The derailment was brought about when the NN engine came to a depression in the rails, causing a further depression owing to its weight in addition, the sleepers required "packing."

Cross-examined by Mr. Cargill, witness maintained that the NN engine had been unbalanced. He said that there had been wear in the rubbing block and excessive wear In the back centre draw plate and draw hook, and that these defects would have a serious effect in the running of the engine and might ultimately result in the wheels of the leading bogie of the tender coming off the rails. All NN engines were unbalanced, because they carried too much weight on the trailing axle.

Franz Otto Johannson, engine driver, said that he had charge of engine 3504 on the express from Newcastle to Werris Creek. Witness said that he thoroughly examined his engine before clearing the loco points.

Arriving at Muswellbrook at 8.30 p.m., witness examined the engine again. The train passed through Aberdeen at a maximum speed of 12 miles an hour, and about a quarter of an hour later the engine "fell into a hole," but immediately righted itself. The train was then travelling at not more than 36 miles an hour. Shortly afterwards an unusual sound alarmed him, and he caught hold of the regulator. He then found himself in the right-hand corner of the cab, on the ground.

Mr. Cargill: What is your opinion of the NN type of engine?

Witness: The best passenger engine on the New South Wales railways.

Asked by Mr. Sayegh whether he had formed an opinion as to the cause of the accident, Johannson said that the accident was brought about by derailment of some vehicle behind, bumping over the sleepers and weakening the bridge. Finally this vehicle fell over, pulling his engine and tender over the abutment of the northern end of the viaduct.

Mr. Sayegh: What do you consider was the cause of the derailment?

Witness: The bad break in the road. [1002]

On the final day of the inquest the coroner read a statement by Fireman Leonard, of the NN engine, that he was still inconvenienced by injuries and could not appear in Court. Leonard mentioned that at the time of the accident he was unable to say whether the brakes were applied or not, otherwise he practically corroborated Driver Johannson.

James Henry Monkley, driver of the leading engine of the train, said that after leaving Aberdeen he did not open his regulator approaching the viaduct. He was going at about 30 miles an hour. He noticed a depression in the track, which caused the engine to roll, and he made a slight application of the brake to steady the train. After applying the brake he had looked and seen what appeared to be some vehicle behind the NN engine rolling out of the line of the train. He had immediately applied the emergency brake, being then somewhere near the middle of the viaduct. The NN engine had then broken away, pulling the draw bar right out, and lifting his tender off the rails. Witness' engine had continued on its way and the NN engine had tilted and fell over the east side of the viaduct.

To Mr. Sayegh (for Mrs. Isaacs, sister of Mr. Mansour), witness said that he did not think it was the speed of the train, but the depression in the road which caused the derailment.

Albert Osmond, driver of the No. 1 express. which passed over the viaduct an hour and a half before the accident, said he did not notice anything wrong with the road.

Norman Charles Harris, assistant chief mechanical engineer of Victoria, said that he thought it highly improbable that the NN engine was the primary cause of the derailment. He believed it was caused by irregularities in the track setting up oscillations and permitting the flange of the wheels to leave the track.

Alfred David John Forster, Assistant Commissioner for Railways, said that he had made exhaustive inquiries in endeavouring to ascertain the cause of the accident. There were some track irregularities, and the exact condition of the track on the south side of the point of derailment was a succession of small but equally spaced waves. Their amplitude increased towards the point of derailment. They might set up in a spring-carried vehicle conditions causing a heavy oscillation. This condition could have caused unloading of the spring-borne weight on the left trailing wheel of the 16 class tender, and at 25 miles the vehicle could lurch severely. At a critical speed it could cause the wheel to leave the rail. It appeared that the trailing wheels of 16 class tender were derailed on the west side of the line. This could have happened without any noticeable indication to the train crews at the middle of the viaduct.

The Coroner, Mr. A. W. Swaddling found that the four victims had died from the effects of injuries accidentally received.

He added a rider to the finding, in which he stated that the condition of the permanent

[1002] Ibid, 2nd July 1926

way at and near the scene of derailment was such that it was not in a condition to carry safely an engine of the NN type at a speed exceeding 35 miles an hour.

In reviewing the evidence, the coroner said that in spite of evidence given as to existence in the bridge of dry rot and white ants, he thought that these factors in no way contributed to the derailment. It was the train that wrecked the bridge, he said, and not the bridge that wrecked the train.

Regarding the NN type of engine, the coroner said that certain evidence had been given as to its effect when travelling at a high speed, but the evidence given did not enable him to make any definite statement. He thought, nevertheless, that it was a matter for inquiry outside the Coroner's Court. [1003]

The Railways Department board of inquiry came to an identical conclusion to the inquest, finding that irregularities in the track had set up oscillations in the engines which had caused the derailment and that the condition of the viaduct had been sound and its wrecked had resulted from the derailment. [1004]

A fifth victim, Mrs Annie Macdonald, 62, of Ashfield, died of her injuries on 6th July. [1005]

The day after the Sevenoaks disaster in England the little Chemin de fer du Montenvers rack railway suffered a disaster in France. The metre gauge line was only just over 5 kilometres long and ran from Chamonix station on the Chemin de fer de Paris à Lyon et à la Méditerranée with gradients varying from 11% to 22% (1 in 9 to 1 in 4½) to the Mer de Glace glacier on the northern slopes of Mont Blanc, a major tourist attraction. It was worked at this date 2-4-2T steam locomotives and coaches. [1006]

On the afternoon of Thursday 25th August 1927 a descending train got out of control. It ran down the steep incline at increasing speed until on a curve the engine left the rails, taking a piece of the rack with it. Unfortunately at the point where the train became derailed the line was on a viaduct, about 20 metres high near the Montenvers Hotel, and the engine and leading coach fell from the viaduct, the engine falling on top of and crushing the carriage. Conductor Ribery in the second carriage was able to apply the brakes and prevent it from following the first over the viaduct. Some of the passengers, sensing the danger, also reportedly jumped from the train, possibly from the first coach. After the accident uninjured passengers assisted in the rescue of their fellow tourists and the injured were distributed between various villas and hotels and the hospital in Chamonix. An initial report spoke of 21 dead and 30 injured, several very seriously, and the final number of dead appears to have been 22, among them Driver Lyard on the engine. Those who lost their lives included three Germans, two English and an American living in Norway. The fireman was thrown from the engine when it left the track and escaped unhurt.

The train involved in the disaster was the second of three to start from Montenvers at short intervals. The regular train left at 16:50 followed by a train *bis* five minutes later (the one which met with the accident) and a train *ter* five minutes after that. There was thus a real possibility of the third train becoming derailed on the section of rack destroyed by the second train and/or colliding with the car remaining on the line. Fortunately, a

---

[1003] Ibid, 3rd Jul 1926

[1004] Ibid, 26th July 1926

[1005] *Evening News*, Sydney, New South Wales, Australia, 7th July 1926

[1006] Wikipedia – Chemin de fer du Montenvers

Mme Mayor, presumably a passenger, was able to make her way up the slope despite foggy conditions and falling snow and warn the driver of the *ter* train in time for him to be able to pull up short of the obstruction. M. and Mme Mayor of Montenvers were then very active in the rescue work.

The parquet of Bonneville opened an investigation into the accident the following morning. [1007] The inquiry disclosed that the accident had its origins in a somewhat complex chain of events which began with a severe snowstorm which resulted in a large number of tourists deciding to return to Chamonix at the same time, hence the departure of three trains in quick succession. Another factor was that there was a short stretch of more or less level track at the summit before the steep descent began. The drivers of descending trains were therefore in the habit of accelerating at the beginning before reversing their engines at the start of the rack section. The driver of the *bis* train stopped to pick up some guides and having done apparently forgot that he had not reversed his engine. As the steep section was reached the speed of the train quickly rose to 10 kilometres an hour, at which point the brakes came on automatically. They also applied with some force, and the conductor on the first coach, who was still belatedly closing the doors, was thrown off. Driver Lyard seems then to have lost his head and still failed to reverse his engine, and the train had run away to disaster. [1008]

The line was not much damaged, and it was hoped to reopen it by the following Monday or Tuesday. [1009]

An accident at Kyautaga in Burma on 28th January 1928 was apparently due to sabotage and is dealt with in the Sabotage chapter.

An accident on the Fort Worth and Denver Railroad near Oklaunion, Texas, United States, on Sunday 3rd August 1930 was not especially serious in terms of casualties but has been thought worthy of inclusion because it was possibly the first ever accident of its type. While the greatest proportion of intersections between railways and roads outside of the major cities in the United States, certainly in 1930, were in the form of level crossings, at Oklaunion an underbridge was provided to carry the Fort Worth line over a road.

On the day of the accident a lorry driven by Horace H. Cummings, of Duncan, Oklahoma, accompanied by his father, and carrying what was described as a huge boiler, attempted to pass beneath the bridge but the boiler was too high and fouled the bridge, displacing the rails of the track above. Before the father and son could do anything a southbound passenger train arrived at the bridge and was derailed. Engineer F.P. Robinson and Fireman A.B. Vance were killed, the engine and five coaches falling from the bridge. Twenty-two people were injured.

After the accident, the two Cummings ran from house to house in Oklaunion to summon assistance. [1010]

[1007] *Le Populaire*, Paris, France, 27th August 1927

[1008] *Railway Gazette*, Vol 47, pg. 392

[1009] *The Times*, London, England, 27th and 29th August 1927

[1010] *The Dubois Courier*, Dubois, Pennsylvania, United States, 4th August 1930

Like the Kyautaga accident in Burma in early 1928 one near Banpur, India, on 17th September 1930, was due to sabotage and is described in that chapter.

CHAPTER 11

TUNNEL ACCIDENTS

In his classic work on British railway accidents, *Red for Danger*, L. T. C. Rolt remarks on the acute trepidation felt by Victorian travellers when they were passing through a tunnel, and their oft repeated pleas to the railway authorities that all tunnels should be artificially illuminated, pleas which, probably for practical reasons if nothing else, fell upon deaf ears. He goes on to point out that these fears were by no means unjustified, especially in the period before the introduction of the absolute block system, when trains were merely spaced by the time interval system. If a train should chance to break down in the open there was a good chance that the crew of a following train might be able to see it in time to pull up, even if there was insufficient time for the guard of the first train to go back and protect its rear. In the smoke- and steam-filled darkness of a tunnel there was no such hope. Furthermore, should a serious collision occur in a tunnel the confined space would be likely to result in the wreckage being hideously compressed, and should a fire break out in a long tunnel provided with ventilating shafts, the latter could act as chimneys. In addition the rescue work would be hampered by the darkness and the difficulty of working in a confined space where the use of breakdown cranes was impossible, and the only resort was to drag the wreckage bodily out of the tunnel.

On the evening of Thursday 9th August 1917 crowded suburban train No. 260 left Paris Orsay terminus of the Chemin de fer de Paris à Orléans at 18:10 for Juvisy. It left slowly and was just 50 metres into the tunnel at the exit from the station when it collided with set of coaches which were being manoeuvred to form the 19:15 to Toulouse. The suburban train struck the rear of the empty stock in a sidelong collision and the front coaches of the Juvisy train and the last of the coaches being shunted overturned.

A rescue operation was of course immediately begun by the station staff. The bodies of three passengers who had lost their lives were taken from the train and 13 other passengers were injured. Two of those who died were Gaston Dalys, aged 18, of Choisy-le-Roi and Alfred Blondel, aged 33, of Athis-Mons. The third victim was a girl aged about 25 who could not immediately be identified. The more seriously injured were taken to Rollin hospital, the others were bandaged and sent home. [1011] While the line was blocked intending passengers were taken by car from Orsay to Austerlitz station. [1012]

An inquiry was opened to determine responsibility for the accident, which was not immediately clear. The most likely theory, according to a newspaper reporter, was that the driver of the train of empty stock had begun to set back prematurely against a stop signal in the path of the departing Juvisy train.

The third victim was later identified as Marie Harang, aged 21, of Orly. [1013]

On Saturday 29th September 1917 on the Great Indian Peninsula Railway a goods train ran into a rock which had fallen from the roof of Parsik Tunnel, India, and was derailed. A following up local train then entered the tunnel and ran into the rear of the goods train, killing four porters riding in the front brake van and injuring two others. Two passengers in the local train were slightly hurt. Due to a failure of the block instruments the trains were running on "Authority to proceed without line clear" rules. The regulations of the railway company permitted trains to follow one another through the Bombay – Kalyan section at five-minute intervals in such circumstances until it could be determined that the line was clear, subject to certain precautions, one of which was not entering a tunnel until it had been ascertained that it was clear. After exhaustive departmental and judicial investigations the driver of the local train was fined 100 rupees. [1014]

Carelessness on the part of a signalman and inaction on the part of a guard, who was possibly dozing, led to a double collision in Redhill Tunnel, England, on the Quarry line of the London Brighton and South Coast Railway, on Thursday 18th April 1918.

The events leading up the accident began with the stopping of the 2.10 a.m. up Eastbourne goods train (2-6-0 engine No. 340, ten loaded and 49 empty waggons and a brake van) at the Earlswood Junction up advanced starting signal, south of the tunnel, while the Earlswood Junction signalman awaited the clearance from Quarry Intermediate box of the previous train. At 4.56 a.m. this was received, the Eastbourne train was accepted, and the advanced starting signal was cleared. The Eastbourne train then restarted but a snatch must have occurred as it did so, and the train became divided. The three waggons and guards' van which made up the detached part of the train followed the front part of the train down the falling gradient into the tunnel but inside the tunnel the gradient changed to a rising one and the rear portion came to a stand.

The front portion of the Eastbourne train passed Quarry Intermediate box at 5.09 a.m. Signalman George Towes failed to see either a tail lamp or side lights and telephoned his colleague at Star Lane and asked him to check that the lights were there. However, he still gave the "Train out of Section" signal to Signalman Scollick at Earlswood Junction and accepted the following up 11.40 p.m. Portsmouth goods train (0-6-0 engine No. 541, 38 waggons and a van). Worse still there was a down goods train, the 4 a.m. down special from Lillie Bridge (engine No. 536, 39 loaded waggons and van) on its way from Quarry

[1011] *L'Humanité*, Paris, France, 10th August 1917
[1012] *La Lanterne*, Paris, France, 11th August 1917
[1013] *La Petit Parisien*, Paris, France, 10th and 11th August 1917
[1014] Railways in India: Administration Report for the Year 1917-18

Intermediate towards Earlswood. This would have been a Government Stores (munitions) train.

Driver Roberts and Fireman Howard of the Eastbourne train did not notice any snatch as they started away and because of the curvature of the line they were unable to exchange signals with their guard in accordance with Rule 171(h) before their train entered the tunnel. After leaving the tunnel a rather foggy atmosphere and smoke beating down from their engine prevented their seeing the rear of their train.

Goods Guard Austin did feel a slight snatch as his train started but believed that his van was following the remainder of the train properly. He told the Board of Trade inquiry into the accident that when his van had come to a stand in the tunnel he had been at a loss as to what to do. After waiting five or seven minutes he had taken his handlamp, gone forward and found that his train had become divided. Returning to his van he had just picked up some detonators and had been about to go back to protect his train when it was run into by the Portsmouth goods. He was lucky to escape without being killed or seriously injured.

The collision derailed the rear vehicles of the Eastbourne train and the engine and leading vehicles of the Portsmouth train, some of which fouled the down line. Shortly afterwards, before it could be stopped, the Lillie Bridge train ran into the wreckage. The result was that 40 feet of the tunnel was blocked to the roof; two engines were badly damaged, and 26 waggons were either destroyed or badly damaged and two guards were injured. Had a passenger train been involved the accident could of course have been much worse.

The Board of Trade inspecting officer, Colonel J.W. Pringle, noted that Driver Sharp and Fireman Keats stated that the tunnel had been full of smoke and steam when they had entered it and they had been able to see nothing of the brake van of the Eastbourne train before they struck it. The same situation was faced by Driver Hunt and Fireman Mapp on the Lillie Bridge train. For the same reason Driver Roberts and Fireman Howard could not be blamed for having failed to realise that there had been a breakaway.

The negligence of Signalman Towes with respect to his non-observance of a tail lamp or side lamps could however not be excused, the inspecting officer remarked. The signalman pleaded that smoke and steam in the cutting in which his box was situated made it difficult to spot tail lamps and also complained that paintwork on the windows of his box (due to the blackout?) had obscured his vision. The fact that the last ten vehicles of the front portion of the Eastbourne train were low waggons rather than box vans clearly demonstrated, said Colonel Pringle, that Signalman Towes had not been carrying out his duties of observation properly.

Guard Austin of the Eastbourne train was, if anything, criticised even more severely than the signalman by the inspecting officer, who found the guard's inaction for probably ten minutes after his van had come to a stand to be 'reprehensible.' It was also noted that there were no entries in the guard's journal since the train had passed Lewes, 28 miles from Earlswood. Had the guard been properly alert there would have been plenty of time to stop the Portsmouth train. Probably the guard had been dozing.

The inspecting officer placed the responsibility for the accident equally upon Signalman Towes and Guard Austin. Concluding his report he praised the efforts of the breakdown gang who had managed to clear the tunnel so that it could be reopened to traffic within only 40 hours. [1015]

An accident occurred on the Central Vermont Railroad on Wednesday 5th June 1918 near Burlington, Vermont, United States, in the tunnel beneath one of the city streets north of Burlington station and between there and the next station at Winooski. A passenger train collided head-on on the single line with a light engine, killing or fatally injuring nine railwaymen and injuring eight railroad employees.

On the morning of the accident engine No. 393 and seven cars had left Burlington for Winooski and probably beyond at 5.45 with Conductor Notemyer, Engineer Eli Griffin and Fireman Fred Benoit. It was engaged in shunting operations at Winooski when a car became derailed in the yard. It proved impossible to rerail the car without further help and so the conductor telephoned the dispatcher at St Albans saying that he needed to return to Burlington for help. At 9.10 a.m. he was ready to go, and the dispatcher issued train order No. 39:

Engine 393 run extra Winooski to Burlington and return to Winooski.

The dispatcher also gave the conductor a message:

If you can make Winooski by 9.40 leave flagman there so you can go ahead and return for No. 72.

The engine left Winooski at 9.20 or 9.25, ran the three miles to Burlington, took water there, picked up the necessary men and tools and was on its way back northwards to Winooski when it collided about 130 feet from the north portal of the tunnel with southbound passenger train No. 72 from Cambridge Junction to Burlington. The latter train consisted of engine No. 333 and six cars, in charge of Conductor Berry (or Barry) with Engineer Frank Andrews and Fireman Clayton Scott. Engineer Griffin, Fireman Benoit, a brakeman and three shopmen on the light engine and Engineer Andrews, Fireman Scott, and a brakeman on train No. 72 lost their lives. [1016]

Dispatcher Earle told the Interstate Commerce Commission (ICC) inquiry of receiving the telephone call from the conductor and of issuing the order and adding the message. He stressed that the message had merely been a suggestion as to what the conductor could do and was not intended to modify the rights of the engine under order No. 39. His idea was that if the engine did not get back to Winooski in front of train No. 72 it would also have to wait for northbound first class passenger train No. 29, due out of Burlington at 9.59 a.m., and an hour's delay would result.

Station Agent Kellogg at Winooski stated that the engine had left for Burlington at about 9.20 and train No. 72 had arrived at 9.26. He did not see the passenger train leave but understood that it had done so on time at 9.30. Soon afterwards he had learned that the engine was returning from Burlington and had advised the dispatcher of the impending collision.

Operator Barlow at Winooski testified that he had handed the order and message to Conductor Notemyer and the latter had remarked that he did not think they could get back to Winooski by 9.40. The engine had left for Burlington about 9.20 and he thought

---

[1015] Board of Trade report

[1016] *Boston Post*, Boston, Massachusetts, United States, 6th June 1918 that train No. 72 had left on time at 9.30. As train No. 72 was passing out of sight he had overheard a telephone message between the Burlington telegraph office and the tower there to the effect that the engine was on its way back and had told the dispatcher.

Operator Perry stated that at the request of the towerman he had registered engine No. 393 as arriving at Burlington at 9.30, and as leaving again for Winooski at 9.35. At the time that he had received the latter message from the towerman he had already been advised by Winooski that train No. 72 had left there, but the towerman had said that it was then too late to stop the engine.

Towerman Lounsboury testified that the conductor of engine No. 393 had asked him to register them in and said that he had left a flagman at Winooski to hold train No. 72 until they returned there, nothing being said about 9.40 or any other time. The engine had taken water, loaded men and materials and set off back at he thought 9.40. He had notified the operator (Perry) that the engine had left, he denied that the operator had said anything to him about stopping it. His duties consisted solely of operating the signals and switches and he knew nothing about the orders. He put the speed of the engine on its return journey at about 35 miles an hour. Before the engine had gone out of sight the annunciator bell had indicated the approach of train No. 72, by then the engine was beyond his signal and unstoppable.

According to Foreman Fitzsimmons at Burlington engine No. 393 had arrived there just before 9.30, had quickly taken water and had left again at 9.32. The engineer had told him that they had left a flagman at Winooski with orders to hold everything until 9.40. He (the foreman) thought that they had had ample time to get to Winooski by 9.40. The engine had left with the engineer, fireman, conductor and eight engine-house men, including himself, riding on various portions of it.

Flagman Desaw of engine No. 393 stated that it had been nearly 9.30 when Conductor Notemyer had told him that the engine was going to Burlington for water, tools, and men and to hold train No. 72 until 9.40. He had then worked on the derailed car. When train No. 72 had arrived he had told Brakeman Foster that the train was to be held until 9.40. and explained why. The train then began to move forward, and he had shouted the same thing to the enginemen. He had thought that the train was just pulling down a little and had not paid much more attention to it, going back to working on the derailed vehicle. He had only noticed later that the train had gone.

Conductor Berry of train No. 72 said that it had left Winooski at 9.31 or 9.32. He had timed the collision at 9.40. He had not seen Flagman Desaw at Winooski and had known nothing of any flagging instructions. His train had been moving at about 8 miles an hour when they collided with the engine.

The ICC concluded that the accident had been due to the failure of Flagman Desaw to carry out his instructions and by Conductor Notemyer and Engineer Griffin allowing engine No. 393 to occupy the main track on the time of a superior train without protection.

Although the flagman's instructions to hold all trains until 9.40 had been definite and positive it was not believed that he had made them clear to the engineer of train No. 72, which had departed on time with no attempt to wait until 9.40.

There was some conflict as to the time that the engine had left Burlington on its return journey, but it seemed evident that the collision had occurred at a time very close to 9.40.

As the point of the collision was still two miles from Winooski it was clear that the engine had occupied the main line on the time of train No. 72 without flag protection.

The ICC pointed out other irregularities and instances of sloppy working and recommended a rule, or a rewording of the present rule, to require a train under similar circumstances to secure a clearance card before leaving a terminal. It also noted that the accident would have been averted had the block system been in use. [1017]

The Chemins de fer de Paris à Lyon et à la Méditerranée suffered a major disaster on Thursday 19th September 1918 in Pacy Tunnel, France, between Lézinnes and d'Ancy le Franc stations. Train No. 12062, which had left Marseille at 12:30 on Wednesday and Lyon, after a pause there of almost three hours, at 22:00, was run into by train No. 12062 *bis* in Pacy tunnel at 04:30. The first train had apparently stopped after a brake pipe had burst in the tunnel. The rear van and two coaches were telescoped and more or less destroyed in the collision. Two crowded third-class carriages were involved, and the number of casualties was high. An early report spoke of 28 dead and 38 injured. Train No. 12062 *bis* was only five minutes behind the first train and was reportedly travelling at 100 kilometres an hour when it ran into the train in front.

Given the early hour it took some time to organise the rescue operation, but relief trains were sent from Dijon and Paris. The general commander of the American troops in the area placed all the motor vehicles that the military possessed at the disposal of the railway company for the removal of the dead and injured. [1018] The American camp by chance was close to the tunnel.

Occurring as it did in the closing days of the First World War, news of the accident soon disappeared from the newspapers which were more interested in stories of the fighting and the pending Allied victory. It has thus not been possible to find the result of any inquiry into the disaster to answer questions such as why the second train was allowed to enter the tunnel when it was still obstructed by the first or why the conductor of the first train did not protect it in the rear with petards (detonators).

Chemins de fer de l'Est passenger train No. 47, which had left Paris Est for Metz and Strasbourg at 20:15 on Wednesday 5th February 1919, its 14 coaches packed with passengers. All went well as far as Ferté-sous-Jouarre station, beyond which the train entered the 940-metre-long Nanteuil-Saacy Tunnel, France. About halfway through the tunnel the emergency alarm sounded, the driver applied his brakes, and the train came to a stand 250 metres short of Nanteuil station. All the passengers in the two leading coaches, one second class, the other third class, and which the French had recently acquired from Germany, opened the doors of their carriages, and got down onto the opposite running line as small flames began to appear beneath the vehicles. Unfortunately at this moment a leave train arrived from the opposite direction. When the passengers saw the lights of the oncoming train it was within a few metres of them. Some threw themselves against the tunnel wall, others managed to perch on the running boards of the coaches, while others panicked in the darkness and began to run, without knowing where

---

[1017] Interstate Commerce Commission

[1018] *La Petit Parisien*, Paris, France, 20th September 1918 they were going. This last group was caught by the leave train and thrown and crushed between the coaches of the two trains, their terribly mutilated bodies littering the track after the leave train had passed.

Nanteuil station was immediately advised of the accident and staff went into the tunnel and despite the flames were able to divide the train behind the second carriage. The engine then went forward with the two burning vehicles and shunted them into a siding before returning for the remaining 12 coaches left in the tunnel. Meanwhile other railwaymen, aided by uninjured passengers, were taking care of the injured, who numbered 18. They were placed in a coach which was used as an ambulance and took them to Château Thierry station. During the journey, an elegantly dressed woman aged about 25, who had no identity papers upon her, died of her injuries. A four-year-old child, Eugène Tanot, also succumbed during the journey, bringing the number of dead up to five. The others known to have lost their lives were two soldiers, Albert Doyen, of the 215th Artillery and Julien Malherbe, of the 48th Infantry, plus a civilian, M. Edouard Barthelemy.

A superficial examination of the damaged carriages by officials of the railway company and the Ministry of Public Works concluded that the fire had been caused by an overheated axle box, such occurrences being common due to the poor quality of the available axle grease. However, a further examination revealed that there was nothing wrong with the axle boxes and a different explanation had to reached as to the origin of the fire. A soldier who had been a passenger in the leading carriage stated that right from the departure from Paris he and his companions had been aware of an unusual amount of heat. They had been wondering where the heat was coming from when gas began to issue from the floor of the carriage, it smelt of petrol and quickly began to suffocate them. Along with some of his companions he had smashed windows to get some fresh air, while others called for help. Hearing the cries passengers in the second carriage had sounded the alarm and the train had stopped. He had then taken a little girl in his arms and jumped down onto the track. He had been able to avoid being struck by the leave train by keeping close to the tunnel wall.

This account would tie in with the fact that the German coaches were gas lit, using coal gas, the smell of which closely resembled the smell mentioned by the soldier. However, a report by railway company engineers the following day stated that the fire had not originated from either the lighting or heating equipment, nor from a hot axle box. The number of dead had now climbed to eight following the discovery of the bodies of three more victims in the tunnel, while there were 16 injured in Château Thierry Hospital, eight of them suffering from gas poisoning.

Rumours then began to circulate that the gas outbreak and fire had results from the use of some novel weapon employed by the Germans and it was said that the symptoms exhibited by the Tanot boy had been identical to those suffered by soldiers in the trenches from poison gas attacks. There were also reports that the remains of an 'infernal machine,' that is a bomb, had been discovered in the wreckage, but M. Gaudel, the public prosecutor, rejected suggestions that the fire had been an act of German sabotage. [1019]

The origin of the fire remained a mystery, and the supposed infernal machine was later identified as a gramophone player. A new and more likely theory was that this

---

[1019] *Le Petit Journal*, Paris, France, 7th February 1919

player, together with some records, had been placed by a passenger out of the way beneath a seat, and the records had accidentally caught fire by coming into contact with a heater. [1020]

The final number of fatalities appears to have been 10.

A collision occurred in the St Gothard Tunnel, Switzerland, on Friday 9th January 1920. 'Due to a misunderstanding and smoke in the tunnel' a service train collided with three ballast waggons in the tunnel. A number of workmen were sat on the waggons when they were struck. They were thrown to the ground, and one was killed outright while others were injured. The six most badly hurt, who had suffered head injuries or broken limbs, were taken to hospital at Altorf. It was hoped that they would all recover. [1021]

On Monday 21st March 1921, a train was derailed and caught fire in a tunnel at Niigata, Japan. Ten people reportedly died and 50 were injured. [1022] [1023]

In 1921 just outside of the Saint Lazare terminus of the Chemins de fer de l'État in Paris, France, the various tracks entered Batignolles Tunnel, which because of the number of lines actually consisted of four parallel bores: No. 1 (St Germain), No. 2 (Argenteuil), No. 3 (Versailles) and No. 4 (Auteuil). The bores were 330 metres long. The Batignolles tunnels had long been a bottleneck and a nuisance and there had been proposals to open them out in the 1890s but nothing had been done due to the cost and the need to demolish many properties on the land above them. A new scheme to remove three of the bores was considered in 1912 after several years of studies and tenders for the work were finally approved by the French government on 12th October 1914. By then, of course, the First World War was in progress and the project was again shelved. [1024]

On Wednesday 5th October 1921 train No. 333 left Saint Lazare station for Puteaux and Versailles. It consisted of an engine and 21 old wooden coaches, some of them double-decker, packed with passengers going home from work. The train entered the Versailles bore of Batignolles Tunnel and was about in its centre when the air brake connection between the fifth and sixth coaches separated, the brakes applied automatically, and the train came to a stand. Driver Blanqui got down from his engine and went back along his train to find the cause of the stoppage and Conductor Vilemer left his van to do the same. Neither considered that their train was in any danger of being run into because the lines through the tunnels were protected by a semi-automatic signalling system.

The damaged pipe was soon discovered and while the conductor remained beside the broken connection the driver returned to his engine to get the materials to make a repair. Suddenly the unmistakeable sound of another train coming up behind them could be heard. It was train No. 253 for d'Issy-les-Moulineaux, which had left Saint Lazare four

---

[1020] *The Times*, London, England, 10th and 12th February 1919

[1021] *Journal de Geneve*, Switzerland, 11th January 1920

[1022] *The Chicago Eagle*, Chicago, Illinois, United States, 26th March 1921

[1023] *Nebraska State Journal*, Lincoln, Nebraska, United States, 22nd March 1921

[1024] Wikipedia – Batignolles Tunnel

minutes after train No. 333. As the second train entered the tunnel under clear signals Driver Huvey saw the tail lamp of the train in front and jammed on the brakes but the distance was much too short to stop. [1025] A collision ensued and the engine of the colliding train demolished the rear van of the front one and threw the two preceding coaches to one side. A panic naturally ensued among the passengers but the collision itself probably only resulted in some fairly minor injuries and no deaths. What made a relatively minor accident into a disaster was the fact that the old coaches of train No. 333 were gas lit. The force of the collision punctured one or more of the gas tanks beneath the coaches, the gas ignited and the soon spread along the train. No less than 16 of the coaches of this train were reduced to ashes; the engine managed to draw the front five coaches clear of the fire and they and their occupants escaped, thanks to the presence of mind of Conductor Vilemer. The coaches of train No. 253 were also involved in the fire, which this time spread from front to back along the train, but the engine of a Rueil train was coupled to the last of them and was able to draw the ten rearmost vehicles back clear of the fire. The others were reduced by the fire to mere shells.

Describing the scene a reporter wrote, 'The spectacle then was sinister. The tunnel, violently illuminated by the gushing flames, vomited acrid, glowing smoke and fumes. One would have said the face of a burning oven. People escaped, blackened, burnt, their clothes ragged. Women were running, their hair scattered; children, their eyes staring and wide-open with fear, were fleeing without knowing where they were going…' [1026]

As well as the passengers who were able to exit the tunnel via one or the other of its portals some were able to escape via connecting doors to the Argenteuil and Auteuil bores.

Rescuers were soon on the scene, the proximity of the Saint Lazare station being perhaps the only fortunate aspect of the disaster. The firefighting efforts of the station staff were soon supplemented by firemen from the Rue Blanche and then from four other fire stations. Hoses were brought into the tunnel to directly attack the blaze. More help came from the far end of the tunnel. The driver of a train from Colombes had stopped his train on seeing the fire ahead and the passengers on the train included the members of a police football team returning from a training session.

Posters stating that traffic had been interrupted due to an accident in Batignolles tunnel were hastily displayed and the station was evacuated with the help of soldiers from the 5th infantry regiment, stationed at the Pépinière barracks, and then closed. Intending passengers were forced to use trams or the Metro instead, though Metro line 2, which passed over the tunnel, was taken out of use for some hours for fear that the heat had damaged the tunnel roof and might cause it to collapse. Other passengers took taxis or simply walked home. Trains coming into Paris were terminated at Batignolles station, on the far side of the tunnel. Traffic from Saint Lazare station was resumed at 21:30 but the lines through the Versailles bore were of course still out of use and the reduced capacity and diversions meant severe delays.

By 23:00 the firemen had extinguished the fire and the remains of the first victims of the accident had been discovered. More remains would be discovered during the process of clearing away the wreckage. Most of the bodies were very badly burnt. They were

[1025] *Le Matin*, Paris, France, 6th October 1921
[1026] *La Petit Parisien*, Paris, France, 6th October 1921

placed in temporary coffins and put on display in the hope that family members or friends might recognise them, then those which had not been identified were taken to the mortuary for possible identification by a forensic pathologist. On 31st October those who had still eluded identification were interred in the Ivry cemetery after a funeral service in the Notre Dame cathedral. In all 28 passengers were believed to have perished in the accident, about 40 were seriously injured and about 60 less badly hurt. In addition one of the firemen, named Lafon, had to be admitted to the Beaujon hospital suffering from asphyxiation. Those who lost their lives included Sabran de Rohan-Chabot, son of the Countess of Rohan-Chabot, who had been on his way to the Sainte-Geneviève college at Versailles, and Lieutenant Aviator Grange, a hero of the First World War, whose charred body could only be identified by his wife by a ring upon one of his fingers.

Less than an hour after the disaster. M. Yves Le Trocquer, the Minister of Public Works, and various other senior officials were on the scene, and investigations into how the accident could have occurred were opened by the railway company and by the public prosecutor, M. Edmond Scherdlin.

The lines through the tunnels were provided with a semi-automatic signalling system incorporating Aubine treadles. The signal boxes on the Versailles line at the Saint Lazare and country ends of the tunnel were posts Nos. 4 and 6 respectively. When train No. 333 passed the signal at post No. 4 at the entrance to the tunnel the Aubine treadle should have been activated and automatically placed that signal at danger until the train passed over a further treadle located at a signal at post No. 6 at the far end of the tunnel. Driver Huvey of train No. 253 stated however that he had received a clear signal, and this was apparently not disputed; it was not a case of a driver overrunning signals. Nor was there any question of the equipment itself malfunctioning, so the cause of the accident must have been a mistake by one or other of the signalmen in overriding the safety apparatus and clearing the signal prematurely.

Suspicion fell at first on Signalman Édouard Lohazic, at post No. 6, and he was arrested and confined in the Santè prison, charged with homicide by negligence. It was alleged that he had cleared back train No. 333 prematurely and his answers to questions posed by the examining magistrate were said to be contradictory. Meanwhile the four signalmen at post No. 4 (clearly a very busy box); Jean Auvray (Head Signalman), Chapin, Maledant and Pigache, remained at liberty but were asked to elaborate upon their version of what had happened to the examining magistrate. [1027] Auvray and Maledant had initially claimed that Lohazic had advised them by telephone that although the signal at post No. 4 was still locked the tunnel was clear and that they could permit train No. 253 to proceed.

Tests soon showed, however, that the signal at post No. 4 could be placed at clear by manipulating the apparatus to get a release without any involvement of the signalman at post No. 6 and that the version of events given by the men at post No. 4 was not entirely credible. Signalman Lohazic, who had throughout declared his innocence, was released on bail.

Subsequently a report prepared by three experts indicated that given the interlocking and the positions of other signals controlled by posts Nos. 6 and 7 (the next post beyond No. 6) it would not have been possible for Signalman Lohazic to have released the signal

[1027] *L'Homme Libre*, Paris, France, 11th October 1921

at post No. 4 as claimed by the signalmen there, and it appeared that the signal had been wrongly released by the men at post No. 4, who had then attempted to throw the blame upon their colleague for their own mistake. The charge against Signalman Lohazic was dropped and instead Head Signalmen Auvray and Signalman Maledant were charged with causing death and injury through negligence. They were tried on 30th March 1922, found guilty and Signalman Maledant was sentenced to six months' imprisonment and fined 300 francs and Head Signalman Auvray was awarded 4 months' imprisonment and fined the same amount.

The two signalmen appealed against their sentences and the Paris Court of Appeal, having visited the scene of the accident, decided to give them the benefit of the doubt and quashed their convictions. This decision was welcomed by the left-wing newspaper *L'Humanité*, which wrote that, 'Now that justice is finally done for two innocent people, are we going to look for the real culprits?' The paper claimed that the causes of the disaster must be looked for at the highest levels of the State. [1028]

A sequel to the accident was the issuing of a strongly worded circular on 14th October by M. Le Trocquer, the Minister of Public Works, addressed to all of the railway companies, instructing them that they must take immediate steps to abolish the use of gas for train lighting, and that it must disappear on all express trains by 1st January 1923 and on all suburban trains from 1st January 1924. Returns were be submitted showing how quickly the abolition of gas lighting was progressing. The use of coal fires for train heating was also to be dispensed with within two years. [1029]

Two weeks after the accident work was begun on installing lighting in the tunnel, something which had been demanded in newspaper editorials, though most tunnels worldwide are not lit. Work on opening out the St Germain, Argenteuil and Versailles bores began in 1922 and was completed in 1926; at the same time, the six tracks through the three tunnels were increased to eight. The Auteuil bore was left intact.

Four men were initially reported to have lost their lives in an accident in Palliser Tunnel, just east of Palliser station, British Columbia, Canada, on the Canadian Pacific Railway, at 11 p.m. on Thursday 20th October 1921. An eastbound freight train in charge of Conductor John Beck had left Revelstoke in the afternoon. It had two engines at the front together with a third engine at the rear. Engineer Palmer Dean and Fireman George Goodwin were on the leading engine and Engineer Ole Petersen and Fireman Rocksford Richmond were on the train engine. As the train was passing through the tunnel part of its roof, which was being relined with concrete at the time, fell on it, killing engineer Dean, Firemen Goodwin and Richmond and a man stealing a ride identified as Lance McCarrigan, of Kincardine, Ontario. In addition Engineer Petersen and a trespasser named Robert Watson were injured. Oil escaping from the tender of one of the damaged engines subsequently caused a fire which destroyed four freight cars.

The railway company announced that repairs to the tunnel would take several days and meanwhile a temporary line was being constructed around the obstruction, but this would involve the construction of three bridges, one across the Kicking Horse river, so it would be two or three days before any traffic could be handled between Golden and

[1028] Wikipedia – Accident ferroviaire du tunnel des Batignolles
[1029] *The Times*, London, England, 15th October 1921

Field. Trains could, however, be diverted over the Kootenay central branch and the Kettle Valley railway with little or no delay. As to the cause of the accident, the company stated that, 'The tunnel, which was constructed in 1905, is three hundred and ten feet long, and lined throughout with concrete. The trouble was apparently due to heavy rains causing a movement of the mountain above in the direction of the Kicking Horse river, which resulted in a pressure on the roof and sides of the tunnel, which proved overwhelming. General Manager Charles Murphy, Chief Engineer Fairbairn and Assistant Chief Engineer James are on the way to Pallister in order to complete a thorough investigation.' [1030]

Engineer Petersen died of his injuries two days after the accident and the finding of the body of another, unidentified, trespasser in the wreckage increased the number of fatalities in the accident to six. An inquest was subsequently held at Golden upon Engineer Palmer as representative of those who had lost their lives. At the hearing surviving members of the train crew testified that a "slow order" had been issued for the tunnel and the train had not been travelling at more than 12 miles an hour.

Bridge and Building Master W. Miller explained that no watchman had been posted at the tunnel because he had not considered that there was any danger of a cave-in.

Foreman Joe Peressini, who had been in charge of the gang which had been re-concreting the tunnel, stated that the work had been completed at 4.30 p.m. on the day of the accident. He considered that the job had been perfectly safe when they had left it.

The crew of the pusher engine (banker) testified that they considered that it would have been almost impossible for the engineer of the leading locomotive to have spotted any obstruction due to the curvature of the tunnel and escaping steam from the engine restricting the view.

After an hour's absence the inquest jury returned a verdict, 'That Palmer Dean came to his death on October 20th, about 11 p.m., through the engine of a freight train running into a cave-in in tunnel No. 21.45, about half a mile east of Palliser on the Canadian Pacific Railway. And it is the opinion of this jury that when a tunnel is being repaired a watchman should be stationed at the tunnel so that warning could be given to approaching trains.' [1031]

Westbound Utah Railway freight train extra No. 102 on Monday 28th August 1922 consisted of engine No. 102, 3s loaded car cars and a caboose, in charge of Conductor Nickerson and Engineer Johnson. At Hiawatha it received "31" train order No. 306, which read:

"Engine 102 run extra Hiawatha to Martin, not pass Jacobs until two thirty five 2.35 p.m. No. 606 engine 2 wait at Martin until three fifteen 3.15 p.m. for Extra 102 west."

The train was given a slow order and a clearance card, and departed from Hiawatha at 1.10 p.m.

Meanwhile Eastbound freight train No. 3, consisting of a caboose and seven empty coal cars, being propelled by engine No. 3, in charge of Conductor Burke and Engineer

---

[1030] *Manitoba Free Press*, Winnipeg, Manitoba, Canada, 22nd October 1921
[1031] *Okanagan Commoner*, Enderby, British Columbia, 27th October 1921

Ratliffe, had received "31" train order No. 309 at Martin, which read:

"Engine No. 3 run extra Martin to Jacobs. This order annulled three fifteen 3.15 p.m. No. 606 Eng. 2 run two 2 hours and thirty 30 minutes late, Martin to Jacobs. No. 612 Eng. 2 run two 2 hours thirty 30 minutes late Jacobs to Standardville."

This train left Martin at 2.45 p.m., entered the single line and collided head-on with Extra No. 102 in No. 1 Tunnel. The caboose of extra No. 3 was demolished and the end of the first coal car crushed in. Engine No. 102 was derailed and its front end damaged. After train No. 3 had been pulled clear the wreckage in the tunnel caught fire and locomotive No. 102 and the six cars behind it were consumed by the fire. Brakeman Roy Ranson was killed [1032]

Conductor Nickerson of Extra No. 102 told the Interstate Commerce Commission (ICC) inquiry that order No. 306 had been handed to him personally by Dispatcher Gutzler, whose office was at Hiawatha.

Engineer Johnson of Extra No. 102 stated that the headlight on his engine had not been lit as he had twisted the bulb and had forgotten to replace it.

Conductor Burke of Extra No. 3 said that he had been in the office at Martin when Operator Opperman had received the order and had read it over his shoulder. He told the operator to write his (Burke's) name on the train order while he made out the clearance card and signed the operator's name on it.

Operator Opperman testified that Conductor Burke had been in a hurry to get away, so he had signed the conductor's name and vice versa.

Dispatcher Gutzler said that he had also been doing clerical work when he wrote out train order No. 309 and had not referred to any of the previous orders which he had written except the last words of order No. 306 and had taken the time 3.15 for order No. 309 when he should have used 2.35. He had thus made the worst mistake that a dispatcher could make by issuing a "lap order," one giving both trains westbound No. 102 and eastbound No. 3 authority to be on the single line between Martin and Jacobs between 2.35 p.m. and 3.15 p.m., with neither train crew knowing of the other's existence. The dispatcher said that he had realised his mistake about a minute and a half after the order had been sent and had tried to contact Martin by both telephone and telegraph but had not been able to get through until after the collision. He said that since he had begun as dispatcher it had been the practice to use the single train order system, issuing orders giving trains rights over opposing trains without issuing them to the other train.

Trainmaster Thompson stated that he had been detailed to watch the work of the train dispatchers. He had checked the train dispatchers' books three times. He was aware that the single train order system was in use, thought it unsafe and intended to change it but had not got around to doing so.

As might be expected the ICC found much to criticise. In its conclusions it wrote, 'The investigation of this accident disclosed that the train order system in use is what is commonly called the "single order system." In this system each train is given an individual order and no attempt is made at issuing the order in the same words to all

[1032] *The Deseret News*, Salt Lake City, Utah, United States, 31st August 1922 trains affected. Early in the days of railway development this system was abandoned as unsafe on account of the errors likely to made on the train dispatcher and the possibility of misinterpretation on the part of those who were to execute them. In its stead was adopted the "double order system" by which the order for all trains affected is transmitted simultaneously and each train holds a duplicate of that held by the others. Not only was there an antiquated and unsafe system of issuing train orders on this railroad, but from the statement of Dispatcher Gutzler and an examination of the train dispatcher's order book it appears that it was the practice to issue train orders restricting the rights of trains without even furnishing a copy of such orders to the trains thereby restricted. Under such a practice the protection furnished train movements depends entirely upon the alertness, carefulness, and memory of the train dispatcher. In case of error on the part of the dispatcher as in this case, or of failure of communication, a condition would be created such is almost certain to result in disaster.

'Under the rules adopted by this company it was not permissible to use the single train order system, the prescribed forms must be used when applicable, orders must be given in the same words to all trains addressed, they should be sent to all addresses simultaneously, and when not so sent they must be sent to the superior train first, and complete must not be given to the inferior train until the order for the superior train has first been acknowledged by the operator receiving the same. Dispatcher Gutzler customarily disobeyed all of these rules. He used the single order system, he used a form of his own which was not prescribed by the rules, and which did not provide for any clearance time, the order was not transmitted simultaneously to all trains affected and in many instances the orders were not transmitted to the superior train at all. The only excuse offered by Dispatcher Gutzler was that it was the custom when he came there. As far as his other work was concerned his duties as train dispatcher were paramount and nothing whatever should have been allowed to interfere with them.'

The ICC continued that although Dispatcher Gutzler's mistake was the direct cause of the accident the underlying cause was the use of the single order system, for which the higher officials of the company were responsible. Trainmaster Thompson's attitude it described as amazing. He had been sent to the district to supervise the dispatchers, had checked the train dispatchers' books three times without even discovering such glaring faults as that restrictive orders were being issued without a copy being given to the trains so restricted, had seen that the single order system was in use, thought it unsafe, was not sure that it was authorised by the rules, yet had done nothing about it. The ICC considered that, 'Such negligence on the part of a supervising officer whose primary duty is to keep in touch with the operating details so as to insure safety of operation is inexcusable.'

The ICC also pointed to other practices not directly involved in the accident but which it said indicated loose operating methods and lack of supervision, such as the crew of train No. 102 not comparing watches, the engineer of that train not comparing his watch with the standard clock, indeed he did not know where the standard clock was located (it was at Hiawatha), the conductor and operator at Martin forging one another's signatures and so on. It remarked that, 'The train order system of train operation in its highest state of efficiency leaves much to be desired in the way of safety, but where such a system is operated as it is on this railroad, even a reasonable degree of safety is lacking.' [1033]

At 11.59 a.m. on Saturday 1st September 1923 Japan was struck by the Great Kantō Earthquake, a massive quake with a magnitude of 8.0 on the Richter scale. The cities of Tokyo and Yokohama were devastated, and it was believed that almost 143,000 people lost their lives. [1034] A number of railway accidents were a direct result of the earthquake and in particular it was widely reported that 600 people had died when the roof of a tunnel at Sasako collapsed. [1035] A Japanese site dealing with railway accidents through the years states however that the worst accident to occur as a result of the earthquake was one at Nebukawa (see Miscellaneous Accidents chapter) and makes no mention of a Sasako Tunnel disaster, casting doubt on the assertion that 600 people had died in such an accident.

On Wednesday 1st October 1924 train D 670 from Köln to Basel set off from Mainz Hbf, Germany but had only gone a short distance when its brakes came on and it ground to a halt in the tunnel between there and Mainz Sud, Germany. Its crew then attempted to determine, in the poor visibility and atmosphere of the smoke-filled tunnel, the reason for the undesired brake application and to release the brakes. Before this could be done the train was run into from behind by train P 682 from Mainz to Worms, due to depart from Mainz five minutes after the Basel train. Fourteen passengers died and many were hurt. The conditions in the tunnel made the rescue operation difficult and some of the fire brigade, medical staff and military involved had to wear gas masks.

Mainz was in the part of Germany under Franco-Belgian occupation at this time, and their administration placed the blame for the accident upon the crew of the Basel train, for failing to protect it in the rear. The facts of the matter were, however, somewhat different. Soon after the new administration had taken over, following a problem with the signalling, the signal-dependent block system through the tunnel had been suspended, and it was ruled to be no longer necessary for a preceding train to be reported back clear of the tunnel for another to be allowed to enter it. That an accident should occur were a train to stop unexpectedly in the tunnel was thus predictable. [1036] [1037]

On the afternoon of Tuesday 28th December 1926 a train from Concepción stopped in a tunnel near Lota, Chile, after colliding with an unspecified animal. Another train coming from Lota then ran into it. Oil lamps smashed in the accident set fire to the wreckage and the passengers were unable to escape. Their charred bodies were recovered the next day. Between 10 and 15 people were believed to have lost their lives. [1038]

A serious accident occurred in England on the Settle and Carlisle section of the former

---

[1033] Interstate Commerce Commission report
[1034] Wikipedia – 1923 Great Kantō Earthquake
[1035] *The Times*, London, England, 3rd September 1923
[1036] *Eisenbahn-Katastrophen in Deutschland* by Hans Joachim Ritzau, pg. 73 (in German)
[1037] Wikipedia – Eisenbahnunfall von Mainz
[1038] The Kingsport Times, Kingsport, Tennessee, United States, 29th December 1926

Midland Railway on Thursday 6th March 1930, when there was a collision in Waste Bank Tunnel, between Culgaith and Langwathby stations on what was by then the London Midland and Scottish Railway.

A ballast train from Carlisle had arrived at Culgaith on the up line at 8.47 a.m. in connection with work on the renewal of Waste Bank Tunnel, specifically to load up spent ballast from the down line. The train was duly crossed over onto the down line and thence propelled to the tunnel, Signalman Wileman at Culgaith having had it accepted by his colleague at Langwathby. Before leaving Culgaith, Flagman Taylor had explained to Signalman Wileman what was to be done and enquired as what down trains were due. The procedure was to be that when the down line had to be cleared for a through train the ballast train would proceed to Langwathby, cross over to the up line and return to Culgaith, thence to be propelled to the tunnel again. The next timetabled train was the 5.10 a.m. down Skipton to Carlisle mineral train, but this train was often overtaken at some point by the 8.05 a.m. down Hellifield to Carlisle stopping passenger train.

Flagman Taylor agreed with Signalman Wileman that the latter would shake the up distant signal as an indication that was time for the ballast train to clear the line. The flagman then proceed to protect the ballast train by placing three detonators on the line at each end of Culgaith Tunnel and stationed himself at the south end of Culgaith Tunnel with a red flag, in accordance with the regulations.

Guard Proudfoot of the ballast train should have posted an inner flagman with detonators and a red flag a quarter mile in the rear of the ballast train, but he failed to do so.

The down mineral train duly arrived at Culgaith at 9.58 a.m. and Signalman Wileman shunted it onto the up line out of the way of the passenger train. Flagman Taylor was close enough to see the shunting of the mineral train and realising that the passenger train must be close at hand he made his way to the ballast train in Waste Bank Tunnel to tell them it was time to proceed to Langwathby. As he did so he picked up the two lots of detonators that he had put down.

When Signalman Wileman was offered the passenger train at 10.07 a.m. he rang Langwathby and asked Signalman Kirkpatrick there if there was any sign of the ballast train. Told that there was not, he kept his home and starting signals and placed a collar on the lever of the starting signal. When the passenger train whistled he cleared the home signal, possibly a little prematurely. The train stopped at the station at 10.09 a.m., a small amount of station work was done, the coach doors were closed and, without looking at the starting signal, Guard Brogden gave his driver the "Right Away" signal. He then got into his van and began arranging the parcels for Langwathby as the train moved off.

As soon as the train had arrived at the station platform and cleared the level crossing adjoining his cabin Signalman Wileman opened the crossing gates for a vehicle and then was making entries in the train register when Fireman Armstrong of the mineral train, who had gone to the cabin to carry out Rule 55, alerted him that the passenger train was setting off against the starting signal. As the train passed the signal and continued towards Waste Bank Tunnel Signalman Wileman sent the "Train Running Away on Right Line" emergency signal to Langwathby and told Signalman Kirkpatrick on the phone what had happened. There was nothing more that he could do.

The passenger train consisted of 4-6-0 engine No. 5971, four coaches and a van, with

Driver John Bowser and Fireman John Stone on the engine. Driver Bowser was killed in the accident. Fireman Stone told the Ministry of Transport inquiry that it was he who had actually been driving the engine at the time. The fireman, who was quite badly hurt in the collision, stated that as far as he could recall the Culgaith down home signal had been clear when he first saw it. They had stopped at the station and when the guard gave the departure signal he had set off, assuming that the starting signal was lowered. His driver had done nothing to stop him, and he had not seen the ballast train before the passenger train ran into it while travelling at about 35 miles an hour, with the engine regulator open and its brakes unapplied.

Flagman Taylor, standing by the engine of the ballast train in the tunnel, saw the passenger train coming and had time to warn the crew of the ballast engine to jump clear, but in addition to Driver Bowser one passenger was fatally injured, three passengers were seriously injured and two were slightly hurt. Both engines were damaged, and the second coach of the passenger train and three ballast waggons were destroyed.

The inspecting officer, Colonel A.C. Trench, decided that there could be no doubt that the Culgaith down starting signal had been at danger. He did not criticise Driver Bowser for permitting his fireman to drive the engine, pointing out that this was the only way that a fireman could obtain driving experience, but he stressed that the driver remained fully responsible for working his engine, and that the deceased Driver Bowser was primarily responsible for the starting signal having been passed at danger, though the fireman was little less to blame. With the reverse curvature and smoke and steam lying about there was little chance of the crew of the passenger train seeing the obstruction in the tunnel.

In the circumstances Flagman Taylor was justified in returning to his train before being recalled but he was seriously at fault for picking up the protecting detonators in contravention of Rule 251(a). The flagman could only say that this was frequently done. The failure of Guard Proudfoot to put down detonators a quarter mile from the train had no practical bearing on the collision but called for adverse comment.

The arrangement between Signalman Wileman and Flagman Taylor of the former shaking the up distant signal as an indication that it was time for the ballast train to move was, the inspecting officer considered, 'most undesirable' and 'much to be deprecated.' It might mislead the driver of an up train and as the line between Culgaith and New Biggin was worked by rotary interlocking block would entail the Culgaith signalman getting a release for an imaginary train in order to clear his signals. A hand signal could have been employed but in any case Guard Proudfoot should have seen that the ballast train was taken forward to Langwathby before the booked time of the passenger train.

The evidence as to whether Signalman Wileman had complied fully with Rule 40 and not cleared the home signal too early given that the section ahead was blocked was contradictory. Fireman Stone recalled that it had been clear when he first saw it, on the other hand Driver Blaylock of the mineral train was confident that the passenger train had been checked.

It was very regrettable that Guard Brogden had not observed the starting signal but strictly speaking the responsibility for its observance did not rest with him.

In conclusion Colonel Trench remarked that the accident had resulted in three separate failures:

(a) The failure of Driver Bowser and Fireman Stone to observe the Culgaith starting signal at danger.

(b) The failure of Flagman Taylor to leave the detonators in place to protect the ballast train.

(c) The failure of Guard Proudfoot to move the ballast train to Langwathby before the passenger train was due.

Finally the inspector pointed to the value of the shock-absorbing buffers fitted to the leading and third carriages of the passenger train in minimising the effects of the collision.[1039]

[1039] Ministry of Transport report

CHAPTER 12

LEVEL CROSSING COLLISIONS

Wherever roads and railways intersect they must do so either by means of a bridge or on the level. Bridges have the advantage of allowing road and rail traffic to operate entirely without interference from one another, but they are much more expensive to provide, especially in flat countryside, due to the need to provide extensive earthworks for approach embankments or cuttings as well as the cost of providing the span itself. Where underbridges are provided there is often also the need to build retaining walls and there can be problems with drainage. [1040] The provision of level crossings, especially if, as in many parts of the world, they are unstaffed, is the much cheaper alternative, but where they exist, they introduce an additional source of danger to both road and rail users. Robert B Shaw in his outstanding book on American railway accidents, *A History of Railroad Accidents, Safety Precautions and Operating Practices*, writes that he has excluded ordinary grade (i.e. level) crossing collisions from consideration on the grounds that they are not proper train accidents, inasmuch trains usually suffer much less harm from such accidents than the road vehicle involved. This series of volumes, on the other hand, will attempt to include, amongst others, all such accidents which resulted in five or more fatalities, even if none of those who died were train passengers or railwaymen. The First World War provided a great stimulus for the development of the internal combustion engine and the number of motor vehicles in use, and consequently of level crossing collisions involving them, greatly increased following the war.

The westbound *Twentieth Century Limited* express of the New York Central Railroad gave a driver and five small schoolchildren what was described as an awful fright on the morning of Friday 28th January 1916. The children had been riding to school in the Jamestown school waggon, in charge of a man names Jas Holderman, who also owned the waggon and the team drawing it. The driver and all the children claimed that at a level crossing near Osceola, Indiana, United States, no warning signal had been given by the engineer of the *Limited* until the train had been almost upon them. The train struck the

[1040] *Backtrack*, Vol 24, pg. 88.

two horses pulling the waggon, valued at $300, carrying one of the animals a quarter of a mile and killing both. The waggon itself was not struck and the driver and his passengers thus escaped. [1041]

A mistake by a crossing keeper led to a collision with a fire engine on Talbot Street level crossing, London, Ontario, Canada, on Sunday 18th October 1916. The fire truck and a car carrying Fire Chief Aiken and his driver were responding to what proved to be a false alarm from South London. When the two vehicles reached the crossing in Talbot Street they found the barriers lowered and a westbound freight train blocking the road. The fire truck and the light roadster carrying the fire chief stopped alongside one another and waited two or three minutes. The westbound Grand Trunk Railway freight then moved off the crossing and Watchman John Primo, seeing the fire vehicles waiting impatiently, raised the gates. The car carrying the fire chief accelerated over the crossing and just escaped being struck by an approaching eastbound freight train. The heavier fire truck was slower to move, it was struck squarely by the engine of the freight train and carried 100 yards down the line. Fire Captain John Case was killed, three other firemen were seriously hurt and two slightly injured. The crossing watchman was arrested. He said that in his anxiety to let the fire vehicles proceed he had failed to notice the oncoming freight train. [1042]

Two people died and 11 were reportedly fatally injured when a train collided with an omnibus full of passengers at Minho, Portugal, on Tuesday 4th July 1916. [1043]

A party of six people were on their way to church in Jacksonville, Illinois, United States, early on the morning of Sunday 16th July 1916. Mr William Groves was driving a Case car which he had only purchased a fortnight before. Beside him in the front passenger seat was Miss Margaret Allen and on the back seat were Mrs Mary McSherry, Mrs Anna Bernard with her three-year-old son Allen on her lap and Mr Groves' 19-year-old daughter Ellen. At about 7.18 a.m. the car reached the ungated level crossing at Spaulding. Mr Groves asked Miss Allen if she could see or hear a train. She replied that she could not. Mr Groves reduced speed and listened himself but could not hear anything either. Visibility at the crossing was obscured by vegetation growing alongside the road.

Mr Groves set off up the slope to the crossing, completely unaware of the proximity of Wabash Railroad passenger train No. 3. The train struck the car, threw it against a cattle fence which was demolished, and the car was pushed 30 feet. Four of the passengers in the car were killed outright and Mrs Bernard was fatally injured. Mr Groves survived.

The inquest on the victims of the accident was quickly held and its jury found the Wabash Railroad negligent for leaving the crossing so obscured by weeds and vegetation that road users could not see approaching trains and the engineers of trains could not see

---

[1041] *Goshen Daily Democrat*, Goshen, Indiana, United States, 28th January 1916
[1042] *The Brandon Weekly Sun*, Brandon, Manitoba, Canada, 29th June 1916
[1043] *The Times*, London, England, 6th July 1916

approaching automobiles. [1044]

Charles H. Buttenbender, real estate agent, builder and prominent citizen of Elyria, Ohio, United States, was one of five people killed in a level crossing collision on the Chestnut Street Crossing in that place on the evening of Wednesday 16th August 1916. The other victims were his wife, Mrs Adaline Conaway (his sister), Mrs Mary Emmert, and Mrs Wise. The party had been returning from the Municipal picnic at Cascade Park, where Mr Battenbender had captained a ball team which had played against the banker's team.

Mr Battenbender had apparently caught sight of the eastbound New York Central Railroad passenger train at the last moment and had made a desperate and unsuccessful attempt to get over the crossing in front of it. The crossing was protected by gates (lifting barriers) under the control of a gateman. The latter, Thomas Meagher, admitted that he had not lowered the gates and pleaded that he had not heard the train coming, an omission that he reportedly blamed on his deafness, nor had he seen it. He also complained that the gates had not been working properly all day and had been slow to lower. There was apparently no bell or indicator to warn crossing keepers of the approach of trains. Mr Meagher was arrested pending an investigation.

Following the accident city officials urged the railway company to speed up the elimination of the six most dangerous level crossings in the city: at Cedar Street, Cleveland Street, Mill Street, Washington Avenue and West Avenue as well as at Chestnut Street. The work was scheduled to take three years to complete and to cost $2,000,000. [1045]

A collision on Fairview Road Crossing, a quarter mile from Quakertown station, Pennsylvania, United States, wiped out all but one of two families from Bethlehem, Pennsylvania, on the afternoon of Sunday 17th September 1916. Mr William Hunsziker and his wife, Mr Andrew Kneff and his wife, and the Kneff's two boys, Charles, aged six, and George, aged 12 all lost their lives in the accident. The two men were foremen at the Bethlehem steelworks, and the two families had set off from Bethlehem at noon on a pleasure trip, leaving two-year-old Gertrude Hunsziker in the care of neighbours.

A freight train had just gone into a siding near the crossing, and it was supposed that Mr Kneff, who had been driving, had assumed that there were no other trains in the vicinity. The car was struck by a Philadelphia and Reading Railroad passenger train from Niagara Falls to Philadelphia. Mrs Kneff and her son Charles was not killed outright and a graduate nurse, Elizabeth Klauder, who had been riding on the special, made a heroic attempt to save their lives, but in vain.

The crossing concerned was provided with a warning sign, but nothing else. [1046]

Engineer Ernest Trout, Fireman Walter Knox and Conductor Thomas Brunner all gave evidence at the inquest on the victims of the accident, the engineer telling the jury

---

[1044] *Jacksonville Daily Journal*, Jacksonville, Illinois, United States, 18th July 1916

[1045] *Elyria Evening Telegraph*, Elyria, Ohio, United States, 17th August 1916

[1046] *The Philadelphia Inquirer*, Philadelphia, Pennsylvania, United States, 20th and 21st September 1916 that his train had been running at 40 miles an hour and he had not seen the automobile until his train was 35 feet from the crossing. The jury specifically exonerated the engineer but severely censured the railroad company for maintaining such a dangerous crossing and recommended its early elimination. [1047]

On the evening of Sunday 1st October 1916 a street car filled with over 90 passengers was struck by a switch engine pulling two freight cars on Forest Avenue Crossing, in Detroit, Michigan, United States. The car was struck almost at its centre and panic-stricken passengers jumped from the front and rear doors or climbed out through windows, only for many of them to be killed by falling under the wheels of the still-moving freight cars. Ambulances from every hospital in the city rushed to the scene. Many of the passengers had been returning from the theatre. Ten people died and about 30 were injured, several it was thought fatally. The accident was the third crossing accident within a year and the second at the Forest Avenue Crossing.

After the accident witnesses stated that the gates had not been lowered and the gateman, 65-year-old Valentine Scabinski, was arrested on a tentative charge of involuntary manslaughter, together with Alex Cliske, the conductor of the street car, while Conductor Robert O'Keefe of the freight train and Motorman William Carter of the street car were held as witnesses. [1048] There were allegations that there had been no lights on the train and also questions as to why a derailing device at the crossing had not operated. No less than five investigations were opened into the accident by the State Railway Commission, the coroner's office, the prosecuting attorney, the Grand Trunk Railway, and the Detroit United Railways. [1049]

Engineer Gibson and Conductor O'Keefe alleged that they had the right of way and that there had been lights displayed on the ends of the freight car. The conductor added that he had been standing on top of the first freight car swinging a lighted lantern. He did not see the street car until the last moment as Forest Avenue is closely built up right up to the tracks. The switching train was reportedly moving at about 20 miles an hour when it reached the crossing.

Three weeks after the accident the coroner's jury found the Grand Trunk Railway negligent for employing an incompetent gatekeeper. Evidence was given that the crossing gates had not been lowered (they had been undamaged after the collision) and the street car employees alleged that Scabinski, a cripple, had been asleep at his post. [1050]

Five residents of Oakland, California, United States, riding in one of three cars on a Sunday picnic trip, were killed when the *Comet* express of the Oakland, Antioch and Eastern Railway struck their vehicle on Moraga Crossing, California on 27th May 1917. Those who lost their lives were A.J. Hawkins, aged 30, a shipyard worker; Miss Gladys Mortimer, aged 20, a stenographer; Mrs F.C. Cannon, 65, her grandmother; Miss Eva Walker, 16, and Mrs Richmond, 24.

[1047] Ibid, 23rd September 1916

[1048] *The Yale Expositor*, Yale, Michigan, United States, 5th October 1916

[1049] *The Topeka Daily State Journal*, Topeka, Kansas, United States, 2nd October 1916

[1050] *The Fort Wayne Sentinel*, Fort Wayne, Indiana, United States, 20th October 1916

The automobile was thrown against a telegraph pole with so much force that it snapped the pole off, while the three-car train was derailed and bumped along on the sleepers for 500 feet before coming to rest. There were 150 passengers on the train and at least some of them reportedly panicked; women fainting and men attempting to fight their way out of the train, while Conductor T.G. Deyo attempted to calm them.

A warehouse 300 feet down the track obscured the view of the train from a driver approaching the crossing, but the people in the following two vehicles could see the danger and tried to warn Mr Hawkins, who was driving. He apparently then caught sight of the train, travelling at perhaps 45 miles an hour, but lost his head and instead of pulling up, tried to beat the train over the crossing. Motorman Al Cooke did all he could to stop the train, but its momentum was too great, and it was too close to the crossing. Mr Richmond and his little daughter, riding in the second car, saw their wife and mother killed in front of them. [1051]

Terre Haute, Indiana, United States, was the scene of a tragedy early on Thursday 21st June 1917. Mr Timothy Wrenn, aged 44, had taken his two daughters, Nellie, aged 9, and May, aged 7, out for a ride together with two nieces, Alma Vaughn, aged 8 and Frances Waggoner, aged 7. As the party were returning home they had to pass over the Big Four (Cleveland Cincinnati Chicago and St Louis Railroad) at Fort Harrison Crossing. According to witnesses Mr Wrenn was distracted by the children and failed to notice the fast-moving train close at hand at a crossing which was not regarded as being dangerous as there was a good view in each direction. The car was carried 20 yards by the locomotive. Mr Wrenn and three of the children were killed outright and the other child died a few minutes later in the arms of a woman who had seen the accident and had hurried to the scene. [1052]

A car being driven by Mr Charles Wiley and containing in addition to himself his wife, their three-year-old daughter, Mr Samuel Kirkwood and his wife and Mrs Rebecca Anderson, was struck at about 8.30 p.m. on Friday 22nd June 1917 by the double-headed southbound Pittsburgh express of the Pennsylvania Railroad. The accident happened on a level crossing near Whitehall station, Maryland, United States. Having enjoyed supper at the Wiley home the party had gone for a drive before heading for the Underwood home. All were killed outright or fatally injured. There were no witnesses to the accident, so it was unclear if Mr Wiley had tried to beat the train over the crossing or if his vehicle had stalled. [1053]

On Tuesday 24th July 1917 Mr and Mrs Clarence Joseph were on their honeymoon, having been married ten days earlier. They embarked on a trip to Camp Perry in a Buick Six limousine, the property of Mr Fred Sherman, a travelling man. Mr Sherman was not in the car, being in Michigan on business, but his wife was, as were Mrs Frank Joseph, Clarence's mother; Miss Esther Joseph, Clarence's sister; Herman Hohls, Esther's fiancé,

[1051] *Oakland Tribune*, Oakland, California, United States, 28th May 1917
[1052] *The Evening News*, Connersville, Indiana, United States, 22nd June 1917
[1053] *The Washington Post*, Washington, D.C., United States, 23rd June 1917

and Mrs Mary Notaker, making a party of seven, all of them from Pemberville, Ohio, United States. Having enjoyed a picnic at Camp Perry the party were heading for Catawba Island, and the journey involved crossing the New York Central Railroad Cleveland – Toledo main line at Port Clinton, Ohio. The crossing was reached just after 3 p.m. Road and railway ran alongside one another for some way before the road swung right to cross the line, and it was unlikely that any of the occupants of the car had noticed eastbound fast mail train No. 4 coming up behind them. As the car got onto the crossing it stalled; and eyewitnesses on the nearby beach could see the driver (either Clarence Joseph or Herman Hohls) desperately working at the controls. Then there was what was described as a terrific crash and the car was carried a quarter mile before the train could pull up.

Automobiles and physicians were soon on the scene. Five of those in the car were dead and Herman Hohls was fatally injured and died an hour later. Esther Joseph was seriously injured but it was thought that she would recover. The finding of Mr Hohl's military registration card and Mrs Notaker's pocket book led to identification of the victims and word was sent to Pemberville. Frank Joseph, a wealthy retired farmer, went from there to the scene and was able to identify all of those killed, who included his wife, his son, and his daughter-in-law.

The crossing was described as a dangerous one. Previously, after a request from local authorities, a watchman had been stationed there, but he was later withdrawn. [1054]

A forerunner of accidents to come was a collision between a southbound Central of New Jersey Railroad passenger train and motor truck of the Gulf Refining Company just north of Galilee station, New Jersey, United States, at 9 p.m. on Monday 30th July 1917. It appeared that Russell Ferry, the driver of the lorry, had assumed that the train was going to stop at Galilee station and that he therefore had time to cross the line in front of it. Instead the train was not due to stop until Monmouth Beach station, a half mile further on. The locomotive struck the truck and an explosion and fire followed. Driver Ferry was killed outright, and Engineer William Miles and Fireman James Fox were both burned, the engineer fatally so. Both Galilee station and the two coaches of the train caught fire, though the 40 train passengers escaped injury. [1055] Baggagemaster H.E. Harrison was just about able to save the mail from the coaches at considerable risk to himself. The Byton fire company eventually extinguished the blaze. [1056]

The partial disability of a taxi driver was put forward as a possible explanation for a level crossing accident at Rockford Crossing, just north of Seymour, Indiana, United States, on the evening of Sunday 29th July 1917. A party of six young people had been spending the day on the White River and had booked a taxi driven by the one-armed Lon Prewitt to take them home. The taxi approached the level crossing with the Pennsylvania Railroad at the same time as a fast southbound train. There were no eyewitnesses to the collision, but it was supposed that the taxi driver, perhaps because of his having only one arm, was unable to pull up when he realised that he was not going to be able to get across

---

[1054] *Sandusky Star-Journal*, Sandusky, Ohio, United States, 25th July 1917
[1055] *Trenton Evening Times*, Trenton, New Jersey, 31st July 1917
[1056] *Red Bank Register*, Red Bank, New Jersey, United States, 1st August 1917

in front of the train. Instead the taxi struck the engine at the junction of its cab and tender. Mr Prewitt, Julius Orgstadt, teacher, aged 23; William Steinker, cashier, aged 26; and Clara Windhorst, aged 22, were killed more or less outright and Edna Zimmerman, clerk, aged 21, died soon afterwards. Two others, Ola Droege, cashier, aged 22 and Martin Steinwedel, businessman, aged 22, were seriously injured but it was expected that they would recover.

Engineer George Stofer told the inquest on the victims of seeing a light near the crossing but not realising that it was a car going to cross until the last moment when he had braked. A verdict of "Accidental Death" was returned. [1057]

A passenger train on the Tennessee Central Railroad struck a car on a level crossing at Crossville, Tennessee, United States, on Saturday 4th August 1917. Merchant Jack Highfield, who was driving, Mrs Rosa Cox, Miss Mabel Dayton, Mrs Hedgecough and Mrs Vanhoy were killed outright, and Miss Dorothy Dayton was so badly hurt that she died shortly afterwards. Miss Ethel Hughes was thought to be fatally hurt. No details were given as to how the accident happened. [1058]

Shortly after midnight on Monday 20th August 1917 a New York New Haven and Hartford Railroad light engine, on its way from New Haven to New London, Connecticut, collided with a car on Ferry Crossing, Old Saybrook, Connecticut, United States. There were no witnesses to the accident but Towerman Edmund Bird, stationed nearby, heard the crash and alerted Watchman Walter Buckley, and the two of them found the wreckage of the car scattered along the track. There had been nine people in it, and eight of them were killed outright or fatally injured.

Those who lost their lives were Mrs Asterino and her children aged five, three and four months; Robert Rohloff, Mabel Wagner, and Gustave Weyer and his wife. The Weyers' daughter, Louise, aged nine, was the only survivor. The party were returned from Niantic, where they had been visiting Louise's brother, William, who was a member of the field hospital there. Mrs Asterino was the sister of Louise and William Weyer and Miss Wagner was described as a 'close personal friend' of William Weyer.

Engineer O'Dea, Fireman Horace Andrews (named as a foreman in the newspaper report but presumably the fireman) and Flagman J.E. Champlin on the engine stated afterwards that they had seen nothing of the automobile. The gong at the crossing was still sounding 15 minutes after the accident. Gates were provided at the crossing but they the crossing was not manned after 9 p.m. Other accidents had happened previously there. [1059]

In contrast to the frequent crossing accidents involving Sunday afternoon picnickers or other pleasure seekers a collision on Short Lane Crossing, just short of Aberdeen, Maryland, United States, involved a group of corn pickers (or tomato pickers according to another source) on their way to work in a waggon drawn by a pair of horses. As the

---

[1057] *Brownstown Banner*, Brownstown, Indiana, United States, 1st August 1917

[1058] *Alexandria Gazette*, Alexandria, Virginia, United States, 6th August 1917

[1059] *Bridgeport Evening Farmer*, Bridgeport, Connecticut, United States, 21st August 1917 waggon neared the crossing on Thursday 30th August 1917 an express train thundered past. Thinking that it must now be safe to cross the driver whipped up the horses and drove onto the crossing just as a light engine, No. 5264, came along on the other track. The waggon was smashed, and the 14 riders were hurled in all directions. Seven men were killed outright: Joseph Berry, Lee Cassidy, Frank Fanning, Albert Horny, R.S. Reckson, Barry Shane and Thomas Wrightson. Three of the others; W.S. Blake, Sheridan Morris and F.C. Smeltzer, were seriously injured; the other four escaped with slight injuries. [1060]

Ironically, the accident would not have occurred had the waggon arrived only four minutes later, for it happened at 5.56 a.m. and the crossing watchman came on duty at 6 a.m. No protection was provided at the crossing outside of the hours when the watchman was on duty and the view of the track was obscured by a cornfield. [1061] [1062]

The westbound *Black Diamond Express* of the Lehigh Valley Railroad, en route from New York to Buffalo, collided with a car which was on its way from Wayland to Rochester on the early evening of Sunday 9th September 1917. Mr Joseph Geisler, grocer, aged 56; his wife Amelia; their daughter, Teresa, aged 18; their son Edwin, aged 13, and Mrs Florence Heldt, aged 26, a married daughter, were all killed outright when the train hit the car while it was in the middle of the track. Another son, Joseph Geisler Junior, aged 16, survived only long enough to be taken to Rochester Hospital. The accident occurred near Rush, New York, United States. The crossing was unprotected but there was a fairly good view along the line in both directions. [1063]

At 08:30 on Thursday 20th September 1917 a bus travelling from Uzès to Nimes was struck by train No. 11.019 going from Martinet to Tarascon on a level crossing three kilometres from Uzès, France The bus was wrecked, eight bus passengers were killed and another ten were seriously injured. [1064]

Seven residents of Detroit, Michigan, United States, lost their lives on a level crossing at Hartsig, 11 miles north of Detroit, on the afternoon of Sunday 14th October 1917. The Pohl and Harvey families had set off from Detroit in the early afternoon to purchase poultry and were returning to the city in a car driven by Mr Joseph Pohl. Charles Van Dam, a farm lad who witnessed the accident, stated that the occupants of an automobile which had gone over the crossing in front of the Pohl car had attempted to warn Mr Pohl of an approaching Michigan Central Railroad passenger train but Mr Pohl had continued onto the crossing at a speed of 20 miles an hour, right into the path of the flyer. Supervisor Hartsig, who lived close to the crossing, said that the side curtains of the car were drawn. He rushed to the spot after the accident and finding that a woman still showed signs of life returned to his house for some water. By the time that he had got back to the crossing she had died.

1060 *Evening Illustrated Ledger*, Philadelphia, Pennsylvania, United States, 30th August 1917
1061 *Gettysburg Times*, Gettysburg, Pennsylvania, United States, 31st August 1917
1062 *Cumberland Evening Times*, Cumberland, Maryland United States, 30th August 1917
1063 *Wellsville Daily Reporter*, Wellsville, New York, United States, 10th September 1917
1064 *L' Œuvre*, Paris, France, 21st September 1917

Killed in the accident were Mr and Mrs Pohl; their children Helen, 5, and Ralph 3; Mrs Harvey and her children Thomas, 3 and Mary, 1.

There was a clear view of trains from either direction at the crossing, but it appears to have been a dangerous one, as only three days earlier a train had struck a lorry at the same spot, seriously injuring two men in the truck. [1065]

A school bus featured in a tragic accident near Barnum, Minnesota, United States, on the evening of Friday 1st February 1918. Seven children lost their lives in the collision and two cars of the southbound Northern Pacific Railroad train were derailed, but no train passengers were hurt. The children who died were William Fogarty, Alpha Harrington, Mary Snooks, Homer Staller, and three brothers, Charley, John, and Karl Kalvo. All the other children and the driver, Carl Mueller, were injured, five of the children seriously. After the accident, the train passengers alighted and helped to care for the injured and relief trains soon arrived with doctors and nurses from Carlton and Willow River.

According to the police authorities the bus should have been flagged over the crossing or its driver have ensured in some other manner that it was safe to cross but witnesses on the evening in question said the driver had failed to do so. The bus itself was described as a box-like affair without windows. [1066] [1067]

The day after the accident it was stated that Marion Fogarty, the most serious of the injured, was still unconscious and might well die. [1068]

Only two days earlier, on 31st January 1918, a Great Northern Railway train had struck a school bus at Nelson, Minnesota. Two children were reportedly killed and 16 injured in this accident, including the daughter of Swan Nelson, the driver of the bus, who subsequently suffered a nervous breakdown. [1069]

At 7.15 p.m. on Tuesday 5th March 1918 a tram and trailer en route from Barcelona to San Andrés de Palomar was struck by a goods train on a level crossing with the Norte line in Barcelona, Spain. The tram driver, Seville de Aipellido, and four tram passengers were killed and 24 were injured. What exactly happened was not immediately clear, but it was initially alleged that the crossing keeper had opened the gates after a train had passed, only to close them immediately upon the approach of a goods train on the other line, trapping the tram on the crossing. The crossing keeper, José Regolés, and the conductor of the tram were arrested. It would seem that the crossing keeper was not, however, to blame, for he was released without charge the following day. [1070]

A large touring car carrying the members of three families was struck by a Wabash Railroad train on Winter Street Crossing, Fort Wayne, Indiana, United States on the

[1065] *Pinckley Dispatch*, Pinckley, Michigan, United States, 18th October 1917
[1066] *The Ogden Examiner*, Ogden, Utah, United States, 2nd February 1918
[1067] *The Ogden Standard*, Ogden, Utah, United States, 2nd February 1918
[1068] *The Winona Republican-Herald*, Winona, Minnesota, United States, 2nd February 1918
[1069] Ibid, 1st February 1918
[1070] *La Vanguardia*, Barcelona, Spain, 6th and 7th March 1918

evening of Sunday 7th April 1918. Henry Stegman, a dairyman and the driver of the car, Mrs Stegman, Rudolph Alexander, Mrs Louise Brand and two children, Otto, and Mary Elizabeth Fatzer, aged 8 and 4 respectively, were all killed outright. Otto Fatzer, the father of the two children, was possibly fatally injured and his wife, daughter Ida, aged 3 and Albert Brand, son of Mrs Brand, were less badly hurt.

The accident occurred at 7.53 p.m. Visibility at the crossing was severely limited by buildings and box cars and the crossing was stated to be a constant menace to incautious drivers and pedestrians. During the daytime it was manned by a watchman, but he went off duty at 7 p.m. Gates were provided at the crossing, and it was believed that Mr Stegman, who knew that there was watchman on duty during the day, forgot that he was not present after 7 p.m., and seeing the gates raised had assumed that all was clear and driven onto the crossing without looking. The top was up on the car and its tonneau was drawn but the front of the vehicle was open. Otto Fatzer saw the headlight of the train at the last moment and shouted a warning a moment before the impact. A witness confirmed that Engineer Charles Christy had whistled and said that had expected the car to pull up but it just kept going.

A local newspaper wrote, 'A bachelor brother of Fatzer's was home waiting for his brother and his family to come home for the evening meal. "I had begun to wonder what the matter was and had just put on my coat to go out and see what had become of everybody when I heard people coming to the house," he said. "They told me what had happened. I had heard the crash but didn't pay any attention to it at the time. You know the thing happened just a little ways from our house."' [1071]

At about 3.15 p.m. on Thursday 18th April 1918 a Chicago and North Western Railway passenger train collided with a car containing six men who were on their home to Milwaukee on a level crossing at Carrollville, Wisconsin, United States. The men all worked at the Newport Chemical Company in Carrollville. Five of them were killed. They were Emmett Doddy, John Crock, George McCoy, Louis Urban and Frank Worm. Charles Schmidt, the driver of the car and only the survivor, was injured. He said that the car had stalled on the track and the train had been too close for anyone to get out of the vehicle. [1072]

Chelten Hills station, Pennsylvania, United States, was the scene of a bad accident on Monday 3rd June 1918. The level crossing concerned, with the Philadelphia and Reading Railroad, was situated at the bottom of steep hills on either side and with the view obstructed by trees, shrubs, and walls. Many accidents had happened at the crossing, and it was provided with gates and a watchman. At the time of the accident the gates had been lowered for a southbound express, running at 40 miles an hour towards Philadelphia. An accident nevertheless occurred when a car occupied by six women came down the hill and failing to stop crashed through the lowered crossing barrier and directly into the path of the express. Five of the women were killed outright or died shortly afterwards. They were Mrs Margaret Ewell, Mrs Longaker, Mrs Mahoney, Mrs Phillips, and Mrs Ryan. The sixth woman, Miss Margaret Taylor, had, it was said, not a chance of surviving.

[1071] *The Fort Wayne News and Sentinel*, Fort Wayne, Indiana, United States, 8th April 1918

[1072] *The Racine Journal-News*, Racine, Wisconsin, United States, 19th April 1918

It was not clear whether some problem with the car, such as a failure of its brakes, had prevented its driver from stopping or whether it was just being driven too fast. [1073]

In a typical level crossing accident scenario a car carrying two families on their way to a picnic was involved in a collision four miles from Waterloo, Iowa, United States, on Sunday 28th July 1918. Those who lost their lives were Mr and Mrs Milton Brown, Irene Brown and Clifford and Edith Bartlett. Three other members of the Bartlett family were riding in another car which had been just ahead of the one involved in the collision. The train involved was an interurban car of the Waterloo Cedar Rapids and Northern Railway travelling at 35 miles an hour. The motorman's view of the crossing was obscured by trees. He said that the driver of the car did not appear to see the train and had "killed his engine" on the crossing. The car was carried 200 yards along the track. Its occupants were not thrown out but were badly mangled. A sixth occupant of the car survived with injuries. [1074]

Alleged negligence on the part of the conductor and motorman of a street car led to an accident near 63rd Street and Archer Avenue in Argo, Illinois, United States on Sunday 4th August 1918. The two men were supposed to have seen an approaching freight train but attempted to cross the line in front of it anyway. They got halfway across when the street car stalled. The motorman and conductor were said to have jumped off and escaped, leaving the passengers to their fate. The Belt Line of Chicago Railroad train hit the trolley car full on, throwing it 20 feet from the track and overturning it. Five passengers lost their lives, only one of whom, the baby daughter of Mrs Grann, of South Chicago, could be immediately identified. Fifteen other trolley car passengers were injured. The conductor and motorman were arrested, together with the engineer of the train. [1075]

The Rev. Solomon Duma, pastor of the Romanian Greek Orthodox Church, had a number of visitors at his home in Canton, Ohio, United States, in the early evening of Sunday 17th November 1918 when he received a telephone call to say that a member of his congregation was dying and could he visit them. Grocer Mike Musat, offered to take the pastor there in his car, an eight-seater Buick, and the pastor's wife and remainder of the visitors went along for the ride.

The route taken by the car led over the Eighth Street level crossing. As Mr Muscat, who was driving, reached the crossing a westbound train consisting of a locomotive and a passenger car was approaching. He waited for this train to pass but then set off over the crossing with the double track line without any thought of a second train and the car was struck by an eastbound passenger working. Mr Musat, Mrs Duma, 12-year-old John Gherman, the Duma's adopted son; Mrs Annie Crisan, her son John Junior, aged 9, and Matthew Cozma were all killed, and John Crisan Senior was probably fatally injured. A daughter of the Crisan's escaped with a broken leg and the pastor and 12-year-old

---

[1073] *The Philadelphia Inquirer*, Philadelphia, Pennsylvania, United States, 4th June 1918
[1074] *The Omaha Daily Bee*, Omaha, Nebraska, United States, 29th July 1918
[1075] *Centralia Sentinel*, Centralia, Illinois, United States, 5th August 1918

Cornella Crisan were unhurt. None of the occupants of the car had apparently seen the train. A warning bell at the crossing was still ringing when a relief train arrived at the scene.

Westbound Grand Trunk Western Railroad mail train No. 9 collided with a school waggon just east of the town of Kingsbury, Indiana, United States, some time after 4 p.m. on the afternoon of Friday 18th April 1919. Six children were killed and seven injured, some seriously. The victims who lost their lives were Winifred Condon, 14; Evelyn Harris, 12; Hazel Kregel, 15; Inez Mitchell, 14; William Smitz, 6 and Irene Wildt, 13. The pupils had attended the consolidated school at Kingsbury and were being taken home after their lessons by Mrs Walter Johnson, whose husband usually drove the waggon. The waggon was squarely across the track when the train was first spotted and was splintered by the train. Mrs Johnson and her 5-year-old son escaped unhurt, as did the horses; her other son was badly hurt. Another pupil, William Condon, 12, heard the engine whistle, jumped, and saved himself.

Mrs Johnson said afterwards, "I stood up three times on approaching the crossing to look for a train but did not see it. I did not hear any whistle blown and was on the track before I noticed our danger. I whipped up the horses in an effort to escape, but it was too late."

An overbridge carried the Wabash Railroad across the road just short of the crossing and interfered with the view.

Conductor H.G. Walworth said that the train, due in Kingsbury at 4.5 p.m. It was fifteen minutes late and making up time. [1076]

The First World War had of course ended by May 1919 but not all the foreign Allied troops had yet gone home from France and at 22:00 on Thursday 1st May 1919 a lorry carrying American soldiers was struck by a Paris – Toulouse train on a level crossing near La Ferté-Saint-Aubin. Three of the men were killed and eight were seriously injured. [1077] [1078]

Six people were killed instantly when a Chicago Burlington and Quincy Railroad train crashed into their car on Hiawatha Avenue Crossing, Berwyn, Illinois, United States, on Monday 9th June 1919. The victims were Mrs William Wigand, of Berwyn; her sons Willard, aged 6, and Raymond, aged 3, her daughter Mildred, aged 3 months; her brother-in-law, William Steinberg, of Gary, Indiana; and his brother-in-law, William Eick, of Hadley, Michigan. [1079]

The Sunday tragedies in the United States continued Sunday 29th June 1919 with a level crossing collision a mile west of Oxford, Nebraska. Mr Fred Flohr, the driver of the

---

[1076] *The Rock Island* Argus, Rock Island, 19th April 1919

[1077] *La Croix*, Paris, France, 4th May 1919

[1078] *La Dèpêche*, Paris, France, 3rd May 1919

[1079] *Wanatah Mirror*, Wanatah, Indiana, United States, 12th June 1919 car, who was seriously injured, stated afterwards that he had been unaware of the approaching Chicago Burlington and Quincy Railroad train until he was on the crossing. He had then tried to accelerate out of the way but instead had stalled the vehicle. His wife, their four children, Dorothy, 5; Velma, 4; Frances, 3, and Corraine, 1, were all killed, together with Mildred Ferguson, aged 12. An eighth occupant of the car, Dorothy Glenn, was seriously injured. The engineer of the train stated that he had been travelling at 30 miles an hour, had whistled for the crossing, and had not seen the car until too late to stop. [1080]

Also on Sunday 29th June 1919 at Flint, Michigan, United States a car containing the Bronson family was struck by an interurban train, killing Mr and Mrs Worth Bronson, Frank Bronson, Mr Worth Bronsons's father, Frankie, aged 4 and Lawrence, aged 5 months, the Bronson's children. Mr Worth Bronson apparently turned his car directly into the path of the interurban, which struck it and carried it two blocks. [1081] [1082]

Finally on Sunday 29th June 1919 seven persons were reportedly killed and two seriously injured in a level crossing collision with an Illinois Central Railroad train near Morgantown, Indiana, United States. [1083]

On about Tuesday 1st July 1919, a collision occurred on Rideview Park Crossing, several miles east of Derry, Pennsylvania, United States. A car containing Otto von Wege, head of the Eller Lumber and Mill Company, of Pittsburgh, and his family, was struck by the westbound *Metropolitan Express* of the Pennsylvania Railroad. Mr von Wege, his wife Bertha; their daughter Mildred, aged 29; their son Walter, aged 14; and Walter Schultz, aged 22, were all killed. As in the Canton accident the previous November Mr von Wege had stopped his car and waited for an eastbound train to make a station stop, then started across the line behind the rear coach of the eastbound train into the path of the *Metropolitan Express.* The gasoline (petrol) tank of the car was ruptured by the impact of the collision and the remains of the car became a mass of flames, though railroad employees were able to extinguish the blaze. [1084]

The 4th of July 1919 Independence Day celebrations in Freeport, Maine, United States were also used as an opportunity to celebrate the return of troops from Europe. Unfortunately, the festivities were marred by a series of road accidents, the most serious of which was a collision on Hunter's Crossing, a mile west of Freeport station, which claimed six lives. The accident happened at 10.35 a.m. when a Maine Central Railroad special train of empty passenger cars running from Brunswick to Portland collided with an automobile driven by Mr Patrick Hunter, less than 200 from the latter's home at Spring Brook Farm. In the car with Mr Hunter were his sister Margaret, their nieces Lois

---

[1080] *The Lincoln Daily Star*, Lincoln, Nebraska, United States, 30th June 1919
[1081] *The Indianapolis Star*, Indianapolis, Indiana, United States, 30th June 1919
[1082] *Van Wert Daily Bulletin*, Van Wert, Ohio, United States, 1st July 1919
[1083] *Middlebury Independent*, Middlebury, Indiana, United States, 4th July 1919
[1084] *New Oxford Item*, New Oxford, Pennsylvania, United States, 3rd July 1919

and Elizabeth Quincannon, aged 8 and 4 respectively, and two women who had accepted a lift into Freeport: Mrs Henrietta Reed and Mrs Maud Ausher. All six were killed outright or fatally injured. The crossing was unguarded, but it was a mystery why Mr Hunter had failed to see the train. One theory was that he had been changing gear as he had negotiated the crossing and had stalled the motor. [1085]

An attempt to beat a speeding Illinois Central Railroad excursion train over a crossing at Del Rio, Illinois, United States, on or about Saturday 5th July 1919 resulted in the death of five persons. They were William Huebner Jr., Gertrude Huebner, Joseph Burns, Agnes Fitzmaurice, and Edna Peet, all of Chicago. [1086]

A level crossing with the Chesapeake and Ohio Railroad just west of Merrillville station, Indiana, United States, was described as inherently dangerous. The railroad at this point ran on an embankment at a considerably higher level than the road and visibility for road vehicles approaching from the south was further restricted by sand banks which made it impossible to see trains until they were 100 feet away. At 8.15 p.m. on Saturday 26th July 1919 a car containing seven members of the Ebert and Piske families and being driven by Mr Gust Piske was struck by a pair of engines and a caboose. Mr Emil Ebert, his daughter Esther, aged 9; Mr and Mrs Gust Piske and their son Arnold, aged 6, were all killed. Mrs Ebert was seriously and possibly fatally injured, and Ruth Piske, aged 11, was slightly hurt.

Neither of the survivors was able to explain how the accident had happened but eyewitnesses stated that Mr Piske had approached the crossing cautiously, then, seeing no train, had moved onto the crossing and was directly over it when his automobile was struck. The train was allegedly travelling at 40 miles an hour and did not whistle until it was almost on the crossing. A valuable diamond ring owned by Mrs Piske was missing after the accident. [1087]

On Thursday 14th August 1919 at Parkersburg, West Virginia, United States, the regular street car from Marietta of the Monongahela Valley Traction Company was being followed by a second car carrying about 40 small children from the Reno, Ohio, Sunday school, accompanied by a few adults. As the two cars arrived at a level crossing with the Baltimore and Ohio Railroad the first car passed over the crossing and Motorman Barnes, driving the second car, attempted to follow, having failed to notice an approaching switch engine. When he did see the engine at the last moment he just had time to apply the brakes and jump, and the train struck the crowded street full on. Both the car and the engine were considerably damaged but what made the accident much worse was the breaking of a steam pipe on the engine which allowed scalding steam to enter the street car. By the end of the day, six of the children and one of the adult carers were dead and an eighth fatality was Mr. E.E. Anderson, who collapsed and died as he was assisting in the rescue work. The City and St Joseph's hospitals were filled with many other sufferers,

---

[1085] *Boston Daily Globe*, Boston, Massachusetts, United States, 5th July 1919
[1086] *The Kingston Daily Freeman*, City of Kingston, New York, United States, 5th Jul 1919
[1087] *Hammond Lake County Times*, Hammond, Indiana, United States, 29th July 1919

a number of whom could not be immediately identified as they were too frightened or too young to give their names. The crew of the engine said that they had done all they could to warn the motorman but in vain.

The street car was said to have been of an obsolete wooden type unfit for active service. It was also pointed out that Section 61 Chapter 54 of the West Virginia state code decreed that a watchman must be provided at all crossings of two railroads, at the joint expense of the companies concerned, and that all cars and trains must come to a dead stand at least 200 feet from the crossing and not proceed until signalled to do so by the watchman. No watchman was provided at the crossing. [1088]

County Prosecutor James Ward announced that there would be a rigid inquiry into the accident. [1089]

By the end of the following day the number of dead had risen to 11 and others were said to be dying. [1090]

Seven members of a Cleveland, Ohio, family lost their lives in a level crossing collision at West Springfield, Pennsylvania, United States, on the evening of Saturday 16th August 1919. They were Mr John Trainor Senior, his wife Penella, their sons John Junior, aged 23, Thomas, aged 14 and Vincent, aged 10, and their daughters Margaret, aged 25, and Penella, aged 8. An unnamed female relative was an eighth fatality. The party had been on their way home from Erie, Pennsylvania. Heavy rain was falling at the time of the accident and the side curtains of the car were down, further restricting visibility. The driver of the automobile had waited for an eastbound New York Central Railroad freight train to pass, then pulled onto the track into the path of a westbound passenger train.

The only surviving member of the family was another 18-year-old daughter, Sadie, who had been away on vacation. She returned home just as news of the accident was received. [1091]

Late in the evening of Saturday 1st November 1919 members of the Glen Social Club of Philadelphia were on their way to a midnight masquerade party at the Rosebud Boat House at Billingsport, New Jersey. The party comprised 28 men and four women, a number of other women having dropped out because it had begun to rain. They were travelling in a three-ton open lorry, fitted with wooden seats along each side and with a tarpaulin stretched across the framework to keep out the rain. Many of the men were veterans of the war who had only just returned from France. August Brehmyer, the owner of the lorry, was driving with Julius Hahn sat alongside of him. The Salem pike along which the lorry was running crossed the line of the West Jersey and Seashore Railroad at Clarksboro, New Jersey, United States. The crossing was unprotected with either a watchman, gates, or a warning bell. The view of the line was blocked by the presence of two houses and a small orchard near the crossing. Hahn, who was not badly hurt in the

---

[1088] *The Bradford Era*, Bradford, Pennsylvania, United States, 15th August 1919
[1089] *Fort Wayne Journal-Gazette*, Fort Wayne, Indiana, United States, 15th August 1919
[1090] *Alexandria Gazette*, Alexandria, Virginia, United States, 15th August 1919
[1091] *The Commercial Tribune*, Cincinnati, Ohio, United States, 17th August 1919

accident, said that there had been a drizzling rain. He and the driver had listened as the lorry approached the railroad travelling at eight miles an hour, but they could not hear any train, nor see one, and according to him there was no whistle sounded or warning bell ringing on a locomotive. Brehmyer began to take the truck over the crossing when they were horrified to see the headlights of a train at close hand. "Brehmyer crashed his foot down on the brake and put on the emergency," Hahn related afterwards. "We could both see we were going to be hit. In an instant I yelled 'jump' to the crowd inside and then I jumped myself. I was stunned for a few minutes and then I saw what had happened. The train was at a standstill about 100 yards away from the crossing. The bodies of the boys and girls were strewn along the cornfield and tomato patch right from where we had been hit to where the fragments of the truck still remained wedged fast against the cowcatcher. We never had a chance. The train did not have its bell ringing nor did the engineer blow his whistle. We just saw the light and then we were hit."

Describing what happened then, a newspaper reporter wrote, 'Clad in their masquerade costumes, some of the boys dressed as girls and one of the girls dressed in a sailor suit, twelve lost their lives the instant the train hit the truck. Of the whole thirty-three in the party, only Hahn, Kilbride, Sautter, Daunback, Heinbecker, Bach and Lehman escaped serious injury. Bruised and battered themselves, they helped the trainmen extricate the dead and injured. All the killed were horribly mutilated. Sautter had helped to carry and lay to one side three of his dead pals. Then the pain of his own injuries became too much for him and he fell in a dead faint just as he was about to light a cigarette.

'The habit of the wounded soldier in France came to the top in McCready, Hartung and Anderson, who, badly battered and sorely hurt, asked their rescuers for a "smoke."

'The occupants of the few houses in the vicinity were roused from their beds. Hurried calls for assistance were sent to Clarksboro, Woodbury and Swedesboro. Dr Downes, the Coroner of Gloucester county, was among the first to arrive, and along with County Prosecutor Oscar B. Rediow directed the work of rescuing the unfortunates pinned under the shattered truck. One by one they were dragged out, and the living sorted from the dead. It was impossible to move the train. The wrecked truck was wedged firmly on the cowcatcher. A flagman was hurried over to the grade crossing and with his red lantern stopped all the passing automobilists and commandeered their cars.

'It was seen at a glance that two of the girls were dead, killed instantly. There was still a spark of life in the other two and their crushed bodies were lifted into a machine, one of the first to reach the crossing, and they were hurried to the Underwood Hospital at Woodbury. Soon more cars came up and as fast as they did so the sorely injured were placed in them and whisked along the Salem pike to the Cooper Hospital in Camden. Eleven were sent there, but two died a few hours after they reached the institution.

'Edward Clark, the Clarksboro undertaker, was summoned and the bodies of the twelve dead were removed to his establishment. There in what had once been a stable under the flickering light of dim oil lamps he and his assistants worked. Meanwhile a wrecking crew was busy removing the debris from the tracks and cleaning up the line. The traffic was resumed. The truck was thrown to one side and the kegs of beer were rolled alongside.

'Nothing more could be done. The dead were beyond help. The injured were in the hands of the doctors.

'A narrow escape was that of Harry Kilbride. He was sitting near the centre of one of the seats inside the truck. "We were going along fine enough," he said. "I should say about eight miles an hour, and we were all wishing we were at Billingsport, because Hahn would not let us have any of the 'near beer' on the road. Suddenly I heard someone yell 'jump.' I didn't have time to jump anywhere, for less than a second afterwards something hit us. I was stunned for a moment, and when I came to I was lying on the railroad track, my feet just a few feet away from the rails.

"I jumped up and, naturally, was dazed. I could not realize what had happened and then I saw the bodies. Then I knew. I helped the others get some of the party out and did what I could."' [1092]

The Volstead Act (Prohibition) had come into effect a couple of days earlier and local officials and revenue agents seized the kegs of beer and flasks of whisky that they found on the bodies of some of the dead men. [1093] Two witnesses at the inquest stated that people on the lorry had been intoxicated but this was denied by physicians who stated that they had, however, given whisky and brandy to the injured after the accident.

Frank Henry, one of the most seriously injured, died on 4th December, bringing the number of dead up to 15.

The jury at the inquest on the victims of the accident exonerated the train crew from blame and recommended that the Greenwich Township Committee take action to have the New Jersey State Board of Public Utility Commissioners order better protection at the crossing. [1094] Engineer Henry Githens, Fireman William Fraser and Conductor Thomas Walker had all sworn at the inquest that the whistle had been blown 300 yards from the crossing and a second time 100 yards from it. People on the lorry all agreed that it had not been, or at least that they had not heard it. For reasons which the newspaper account does not make entirely clear (to the author at any rate) the relatives of the victims were reportedly very unhappy at the inquest verdict, presumably because it had not found fault with the train crew. [1095]

On the morning of Monday 17th November 1919 a lorry taking 35 workmen to their employment at Hog Island, Philadelphia, United States, was hit by Philadelphia and Reading Railroad train No. 307, also bound for Hog Island, just before 7 a.m. There were no gates at the crossing but there was a warning gong, which was sounding at the time of the accident. The driver of the lorry, William Astlin, said that the gong was always ringing but he 'had never been in any danger before.' There was a thick fog on the day of the accident. He was almost over the crossing, not having taken any notice of the gong when his truck was struck by the train. Six of the workmen were killed instantly or fatally injured and nine were injured, one probably fatally. Astlin, who escaped unhurt, was arrested after the accident but later allowed to go home, pending an appearance before the magistrate.

[1092] *The Philadelphia Inquirer*, Philadelphia, Pennsylvania, United States, 3rd November 1919
[1093] *The New York Times*, New York, United States, 3rd November 1919
[1094] *The Philadelphia Inquirer*, Philadelphia, Pennsylvania, United States, 5th November 1919
[1095] Ibid, 4th November 1919

Neither Engineer George Hill nor Conductor C.W. Tell was aware of the presence of the lorry before their train hit it. [1096]

A newspaper reported that Astlin basically admitted that the accident had been his fault. '"I took a chance," Astlin, the driver of the truck, sobbingly told Lieutenant Bausline, of the Sixty-fifth street and Woodland avenue station, yesterday. "The fog caused the accident, and I couldn't see five feet in front of me. I heard the bell ring several times and I thought the train had passed. I took a chance and I thought I could get over. Then the crash came before I could realize the danger."'

A survivor of the accident also spoke to a reporter. '"I've been under some fierce shellfire in my time, but I never heard anything like the crash when that engine hit our truck." This was the manner in which O. Nichelson, a veteran of the World War and employed in ship construction at Hog Island, described the crash between a train and a motor truck as he lay on a cot in the Hog Island emergency hospital. His chief concern was for his wife, an English girl, who came to this country to marry him. "Gee! Wouldn't it have been tough if I had been killed?" he exclaimed. "This little girl doesn't know a soul in this country but me." Nichelson received his injury as he dived out of the rear of the truck when he saw the accident was inevitable. His foot was crushed when a wheel of the engine virtually tore a shoe from his foot.' [1097]

The day after the Hog Island accident, Tuesday 18th November 1919, seven people lost their lives in a level crossing accident at Cheektowaga, New York, United States, just outside of Buffalo. A car driven by William Mummery, of Buffalo, and carrying six young women, all present or former nurses at the Buffalo Homeopathic Hospital, was struck by a New York Central Railroad train on the Harlem Street Crossing in Cheektowaga. Everyone in the car was killed. Apart from Mr Mummery they were his daughter, Doris; Margaret Breese, of Buffalo; Mabel Hodgins, of St Catherine, Ontario, Canada; Jessie McMillan, of Tavistock, Ontario; Myrtle Nunn, of Fenwick, Ontario, and Jean Scott, of Galt, Ontario.

The nurses had been to a reception at Lancaster in a car driven by a brother of Margaret Breese but on arrival at Lancaster the car had broken down. Miss Mummery had then telephoned her father explaining what had happened and asking him to come and pick them up and bring them back to Buffalo.

When Mr Mummery had reached the crossing a freight train had been passing over it. As soon as that train had cleared the crossing he had set off and his car had been struck by the Chicago flyer travelling in the opposite direction. It took the train half a mile to stop. [1098]

Fog was a significant factor in an accident at Homewood, Illinois, United States on Wednesday 31st December 1919. Mr John Kasper, a carpenter of Chicago Heights, had set off in the morning with his wife and three other people for Chicago, but for some reason turned back after going part of the way. At the level crossing of the Illinois

[1096] *Evening Public Ledger*, Philadelphia, Pennsylvania, United States, 17th November 1919
[1097] *The Philadelphia Inquirer*, Philadelphia, Pennsylvania, United States, 18th November 1919
[1098] *Syracuse Herald*, Syracuse, New York, United States, 18th November 1919

Central Railroad and the Dixie Highway at Homewood, Illinois, United States his car was struck by southbound train No. 23, travelling at 60 miles an hour, thrown onto the opposite running line and struck again by northbound train No. 10. Mrs Elizabeth Kasper, John Kasper's wife; Mrs Grace Buknis, Mrs Adeline Slushnis and her three-year-old son were all killed outright, and John Kasper died later in St James' Hospital, Chicago Heights. There was a watchman at the crossing, but the fog prevented him from seeing the car until it was too late.

Several years earlier, when a new yard was being constructed at Homewood, the Village President, Fred Gold, had insisted that the railroad company build a subway in place of the crossing at the same time. 'The subway is under construction and will save many lives in the future,' a newspaper added. [1099] [1100]

Tragedy struck the town of Ardmore, Oklahoma, United States, early on the morning of Wednesday 3rd March 1920. At 1.45 a.m. a party of ten young people, all but one of whom lived at nearby Wilson, were returning there from a dance at the Convention Hall in Ardmore, riding in a Ford truck. Meanwhile Oklahoma Pacific and New Mexico Railway local freight train No. 88, due in Ardmore at 12.30 p.m., was running 12 hours late and did not arrive there until after midnight. When the party of dance-goers set off homewards the local freight was busy shunting cars across McLish Avenue Crossing, and a collision with the Ford truck ensued. The cars were being pushed over the crossing by the engine and the truck was ground beneath a Missouri-Kansas-Texas Railroad stock car. Miss Helen Bridges, aged 20, of Ardmore, and Bailey Morris, aged 18; Miss Lucille Kimbrell, aged 16; Ike Kimbrell, aged 23 and Miss Iva Beall, aged 24, all of Wilson, were killed. Two of the others were injured and the remaining three escaped unhurt.

Winston Brown, who was slightly injured, said that no one on the truck had been aware that a train was anywhere near. They did not hear it, probably because the truck itself was making a considerable amount of noise; nor did they see any lights on the train or any warning signal.

Gus Key, who lived on McLish Avenue near the crossing, had just returned home in his car and had put it away when he heard screams from the crossing. He went to the scene, saw what had happened and went into town to fetch Sheriff Garrett. The latter shortly arrived and found the members of the train crew surveying the scene but making no effort to remove the dead or injured. They allegedly told the sheriff that they could not do so until an inquest had been held. (To the author it seems preposterous that the crew would not assist the injured because there had been no inquest, but this is what was alleged in the local newspaper.) The sheriff told the crew that an inquest was not necessary as there was no doubt as to the cause of death and told them to get jacks and remove the victims from under the cars, which the crew did.

The crew of the train were Conductor C.F. Vaughn, Engineer R.L. Bell, Fireman Elmer Webb and Brakemen W.F. Wright and T.C. Walsh.

Those killed were reported to have all been well-known and prominent in Wilson

[1099] *The Pointer*, Riverdale, Illinois, United States, 2nd January 1920
[1100] *The Daviess County Democrat*, Washington, Indiana, United States, 2nd January 1920

society.[1101]

Two days after the accident County Attorney Brown charged the whole crew of the train with second degree manslaughter, alleging that they did 'wilfully, negligently and unlawfully operate and move said engine along and upon a certain switch line of said railway company passing through the western portion of the city of Ardmore, Oklahoma and across the streets and highways of said city without giving any signals whatever when approaching railroad crossings and without equipping said string of cars with a brakeman or other person at the rear end of said string of cars when said train was in backward motion and at the point of intersection of said switch line with McLish Avenue in said city and while so approaching the same and when said string of cars was in backward motion along and upon said switch line said defendants did so wilfully, negligently and unlawfully as aforesaid and in the manner aforesaid, collide with a certain automobile truck then proceeding along and upon McLish Avenue, thereby causing the death of…'[1102]

It has not been possible to determine if the case ever came to trial.

Carelessness on the part of the driver of a jitney bus was blamed for an accident which cost the lives of at least six people at 3.30 p.m. on the afternoon of Sunday 7th March 1920. A party from South Bend, Indiana, United States, had been on a visit to Bremen, a few miles to the south and had boarded the bus for their return journey to South Bend. Witnesses said that the engineer of a Baltimore and Ohio Railroad express had whistled for the crossing. A box car on a siding obstructed the view and the jitney driver, Harry Wyrouth, apparently neither heard the whistle nor stopped to check that the line was clear before entering the crossing. He paid for his negligence with his life.

The passengers killed were Mrs Carrie Bager, Mrs Berger, Mrs Borts and Mrs Lena Rospzir, all of South Bend, and a man who could not immediately be identified but was believed to be a resident of Elkhart, Indiana. Mrs Borts' husband, Walter, and their infant daughter were said to be dangerously injured.[1103]

Two bad level crossing collisions occurred in the United States on Sunday 27th June 1920, both in the state of Indiana. The first and more serious in terms of the number of casualties happened at 9.45 a.m. An automobile truck loaded with picnickers was struck by Southern Railway passenger train No. 5 on a crossing half a mile west of Huntingburg. Ten people, adults, and children, were killed outright, one more had died by the next day and about eight were badly hurt. Only two, man and wife, escaped injury altogether, one of them having jumped from the truck when they saw that an accident was inevitable.

Mr Jacob Bretz, the picknicker who had jumped, spoke to a newspaper reporter afterwards. "We had started across the track when I saw the train approaching, about 150 feet away. I think every occupant of the truck, as well as myself, saw the train, but not one word was spoken. Everyone seemed to realize the predicament we were in – the one

---

[1101] *Daily Ardmoreite*, Ardmore, Oklahoma, United States, 3rd March 1920

[1102] Ibid, 5th March 1920

[1103] *The Tipton Daily Tribune*, Tipton, Indiana, United States, 8th March 1920 hope that the truck might get across before the engine arrived. In a frantic effort to make haste Rehl attempted to make speed, causing his engine to go dead. Realizing that the train was upon us I jumped, worried with the fear that my wife would be killed in the crash. Hardly had I hit the ground until my wife landed on the ground near me, Except for the nervous shock we both escaped until."

Bretz had been sitting on one side at the rear of the truck with his wife sat opposite. The truck was struck by the train about three feet from its rear and Mrs Anna Bretz had been thrown backwards and off the vehicle by the impact. Bretz afterwards reportedly worked 'like a Trojan' aiding the injured. There had been three truckloads of picknickers in all and he was assisted by Amos Martin, the driver of the second truck, while Joseph Riglesberry, the driver of the third, went off to get assistance. Bretz and Marin carried the bodies of the dead and injured to a more comfortable location and tended the injured but were so shocked by what they saw that as soon as more helped arrived they had to leave the scene.

It took the train a full 1,000 yards to pull up and it was supposedly half an hour before it set back to the crossing. Most of the victims were then loaded onto it to be taken to the Methodist Hospital in Princeton, but some of the less seriously injured were taken in the two other trucks to a doctor in Huntingburg. Another wreck prevented the train reaching Princeton and it was forced to return to Huntingburg and take them to Evansville instead, reaching there at 3.30 p.m. [1104]

The driver of the truck, Ed Rehl, was one of those killed, together with his wife Carrie and their son Carl. It was theorised that as he was familiar with the times of the trains passing through Huntingburg he had assumed that the train involved had already gone, which it would have done had it not been 40 minutes late. Visibility at the crossing was reduced by a high embankment. As he reached the track he probably changed into a higher gear and then seeing the train he had probably pushed down the accelerator, flooding the engine with gas (petrol) and causing it to stall. Then, before he was able to restart it, the train was there. The engineer of the train, Al Lewis, declared that he had not seen the truck until he was right upon it. His train was reportedly travelling at 50 miles an hour.

The locomotive involved in the collision had a bad reputation, having previously collided with some freight cars in 1904, severely injuring its fireman, and struck a truck load of picknickers, killing eight, at Maud, Illinois, in 1912. [1105] The author has not been able to trace a record of the latter accident.

At 5.30 p.m. on 27th June a car being driven by Mr William Litherland, a garage owner, of Nashville, Indiana, was struck by an Illinois Central Railroad train on a level crossing near Morgantown, Indiana. Mr Litherland had been following another car driven by a farmer named Colon Thompson. The Thompson car had got over the crossing safely and the Litherland car had followed and been hit by the train. Mr Thompson said that the train had whistled but Litherland either did not hear it or panicked. The crossing was said to be particularly dangerous as the track ran through a cutting approaching it, so approaching trains remained invisible until they were close at hand. The car was carried

---

[1104] *Madison Daily Herald*, Madison, Indiana, United States, 28th June 1920

[1105] *The Princeton Daily Democrat*, Princeton, Indiana, 28th and 29th June 1920

50 feet down the track. Mr Litherland and his four passengers lost their lives. A 17-year-old daughter, who had chosen to stay home, escaped the accident. [1106] [1107]

Sunday 8th August 1920 saw another two serious level crossing collisions in the United States. Three generations of a family who lived near Pontiac Michigan, were wiped out in the afternoon when their car was struck by an interurban car near Orion, Michigan. They were Mr and Mrs Edward Spies, their son Arthur and his wife, and their grandson. The victims were reportedly thrown 200 feet and all but one were killed outright. The motorman of the interurban said that he had whistled repeatedly as he had approached the crossing but Arthur Spies, who had been driving, apparently failed to hear the warnings or see the danger. [1108]

At 9.30 p.m. on Sunday 8th August a car carrying six 'Salem Roumanians,' presumably new immigrants, was struck by Pittsburgh Fort Wayne and Chicago Railroad eastbound passenger train No. 28 on New Garden Street, Salem, Ohio, United States. Emil Bolta, night foreman at the Buckeye Engine Company, Salem, was driving his new car with five Romanian passengers: George Balon, John Beduna. John Danella, George Nan and John Popa, all employees of the American Steel and Wire Company at Salem. The crossing here had gates, but they did not close simultaneously. Mr Bolta drove onto the crossing before he noticed that the west gate was down. It must have been a busy crossing, for he was then prevented from reversing by three automobiles which had stopped behind him. Attempting instead to go forward and drive through the gate he stalled the car, then the fast-moving train struck it. All six occupants of the car lost their lives. [1109]

At 8.30 p.m. on the evening of Sunday 22nd August 1920 a bus with 15 passengers was struck by a train on a level crossing at South Camden, New Jersey, United States. The bus, which was full, was en route from the Federal Street Ferry in Camden to Yorkshire Village and had travelled the five miles to the Morgan Street Crossing in South Camden without incident. Here the road crossed the West Jersey and Seashore Railroad by a crossing which had been the site of numerous previous accidents, though it was protected by a watchman. As the bus approached the crossing the watchman signalled for it to wait as a southbound train was close at hand. The bus waited, the southbound train went by, and the bus driver started to cross. "Go back," the watchman shouted and frantically waved his arms for the bus to stop, but it came on, into the path of a northbound electric train. It was theorised that the noise made by the southbound train might have drowned out the sound of the northbound one. The electric train tore the bus to pieces, it was said. By the following day nine passengers and the bus driver, known as Joseph Stratton, but whose real name was Vincenzo Stratoni, had succumbed.

[1106] *The Bedford Daily Democrat*, Bedford, Indiana, United States, 28th June 1920
[1107] *The Commercial Tribune*, Cincinnati, Ohio, United States, 28th June 1920
[1108] *The Sedalia Democrat*, Sedalia, Missouri, United States, 13th August 1920
[1109] *The Evening Review*, East Liverpool, Ohio, United States, 9th August 1920

The following day Coroner David Bentley of Camden stated that, "The accident was the fault of the driver, according to the testimony of witnesses I have examined. The watchman at the crossing swung his light and called for Stratoni to stop. Stratoni failed to heed the warning and drove directly into the path of the express. It is a bad crossing, and drivers of vehicles should be more careful when crossing at that point." [1110]

It was ruled that an inquest on the victims was unnecessary but separate investigations were opened by the Camden county and the Camden city authorities. [1111]

An eleventh victim, John Myers, aged 24, died of his injuries on Tuesday 24th. [1112]

On Sunday 29th August 1920, a car driven by Mr Robert Yust, a farmer residing near Sylvia, Kansas, United States, was struck by Atchison Topeka and Santa Fe Railway passenger train No. 2 a short distance from his home. Mr Yust was driving his family in a Dodge sedan the short distance into town to attend the Chautauqua (a popular adult education and social movement of the time) [1113] which was in session there. As his car reached the crossing west of town Mr Yust reportedly "killed" its engine just as got to the track. Visibility was impaired by a tall cornfield and the car was on the crossing before he saw the train. He tried to reverse but "killed" the engine again and the train then struck the car. Mrs Yust and their children Howard, aged 14, Adelaide, aged 11 and Victor aged 5 were killed outright. Mr Yust and his daughters Margaret, aged 12, and Grace, aged 3, were taken to hospital in Hutchinson on the train, where Grace also died. No hope was held out for Margaret either, but it was thought that Mr Yust would survive. The oldest Yust son had remained at home.

Coming into Emporia, Kansas, in the evening, the same train ran down a pedestrian named John Clark, who was walking along the track at the station, but he was not seriously hurt. [1114]

At 22:00 on Tuesday 8th February 1921 a Saint Etienne – Paris express struck a bus carrying 11 conscripts from Feurs to Panissières on a level crossing close to Feurs station, France. The conscripts had just been seen by a review board and had hired the bus to take them to Panissières. The bus driver had found the barriers at the crossing open and had driven onto the crossing only for his vehicle to be struck by the train.

After the accident four conscripts named Berthaud, Gubien, Marin and Paccard were found to be dead and four others named Didier, Dourgin, Giroud and Fetinet were grievously injured. The three others were missing. One of them, Vial Georges, was found on the front of the locomotive of the express when it stopped at Roanne; he was still alive and was taken to hospital but died shortly afterwards. The bodies of the other two missing conscripts, Jay Stéphane and Pétrus Cuissard, were found by the lineside 30 kilometres from the scene of the accident. By the following morning Didier, Dourgin and Giroud

[1110] *Evening Public Ledger*, Philadelphia, Pennsylvania, United States, 23rd August 1920
[1111] Ibid, 24th August 1920
[1112] *The Philadelphia Inquirer*, Philadelphia, Pennsylvania, United States, 25th August 1920
[1113] Wikipedia – Chatauqua
[1114] *The Emporia Gazette*, Emporia, Kansas, United States, 30th August 1920

had died of their injuries as well. [1115]

On Monday 28th March 1921 M. Antoine Payra, a Perpignan liquorist, took his family and a friend in his car to visit friends at nearby Saint-Hippolyte. After a pleasant spring day there they set off back home, but at Pia, France, their car was struck by a Pia – Rivesaltes train at a level crossing on the edge of the village. A lineside house prevented M. Payra from getting a good view of approaching trains, the noise of the engine of his car prevented him from hearing the warning whistle of the train and chains which should have been stretched across the road were not in place.

All five of the occupants of the car were killed instantly. M. Payra was thrown under the wheels of the locomotive and his body could not be released even after five hours' effort. The others were cut to pieces, Mme Suzanne Payra being recognisable only by her handbag. The others who lost their lives were the couple's daughters Lucienne and Simone, and Elise Alabert, presumably a family friend.

The Perpignan parquet (examining magistrate) went to Pia in the morning to open an investigation. [1116]

From the brief newspaper report the accident would appear, like that at Feurs the previous month, to have been due to negligence on the part of the crossing attendant.

Early in the evening of Wednesday 14th June 1921 a Central of New Jersey Railroad Lakewood express bound for Barnegat City collided with a fire engine of the Eagle Hose and Chemical Company on the Market Street Crossing in Perth Amboy, New Jersey, United States. The results were very serious; six firemen were killed outright and three more had died of their injuries by the following afternoon. There was something of an outcry locally after the accident and the next day the city aldermen held a special meeting regarding the accident, at which, among other measures, it was demanded that the Central of New Jersey, Lehigh Valley, New York and Long Branch and Pennsylvania Railroad Companies limit the speed of all their trains through the city to ten miles an hour while passing over Market Street, New Brunswick Avenue, Smith Street and Washington Street level crossings. There were allegations not only that the gateman had failed to lower the barriers but that the train had been travelling at excessive speed. More reasonably the aldermen called upon the New Jersey Public Utilities Commission to call a meeting to discuss the abolition of the level crossings. [1117]

Two of the train crew, Conductor William Ridgeway and Engineer Theodore Brown, were arrested after the accident and charged with manslaughter before being released on bail. Engineer Brown stated that had not seen the fire truck until his train was close to the crossing. He said, "When I got within the required distance of the crossing I blew the whistle. I saw nothing of the truck until just a second before the crash. I jammed on the air and reversed the throttle. That was the best I could do then. The big truck rammed the side of my engine directly alongside me, but fortunately I escaped injury." [1118]

[1115] *Le Radica*, Paris, France, 10th February 1921

[1116] *Le Petit Journal*, Paris, France, 30th March 1921

[1117] *Perth Amboy Evening News*, Perth Amboy, New Jersey, 16th June 1920

[1118] *Boston Daily Globe*, Boston, Massachusetts, United States, 16th June 1920

The crossing gates had reportedly not been lowered because they were temporarily out of order, but the gate tender, Andrew Thomas, 64, had been stood on the crossing waving his red flag when the fire truck appeared, overtook a line of automobiles which had stopped in response to the red flag and entered the crossing just as the express got there. Departure of the fire engine from its station after an alarm had been received was apparently delayed several minutes because of problems with its carburettor. This not only meant that more fireman had time to board it – there were 14 on it in all – but probably meant that they were in more hurry than usual to get to the fire. The gateman was injured, presumably by being struck by either the fire truck or the train, and he was taken to hospital. Unlike the conductor and the engineer he was apparently not arrested.

The accident was the subject of an exhaustive investigation by a Grand Jury which published its findings towards the end of July. It recommended as a temporary measure the employment of two watchmen at each of the Perth Amboy level crossings and as a permanent solution either the elevation of the railroad tracks and/or the depression of the roads so as to eliminate the crossings altogether.

The Grand Jury noted that the dangers of the level crossings in Perth Amboy were greatly accentuated by:

(a) 'The excessive rate of speed at which trains operate through Perth Amboy.

(b) The neglect on the part of the railroads to use proper care in the selection of employees who are physically and mentally equipped to perform their duties and carry out the high responsibility resting upon them.

(c) The failure of the railroads to employ help in sufficient numbers to properly protect the grade crossings.

(d) The frequency with which accidents occur at the grade crossings within the city of Perth Amboy is a forensic argument, in impelling one to the conclusion that these crossings are a danger and menace to human life and property.' [1119]

The rashness of a lorry driver resulted in his own demise and the death of four of his passengers in a level crossing accident a mile north of Anguilla, Mississippi, United States, at 3 p.m. on Thursday 11th August 1921. A construction company were in the process of building a public road in Sharkey County and the General Motor Company lorry, towing a trailer in which some of the passengers were riding, was on its way to Nitta Yuma to get gravel. The road along which the lorry was being driven ran alongside the line of the Yazoo and Mississippi Valley Railroad for some distance before crossing it at an angle of 75 degrees. The truck and trailer and the northbound train which struck them were travelling in the same direction. All of those killed had been riding in the truck, those in the trailer being injured when it overturned following the collision. Apart from Charles Brown, white, who was driving, those killed were John Vance, white, and three negroes named James Murray, Will While and Percy Williams. Foreman Bud Crowder, white, and nine negroes were injured. [1120]

[1119] *Perth Amboy Evening News*, Perth Amboy, New Jersey, 22nd July 1920

[1120] *The Daily Herald*, Mississippi Coast, United States, 12th August 1921

A party of six people returning from a camp meeting at Portage were killed instantly on the evening of Sunday 14th August 1921 were killed when their car was struck by a Baltimore and Ohio Railroad passenger train, said to have been running at 50 miles an hour, on a level crossing two miles north of Weston, Ohio, United States. Those who lost their lives were Joe Oberla (or Oberly), a Civil War veteran; his granddaughter Opal, aged 12; Mr and Mrs George Burditt and their son Ray, and Mrs Miller, a widow. The crossing was unobstructed, but the car reportedly stalled on the track. [1121] [1122]

Five people, Mr and Mrs J.A. Miller, of Bettendorf, Iowa, Mr and Mrs Charles Ege, of Albany, Illinois, and a Mrs Blean, were killed when their car was struck by a westbound Toledo Peoria and Western Railroad passenger train on a level crossing at La Harpe, Illinois, United States, at noon on Saturday 27th August 1921. Mr Miller, who was driving, had apparently entered the level crossing before he saw the train, then panicked and stalled the engine of the car. [1123]

On Thursday 20th October 1921, a party of five Maine women set off from East Hiram in a small sedan car either for a day's shopping in Portland or according to another account to attend a convention of the Pythian Sisters [1124] (a fraternal organisation, the female auxiliary of the Knights of Pythias) [1125] at Steep Falls. Mrs Edith Howard was driving. The other women were Mrs Nettie Foss, Mrs Irma Hybert, Mrs Zola Small and Mrs Frank True. It was raining hard when they approached the level crossing at Mattocks station, Maine, United States, on the Maine Central Railroad at about 9.30 a.m. Mrs Howard was said to have been a very good driver and it was theorised that rain on the windscreen had obscured the approaching train. The westbound late forenoon passenger train struck the car, killing Mrs Hybert and Mrs True outright. A special train, made up from a freight train in the yard at Mattocks, took the other three women to the Maine General hospital in Portland, but Mrs Small died on the journey and Mrs Howard shortly after arrival. [1126] [1127] Mrs Foss clung to life a little longer before dying two days after the accident. [1128]

Five members of the Sales family died in a level crossing accident at Los Medanos, California, United States, on Sunday 6th November 1921 when their car was struck by Southern Pacific Railroad train No. 109. The victims were Mr and Mrs W.D. Sales Junior, Mrs W.D. Sales Senior, Mr R.H. Bradley, father of Mrs W.D. Sales Junior, and Mr and Mrs Sales' daughter Marjorie, aged 10. A son, Donald, aged 9, was injured. The accident happened close to the Standard Oil Company pumping station where Mr Sales worked, and the car had just left the pumping station and turned onto the highway when it

---

1121 *The Lima News*, Lima, Ohio, United States, 12th August 1921
1122 *Norwalk Reflector-Herald*, Norwalk, Ohio, United States, 12th August 1921
1123 *Free Trader-Journal*, Ottawa, Illinois, United States, 29th August 1921
1124 *The Oxford Democrat*, South Paris, Maine, United States, 25th October 1921
1125 Wikipedia – Pythian Sisters
1126 *Boston Daily Globe*, Boston, Massachusetts, United States, 21st October 1921
1127 *Biddeford Daily Journal*, Biddeford, Maine, United States, 20th October 1921
1128 Ibid, 22nd October 1921

was struck. The two women and the children were carried over a quarter mile before the train could be pulled up. Doctors Earl Gregory and H.E. Peters chanced to witness the accident and stopped their automobile to offer assistance. They said that Mr Sales had not attempted to beat the train but rather had evidently been unaware of its presence. [1129]

Fog and the inexperience of the bus driver were features of a disastrous collision at Proberta, California, United States, between Southern Pacific Railroad train No. 15 just before noon Wednesday 30th November 1921. Despite its magnitude the accident does not seem to have received a great deal of coverage in the American press, perhaps because it happened in the far west of the country. The main facts of the accident were, however, brought out in a newspaper report of an investigation conducted by a Board of Inquiry, which board was composed of local residents of nearby Red Bluff, chosen because they had no connection with the railroad company and hence could be considered impartial.

The events leading up to the disaster began when the bus which was involved was delayed by engine trouble. A phone call was made to Red Bluff high school and a relief bus was sent out from there but the problem with the engine was then rectified and the regular bus was proceeding towards the school, keeping close behind the relief bus in what was described the next day as a thick fog, though the thickness of the atmosphere does not seem to have been mentioned in the report of the investigation, and was, it seems, much exaggerated. Much more significantly, no doubt, the driver of the bus, Charles Bosworth, was himself one of the students, and only 16 or 17 years old. [1130]

According to the report Engineer V. Greer gave the usual warning blasts on his whistle as his train neared the crossing and whistled again when he saw the first bus going over the crossing close in front of him. Driver Bosworth, however, perhaps intent upon not falling so far behind the relief bus as to lose sight of it in the fog, failed to notice the whistles or to take any precautions at the crossing, and drove onto it directly in front of the train. It was about 10.30 a.m. The impending accident was witnessed by a man named W.H. Mackenzie, who shouted and waved his arms to warn Bosworth, but the bus was already between Mackenzie and the crossing, and he was not seen. Travelling at between 45 and 50 miles an hour the train slammed into the bus, throwing it against a lineside signal. The bus was wrecked and 14 students, including Charles Bosworth, were killed outright or fatally injured. One lucky student who should have been on the bus had stayed off school sick, three who had been on it were given a lift to school in a car when the vehicle broke down and only one who was actually on the bus at the time of the accident survived. Opal McNaughton suffered a fractured pelvis and compound fractures of both legs and was a year in hospital, but she survived to marry and raise a family. [1131]

The Board of Inquiry found that the blame for the disaster lay entirely with Charles Bosworth, and that no blame attached to Engineer Greer or to the railroad company. It recommended that a law be enacted requiring the drivers of automobiles, both private vehicles and public buses, to stop before attempting to negotiate any railroad level

[1129] *The Bakersfield Californian*, Bakersfield, California, United States, 7th November 1921
[1130] *The Bisbee Daily Review*, Bisbee, Arizona, United States, 1st December 1921
[1131] Webpage – https://disasteroushistory.blogspot.com/2016/03/proberta-california-train-bus-crash.html

crossing. In addition it recommended that the drivers of Red Bluff high school buses be instructed not only to stop before negotiating level crossings but for one of the passengers to alight and look up and down the track before signalling to the bus driver to proceed. [1132]

Less than two months after the Proberta disaster there was another bad school bus crossing accident in the United States. Again fog as involved, and it probably played a more important role than at Proberta. The scene this time was Dolby's Crossing, four miles east of Van Wert, Ohio. At 8 a.m. on Tuesday 10th January 1922 a school bus carrying 28 children to Ridge Township Consolidated School was struck on the crossing by a Pennsylvania Railroad freight train, There was a thick fog at the time. Four children, sisters Lucille Fackler, 13, and Jessie, 11; Roland Imler, 8, and his sister Marietta, 7, were killed more or less outright. Fifteen other children were taken to hospital, five of whom it was thought at the time were fatally hurt. In the following days this turned out to be somewhat pessimistic, but August Stegman, 12, died two days after the accident [1133] and his sister Josephine, 13, passed away a day or so later. [1134]

Unlike his counterpart at Proberta bus driver Joseph Overholtz had stopped his vehicle short of the crossing, alighted and gone forward on foot to check if all was clear. He said afterwards that he had been able to hear and see nothing. Satisfied that it was safe to cross he had returned to his bus and started it up the slight incline to the crossing. Just as the bus reached the line the dim black outline of a locomotive loomed out of the fog. The left side of the pilot of the engine struck the bus and threw it 75 feet. Driver Overholtz was somewhat hurt himself and his own son Lawrence, aged 10, was one of the children who were seriously injured.

As soon as news of the accident was received three ambulances and all the physicians of Van Wert were rushed to the scene and Van Wert County Hospital prepared to receive the injured. In addition the Chas Siferd ambulance was ordered from Lima and all available invalid coaches from Delphos, and the Pennsylvania ran a special train with medical aid from Lima.

As might be expected, 'News of the terrible occurrence spread like wildfire over the countryside. Frantic mothers and agonized fathers whose little ones were in the bus set out for the scene. Some did not wait for a conveyance but ran to the place. Several mothers are said to have run for distances as great as two miles. They began searching among the moaning children for their own. Many of the women were prostrated. Some fainted on being told the tragic news. It was at first reported that 20 people had died.'

Engineer J.D. Trimble told the authorities that he had whistled for the crossing as usual, but this was denied by witnesses, and there were also allegations that the headlights of the engine had not been burning. [1135] Van Wert co-prosecutor Carl Dunifon declared that Engineer Trimble, Fireman W.H. Koons and Conductor James Raines would be charged with manslaughter. Later in the day just the engineer was charged with

---

[1132] *Berkeley Daily Gazette*, Berkeley, California, United States, 7th December 1921

[1133] *The Evening Review*, East Liverpool, Ohio, United States, 12th January 1922

[1134] *The Daily Union-Herald*, Circleville, Ohio, United States, 14th January 1922

[1135] *The Star Journal*, Sandusky, Ohio, United States, 11th January 1922

manslaughter by Ray Imler, the father of the two little girls who had died. [1136]

A week after the accident the Grand Jury threw out the indictment against Engineer Trimble, stating that there was insufficient evidence against him. They recommended that more protection be provided at level crossings in Van Wert country. [1137] A few days later and inquest upon the six victims found a verdict of Accidental Death, Coroner Keyser finding that the thick fog had obscured the view of the bus from the train engineer and of the train from the bus driver. [1138]

A car carrying seven people was struck by a military train on a level crossing at Los Barrios station, Spain, at 07:00 on Friday 20th January 1922. The barriers at the crossing had reportedly been left open. Five of the occupants of the car lost their lives and the other two were injured. [1139]

The American level crossing collisions involving buses continued the evening Friday, 3rd March 1922 with one at the St Clair Street Crossing, Painesville, Ohio, though this time it was an ordinary public bus running between Painesville and Fairport and not a school bus. Many of the bus passengers had just left a dance hall, others had attended a lodge meeting. The bus, which was making its last trip of the day, had stopped just south of the railroad track to pick up two passengers. When it started up again the crossing watchman flagged the bus to stop as a train was approaching but the driver kept going. The train, travelling at an estimated 60 miles an hour, struck the bus, wrecking it and rupturing its gasoline tank, starting a fierce fire which badly burnt some of the bodies of the bus passengers. To complete the destruction westbound train No. 3 ran into the wreckage before it could be stopped.

Chester Parrish, the manager of the bus company, stated that a large oil tank had obscured the view of bus driver Andrew Adams, both of the watchman's lantern and of the oncoming train. One of the bus passengers, named Charles Sullivan, discounted this and said, "I saw the train before we reached the tracks. We had stopped to pick up two passengers just south of the railroad or we would have been across. People near me screamed as the engine came on. I saw the driver reach down and pull something, I don't know whether it speeded us up or slowed us down. It was dark in the machine. I thought we had cleared the track. Then it hit us."

The train was eastbound express No. 600 of the New York Central Railroad, en route from Cleveland to Buffalo. Engineer Frank Fernandez declared, "I blew the whistle just as we reached the limits of town. An instant after I saw the automobile on the track immediately ahead of me. The space before we struck it was so brief I could not even know whether the automobile was standing or in motion. As I threw on the brakes I felt the collision and fragments of the wreckage flew past the engine cab. I am sure we stopped as quickly as possible." Fireman Milo Cornell concurred with his driver.

1136 *The Lima News* and *Times Democrat*, Lima, Ohio, United States, 10th January 1922
1137 Ibid, 19th January 1922
1138 *The Portsmouth Daily Times*, Portsmouth, Ohio, United States, 25th January 1922
1139 Webpage: Federación Castellano Manchega de Amigos del Ferrocarril – Accidentes Ferroviarios en España

Assistant General Manager D. R. McBain of the New York Central also issued a statement, which in part read, "The crossing is protected by a watchman, who was on duty at the time of the accident and who endeavoured to stop the bus and warn the driver of the imminent danger by swinging a red lantern. We have several witnesses to verify this action of the watchman. Evidently the bus driver wanted to beat the train which resulted in the disaster which followed. After striking the bus the train ran about twenty car lengths before it was stopped. Witnesses verify the declaration of the engineer that the electric headlight, which can be seen for a distance of five miles when unobstructed, was burning brightly, the whistle was blowing and the bell ringing. The crossing is unobstructed, and a clear view west can be had."

Painesville City Manager Thomas Wyman promised a thorough investigation into the accident. He stated, "This has been one of the most notorious crossings in the state, and all plans have been made to have it eliminated, either by overhead street bridges or a subway. It is unfortunate that this work had not been done prior to this accident. There have been several narrow escapes at this crossing during the last two years, and one accident in which several people were injured. The railroad company has a watchman at this point, and I understand he was on duty at the time the accident occurred. At least he was there when the officers arrived on the scene. At present I am unable to fix the blame for the accident, but if my investigation shows negligence on the part of anyone, a rigid prosecution of the person or persons will be demanded."

By the next day 14 people were dead and several others, including Driver Adams, were seriously injured. [1140]

At the close of the inquest on the victims of the accident Coroner M.H. Bradley found Flagman A.W. Barnes guilty of negligence, stating that seven different witnesses had testified to this effect. [1141] At the opening of the inquest Barnes had testified that he had seen the headlight of the express fully four miles away, that he had immediately got out a white lantern and swung it in the middle of the street, had seen the bus coming down the street two minutes before the accident, had seen it stop to pick up passengers and then restart, and that he had swung his lantern at it, but it had failed to stop. [1142]

Despite the coroner's verdict the Ohio State Utilities Commission announced on 27th April that it was powerless under the law to fix responsibility for the disaster as it could not determine whether the watchman had flagged the bus in time for it to stop short of the crossing. It declared that this was a matter for a jury to decide. It added that it considered protection at the crossing inadequate and would take steps to ensure that the railroad company installed gates there.

The number of dead had by now risen to 18. [1143]

Late on Sunday 7th May 1922 a westbound Canadian Pacific Railway train on the Kettle Valley line collided with a car at Nicomen Crossing, near Mission City, British Columbia, Canada. All six occupants of the car were killed. They were Oliver Deroche, a

[1140] *Steubenville Herald-Star*, Steubenville, Ohio, 4th March 1922
[1141] *Connersville News-Examiner*, Connersville, Indiana, United States, 14th March 1922
[1142] *The Commercial Tribune*, Cincinnati, Ohio, United States, 5th March 1922
[1143] *The Canton Daily News*, Canton, Ohio, United States, 28th April 1922

prominent local dairy farmer who had a town named after him; Mrs Rose MacDonald and her son, Alex; Bert Gregory, a farm hand; Miss Margaret Pentland, of Mission City and a man identified only as "Shorty." As a result of the collision the automobile was flung against a box car standing in a siding.

The Hon. Dr. W.H. Sutherland, Minister of Public Works for British Columbia, was riding on the train with the Premier, John Oliver, and he attended to the victims after the collision, though there was little that he could do. He said that five of them had been killed outright, while Miss Pentland lived for a few minutes. [1144]

Fifteen tinplate workers employed at Baldwin's had received their pay and were being taken home to Morriston in a lorry when it was struck by a light engine and cut in two on a level crossing at King's Docks, Swansea, Wales, on Friday afternoon, 16th June 1922. Six of them were killed and others badly hurt. A jack had to be used to get some of the bodies from beneath the engine. Doctors were quickly on the scene and the injured were taken to hospital. The men killed were J. Griffiths, Ben Haste, D.T. James, Robert John, Tom John, and D.T. Williams. There could have been more men on the lorry but some of them had chosen to walk home. [1145]

The family of Mr El Smilak, an oil operator and junk dealer of Hartford City, Indiana, United States, were out on a pleasure trip on the afternoon of Thursday 13th July 1922 when their car was struck at about 4 p.m. by a westbound Pennsylvania Railroad passenger train at a level crossing a mile from that place. Mr Smilak, his daughters Cecilia, 14 (the driver of the car), and Sophia 13, his sister-in-law Mrs Karklin and his nephew Robert Karklin, 8, were killed; and his wife, his son Howard, 7, and his niece, Mrs Levi, were injured.

Engineer Pittman stated that he had whistled as usual for the crossing and had been slowing down to stop at the station a mile beyond. It was thought that Miss Smilak had been looking out for the eastbound train which usually passed at that hour and had failed to see the westbound one until the last moment. Mrs Smilak, in a post-accident statement, blamed the train crew for the accident, saying that they had not whistled.

Mr Smilak had been a prominent local businessman, and his daughters were talented young musicians. [1146] [1147]

No trains normally ran on Sundays on the Belfast and Moosehead Lake Railroad between Burnham Junction and Belfast, Maine, United States, and doubtless because of this Mr Evart Varney, a prominent Burnham resident who was road commissioner for Waldo country, did not bother to look along the line before driving onto the track at Unity Crossing at 7.30 a.m. on Sunday 16th July 1922. Unbeknown to him on this particular Sunday a special troop train was being run from Belfast to Camp Devens, Massachusetts, carrying F company of the 103rd Maine Infantry. With Mr Varney in the

---

[1144] *The Lethbridge Daily Herald*, Lethbridge, Alberta, Canada, 8th May 1922
[1145] *The Times*, London, England, 17th June 1922
[1146] *The Kokomo Daily Tribune*, Kokomo, Indiana, United States, 14th July 1922
[1147] *Decatur Daily Democrat*. Decatur, Indiana, United States, 14th July 1922

car were his wife, their son Everett, their 18-month-old adopted daughter Muriel, Lawrence Burrell, a half-brother of Mrs Varney, and two family friends: Miss Clara Hinckley, a Boston, Massachusetts, teacher, and Mrs Margaret Marshall, of Chelsea, Massachusetts. The party were on an outing to Searsport Beach.

The three women, riding in the back of the small touring car were killed instantly and Everett died within a few minutes. As soon as he saw the car the engineer of the troop train had applied his brakes in emergency, but the train still slid 150 feet with locked wheels after hitting the car. When it was realised what had happened the soldiers formed rescue squads and the ambulance detail went with stretchers to the roadside. It was clear that nothing could be done for the poor women, so they concentrated on the men. Though there were no doctors on the troop train Drs P.W. Whittaker and H.L. Trueworthy from Unity did what they could, and the two injured men were taken on the troop train, running at high speed under special orders, to Waterville station, where ambulances were waiting to take them to the hospital. Unfortunately, Messrs Varney and Burrell died shortly after their admissions. The one survivor of the accident was little Muriel, who was thrown clear and escaped with a few cuts and bruises. Having been checked over by the surgeons she was later taken to Pittsfield to be cared for by relatives of the Varney family. [1148]

At about 1 p.m. on Thursday 20th July 1922 a Toronto Hamilton and Buffalo Railway train from Toronto, Ontario, to Buffalo, New York, collided with a car at Stevensville, Ontario, Canada. Eight people were riding in the car, which was driven by Mr Daniel Fretz, at the time of the accident living with his family in Colorado but formerly a farmer of Bertie Township, Ontario. The Fretz family had returned to Canada for a holiday and shortly before the accident had left the residence of Deputy Reeve James Baker, close to Stevensville station. Mr and Mrs Fretz, their children Ethel, Harold and Carl, and Robert Baker, aged 10, were killed outright. Mrs Edith Baker, the wife of the Deputy Reeve, was taken from the wreckage alive but died before reaching Buffalo. Thelma Baker was taken to hospital in serious condition, but it was thought that she would recover. Robert and Thelma were Mr and Mrs Baker's son and daughter.

The train had been travelling at about 50 miles an hour. Engineer Humphreys declared that he had whistled for the unguarded crossing and had not seen the car at all. His fireman said that he had seen the vehicle at the last moment, but too late to warn his engineer. The injured were put on the train and taken to Buffalo. The car had been torn to pieces and the bodies of those killed were badly mangled. [1149] [1150]

In the great majority of level crossing collisions the principal and often the only victims are the occupants of the road vehicle struck by the train, but there are always exceptions. Thus of the ten persons killed in the accident at Annandale, Minnesota, United States, on Saturday 12th August 1922 six were train passengers and two were railroad employees.

[1148] *The Lowell Sun*, Lowell, Massachusetts, United States, 17th July 1922

[1149] *Manitoba Free Press*, Winnipeg, Manitoba, Canada, 21st July 1922

[1150] *The Evening Huronite*, Huron, South Dakota, United States, 20th July 1922

Westbound Minneapolis St Paul and Sault Ste Marie Railroad passenger train No. 107 left Minneapolis, Minnesota, at about 12 noon. It consisted of engine No. 727, a mail and express car, a baggage car, a smoker, two coaches, two tourist cars, a dining car and five sleeping cars in charge of Conductor Matthews with Engineer Whitchurch and Fireman Inman. An hour and a quarter late and travelling at about 45 miles an hour, it collided with an oil tanker a quarter mile east of Annandale station. The lorry was thrown violently against a switch stand, opening the points, so that the whole of the train apart from the engine, tender and leading mail car was diverted onto the siding. The baggage car smashed up against the engine of eastbound freight train No. 2 standing on the siding and was demolished and the next five vehicles were more or less damaged. The baggageman of train No. 107 and Engineer Chris Wallace of train No. 2 were killed. The other two to lose their lives were Fred Lamer, the lorry driver, and Nels Hanson, his passenger. [1151] Another 35 train passengers and four railwaymen were injured.

No inquest was held on the victims of the accident, Coroner Moffat deciding that one was not necessary, [1152] but this time it was investigated by the Interstate Commerce Commission (ICC). Brakeman Rahm of train No. 2 was the first person to give evidence. He stated that in accordance with the usual practice, although there was no written rule to that effect, when his train had stopped close to the crossing he had got down to flag. He recalled that when train No. 107 had come in sight he had walked towards the east switch, which had been securely locked for the main track, and he had heard the express give both the normal whistle at a passing place and the whistle for the highway crossing. He had then seen the oil truck approaching and had moved to the middle of the highway and flagged it to stop. He had waved his flag four or five times, but the lorry driver and his passenger appeared to be talking and laughing and took no notice of him. The lorry was travelling at 12 or 15 miles an hour. Only when the truck was 40 or 50 feet from the crossing did its driver look up, see the train and swerve to the left. Brakeman Rahm said that when he had seen that there was likely to be a collision he had given emergency stop signals to the train engineer, then jumped out of the way to avoid flying debris, hearing the brakes of the train being applied as he did so.

Engineer Whitchurch stated that his train was travelling at between 40 and 50 miles an hour as it approached the crossing. He had a heavy train and although they were late he was not trying to make up time. He had whistled for the crossing and having done so he had seen the brakeman giving violent signals to someone on the opposite side of the track, then his fireman had shouted, "You are going to hit a truck." He had made an emergency brake application, then almost at once he and his fireman had been temporarily blinded by gasoline which had splashed from the wrecked truck, which he had not seen at any point before the collision. The brakes of his engine, which was only slightly damaged, had been working properly and its bell, which worked automatically, had been ringing all the way from Maple Lake, 6¼ miles back, and had continued to ring after the accident.

Fireman Inman said that he had not seen the brakeman, who had been on the engineer's side of the engine, but he had seen the yellow oil truck when it was between 50 and 100 feet from the crossing. It was moving slowly but when he saw that it was not stopping he had shouted to his engineer, who had applied the brakes at once.

[1151] *The Bisbee Daily Review*, Bisbee, Arizona, United States, 13th August 1922
[1152] *The Wisconsin State Journal*, Madison, Wisconsin, United States, 14th August 1922

The ICC reached the obvious conclusion that the accident had been the fault of Driver Lamer, who paid for his negligence with his life. The commission added that, 'There can be no possible extenuation for such gross carelessness as was shown by the truck driver in failing to heed the signals given by Brakeman Rahm as well as in neglecting to look toward the east when approaching the crossing. Unless automobile drivers heed warning signs and flagmen's signals at railroad crossings, other accidents of this character may be expected to occur in the future.' [1153] A prescient comment.

The only fortunate aspect of the accident was that there was no fire.

Only 10 days after the Annandale accident, on Tuesday 22nd August 1922, a Michigan Central Railroad passenger train, the northbound Toledo – Detroit Flyer, collided with a truck loaded with turpentine and gasoline at a level crossing with the Dixie Highway on the edge of the village of Rockwood, Michigan, United States. This time there was no derailment but there was a fire, both locomotive and truck being engulfed in flames. Engineer George Engle was killed was killed more or less outright, Fireman E. Webb died soon after reaching hospital and the driver of the lorry, W.J. Leinders, was badly burned and died later in hospital. The coaches of the train were scorched but none of the train passengers was hurt. Engineer Engle had had time to apply the brakes and the train stopped some distance down the track. [1154] [1155] The brief newspaper reports on the accident do not indicate what sort of protection, if any, there was at the crossing.

A Chevrolet touring car carrying eight members of a family who were returning from a fairground at Luxemburg, Wisconsin, United States, at about 4.45 p.m. on the evening of Wednesday 6th September 1922. Their journey had only just begun when the car was struck by Kewaunee Green Bay and Western Railroad passenger train No. 14 on a level crossing two blocks from Luxemburg station. Mr and Mrs Frank Guradnik, their daughter Adeline, aged 20, and Maron Lasnoski, their son-in-law, were killed instantly and their daughter Erne, aged 17, was fatally injured. Daughters Hilda, 15, Armilla, 11 and Olga Lasnoski (nee Guradnik) escaped with slight injuries. Mr and Mrs Lasnoski had been on their honeymoon.

The bell of the engine, in charge of Engineer Steve Slattery, had been ringing as the train approached the crossing, but the view of the track was obscured by the Krause cheese warehouse and other buildings. Mr Guradnik had approached the crossing slowly but according to eyewitnesses had failed to notice the train. The car was dragged 300 feet before the train could be pulled up. The train was stopped for an hour while the injured were cared for, the bodies removed, and the wreckage cleared away. [1156] [1157] [1158]

At the beginning of October 1922 the Espen and McGee families, who were related,

---

[1153] Interstate Commerce Commission report

[1154] *The Commercial Tribune*, Cincinnati, Ohio, United States, 23rd August 1922

[1155] *Bluefield Daily Telegraph*, Bluefield, West Virginia, United States, 23rd August 1922

[1156] *Door County Advocate*, Sturgeon Bay, Wisconsin, United States, 8th September 1922

[1157] *The La Crosse Tribune*, La Crosse, Wisconsin, United States, 7th September 1922

[1158] *The Escanaba Daily Press*, Escanaba, Michigan, United States, 10th September 1922 had adjoining farms two miles south of Exeland, Wisconsin, United States, but since the early summer the McGees had been living in St Paul, Minnesota, and Mr Espen, brother-in-law of Mr William McGee, had secured a job in the Minnesota city. On Monday 2nd October, the two families were moving to St Paul, eight of them being packed into a small Ford touring car. As the car, loaded with household effects in addition to its passengers and driven by Mr McGee, reached St Paul, it reached the tracks of the Chicago St Paul Minneapolis and Omaha Railroad at Atlantic Street. Atlantic Street was little more than a driveway, little used by motorists, [1159] and it was thought that the party had got lost in attempting to get to Seventh Street, where the McGees were residing. On the approach to the level crossing the narrow street was bordered by piles of lumber (dressed timber) on either side so that the view of the track was much obscured. As the car reached the crossing a northbound train from Chicago was coming down the grade at 35 miles an hour. Engineer Albert McKay saw the car at the last moment, applied his brakes and sounded the whistle, then came a terrific impact as the locomotive struck the car.

Mr and Mrs William Espen, their children Francis, 10, Lloyd, 9 and baby Glen Earl, 4 months; together with Mr and Mrs R.W. McGee, were killed outright. Seven-year-old Marvel Espen, the other passenger in the car, was critically injured and died in the City hospital the following afternoon. No one in the car survived the accident.

A newspaper reporter described in considerable detail the aftermath of the accident, 'Identification was made by certificates of the Exeland school which were found in the pockets of some of the children. The certificates arranged the entrance of the children to St Paul schools. A bank book showing a deposit in a Texas bank gave rise to a report circulated in the large crowd that gathered at the scene of the accident that the victims were tourists. Identification was made positive by Frank McGee, 549, Canada street. Espen, while in St Paul, had roomed with Frank McGee.

'A litter of twisted metal, feathers, home-made jelly and broken dolls, over which a policeman keeps vigil and tells the story of the accident to curious loiterers, is all that remains to bear witness to the hopes of William Espen and his family. The dolls were the companions of the four Espen children. Of the children only one remains, and Monday night it was questionable whether she would survive to repair the damage to doll limbs and rents in doll attire.

'Espen had arranged with high spirits to bring his family to St Paul. He had found a home, he had remunerative work, and the children were to start school next week.

'Although eight persons were crowded into the little automobile, there was room apparently for many trifles to please the children and many articles of household equipment. In the tangle of iron one witness found two rabbits, dead. They were pets of nine-year-old Lloyd, and in the crash they met the same fate as their master. Over all the mass were feathers, and in the conglomeration of junk were found the carcasses of chickens. Two rifle stocks, but the barrels were wrenched off, and were lost to view in the pile of broken machinery. Intermingled were bits of cloth, torn from clothing and a pile of blankets. The car was so wrecked, and many studied it for minutes before they were able to determine its make.

'Mrs Espen, in the moment of death, tried instinctively to shield her baby, witnesses declared. The child was in the mother's lap, and they were seen hurled from the machine

[1159] *The Sedalia Democrat*, Sedalia, Missouri, United States, 3rd October 1922 together. The mother lost her hold, however, but her arm was outstretched to the prostrate form of the baby when she was picked up by Patrolman C.J. Watson.' [1160]

In a strange repetition of the St Paul accident only a week earlier on Sunday 8th October 1922 a car containing a family moving from Chicago, Illinois, to a farm they had rented near Plymouth, Indiana, was involved in a level crossing accident at the Yellowstone Crossing, about three miles west of Valparaiso, Indiana, United States. The car was following a truck loaded with the family's furniture. Just as the car reached the crossing it stalled. Mr Rene Hargot Senior, who was driving, attempted to push the vehicle off the track assisted by Charles Misholan, a family friend who was travelling with the family. They were unsuccessful, and shortly a Pennsylvania Railroad train was upon them. Mr Hargot just had time to grab his 14-year-old son Rene Junior, who had been riding in the front seat, and jump clear, but the rest of the family had no chance to escape. It was raining and the curtains of the car were drawn so they probably did not even realise their danger. Mrs Harriet Hargot and their children Raymond, 12; Irene, 10; Richard, 8; Susan, 6; Margaret, 3; and Edmund, 18 months, were all killed. [1161] [1162] Mr Hargot was later reported to be in a critical condition, suffering severely from shock. [1163]

Mr Hargot was clearly one of those people who are just unlucky. It was said that he had lost over $200,000 in France due to the war and had moved back to Chicago after the armistice. [1164]

Only briefly mentioned in newspaper reports was a level crossing collision at Waggoner, near Lichfield, Illinois, United States, on Sunday 22nd October 1922. Mr H. Ray Keefe, his wife, and their baby; Miss Lucille Scott, aged 20, and Miss Ruth Boyd, aged 19, lost their lives when the automobile in which they were riding was struck by an Illinois Central Railroad train. The same day a Wabash Railroad train killed three people in a level crossing collision at Bement in the same state. [1165]

Five teenagers from New Bavaria, Ohio, United States, were killed in a level crossing accident at nearby Holgate at about 10.30 p.m. on Thursday 26th October 1922. The victims were Virgil Brown, 18, Thelma Stimet, 18, Larry Schwab, 16, Cecelia Royal, 15 and her brother John, 14. The girls, or possibly all five, were students at Holgate High School and the teenagers were probably going home after a party. Their car was struck on an unguarded crossing by a Baltimore and Ohio Railroad train which was performing switching operations. [1166] [1167] [1168]

[1160] *Ladysmith News-Budget*, Ladysmith, Wisconsin, United States, 6th October 1922
[1161] *Connersville News-Examiner*, Connersville, Indiana, United States, 9th October 1922
[1162] *Linton Daily Citizen*, Linton, Indiana, United States, 9th October 1922
[1163] *The Tipton Daily Tribune*, Tipton, Indiana, United States, 9th October 1922
[1164] *The Indianapolis Times*, Indianapolis, Indiana, United States, 9th October 1922
[1165] *Fremont Eagle*, Fremont, Indiana, United States, 26th October 1922
[1166] *The Lima News*, Lima, Ohio, United States, 27th October 1922
[1167] *The Commercial Tribune*, Cincinnati, Ohio, United States, 28th October 1922
[1168] *Hamilton Evening Journal*, Hamilton, Ohio, United States, 27th October 1922

The classic and all too common mistake by motorists of misjudging the speed and closeness of an approaching train and trying to cross the line in front of it was responsible for the death of four adults and two children at Oklahoma City, Oklahoma, United States. The adults: Opal Darnell, Ruth Howard, Etalia McFadden and Cora Phelan, were all schoolteachers, and the two children: Frank Davis, 10, and Henry Maloney, 9, were, it was presumed, being given lifts home. The accident happened at 4.45 p.m. on Monday 15th January 1923. Miss Howard owned the Ford car which was going west when it was struck by an Atchison Topeka and Santa Fe Railway northbound train travelling at an estimated 50 miles an hour and carried three-quarters of a mile. [1169] [1170]

A collision on Oakland Park Avenue Crossing, Columbus, Ohio, United States, on Friday 30th March 1923 was another of those in which the train suffered badly as well as the road vehicle. Southbound Cleveland Cincinnati Chicago and St Louis (Big Four) Railroad passenger train No. 11, otherwise known as the *Southwestern Limited*, collided with a car on the crossing and was badly derailed. Three occupants of the car, three railroad employees, a train passenger and a trespasser lost their lives and 43 passengers, and 5 railroad employees were injured. The Interstate Commerce Commission (ICC) investigated the accident.

The train consisted of engine No. 6480, a club car, six Pullman cars and a dining car in charge of Conductor McGraw with Engineer Andrew Green. The latter told the inquiry that he had whistled for the crossing and his engine bell had been ringing for some time. His train, which was an hour late, was running at 70 miles an hour at the time. He said that he had not seen the car until immediately before the collision, when he had applied the brakes in emergency. He thought that some part of the wrecked car had got under the wheels of his engine and derailed it. The three railwaymen killed were his fireman and two assistant fuel supervisors who had been riding on the engine.

Conductor McGraw, riding in the fifth vehicle of the train, stated that his first warning of anything wrong had been hearing a grinding noise and feeling the brakes coming on. From his position in the train he could not hear either the engine bell or its whistle.

Operator Butler had been standing at the door of the Clintonville telegraph office, situated 300 yards north of a crossing. He said that he had seen train No. 11 approaching and the radiator of the car involved appear from behind a box car parked on a siding. He had assumed that the car would stop. It did not and he had witnessed the collision. He thought that the train was travelling at 85 to 90 miles an hour. The engine whistle had been blown and its bell was ringing.

Brakeman Dumont of a Pennsylvania Railroad train switching nearby testified that he had seen the automobile nearing the crossing. Like the operator he had imagined that the car would stop and did not take much notice of it until he saw it enter the crossing in front of the train. The car was only moving at a moderate speed, about 15 miles an hour. He, too, heard the train whistle for the crossing. Neither he nor anyone else had made any signal to the car driver.

[1169] *The Ada Evening News*, Ada, Oklahoma, United States, 16th January 1923
[1170] *The Ada Weekly News*, Ada, Oklahoma, United States, 18th January 1923

Other witnesses agreed that the car had approached the crossing at a moderate speed and had not tried to race the train over the crossing, it had just failed to stop. The car driver, Mrs Hemminger, the wife of an attorney, had been driving the car or a similar model for six years, had lived a mile from the crossing for several years, and was considered by a neighbour to be a competent and careful driver. Mrs Hemminger and her children, five-year-old Robert and an unnamed baby, were all killed. [1171]

The ICC concluded that the accident had been caused by the car driver, who after passing some buildings which had obscured her view, had failed for some reason to see train No. 11, perhaps because her attention had been distracted by the shunting Pennsylvania train. The evidence indicated that the whistle had been blown by the engineer and that the speed of the train had probably not exceeded the line limit of 20 miles an hour. The commission remarked that, 'This accident again calls attention to the vital necessity for automobile drivers to approach railroad crossings with caution and to make certain before proceeding over them that no train is approaching.'

The only protection provided at the crossing were crossbucks and signs reading 'RAILROAD CROSSING' and 'TWO RAILROAD CROSSING DANGER.' A 24-hour survey carried out by the ICC showed 495 motor vehicles, 53 horse-drawn vehicles, 75 pedestrians, 11 equestrians and 6 livestock passing over it during that period, plus 60 train movements. The ICC considered that given the volume of road and rail traffic over the crossing more protection that mere signs was necessary. Finally it noted that the all-steel rolling stock of the train had probably minimised the loss of life to train passengers. [1172]

On Sunday 1st April 1923, a five-passenger touring car driven, according to one report, by Mr J.R. Saunders, of Normal, Kentucky, United States, drove onto Bellefont Crossing, two miles west of Ashland and about four miles from the Saunders' home, into the path of a Chesapeake and Ohio Railroad passenger train. The car was not very old, dating from 1919, but according to an eye-witness it had covered thousands of miles and when being driven in intermediate gear was very noisy. It would appear that the eye-witness, Homer Fuller, had been stopped at the crossing waiting for the train when the Saunders' car overtook him. He said that its engine had been labouring and its driver appeared oblivious of the train, which was train No. 2, the *Fast Flying Virginian*. Mr Saunders, Mr. Ernest Tinsley, his wife Stella, their 6-year-old daughter, and Miss Carrie Keebler, aged 21, were all killed. Mrs Saunders and the Saunders' children, Pearl, aged 2, and Opal, aged 1, were thrown from the vehicle and not badly hurt. They were taken to the home of her uncle, Mr Alva Saunders, of South Ashland. It was noted that Mr Saunders had been driving a car for some time and was considered a good driver. [1173]

A report in another newspaper put a somewhat different complexion on the accident. To begin with it stated that it was Mr Tinsley rather than Mr Saunders who had been driving the car. It wrote that the accident, which had happened at 3.15 p.m., had been due not to the car driver failing to see the train but to his attempting to beat it over the crossing and misjudging its speed, adding that Mr Tinsley was an inexperienced driver

---

[1171] *Steubenville Herald-Star*, Steubenville, Ohio, United States, 30th March 1923

[1172] Interstate Commerce Commission report.

[1173] *The Portsmouth Daily* Times, Portsmouth, Ohio, 2nd April 1923 and had he accelerated he could probably have cleared the crossing in time. The train had been travelling at 40 miles an hour. The view at the crossing was unobstructed in both directions. The party had enjoyed Easter dinner at the Tinsley home and were going for a ride, it was believed to Ironton. They had only been on the road 15 minutes. [1174]

A train ran into a car on a level crossing at Mantua in northern Italy on the morning of Tuesday 3rd April 1923, killing seven of its eight occupants. [1175]

The French occupied region of the Saar, Germany, was the scene of a serious level crossing accident on Sunday 8th April 1923. A car broke through a lowered crossing barrier on the road between Dillingen and Bothen and was struck by a train. Six of the passengers in the car were killed outright and its driver died later in hospital. [1176]

Undue haste by a man anxious to get his expectant wife to hospital led to five deaths in a level crossing accident at Alhambra, California, United States, on the evening of Sunday 29th April 1923. Mr Daniel Durate of San Bernardino was heading for Los Angeles with his wife and Mr and Mrs Francisco Morales. The car collided with a Pacific Electric Railway interurban train and the four were killed almost instantly. As a result of the accident the dead or dying Mrs Durate gave birth to a baby boy, but the child lived only a short time. The victims were all Mexicans. [1177] [1178]

Saginaw Avenue Crossing, a mile north of Selma, California, United States, was the scene of a bad accident at about 6.45 p.m. on Sunday 5th May 1923 when a car containing six members of a family was struck by Southern Pacific Railroad passenger train No. 107, the *Valley Flyer*, which had left Bakersfield at 3.40 p.m.. The train was in charge of Conductor Charles Fowlie with Engineer R.O. Quakenbush. It was travelling at about 45 miles an hour and took 32 rail lengths to pull up. The Ford automobile was demolished, and all its occupants were killed. They were Mr and Mrs John De Motz, of Orange Cove; Edward De Motz, brother of John, also of Orange Cove; Mrs Jongewaard, daughter of the De Motzes, of Merced; and Dr and Mrs Ogg, of Orange City, Iowa. Mrs De Motz and Mrs Ogg were sisters. The party had been returning home from the nearby ranch of Julius Christensen. The crossing was unprotected but there was a clear view from it along the line in both directions and with the death of all those involved the accident remained unexplainable. [1179]

A lorry carrying the possessions of the Williams family, who were moving from Franklin to Germantown, was struck by Baltimore and Ohio Railroad passenger train No.

[1174] *The Commercial Tribune*, Cincinnati, Ohio, United States, 2nd April 1923
[1175] *The Times*, London, England, 4th April 1923
[1176] Ibid, 10th April 1923
[1177] *Reno Evening Gazette*, Reno, Nevada, United States, 30th April 1923
[1178] *Berkeley Daily Gazette*, Berkeley, California, United States, 30th April 1923
[1179] *The Bakersfield Californian*, Bakersfield, California, United States, 7th May 1923

64, travelling at 50 miles an hour, on a level crossing at Carlisle, Ohio, United States, a mile from Franklin on Monday 21st May 1923. Riding in the back of the lorry, sat on a sofa, were Mrs Margaret Williams, 53, her daughter Sarah, 13, her daughter-in-law, Mrs Lelia Williams, 20, and the latter's children, Mary Louise, 3, and William, 4 months. Majess Gross, the driver of the lorry, was seriously if not fatally injured and a man named Solen Kindred escaped with cuts and bruises. As the lorry approached the crossing with the main street of Carlisle a freight train was passing and although the crossing bells continued to ring Gross drove onto the line behind it to be struck by the passenger train coming in the opposite direction.

Bert Williams, husband of Margaret, was following the lorry riding in a bus and initially unaware of the accident. When he found the body of his wife he fainted and after being revived was taken to the home of friends in Franklin. Mrs Lena Williams' husband, from whom she had been estranged for some months, was believed to be in Canada. [1180] Gross stated after the accident that a building and a string of cars had obscured his view of the track and Engineer S.J. Bennett stated that the same cars had obscured his view of the lorry. [1181]

A train collided with a lorry converted into a bus by installing benches along each side of its body on a level crossing at Kärra, between Kållered and Fässberg, Sweden, on Sunday 22nd July 1923. Five bus passengers were killed outright, four more died of their injuries. All the other bus passengers, who numbered 32 in all, were more or less hurt. The vehicle was reportedly 'smashed to kindling wood.' [1182] [1183]

Early on Sunday 29th July 1923 Gateman Edwin Scott closed the gates at the Merrick Road Crossing, Rockville Center, New York, United States, for the last Long Island Railroad Babylon theatre crowd train from New York. That the gates had been down was confirmed by Engineer F.X. Yaeger in charge of the train. Despite this a car appeared put of the gloom, smashed through the gates without having reduced speed and was immediately hit by the train. The car driver, William Gottlieb, President of Gottlieb and Rabinowitz Inc., garment manufacturers, and his four young women passengers were all killed. Four of the victims were killed outright and one of the young women died as she was being removed from the wreckage. The party had been returning from an outing of the National Merchandise Fair arranged by the Retail Dry Goods Association.

Although the testimony of the gateman was supported by the engineer he was arrested by the Rockville Center police, presumably on the basis of a statement from William Keller, a motorist in a following car, who said that he believed that the gates had been open, although he later admitted that the night had been misty, and he could have been mistaken. In any case the fact that one gate and its lanterns was demolished, and the other was damaged surely proved that they had been closed as the gateman stated.

The four women were later identified as Miss Margaret Lewis, a stenographer, of

---

1180 *The Commercial Tribune*, Cincinnati, Ohio, United States, 22nd May 1923
1181 *The Evening Gazette*, Xenia, Ohio, United States, 22nd May 1923
1182 Webpage – Järnvägsolyckor I Sverige
1183 *The Times*, London, England, 23rd July 1923

Manhattan; Miss Verna Totter, a writer, of Manhattan; Mrs Gladys Miller, an actress, of Brooklyn; and Miss Thelma Keough, an actress and sister of Mrs Miller. [1184]

Conductor C. S. Hornshu, Engineers Glen Duffy and O. A. Black and other members of the crew of eastbound Pennsylvania Railroad passenger train No. 30 were stated to have been on the point of nervous collapse when there train finally arrived at Terre Haute, Indiana, United States, on the evening of Sunday 29th July 1923. Firstly, at Highland, Illinois, their train struck a car which had apparently stalled on a level crossing, killing four of its occupants. Continuing its journey two hours late after that accident the same train had then collided with another car at Liggett, eight miles west of Terre Haute. G.G. Gillishie, the operator at Liggett tower, who witnessed this second accident, said that the car driver had apparently attempted to beat the train, which was travelling at 50 miles an hour, as he had stopped momentarily before proceeding onto the crossing. This time no less than nine people in the car lost their lives, the vehicle being 'torn into small bits of junk.'

The victims of the Liggett accident were Mr Raymond Thompson, the driver of the car; Mrs Velta Bostwick; her children May, 18; Richard, 16; Clarence, 13; Trilla, 13; and Etta, 3 and Mrs Ethel Slavens and her daughter Anna, 3 months. The party, from Danville, Illinois, were on their way home after visiting friends in Terre Haute and Liggett. They had just left the home of John Roberts, only 200 yards from the crossing. [1185]

On Tuesday 7th August 1923 nine members of the Kendall and Spalding families from Kingsville, Ohio, were on their way to Chicago, Illinois, to attend the Golden Wedding anniversary of Mr and Mrs James Hutton. Railroad officials surmised that the party were unfamiliar with the situation at Griffith, Indiana, United States, where the Chesapeake and Ohio, Elgin Joliet and Eastern, Erie, Grand Trunk Western and Michigan Central Railroads met, and their large touring car was struck on a level crossing by an Elgin Joliet and Eastern train. Mr O. L. Kendall, his son, Robert, 9 and daughter Dorothy, 3, together with Mr Charles Spalding and his daughter Leola, 15, were killed. The other four family members were injured, Mrs Kendall it was feared fatally. The car had tried to cross in front of the train. [1186]

Erie Railroad train No. 6, the eastbound Cleveland – New York flyer, travelling at 60 miles an hour, struck a car on Gray's Crossing, Binghampton, New York, United States, on the evening of Sunday 26th August 1923. The victims of the accident had been to a party at Binghampton at the home of Katy Twaryones, one of those killed, but four of them – Karto and Mary Huberis, Matrys Rudis and Tony Kamlusky – lived at Endicott and were being taken home there. The sixth person to lose his life, Frank Hardick, lived at Binghampton. Mary Huberik's daughter, Helen, aged 6, was taken from the pilot of the engine a quarter mile from the crossing and rushed to hospital with little hope of recovery.

[1184] *New Britain Herald*, New Britain, Connecticut, United States, 30th July 1923

[1185] *The Brazil Daily Times*, Brazil, Indiana, United States, 30th July 1923

[1186] *Steubenville Herald-Star*, Steubenville, Ohio, United States, 7th August 1923

Coroner Frank Goodwin, who investigated the accident, stated that he would interrogate the engineer of the train, but he did not think that an inquest was necessary. There was a thick fog at the time of the accident and the car driver ignored the warning signals at the crossing, perhaps because they were not seen in the fog.

Two level crossing collisions with buses occurred in the United States on Thursday 11th October 1923. In the less serious accident a Pennsylvania Railroad express collided with a motor bus running from Williamsport to Blossburg and carrying ten passengers near Hubbensville, Pennsylvania. The bus had stopped 25 feet from the crossing to let passengers alight. According to witnesses the bus driver then changed into a higher gear and set off over the railroad without thinking to check if a train was coming. The train struck the bus squarely in the middle, killing three passengers outright and possibly fatally injuring two more.

The other accident was different in that it involved a horse-drawn vehicle, and it was a school bus rather than an ordinary passenger service. Again a Pennsylvania Railroad express was involved, this time it was the westbound *Clevelander* from New York to Cleveland. Ohio. The scene of the accident was Lambert's Crossing, Ohio, between Atwater and Ravenna and only a short distance from Rootstown school, the destination of the bus.

Pupil Andrew Kazimer, 14, had been riding at the very rear of the bus, heard the whistle of the express and saw it coming around the corner. He said afterwards that he had screamed a warning to the other children and jumped off the back of the bus. Five of the other pupils followed suit and escaped injury. Andrew said that the bus driver, like his colleague at Hubbensville, had failed to check that the line was clear before he began to cross. When he saw the train he made a frantic effort to get the bus clear as the engine of the train, its brakes grinding, came up at between 50 and 60 miles an hour, but only the horses and the front of the bus got clear in time.

Eight children died in the accident, three of them, Mildred, 11, Harold, 9 and Verna Shaw, 2, being the offspring of farmer C.G. Shaw. A fourth Shaw child, Lawrence, 8, was injured. The other children who died were Thelma Benshoff, 9; Margaret Kuntz, 12; Richard Silvafi, 8; Ella Staincke, 14 and Julia Wancik, 13. Julia's brother, Steve, was seriously injured as was the bus driver, Lewis Klein. [1187] [1188]

Engineer G.G. McCaslin of the express also placed the blame for the accident on the bus driver. The road crossed the line at an angle but there was a clear view of approaching trains and it would seem that as at Hubbensville the bus driver had just allowed his attention to wander and simply forgot to look for trains.

Nine lives were again lost in a level crossing collision at Brookfield, Indiana, United States, on the morning of Sunday 14th October 1923. A car containing ten persons was struck on a level crossing by Cleveland Cincinnati Chicago and St Louis (Big Four) Railroad train No. 1, killing or fatally injuring Mr James Means, his wife Jennie, their son Byron, 13; their married daughter, Madge Britton, their granddaughter, Mary, 7; Mrs

---

[1187] *The Star-Journal*, Sandusky, Ohio, United States, 11th October 1923

[1188] *The Lima News*, Lima, Ohio, United States, 11th October 1923

Jennie McGuire, sister-in-law of Mrs Means, Marjorie McGuire, 6, daughter of Mrs McGuire and Golda and Martha Gaither, 20 and 18, sisters of Mrs McGuire. The only survivor of the party was Mr Roy Britton, who escaped with slight injuries.

Mr Britton said that going up the incline to the crossing Mr Mears had kept the car in a high gear and when the vehicle got onto the crossing it had stalled when the train was close at hand. After the accident he had found Mr Mears lying beside the track a short distance from the crossing. Mr Mears was conscious and asked Mr Britton if all the others had been killed, then had added, "It was all my fault." Shortly afterwards he had died. [1189]

An inquest was held into the accident, which heard the evidence of Mr Samuel Games, the only eye-witness to the tragedy. Mr Games stated, 'I was standing on my porch which is about 100 feet from the railroad crossing in Brookfield. I heard the train coming and saw a machine coming slowly toward the track. I thought he would stop, and I did not yell at him. I saw him drive on the track and the car go dead. He would have gotten on across if his car had not died on the track. When the car went dead I jumped off the porch and started towards the crossing. The train gave two short hoots as a warning blast but did not whistle at the whistling post. I saw the train strike the machine and the car and bodies thrown about. I went to all the parties except one child that was thrown farthest down the track. I found Mr Means living. Mrs Means, Byron Means, Mrs Britton and Mary Rosamund Britton were all dead. Mrs McGuire and one of the Gaither girls were dead. One of the Gaither girls was living. I did not see the little McGuire girl.

'I picked up some money and Mr Gants gave me a bill he had found. The conductor demanded it and I turned it over to him. I never counted it, but I know there was $40 or $50. The crossing is a grade crossing, and the view is obstructed until a person is within 10 or 12 feet of the main crossing. A person on foot could see further, but one in a machine could not have a clear view until on the switch Track. There is no signal bell nor any warning to the public except the usual R.R. crossing signs.'

Engineer George Diver, Fireman Ray Stevens, Conductor C.E. Kennedy and Brakeman Marion Overting all testified that the train had whistled at the whistle post 1,500 feet from the crossing as well as when the car was seen.

Dr G.I. Inlow, Coroner for Shelby county, where seven of the victims had died (the remaining two had died in Indianapolis Hospital in Marion county) found that the accident was 'due to the failure of J.W. Means to look for an approaching train and his automobile stalling on the track.' He recommended that the railroad company install an electric signal bell at the crossing. [1190]

Three ministers of the gospel were among five persons killed in a level crossing accident a mile north of Winfield, Alabama, United States, late on Sunday 21st October 1923. Those killed were Rev. P.F. Lucas, his wife and two-year-old daughter, of Winfield; Rev. William Otts, of Sulligent, Alabama, and Rev. Mack Hankins, of Fulton, Mississippi. Two others in the car, Rev. D.A. Carter, of Hamilton, Alabama, and Rev Owens, of Fulton, Mississippi, were injured, the latter critically so. The party were

[1189] *The Shelby Democrat*, Shelbyville, Indiana, 18th October 1923
[1190] Ibid, 25th October 1923

returning from a revival meeting when the accident happened. The car had almost cleared the crossing when it was struck by a St Louis-San Francisco Railway train. [1191]

On the afternoon of Tuesday 13th November 1923 a New York Central Railroad local freight train in charge of Conductor Hiram Austin with Engineer Marley and Fireman Hover had called at Craryville station but was not intending to stop at the next station at Martindale, New York, United States, as there was no traffic for it to drop off or pick up. As it neared the latter place Engineer Marley whistled for the level crossing there, then when he saw a car stopped on the crossing, he made every effort to stop his train, but it struck the car and wrecked it. Four of the occupants of the automobile, a Willys-Knight, were killed outright. They were Mr Frank Whiteman, Mrs Alida Miller, his housekeeper, who was driving the car; Mrs Miller's daughter Dorothy, aged 5, and Mrs Dingman. Mrs Dingman's husband, Mr Asbury Dingman, died three hours later in hospital. All lived near Craryville.

Engineer Marley believed that when Mrs Miller had seen the train close at hand she had panicked and stalled the car. William Lockard, the assistant station agent at Martindale, was alerted by the continuous whistling of the engine and witnessed the accident.

The crossing was regarded as one of the most dangerous in the county. The road ran parallel to the track for some distance before turning and crossing it at right-angles and the view of trains offered to car drivers was very limited. An electric bell had been installed at the crossing two years previously following pressure from a local newspaper, but it was thought it was not heard above the noise of the car engine whilst the closed curtains of the vehicle further reduced visibility. There were efforts to either replace the crossing with an underpass or else to re-site it north of the station. [1192]

Six members of the Whittenmeyer family died at 11 a.m. on Sunday 13th January 1924 when their car was struck by a Chicago and Eastern Illinois Railroad passenger train on a level crossing at Emison, Indiana, United States. Mr Claude Whittenmeyer, his wife Lula and their children Helen, 16; Mary, 8; Lorene, 4, and Charles, 3 months. They had been on the way to relatives for Sunday dinner when the accident occurred. Mr Whittenmeyer had waited at the crossing for a freight train to go by, then drove onto the crossing into the path of the fast-moving passenger train on the other running line. The Emerson operator, William Wolfe, had a narrow escape when the wreckage of the car was thrown against the station buildings. The only survivor of the family was a boy named Junior Walker, an orphan who had been adopted by the Whittenmeyers and had gone to Sunday School in Emison that morning. He now became an orphan for a second time. [1193] [1194]

Seven newsboys who had been attending a party at nearby Cudahy were being taken home in a truck to St Francis, five miles south of Milwaukee, Wisconsin, United States, and were almost there when the vehicle was struck by a Chicago and North Western

---

[1191] *The Decatur Daily*, Albany, Alabama, United States, 22nd October 1923

[1192] *The Brewster Standard*, Brewster, New York, United States, 23rd November 1923

[1193] *Linton Daily Citizen*, Linton, Indiana, United States, 14th January 1924

[1194] *The Indianapolis Times*, Indianapolis, Indiana, United States, 14th January 1924

Railway train on a level crossing at 10.10 p.m. on Friday 8th February 1924. The driver of the lorry, Arnold Voight, 26, and newsboys Bailey Clements, 12; Lester Holling, 13; Vilas Kohler, 13; Michael Mochalski, 14, and (probably but not immediately positively identified) two brothers named Loganover, aged about 15 and 16. The remaining boy, Gilbert Kohler, 14, brother of Vilas, suffered a fractured skull and other injuries and it was not clear if he would survive.

A watchman attended the crossing but had gone off duty at 10 p.m. The train, which had left Chicago at 8 p.m. and was due in Milwaukee at 10.15, had been ten minutes late. If the watchman's hours of duty ended at 10 p.m., as seems likely, it is hard to see how he could be strictly blamed for going off duty when he did, especially if he had had no information as to how late the train was, though perhaps a more conscientious person might have remained longer. The engineer had reportedly whistled for the crossing. First reports that the truck had stalled on the crossing could not be confirmed.

The newspaper account of the accident added, 'Pathetic evidence of the joy immediately preceding death was disclosed in the clothing of the dead, when the bodies were removed to the morgue. Crushed apples, oranges and candles were found in the pockets of the victims.

'Voight is survived by a widow and a small baby. Mrs Voight had not heard of the tragedy when a reporter arrived at her home, bearing first news of the crash that cost the life of the husband and father. Hysterical with grief, she hurried to the morgue to identify her husband.' [1195]

Mr John Acee, an Amsterdam, New York, United States, clothing merchant, went for a visit on Tuesday 6th May 1924 to the nearby farm of Mr Thomas De Capio, taking with him his family, together with his sister-in-law Mrs Schahanda Harb and her daughter. In the evening they set off back to Amsterdam, taking with them David Spirono, an Amsterdam mill worker. The farm adjoined De Graff Crossing with the New York Central Railroad main line. As Mrs De Capio was bidding the party farewell she heard the sound of a train, as the car was about to cross the tracks. Running in front of the car, she said afterwards, she had shouted and waved at Mr Acee to stop, but distracted either by adjusting something in the vehicle or by talking to his passengers he had taken no notice of her, and she had had to jump out of the way to avoid being run down. The train was in fact the first section of the westbound *Twentieth Century Limited.* The car almost cleared the crossing, but the train just caught it, swung it around and wrecked it. Mr Acee, his wife Emeline, their daughter Julia, 14; their sons Thomas, 5, and Joseph, ten months; Mrs Harb, her daughter Mary, 3, and Mr Spirone were all killed. [1196]

An effort by passers-by to prevent an accident seems instead to have caused one on the evening of Friday 6th June 1924. A touring car driven by a Mr Adam Pitman and containing most of his family was approaching a level crossing at Dawn, Ohio, United States, at the same time as an eastbound Cleveland Cincinnati Chicago and St Louis (Big Four) Railroad passenger train. Fearing an accident two men who chanced to be at the

---

1195 *The Racine Journal*, Racine, Wisconsin, United States, 9th February 1923
1196 *The Republican Press*, Salamanca, New York, United States, 7th May 1924

crossing, W.A. Hartman of Ansonia, and J.A. Young of Dawn, ran out into the road to stop the car. Confused by the gesticulations of the men Mr Pitman stopped his car, but not before it had got onto the crossing. He managed to restart it, but it was too late, and it was struck by the engine. The gasoline tank immediately exploded, and the occupants of the car were enveloped in flames as they were thrown from the vehicle. Mr Pitman, 60; his wife Phoebe, 54; their children Alden, 25; Edwin, 9; and Effie, 7; and their daughter-in-law Hazel, 20. It took the train a quarter mile to stop. The party had been returning from a day spent with friends celebrating Mr Pitman's 60th birthday. Two-year-old Glenn Pitman escaped the accident as she was ill and had been left with neighbours. [1197]

Just under a fortnight after the Dawn accident another family lost their lives in another Ohio level crossing accident. Shortly before 9 a.m. on Thursday 19th June 1924 a car driven by Mr Andrew Petersen was struck by a Lake Shore Electric Railway interurban train at Hickory Corners, a mile and a half west of Genoa. Mr Petersen, his wife, Mrs Josephine Petersen, Mrs Mary Henrichsen, Mrs Gladys Henrichsen and the latter's sons, Richard, 2 and Robert, 9 months, were all killed. The party had left Sandusky at 6.40 a.m. to visit a daughter at Curtice. The interurban had left Sandusky at 7.45 in charge of Conductor M.J. Denigan with Motorman Tomas McGrath and was due in Toledo at 9.50. Road and interurban ran alongside of one another from Genoa to Hickory Corners where the road turned right to cross the line. Mr Petersen turned right across the line directly in front of the train. The car was wrecked, and the victims were badly mangled; the license plate of the automobile was used to identify the party. The only witness to the accident, Mrs Lucy Hurdelbrinke, said that both the interurban and the car had been travelling at high speed, and the car driver had apparently been unaware of the train. [1198]

Mr John Arnold of Diamond Bluff, Wisconsin and his large family were on their way to visit relatives in Minnesota on Thursday 3rd July 1924 when their car was struck by an eastbound Chicago and North Western Railway passenger train on a level crossing at the Wagoner Street Crossing in Eagle Lake, Minnesota, United States. Mr and Mrs Arnold and six of their children, aged between 2 and 22 years were all killed instantly. Luther Arnold, 19, was taken to Mankato hospital with serious injuries and his recovery was thought doubtful. Another son remained at home on the farm to care for the stock while the family were away. The accident was said to be unaccountable as the crossing was unobstructed. The coroner's jury found that the accident had resulted from Mr Arnold's negligence. [1199] [1200]

The driver of a motor vehicle failed to observe the warning lights at a level crossing at Röhrensee, Germany, on the Bayreuth – Altstadt line on Monday 7th July 1924 and the vehicle was struck by a train. Five people died and two were injured. [1201]

[1197] *The Commercial Tribune*, Cincinnati, Ohio, United States, 7th June 1924
[1198] *The Star Journal*, Sandusky, Ohio, United States, 19th June 1924
[1199] *The Redwood Gazette*, Redwood Falls, Minnesota, United States, 9th July 1924
[1200] *The Wisconsin State Journal*, Madison, Wisconsin, United States, 4th Jul 1924
[1201] *Eisenbahn-Katastrophen in Deutschland* by Hans Joachim Ritzau, pg. 51 (in German)

A private level crossing with the Erie Railroad Bergen cut-off line near Passaic Junction, New Jersey, United States, was approached by quite a severe incline. On Saturday 12th July 1924, a car driven by Mr Andrew Bromelski made two unsuccessful attempts to climb the hill. On the third attempt he succeeded, only to reach the track at the same moment as an eastbound express. The car was demolished and Mr Bromelski, his mother Mrs Julia Bromelski, Mary Yovonipski, 10, Michael Sikarski, 7, and his sister Anna, 6, were all killed. Mrs Bromelski's husband witnessed the accident from the porch of his home nearby. The three children were the elder Bromelski's grandchildren. [1202] [1203]

About 200 people went on an annual outing from Attica, Ohio, to Sand Beach, on the shore of Lake Erie, on Wednesday 23rd July 1924. In the evening 26 of them were returning home in a truck driven by Mr George Spangler. At 6.45 p.m. they arrived at the Locust Street Crossing at Oak Harbor, Ohio, United States. According to eye-witnesses Mr Spangler stopped his track at the crossing, alighted and walked to the tracks to check that all was clear. Having assured himself that it was he climbed back into his vehicle and set off, only for it to be struck by a westbound New York Central Railroad passenger train from Cleveland to Toledo. The truck was wrecked, and its wreckage carried 300 feet down the track. Ten people were killed outright, including Mr Spangler, and an equal number were injured, some seriously. It was thought that more would have lost their lives had not some of them seen the train coming and jumped before the collision. The victims were mainly boys and girls from Rice and Green townships, on their annual outing to Locust Point beach.

An investigation into the accident disclosed that there had in fact been two trains involved. Mr Spangler had waited for a slow-moving eastbound freight train to pass and then, rashly assuming that all was clear, had started over in front of the passenger train.

The accident immediately followed a day-long conference involving representatives of the railroad companies and automobile associations held to discuss measures to prevent just such accidents. [1204]

Mr Roy Fawns and his wife Myrtle, who had been married less than 24 hours, were killed together with three relatives in an accident at McGonigle, two miles south of Oxford, Ohio, United States, at 2.30 p.m. on Sunday 10th August 1924. They had been en route from Hillsboro, Kentucky, to New Castle, Indiana, where they intended to set up home, when their car was struck by a southbound Cincinnati Indiana and Western Railroad passenger train at the crossing of the line and the Hamilton – Oxford pike. The others who lost their lives were Lilia Fawns, 15, a sister of Mr Fawns; Mr J.F. Johnson, an uncle of Mrs Fawns; and Deloris Johnson, 10, his daughter. The train was travelling at between 50 and 60 miles an hour. The small touring car was wrecked and caught fire almost immediately; shortly afterwards its petrol tank exploded. Four of the victims succumbed instantly but Mrs Fawns survived long enough to be taken to hospital, where she briefly regained consciousness and was able to give details of who the occupants of the car were. She then lapsed into unconsciousness and died at 6 p.m.

[1202] *New Britain Herald*, New Britain, Connecticut, United States, 12th July 1924
[1203] *New Castle News*, New Castle, Pennsylvania, United States, 12th July 1924
[1204] *The Times Recorder*, Zanesville, Ohio, United States, 24th July 1924

Visibility at the crossing was obscured by the McGonigle station buildings and some roadside shrubs. Coroner Cook visited the scene and was severely critical of the lack of protection at the crossing. He stated, "This accident was due entirely, in my estimation, to a lack of sufficient safety devices at the crossing. The crossing is one of the most dangerous grade crossings in Butler County." He said that railroad officials had been requested several times to provide additional protection, but nothing had been done. He intended to immediately raise the matter with the railroad again.

The train had been in charge of Conductor Matt Matthews with Engineer Ike Stricker. [1205]

Sunday 24th August 1924 saw a bad accident at the ironically named Cemetery Crossing, near Northumberland, New Hampshire, United States. A 'high powered seven passenger limousine' carrying the Donahue family, who had been visiting relatives in Groveton and were returning to Lancaster, was run into just before 6 p.m. by a northbound Boston and Maine Railroad passenger train, which carried the car 200 feet and then threw it into a ditch. Mrs Maude Donahue and children Mary, 17; William, 16; Charles, 13; and Thomas, 8, were killed instantly. Mr William Donahue, a local contractor, who had been driving, died in Lancaster Hospital less than three hours later. The train had whistled for the crossing and another car driven by Mr Harry Lucas of Lancaster was already stopped waiting for it. Mr Donahoe approached the crossing at some speed, veered to the left to overtake the waiting car and drove onto the crossing. As he did so he either saw or heard the train and stepped on the accelerator, but the engine caught the car broadside anyway. An investigation into the accident was still underway at the time of the newspaper report but the preliminary inquiry indicated that Engineer John Stevens had not been to blame. [1206] [1207]

Earlier on Sunday 24th August 1924 a similar accident had occurred at Dedrick Crossing, two miles west of Sterling, Nebraska, United States. A car carrying Mr Chris Hoehne, his wife Emma and their children Zelma, 14; Arthur, 12; Alfred, 10; and Walter, 7, was struck by a Chicago Burlington and Quincy Railroad train and the whole family was wiped out. The westbound St Louis-Pacific Coast express was travelling at an estimated 45 miles an hour when it struck the car at about 11 a.m. It was supposed that Mr Hoehne, who had been driving, had not seen the train at all, possibly because the view to the east from the crossing was partly obscured by a clump of trees. Engineer Frank Newall said that he had seen the car enter the crossing and then come to a stop as if its driver had lost control when he saw the train. Instead of reversing the car driver had then attempted to turn it around. The train was then right upon it and the car was caught on the cowcatcher and thrown through a cattle guard before being carried about 120 feet along the track.

Mr Hoehne was a wealthy farmer who had come to the United States from Germany. He and his family had been on the way to visit neighbours who lived near the crossing. They witnessed the accident and were said to have been overcome. The train was delayed

[1205] *The Commercial Tribune*, Cincinnati, Ohio, United States, 11th August 1924

[1206] *The Lowell Sun*, Lowell, Massachusetts, United States, 25th August 1924

[1207] *The Portsmouth Herald*, Portsmouth, New Hampshire, United States, 25th August 1924

35 minutes. When it proceeded again it was almost involved in a second crossing accident near Hickman, Nebraska, when a small car just managed to reverse clear of its path.

The inquest into the accident heard evidence from Engineer Newall, Fireman J. Skillet and Conductor Ed Mauck and other witnesses. Its jury returned a verdict that the accident had been unavoidable. [1208]

A landau carrying ten persons was negotiating a level crossing near Cuneo in Piedmont, Italy, on the evening of Monday 25th August 1924 when it was run into by a train. Five of its occupants were killed instantly, three more died on the way to hospital and little hope was entertained for the other two. [1209]

On Saturday 30th August 1924, the parents and some of the children of the family of Mr William Haen left Sheboygan for Oconto Falls, accompanied by Mr Haen's sister Mrs Keene and her husband, Peter. On Monday 1st September they were on their way home. They travelled as far as Brickyard Crossing, a mile north of New London, Wisconsin, United States, when their new grey Dodge touring car was struck by southbound Chicago and North Western Railway passenger train No. 116, in charge of Conductor John Wall with Engineer Frank Ostermeier. Travelling at between 30 and 40 miles an hour the train reportedly smashed the car into an unrecognisable mass, killing Mr and Mrs Haen, Mr and Mrs Keene and two of the Haen children, William, aged 17, and Ruth, aged 3. The one survivor of the seven occupants of the car was son Raymond, aged 16, who apparently was driving the vehicle. He was injured and suffering from shock, but it was thought would survive.

County Road D, upon which the accident occurred, ran parallel to the railroad for about a mile before turning sharply to cross it. From the point where the road turned there was a good view of the track. Possibly the inexperience of the driver was a factor, together with the fact that two cars just in front of him had dashed over the line in front of the train. An eyewitness in a following car said that Raymond seemed to observe the train at the last moment, for the car stopped when almost on the track, then jerked forward. It was surmised that Raymond had attempted to reverse his car but had panicked and put the vehicle into forward gear instead.

For some reason, perhaps insufficient room in the car, four of the Haen's daughters were left at home with their cousins in Sheboygan. Neighbours had the unenviable task of informing the cousins of the tragedy. [1210]

The failure of the car driver to see the train was the probable cause of a level crossing accident near Port Allegheny, Pennsylvania, United States, on the afternoon of Sunday 5th October 1924. The small touring car in which the Fitzoli family were riding was hit by a Pennsylvania Railroad train travelling from Washington D.C. to Buffalo, New York. Mrs Fitzoli, two daughters and four sons were all killed outright and Mr John Fitzoli was

---

[1208] *Nebraska State Journal*, Lincoln, Nebraska, United States, 25th August 1924
[1209] *The Times*, London, England, 28th August 1924
[1210] *Appleton Post-Crescent*, Appleton, Wisconsin, United States, 2nd September 1924

critically injured. The accident happened near the family's home. [1211]

Only briefly mentioned in the American press was an accident at Crown Cotton Mills Crossing, near Dalton, Georgia, United States, on 7th October 1924. The first section of the Nashville Chattanooga and St Louis Railway southbound *Dixie Flyer* collided with a car, killing Mr and Mrs Will Johnson and their son, Mr and Mrs Ellis Quinn and Mr Lester Champlain. Five of the victims were killed outright and Mrs Quinn succumbed after being taken to Salton Hospital. [1212] [1213]

No less than five different investigations were opened into an accident which occurred early on Sunday 2nd November 1924 in Chicago, Illinois, United States. The Coroner, Oscar Wolff; the Chicago Milwaukee and St Paul Railway company, a street car company; the city police, and the Illinois Commerce Commission each opened their own inquiries, which appears a very inefficient way of doing things. The accident was admittedly a bad one. A street car was just negotiating a level crossing when it was struck by a backing freight train, reportedly cut in two and thrown against the adjoining watchman's tower. Ten people, five men and five women, lost their lives and 31 were injured. Most had been attending a Halloween party and many were dressed in masquerade costumes.

Motorman Robert Glenn and Conductor Carl Olson of the street car stated that the crossing barriers had been raised for them to cross and this appears to have been the case. The conductor added that when the gates were raised he had gone onto the crossing to check that all was clear but had not seen the backing train in the darkness as there had been no lights on the rearmost gondola cars.

The crossing watchman had lowered the gates for the train to make a switching movement over the crossing but apparently had not realised that it was the intention of the train to set back over the crossing again and had then raised the gates to let the street car across. Switchman John Gaskett shouted to the watchman to lower the gates again and to the train crew to stop, but it was too late. According to one account engineers Walter Vincent and Walter Wiedel, in charge of the two engines making the switch movement, said that they had heard Gaskett's shout, but the momentum of the train had carried it over the crossing. [1214] Another account stated that they said that they had not seen the lantern signal from the switchman for them to stop art once. [1215]

The crossing watchman, Joseph Brahe, 59, who had previously lost an arm in a railroad accident, was taken into custody following the collision. There were allegations, denied by Brahe, that he had been intoxicated. Two physicians who examined him after the accident believed he had been drinking and detectives found a small bottle smelling of whisky in his tower. Furthermore a saloon keeper whom Brahe had admitted had sold him 'a couple of drinks,' before he went on duty was also arrested.

[1211] *The Titusville Herald*, Titusville, Pennsylvania, United States, 6th October 1924
[1212] *Scott County Democrat*, Benton, Missouri, United States, 16th October 1924
[1213] *The Anniston Star*, Anniston, Alabama, United States, 9th October 1924
[1214] *The Commercial Tribune*, Cincinnati, Ohio, United States, 3rd November 1924
[1215] *Scott County Democrat*, Benton, Missouri, United States, 13th November 1924

Questioned at the inquest Brahe denied knowledge of a rule that the gates should not be raised sooner than three minutes after a movement over the crossing and maintained that he had received no signal to warn him of the backing movement. Switchman Bernard McGraw confirmed that there had been no light at the end of the backing train. [1216]

The results of the outcome of the inquest or other investigations do not seem to have been reported.

The day after the Chicago tragedy, Monday 3rd November 1924, eleven people died in a collision between a fast Chesapeake and Ohio Railroad passenger train and a bus at Lees Crossing, near Hampton, Virginia, United States. The bus was demolished in the collision and some of the victims could only be identified by articles of jewellery or other personal property. A Mrs Dixon was found on the pilot of the engine with her baby clasped in her arms. Both were injured and were taken to Hampton hospital, along with William Rand, the owner and driver of the bus, who was thought to be fatally hurt. [1217]

Conductor Lane and Engineer Catlin of the train were unable to make any statement as to the cause of the accident. The location was known as a "blind crossing," the view of the line being blocked by a row of houses, and several accidents had occurred there. [1218] [1219]

On the evening of Wednesday 12th November 1924 a Pennsylvania Railroad accommodation train from Pittsburgh, Pennsylvania to Dennison, Ohio, in charge of Conductor Elmer Noiles with Engineer L.E. Carver, was nearing the end of its journey when it collided with a touring car on Polen Crossing, Ohio, United States, four miles east of Dennison. Mr Earl Farren, a well driller, had been returning home with his family after purchasing supplies at the nearby Lone Pine Inn general store; they lived not far from the crossing. Mr Farren and children Clyde, 16; Treva, 13; Bernard, 10; and Kenneth, 6, were all killed. Mrs Farren was seriously, possibly fatally, hurt. It could not be exactly determined how the accident had happened but the crossing, with a private road, was obstructed on all sides until the track was reached. [1220]

"I saw the light of a train and cried to step on the gas. Almost on the instant it seemed like a bomb exploded in our car, and the next I remember was seeing bodies lying all around except one which stood upright against the auto. I was loaded into an ambulance and went to sleep. We were out for a ride around town. We had left Miss Schultz's house about half an hour before the accident." Such was the testimony of August Garbe, 20, the only survivor of a party of six teenagers whose car had been struck by a Chicago Burlington and Quincy Railroad fast night mail train on a level crossing three miles south of Aurora, Illinois, United States, at 10.30 p.m. on Thursday 27th November 1924. Those who died were Robert Coughlin, 19; Marian Schultz, 17; Kenneth Murray, 16; Theron

[1216] *The Evening Tribune*, Albert Lea, Minnesota, United States, 7th November 1924
[1217] *The Atlanta Constitution*, Atlanta, Georgia, United States, 4th November 1924
[1218] *The Billings Gazette*, Billings, Montana, United States, 4th November 1924
[1219] *The Daily Illini*, Urbana-Champaign, Illinois, United States, 4th November 1924
[1220] *The Evening Independent*, Massillon, Ohio, United States, 13th November 1924

Marshall, 18; and Lotus Babcock, 18.

August Garbe had been sat on the back seat of the car and was the only one in it to see the headlight of the train. When he had shouted the driver had put his foot down, but the back wheels of the automobile were still on the rails when the train arrived. Although he had suffered two broken legs in the collision it was thought that he would recover. [1221]

The same evening three women died when a car was struck on a crossing at Lombard, Illinois, by a Chicago and North Western Railway train. In this accident the car driver had waited for a freight train to pass and then driven onto the line in front of a passenger train on the other track. [1222]

Wyandotte, Michigan, United States, was the scene of a nasty level crossing accident involving an interurban train and a truck and trailer owned by the Liberty Motor Company of Toledo and loaded with steel drums of turpentine. The accident happened shortly after 6 p.m. on Wednesday 3rd December 1924. Fire broke out immediately after the collision, four passengers and the motorman lost their lives, and ten passengers were injured. The Interstate Commerce Commission (ICC) investigated the accident.

The Detroit United Lines Railway interurban (train No. 226) consisted, according to the ICC report, of cars 8002 and 181 coupled, with Conductor Byers and Motorman Callahan and Conductor Smith and Motorman Rollins respectively.

Conductor Byers told the inquiry that Motorman Callahan (who died in the accident) had sounded the usual crossing whistle, then there a long blast on the whistle, a severe brake application and, immediately before the impact, the car was reversed. After the collision and the outbreak of fire he and Motorman Rollins had attempted to go to the front of the train to fight the fire with extinguishers, but the smoke and heat had been too intense. After assisting the passengers at the rear of the front car he and Motorman Rollins had uncoupled the second car and moved it back out of danger, while Conductor Smith went back to protect the train.

Lorry driver Harry Culver, 20, stated that he had been driving about four years and held both Michigan and Ohio driving licences. He had been returning from Detroit to Toledo. He was familiar with the crossing and knew that he must be nearing it, but his headlights had been dimmed in accordance with a city ordinance. It was only when he had seen the reflection of the truck's headlights on the rails that he had realised how close to the crossing he was, and at the same time he saw the approaching train, moving at 50 or 60 miles an hour. He was only travelling at about 12 miles an hour but was too close to the line then to stop and with the weight of the truck and its trailer he could not go any faster either. [1223]

A newspaper report named one of the interurban passengers, Henry Culp of Norwalk, as the hero of the accident, stating that the lights of the interurban had gone out in the collision but regardless of his own safety he had seized a lantern from the rear car and dashed forward to the front car to aid the passengers. [1224]

[1221] *Moline Daily Despatch*, Moline, Illinois, United States, 28th November 1924
[1222] *Laurel Daily Leader*, Laurel, Mississippi, United States, 28th November 1924
[1223] Interstate Commerce Commission report
[1224] *The Commercial Tribune*, Cincinnati, Ohio, United States, 4th December 1924

The ICC concluded that the accident had been the lorry driver's fault and made the usual comment about 'the vital necessity for automobile drivers to approach railroad crossings with caution and to make certain before proceeding over them that no train is approaching.'

A serious accident on Market Street Crossing, Marcus Hook, Pennsylvania, on the morning of Tuesday 10th February 1925 occurred at a busy level crossing which there had for some time been pressure to abolish. During a dense fog, a light engine collided with both a lorry and a car, resulting in six deaths. The crossing was protected by gates operated by a crossing watchman, named Sabatino Bostilio, who was warned of the approach of a train by the ringing of a bell activated when a northbound or southbound train was 300 yards from the crossing. As an additional precaution railroad policemen, provided with whistles, were stationed north and south of the crossing to give warning to the watchman to lower the gates.

On the morning of the accident a southbound train had just passed over the crossing and the watchman had then raised the gates for the waiting vehicles and pedestrians. Before he had time to lower them again a light engine, No. 1237, driven by Engineer Frank Firestone with Fireman T.C. Smith, loomed out of the fog and collided with a light sedan car and a truck owned by the Sinclair Refining Company, causing the deaths and a number of injuries. Two of those killed were passengers who had just alighted from the southbound train. [1225]

As usual there were several separate investigations into the accident. The day after it had happened Engineer Firestone told the Public Service Commission that he had whistled for the crossing and his fireman had rung the engine bell. [1226] At the railroad company inquiry the engineer explained that he had been returning light after having assisted a heavy freight train to North Philadelphia. His engine had been running at 40 miles an hour. At the close of the railroad inquiry Pennsylvania Railroad Superintendent James Buckalew insisted that none of the railroad employees was to blame for the accident. [1227]

The inquest on the six victims was not held until 25th February. There Engineer Firestone repeated the evidence that he had given before and added that he had been given no special orders on account of the fog. His engine had arrived at Marcus Hook at about 9.55 a.m. Sergeant William Blankenburg stated that he had blown his cyclone whistle to warn the gateman when the engine was about 300 feet away from him. Gateman Bostilio, speaking through an interpreter, said that he had received the warning too late to lower the gates as the truck and pedestrians were already on the track. The coroner's jury found that the accident had resulted from the negligence of the Pennsylvania company in not providing proper protection at the crossing and they recommended that immediate action be taken to abolish the crossing. [1228]

In mid-July there was still argument over how the crossing should be replaced,

---

[1225] *Chester Times*, Chester, Pennsylvania, United States, 10th February 1925
[1226] Ibid, 11th February 1925
[1227] Ibid, 12th February 1925
[1228] Ibid, 26th February 1925

whether by an offset bridge or by elevation of the railroad tracks. [1229]

Just before 10.30 a.m. on the morning of Friday 6th March 1925 Mrs Ruby Bodine set off in a small touring car from her home near Williamsport, Indiana, United States, to visit relatives at Judyville. With her in the car were her children, Elnora, 5; Mabel, 3; Frederick, 19 months; and Ellsworth, one month, together with her cousin, Miss Clara Hurd. All the curtains of the car were up as they reached a level crossing on the outskirts of Williamsport and stopped to allow an eastbound Wabash Railroad passenger train to pass. The train having done so Mrs Bodine set off over the crossing despite the warning shouts of Crossing Watchman John Shamp and Brakeman B. F. Mullins, who was struck by the car as he tried to stop it and slightly hurt. Mrs Bodine had failed to notice a loose cut of waggons bearing down on the crossing on the other line. The trucks collided with the automobile and pushed it along the track against a switch stand. All six occupants of the car were killed or fatally injured. [1230] [1231]

On the evening of Monday 23rd March 1925 a party of eight people was on its way in a motor car to attend an entertainment being held in aid of the Roman Catholic church at Cheltenham when it was struck by the 7.38 p.m. down Melbourne to Frankston electric parcels van on Wickham Road Crossing, near Highett station, Victoria, Australia. All eight occupants of the car were killed or fatally injured. The victims were: Mr Clarence M'Donnell and his wife Pearl, their son Raymond, aged four months, Mr Clarence M'Donnell's father Michael, Mr Edward Shacklock and his wife Josephine, Mr Martin Foley, and Miss Enid Grace.

The crossing was worked by a crossing keeper and his wife. It was provided with four hand-worked heavy wooden gates which in the British fashion swung alternately between the road and the railway rather than lifting barriers as in North America and elsewhere. The gates normally stood closed to road traffic and the south-east and south-west ones were provided with lights showing both to road users and to train crews. The crossing keeper was warned by bell when a down train left Moorabbin, 1,087 yards from the crossing, and a home signal was also provided, 200 yards from the crossing, which should have been replaced to danger before the gates were opened to road traffic.

The crossing keeper, William Resuggan, alleged that on the evening in question the bell had not been rung when he had opened the gates for the M'Donnell car, though Signalman Alfred Read at Moorabbin stated that he had rung the bell for the parcels train and had an acknowledgement, while the bell was tested and found to be working normally after the accident. Resuggan admitted that he had not replaced the home signal before opening the gates and added that when he had seen the parcels train he had shouted to his wife to, 'Throw the stick back,' and to the driver of the slowly moving car, 'For God's sake shift yourselves.'

The inquest into the accident was lengthy; 40 witnesses were examined, including Motorman John M'Knee, Guard Bernard Rose and Assistant Guard Albert Miller of the

[1229] Ibid, 16th July 1925

[1230] *The Evening Tribune*, Marysville, Ohio, United States, 6th March 1925

[1231] *The Logansport Morning Press*, Logansport, Indiana, United States, 7th March 1925 parcels train, Mrs Resuggan, two of the Resuggans' sons and so on, so that even the local newspaper reported that it 'drags on.' It was finally concluded on Friday 17th April, when the Coroner, Mr Berriman, P.M. found Resuggan guilty of manslaughter and committed him for trial. [1232]

The evidence given at the inquest was largely the same at Resuggan's trial. Resuggan repeated that the warning bell had not rung and his wife, Florence, stated that the parcels van had been travelling 'at a terrific speed.' Harold Resuggan, one of the Resuggans' sons, stated that the warning bell had failed to ring numerous times previously. Mr Foster, the defence counsel, pointed out that Resuggan had worked the crossing safely for 16 years and that the parcels train, which had been late, should have passed at the time of the accident (8.15 p.m.) and no more trains had been due for almost three-quarters of an hour.

The jury found Resuggan "Not Guilty," and recommended that the gates and signal should be worked by one lever, the crossing should be better illuminated and the lights on the gates should be improved. [1233]

Gale-force winds which drowned out the sound of the engine whistle of an approaching Chicago and Alton Railroad train were a probable factor in a collision on an open crossing at Gardner, Illinois, United States, on Sunday 29th March 1925. Coal miner Robert Harrop, his, wife and their children Joe, 9; Margaret, 4; Harriet, 2; and Mary, 5 months, were all killed when the family car was struck by the railroad's *Million Dollar Special* passenger train. [1234] [1235]

On Sunday 19th April 1925 Mr Ralph Cottey Senior and his family were on a visit by car from their home in Indianapolis to a cousin living in Ingalls, Indiana, United States. They had come within half a block of their destination when their car was struck on the Alfont Street Crossing by a fast Union Traction Company interurban car bound for Indianapolis and known as the *Hosierland Flyer*. Mrs Mary Cottey, their children Jennie, 8; Dorothea, 6; and Ralph Junior, 3, were all killed, as was the elderly Mr Lincoln Cottey, who was presumably Ralph Senior's father. Ralph Cottey Senior was thrown from the car and suffered a broken hip and other injuries. He was taken to a nearby house and thence to the Methodist Hospital in Indianapolis. He remained conscious and kept asking for his wife and children, but it was not thought wise to immediately tell him that they had all died.

The interurban was kept at the scene until Coroner Earl Sells arrived at 10 a.m. to examine the scene. It appeared, the coroner said, that Mr Cottey had neither seen nor heard the train. [1236]

Fast train D 186 collided with a lorry pulling a trailer carrying a choral society on a

---

[1232] *The Argus*, Melbourne, Victoria, Australia, 24th March and 8th, 9th, 11th, 17th and 18th April 1925

[1233] Ibid, 11th, 12th and 13th June 1925

[1234] *The Evening Tribune*, Albert Lea, Minnesota, United States, 30th March 1925

[1235] *The Evening Gazette*, Xenia, Ohio, United States, 30th March 1925

[1236] *The Tipton Daily Tribune*, Tipton, Indiana, United States, 20th April 1925 level crossing at Rot-Malsh, Germany, on the Bruchsal-Heidelberg line on Sunday 10th May 1925. The results were disastrous; the train was derailed, and 11 people were killed and six injured. The crossing barriers had not been closed. [1237]

Sunday 28th June 1925 saw two more bad level crossing accidents in the United States. Mr C.P. Woodring, his wife Arminta, their daughters Marie, 17, and Bernice, 15, and their son Glenn, 11, were all killed when their car was struck by the second section of the eastbound New York Central Railroad *Interstate Express* running from Chicago to New York. The car was carried half a mile and their bodies were badly mangled. They had apparently been on their way to church and were identified by a Sunday School book found among the remains. West Street Crossing, Ashtabula, Ohio, United States, the scene of the accident, had seen a number of previous accidents. Mr Woodring had apparently stopped and waited for a westbound freight train to pass, then, in the usual scenario, had started across in the path of the eastbound train. A watchman was stationed at the crossing, but he was hidden by the trains from the car driver and the crossing lacked gates. [1238] [1239]

In the other accident on 28th June six members of the Beaver family were killed when their touring car was struck by an Illinois Traction System car near the Elkins stop, seven miles west of Bloomington, Illinois, United States. The car was carried 90 feet down the track and crushed up against a support pole. Mr L. M. Beaver, his wife Zella, their children Loretta, 10; Paul, 8; Jessie, 3 and a baby which was born as a result of the accident all lost their lives. Another daughter, Zora, 9, was seriously injured. The family's pet terrier also died. The interurban car was derailed, and its 20 passengers were shaken up; three were taken to Bloomington hospital. [1240]

On 4th July 1925, a tartana (horse-drawn light carriage) was struck by a Murcia – Cartagena goods train on a level crossing at Avileses, Spain, on the Chinchilla – Cartagena line. The vehicle was carrying seven people aged between 15 and 23. They were coming from the village of Cegarra where they had been at a fiesta and were bound for Balsicas to visit a friend of one of the girls riding in the tartana. Six of them were killed and one hurt. [1241]

On Sunday afternoon, 19th July 1925, Mr Mathew Severin Senior, of Youngstown, Ohio, asked Joseph Babich to drive him, his wife Agnes and their son, Matthew Junior, aged 10, over the state line into Pennsylvania to collect their two other children, Barbara, 7, and Charles, 5, from where they had been staying on their uncle's farm. As the party were setting off they picked up Mary Moreich, the 12-year-old daughter of a neighbour,

---

[1237] *Eisenbahn-Katastrophen in Deutschland* by Hans Joachim Ritzau, pg. 66 (in German)
[1238] *The Commercial Tribune*, Cincinnati, Ohio, United States, 29th June 1925
[1239] *The Times Recorder*, Zanesville, Ohio, United States, 29th June 1925
[1240] Ibid.
[1241] Webpage: Federación Castellano Manchega de Amigos del Ferrocarril – Accidentes Ferroviarios en España

to come along for the ride.

A mile and a quarter north of Brookfield, Ohio, United States, the car reached a level crossing with the New York Central Railroad at the same time as southbound passenger train No. 81, in charge of Conductor John Campbell with Engineer B. W. Wood and Fireman Brown. Engineer Wood saw the approaching car, whistled, and the car seemed to slow down as if it was going to stop. Then, suddenly, the car gained speed. Engineer Wood applied the brakes in emergency, but the distance was too short. The car was crushed beneath the engine and all five occupants were killed instantly. [1242]

The Fleck family had moved from Richardson, North Dakota, to a farm near Santa Rosa, California, but for some reason had decided to sell up their effects there and return to Richardson, where Mr Fleck's parents lived. On Friday 17th July 1925 they were on their way back in two cars when the leading vehicle, carrying Mr Ralph Fleck, his wife and six children stalled on a level crossing a mile from Billings, Montana, United States. A daughter, Rosie, aged 18, had time to jump to safety from the car but its other seven occupants lost their lives when it was struck by a fast moving eastbound Northern Pacific Railroad train. Three other Fleck children were riding in the following car with their uncle, Stanley Sattler. [1243] [1244]

With two other families the Devine family of Boxville, Kentucky, were returning home after a trip to Indianapolis, Indiana, on Monday 27th July 1925. Their car was the middle vehicle of the three when it was struck by a northbound Chicago and Eastern Illinois Railroad passenger train on a level crossing at Busseron, Indiana, United States. Mrs Devine and their sons aged 5 and 3 were killed outright, Mr Jess Devine died in the ambulance taking him to hospital and their 11-year-old daughter succumbed as she was being taken out of the ambulance. A 10-year-old son was in the Good Samaritan hospital at Vincennes and was not expected to live. The crossing was approached by an "S" curve in the road and was regarded as being especially dangerous. [1245]

Heavy rain on the morning of Saturday 8th August 1925 resulted in five men working on the new Broadway – Gowanda boulevard deciding to stop work and drive into Buffalo. As the automobile that they were in reached a crossing known as "Blind Crossing," near Gowanda, New York, United States, the same heavy rain was very likely a factor in their failing to observe an Erie Railroad Jamestown – Buffalo train due in Gowanda at 10.44 a.m. As its name implied the crossing was in any case one with limited visibility of approaching trains. The train hit the car and Henry Ruper and Elmer Tibbets, of Otto; and Calvin Locke and Fenton Wite, of Cattaraugus were killed instantly. William Merow, of East Otto, who had been driving the car, was fatally hurt, and died later in the day in a Buffalo hospital. [1246]

[1242] *The Evening Independent*, Massillon, Ohio, United States, 20th July 1925
[1243] *The Brownsville Herald*, Brownsville, Texas, United States, 17th July 1925
[1244] *The Moore Independent*, Moore, Montana, United States, 23rd July 1925
[1245] *Seymour Daily Tribune*, Seymour, Indiana, United States, 28th July 1925
[1246] *The Republican Press*, Salamanca, New York, United States, 8th August 1925

At the inquest on the five men Engineer Clifford Holland, Fireman Charles Cary and Conductor Arthur Craige testified that the whistle had been sounded for the crossing and the engineer stated that he had stopped the train in 500 feet. Erie Trainmaster W.G. Bucks stated that the crossing bell had been working properly after the accident. Sixteen-year-old Robert Dorey said that he had waved to the car to stop it, but its driver had ignored him. District Attorney A. M. Laidlaw declared that the evidence showed no evidence of negligence on the part of the railroad and the accident was due to the carelessness of the car driver. [1247]

As a newspaper noted at the beginning of September 1925 the annual railroad crossing toll in the United States was increasing by an alarming rate every year. An accident near Lodge Pole, Nebraska, on Saturday 29th August added eight to the number of deaths for the year and wiped out the entire family of tenant farmer Conrad Nein. The Neins lived near Scottsbluff, Nebraska, and were on their way to visit an uncle, Heinhart Nein, in Colorado. As the Neins' Ford touring car reached the crossing with the Union Pacific Railroad it was blocked by a freight train. When that train moved forward Mr Nein moved onto the crossing and the car was struck by passenger train No. 26 running on the other track and travelling at, it was said, over 60 miles an hour. Mr Nein, his wife Elizabeth, their children, Reinhart, 10; Fred, 8; Harry, 6; Irene, 4; and Albert, 6 months, plus Katherine Harris, a niece aged 12, were all killed instantly. The accident was witnessed by a Mrs Bales and her child, of Lodge Pole, who were on foot and had been about to cross as well but had stepped back upon hearing the whistle of the express. [1248]

Southern Railway fast passenger train No. 35 collided with a car on Carter Street Crossing, Reidsville, North Carolina, United States, at about 7 p.m. on Sunday 30th August 1925. It was supposed that the loud ringing of the warning bells of the crossing had somehow confused Mrs Irvin, who was driving, for the car came to a dead stand on the crossing. The bell on the engine of the train was also ringing, it had whistled, and its headlight was burning. Mrs Irvin was killed together with her four passengers: Mr Jack Carter, Mrs Manton Oliver (his sister), Mrs Lilian Oliver and Mrs Nina Cone. [1249]

Mr Claude Warren, a farmer living near Beardstown, Tennessee, United States, his wife, two of their children, Mrs J.C. Cude (his mother-in-law) and Mr Tom Curl were killed when the car in which they were riding was struck on a level crossing by an Illinois Central Railroad train on Sunday 6th September 1925. The car was described as 'splintered' but a two-year-old daughter of the Warrens escaped unhurt. [1250]

Westbound Great Northern Railway fast mail train No. 27 struck a Dodge touring car on Sullivan Crossing, Shelby, Montana, United States, on the morning of Sunday 13th September 1925. Mr Michael Hasquet and his wife Jennie, Mrs O.V. McIntire and her

---

[1247] Ibid, 11th August 1925

[1248] *The Telegraph*, Sidney, Nebraska, United States, 4th September 1925

[1249] *Gastonia Daily Gazette*, Gastonia, North Carolina, United States, 31st August 1925

[1250] *The Commercial Tribune*, Cincinnati, Ohio, United States, 7th September 1925 daughters Jean, 12, and Virginia, 10, had been on the way to church. Less than 50 feet from the crossing Mr Hasquet, who was driving, had stopped the car to pick up a Mrs Lynch, who was also on her way to church. The car then set off again into the path of the train. All the occupants of the car were either killed outright or fatally injured.

The inquest on the victims cleared the crew of the train from blame and found that the accident had been due to the failure of Mr Hasquet to notice the train, largely because his view was blocked by two box cars parked on an industry track (siding) east of the crossing. The jury also criticised the railway company for the lack of all safety devices at the crossing. [1251]

An interurban freight car, in charge of Conductor Fosdick with Motorman Moore, collided with a touring car carrying the Routough family near New Lebanon, Ohio, United States, on the evening of Sunday 20th September 1925. The car had shot out in front of the train while it was raining hard. Mr Noah Routough, his wife Lizzie and their children Virginia, 12; Robert, 10; Paul, 8 and Clinton, 6, were all killed. An infant son escaped unhurt. The car was carried 1,000 feet before the train could be stopped. The family had been visiting friends in New Lebanon and were only two miles from home. [1252]

The Canadian National Railway *International Flyer*, in charge of Conductor Dickson with Engineer T. Gorman, travelling at about 35 miles an hour, collided with a car containing two families at St Clair Crossing, near Scarborough Junction, Ontario, Canada, on Sunday 11th October 1925. The car was demolished and Mr Arthur Nichols, his wife Rose, and their 4-month-old daughter Shirley; Mr Arthur Mould, his wife, and their daughter Ruth, 12, were all killed. Daughter Hilda Mould, aged 4, suffered concussion and other injuries and was taken to hospital but was not thought to be seriously hurt. There was a good view of trains at the crossing, and it was not known why Mr Nichols, who was driving, had failed to give way to the train. [1253]

Late on Sunday 18th October 1925 former Michigan state representative George Long of Pompei, his wife and three girls, the daughters of friends, were killed when their automobile was struck by an interurban car on a level crossing near Lansing, Michigan, United States. [1254]

The beginning of November 1925 saw a tragic level crossing accident involving a school bus at Bennetts Crossing, Nahunta, Georgia, United States. At 8.15 a.m. on Monday 2nd November a school bus carrying 30 children and driven by Elroy Strickland approached the crossing at the same time as Atlantic Coast Line Railroad train No. 81, the *Everglades Limited* express. A string of box cars on either side of the crossing restricted the view; there was a gap between the cars of about 15 feet for vehicles to pass.

[1251] *The Kevin Review*, Kevin, Montana, 17th September 1925
[1252] *The Commercial Tribune*, Cincinnati, Ohio, United States, 21st September 1925
[1253] *The Winnipeg Evening Tribune*, Winnipeg, Manitoba, Canada, 12th October 1925
[1254] *The Frederick Post*, Frederick, Maryland, United States, 19th October 1925

The normal procedure was for the bus driver to stop short of the crossing, alight and walk up to the track to check all was clear. On this morning it was raining so although he reportedly stopped the bus he did not alight but merely listened for a train. Not hearing one, he then started the train over the crossing. Another account stated that the curtains of the bus had been down, and the driver had asked one of the children to look one way while he looked the other. [1255]

Eight children were killed more or less outright, others were badly hurt and four more died in the following days. [1256] Accounts of the accident in local newspapers contained moving details of the victims, thus, 'All the children in the bus, ranging from 7 to 15 years of age, were singing and none of them knew of the fate that was to be theirs. One little boy, who was about eight years old, was picked up when the train stopped and still conscious but in a dazed condition, offered his teacher who had arrived at the scene, some candy he had in a bag and said he still had the penny his mother had given him. At the same time this little fellow was dying with part of his brains exposed to view. He was rushed to a Waycross hospital but died on the way there. His name could not be learned.' [1257]

The train had slowed to 20 miles an hour when it hit the bus, knocking it against the box cars. An official statement from the superintendent of the Savannah district of the Atlantic Coast Line maintained that the train crew had followed every rule and that five witnesses had declared that the school bus had approached the crossing at high speed, this, of course, contradicting statements that the bus had stopped short of the crossing.

The public service commission launched an investigation into the accident and raised the question as to whether it had the authority to prevent railroad companies from side tracking box cars onto sidings adjoining level crossings. [1258]

In another development State Superintendent of Schools Fort Land proposed making it mandatory for the drivers of school buses to stop short of railroad crossings and send someone ahead to flag "all clear," with drivers failing to do so facing instant dismissal. [1259]

No report of an inquest on the victims of the accident has come to light.

Four men and three women died late on Saturday 28th November 1925 when there automobile was struck by an interurban car on a level crossing at Grand Rapids, Michigan, United States. Motorman Donald Peterson, who 'admitted' that his train had been travelling at high speed, said that the collision had disabled the air brake preventing him from pulling up more quickly, and that he had seen the automobile nearing the unobstructed crossing but had assumed that it would stop.

Four of the dead were: John Berrak, Veronica Prawdzik and Walter Rosklewicz, all of Grand Rapids; and Frank Benzer, of Boine Falls. The other three could not immediately

[1255] *Beatrice Daily Sun*, Beatrice, Nebraska, United States, 3rd November 1925
[1256] *The Jackson Herald*, Jackson, Georgia, United States, 12th November 1925
[1257] *Thomasville Times-Enterprise*, Thomasville, Georgia, United States, 3rd November 1925
[1258] Ibid, 4th November 1925
[1259] *The Jackson Herald*, Jackson, Georgia, United States, 12th November 1925

be identified. All had been killed instantly. [1260]

Towards the end of 1925 Mr Lawrence Oxford and his family moved from Malden, Missouri to Archillion, Arkansas, United States, and to a home a quarter of a mile from that of his brother Clem. On Monday 30th November Lawrence's family had spent the day picking cotton at Clem's place and at the end of the day Clem was driving them the short distance home, the car carrying most of the members of both families. The journey home involved crossing the line of the St Louis-San Francisco Railway, and fast passenger train No. 801, with Conductor C.S. Pawkett and Engineer Fred Dierssen, was approaching, having left Blytheville at 4.15 p.m. It was running at 40 miles an hour. Engineer Dierssen had whistled for the crossing. His fireman was the first to see the danger and the brakes were applied but the collision could not be avoided, and the car was thrown 40 feet.

Andrew Smith, his wife, and Mrs Maggie Hawkins, who were busy picking cotton in a nearby field, witnessed the accident and agreed that Clem Oxford had speeded up his car, which was a new one only two months old, immediately before the collision, as if he were trying to beat the train over the crossing.

Clem Oxford, his daughter Mildred, 9, and his sons Omar, 7, and Clem Junior, 5; Lawrence Oxford, his wife Velzie, their son Theodore and their daughter Odyssey all lost their lives. William Oxford, 8-year-old son of Lawrence, was badly hurt but it was thought might survive. He was loaded aboard the train, which returned with him to Blytheville. [1261]

In contrast to the situation in many countries most public road level crossings in Great Britain were protected in the early twentieth century by gates worked either from a signal box or by a crossing keeper. Level crossings collisions did still occur from time to time, but they were relatively rare and major accidents were rarer still. Two bad accidents did occur, however, in 1925 and 1926.

On Monday 7th December 1925, the first of these accidents happened at Fenny Stratford, England, on the London Midland and Scottish Railway line between Cambridge and Bletchley. Although neither the motor coach nor the train involved in this accident was going very fast the results of it were disastrous; the coach was demolished, six people were killed outright, three more died of their injuries and four others were seriously hurt.

The crossing concerned adjoined Fenny Stratford station and consisted of four main wooden gates, two on each side of the line, worked together by a wheel in the signal box and interlocked with the signals so that the signals could not be lowered unless the gates were closed across the road and bolted shut. Paraffin oil lamps mounted upon the gates showed red lights to warn road users when they were closed.

At 8.22 p.m. on the day of the accident Signalman William Souster at Fenny Stratford accepted the 6.30 p.m. up passenger train from Cambridge and at 8.32, the train having

---

[1260] *The Kokomo Daily Tribune*, Kokomo, Indiana, United States, 30th November 1925
[1261] *The Sikeston Standard*, Sikeston, Missouri, United States, 4th December 1925

been accepted by Bletchley No. 5 signal box, he closed the closing gates and lowered his signals for it. He then went to the window to watch the approaching train, and saw the headlights of a motor vehicle, travelling at about 10 miles an hour, about 20 yards from the crossing. It did not seem to be pulling up and he shouted, "Look out, you will be into the gates." The motor coach, then crashed through the southern gate and was stuck by the engine of the train.

Driver Thomas Bazeley of the passenger train explained that it had left Cambridge 15 minutes late awaiting connections and had not regained any time. The Fenny Stratford signals had been clear, and he had been approaching the station at 10 or 12 miles an hour preparing to stop there when he saw the headlights of the motor coach. He had not anticipated it failing to stop and the collision had come as a complete surprise to him.

Fireman George Atkin confirmed his driver's evidence, as did Guard John Chappell.

Describing the aftermath of the accident a newspaper reporter wrote, 'The Rev. W. Nightingale {one of those who lost their lives} was a retired clergyman, who had formerly lived in Nova Scotia. Mrs Garrett, another of the killed, was the wife of a postman. The news of her death did not reach her husband until this morning. When she did not return last night, Mr Garrett concluded that she had decided to spend the night at Bletchley with relatives. This morning, on taking up his duties, he heard of the accident, and on making inquiries learned that his wife was dead. Many pathetic stories were told to-day by those who took part in aiding the injured. Miss Daphne Tansley, the six-year-old daughter of Mrs Tansley, while being treated in the signal box, kept asking for her "mummy," who was severely injured. Despite her suffering Mrs Tansley, while lying in the waiting room waiting to be removed to the hospital, urged that the helpers who had brought her hot water bottles should look after the other injured first.

'Mr C. Lovell, who lives in a cottage about 50 yards from the scene of the accident, told a graphic story of the scene when he arrived. He said he saw a group of three bodies huddled on the ground. One of them was a woman who was alive. She had a very severe wound on the head, and he stayed by her for some time. He asked someone who had by that time arrived to bring him some hot water, and he used that to bathe the woman's head. Bodies, he added, were lying huddled up all over the place. For a time they had to work in almost complete darkness, as the only light that was available was the dim glimmer which came from the signalman's cabin. Eventually people came with all kinds of lamps, and they were able then to work under much better conditions.' [1262]

The motor coach involved was a 14-seater Ford charabanc. It was in private use, and on the day of the accident had been booked to carry a party of 12 people from Woburn Sands to the Wesleyan chapel in Bletchley for a lecture. Mr Nelson Payne, its regular driver, was feeling unwell, and so it was being driven by another employee, Mr Fred Griffin. Mr Payne was injured in the accident and his recollection of the events leading up to it was hazy, while Mr Griffin was fatally injured. The only useful evidence as to the operation of the coach came therefore from a Mr Porter, the only uninjured passenger on it, who told the Ministry of Transport inquiry that as the coach had neared the crossing he had thought that it was not going to stop in time, though the brakes were applied at the last moment. He saw the red light on the far crossing gate (the hood of the vehicle hid the other from his view from where he was sat).

[1262] *The Times*, London, England, 8th December 1925

Major G.L. Hall, the Ministry of Transport inspecting officer, concluded that there was nothing in the crossing equipment or manipulation of the gates which could have contributed to the accident, nor were any of the railwaymen to blame for it. As the driver of the coach had died in hospital without being able to give any statement, the reason for his failure to stop must remain conjecture. The alternatives were a failure of the coach equipment and an error of judgement on the part of Driver Griffin.

Mr Payne assumed that there had been a failure of the coach brakes but there was no evidence to support this theory and the evidence of Mr Porter, an experienced car driver, went against it. In his opinion, Major Hall continued, it was much more probable that Driver Griffin had made an error of judgement than that there had been a brake failure. The locking of the coach wheels on a muddy, slippery road might have been a contributory factor. As Mr Griffin had lost his life in the collision Major Hall considered that no useful purpose would be served in attempting to determine how much, if at all, he was to blame for it. There was no suggestion that Mr Griffin had been drinking.

The inspecting officer in his closing remarks stated that, in answer to letters received regarding the poor visibility of other level crossings, this was not a problem at Fenny Stratford. As regards the coach he was critical of its seating arrangement which had resulted in a woman passenger being sat directly alongside the driver, and he stated that, 'I cannot but think that the presence of the passenger on the driver's seat had some bearing on the tragic sequel.' [1263]

Christmas Day, 25th December 1925 saw an accident on Butterfield Crossing, three miles west of Wisner, Nebraska, United States, which all but wiped out the family of Mr Fritz von Zeggeren* Senior, a prominent Nebraska farmer. Mr Von Zeggeren Senior, his wife, their son Emil, 22 and daughters Anna, 20 and Louisa, 16, all lost their lives when the car was struck by a Chicago and North Western Railway train. Fritz von Zeggeren Junior, who was driving the car, was said to be fighting for his life in a Norfolk hospital. Railroad officials were unable to explain the accident as there was nothing at the crossing to obstruct a motorist's view. [1264] [1265] [1266]

*Or Segern or Seggern according to different newspapers.

In the mid-1920s there were not yet that many motor cars in use in Third World countries but at about 12 noon on Saturday 2nd January 1926 a car belonging to the Patiala Durbar and carrying nine purdah (concealed with covering clothing and veils) women was struck on Crossing No. 123 of the Delhi-Umballa-Kalka Railway between Chandigarh and Ghaggar stations, India, by No. 8 down train. Seven of the women were killed outright or fatally injured and the other two were badly hurt. The car driver (chauffeur?) was also severely hurt. The car had been driven onto the line directly in front of the train. [1267]

[1263] Ministry of Transport report

[1264] *The Evening Gazette*, Xenia, Ohio, United States, 26th December 1925

[1265] *The Wisconsin State Journal*, Madison, Wisconsin, United States, 26th December 1925

[1266] *Nebraska State Journal*, Lincoln, Nebraska, United States, 26th December 1925

[1267] Report by the Railway Board on Indian Railways for 1925-26 Volume 1

A flashing light warning signal apparently failed to prevent the driver of a small coupe from permitting his vehicle to come into collision with a New York Central Railroad train on Secor Road Crossing, five miles west of Toledo, Ohio, United States, at 5 p.m. on Tuesday 19th January 1926. Mr Joseph Morse Senior, his wife Lilian, their eight-month-old son Joseph Junior; Mrs Catherine Dressler (Mrs Morse's mother) and Laura Dressler (Mrs Morse's sister) all lost their lives. It was surmised that the crowded state of the car had prevented Mr Morse from seeing the train. [1268]

On Sunday 28th February 1926, a car driven by Mr W.E. Fortson and containing his wife and five children was struck by a Seaboard Air Line Railroad Atlanta – Washington D.C. passenger train which Mr Fortson knew was due but had failed to observe at Oglesby, near Elberton, Georgia, United States. Mrs Fortson and children Hattie, 11; Lena, 6; Beatrice, 4; and Earl, 2 were killed. Mr Fortson and his son William were seriously injured. [1269]

A car containing Mr J.L. Smith, a farmer living near Bancroft; his mother, Mrs Pillsbury; his children Marion, 7; Billie, 3, and Wayne, 18 months; and Florence Hemingway, 11, was on its way to nearby Owosso on Saturday 27th March 1926 to pick up Mrs Smith, who worked there. As his car reached the level crossing with Shiawassee Street at Corunna, Michigan, United States, Henry Wille, the station agent there, saw that the car's curtains were fastened and that its driver made no attempt to stop before it reached the track. A westbound double-headed Grand Trunk Railway passenger train from Detroit to Grand Haven was also approaching the crossing and its leading engine struck the car, tearing it to pieces and killing its occupants outright. Engineer Listenfeltz said afterwards that he had seen nothing of the car until its wreckage flew past his locomotive. The collision occurred at 3.57 p.m. [1270]

A train ran into a motor car containing three men and three women near Shizuoka, Japan, on Saturday 3rd April 1926. All were killed. [1271]

The Australian state of Victoria suffered its worst level crossing accident up to that time on the evening of Monday 26th April 1926 at Boronia. A Melbourne family party had formed a little club. They all subscribed a small sum of money and hired a van from a local carrier in which they went for picnics at weekends and on public holidays. On the day of the accident a large Ford van, serving as a bus, had thus been hired by them from Mr Samuel Tantanini of Fitzroy to take them to Fern Tree Gully, though as rain had threatened the party was smaller than usual, numbering only 20. The party had duly spent the day at Fern Tree Gully and were returning home to Clifton Hill, a Melbourne suburb.

1268 *The Commercial Tribune*, Cincinnati, Ohio, United States, 20th January 1926
1269 Ibid, 1st March 1926
1270 *The Escanaba Daily Press*, Escanaba, Michigan, United States, 28th March 1926
1271 *The Times*, London, England, 5th April 1926

The crossing at Boronia was unprotected and visibility was impeded to some degree by a dense bank of native shrub, though the crossing was not considered especially dangerous. It was still only early evening when the bus reached it, the party having decided to return a little early to avoid the heavy traffic, and the bus was only travelling slowly, its driver having been warned of the vicinity of the crossing. The children in the party were asleep, while the adults were chatting. Also approaching the crossing was the 5.58 p.m. electric train from Fern Tree Gully to Melbourne Flinders Street. There was ample testimony from local residents as well as the Boronia station staff that the train had whistled more than once as it neared the crossing, but Driver Tantanini failed to hear the whistle, perhaps because it was drowned by the noise of the engine of his vehicle. He also failed to see the train and drove onto the crossing directly in front of it. The train caught the back of the van and pushed it against a fence and the cattle pit. The van was wrecked. Nine people were killed more or less instantly, and 15 others were injured, among them Driver Tantanini, who suffered a broken leg, were injured. One of those injured subsequently died increasing the number of fatalities to 10.

Help was quickly at hand after the accident. The local postmaster opened his office and telephoned to five doctors while troops of Boy Scouts from Fitzroy, Kew and Camberwell were praised for their assistance. Nurse Gale from Melbourne also chanced to be on the train. Ambulances took most of the injured to Melbourne while the bodies of those killed were loaded onto the train. Motorman William Edwards and Guard William Sharp both suffered severely from shock and a relief crew took the train on to Melbourne. [1272]

The inquest on the victims was said to have been one of the most pathetic and difficult ever held by the City Coroner. At its conclusion Mr D. Berriman recorded a verdict of "Accidental Death." He stated, "There is no doubt that the driver of the motor van and two of his passengers knew there was a crossing there, and they should have conceived in their minds certainly greater care than was exercised. The rattle of the van might have stopped the sound of the train and the whistles. However I am going to find that all the ten people were killed through an accidental collision between the van and the train. The evidence is rather towards the conclusion that this is not a safe crossing, where the five roads converge, and with the height of the tea-tree skirting the angle. The tea-tree is a distinct danger. Whether some devices might be conceived is a matter for the Railway Commissioners, but I strongly feel that some measures for safety should be taken. If the tea-tree were removed that would be something." [1273]

Less than a week after the Boronia tragedy, on the evening of Friday 30th April 1926, Mr John Caire, his wife Bertha, son Clifton, daughter Iverna and daughter-in-law Mrs Amy Caire lost their lives in another Australian level crossing collision, this time at Cheltenham in South Australia.

The crossing where the accident occurred adjoined Cheltenham station. It had manually operated gates which were operated during the day by a crossing keeper, but the latter left duty at 5.30 p.m. until 8 a.m. the next morning, during which time they were worked by the station staff. The gates were close to the signal box but were not

---

[1272] *The Argus*, Melbourne, Australia, 27th April 1926
[1273] *The Herald*, Melbourne, Australia, 25th May 1925

interlocked with the signals. As the 30th of April was the last day of the month the stationmaster was busy making up his accounts and an extra porter had been sent from Port Adelaide to act as crossing keeper. [1274]

The inquest on the victims heard evidence from Cheltenham Signalman Stanley Wallace that the next box at Woodville had reported the train involved leaving there at 9.25 p.m. At that point everything had been normal. He had seen the headlights of the train a short distance away and then had heard a click as the north gate was opened. He saw the Caire car enter the crossing and gave a shout but the next moment the train hit the car. He knew it was the first day that the crossing keeper, Gabriel Cohen, had worked there and saw that the keeper wore glasses. There was no system of communication between him and the crossing keeper. It was a wet and windy day.

Stationmaster Arthur Willoughby stated that it was usual to employ a relief crossing keeper at the end of the month.

Crossing Keeper Cohen stated that he usually worked in the Port Adelaide goods yard. He had worn glasses since he was a child and was very short-sighted. The regular crossing keeper had showed him how the gates worked. After dark he had listened to the bells and levers in the box as a guide as to when to close the gates. He had expressed his misgivings as to being able to work the crossing to the signalman, but the latter had assured him that he would be all right. When the headlights of the Caire car came in sight the gates had been closed to the road but as he could neither see nor hear any trains approaching he had opened them to let the car across. The first that he had known of the accident was the sound of an explosion.

Further evidence was given at the inquest that Cohen had not been supplied with a timetable and had no knowledge of the significance of the fixed signals.

At the conclusion to the inquest the City Coroner, Dr. Ramsay Smith, returned a verdict that the death of Mr Clifton Caire had been caused by a train colliding with his car, "due to the responsible officers of the South Australian Railways, in placing on duty Gabriel Cohen, a person who was not qualified to perform the necessary duties of a crossing keeper by means of physical conditions." The death of the deceased had not been the immediate and direct effect of the personal neglect or omission of the officers, so it was not a case of manslaughter. [1275]

The stalling of a sedan on a level crossing at Flint, Michigan, United States, on Sunday 23rd May 1926 led to the deaths of two young mothers and their four small children when it was struck by a Pere Marquette Railroad train. The victims were Mrs Jeffries, her children Ruth, 4 and Flora, 15 months; and Mrs Fay Meade and her children Velva, 6 and Wilfred, 13 months. [1276]

Five occupants of a car, four of them from the same family, lost their lives when it was struck by a northbound Pennsylvania Railroad train on Bucks Crossing, near Wilmington, Delaware, United States, on the evening of Monday 5th July 1926. Those

---

[1274] *The Advertiser*, Adelaide, South Australia, Australia, 1st May 1926

[1275] *Observer*, Adelaide, South Australia, Australia, 22nd May 1926

[1276] *Marshall Evening Chronicle*, Marshall, Michigan, United States, 24th May 1926 who died were Mr Caleb Cannon, his wife Annie, his son Lawrence, his granddaughter Annie, aged 9, and Arthur Tiehler, aged 8. Mr and Mrs Cannon's daughter Selena, mother of the little girl, was seriously injured. Only the train crew witnessed the accident, and they were unable to say how it had happened. [1277]

Another five people died in a level crossing collision south of Lackawanna, New York, United States, on the evening of Friday 16th July 1926. Mrs Anna Jablowski, her son John, 16; Joseph Dylonp and Mrs Ega Marchikowski, all of Lackawanna; and Mr Lous Wolpa, the driver, lost their lives when the automobile stalled on the crossing and was struck by a Buffalo Rochester and Pittsburgh Railroad train. Mr John Storescz was said to be dying, having suffered a fractured skull and other injuries. [1278]

The third bad United States level crossing accident in July 1926 occurred on Whiteside Crossing, five miles south of Lakewood, New Jersey, on Sunday 25th. Mr Sylvester Chuskis, his wife Helen, their son Julius, aged 7; Victoria Matusa and Julius Yarmelowicz, all of Trenton, had been on their way to a summer cottage at Sea Heights. All were killed when their car was struck by a Central of New Jersey Railroad express. The South Lakewood station agent, Walter Buzby, who witnessed the accident from 1,000 feet away, said that the view at the unprotected crossing had been obstructed by the cars of a freight train shunted out of the way of the express, but the engineer of the latter had whistled for the crossing and the bell on its engine had been ringing. [1279]

A train collided with a car on a level crossing near Alessandria, northern Italy, on Sunday 1st August 1926. The car was dragged 150 metres along the line, its petrol tank exploded, and its occupants were engulfed in flames. All five of them were killed, three of their bodies being charred beyond recognition. [1280]

In an all too familiar scenario in the United States, also on Sunday 1st August 1926, a Ford roadster truck stopped at the Gamble Crossing, near Gastonia, North Carolina, for a Southern Railway northbound freight train, then started over the crossing in the path of southbound passenger train No. 45, just over 6 p.m. The accident was witnessed by the driver of a car which had waited for both trains. The driver of the truck, Miss Viola Cloninger, 15, was killed, along with four of her passengers: Miss Bertha Wright, 16; her sister Laura, 20; Miss Inez Jenkins, 16 and her sister Villiard, 19. Miss Viola Bess, 16, was seriously, possibly fatally, injured. Conductor B.L. Ormond, in charge of the train, stated that the accident had been unavoidable as far as the train crew were concerned. [1281]

Five men returning early on Sunday morning, 8th August 1926, from a fraternal

---

[1277] *Salamanca Republican-Press*, Salamanca, New York, United States, 6th July 1926

[1278] *The Sandusky Register*, Sandusky, Ohio, United States, 17th July 1926

[1279] *The Bridgeport Telegram*, Bridgeport, Connecticut, United States, 29th July 1926

[1280] *The Times*, London, England, 3rd August 1926

[1281] *Gastonia Daily Gazette*, Gastonia, North Carolina, United States, 2nd August 1926 meeting, were killed by the westbound *Ohio Limited* train of the New York Central Railroad when their light sedan was struck by it on a level crossing at Williamsville, New York, United States. The young men, who were all single, were Roy Pope and his brother Arthur, of Snyder, New York; and Harold Krieb, Henry Link and Charles Smith of Williamsville. The car had entered the crossing behind an eastbound train stopped at the station. [1282]

A school charabanc coming from Bacău was hit by a passenger train on a level crossing at Letea-Trei Izvoare, Romania, at 11 p.m. on about Thursday 26th August 1926. There were 5 deaths and 4 injuries. [1283]

The second disastrous English level crossing accident in nine months occurred at Naworth station, on the Newcastle – Carlisle secondary main line of the London and North Eastern Railway, on Monday 30th August 1926. Again there were nine fatalities, eight of them the passengers riding in a motor coach and the other the crossing keeper.

Naworth was only a small station and its staff consisted of just three porters, only one of whom was on duty at a time. The crossing was hand worked and only a single pair of 'occupation type' wooden gates was provided, one on each side of the double line. About 10 yards from the gates on the Carlisle and Newcastle sides of the road respectively were the booking office, where there was a telephone communicating with Brampton Junction, and the porter's room, where there were three position block repeaters for both lines showing the indications of the block instruments in the adjoining signal boxes at Brampton Junction and Low Row, each about a mile and a quarter distant. A six-lever ground frame adjoining the porter's room worked distant, home and starting signals in each direction. Unlike at Fenny Stratford these signals were not interlocked with the gates, so the signals could be clear even though the gates were open for road traffic. The gates normally stood closed across the road and according to the regulations they must not be opened for a road vehicle by the crossing keeper if a train was approaching, that is when one or both block indicators showed "Train on Line," and not until the fixed signals had been placed at danger. Also the far gate must be opened first to prevent vehicles having to stop on the line to await its opening.

On the day of the accident the 1.18 p.m. down express from Newcastle to Carlisle passed Low Row at 2.35 p.m., travelling at nearly 50 miles an hour. Driver William Embleton told the Ministry of Transport inquiry that the Naworth distant and home signals had been clear for him, and he had whistled as usual 500 yards from the crossing. His engine had been less than 25 yards from it when the motor coach came into view, and he had braked and reversed his engine a moment before the collision.

Fireman Archibald Kennedy confirmed his driver's evidence as far as the signals and the speed of the train. Busy swilling down the footplate he said he had not seen the coach before they struck it. After the accident he had noticed that the up signals were also off and had gone to the lever frame and replaced them, while Guard John Atkinson had broken into the booking office and telephoned Brampton Junction to report the accident.

[1282] *Cattaraugus Republican*, Little Valley, New York, United States, 11th August 1926

[1283] *The Times*, London, England, 28th August 1926

The motor coach was the property of Mr Sowerby, of Gilsland, a minor holiday resort, and it making an advertised excursion to Alston which had set off at 2 p.m. It was being driven by Mr Fred Lamb and was carrying 15 passengers. [1284] Driver Lamb and a teenager named Nathan Green, who was also sat on the front seat, gave evidence at the inquiry. Lamb said that as he had turned off the main road onto that leading to the crossing Green had blown the vehicle horn repeatedly. As the coach neared the crossing, which was less than 300 yards from the main road, the gates had been closed but the porter, William Oliver, had then appeared and opened the near gate. The coach had been travelling at about 10 miles an hour. Oliver had then walked across and opening the far gate had motioned for them to come on, whereupon the coach had entered the crossing, moving at a walking pace. He (Lamb) had looked to the right and then to the left when he had seen the train almost upon him and had had time to stamp on the accelerator in an effort to get clear.

Green generally confirmed Lamb's evidence and added that Oliver had come from the booking office.

Evidence was also given of a previous near miss involving a bus driven by another of Mr Sowerby's drivers and a car driven by a local schoolteacher. The gates had been opened for the bus which had just got over the crossing in front of a train, while the teacher had just pulled up in time to avoid it. The crossing keeper involved, not Oliver, denied that any such incident had occurred, but admitted that he did not always replace the signals before opening the gates and the inspecting officer, Lieut. Colonel A.H.L. Mount, did not believe his denial.

Porter Oliver had only been working at Naworth a week, having moved there after being made redundant at Edlingham. Porter John Bell, whom he relieved daily, spoke of him as a quiet conscientious man and Oliver's immediate superior, Thomas Arnold, the Acting Stationmaster at Brampton Junction, said that he had been impressed by Oliver's carefulness and intelligence. It was Bell who had first instructed Arnold of his duties on 23rd August, first as regards the booking office and then as to operation of the crossing, telling him always to keep the gates closed except when letting vehicles across and to replace the signals first. Acting Stationmaster Arnold had made his daily inspection of Naworth on the afternoon of 23rd August when Oliver said that Bell had instructed him as to what to do and that he (Oliver) thought that it was a very easy job. Given Oliver's previous railway experience as porter signalman and signalman at Edlingham, Alnwick and Edlingham again, Arnold considered this assurance sufficient, and on a further visit to Naworth he had seen Oliver operate the crossing correctly.

Concluding his report the inspecting officer stated that there could be no doubt that the train had passed the distant signal at clear and was approaching the crossing when the gates were opened and that neither the train driver, Embleton, nor the bus driver, Lamb, was in any way to blame for the accident, for which Porter Oliver was solely responsible. Oliver had failed to carry out three simple and common-sense duties and had clearly been unaware of the risk he was taking as he himself had been killed as a result. Firstly he had failed to check the block indicators, secondly he had not replaced the signals and thirdly he had opened the gates in the wrong order. Had he checked the indicators he would have seen that a train was approaching and not opened the gates at all; had he replaced the

[1284] *The Times*, London, England, 31st August 1926

signals even after the train had passed the distant signal Driver Embleton might have been able to reduce speed sufficiently to enable to coach to clear the crossing, and had the far gate been opened first the train would probably have arrived before the coach entered the crossing.

There were, however, no safeguards at the crossing against forgetfulness or carelessness, or due to distraction from preoccupation with booking duties or the persistent blowing of car horns by impatient road users.

Turning to remarks and recommendations, Lieut. Colonel Mount noted that the line had opened when railways were in their infancy and that for new works it had been a requirement for more than 50 years for the crossing gates to close alternately across road and railway and for the gates and signals to be interlocked had been a requirement on new lines for over 40 years (in passing, it may be noted that regardless of this the author is aware of a level crossing close to him where the gates never closed across the line up to its abolition in the early 1970s). He therefore recommended that the crossing should be modernised, with gates closed across both road and railway and interlocked with the signals. Given that the crossing keeper had other duties to perform a gong or track circuit to warn him of the approach of trains would be useful, or alternatively an arrangement by which the crossing keeper must obtain definite permission from the signalman before opening the gates, Also the block indicators, lever frame and telephone should be brought together in one cabin.

Finally, although it had no bearing on the accident, he commented like Major Hall on the objectionable arrangement of having a passenger sat beside the driver and on the excessive use of the coach horn. [1285] [1286]

Chicago attorney C. Paul Tallmadge, his wife and three children were killed on the evening of Friday 3rd September 1926 when the car that he was driving was struck by a Cleveland Cincinnati Chicago and St Louis (Big Four) Railroad passenger train on a level crossing at Woodside, Ohio, United States. The bodies of the victims were scattered along the right of way as the car was carried a quarter mile along the track and when the train stopped it burst into flames. The track was damaged at one point and the train crew said it was a miracle that the train had not been derailed. The family had been returning home from a vacation. The view of the track at the crossing was unobstructed. [1287]

Six people died when the Southern Pacific Railroad's westbound *Sunset Limited* passenger train, travelling at 45 miles an hour, collided with a train on a level crossing at Puente, California, United States, late on Sunday 5th September 1926. The victims were Mr W.H. Hoag and Mrs Mary Hoag, of Baldwin Park, California; Miss Grace Hoag, 21, of Vandalia, Illinois; Mrs L. Meader, of Vandalia; Donald Smith, aged 7, of Baldwin Park; and Daniel Opperman, of Lodi, California. Mr and Mrs R.B. Smith of Baldwin Park were injured, Mrs Smith probably fatally. [1288]

[1285] Ibid, 1st and 3rd September 1926

[1286] Ministry of Transport report

[1287] *The Commercial Tribune*, Cincinnati, Ohio, United States, 4th September 1926

[1288] *Oakland Tribune*, Oakland, California, United States, 6th September 1926

Heavy rain interfering with his view probably caused Mr Henry Astill to fail to see an approaching goods train as his car reached Dean's Crossing, near Molong, New South Wales, Australia, at 7.45 a.m. on Thursday 30th September 1926. He drove onto the track in front of the train and its engine smashed the car, killing him and his passengers: Mrs Mary Staines, her children Joyce, 6, and Norman, 2; and May Walker, 13. The bodies of the victims were badly mutilated. The train was bound from Parkes to Orange and was in charge of Guard W. Considine with Driver H. Gouge and Fireman Brown. Driver Gouge told the police that he had seen the car and applied the brakes, but the car had tried to cross in front of his engine. He and his fireman were badly affected by shock and had to remain for some time at Molong. The road approaching the crossing was straight, the view of the line unobstructed and the engine had whistled. The road also climbed to the crossing, precluding the possibility of a brake failure as a cause of the accident. [1289]

The *Transatlantic Limited* passenger train of the New York Central Railroad, in charge of Conductor George Ekins, with Engineer George Burns and Fireman George Whelan, struck a small sedan car carrying five men on a hunting trip on a level crossing at Syracuse, New York, United States, on Sunday 10th October 1926, killing them all. They were John Ciccihilli, Phillipa De Mascio, Mike Montreal, Angelo Serio and Fillipo Sioblia. The view of trains at the crossing was unobstructed. [1290]

At 8 p.m. the same day another New York Central Railroad passenger train struck an automobile truck at Crissey, Ohio, United States, 14 miles west of Toledo, killing Mr Eli Orner, his wife, their daughter Eleanor, 15, their son Simon, 3 months, and their granddaughter, Lavine, 4. The truck was carried nearly half a mile. [1291]

Max Wolf, a pedestrian, saw that a car was likely to be struck by a train on the unguarded level Lemon Street Crossing in Kenosha, Wisconsin, United States, on Saturday 23rd October 1926. He gesticulated to the car driver to stop but his efforts were in vain. The southbound Chicago and North Western Railway passenger train struck the car, killing Mrs Mary Maxim, Zigmund Maxim, aged 5, Joseph Zeindewich, Frank Wienestosky and Mrs Tuskowski. An infant was missing and was believed to have also died.

Witnesses agreed that Engineer W.R. Kennedy had whistled several times for the crossing, but it was thought that with the windows of the sedan being closed because of the cold the whistles had not been heard. The occupants of the car had been on an outing to nearby woods. [1292]

Nine people died in different level crossing accidents in Ohio, United States, late on Saturday evening, 30th October 1926. Five of them were teenagers who had been celebrating Halloween who lost their lives when their car was struck by an Erie Railroad train at Warren. Their names were William Faber, 19, Frederick Jaglelsky, Joseph Johim

---

[1289] *The Leader*, Orange, New South Wales, Australia, 1st October 1926

[1290] *Syracuse Herald*, Syracuse, New York, United States, 11th October 1926

[1291] *The Commercial Tribune*, Cincinnati, Ohio, United States, 11th October 1926

[1292] *The Sheboygan Press*, Sheboygan, Wisconsin, United States, 23rd October 1926 and Louise Missu, all of Newton Falls and Dosetta Thompson of Ravenna. [1293]

Five cotton compress labourers were killed, and 14 others were injured, 12 of them seriously, in a level crossing collision near Houston, Texas, United States, on Tuesday 9th November 1926. A Southern Pacific Railroad accommodation train carrying workers to an oil refinery several miles down the Houston ship channel collided with a truck pulling a large passenger trailer. A large cotton warehouse obscured the view of the train from the truck driver; when he did finally see it he accelerated in an attempt to get across the line in front of it. The victims were all negroes and Mexicans. [1294]

A crossing keeper was blamed for a collision between No. 89 up passenger train from Aligarh to Bareilly and a camel cart on Gate No. 87 crossing, near Aligarh, India, at about 1.15 a.m. on Monday 22nd November 1926. The cart was completely destroyed. Four of its occupants were killed outright and eight others were seriously injured, two of them dying later in hospital. The gateman was blamed for leaving the gates open, unattended and without lamps. He was prosecuted and sentenced to 18 months' rigorous imprisonment. [1295]

1927 was only a few days old when there was a serious collision involving a train and a one-man street car on the 21st Street Crossing in Superior, Wisconsin, United States. At about 8.15 p.m. on Thursday 6th January 1927 the *Gopher Limited* passenger train of the Great Northern Railway struck a street car carrying 24 passengers from suburban Billings Park. Street car passengers Alice Isaacson, William Manning, Alvira Nelson and Fred Ryan were initially reported as having been killed, as was Fireman George Lindberg, who was thrown from the engine of the train by the impact. All the other street car passengers were injured to a greater or lesser extent, and two more were to die in the following days. Motorman N.J. Pellman of the street car, Engineer Fred Benedict and Watchman Antone Johnson, whose shanty was demolished by the street car and caught fire, were also hurt.

The exact sequence of events leading up to the accident was unclear. It was reported that the street car had stopped short of the crossing, then started forward across the tracks, when the crossing gates had descended upon it, knocking the trolley pole off the car, and stopping it.

Engineer Benedict stated, "I could see the street car stopped on the right side of the crossing when I was two blocks away. I was travelling about 25 miles per hour and had increased my speed after signalling the crossing. Of a sudden the car moved ahead, stopped directly on the tracks, backed up again and was clear. My train was pounding down on the crossing and when I was about fifty yards away the car again started across. God, how I tried to stop that engine! I gave her everything I had in brakes, sand even, but the crash I heard sickened me. The next I remember was listening to the cries of the dead and dying above the hiss of steam."

1293 *The Evening Independent*, Massillon, Ohio, 1st November 1926
1294 *The Sedalia Democrat*, Sedalia, Missouri, United States, 9th November 1926
1295 Report by the Railway Board on Indian Railways for 1926-27 Volume 1

Gateman Johnson said, "I tried to stop that car by putting down the gates. But he kept coming. He did not know what to do, I guess. The last I remember I was falling with the shanty." [1296]

The inquest on the seven victims of the accident was held at the end of January. Its jury exonerated the crossing watchman, the street car motorman and the train engineer, but found the Duluth-Superior Street Railway, the Great Northern and Northern Pacific Railroads (the Great Northern train was running on Northern Pacific tracks) and the Wisconsin railroad commission jointly negligent. The street car company were blamed for operating one-man cars, the Great Northern for permitting their trains to run at excessive speed, the Northern Pacific for having faulty crossing gates and the railroad commission for refusing to consider a petition calling for a viaduct to replace the crossing. [1297]

A bus carrying the Baylor University baseball team from Waco to Austin, where they were due to play against a University of Texas team, was struck by an International-Great Northern Railroad passenger train No. 2, the *Sunshine Special*, on a level crossing at Round Rock, Texas, United States, at noon on Saturday 22nd January 1927. The results were disastrous. Ten of the students, including five members of the baseball team, were killed outright and 13 others were injured.

A newspaper reporter described the accident. 'The accident was the worst that has ever happened in the history of any Texas college. The boys were laughing and talking about the prospects for the game tonight as they passed through the business section of Round Rock without stopping. Almost before they knew what happened, the crash came.

'Joe Potter, a student, was driving. "He could not see the train for a string of box cars," said John Kane, one of the survivors. "Until we got right on the track. Then several of us saw the train coming. Joe automatically tried to stop, but realizing it was too late, he turned down the track. Nobody had time to say anything." Kane said his first feeling was that of coming out from under an anaesthetic, or of whirling in the air, and then he looked and saw that the body of the big bus was being hurled towards him. "So I shut my eyes and crawled away," Kane said. He opened his eyes after crawling a few feet, he said, and saw Keifer Strickland and Lewis Slade standing in the road. Blood was gushing from Strickland's head. He looked the wreck and saw the dead and dying, and he found that he, himself, was hurt, the blood oozing from his head and nose. Then someone led him away.

'Abe Kelley, elected football captain for next year, died a hero, Kane said. "He and Washam were inseparable pals," he said. "They were sitting together. Just before the impact Abe shoved Washam out of the car and clear of the wreck. Washam was saved. Abe was killed."

'Ed Gooch saved his life through an act of kindness. At Temple when the bus passed along, the student saw Ivey Foster Jr, of Taylor. "He will freeze out there, and he wants to go to the game," Coach R. Wolfe said, according to Kane. "So let's take him along."

1296 *The La Crosse Tribune* and *Leader Press*, La Cross, Wisconsin, United States, 7th January 1927

1297 Ibid, 28th January 1927

The bus stopped and picked up Foster and Gooch got up and gave him his seat. Foster was killed, and Gooch was not.

'Ivey Foster Sr, Taylor business man, was helping unload the bodies from the train at Taylor upon its arrival there and as he was helping lift one of the bodies from the baggage car the covering slipped from the face. It was Ivey Foster Jr, his son. Mr Foster collapsed.'

Immediately after the accident R.L. Bobbitt, speaker at the House of Representatives, said that he would advocate a law requiring railroad companies to install automatic gates at all grade crossings, while it was stated that the Texas legislature would introduce a bill immediately to compel all railroads and interurban companies in Texas to eliminate grade crossings by replacing them with underpasses or overhead crossings. [1298]

An inquiry into the accident opened the following Wednesday and heard evidence from a considerable number of witnesses, including Engineer Sid Crewes and Fireman Ben Westrup, though not conductor T.D. Barnes, who was in hospital suffering from a nervous breakdown. The engineer stated that he had whistled for the crossing and the engine bell had been ringing continuously. Plenty of witnesses agreed that the whistle had been sounded. The train had been brought to a stand 800 or 900 feet beyond the crossing. Evidence was also given the weather had been cloudy, misty, and foggy, though whether the fog had anything to do with the accident seems debatable. The inquiry was adjourned without responsibility for the accident having been determined. [1299]

Legislation was in fact passed for the construction of underpasses or overbridges at all intersections of state highways and railroads and the first overpass in Texas was constructed at Round Rock, but not until 1935. Other legislation required all school buses in the United States to stop dead before entering any railroad crossing. The students who lost their lives became known as the *Immortal Ten.* [1300]

All but one of the Knight family, who had only just moved to Illinois from Red Boiling Springs, Tennessee, were killed in an accident at Blackhawk Park, Rockford, Illinois, United States, at 4.30 p.m. on the afternoon of Sunday 23rd January 1927 when the car they were in was struck by a passenger train on a level crossing. Mr I. F. Knight, his wife Martha, and their children Eloise, 11; Clinton, 8; Dola, 5; and Nellie, 4 were killed. Another daughter, Zella, aged 13 or 15, escaped with slight injuries, but was left, as a newspaper pointed out, 'alone and virtually penniless.' The view at the crossing was unobstructed but Mr Knight apparently drove onto the track without noticing the train. [1301] [1302]

Rather confusingly a grade (level) crossing with the Baltimore and Ohio Railroad line at Kensington, Maryland, United States, was provided with lifting gates, but these were only operated between 6 a.m. and 6 p.m. daily. Outside of those hours the gates were left

---

1298 *San Antonio Sunday Light*, San Antonio, Texas, United States, 23rd January 1927

1299 *San Antonio Express*, San Antonio, Texas, United States, 27th January 1927

1300 Webpage: Williamson-county-texas-fistory.org/The_Immortal_Ten_Crash_in_Round_Rock

1301 *Evening Telegraph*, Dixon, Illinois, United States, 24th January 1927

1302 *The Harvard Herald and Independent*, Harvard, Illinois, United States, 27th January 1927 raised regardless of whether or not a train was due, though an automatic electric bell was also provided which operated throughout the 24 hours. Mr Walter Twombly, a local man, was presumably aware of the conditions at the crossing, but for some reason early on Friday 4th February 1927 he drove onto the tracks despite the ringing of the bell and the whistling of the engine of an eastbound St Louis, Missouri – Washington, D.C. express. The light touring car was rolled into a ball and tossed aside; its five occupants being killed instantly. As well as Mr Twombly they were his wife, Elizabeth; their son Paul, 20; their daughter Elizabeth, 18; and Mrs Ellen Holland, a neighbour. The party had been returning from a card party at Forest Glen and had passed the Twombly home to take Mrs Holland home. [1303]

Five people died in an accident near Wellsville station, Kansas, United States, at 9.40 p.m. on the evening of Thursday 17th March 1927. *The Chief*, the crack transcontinental train of the Atchison Topeka and Santa Fe Railway, collided with a Ford coupe driven by Mr William Hayes, a local farmer, and containing Mrs Petrel, the wife of the second trick Wellsville operator, who was on duty at the time, 4-year-old Dale Gregory (Mrs Petrel's son by a previous marriage?), Mrs Bessie Garrett and the latter's infant son.

Engineer C.E. Burr and Fireman A.E. Sparrier were unaware that they had struck a car until they reached Kansas City Union station, though they knew that their engine had hit some obstruction and reported the fact on their arrival so that the line could be examined.

The accident was witnessed by Mr H.M. Brehelsen and his son Howard, whose car had been stopped at the crossing waiting for a westbound fast mail train and the eastbound *Chief*. They believed the other people in Mr. Hayes' car had prevented him seeing the train. [1304]

On the evening of Wednesday 15th June 1927 the Whippel family set off from their farm two miles from Sunman, Indiana, United States, to go into town to watch the movies. At the Meridian Street Crossing with the Cleveland Cincinnati Chicago and St Louis (Big Four) Railroad their automobile stopped to allow a freight train to pass. It then set off into the path of a Chicago – Cincinnati passenger train running on the opposite track. Seeing what was about to happen Engineer Pat Hickey had applied his brakes, but his train ran a quarter mile before it could be pulled up, having carried the car 100 feet. Mrs Whippel and her daughter Romelda, 19, were killed instantly, her son Wilson passed away soon after being taken to a doctor's office and her son Melvin, 17, and Mr Frank Whippel, the last survivor of the family, died during the night. [1305]

Five people died in a collision between a train and a car near Lübeck, Germany, on the Hamburg – Lübeck line on Saturday 25th June 1927. Four of the victims were Herr Debald, the head of a Hamburg export firm, his wife, and his parents. The crossing keeper had failed to close the barriers. He was arrested. [1306] [1307]

[1303] *The Morning Herald*, Hagerstown, Maryland, United States, 5th February 1927

[1304] *The Emporia Gazette*, Gazette, Kansas, United States, 18th March 1927

[1305] *The Brazil Daily Times*, Brazil, Indiana, United States, 16th June 1927

[1306] *Eisenbahn-Katastrophen in Deutschland* by Hans Joachim Ritzau, pg. 94 (in German)

A seaside excursion ended in tragedy early on Sunday 3rd July 1927 at Belmar, New Jersey, United States. The morning before Mrs Catherine Damiano and her six children had set off with three of her neighbours from South Orange to the seaside, promising her husband Bonificio that they would be back on Sunday in time for him to go to look over a farm he was thinking of purchasing for his family. One of the neighbours, James de Cicco, an experienced driver, was at the wheel.

The party remained at the seaside until sundown, with the children playing in the sand, and then deciding that it was too late to make to full return trip they parked the car by the roadside and slept most of the night. About 5 a.m. they set off again. Mrs Damiano, who had only begun to drive a week earlier, decided that she would like to take the wheel and did so. As they arrived at the level crossing with the New York and Long Branch Railroad at Belmar, following another sedan, a train of empty cars was approaching. This was another crossing provided with a watchman but without gates. The watchman, James Whidbee, signalled to the two cars to stop but the first sedan ignored him, drove over the line, and went on its way. That driven by Mrs Damiano at first appeared to be going to do the same, then pulled up just short of the tracks. Then Mrs Damiano for some reason, probably related to her lack of experience, restarted the car, only to stop it again squarely across the tracks. The car had now stalled, and she frantically tried to get it to restart, but it would not. De Cicco, sat alongside her, then jumped out and attempted to push the sedan off the tracks, assisted by Whidbee, while the children were asleep in the rear of the vehicle and the other adults were apparently too paralysed with fear to do anything. As Mrs Damiano was still struggling to get the car to go the train stuck it, killing her, her children Anna, 13; Alexander, 8; Ames, 6; and Nicolina, 14 months, together with her neighbours Santo de Filite and Mildred Ballentino. Rose Damiano, 10, and her brother Anthony, 3, were injured.

Pending an investigation the black crossing watchman was held on $10,000 bail, though he had done the job for many years and had never previously been found negligent. Whether his colour had anything to do with his being arrested it is hard to say; the accident was clearly not his fault. [1308]

The Pawloff family of New York chose to spend the United States Independence Day holiday at home until they went for a drive in the evening of Monday 4th July 1927. They had been away from home less than an hour when their car arrived at the Sugar Street Crossing at Echota, a suburb of Niagara Falls. Nearing the crossing at the same time was a New York Central Railroad passenger train, the engineer of which saw the automobile, which appeared to be slowing down, but when he saw that it was not going to stop he could do nothing to avoid hitting it. There was a flagman at the crossing even at night-time, though it had no gates, and the car driver took no notice of the flagman's frantic signals to stop. Caught by the locomotive the car was rolled down the track about 100 yards and its occupants were all killed or injured. Killed outright were Mrs Mary Pawloff, her sons Andrew, 16; Theodore, 9; Walter, 6; and Carl, an infant, together with her daughter Margaret, 5. Mr Samuel Pawloff and another son, Michael, 8, were seriously

---

1307 *The Times*, London, England, 27th June 1927

1308 *The Commercial Tribune*, Cincinnati, Ohio, United States, 4th July 1927 injured. Police and ambulances called to the scene were delayed by the traffic jam caused by the accident. [1309] Michael died in hospital the next day. [1310]

Only briefly mentioned in the newspapers was a level crossing collision at Brady, Kentucky, United States, on Monday 18th July 1927. Chesapeake and Ohio Railroad passenger train No. 26 struck an automobile, killing Mr and Mrs Alec Patton, their 8-month-old daughter Alla, Grace McLain, 16, Albert McLain, 4, and Elmo Caudill, 6. All lived in nearby Morehead. [1311]

A Sunday morning drive ended in disaster almost as soon as it had begun for the family of Mr W.B. Wrigley, of Yazoo City, Mississippi, United States, on 7th August 1927. His view obscured by a freight train, Mr Wrigley drove onto the track at a level crossing with Eighth Street, only a short distance from his home, in front of a Clarksdale – Jackson passenger train at about 10 a.m. Engineer Kelly said that he had not seen the car until the last moment, when it was apparently at a standstill on the line and attempting to back off. The touring car was demolished, killing Mr and Mrs Wrigley, a daughter and son aged 10 and 8, and a deaf and mute man named Robert Laney. Mr W.B. Wrigley's father, Ben F. Wrigley and an infant of 6 or 8 months were probably fatally injured. [1312]

The *Sunshine Special* passenger train of the Missouri Pacific Railroad collided with a small sedan car on a level crossing at Summit, Missouri, United States, on Sunday 11th September 1927, killing miner John Burlbaw, his wife and two children and his mother, Mrs Ellen Bloodsaw. The train, travelling at 50 miles an hour, took a quarter mile to stop. [1313]

At 14:10 on Saturday 1st October 1927 No. 82 down passenger train collided with motor bus No. 2180-A at a level crossing at Verka station, India, on the Amritsar-Pathankot section of the North Western Railway. The bus was badly smashed and carried over 400 feet on the cowcatcher of the engine. Ten bus passengers died and 18 were injured. The accident was blamed on the rash and negligent operation of the bus by its driver. [1314]

Late on Friday 7th October 1927 the Booker family were returning to their home in Ripley, Mississippi, from Memphis, Tennessee. At the Wise Crossing with the St Louis-San Francisco Railway, near Hickory Flat, Mississippi, United States, Mr John Booker waited for train No. 6, the fast mail, to pass and then started over the crossing and his car was struck by a freight train travelling in the opposite direction on the other running line.

[1309] *Syracuse Herald*, Syracuse, New York, United States, 5th July 1927

[1310] *Salamanca Republican-Press*, Salamanca, New York, United States, 5th July 1927

[1311] *The Commercial Tribune*, Cincinnati, Ohio, United States, 19th July 1927

[1312] *The Daily Herald*, Biloxi, Mississippi, United States, 8th August 1927

[1313] *The Commercial Tribune*, Cincinnati, Ohio, United States, 12th September 1927

[1314] Report by the Railway Board on Indian Railways for 1927-28 Volume 1

The car caught fire as a result of the collision as well as being badly damaged. Mr Booker, his daughter Lottie, 17, his nephew James Booker and the latter's wife and infant son all lost their lives. Mr John Booker's son J.T., 13, and son-in-law Clayton Nix escaped with slight injuries. [1315]

The failure of the road user to see or hear an approaching train is probably the most common cause of level crossing collisions and such was apparently the case in a major accident in the north-east suburbs of Indianapolis, Indiana, United States, on the evening of Saturday 14th October 1927. What were described as an automobile truck and trailer, the latter carrying 65 men and women on their way to a barn dance, was struck by Union Traction Company interurban car en route from Indianapolis to Muncie. Sixteen people were initially reported killed and about 20 injured, four it was thought fatally. All the victims had been riding on the trailer except the motorman and the conductor of the interurban, who were injured. Motorman W.W. Merrill suffered a broken leg, and the car ran for half a mile before it could be pulled up. Conductor Hal Titus was briefly knocked unconscious. The men who died had all been members the Mystic Order of Veiled Prophets of the Enchanted Realm, described as 'the fun organisation of the Masonic Lodge.' The members of the Santa Grotto drill team which was giving the dance on the trailer were playing musical instruments and the whole party, it was said, was singing and laughing at the time of the accident.

The truck had cleared the line at the moment of impact and its driver, Harry Stewart, was unhurt. He was taken into custody by the police pending an investigation and admitted that he had neither seen nor heard the interurban car, adding that the first that he had known of the accident had been when his machine had suddenly leapt forward. Motorman Merrill told the police that he had whistled repeatedly for the crossing. [1316] [1317]

By the afternoon of the following day the number of dead had risen to 18 and it later increased to 20.

On 2nd November Coroner C.H. Keever, who held the inquest on the victims of the disaster, stated that he would recommend that both Driver Stewart and Motorman Merrill be bound over to appear before the grand jury. Meanwhile the public services commission investigation blamed Driver Stewart, the dim headlight of the interurban (required by law within the city limits), members of the party on the trailer for not noticing the interurban and the public in general for requiring speedy transport (the interurban had reportedly been travelling at 40 miles an hour). It recommended that such trips as the party's should be flagged over crossings as school buses were. [1318]

No report of any further action against the truck driver or motorman has been found.

On the afternoon of Wednesday 19th October 1927 Russell Cox, cashier of the Byesville state bank, Ohio, United States, drove his Ford sedan to the school to collect his 7-year-old son, Ralph. It was raining hard, and he picked up his son's classmates,

---

[1315] *The Anniston Star*, Anniston, Alabama, United States, 8th October 1927
[1316] *The Evening News*, Jeffersonville, Indiana, United States, 15th October 1927
[1317] *Sarasota Herald*, Sarasota, Florida, United States, 15th October 1927
[1318] *The Indianapolis* Times, Indianapolis, United States, 2nd November 1927

William Breeden, Stephen Gold, William Gooden and Raymond Kostelnik, all also aged 7, to take them home as well. When his car reached the Sixth Street level crossing at Byesville at 3.30 p.m. witnesses reported that he brought his car almost to a full stop at the crossing, then, believing all was clear, he had proceeded, and the car was struck by a southbound Pennsylvania Railroad passenger train. Engineer John Bean saw the car enter the track at the last moment but with slippery rails could do nothing much to lessen the speed of his train. The car was reportedly 'crushed like an egg shell.' Stephen Gold, the only survivor of the accident, was in hospital with a fractured skull and 'extremely critical.' The crossing was a 'blind' one with limited visibility but was not provided with gates, a crossing watchman or even a warning bell. [1319]

Five residents of Tiffin, Ohio, United States, died in an accident at nearby Adrian at 3 p.m. on Saturday 22nd October 1927. Peter Brittner and his wife Alice had travelled to Adrian to look over some property and were on their way home. With them in the car, coming along for the ride, were neighbours Grover Sours and his son Calvin, 8, and Margaret Elens, 16. At the Main Street Crossing in Adrian Mr Brittner failed to notice a fast approaching Cleveland Cincinnati Chicago and St Louis (Big Four) passenger train. His small touring car was caught by the cowcatcher and carried 150 feet down the track, bursting into flames as the train came to a halt. It was believed all five occupants were killed instantly. [1320]

Towards the end of 1927 work was in progress on the Puddingstone Dam in California, United States. On Friday 28th October 1927, in connection with the dam construction, a large dump truck with 12 workmen was sent to load a consignment of cement. Ten of the men were riding stood in the back of the truck, the other two being its driver, W.S. Powers, and a man sat beside him. As the truck reached a level crossing near San Dimas station Driver Powers either failed to spot or tried to beat an Atchison Topeka and Santa Fe Railway local train, due at San Dimas at 8.38 a.m. and which reached the crossing at 8.42. San Dimas was only a request stop and the train was not going especially slowly as there was no custom for it that morning. The train struck the rear of the truck, and seven men were killed or fatally injured, the other five being more or less hurt. Identification of the victims proved difficult as only one was local, the others living at a camp by the dam. They each had pay off numbers, carried in metal figures on their caps or blouses, but the caps were all scattered in the collision and mixed up.

An inquest held in Pomona the day after the accident brought in a verdict of negligence against Driver Powers, who had escaped with slight head injuries from the accident. [1321]

A disastrous level crossing accident happened in France at 18:30 on the evening of Thursday 9th February 1928. A loaded mineral waggon ran away down an incline at Thionville in the department of Moselle. At the level crossing with the road from Daspich

[1319] *The Times Recorder*, Zanesville, Ohio, United States, 19th October 1927
[1320] *The Morning Republican*, Findlay, Ohio, United States, 24th October 1927
[1321] *The Covina Argus*, Covina, California, United States, 4th November 1927

to Terville it struck and cut in two a tramcar running between Hayange and Terville, killing 18 of the tram passengers. Most of the victims were young men and women on their way home at the end of their day's work at the Weymerskirch sawmill. Four other seriously injured passengers were taken to hospital.

The following afternoon the Merz parquet (examining magistrate) went to the scene to open an investigation into the accident and to determine if the man stationed by the Angevilliers mine at the crossing to protect road traffic had done his duty. At the same time M. Andre Tardieu, the Minister of Public Works, dispatched M. David, Assistant General Inspector of Railways, to carry out a separate investigation. [1322]

An interurban car of the Terre Haute Indianapolis and Eastern Traction Company, in charge of Conductor Lawrence Hensley with Motorman Ralph Raridan, collided with a car driven by Mr John Buckner on a level crossing at Centerton, Indiana, United States, late on Saturday 25th February 1928. Mr Buckner, his wife, their children Genevieve, 10, and Katherine, 9, Mr Buckner's mother Amma Thacker and his stepsister Hazel, 13, were all killed. The party were returning to nearby Brooklyn after visiting a sister of Mrs Thacker at Martinsville, a few miles to the south. Visibility at the crossing was restricted by a store and it was known as 'Death Crossing,' almost 30 people having lost their lives there over the years. [1323]

An interurban car of the Illinois Traction System forming a special working carrying 46 students of Chaminade College on Sunday 26th February 1928 collided with a car carrying four coal miners and the wife of one of them on a level crossing north of Granite City, Illinois, United States. All five occupants of the car were killed. They were Robert Craig, Durant Douglas and his wife Anna, Louis Hubbard, and Frank Oettle, all of Glen Carbon, Illinois. Motorman William Clutter said that he had seen the car 300 feet from the crossing, whistled and the car had slowed down, but when it neared the crossing it had suddenly accelerated and gone into the path of the train. It seemed as if the driver had attempted to beat the train over the crossing. [1324]

The 10 a.m. special train from Mildura had not proceeded very far when it collided with a truck carrying nine people on Koorlong Avenue Crossing, Irymple, Victoria, Australia, on the morning of Monday 26th March 1928. The driver of the truck, William Beauman, manager of Mrs Lamont's horticultural block at nearby Redcliffs, was killed outright, as were Ivy Moss, 18; Colin Tickell, 15; and Josephine Lamont, 12, all of Redcliffs. The other passengers on the truck were all injured. They were Mrs Lamont and her daughters Gladys, 12; Nancy, 10; Peggy, 8; and Lorna, 3. Mrs Lamont died of her injuries three days after the accident and Nancy the following day. The party had been going to the Mildura carnival.

Beauman reportedly failed to reduce speed at the crossing, at which there was limited visibility. The train driver, John Giles, stated that he had been travelling at 30 miles an

---

1322 *Excelsior*, Paris, France, 11th February 1928
1323 *The Logansport Press*, Logansport, Indiana, United States, 26th February 1928
1324 *The Commercial Tribune*, Cincinnati, Ohio, United States, 27th February 1928

hour. He had whistled when he was 200 yards from the crossing and continuously from a distance of 100 yards from it. He said that, "I just saw a motor vehicle flash past the engine." [1325]

At the inquest on four of the victims the deputy coroner, Mr. J. Patterson, J.P., returned a verdict simply that their deaths were due to a collision between the truck driven by William Beauman and a railway engine. He found that no blame attached to the engine driver. The coroner added that he hoped that some invention might be discovered, apart from notice boards and the whistling of engines, to prevent drivers of vehicles proceeding when a train was approaching. [1326]

Reckless driving by two motorists led to at least eight deaths and several cases of injury around noon on Sunday 13th May 1928 in a level crossing collision with two automobiles at Espyville, Pennsylvania, United States. One car was occupied by the Hayes family, who were making a short trip from their home to Espyville to attend a Mother's Day event. The other car carried the Howard family, on their way to visit relatives at visit relatives at Kinsman, about 10 miles away. As the cars neared the level crossing at Espyville station, half a mile from the town, they were racing one another, and 300 yards from the crossing the Ford sedan driven by Mr Robert Hayes overtook the Chevrolet sedan driven by Mr William Thomas. Also approaching the crossing and activating the electric warning bell at the otherwise unguarded crossing, was a Pennsylvania Railroad passenger train which was not booked to stop at the station.

An eyewitness described what happened next. '"I saw the two cars racing and also heard the crossing bell," said Mrs Margaret Rhodes this afternoon, "and I rushed to my front door and saw exactly what I feared I would see – both cars crashing into the train."'

The Hayes' car struck the front of the locomotive of the train and the Thomas' car struck the engine just under its cab. Both cars were crumpled up and almost destroyed. When the train came to a stand the bodies of the victims were scattered all along the track. The bodies of the dead and injured were loaded into the baggage car of the train and it eventually proceeded to Erie, about two hours late.

Killed in the accident were Mr and Mrs Hayes, their children Everett, 10; Melvin, 8; Delbert, 5; and Gary, 3, and Mr Thomas and his father, Ward. Mrs Ward Thomas was probably fatally injured while Donald Hayes, 1, was the only survivor of his family.

Coroner W.E. Byham took the evidence of Mrs Rhodes and other witnesses, all of whom agreed that the two cars had been racing one another at high speed as they neared the crossing. He did not interview Engineer James Pilmer because the engineer had gone on with his train, but the latter had stated to others that he had been running fast and had whistled for the crossing. The inquest had not been completed at the time of the newspaper report on the accident; it was obvious, however, that its verdict would be that the accident had resulted from the stupid driving of Messrs Hayes and Thomas. [1327]

A few minutes before the Espyville accident, at 11.45 a.m. on Sunday 13th May 1928,

[1325] *The Argus*, Melbourne, Victoria, Australia, 27th March 1928

[1326] Ibid, 7th April 1928

[1327] *The Titusville Herald*, Titusville, Pennsylvania, United States, 14th May 1928 what was described as an open truck carrying 11 people home from a Sunday School meeting at Corinth Baptist Church was struck by an Atlantic Coast Line Railroad passenger train on Braswell Crossing, four miles west of Chadbourn, North Carolina, United States. Four people were killed outright and Mr W.C. Bullard, the superintendent of the Sunday School, died of his injuries while he was being taken to hospital. R.J. Bullard, aged 16, the driver of the truck and presumably the son of Mr W.C. Bullard, was one of several seriously injured, at least one of whom probably subsequently died. The truck was described as having been completely demolished and the bodies of the victims were scattered along the right of way for 300 yards. The two least seriously injured victims, Donnie Mae Kellyhan and Mattie Lee Ellis, walked to the home of Mr D.L, Bullard, 200 yards away, to report the accident. 'Their shoes and clothes were in shreds, and they were dazed.'

How the accident had happened was said to be a mystery. The train engineer was said to be particularly careful to whistle for the crossing. Visibility at the crossing was, however, partly obscured and it seems that the truck driver drove onto the line without checking that it was safe to do so. Possibly his inexperience was a factor. [1328]

Cleveland Cincinnati Chicago and St Louis (Big Four) train No. 11, in charge of Conductor George Clark, with Engineer Tom Malia and Fireman L. W. Christy, collided with a car near Mackinaw, Illinois, United States, at 3.37 p.m. on Thursday 7th June 1928. The car was carrying the Hasty family, who were returning from visiting the graves of relatives at Mount Pleasant cemetery. Six of the occupants of the car, Mr Ransom Hasty, his mother Mrs Frances Hasty, three of his children, Kilby, 8; Teddy, 7 and Lawrence, 5; and Mr Roger Watson, were killed. Another daughter, Nellie Hasty, 10, suffered a fractured leg. The road on which the car was travelling was a little used one and the view at the crossing was obstructed.

Speaking after the accident Engineer Malia said, "It all happened very quickly. I was sounding my whistle and the bell was ringing. I never saw the car, only felt the jar when the engine struck it. That was directly at the crossing, and it was at that moment that I first applied the brakes and brought the engine to a stop." [1329]

Ten people were killed instantly and 18 were seriously injured when a train collided with a bus on a level crossing near Bialystok, Poland, on Saturday 14th July 1928. [1330]

At about 14:00 on Wednesday 25th July 1928 a light engine running between Trichinopoly Junction and Trichinopoly Fort, India, on the South Indian Railway, and travelling at about 12 miles an hour tender first, was in collision with a motor bus on the level crossing at mile 252/10. All eight occupants of the bus died, some killed instantly, and others succumbed later in hospital from their injuries and burns. An inquiry by the Government Inspector of Railways attributed the accident to the recklessness of the bus

---

[1328] *The Robesonian*, Lumberton, North Carolina, United States, 14th May 1928
[1329] *The Urbana Daily Courier*, Urbana, Illinois, United States, 8th June 1928
[1330] *The Evening Tribune*, Marysville, Ohio, United States, 14th July 1928

driver, who had tried to cross in front of the engine. [1331]

Two taxis returning from the Saskatoon Exhibition, a summer festival held on the outskirts of Saskatoon, Saskatchewan, Canada, were in collision with a Canadian National Railway yard engine on a level crossing at 9.35 p.m. on Saturday 28th July 1928. The initial collision, with a taxi driven by Mr Adam Hopfinger, was by far the most serious, resulting in the death of five of the occupants of the taxi and the injury of five others. Mr Hopfinger drove onto the line in front of the engine, which he had apparently failed to notice. No one was hurt in the collision of the second cab with the engine.

The victims seem to have been a strange assortment of people. Killed outright were Professor H.R. Karstens, a retired theology professor; an unidentified man, probably a farmer; and June Nelson, 3½, daughter of a member of the Nelson troupe of acrobats; while Steve Dzuba, a farm labourer from Hubbard, Saskatchewan, and Hans Hartig, 16-year-old theology student, died in hospital. Those injured were Mr Hopfinger and his wife, Mrs Nelson, mother of June; Gilbert Morrison, a local carpenter; and Yard Foreman Charles Burns, who had been riding on the engine. The condition of Mrs Hopfinger was said to be critical. [1332]

Five people died when a train collided with a car on a level crossing at Curia station, Portugal, on Friday 10th August 1928. [1333]

A car containing six American tourists from Cleveland, Ohio, was struck by an eastbound Canadian Pacific Railway boat train travelling at 45 miles an hour on a level crossing near Yamachiche, Quebec, Canada, on Wednesday 22nd August 1928. Mr George Wise and his wife, Mrs Vanderwerf and her two small children, Jack and Jean, were all killed. Mrs Mart Dotty was taken to the Norman Cross hospital in Three Rivers, Quebec, in a critical condition. The car was travelling in roughly the same direction as the train and Mr Wise apparently only became aware of the train at the last moment when he failed in his attempt to pull up short of the crossing. [1334]

Tragedy befell the McCain family at 2.16 p.m. on the afternoon of Sunday 9th September 1928 when their car was struck by an eastbound Canadian Pacific Railway passenger train from Sault Ste Marie at Blind River, Ontario, Canada. Engineer B Allen said that he had been right on top of the car before he had seen it and there was nothing that he could do to prevent the accident. He had whistled for the crossing and the automatic bell there was ringing. Mr W. McCain, who had been driving the car; his daughters Margaret, 19, and Marie, 13; his sisters-in-law Mrs Frank McCain and Mrs Bell, and Miss Beulah Hannan, 27, were all killed or fatally injured. Another daughter, Bernice, 17, was critically injured and it was feared not likely to survive. She remained conscious and explained afterwards that when her father had belatedly seen the train he

---

[1331] Report by the Railway Board on Indian Railways for 1928-29 Volume 1
[1332] *Manitoba Free Press*, Winnipeg, Manitoba, Canada, 30th July 1928
[1333] *Excelsior*, Paris, France, 11th August 1928
[1334] *Manitoba Free Press*, Winnipeg, Manitoba, Canada, 23rd August 1928

had thought of trying to stop short of the track but doubting if his brakes were capable of stopping the car that quickly he had tried to get across the line in front of the train instead.[1335]

An entire family of seven was wiped out in a collision between Northern Pacific Railroad passenger train No. 11, en route from St Paul and Brainerd, and a car on an unprotected level crossing near Rice, Minnesota, United States, on Wednesday 12th September 1928. Mr Morgal and his family were returning from a visit to his sister Mary. The victims were Mr Morgal, his unnamed wife, and their children Alois, 5; Veronica, 4; Sophia, 3; Gertrude, 2; and a 5-month-old baby. The crossing was unobstructed, and the only explanation offered for the tragedy was that it was somewhat foggy. An alternative possibility would seem to be that Mr Morgal was distracted at the crucial moment by one or other of his young children. The county coroner, Mr J.S. Friesleben, decided that as the accident was clearly the fault of the car driver no inquest was necessary.[1336] [1337]

At 1.50 a.m. on Thursday 13th September 1928, another seven lives were lost in an accident on the South Broadway Crossing with the Grand Trunk Railway just outside of the city limits of Gary, Indiana, United States. Westbound fast freight train No. 6040, in charge of Conductor John James with Engineer Fred Flisher was a long train of 100 cars. Engineer Flisher stated afterwards that he had seen the automobile stop at the crossing and then suddenly jerk forward as his locomotive neared it. He had then immediately whistled and applied the brakes, but his train had travelled nearly a mile before it could be stopped. The engineer theorised that the driver of the car, a Chevrolet sedan, had let his foot slip off the clutch with the engine running, causing the vehicle to lurch onto the crossing.

Those who lost their lives were Mr. Howard Mainord, his wife Betta, a four-month-old baby, believed to have been their son; Mrs E.L. Jones, a year-old-boy believed to be her son; Mr D.D. Pollock and Mr Howard Smith. Seven-year-old Freddie Jones, another son of Mrs E.L. Jones, was taken to the Mercy hospital in Gary with a fractured skull and internal injuries and was not expected to recover. The crossing was protected only by a crossbuck.[1338]

A car containing five members of a vaudeville troupe, *The Violin Mania*, was driven into the side of Chicago Milwaukee St Paul and Pacific Railroad No. 21 at a level crossing with Highway 21, a mile southeast of Elkhorn, Wisconsin, United States, at 9.55 a.m. on Monday 17th September 1928. Mr Arthur Haggerty, his wife Hazel, Maurice Kaplan, Julia Lathowens and Isadore Platkin were all killed. The troupe had performed at La Crosse, where they had performed the night before, to Chicago. Eye-witnesses said that the car had approached the crossing travelling at high speed. When its driver had realised that he could not get over the crossing in front of the train he had swerved off the highway, his car had struck a guard rail and it had been thrown against the front of the

[1335] *Manitoba Free Press*, Winnipeg, Manitoba, Canada, 10th September 1928
[1336] *The Brainerd Daily Despatch*, Brainerd, Minnesota, United States, 13th September 1928
[1337] *The Winona Republican-Herald*, Winona, Minnesota, 13th September 1928
[1338] *The Times*, East Chicago, Indiana, United States, 13th September 1928

train between the engine and its tender. The car was carried about 12 feet with its right side demolished. [1339]

Westbound Northern Pacific Railroad passenger train No. 7 collided with a car on a crossing near Yates, Montana, United States, just west of the state line with North Dakota, on the evening of Saturday 22nd September 1928. Five members of a family from Sentinel Butte, North Dakota, lost their lives. The victims were Mr Joe Cook, his wife and infant daughter; Mrs Gertrude Northorpe, Mrs Cook's mother; and Mrs Northorpe's youngest daughter, Geraldine. The adults were all killed outright; Geraldine died later in hospital. The family had been to Glendive, Montana, where Mr Northorpe was recovering from an operation. It was thought that the car driver had failed to see the train. [1340]

Grocer Fred Kelly of Geelong West, Victoria, Australia, took his family of four, a Mrs Smallicombe and her four children and a man named Fred Leary from Geelong West to a function at Corpus Christi College, Werribee, on Sunday 30th September 1928, the party travelling in his tradesman's van (described as a motor waggon in the newspaper). On the return trip in the evening he also picked up a man named Leslie Hargreaves, whose car had run out of petrol. Proceeding along the Melbourne – Geelong road the party duly reached the level crossing at Corio, Victoria, and it was here that disaster struck. Also returning from Werribee to Geelong and reaching the crossing at the same time as the van was a special passenger train in charge of Guard Kenshole with Driver Nitchie and Fireman Myerscroft. Witnesses agreed that Driver Nitchie had whistled for the crossing four times, but Mr Kelly misjudged the speed of the train and believed that he could get across in front of it. He almost made it, but the train struck the rear of his van and reduced it 'to a heap of splintered wood and tangled iron.'

Leslie Hargreaves, Fred Leary, Eileen and Bessie Smallicombe, aged 16 and 15 respectively; and Helen Kelly were killed and James Kelly, his wife Margaret, their daughter Evelyn, 11; Mrs Smallicombe and her sons Louis, 17, and Alfred, 12, were all injured. [1341]

At the conclusion of the inquest into the accident the coroner, Mr D.W. O'Grady, simply found that the victims had met their deaths through the train colliding with the van. He remarked that the train crew were blameless and that the fact that the engine was running tender first had not been significant. He did not cast any blame either on Mr Kelly for trying to beat the train over the crossing. [1342]

Six level crossing accidents occurred in the eastern United States on Sunday 30th September 1928, killing 19 people. Near Massapequa, New York, a car driven by Mr George Schwartz and carrying a number of children either to or from a birthday party was struck by a Long Island Railroad electric train travelling at 40 miles an hour. Contact

---

[1339] *Janesville Daily Gazette*, Janesville, Wisconsin, United States, 17th September 1928
[1340] *The Bismark Tribune*, Bismark, North Dakota, United States, 24th September 1928
[1341] *The Argus*, Melbourne, Victoria, Australia, 1st October 1928
[1342] Ibid, 15th December 1928

with the third rail after the collision set fire to both the automobile and the leading coach of the train. Mr Schwartz; Madeline Doeltel, 12; Mildred Stenholm, 16; Dagmar Stenholm, 10; and Eleanor Rohling, 17, all of Hempstead, Long Island, were killed. Alice Doeltel, 14; Priscilla Doeltel, 9; and Valborg Stenholm, 18, all of Hempstead; and Raymond Miller, 18, of Brooklyn were injured. The current had to be switched off during the rescue operation and train services were at a standstill for an hour. [1343]

Six members of a wedding party on their way from Bayonne, New Jersey, to Philadelphia, Pennsylvania, lost their lives when their car was struck by the southbound Pennsylvania Railroad Crescent Limited express from New York to New Orleans, travelling at perhaps 70 miles an hour. The accident happened at 10.12 p.m. on 30th September 1928 on Green Lane Crossing, near Bristol, Pennsylvania, United States. The crossing watchman had left duty a few minutes earlier and it was believed that the car driver had misjudged the speed of the train. The bodies of some of the victims were carried a mile down the track. The engineer of the train, E.R. Stebbing, was arrested for manslaughter after the accident and released on bail of $2,000 pending the coroner's inquiry.

The victims were Mr. L. Walker Lees, Mr Lester Seidel, Mrs Catherine Riley, her daughters Marie, 19 and Anna, 17 and Mrs Sarah Hasbroucke, Mrs Riley's sister. [1344] [1345] [1346]

Nine tourists were killed and another 19 seriously injured when a train struck their bus near Kaschau, Czechoslovakia, on Sunday 7th October 1928. [1347]

Sunday 7th October 1928 also saw another level crossing accident which wiped out a whole family. Mr Orson Erickson was en route from his farm north of Delta, Utah, United States, to that place to attend the funeral of a friend, Abner Johnson, who had himself died in an accident a few days earlier. The road along which Mr Erickson was driving ran alongside the line of the Los Angeles and Salt Lake Railroad before turning abruptly to cross it. As Mr Erickson neared the crossing he slowed up, then for some unaccountable reason he appeared to lose control of his vehicle, which shot onto the crossing in the path of southbound train No. 25, running into Delta at 40 miles an hour. The pilot of the locomotive caught the car and carried it along the track for several hundred feet, while its occupants were dropped out one by one to be crushed beneath the engine wheels.

Mr Erickson, his wife Eliza, and their children Ione, 17; Hazel, 14; Lucille, 12; Nelda, 9; Keith, 8' and Ralph, 1½ were the victims. [1348]

[1343] *Albuquerque Journal*, Albuquerque, New Mexico, United States, 1st October 1928
[1344] *The Evening Sun*, Hanover, Pennsylvania, United States, 1st October 1928
[1345] *New Castle News*, New Castle, Pennsylvania, United States, 1st October 1928
[1346] *Bucks County Independent*, Bristol, Pennsylvania, United States, 5th October 1928
[1347] *The Salt Lake Tribune*, Salt Lake City, Utah, United States, 8th October 1928
[1348] Ibid.

Southbound Chicago and Alton Railroad motor train M15, in charge of Conductor Edward Carpenter with Engineer Ollie Cole, collided with a car on Husted Crossing, Roodhouse, Illinois, United States, at 3.30 p.m. on Sunday 21st October 1928. The accident killed Mr Joseph Rook, a local farmer and landowner and his wife Sarah, their daughter Mrs Hazel Erickson, and Mrs Erickson's children Ronald, 15; Vivian, 13; and Richard, 2. Mr Harry Erickson, the driver of the car, was grievously injured. The automobile had almost cleared the crossing when the train caught its rear end. The family had been visiting friends nearby. The vehicle was a touring car with curtains and the accident happened at the north end of the Chicago and Alton yards. It was surmised that the curtains in the car and freight waggons in the yard might have combined to hide the train from the view of the car driver. [1349]

In rural Montana, United States, Mrs Ralph Abbott, the assistant principal at the Howard high school, 10 miles west of Forsyth, was in the habit of taking several other children together with her own to and from school in her small touring car (referred to in some newspaper reports as a school bus). On the afternoon of Thursday 25th October 1928 the car had only gone a short way when it reached a level crossing with the Northern Pacific Railroad. The road over the line must have been a narrow one for Mrs Abbott stopped at the crossing to give way to a car coming in the opposite direction. The curtains of her car were up, and her vision obscured she then set off over the track in the path of a local freight train in charge of Conductor J. B. Flanigan with Engineer William Cavenger. The car was struck squarely, carried 300 yards and 'reduced to a mass of twisted steel and wood splinters.'

Mrs Abbott, her son Emery, 14; Richard Robinson, 8; Marguerite Virgin, 8; Mildred Simmons, 7; and Betty Lou Ferguson, 6, were all killed. Mrs Abbott's daughter Georgia, 12; Mary Sian, 17; and Margaret Robinson, 9, were injured. [1350]

Seven people died on Sunday 28th October 1928 when an automobile stalled on a private level crossing at Harwood, near Hanover, Maryland, United States, and it was struck by a Baltimore and Ohio Railroad passenger train. The gasoline tank of the car was broken in the collision, setting it on fire. The victims were Mr Emil Aswendt, his wife Mildred and their 15-month-old son Louis, Mrs Mary Kirby, Mrs Sally Kirby, Mrs Mary Shriver and her 22-month-old daughter Catherine. [1351]

Also on Sunday 28th October Joseph Tschak, living in the mining village of Charles, Pennsylvania, United States, held a celebration of the christening of one of his of his four small children. At the end of the celebrations the guests were on their way home in two cars, the first driven by John Momola and the second by Tschak himself. At 12.40 a.m. on Monday 29th October the two vehicles reached the four tracks of the Pennsylvania Railroad at Seward. There was no watchman on duty at that hour. Momola's car crossed the lines and Tschak, clearly not anticipating any danger, followed, to be hit by night

[1349] *Alton Evening Telegraph*, Alton, Illinois, United States, 22nd October 1928

[1350] *The Montana Standard*, Butte, Montana, United States, 26th October 1928

[1351] *The Salisbury Times*, Salisbury, Maryland, United States, 29th October 1928 express train No. 42, the crew of which had no chance of avoiding the accident. Tschak, Mr George Zahorchok, his wife Anna, their daughter Josephine, 8; Mr Momola's wife Bertronella and their daughter Ellen, 7, were all killed. [1352]

Six people lost their lives on Sunday 18th November 1928 when a Chicago Rock Island and Pacific Railroad freight train struck a motor truck on a level crossing at Gem, Kansas, United States. Those who died were Mr Otto Borland, his wife and their daughters Ida and Pauline, of Bloomington, Nebraska, and Mrs Ayers and her daughter Beulah May, aged 3. Mr Don Ayers was injured. Mrs Ayers was the sister of Mr Borland and the Borland family had arrived on a visit only the previous evening. [1353] [1354]

Pennsylvania Railroad train No. 216 from Logansport, Indiana to Cincinnati, Ohio, in charge of Conductor W.E. Dempsey and Engineer Walter Eagen, collided with a car carrying eight members of one family on the South Independence Street Crossing at Windfall, Indiana, United States, at 2.23 p.m. on Sunday 16th December 1928. Mrs Mary Hinds, her daughters Mrs Della Carlisle and Mrs Irene Richey, her son-in-law Mr Lloyd Richey and her grandson Eugene Richey were killed outright and another grandson, Frederick Elliot, 5, died on the way to hospital. Mr Frank Carlisle, the driver of the car, and Mrs Elliot, another daughter of Mrs Hinds, were injured. The party all lived in Kokomo, Indiana, and were travelling between different relatives on Sunday visits. Mr Carlisle apparently failed to observe the train, which was 30 minutes late and running fast. The car was demolished, and its parts were scattered over 200 feet. [1355] [1356]

Early on the morning of Thursday 27th December 1928 a Minneapolis express of the Minneapolis St Paul and Sault Ste Marie (Soo) Railroad, running on Baltimore and Ohio Chicago Terminal Railroad (BOCT) line, collided with an automobile on a level crossing at Oak Park Avenue, Illinois, United States, on the outskirts of Chicago. Riding in the car were members of the Henry Olsens of Berwyn and their cousins, the Halvar Olsens of Chicago; seven persons in all. Mr Henry Olsens and his wife were taking the Halvar's home from a Christmas celebration. Mr Henry Olsens and his wife, Mr Halvar Olsens and his wife and their children Virginia, 11 and Arnold, 9, all lost their lives. Their third child, Howard, was injured but expected to recover while two children of Henry Olsens had remained at home with their grandfather.

The busy level crossing was provided with gates but the crossing keeper, Joseph Boharty, had failed to lower them for the Minneapolis train, thus misleading Henry Olsens into thinking it was safe to cross. The watchman was taken into custody after the accident. [1357]

At the inquest, held at the beginning of January 1928, Watchman Boharty, who had

---

[1352] *Altoona Mirror*, Altoona, Pennsylvania, United States, 29th October 1928
[1353] *Lawrence Daily Journal-World*, Lawrence, Kansas, United States, 19th November 1928
[1354] *The Constitution-Tribune*, Chillicothe, Missouri, United States, 20th November 1928
[1355] *The Tipton Daily Tribune*, Tipton, Indiana, United States, 17th December 1928
[1356] *The Kokomo Daily Tribune*, Kokomo, Indiana, United States, 17th December 1928
[1357] *The Oak Parker*, Oak Park, Illinois, United States, 28th December 1928

little English and spoke through an interpreter, testified that he was responsible both for the crossings with both the BOCT line and the parallel Garfield Park branch of the Chicago Rapid Transit Company. An electric bell warned him of the approach of electric trains on the latter line, but he had no warning of BOCT trains other than by seeing their approach. On the day of the accident he had lowered the gates for a Rapid Transit train, then saw the Minneapolis train approaching. Two cars on the crossing had delayed his lowering of the BOCT gates, and the Olsens car had entered the crossing when they were only partly down.

At the conclusion of the inquest the jury found the Baltimore and Ohio to blame for not employing at Oak Park Avenue and other crossings on the line a separate watchman from the man employed to protect the Rapid Transit line. The Rapid Transit was blamed for employing an incompetent watchman and the Soo line was found fault with for taking the crossing protection for granted. Boharty was found not criminally responsible, and the jury recommended that he be released from custody. Finally the jury recommended the elevation of the level crossings at Oak Park Avenue. [1358]

What was described as 'the social event of the season' for the teenagers of Kenosha, Wisconsin, United States, was a party at the home of Mr James Blair, the cashier of the First National Bank. It would end in tragedy.

The party, on the evening of Saturday 29th December 1928, was to begin with a treasure hunt for the 36 teenagers in which they were to drive around the area in eight cars looking for clues. The procession of cars set off and when the leading car, driven by Don Travis Junior, son of the general manager of the Simmons bed company, reached a level crossing with the Chicago and North Western Railway, a northbound freight train was passing. When it had cleared the crossing, which was protected only by wig-wag signals, Travis set off, doubtless slightly irritated by the wait, and oblivious of southbound passenger train No. 120. His car cleared the track and the second car, driven by Ferrin Alford, son of the vice-president of the Nash Motor Company, almost cleared the crossing but the train struck its rear. The third car, driven by Richard Hastings, son of Dr Hastings, was struck squarely by the train. Four teenagers – Ferrin Alford, Alice Judd, Lowell Smith and Mary Slater – were killed outright, Richard Hastings died the next day and doctors held out no hope for a sixth teenager, Guildford Hagmann. Jane Blair, the organiser of the party, and Alice Gephart, were injured but expected to recover. [1359] [1360]

The last major level crossing accident of 1928 happened in Canada on New Year's Eve, Monday 31st December. A packed car was struck by a Canadian National Railway passenger train travelling at 50 miles an hour at Cataraqui, near Kingston, Ontario, killing all eight of its occupants. The victims were Mr Edgar Rivet and his wife, Mr Alexander Dupuis and his wife and four unnamed Dupuis children. The two families were reportedly in the process of moving from Cornwall to Peterborough and the car was also loaded with dishes and other light household goods as well as Christmas gifts.

[1358] *Oak Leaves*, Oak Park, Illinois, United States, 5th January 1929

[1359] *Wisconsin Rapids Daily Tribune*, Wisconsin Rapids, Wisconsin, United States, 31st December 1928

[1360] *The Wisconsin State Journal*, Madison, Wisconsin, United States, 31st December 1928

The train crew failed to notice the car until the last moment but Mr Thomas Burtch, who lived near the crossing, witnessed the accident. He said that the car had been braked at the last moment, presumably when its driver had belatedly seen the train, but it had still been moving at about 15 miles an hour at the moment of impact. Kingston city council had recently recommended better protection for the crossing, the scene of another fatal accident less than two years earlier. [1361]

What was described as a blinding snowstorm was a factor in a disastrous collision at Monroeville Crossing, near Bellevue, Ohio, United States, involving a Lake Shore Electric (LSE) Railway interurban car and a Greyhound bus, early on the afternoon of Tuesday 22nd January 1929.

Lake Shore train No. 213 had left Cleveland at 10.30 a.m. for Toledo. It consisted of car No. 164 with Conductor Fred Truman and Motorman Thomas Schuyler and Express Messenger Robert Boggs. The events leading up to the accident were described by a newspaper reporter: 'When about 700 or 800 feet from the Route 20, Monroeville Road crossing, Schuyler blew the road crossing whistle. While it was snowing at the time, the window wiper on the front of the interurban car's vestibule was going and Schuyler could see for a distance of 800 or 900 feet. Schuyler saw the eastbound bus approaching the crossing when he first blew the road crossing whistle and noted that it was slowing down as it approached the crossing. The bus came to a stop about 10 feet from the tracks when the car was, according to Schuyler's estimate from the crossing, and travelling at a speed of 40 miles an hour. When within 150 feet of the crossing, Schuyler saw that the bus was again starting up. Schuyler says he blew short blasts of the whistle, set the brakes in emergency, and reversed the motors of the car. Then came the crash.

'The interurban struck the bus in the middle, taking the body of the bus completely off the chassis. The car left the tracks and plunged over the body of the bus. Before the car was stopped it was 70 feet beyond the crossing.

'The interurban car carried 11 passengers at the time of the accident. So far as officials of the LSE were able to determine, no passengers or crew members were seriously hurt. According to the reports given to the investigation after the wreck, all passengers and crew members were able to leave the car without assistance, except for one man, who complained of some injuries. A number were bruised, and all badly shaken up.'

Motorman Schuyler just had time to move back into the interior of the car before the impact. He and Express Messenger Boggs, who had been riding in the baggage compartment, were thrown amongst barrels and boxes and badly bruised. As soon as the train stopped Schuyler walked to a nearby farm house to report the accident but finding that they had no telephone he had to continue to a second farm to send his message.

As for the bus driver, the newspaper reporter wrote: 'Interviewed at his bed in Bellevue Hospital, the story of the driver, Edward Butler, Cleveland, corroborated Schuyler's version in many details. Butler said he stopped 20 feet west of the Lake Shore track and got out of his bus. He claims he walked to the track and looked in both directions but believed the track to be clear. He got back into his bus and drove onto the track and the crash came before he was able to clear the crossing.

[1361] *Medicine Hat News*, Medicine Hat, Alberta, Canada, 3rd January 1929

'Butler had driven this stretch of road many times and he was well known to Bellevue bus drivers and taxi men. His condition is perhaps the least serious of any of those injured.' [1362]

Seventeen bus passengers were killed more or less outright in the accident and 16 injured, of whom two more would die by the following day.

Announcing his verdict at the close of the inquest on the victims of the accident, the coroner, Dr Charles Bell, found that Motorman Schuyler and Driver Butler were not criminally responsible for the accident, which he blamed on the public utilities commission and the Lake Shore company. His verdict read: "After having heard the evidence, examined the bodies, and considered the facts and circumstances and after due deliberation, I do find that the death of the 19 people on Jan. 22, 1929, at the Monroeville road crossing, two miles east of Bellevue, O., was accidental, no one being criminally responsible. Both crews of the companies, namely the Lake Shore Electric and the Greyhound Bus Company, were complying with the orders of their respective companies. The bus driver, because of the heavy snow storm, was over-cautious. Had he not left the bus and gone to the track and looked he would have had plenty of time to cross the track in safety.

"The accident was so terrible and the loss of life so great that the public utilities commission of Ohio made a thorough investigation and then ordered the Lake Shore Electric company to reduce speed when approaching the crossing, then later ordered them to install flash signals at that crossing and gave the company 90 days to comply with that order.

"If the public utilities commission has the right to make such an order it had the right to do so before the accident, and therefore is primarily responsible for the accident. Had the public utilities commission been functioning properly and had it been up to the minute in efficiency, with the safety of the travelling public in mind, the order for the flash signals at the crossing would have been issued long ere this.

"But the fact that the public utilities commission did not issue such an order to the Lake Shore Electric Co. before the accident in question does not render the company less liable for the accident.

"All railroad companies, electric or steam, should employ every up-to-date and reliable method for the safety of the people that travel the public highways. Neglect to do so renders them liable for any accident upon the crossing in question and all other crossings where such flash lights or safety devices are needed." [1363]

A car driver ignored or failed to observe the warning signals made by the crossing keeper, smashed through the barrier protecting a level crossing and his vehicle was hit by a train at an unstated location between Eberbach and Hanau, Germany, on Monday 7th April 1929. Five people died. [1364]

[1362] *The Sandusky Register*, Sandusky, Ohio, United States, 23rd January 1929

[1363] *The Star Journal*, Sandusky, Ohio, United States, 15th February 1929

[1364] Katastrophen der Deutschen Bahnen Teil II by H.J. Ritzau, pg. 238

Five people lost their lives in a level crossing accident near Mokena, Illinois, United States, late on Sunday 14th April 1929. The victims were a man believed to be Bud Zuronowsky, the driver of the car, Mr Anton Funik and his wife, and a woman and boy aged about 8 who were not immediately identified. Mr Zuronowsky had stopped his car at the crossing with the Chicago Rock Island and Pacific Railroad line for a westbound freight train to pass. As the caboose of the freight train cleared the crossing he started his car over the line at low speed directly into the path of an eastbound Omaha – Chicago passenger train. The car was carried 150 feet and the engine of the train was derailed. [1365]

Driver inattention cost the lives of six people at the Modena Crossing, Gastonia, North Carolina, United States on Sunday 21st April 1929. A car driven by Mr J.D. McCurry was approaching the crossing at the same time as the southbound Southern Railway *Crescent Limited* express. The crossing lacked gates, but it was provided with a watchman, named J.B. Hollyfield. Apparently oblivious of both the watchman, who was frantically waving his red flag and shouting for him to stop, and of the speeding train McCurry drove his car onto the crossing. Seeing the danger at the last moment McCurry then attempted to reverse his vehicle but in his panic stalled its engine and the car stopped in the middle of the crossing. It was then too late for anyone in the vehicle to escape. The occupants were all thrown from the car as the locomotive of the train hit and demolished it, setting it on fire as well.

Mr McCurry, Mr Fred Parton, Mr Parton's wife Maggie, and their children James, 6; Bud, 4; and Ring, an infant, were all killed. Mrs McCurry and the McCurry's infant son, J.D. Junior, were only slightly hurt; it was believed that Mrs McCurry had thrown her child from the car immediately before the impact. The accident happened at 12.30 p.m., a few minutes after services had ended at the nearby East Baptist church and it was witnessed by many churchgoers. [1366]

A train ran into a bus full of passengers on a level crossing at Artigas, near Flores, Argentina, on about Monday 22nd April 1929. Ten of the passengers were reportedly 'cut to pieces' and the same number were seriously injured. [1367]

An accident at New Hampton, New York, United States, on 28th April 1929 had several very familiar features. For one thing it involved a family out on a Sunday pleasure jaunt. For another, almost the whole family was wiped out, with the death of Mr Antonio Vianchi, his wife, and their children Lucy, 16; Marian, 14; Minnie, 10; George, 8; Louise, 6; Rose, 3; and Helen, 10 months. Only Frank, 12, escaped with serious injuries, doctors giving him an equal chance of survival. The mechanism of the accident was also all too familiar. Mr Vianchi, a railroad labourer, had waited at the crossing for an eastbound Erie Railroad express to pass, then pulled onto the tracks in the path of a westbound New York – Chicago express. [1368]

[1365] *Dixon Evening Telegraph*, Dixon, Illinois, United States, 15th April 1929
[1366] *Gastonia Daily Gazette*, Gastonia, North Carolina, United States, 22nd April 1929
[1367] *The Times*, London, England, 23rd April 1929
[1368] *The Olean Herald*, Olean, New York, United States, 29th April 1929

The day after the New Hampton accident, Monday 29th April 1929, another occurred at Leslie, Maryland, United States. Again two trains were involved. A truck driven by Mr Frank McCumsey, of Quarryville, Pennsylvania, and carrying his wife and family, waited at the crossing for an eastbound freight train to pass and then set off over the crossing in front of a westbound express, despite the efforts of Crossing Watchman J.J. Tulley to warn him by shouting and waving his flag. Mr and Mrs McCumsey, their sons aged 11 and 9 and a daughter of 4 were killed almost instantly. An 18-month-old baby was thrown from its mother's lap and landed on a grassy bank, not seriously hurt but was of course suddenly orphaned. [1369] [1370]

At the beginning of May 1929 Morgan Chamberlain, a boarder with the Bungle family at Riverside, New Jersey, United States, purchased a new automobile. On Sunday 5th he offered to take the family for a ride in his brand new car. The party were only a short way from home in the evening when they came to a level crossing with the Pennsylvania Railroad. An excursion train from the Jersey Shore resorts was approaching but a witness to the accident told police afterwards that Chamberlain, ignoring several whistles from the train, had tried to race it over the crossing, perhaps because he was anxious to demonstrate the capabilities of his new acquisition. He failed to get across and the ensuing collision killed him, John Bungle, Mrs Clara Bungle, their children Ada, 6, and John Junior, 5, and John Woller, 4, of Philadelphia, a relative of the Bungles. Carl Woller, John's father, and Clara Woller, 11, were seriously injured. (Some newspapers state that the younger Clara was a Bungle rather than a Woller.) [1371] [1372]

A family of six were all killed on the evening of Sunday 19th May 1929 when their small touring car was struck by the eastbound *Wolverine* express of the Michigan Central Railway, running at 60 miles an hour, on a level crossing near Wayne, Michigan, United States. Engineer F. Hessmer said that the driver of the car had tried to beat his train over the crossing despite his whistling and frantic gestures. Mr Joseph Kando and his wife were carried nearly a mile on the pilot of the locomotive before the train could be stopped while their sons, Joseph Junior, 10; Wayne, 8; George, 6; and Warren, 3, were thrown from the car. An eyewitness, Bud McCarty, spoke of seeing what looked like found bundles of paper tossed into the air, which he subsequently discovered were the bodies of the children. McCarty just managed to prevent a further tragedy by pulling two bystanders from the path of a westbound train after the accident.

The crossing was completely unprotected but there was a clear view of almost a mile in each direction in clear weather. Trains passed over it daily at a mile a minute. [1373] [1374]

[1369] *The Gettysburg Times*, Gettysburg, Pennsylvania, United States, 30th April 1929
[1370] *The Charleston Gazette*, Charleston, West Virginia, United States, 30th April 1929
[1371] *Indiana Evening Gazette*, Indiana, Pennsylvania, United States, 6th May 1929
[1372] *Altoona Mirror*, Altoona, Pennsylvania, United States, 6th May 1929
[1373] *Mt Pleasant Daily Times*, Mt Pleasant, Michigan, United States, 20th May 1929
[1374] *The Titusville Herald*, Titusville, Pennsylvania, United States, 20th May 1929

The Dominion Atlantic Railway's *New Yorker* express collided with a car on Aikens Crossing, three miles from Windsor, Nova Scotia, Canada, on Tuesday 16th July 1929. The five occupants of the car were killed instantly. They were Mr Charles Wright, his daughter Jean, 16; his son Graham, 11; his father-in-law, Mr N.W. Eaton; and Mr Eaton's daughter, Mrs Heuston. An inquest jury found a verdict of Accidental Death, with no blame attachable to the train crew. [1375]

Four people were killed more or less instantly and a fifth died shortly after admission to hospital as a result of their car being involved in a collision with the section of New York Central Railroad passenger train No. 139 on Long Crossing, near Cambria, New York, United States. The accident happened early on Wednesday 17th July 1929. The car was carried 300 by the engine, then fell alongside the track and caught fire. Engineer J.J. Clapp said that it appeared that the car driver had failed to see his train until it was too late and there was insufficient distance for him to stop the train. The victims were provisionally identified as Mr Nathaniel Kaufman Senior, his wife Louisa and their son, Nathaniel Junior; Mrs Clara Unther and Sarah Kaufman. Sarah survived long enough to be taken to Lockport General Hospital. All were from Philadelphia, Pennsylvania. [1376] [1377]

Messrs Frank Clark and Ray Holman, two employees of the International Harvester Company in Richmond, Indiana, United States, decided on Sunday 21st July 1929 to go for a picnic with their families to a spot over the state line near Dayton, Ohio. The party were returning to Richmond via the Dayton – Richmond pike, described as 'one of the most heavily travelled roads in western Ohio,' when their car, which was being driven by Mr Clark, was struck by the Pennsylvania Railroad's *Chicago Daylight Express* on a level crossing two miles west of Eaton, Ohio. Mr Clark and his wife, Mr Holman, his wife (who was Mr Clark's sister), and their children Cecil, 4; Odell, 2; and Juanita, 2 months, all died. The train carried the car 400 yards down the track. The railroad crossed the highway at an angle and the crossing was equipped with unspecified warning signals.

Conductor William Rant stated that his fireman had told him that he had seen the car run directly onto the tracks in front of the engine from his (the fireman's) side when he had leapt for the brake lever but too late, of course, to be of any use. Three investigations into the accident were begun by Sheriff Sam Walsh, Marshall Willard Armstrong, and Acting Coroner S.S. Lander. [1378] [1379]

The *Sunshine Special*, crack passenger train of the Texas and Pacific Railway, collided with a gravel truck carrying 14 members of two Texas families, killing all but one of them, on Sunday 11th August 1929. The accident happened at Mesquite, Texas, United States, about 20 miles east of Dallas. The Badgett and McHenry families had visited the zoo at Marsalis Park in Dallas so that the children could see the animals and the party were almost home, the crossing being only 150 yards from the Badgett

---

[1375] *The Lethbridge Herald*, Lethbridge, Alberta, Canada, 17th July 1929

[1376] *The Cattaraugus Republican*, Salamanca, New York, United States, 17th July 1929

[1377] *Dunkirk Evening Observer*, Dunkirk, New York, United States, 17th July 1929

[1378] *The Star Journal*, Sandusky, Ohio, United States, 22nd July 1929

[1379] *Defiance Crescent-News*, Defiance, Ohio, United States, 22nd July 1929 residence. Farmer Sherwood Badgett, the father of the Badgett family, who did not go on the trip, reportedly witnessed the accident as he was drawing water from a well, though he denied having done so when spoken to afterwards. The only other witness was Claude Chapman of Mesquite, who stated that the truck had stopped short of the crossing and had then suddenly jerked forward into the path of the train. The truck was smashed squarely in its middle and smashed to pieces.

Those who died were Mrs Ivory Badgett, her children Eva and Vera (twins), 23; Texas, 16; Emma, 16; Birdie, 7; Jessie, 6 and Mary Joe, 2; Mr E.F. McHenry, his wife Annie, and their children Beulah, 14; Emma, 10 and Billie, 4. Mary Lee McHenry, 14, was seriously injured. [1380] [1381]

An accident at an unstated location between Soltau and Buchholz, Germany, on Sunday 25th August 1929 led to five deaths. The car driver had failed to observe a train at an unprotected level crossing. [1382]

A car being driven by Mr Frank Zinda, of Stevens Point, Wisconsin, United States, accompanied by his family had been followed for some distance along U.S. Highway 51 on the afternoon of Tuesday 27th August 1929 by a car occupied by five women from Freeport, Illinois, who were returning from a vacation in northern Minnesota. When they arrived at a level crossing with the Chicago and North Western Railway near Lone Pine, Wisconsin a westbound passenger train, consisting of a gas-electric locomotive and a single coach, was approaching. Mr Zinda thought he could have crossed in front of the train, but he decided not to risk it and stopped. As he did so his car was overtaken by that containing the Freeport women which continued onto the crossing.

Mr Zinda told a reporter, "The car swerved around us, and seemed to try and go around the locomotive. It looked as though it tried to circle around the engine. I was afraid that we would be hit by them when the train struck. Then there was a cloud of dust, and we could not see much more because of the train between us and the car. I jumped out and went up the track because I thought that there might be someone alive. I looked at each of the people as I went along but I could see they could not have lived a minute. Parts of a human body lay on the track."

Engineer C.B. Smith said that he had been too busy applying the brakes to see what was happening on the crossing, then the cloud of dust had blocked his view. Immediately after the impact the car had caught fire. His train had been travelling at 45 miles an hour. He said that he could not understand how the car driver could not have seen his train.

The dead women were Mrs Anna Grandsen, Secretary-Treasurer of the Wagner Printing company; Mrs Hance, wife of the chairman of the board of the Guarantee Trust and Savings Bank; Mrs Kuehner, wife of a furniture dealer and undertaker; Mrs Alice Reitzell, a mathematics teacher; and Miss Marth Young, a nurse (sister of Mrs Grandsen).

The front trucks of the engine were derailed in the collision. When it was rerailed at 5 p.m. a steam locomotive pushed the train to Wisconsin Rapids.

[1380] *The Vernon Daily Record*, Vernon, Texas, United States, 12th August 1929

[1381] *Fredericksburg Standard*, Fredericksburg, Texas, United States, 17th August 1929

[1382] Eisenbahn-Katastrophen in Deutschland by Hans Joachim Ritzau, pg. 94

District Attorney Walter Murat decided that an inquest into the accident was not necessary. [1383]

A Michigan family and a family friend who were returning from a pleasure trip were killed on the evening of Sunday 22nd September 1929 at La Plaisance Crossing, near Monroe, Michigan, United States. The victims were Mr Leo Colebert, a Monroe paper mill worker; his wife Lena; their children Stufco, 5; Fisco, 3 and Mary, a baby; plus Oeiel Sabelbien, a La Salle township farmer. It was theorised that Mr Colebert had been unaware that he was approaching the unlit crossing, where there was also a sharp curve in the tracks. [1384] [1385]

About midday on Sunday 27th October 1929 a car carrying five people who were returning from a baptism was struck by a Lausanne – Milan electric express train on a level crossing at Saint Léonard station, Sion, Valais canton, Switzerland. The car was thrown into a stream and the baptised child, its father and brother were drowned. The other occupants of the car, a godmother, and a midwife, were thrown by the engine one onto the embankment and the other onto the opposite running line. They were also killed. [1386]

Nine people, all members of the same family, died late on Friday 1st November 1929 in a level crossing accident at Osceola, Indiana, United States, when their car was struck by a New York Central Railroad passenger train. The victims were Mr John Flatt and his wife, their daughters Mary Flatt and Mrs Brennan (Christian name not stated), their son-in-law Dewey Brennan and their grandchildren Virgie, 5; Lawrence, 3; and Robert, 10 months; Brennan and Richard Vaugh, 7 months. Betty Bell Vaugh, a granddaughter, 4, was critically injured.

The crossing warning bell was ringing and two schoolgirls, Lucille Davis, and Kathryn Minegar, who saw the accident about to occur, screamed a warning to the car driver, but the vehicle was driven directly in front of the train. [1387] [1388]

Mr W.H. Metzler, a coal miner, his wife, their two daughters and a son were all killed when their car was struck by a Louisville and Nashville Railroad passenger train en route from Nashville to Corbin on a level crossing two miles east of Jacksboro, Tennessee, United States, on Thursday 21st November 1929. The family had been on their way from their home at Turley to look at a farm at La Follette. The railroad passed through a cutting close to the crossing which presumably hid a train until it was close at hand. Parts of the wrecked automobile were scattered as far as a quarter mile from the collision point. [1389]

[1383] *Stevens Point Daily Journal*, Stevens Point, Wisconsin, United States, 28th August 1929.
[1384] *South Haven Daily Tribune*, South Haven, Michigan, United States, 23rd September 1929
[1385] *Defiance Crescent-News*, Defiance, Ohio, United States, 23rd September 1929
[1386] *Le Petit Journal*, Paris, France, 28th October 1929
[1387] *The Indianapolis Times*, Indianapolis, Indiana, United States, 2nd November 1929
[1388] *The High Point Enterprise*, High Point, North Carolina, United States, 2nd November 1929
[1389] *The Kingsport Times*, Kingsport, Tennessee, United States, 21st November 1929

The first snowstorm of the year was indirectly responsible for the death of five prominent Texas women in a level crossing collision near Dallas, Texas, United States, on the morning of Thursday 21st November 1929. The negro chauffeur of the car, Robert Guess, who was badly hurt, said later in Baylor hospital that he had failed to see the Texas Santa Fe and Pacific Railroad train through the falling snow. He was also unfamiliar with the road. The Chevrolet sedan was demolished by the train, and he was lucky to escape the fate of his passengers. The women who lost their lives were Mrs Clark, Mrs G.R. Gryder, Mrs Arthur Gryder (her daughter-in-law) and Mrs Prendergrass, all of Leonard, Texas, and Mrs King of Fort Worth, the destination of the party. [1390] [1391]

Six young adults, two of them a couple on their honeymoon, died in a level crossing accident at Santa Fe Springs, near Whittier, California, United States, early on Sunday 1st December 1929. The victims were Edwin Bike, Mrs Audna Day, William De Ford, Miss Arlene Herr and Mr Joseph McCann and his wife, on their honeymoon from Centralia, Washington. The party had been to a late Saturday night barn dance at the John S. Baker ranch. De Ford, who had been driving, had tried to beat an Atchison Topeka and Santa Fe Railway train over the crossing. He almost succeeded, but the train struck the rear of the small car and threw it against the concrete base of a wig-wag signal. Five of the occupants were killed outright and De Ford died soon after being taken from the demolished car. [1392]

Six people died and 10 were injured on the evening of Sunday 8th December 1929 when a train on the central line collided with a motor bus on a level crossing between Catalina de Guines and Madrugada, Cuba. However, newspapers in the United States were able to reassure their readers that 'There were no Americans among the victims.'[1393]

At 9.10 a.m. on the morning of Thursday 19th December 1929 the regular bus from Villafranca to Barcelona left San Sadurni de Noya for Barcelona, carrying 27 passengers, most of whom had visited the market at San Sadurni. It had only covered two about two kilometres when it reached a level crossing at Torre Ramona, Catalonia, Spain. The barriers of the crossing were raised, and the bus driver, Magin Boqueras, naturally assumed therefore that it was safe to cross, but as his bus entered the track it was struck by a passenger train bound for Barcelona, carried 100 metres down the track and destroyed. Seventeen bus passengers were killed outright or died shortly afterwards, and a number of others were badly hurt. Driver Boqueras escaped with slight injuries.

The crossing keeper Arturo Benaiges, aged 21, was arrested after the accident. The explanation for the accident appeared to be that the passenger train which collided with the bus normally crossed a goods train from Barcelona at San Sadurni. Being aware of this and of the fact that the goods train had not yet reached him, the crossing keeper had

---

[1390] *Waxahachie Daily Light*, Waxahachie, Texas, United States, 21st November 1929
[1391] *Miami Daily News-Record*, Miami, Oklahoma, United States, 21st November 1929
[1392] *The Salt Lake Tribune*, Salt Lake City, Utah, United States, 2nd December 1929
[1393] *The Evening Independent*, Massillon, Ohio, United States, 9th December 1929

assumed that the passenger train must be still waiting in San Sadurni station and that as he could not see the goods train approaching it must be safe to let the bus across. The crossing point of the two trains had presumably been altered because the goods train was late. [1394]

A truck carrying the family of Mr Fay Glass of College Springs, Iowa, United States, was struck on the evening of Friday 20th December 1929 by a Chicago Burlington and Quincy Railroad gasoline railcar on the outskirts of Clarinda, Iowa. The family, together with Mr Charles Atkins, had been in Clarinda doing their Christmas shopping. All were killed. Apart from Mr and Mrs Glass and Mr Atkins the victims were the Glass children – Frank, 13; Dee, 9; Mary, 8; and Meza, 4.

A reporter wrote, 'Debris from the wrecked car, Christmas presents, and articles of clothing littered the right of way of the Burlington road. Toys for the children, all of whom were small, candles and decorations for the tree which was to have made Christmas a merry one at the Glass home were found in the wreckage, tangled with metal and wood from the automobile and the gasoline coach.

'Candles and fruits, intended by Glass and his wife and Atkins for filling stockings next Wednesday morning were mixed with blood-covered clothing strewn along the road for many feet where it had been scattered as the crew of a speeding motor train sought to halt its progress.'

It was impossible to determine why Mr Glass had failed to observe the approaching train. [1395] [1396]

Farmer William Pennywell and several members of his family had been on a visit to his son, Arthur, who was recovering from an operation in Mauston, Wisconsin, United States, on Sunday 22nd December 1929. On leaving the hospital they intended to sell some honey produced on the farm. Their light sedan had only gone a short way when it was hit by an eastbound Chicago Milwaukee St Paul and Pacific Railroad mail express on an open level crossing. The car was thrown 200 feet by the train and Mr Pennywell, his twin daughters, aged 12, his brother-in-law William Hartwell, and a cousin, Tom Van Skoik, were all killed. Another of Mr Pennywell's sons, Frank, was very badly hurt and doctors gave him no chance of recovery. [1397]

C. J. Triplett, a state highway patrol officer, chanced to be standing at a level crossing of the Golden Gate Highway and the Atchison Topeka and Santa Fe Railway, at Calwa, California, United States, on the afternoon of Sunday 22nd December 1929 as a railcar bound for Corcoran and an automobile approached the crossing at the same time, both travelling at a moderate speed. He signalled to the car for it to stop but its driver ignored

[1394] *La Vanguardia*, Barcelona, Spain, 20th December 1929

[1395] *The Telegraph-Herald* and *Times-Journal*, Dubuque, Iowa, United States, 22nd December 1929

[1396] *The Ames Daily Telegraph* and *Evening Times*, Ames, Iowa, United States, 22nd December 1929

[1397] *The Capital Times*, Madison, Wisconsin, United States, 23rd December 1929 him. Train and car collided, and six of the eight occupants of the car were killed outright and a boy of 5 died four hours later. A girl of about 6 was still alive the next day but it was not certain if she would recover. During a brief period of consciousness after the accident she identified herself as Josefine Ramirez. The of the other victims were identified as Tomez Ramirez of Fresno and Antonio Tenorio of Los Palos. Both were general delivery drivers. A bottle of liquor was found in the car, and it was assumed that the driver had been drinking. [1398]

An electric train of the Ferrocarril Central Argentino collided with car on a level crossing in the Olivos suburb of Buenos Aires, Argentina, on New Year's Day, Wednesday 1st January 1930, killing nine occupants of the vehicle. [1399]

At 10.50 p.m. on Saturday 4th January 1930 a bus carrying 20 students from Burbank High School, who were returning from basketball matches at Big Prairie, where the boys had been victorious, though the girls had lost, was struck by a Pennsylvania Railroad mail and express train travelling at between 45 and 50 miles an hour on a level crossing at Shreve, Ohio, United States. The results were very serious; seven students were killed and eight students, the school principal and the bus driver were injured. None of the surviving students could explain how the accident had happened and the bus driver, Joseph Baker, was said to be incoherent from shock, though he attempted to make a statement to Wayne County Coroner J.B. Patterson, who reported that, "He doesn't remember how it happened," and added, "From the partial investigation that I have made it appears that the driver did not come to a full stop before crossing the tracks. He did slow up according to three eye-witnesses. The engineers whistled and warning bells were ringing at the time."

Several car loads of fans had preceded the bus homewards and learnt of the accident upon their arrival. Mr and Mrs H.M. Grubb found that their sons Forrest, 15, and Wilbur, 14, had both died, while Mrs Repp, whose husband was away in Indiana, found that her son Claude, 16, had been killed and her daughter, Edith, 15, was injured. Others were luckier: 'Frank Whitmore was one of the first Burbank residents to speed to the scene. His daughter had gone to the basketball game, and he had urged her to ride in the bus, feeling secure in the carefulness of Joe Baker's driving. He found wreckage of the bus piled in front of his sister's home. She had sought news of her niece too. When the bodies were identified, and the injured listed, Whitmore sped back home, to find his daughter safely in her mother's arms.'

The principal of Shreve High School, Joseph Weltner, witnessed the accident from the lobby of the Carl Hotel. "I ran out of the lobby to see what had happened. My eyes fell on a young girl, mangled and bleeding, lying across the right of way near the crossing. Down the track, as far as I could see, were bodies strewn along the tracks. The bus was lying beside the station platform, so badly battered it was hardly recognisable. I carried the girl into the lobby of the hotel and ran back to aid others. I picked up three others and brought them in. Two, I think, were dead. They were later taken to Wooster in private autos and ambulances. The most horrible scene of all was at the engine where five bodies

---

[1398] *Berkeley Daily Gazette*, Berkeley, California, United States, 23rd December 1929

[1399] *The Times*, London, England, 2nd January 1930

had been impaled on the cow-catcher. I helped the train crew remove the bodies. Among them was Joe Baker, the driver. He seemed dead. We had just laid him beside the track when he suddenly moaned and struggled to his feet, staggered a few feet and then collapsed." [1400]

Apart from the Grubb brothers and Claude Repp, the other students who lost their lives were Willard Baker, 14; Wayne Lehman, 15; Eugene Talley, 17; and Emil Timic, 16.

It was suggested that the shouts and singing of the students might have drowned out the noise of the crossing bell and the engine whistles and that frozen snow on the windscreen of the bus might have obscured the view of the train, [1401] though it was dry at the time of the accident. [1402]

The report of the inquiry into the accident carried out by the public services commission was published less than a week after the accident. It stated, "The prime cause of the accident was due to negligence on the part of the bus driver who failed to bring his bus to a stop at a point where he could see and hear the approaching train. In addition to this he failed to carry out the ordinary rules of safety required of drivers to look and listen before proceeding over the crossing. If this had been done, we feel that the accident would have been averted."

The commission completely exonerated the railroad company and its employees from blame for the accident as the warning bell had been ringing and the engine whistle had been blown. However, in view of the fact that a fast-moving train could only be seen for 10 seconds before it reached the crossing it recommended the employment of a watchman at night and the replacement of the warning bells by light signals if the latter were found to be more effective. [1403]

A mere 19 days after the Shreve tragedy a school bus was demolished by a train on Sheldon Road Crossing, near Berea, Ohio, United States, killing nine small children and the bus driver. The children were all pupils at Brook Park village school. On the morning of Wednesday 22nd January 1930 the bus had picked up 23 pupils and dropped off 13 of them at Berea High School before continuing with the younger children towards the village school. It was struck by New York Central Railroad mail train No. X-19 from Buffalo, New York, to Chicago, Illinois, in charge of Conductor Curtzweiler with Engineer John Hand. This was reportedly the fastest train on the line, faster even than the *Twentieth Century Limited,* with a maximum speed of 70 miles an hour. The Ford bus was almost entirely destroyed. Peter Ovak, a trackman, was working a short distance from the crossing and although he did not actually see the collision he heard it and rushing to the scene, attempted to pick up some of the children. Stopping passing motorists, he got them to take any of the victims who still showed signs of life to Berea community hospital. All available police, physicians and locals soon joined him in the rescue operations and as news spread mothers and fathers whose children rode the bus daily hurried to the scene as well.

[1400] *The Sunday Messenger*, Athens, Ohio, United States, 4th January 1930
[1401] *The Evening Independent*, Massillon, Ohio, United States, 6th January 1930
[1402] *East Liverpool Review*, East Liverpool, Ohio, United States, 4th January 1930
[1403] *The Star Journal*, Sandusky, Ohio, United States, 9th January 1930

Coroner A.J. Pearse was also shortly on the scene and took charge of the investigation. He decided that to all appearances the bus driver was to blame for the accident, stating, "I do not see how the driver could fail to see the approaching train." The crossing with the quadruple-track main line was unprotected except by a pair of warning signs. It was believed that John Taylor, the driver of the bus, had waited for a freight train to pass on another running line and then proceeded to cross, failing to see the fast-approaching mail train. The train stopped for some time at the accident site, then proceeded to Toledo, where its crew were changed as usual. It was only carrying a few passengers in one car, the remainder of it consisting of mail vans.

The children who lost their lives were William Davidson, 14, and his brother Vernon, 7; Evelyn Kaltenbach, 7; William Pastorek, 10; Jacob Walter, 12 and his sister Juanita, 11; and Rita Zielenski, 8 her sister Dorothy, 6, and brother Vincent, 6. Ethel Mary Davidson, sister of William and Vernon, was seriously injured. She lived near the crossing and her mother witnessed the accident. 'Mrs Will Davidson, mother of five children, two of whom were killed, and one seriously injured in the crash, saw the speeding train bearing down on the bus – and turned her head away. Her two dead children are William, 14, who was instantly killed, and Vernie, 9, who died after arrival at Community hospital, Berea. Ethel Mary Davidson, 12, suffered fractured legs and internal injuries.

'The Davidson home is a quarter of a mile from the grade crossing. "I walked to the bus this morning with the three children," Mrs Davidson said, as sobs shook her body. "They were carrying their lunches and school books and laughing. As the bus pulled away from our home, I stood in the road and watched the machine until it reached the railroad tracks. It has been my custom to watch the bus until it safely crosses the tracks. As the bus reached the crossing I saw the train bearing down on the bus. I turned my head away. I knew a crash was inevitable. Then I heard the horrible, grinding noise of the impact. I ran into the house, frantic.

'"People gathered in the road immediately. They were hurrying towards the tracks. My husband, a track man for the Big Four railroad, was notified. He came home and got me, and we hurried to Community hospital. We found our three children there. Vernie had died after reaching the hospital. William was instantly killed. I hope Ethel Mary lives."'

The newspaper report added that, 'Taylor, the bus driver, had two children of his own. The Taylor home is directly across the street from that of J.C. Walter, father of two of the victims. The Taylor and the Walter children played together, and Jacob Walter often rode on the front seat of the bus with the driver. "Even though I have lost two children in the accident," said J.C. Walter, "I cannot help but feel that Taylor was not to blame."' [1404]

The inquest into the accident was quickly held but the formal verdict of the coroner was withheld until he had questioned Engineer Hand, who was reportedly near collapse, and his fireman, John Deal. [1405] The main points elicited at the inquest were that the crossing was unprotected except by 'Stop, Look and Listen' signs, that Driver Taylor had stopped before entering the crossing and that Engineer Hand had whistled.

Before the formal reports of either the coroner's inquest of the investigation by the

[1404] *Defiance Crescent-News*, Defiance, Ohio, United States, 22nd January 1930
[1405] *The Daily Times*, New Philadelphia, Ohio, United States, 23rd January 1930

public utilities commission had been published there was a crackdown on the employment of incompetent or physically unfit or disabled school bus drivers. Thus on Saturday 25th January a newspaper announced, 'A movement was begun today to remove all Ohio school bus drivers who may be judged as unfit, following a statement by Attorney General Gilbert Bettman that all drivers of school buses will be required to pass a physical examination and be licensed as chauffeurs, as required by state law.'

The newspaper noted that previously, 'Enforcement of the chauffeurs licence law has heretofore been left in the hands of local authorities, and crippled and almost totally blind men have been given licences according to Secretary of State Clarence J. Brown. Drivers of school buses, in addition to being licensed as chauffeurs, are also required to possess a certificate as to their moral character and qualifications for the position, Bettman said.' [1406]

The report of the public services commission was approved on Monday 27th January. It found that Driver Taylor had failed to look in the direction from which the express was approaching and so had caused the collision. It absolved the railroad company from blame for the accident but recommended that the warning signs at Sheldon Road Crossing be replaced (supplemented?) by flashing light signals. [1407]

The formal report of Cuyahoga County Coroner A.J. Pearse was announced on Tuesday 28th January and as previously reported in the newspapers and in agreement with the findings of the public utilities commission it stated that the dead and injured children had been "victims of carelessness on the part of John Taylor, the driver." The coroner cleared the train crew of all blame for the accident, but he noted that it probably would not have occurred had there been a watchman at the crossing. He recommended the elimination of all level crossings and failing or pending this ambitious proposal the installation of automatic flashing light signals at all crossings. As for the buses he recommended that each should be provided with a two-man crew, with one man responsible for negotiating crossings safely. [1408]

On the evening of Saturday 8th February 1930 a bad accident occurred on a level crossing at Port-à-Binson, France, on the Paris – Epernay line of the Chemins de fer de l'Est. At about 19:50 a car arrived from the direction of Chatillon, smashed through the lowered crossing barrier and came onto the track directly in front of fast train PA 45, which had left Paris at 17:55 and was running at 80 kilometres an hour. The engine of the train struck the car, threw it a considerable distance, and smashed it to pieces. All six occupants of the vehicle, two men, two women and two children, were killed and badly mutilated.

The investigation into the accident disclosed that the car was the property of the Œilly Lime Company and had been being driven by one of their employees, M. René Parmentier, accompanied by his wife and their 2-year-old daughter Janine. The couple had been returning from Reims where they had spent the day with friends. With them were their friend M. Roland Poisse, his wife, and their little boy, aged 4. The powerful car had just passed over a temporary bridge over the Marne, 50 metres from the crossing

---

[1406] *The Coshocton Tribune*, Coshocton Ohio, Unites States, 23rd January 1930.

[1407] *The Athens Messenger*, Athens, Ohio, United States, 25th January 1930

[1408] *The Star Journal*, Sandusky, Ohio, United States, 28th January 1930

and it was not clear why the car driver had failed to stop. The barriers were provided with clearly visible warning lights and the car should not have been moving that fast having just negotiated the suspension bridge across the river. It seemed likely that the driver had accidentally put his foot on the accelerator instead of the brakes. The train driver had reversed his engine when he saw what was happening but was then only 50 metres from the crossing. [1409]

A Dodge touring car containing ten persons belonging to three families was struck by the 6.50 p.m. up passenger train from Geelong to Melbourne Spencer Street on Hopper's Crossing, 1½ miles on the Melbourne side of Werribee, Victoria, Australia, at 7.45 p.m. on Sunday evening, 16th February 1929. The crossing was an open one, and reportedly had been the scene of previous mishaps. Mr Edward Crowley failed to notice the train until the last moment. He then stopped his car, but its bonnet was foul of the running line, and it was struck by the train. Mrs Marie Ferguson, her daughter Mrs Vivian Warner and her grandchild Cranleigh Ferguson, 3, were killed outright and another grandchild, Jean Ferguson, 5, died in the guard's van of the train before it reached Melbourne. Edward Crowley, his wife Ida, and his daughters Lois, 6, and Joan, 5, were seriously injured. Lois died of her injuries the next day. Stuart Crowley, 2, was less badly hurt and Mr Ferguson was almost unhurt, having jumped from the car at the last moment. He had Jean Warner in his arms, but she was unfortunately still struck by the train. [1410] [1411]

In the vast majority of level crossing accidents the fatalities, if any, are restricted to the occupants of the road vehicles or perhaps passers-by, but there are always exceptions. Thus on 23rd February 1930 an accident on Burlington Road Crossing, near Kenosha, Wisconsin, United States, resulted in 14 deaths, of which 12 were train passengers. Another 122 passengers, five railwaymen, one off duty, and one other person were injured. The accident was the subject of a full Interstate Commerce Commission (ICC) investigation.

The accident occurred on the Chicago North Shore and Milwaukee Railroad, a double-track electric line. Burlington Road Crossing was protected by a wig-wag warning signal, consisting of a bell and a swinging banner with a red light in its centre. The accident happened at 10.45 p.m.

The car involved in the accident was a Buick touring car being driven by Frank Tomczak, with passenger Norman Shinners, both from Milwaukee, and both of whom died in the accident. The train which struck their car was southbound passenger train No. 436. It consisted of five motor cars and was in charge of Conductor Durr and Motorman Hall. Travelling at a speed of between 35 and 60 miles an hour according to different estimates this train struck the car, was derailed fouling the opposite running line, and was then run into by northbound freight train extra No. 546. The latter train consisted of electric motor No. 456, an empty car, 12 loaded cars and a caboose, in charge of Conductor Harris and Motorman Norris.

1409 *Le Peuple*, Paris, France, 10th February 1930

1410 *The Age*, Melbourne, Victoria, Australia, 17th February 1930

1411 *The Herald*, Melbourne, Victoria, Australia, 17th February 1930

The crew of the passenger train could not help the inquiry much. Conductor Durr was unaware of any problem until his train was derailed and Collectors Hickey, Kindel, Searle and Wade could only add that they had felt the brakes applied in emergency. Motorman Hall was too badly hurt to be interviewed.

Motorman Harry Norris testified that he whistled for the crossing and had seen the wig-wag operating. He saw a car stopped on the east side of the crossing, and another automobile approaching at a speed he put at 45 miles an hour. The second car overtook the stationary one and entered the crossing. He had applied his brakes in emergency as soon as he saw that the second car was not stopping and had reduced the speed of his train from 30 or 35 to 15 miles an hour at the crossing. The headlights of both his and the passenger train were burning brightly. Head Brakeman Grooms and Brakeman Anderson, who had been riding in the motor, more or less corroborated his testimony.

The driver of the stationary car, Mr Algot Lindquist, also gave evidence. He stated that he had seen the freight train approaching, had seen the wig-wag operating, had heard the bell, and had stopped his Ford sedan short of the crossing. His passenger, Miss Irene Sand, had then drawn his attention to the passenger train coming from the opposite direction, moving twice as fast as the freight train. Mr Lindquist confirmed that the headlights of both trains had been burning and he had heard continuous whistling from one of the trains. While they were waiting the Buick had approached at high speed, passed their car and driven onto the crossing in front of the passenger train.

The ICC concluded that the accident had been due to the driver of the Buick entering the crossing in front of the passenger train, probably because he had only seen the freight train which he believed he could get across in front of. The passenger train could have been seen almost half a mile away and there was no reason for the accident to have occurred had the car driver exercised reasonable care. The report made no recommendations. [1412]

The inquest jury returned a verdict that the two occupants of the car had died from it being struck by a North Shore train and that the train passengers had died from their train having been derailed. [1413]

The dedication on Monday 24th February 1930 of a new airport at Machado to serve Havana, Cuba, was marred by two accidents. In the first and less serious of these two pilots, Lieut. Rolando Caballero and Sergeant Gonzales of the Army Air Force, lost their lives when their plane crashed. Worse was to follow for a crowded bus taking passengers back to the Cuban capital was struck by an interurban electric train on a level crossing, killing ten of the occupants of the bus. [1414]

George Frey, his wife Dorothy, Mr Albert Holden and his wife (George Frey's father-and mother-in-law), and Mr John Miller and his wife, were all killed when their car was struck on the Plum Street Crossing in Florin, Pennsylvania, United States, on the evening of Sunday 9th March 1930. The Holdens had been visiting Evans Carpenter, who lived

[1412] Interstate Commerce Commission report
[1413] *The Sheboygan Press*, Sheboygan, Wisconsin, United States, 28th February 1930
[1414] *The Morning Herald*, Uniontown, Pennsylvania, United States, 25th February 1930

only 50 feet from the crossing, and Frey had come to bring them back to Lancaster, ten miles away. The Millers had come along just for the ride. The car was struck by the westbound Pennsylvania Railroad *Airyway Limited* express. Buildings obstructed the view of the crossing, which was little used as there was an adjoining crossed protected by a watchman a short distance away, though Plum Street Crossing offered a slightly more direct route to the main road. No one witnessed the accident. [1415] [1416]

A level crossing which was already due shortly to be eliminated because of its previous bad safety record was the scene of a collision which killed eight people on Sunday 23rd March 1930. It was already proposed to replace Shoemaker Crossing, New York, United States, on the Auburn branch of the New York Central Railroad and five miles from that place, with an overbridge in the summer.

On the day of the accident a small car ignored a red flashing danger signal and sped at high speed into the side of a gasoline railcar. The light sedan was hurled into the side of a house 50 feet away while the railcar, which had also been travelling at speed, took almost three-quarters of a mile to come to a stand, its front end badly damaged. The bodies of the occupants of the car were scattered about the track and badly mutilated so that they were hard to identify, but seven of them were named as Darius Auletta, 25; Earl Besaw, 24; Edward Chelgren, 20; Ruth Dunlavey, 17; Bertha Kierks, 19; Dorothy Klink Parks, 25; and Frank Straus, 23. None of the 25 or 30 passengers on the train was more than shaken. Visibility at the crossing was unobstructed. [1417]

Conductor William McGann and Engineer W.R. Wolfe gave evidence at the inquest into the accident, the latter stating that his railcar had been travelling at 50 miles an hour and he had sounded his whistle 1,500 feet from the crossing. Two girls, Frances and Onita Smith, who had witnessed the accident from the porch of their house, also gave evidence. They testified that the wig-wag had been flashing and two cars had already stopped at the crossing when the third car failed to pull up and was hit by the train. [1418]

Coroner Leroy Corey reserved his decision at the close of the inquest and it does not appear to have been recorded, but it was presumably that the accident had been due to the recklessness of the car driver.

What was, at the time that it occurred, the worst level crossing in the country's history, happened near Hikurangi, New Zealand, on the evening of Saturday 5th April 1930. A bus carrying 11 teenagers and young adults going to the pictures at Whangarei, was struck by a special race train carrying passengers from Whangarei to the Bay of Islands. Due to a delay at the racecourse the train was running 13 minutes late. Six bus passengers were killed. They were Irene Darwin, 15; Gavin Davidson, 16; Andrew Frew, 17; Sydney Johnson, 21; James Rothwell, 16; and Edna Wilson, 23. The condition of another passenger, Philip Sparks, 15, was reported to be precarious. The other passengers and the driver, Noel Wilkinson, 19, were less badly hurt. Wilkinson was the son of one of

---

[1415] *The Evening Sun*, Hanover, Pennsylvania, United States, 10th March 1930
[1416] *Somerset Daily Herald*, Somerset, Pennsylvania, United States, 10th March 1930
[1417] *The Daily Messenger*, Canandaigua, New York, United States, 24th March 1930
[1418] *Syracuse Herald*, Syracuse, New York, United States, 30th March 1930

the proprietors of the bus service. It was very dark at the time of the accident and drizzling and it was suggested that the rain might have obscured the driver's vision. [1419] [1420]

Considerably worse than the New Zealand accident in terms of casualties was an accident less than a week later near Isleta, New Mexico, United States, on Friday 11th April 1930. Atchison Topeka and Santa Fe Railway fast mail train No. 7 collided with a Pickwick-Greyhound Lines bus, killing 21 bus passengers and its driver, and injuring seven more passengers. Like the Kenosha accident less than two months earlier the collision was investigated by the Interstate Commerce Commission (ICC).

The Atchison Topeka and Santa Fe train consisted of engine No. 3712, four mail cars and five express cars, in charge of Conductor Irwin with Engineer Davis. The bus had a seating capacity of 25 persons. It was being driven by Mr. F.D. Williams and was on a schedule route from Los Angeles, California, to Denver, Colorado.

The road crossed the track at an angle of 39 degrees. Five hundred yards from the crossing from the direction from which the bus was approaching was a sign marked "R. R.", followed 281 yards from the crossing by a sign marked "STOP", and a standard cross-buck on the opposite side of the crossing. Both a State Railroad Commission regulation and the rules of the bus company required buses to stop and their drivers to check that no train was approaching before crossing the line.

Engineer Davis stated that the bell of his engine had been ringing continuously and he had been sounding the whistle from the whistle board. He estimated the speed of his train at 45 miles an hour. Because the bus had approached from the opposite side of his locomotive he had not seen it before his fireman had shouted a warning.

Fireman Oldham said that when he had first seen the bus he had been satisfied from its speed that it would stop. When the engine was 200 feet from the crossing he had realised that it was not going to halt and had shouted to his engineer to apply the brakes.

Conductor Irwin testified that he had felt the brakes apply in emergency and had realised that something must be wrong. After the accident he had found the wreckage of the bus burning under the engine.

Brakeman Jones and Flagman Reeder confirmed that they had heard the engine whistle sounded approaching the crossing.

Four of the surviving bus passengers gave evidence at the ICC inquiry. Ellis Huff was uncertain if the bus had stopped before reaching the crossing as he had been dozing. He said that a woman passenger had been sat beside the driver. He had not noticed anything wrong with the bus brakes, but the bus had a tendency to stall when running idle.

W.R. Perkins agreed that the bus had stalled several times since he had joined it at Gallup, New Mexico. The bus had approached the crossing slowly and he thought that it had stopped but was not sure.

Conrad Heuvers said that he had seen the train but had not heard it whistle. The bus had slowed but not stopped short of the crossing and at the last moment its driver had

[1419] *Sydney Morning Herald*, Sydney, New South Wales, Australia, 7th April 1930
[1420] *Brisbane Courier*, Brisbane, Queensland, Australia, 7th April 1930

swerved to the left to avoid a head-on collision with the train. He agreed that a woman passenger had been sat beside the driver.

Homer Stein said that the bus had been travelling at 20 to 25 miles an hour, then slowed down near the crossing. Its driver looked to the left but not to the right until he was almost on the crossing, when he had swerved the vehicle to the left, and it had stopped with its front wheels between the rails. Very shortly the train had hit it. He did not hear the train whistle and could not have heard the sound of its bell anyway from within the bus.

Marcelina Zuni lived near the crossing. He stated that he had seen the bus first, then had his attention attracted by the train whistle. The bus had drifted towards the crossing, then suddenly turned to the left, stopped on the track 15 or 20 feet from the crossing, and stalled, upon which its driver had tried to open the door and then to jump through a window.

The ICC concluded that Driver Williams was responsible for the accident by driving his bus onto the crossing directly in front of the train. It noted that there were clear signs warning of the crossing and two different sets of rules and regulations required the bus driver to stop short of it and check that it was safe to cross. 'Under these circumstances, in daylight, and with the sun shining, and in view of the fact that the driver of the bus, when 472 feet from the crossing, could have seen a train more than 1,800 feet distant, there would appear to be no excuse for his action in failing to prevent the accident.' One of the passengers on the bus said the driver looked to the left while approaching the crossing but did not look to the right until the bus had nearly reached the crossing, and the fact that it was not until then that the driver finally realised that a train was approaching from the right, is evident from his action at the last moment in turning the bus off the roadway.

The commission noted that another driver who had operated the bus the previous day stated that it had been in good condition and that Driver Williams had operated the bus for several months and was considered an experienced man. The presence of a woman passenger on the seat beside the driver and the several instances of the bus stalling might have been factors in the accident, though it was impossible to be certain.

In conclusion the commission noted that approximately 2,000 people were killed and 6,000 injured annually in level crossing accidents. Only a very few of these involved buses but when they did the results were likely to be very serious. Every possible precaution should therefore be taken to prevent their repetition and 'this accident calls for an ever increasing vigilance on the part of supervising officials of bus companies to see that even more care and attention are given to the maintenance of equipment to see that employees have ample opportunity for rest and are fit to resume duty, and to make sure that the highest possible spirit of discipline and training is maintained among their employees. It is only by continually emphasising these factors that the occurrence of accidents of this character, or other accidents, at various points en route, can be prevented, or at least reduced to a minimum.' [1421]

Unlike the ICC the jury at the inquest on the victims of the accident was reluctant to blame Driver Williams for it. Its verdict merely stated that the victims, "came to their death by reason of wounds, burns, etc, received and inflicted in a crash and collision

[1421] Interstate Commerce Commission report

between a Pickwick company bus and southbound train No. 7." [1422]

At 13:30 on the afternoon of Thursday 17th April 1930 a bus of La Segorbina company left Valencia, Spain, for its daily journey to the town of Segorbe, carrying 30 passengers. It got as far as Gilet when it was struck by a train and demolished, killing outright or fatally injuring 12 of the passengers, whilst the remainder were injured. According to a Spanish newspaper account of the accident the train was owned by the Compañia de Ojos Negros a Sagunto. It was in charge of Conductor Rafael Tormo with Driver Miguel Esteban, Fireman Pablo Serrano and Brakemen José Pastor and Leoncio Ruiz. The exact nature of the accident is not entirely clear from the newspaper report, but it appears that the woman crossing keeper, Teresa Peñalver Pérez, had failed to put the chain barrier across the road until both bus and train were close at hand. She had then belatedly tried to do so and became the first fatality of the accident when she was run down by the bus, which was unable to stop. [1423] A report of the accident reproduced in various American newspapers states that the crossing keeper committed suicide after realising that she was responsible for the collision, but this does not seem to be correct. [1424]

A family returning from a Victoria Day trip to Montreal, Quebec, to their home in Hamilton, Ontario, was involved in a level crossing accident near Coburg, Ontario, Canada, early on Monday 26th May 1930. (Victoria Day is a Canadian public holiday, originally celebrating the birthday of Queen Victoria of England on 24th May.) [1425] The gates at the crossing were reportedly 'out of repair' and the crossing keeper's efforts to warn the car driver of an approaching Canadian National Railway train by waving a flag and a lantern were unsuccessful. Killed were Mr and Mrs H.J. Johnston Senior, Mr and Mrs H.J. Johnston Junior and 18-month-old Robert Johnston Junior. James Johnston, 5, who had been sat in his mother's arms, was seriously injured. Almost unhurt was Robert Johnston Senior, who had jumped from the car a moment before the impact. [1426] [1427]

The same morning as the Coburg accident, 26th May 1930, another five people died in a level crossing collision at Crittenden, New York, United States. The victims of this accident were Elmer Christner, 20, and Samuel Miller, 17, both of Corfu, New York; Mamie Miller, 21, of Clarence, New York; and sisters Eleanor and Kate Miller, ages unstated, of Nappanee, Indiana. All five were members of the Mennonite religious sect. There was an unobstructed view of the track from the crossing, it was protected by wig-wag signals and the New York Central Railroad train had whistled. It was impossible to say why Elmer Christner, the driver of the car, had missed all the warnings. [1428]

[1422] *Albuquerque Journal*, Albuquerque, New Mexico, United States, 15th April 1930
[1423] *La Vanguardia*, Barcelona, Spain, 19th April 1930
[1424] *Nebraska State Journal*, Lincoln, Nebraska, United States, 18th April 1930
[1425] Wikipedia – Victoria Day
[1426] *The Brandon Daily Sun*, Brandon, Manitoba, Canada, 26th May 1930
[1427] *The Havre Daily News*, Havre, Montana, United States, 27th May 1930
[1428] *Olean Evening Times*, Olean, New York, United States, 26th May 1930

Six people died and a number were injured when a train struck the motor truck on which they were riding at a level crossing at Vina Del Mar, Chile, on Friday 30th May 1930. [1429]

Five of six occupants of a truck died when it was struck by a train on Skinkle's Crossing, near Beatrice, Indiana, United States, on the afternoon of Sunday 8th June 1930. The party had enjoyed dinner at the home of Edward Peterson. After the meal one of them, Virgil Planalp, had suggested that they try the fishing at Ludington ditch, near Lake Eliza and about six miles away, for minnows and fish. They had set off at about 4.30 p.m. and had only gone about a mile when they reached the crossing. The truck was struck by a Chesapeake and Ohio Railroad passenger train, probably a gasoline railcar, in charge of Conductor H. Camack with Engineer C.L. Fisher. Clyde Henley Senior, brother-in-law of Edward Peterson, his son Clyde Junior, aged 8; his daughter Margery, aged 6; Harley Parks, aged 10 and Virgil Planap were killed. Melvin Parks, Harley's father, jumped at the last moment and escaped with injuries. [1430]

The inquest into the accident, held at the end of the week, heard evidence from Calvin Skinkle and J. A. Widner, who had witnessed the accident from the front porch of Skinkle's home. Skinkle testified, "We stood there and wondered whether the truck was going to stop. It didn't and then the crash occurred. It sounded like a dead bump. The truck was picked up and thrown about twenty feet over the side of a bank." Engineer Fisher stated that he had begun whistling when he had seen the truck approaching the crossing and had applied the emergency brakes when he had seen that the truck was not stopping. The inquest was then adjourned pending Melvin Parks recovering to a point where he was fit to give evidence. [1431]

The inquest was resumed a month later. Parks was not able to explain how the accident had happened but emphatically denied that any of the party had been drinking, a possibility raised by the coroner because several bottles supposedly containing alcohol had been found in the truck.

Deputy County Coroner Carl M. Davis returned a verdict of "Accidental Death." [1432]

Just over two months after the Hikurangi accident New Zealand suffered a second one of similar magnitude. At 5 p.m. on the evening of Monday 16th June 1930 three nurses and a matron of what was variously described as the Templeton Home for Children and the Templeton Farm Colony, which was a state-run 'school' for so-called 'mentally defective' children, [1433] were returning from a trip to Christchurch on their day off in a car driven by Mr Ralph Smith. As the vehicle approached the level crossing at Sockburn a mixed train was also approaching it. The crossing was equipped with a crossing keeper but lacked gates. The keeper waved his hand lamp to stop the car, but his signal was not seen until the last moment by Driver Smith. When he did finally notice it and the train he

---

[1429] *The Bee*, Danville, Virginia, United States, 31st May 1930

[1430] *The Vidette-Messenger*, Valparaiso, Indiana, United States, 9th June 1930

[1431] Ibid, 13th June 1930

[1432] *The Lake County Times*, Hammond, Indiana, United States, 8th July 1930

[1433] www.core.ac.uk The Closure of the Templeton Centre.

swerved his car to the left in an attempt to avoid the engine, but the car was caught and carried to a cattle stop. The sudden resistance was sufficient to derail the engine of the mixed train and the seven following waggons were telescoped and piled up in a heap. The wreckage then caught fire, probably from the burst petrol tank of the car. Ralph Smith, Nurses Isabel Bensell, Mary Cameron and Jane Field-Palmer, Matron Isabella Brand and the train driver, C.W. Smith, were all killed. The train driver was pinned beneath his engine and had not been extricated at 10.30 p.m. Fireman Carson was thrown clear but was admitted to hospital suffering from scalds and shock. [1434]

A very lucky escape from what could have been a very serious accident was had at 10 a.m. on Monday 30th June 1930 at Bearmouth, Montana, United States. Carl Sitia, a forest ranger at Bonita, was driving a truck loaded with high explosives when he arrived at a level crossing with the Northern Pacific Railroad. He stopped at the crossing, listened for a train, and not hearing one proceeded, when his vehicle stalled on the crossing. Before he could restart the engine the truck was struck by what was described as a 25-car carnival train and cut in two, 'hurling the explosives about the railroad like so much kindling wood.' It was said that the quantity of explosives was sufficient to have destroyed the train. No explosion occurred, however, and Sitia escaped with a slight cut to the head. The main casualty of the accident was therefore the truck, which had been driven less than 500 miles. [1435]

Shortly after 8 a.m. on Friday 1st August 1930 a 1923 Ford light touring car containing ten people was struck by *The Fox*, a diesel-electric hauled limited train of the Chicago and North Western Railway, on a level crossing half a mile south of Clover, Wisconsin, United States. Four of the occupants of the car were killed outright, one died soon afterwards and a sixth was critically injured. The victims, who were all related, were on their from nearby Manitowoc to the farm of Frank Frozena at Newton, to assist him in harvesting beans. The three adult women in the car were all sisters of Mr Frozena and the other occupants were their children. Killed in the accident were Mrs Mrotek and her daughter Gertrude, 5; Mrs Powlisz (also spelt Povalish) and her daughter Marion, 2½; and Martha Vandebusch, 12. Mrs Vandebusch, the driver of the car; her daughters Leona and Dorothy, her son Donald and Eleanor Mrotek were injured. The Vandebusch girls were rendered unconscious by the accident, and it was feared that Leona would not survive.

The little used road crossed the line on the skew. The crossing was unprotected except for a sign reading "Look out for the Cars." A newspaper report explained that, 'No houses or buildings obstruct the view but with the top of the car apparently up it would be difficult to look back as the tracks approach at an angle of almost 45 degrees, going to an apex where the road and tracks meet.'

Mrs Vandebusch was not badly hurt but was stated to have been overcome by the tragedy, for which she felt herself partly to blame.

There were no witnesses of the accident other than Engineer Kitzerow, who

[1434] *Sydney Morning Herald*, Sydney, New South Wales, Australia, 17th June 1930
[1435] *The Helena Daily Independent*, Helena, Montana, United States, 2nd July 1930

afterwards had set his train back to Manitowoc with the dead and injured aboard. Charles Sherman was the conductor of the train, which consisted of only the locomotive and three coaches. [1436]

At least five and according to some reports eight people died and others were badly hurt as the result of a collision between a Lisbon express and a bus at Braga, Portugal, on Saturday 2nd August 1930. The collision was apparently the fault of the gatekeeper at the crossing who had failed to close the gates. According to differing reports he was arrested after the accident or was one of those who had lost their lives in it. Several children orphaned by the accident were reportedly immediately adopted by local families. Local theatres were closed following the accident. The bodies of the victims were buried in a common grave, the funeral expenses being paid by the state. [1437] There had been 19 passengers on the bus. [1438]

A collision between a Southern Railway freight train and a car killed six people near St Anthony, Indiana, United States, late on Sunday 24th August 1930. Those who lost their lives were William Heddinger, the driver of the car; his daughter Irene, 15; his son Theodore, 12; his mother-in-law, Mrs Bieker; and his nieces Bernice and Leola Bieker, aged 10 and 8. Mrs Heddinger, and her father John Bieker were critically injured and Mary Bieker, 3, was less badly hurt. A severe drought was raging at the time and the party had made the journey from the Heddinger farm to St Anthony to obtain water. Visibility at the crossing was restricted by a long string of box cars on a siding. [1439] [1440]

At about 6 p.m. on Wednesday 27th August 1930 a Baltimore and Ohio Chicago Terminal Railroad freight train collided with a sedan car moving east on 111th Street Crossing, north of Morgan Park station, Indiana, United States, killing five of its occupants. Grocer Charles Cecenas, his wife Sophie, Mrs Magdaline Grozinski and two unidentified girls, aged about 14 and 5, lost their lives. Grocer Joseph Grozinski, husband of Magdaline, was critically injured. The couple's son Frank, 11, and Ruth Sawicky, 8, were less badly hurt.

The railroad approached the crossing through open fields and trains were visible for half a mile in each direction. A wig-wag flashing signal at the crossing was working correctly. The engineer of the train told the police that he had seen the sedan approaching the crossing for several minutes and it had appeared to be under full control. He had whistled three times for the crossing. When he had realised that the car was not stopping he had applied his brakes in emergency but with 60 loaded cars behind his locomotive his train had travelled a quarter of a mile before it could be pulled up. The train was stopped half an hour while the bodies of the dead were removed from the wreckage. Motorists

---

[1436] *The Manitowoc Herald-News*, Manitowoc, Wisconsin, United States, 1st August 1930

[1437] *The Morning Herald*, Hagerstown, Maryland, United States, 4th August 1930

[1438] *Le Matin*, Paris, France, 4th August 1930

[1439] *The Lake County Times*, Hammond, Indiana, United States, 25th August 1930.

[1440] *The Newark Advocate*, Newark, Ohio, United States, 25th August 1930 took the injured to hospital. [1441]

Train No. 16 of the Southern Pacific Railroad, the *West Coast Limited*, collided with a small car at the northern edge of Turner, Oregon, United States, eight miles east of Salem, shortly after 6 a.m. on Monday 1st September 1930. All five women in the car were killed. They were Mrs Robertson, who had been driving; her daughter Ailene, 17; her niece Inez Barnett, 15; Mrs Lois Briggs and her mother-in-law Mrs B.C. Briggs. All lived in Turner except Mrs Lois Briggs who was from Portland. They had been en route to the house of Mrs T.T. Palmer and thence to the Salem cannery where they worked. The train was running a few minutes late. [1442]

Brothers Ambrosio, Claro and Pedro Bocsir, Pedro Abbonag, Namerlano Acoro, Felix Avecilla, and Estoben Codianat, all Filipinos, were killed when the truck in which they were riding was struck by the *Coaster* express of the Southern Pacific Railroad on a level crossing between Spreckets Junction and Spence and five miles south of Salinas, California, United States. The accident happened on Thursday 11th September 1930. Felix Avecilla owned the truck and Estoben Codianat was driving it. Engineer W.H. Thompson said afterwards that the truck had stopped at the crossing to give way to his train and had then suddenly plunged forward onto the line. His locomotive carried it 50 feet down the track before it was tossed to one side.

It was theorised that believing that the truck was out of gear Codianat had released the clutch. It was not and when he had done so the machine had lurched forward as the pressure was removed. [1443]

The train which had departed from Lyon-Est at 18:40 was due to pass over the level crossing at Lentay, France, at 20:35 on Sunday 12th October 1930, a couple of minutes before arriving at Morestel. The crossing keeper, M. Berthier, had placed ropes (chains?) across the road and lit two lanterns as a further precaution. The train approached at considerable speed around the long curve preceding the crossing and at the same time the service bus running between Lyon and Saint-Genix-d'Aosre arrived. The crossing keeper thought at first that the bus would stop but it did not slow up. M. Berthier jumped out into the road and waved his two lanterns at the bus, but the collision was unavoidable. He was the first victim. The bus smashed through the barrier and entered the crossing at the same moment as the train. The shock was terrific, and the body of the bus was separated from its chassis and thrown some metres into a ditch.

Driver Louis Thianet immediately stopped the train and as soon as he had done so Conductor Melchior went to assist the victims, aided by some of the train passengers. The staff at Morestel station were alerted and asked for lanterns and jacks and notice was sent to the local police.

Four of the bus passengers were killed outright, one of whom was decapitated. The other three were identified as Mmes Cœur, David and Juppet, all from Morestel, so they

[1441] *The Winona Republican-Herald*, Winona, Minnesota, 29th August 1930

[1442] *The Athena Press*, Athena, Oregon, United States, 5th September 1930

[1443] *Oakland Tribune*, Oakland, California, United States, 11th September 1930

had almost reached home when the accident happened. Eight other bus passengers, all from the local area, were injured, two fatally, and M. Granger, the owner of the bus, also succumbed after having being taken to the Hôtel Dieu in Lyon. The newspaper report stated that this brought the total number of dead in the accident to seven so M. Berthier, the crossing keeper, presumably survived being run down by the bus. [1444]

Eleven people died and nine more were seriously injured in collision between train and motor bus at Ploesti, Romania, on the night of Monday 10th November 1930 [1445]

Five Filipinos, described as 'ranch hands' and whose surnames were given as Amelino, Caridad, Conte, Matea and Steav, were killed when their car was struck by an Atchison Topeka and Santa Fe Railway gas-electric railcar on a level crossing at Lucca, near Lindsay, California, United States on Tuesday 11th November 1930. A brakeman who could see that the car was not going to stop had tried vainly to flag down its driver, Steav. [1446] [1447]

A private motor bus was in collision with a train on a level crossing between Mhow and Indore, India, on Sunday 16th November 1930, killing 5 people and injuring 14. 'All the killed, who were Muslims, were occupants of the omnibus.'

Nine people died in an accident at Butterworth Crossing, near Dewitt, Virginia, United States, on the evening of Sunday 16th November 1930. Eight of the party had been returning from a funeral of a relative at Petersburg. The victims were three women named Johnson, four Johnson children and a woman named Ruth Williams. They were all negroes, the newspapers informed their readers, whilst the ninth victim, J.C. Flippin, the driver of the car, was a white man 'well known in the Clarksville section.' The accident happened when Flippen tried to cross the line directly in front of a Seaboard Air Line Railroad passenger train. [1448]

In a very serious level crossing accident in Russia on Wednesday 3rd December 1930, 28 people died and a further 19 were injured when a train collided with a tram on a level crossing in Leningrad. The operator of the tram reportedly failed to hear the warning alarm at the crossing and drove onto the line in front of the train. [1449]

A family of six on their way to church to attend a Christmas service were killed in a level crossing collision on Church Street Crossing, the edge of Lomira, Wisconsin, United States, on Christmas Eve, 24th December 1930. Harry Schultz had picked up his

---

[1444] *L'indépendant des Basses Pyrénées*, Pau, France, 15th October 1930
[1445] *Nebraska State Journal*, Lincoln, Nebraska, United States, 11th November 1930
[1446] *Oakland Tribune*, Oakland, California, United States, 12th November 1930
[1447] *Nevada State Journal*, Reno, Nevada, United States, 12th November 1930
[1448] *The Monroe News-Star*, Monroe, Louisiana, United States, 17th November 1930
[1449] *Reno Evening Gazette*, Reno, Nevada, United States, 3rd December 1930

parents, Mr Herman Schultz and his wife, and the three, with Harry's children, Esther, Raymond and Anita, were the victims of the accident. The car was demolished, and the bodies of the victims were scattered along the tracks. Harry Schultz's wife and an older child had remained at home to make preparations for the Christmas celebrations. The car was struck by southbound Minneapolis St Paul and Sault Ste Marie Railroad (Soo line) express No. 2 in charge of Conductor John Anderson with Engineer W.H. Benedict. The accident happened at 6.50 p.m. The crossing with Highway 67 was protected by a wig-wag signal and there was a fair view along the track in each direction. There had, however, been fatal accidents at it in the previous few years. [1450]

[1450] *The Sheboygan Press*, Sheboygan, Wisconsin, United States, 26th December 1930

CHAPTER 13

RAIL/RAIL CROSSING COLLISIONS

At locations where railways and roads intersected the idea of having them cross on the level had obvious advantages in terms of economics over the cost of building an over or underbridge, especially if the level crossing were unmanned and the construction of a bridge would require the building of substantial approach embankments or possibly approach cuttings leading to an underbridge. With road users being limited in the early days to pedestrians, cyclists, and horse-drawn vehicles the dangers to a train from a collision on such a level crossing were also likely to be minimal. For railways crossing other railways on the level, though, the situation was very different. A collision with another train on a crossing was likely to be far more serious than a collision with, say, a horse and cart, and furthermore a train driver approaching a level crossing and seeing another train nearing or on it could not pull up in anything like the distance that a road user could. It is no surprise therefore that in most countries, where the railway system was either built directly by the state or under state supervision, that rail/rail crossings were not provided. It was only in countries where the railways were built by competing private companies without state interference that such crossings existed, most especially in the United States, where lines were often built as cheaply as possible and the economy of not needing to construct a bridge was doubtless an important consideration. Thus, the United States had many rail/rail level crossings, Great Britain at one time had about 20 or so and its colonies, such as Ireland and Canada, had a handful. One or two did also exist elsewhere, such as at Istravoz (later Hilal), Turkey, where the Oriental Railway crossed the Smyrna – Cassaba Railway. [1451] As has been pointed out, the American companies also seemed to have a perverse proclivity for building rail/rail crossings where they could easily have been avoided, perhaps because in the early days train speeds were low, trains infrequent and night operation unusual so that the dangers were not fully appreciated. [1452] By 1916 every one of the 15 or so remaining rail/rail crossings in Great Britain was fully

[1451] Wikipedia – Hilal railway
[1452] *A History of Railroad Accidents, Safety Precautions and Operating Practices* by Robert B. Shaw, pg. 99

signalled, and there were hardly any further collisions on them. In the United States, on the other hand, where many crossings remained with little or no signal protection, collisions on them continued to occur regularly.

Barely mentioned in the press but the subject of an Interstate Commerce Commission (ICC) investigation was a side collision near Fostoria, Ohio, United States, between lines of the Baltimore and Ohio and Hocking Valley Railroads, the crossing being known as "H.V. Crossing" in the timetable of the former company and as "B & O Crossing" in the timetable of the latter company. The crossing was protected by a movable crossing board target and fixed stop boards, the latter placed 200 feet from the crossing.

On the morning of Thursday 15th June 1916 eastbound Baltimore and Ohio express No. 14 consisted of engine No. 1436 and 11 baggage and express cars, in charge of Conductor Hollinger and Engineer Mitchell. It reached Fostoria at 7.06 a.m., 26 minutes late, stopped at the N.Y.C. & St L Crossing, proceeded as far as the stop board for H.V. Crossing, stopped again, the crossing board was cleared, and it proceeded again. At 7.10 a.m. its fourth car was struck by a Hocking Valley freight train.

Northbound Hocking Valley freight train second No. 65 consisted of engine No. 157, 61 loaded cars and a caboose, in charge of Conductor Jones with Engineer F.E. Michael. This train cut right through the fourth and fifth cars of train No. 14, a baggage car and refrigerator express car, both of wooden construction, while the three leading cars of train No. 65 were destroyed. Express messenger Earl Lindsay was killed, and Engineer Michael was slightly hurt.

Conductor Hollinger of the express told the inquiry that he had seen the crossing board cleared for his train. His view of the Hocking Valley line at the stop board was obstructed by a building and stabled cars. When he first saw train No. 65 it was moving at 12 miles an hour, reduced to 6 miles an hour when they met.

Engineer Michael stated that the brakes of his train had been inspected before leaving Columbus and reported to be in good order. There had been no problem with the brakes until he reached the B & O Crossing where he had seen the crossing board against his train. The first brake application had reduced the speed of his train to 15 miles an hour, but a second application had made little difference. He had then made an emergency brake application and whistled in the hope that the operator at the crossing might give him the road. He had not used the sanders as he did not think they were working, and company rules forbad his reversing the engine. After the accident he had found an angle cock closed at the head of the thirteenth car, which he thought must have been why he could not stop.

Conductor Jones stated that he had not anticipated any problems in stopping his train and when he felt a jar at the crossing he had not, until he had gone forward, realised that there had been a collision. At different points on the journey he had seen a man, whom he recognised as a former brakeman, riding the train but had paid no attention to him. Half a mile from the crossing there had been 50 lb of air and he did not think that much could leak off in such a short time. He did not know the cause of the accident.

Fireman Janette of second No. 65 generally corroborated what his engineer had said. He added that most trains stopped at the stop board, but some did not stop but barely crept along.

Flagman Cooley of the freight train generally corroborated the statement of his conductor. He added that Engineer Michael was one of those engineers who always came to a dead stand at the stop boards, even if the crossing signal was clear.

The ICC dismissed the inference of Engineer Michael that a closed angle cock had prevented him stopping his train and concluded that the accident had been due to his failing to have it under proper control. [1453]

A Southern Pacific Railroad freight train collided with a Western Pacific Railroad hand car on a crossing close to Lyoth and just south of Tracy, California, United States, on the early evening of Friday 15th September 1916. There five men of them riding on the car and the foreman, O'Shaughnessy, and three of his men will killed outright. The other man, a Mexican section hand, was badly hurt, and taken to hospital where he died later in the night. Two other men, who must have been the engine crew of the Southern Pacific train, were only slightly hurt. [1454]

Minneapolis and St Louis Railroad (M & St L) train No. 3, which left Albert Lea, Minnesota, United States, at 6.06 a.m. on Wednesday 14th March 1917 with Engineer W.H. Mace and Fireman Monshaw, collided with a Chicago St Paul Minneapolis and Omaha Railroad train on a crossing at Merriam Junction, Minnesota, United States. The locomotive of the M & St L train was overturned, and Engineer Mace and Fireman Monshaw were badly scalded. They were widely known in local railroad circles. [1455]

On Thursday 31st January 1918, a bad collision occurred at Sedro Woolley, Washington, United States, at the crossing of the Great Northern and Northern Pacific Railroads. Trains of the two companies met at 8.20 p.m., killing at least five passengers, and injuring 15 passengers and three railwaymen. Stop boards were provided on the approaches to the crossing and Washington state law and Rule 98 of the Northern Pacific required all trains to stop short of the crossing. Buildings restricted visibility at the crossing and bright arc lights made locomotive headlights harder to distinguish at night.

Eastbourne Great Northern train No. 280 consisted of engine No. 103, a combination mail and baggage car, a smoking car, and a first-class coach, all of wooden construction, in charge of Conductor Corrigan and Engineer Naser. It was en route from Anacortes to Rockport.

Southbound Northern Pacific freight train No. 676 consisted of engine No. 1673, 29 loaded cars, an empty car, and a caboose, in charge of Conductor Heald and Engineer Hall, and was bound from Sumas to Seattle.

In the collision the engine of the freight train cut the rear coach of the passenger train in two, uncoupled it from the smoking car and carried it forward until its side was torn away by striking the buildings of the Northern Pacific depot.

Conductor Heald told the Interstate Commerce Commission (ICC) investigators that a

---

[1453] Interstate Commerce Commission report.

[1454] *The Bakersfield Californian*, Bakersfield, California, United States, 16th September 1916

[1455] *Freeborn County Standard*, Albert Lea, Minnesota, United States, 15th March 1917 brake test had been made before the train had left its last stop at Wickersham. Speed had been reduced from 25 or 30 miles an hour to 15 at Delavan, 2½ miles short of Sedro Woolley. There was another application a mile further on at Wild's Spur and the last one had been at Sedro Woolley. Like Conductor Jones at Fostoria he had not realised at first that there had been an accident. He had heard no whistle from the engine for the hand brakes to be applied, though he had heard whistles for level crossings.

The statement of Rear Brakeman MacDonald corroborated that of his conductor. He said he had believed that the engineer had the train under control.

Brakeman Ringer remembered the brakes being applied at Delavan and on the approach to Sedro and stated that he thought that the engineer had the train under control. He had been standing on the gangway ready to alight at Sedro when the engineer had called out, "There is the Great Northern," sparks had begun to fly from the wheels and the engineer had whistled for brakes. The train was going at 15 or 20 miles an hour. He had still thought that the train was under control and would stop at the stop board until the last moment, when he had jumped.

Engineer Hall stated that he had applied the brakes north and south of Delavan. At the next application approaching Sedro the brakes had not taken hold properly and he had made an emergency application, but without the desired effect, though he could see from the sparks that the engine brakes were working. He then began whistling for brakes and about the same time saw the Great Northern train on the crossing. He had not reversed his engine as he feared it would simply make the wheels skid. After the accident he had found the angle cock on the 22nd car from the front partly turned by a hobo, and he thought that was the cause of the accident (shades of Fostoria again).

Fireman Livesley stated that he had thought that the brakes of the freight train had worked all right before the accident. He put the speed of the freight train at not more than 12 miles an hour at the moment of the collision.

Trainmaster Fitzsimmons stated that he had examined the freight train after the accident and that although there appeared to be an angle cock partly turned closer examination showed this not to be the case.

The ICC concluded that the direct cause of the accident had been the failure of Engineer Hall to bring his train under proper control a sufficient distance north of the stop board. [1456]

Although the ICC stated that only five had died a newspaper report listed six deaths with very likely two more fatal outcomes to follow. [1457]

Unusual circumstances surrounded a fairly minor collision at Racine Junction, Wisconsin, United States, on Saturday 20th April 1918. Here the double-track Chicago and North Western Railway was crossed at right-angles by the single-track line of the Chicago Milwaukee and St Paul Railway. Protection at the crossing was more sophisticated than at many in the United States at this date, for it was provided with fixed signals worked by wire from a tower (signal box) as well as derails to stop trains which

[1456] Interstate Commerce Commission report.
[1457] *The Daily Capital Journal*, Salem, Oregon, United States, 2nd February 1918

ran past the signals. Late on the afternoon of the day in question these failed to prevent a collision between a Milwaukee freight train and a North Western switching engine which injured two employees.

At 5.10 p.m. the North Western switching engine, No. 366, in charge of Foreman Birkem with Engineer Lalonde, was engaged in shunting operations and had moved forward over the crossing on the northbound track, pulling three cars, and come to a stand north of a crossover, preparatory to pushing the cars back over the crossover to the southbound track and thence to sidings on the east side of the line. This backing movement was protected by dwarf signal No. 3 and 57 feet south of the dwarf signal was a Hayes-type derail. After a couple of minutes the crew saw that the dwarf signal had been cleared and began the backing movement, whereupon the leading car was stuck by the Milwaukee train as it reached the crossing.

Milwaukee eastbound extra freight train No. 2101 consisted of engine No. 2101, running tender first, 8 loaded cars and a caboose in charge of Conductor Victor and Engineer Burns. It had left Corliss at 4.50 p.m. and came to a stop at the signal protecting the crossing at 5.09 p.m. The signal cleared after a few minutes, the train set off and collided with the front car of the shunt move as stated above. The struck car was destroyed, engine No. 2101 was derailed, and its tender was badly damaged.

The accident did not get into the newspapers, but the Interstate Commerce Commission (ICC) decided that its circumstances made an investigation necessary.

Foreman Birkem told the inquiry that he had got off the cars as engine No. 366 was pulling them northwards over the crossing. After the collision he had found the derail in the 45 degree position and the derail wired down and inoperative.

Switchman (shunter) Diesness agreed that the dwarf signal had been in the 45 degree position after the accident though he was positive that it had been in the stop position when the shunt move had first stopped north of the crossover.

Engineer Lalonde of engine No. 366 said that he had begun backing up when the switchman had signalled for him to stop, and his fireman had called his attention to the Milwaukee train. His train had almost stopped when it was hit. Afterwards he had seen the dwarf signal in the 45 degree position and the derail was off the track.

Fireman Luhn of engine No. 366 stated that after they had come to a stand north of the crossover he had seen the dwarf signal in the 45 degree position, the switchman had signalled to them, and they had started to back up. After the collision, the dwarf signal had still been "off," and he had seen that the derail was wired up and inoperative. He was sure that the dwarf signal had been "on" when they had first moved northwards.

Engineer Burns of the Milwaukee engine said that he had been stopped two minutes at the interlocking signal when it had cleared, and he had begun to move forward over the crossing. Passing the east end of a freight house blocking his view at 7 or 8 miles an hour he had seen the North Western cars being pushed over the crossing. He had made an emergency brake application but had not been able to avoid the collision.

Fireman Hall of engine No. 2101 said that his engineer had whistled for the signal protecting the crossing, but it had not been cleared. The North Western train had then moved northwards over the crossing, the signal had been cleared and they had started across.

Conductor Victor said that after the collision he had found the wire operating dwarf signal No. 3 broken, and the derail wired up and inoperative.

Operator White stated that after the shunt move had stopped north of the crossover he had cleared the signal for the Milwaukee train to proceed. He could not account for the dwarf signal being clear; he had not worked its lever. Operator Ammon, whom he had relieved, had noted in the daily transfer report that the wire to dwarf signal No. 3 had broken about noon, though he did not remember Operator Ammon mentioning it to him. About an hour before the accident the derail had stuck so that he had had to disconnect it and he had therefore not operated the derail after the northbound shunt move. Nor had he taken steps to ensure that the dwarf signal was secured in the stop position after the derail became inoperative. There had been trouble with the derail and signal over the previous 10 days, which had been attended to by the signal maintainer.

The ICC noted that the back wire of dwarf signal No. 3 had been found to broken after the accident and the spring provided was evidently not strong enough to restore the signal arm to danger by itself. The commissioners concluded that Switchman Diesness and Fireman Luhn must have been mistaken in their assertions that the dwarf signal had initially been "On" after the northward shunt move.

The ICC blamed Operator White for the accident. He had disregarded rules 727, 628 and 638 of the company.

Rule 627 stated:

If a signal fails to work properly, its operation must be discontinued, and the signal secured so as to give the normal indication until repaired."

Rule 628 stated:

Signalmen must observe, as far as practicable, whether the indications of the signals correspond with the positions of the levers.

Rule 638 stated:

If necessary to discontinue the use of any fixed signal, hand signals must be used, and the division superintendent and operating repairman notified.

Operator White had failed to get the crew of engine No. 366 to wait for a hand signal before using the crossover and had failed to check that the position of the dwarf signal corresponded with the position of its lever. Furthermore, although he claimed Operator Ammon had not told him that the back rail to the dwarf signal was broken, the fact was noted in the transfer book where he himself had made entries.

The ICC made no comment regarding the apparently poor maintenance of the signalling equipment. [1458]

In 1919 the crossing of the Central of Georgia and Louisville and Nashville Railroads at Sylacauga, Alabama, United States, had no such fancy protection as that provided at Racine Junction, merely stop boards on each line on the approach to the crossing. The state law required all trains to come to a stop before using such crossings but neither it nor any equivalent rule was included in the rule books of either company.

[1458] Interstate Commerce Commission report.

On Monday 24th February 1919 southbound Louisville and Nashville Railroad passenger train No. 81 consisted of engine No. 361, a combination baggage and mail car and three coaches of wooden construction, two with steel underframes, in charge of Conductor Meigs with Engineer Talford. It arrived at Sylacauga at 8.10 a.m., one minute late, and had not cleared the crossing when the head end of its rear coach was struck by the Central of Georgia train. The struck car was derailed and partly overturned, and the front of the colliding engine also left the track. One unlucky passenger, Mrs Mary Thornton, of Weogufka, Alabama, was killed and 17 passengers and three railwaymen were injured.

Central of Georgia freight train second No. 38 consisted of engine No. 1801, 20 loaded and two empty cars and a caboose, in charge of Conductor Lindsay and Engineer Sweatt. It had arrived at Sylacauga at 7.50 a.m., three hours and 50 minutes late, stopped to take water, then proceeded and collided with the Louisville and Nashville train.

The Interstate Commerce Commission (ICC) investigated the collision and took evidence from many witnesses, beginning with Engineer Sweatt, who stated that he had stopped at Sylacauga about 20 minutes, during which time he had taken water at the west water plug, detached his engine and gone over the crossing to the sand house, taken sand, gone back over the crossing and recoupled his engine to the train. The flagman at the crossing had signalled him to come on and he had whistled twice. He had then answered a proceed signal from his conductor with another two whistles and asked his fireman, "How about the crossing?"

"All right," was the answer.

As he set off the flagman was walking back to rejoin the train. Fifty feet from the crossing, when his train was travelling at 10 miles an hour, he saw the engine of train No. 81 enter the crossing. He had not heard that train whistle. He applied the brakes in emergency and opened the sanders but could not pull up in time.

Fireman Doolittle of the freight train said that his engineer had twice asked him if the crossing was clear and seeing nothing and assuming from the flagman's signal that it was, he had replied that it was clear and gone back to his firing. The engineer had then applied the brakes and looking up, he had seen the rear coach of the Louisville and Nashville train. He was certain that that train had neither whistled nor stopped before entering the crossing.

Flagman Kirven stated that he had heard the Louisville and Nashville train whistle for the station but had not been able to see it. Taking it for granted that that train would stop and as his own train had stopped first he had given it the signal to proceed. He had not considered it necessary to remain at the crossing while his train went over it. He was sure that the Louisville and Nashville train could not have stopped before entering the crossing.

Brakeman Coleman said that he had signalled to his engineer to stop when he had seen the Louisville and Nashville train. He agreed that that train had not stopped short of the crossing.

Engineer Talford of train No. 81, on the other hand, testified that he had stopped just short of the stop board. He had seen some smoke on the Central of Georgia line but no train and checking with his fireman that it was clear, he had started. Whistling to warn some children about to cross the line he had then heard the other train whistle but had not seen it until he was on the crossing. He had then given his train full steam in the hope of getting across in front of the Central of Georgia train.

Fireman Vernon of train No. 81 corroborated his engineer's testimony.

Conductor Meigs of train No. 81 stated that it had stopped for about 15 seconds.

Baggageman Martin and Flagman Jarvis of train No. 81 also agreed that it had stopped dead, as did Chief Dispatcher Brock, Track Supervisor Porterfield, and Car Inspector Streip, all of whom had been passengers on the train.

Central of Georgia Operator McClain, however, was positive that although train No. 81 had slowed down it had not stopped, though it had whistled.

Operator Norris, who was just about to relieve McClain, was equally positive that train No. 81 did not stop.

The ICC concluded that the primary cause of the accident was the failure of Flagman Kirven to properly protect his train after signalling for it to proceed. Contributory causes were the failure of one or both engineers to stop short of the crossing and of the railroad companies to provide adequate safeguards.

It was not within the flagman's ordinary duties to flag trains over crossings but having done so, Flagman Kirven should have remained at the crossing until his train had cleared it, especially as he was aware of the proximity of the Louisville and Nashville train. Had he remained at the crossing he could have flagged and stopped one or other train.

Had Engineer Sweatt stopped at the stop board he would have seen the other train. In relying on the flagman when the latter had not remained at the crossing he had exercised poor judgement.

Whether or no train No. 81 had stopped at the crossing was open to question. The statements of the railwaymen were conflicting and of ten independent eye-witnesses (presumably passengers on the train), six maintained that it had stopped and four that it had not. In view of the statements of the operators and others it seemed doubtful if it had. Considerations of the whereabouts of the two trains just before the collision also made it unlikely that train No. 81 had stopped. If Engineer Talford failed to stop then he had been grossly negligent and was largely responsible for the collision.

Regardless of the failure of the railwaymen involved, the ICC found both companies at fault for not providing more protection at the crossing than stop boards and not even having a rule or regulation regarding its use. The crossing was used by 14 scheduled Central of Georgia and 8 scheduled Louisville and Nashville trains daily. [1459]

Gibbsland, Louisiana, United States, was the scene of a collision on Wednesday 18th February 1920 involving trains of the Louisiana and North West and Vicksburg Shreveport and Pacific Railroad companies. The Vicksburg line crossed the Louisiana line at an angle of 45 degrees. The only protection provided was stop boards, together with special timetable instructions that trains were to stop not more than 200 feet from the crossing.

Southbound mixed train No. 22 of the Louisville and North West Railroad consisted,

[1459] Ibid.

on the day of the accident, of engine No. 22, 17 freight cars and a coach, in charge of Conductor Weaver and Engineer Hattaway. It arrived at Gibbsland at 1.25 a.m. and stopped at the station platform south of the crossing. A few minutes later it was backed onto the crossing and came to a stand on it while the rear coach was detached and coupled to a caboose. While in this position with a gondola car on the crossing it was struck by the Vicksburg train.

Westbound Vicksburg Shreveport and Pacific passenger train No. 11 consisted of engine No. 308, a baggage car, a mail and express car, a combination car, 2 coaches, a sleeping car, and a dining car in charge of Conductor Berkley and Engineer Porter. It had left Monroe, Louisiana, at 11.30 p.m. the previous day and reached Gibbsland at about 1.35 a.m., colliding with train No. 22 at a speed variously estimated at between 8 and 40 miles an hour. As a result of the collision the gondola car of train No. 22 was pushed along the station platform and into the station building, while the coach and caboose of that train were damaged. The engine and tender of train No. 11 were damaged, and the mail and express car was derailed. Two persons (passengers?) on the station platform were killed and 15 of them plus three railwaymen were hurt.

Engineer Hattaway of train No. 22 told the Interstate Commerce Commission (ICC) inquiry that after stopping his train at Gibbsland about two minutes he had been signalled to back up. He was then looking for signals from the rear of his train and had not seen train No. 22 approaching, but he did not believe that Engineer Porter had stopped before using the crossing.

Fireman Roman said he had been attending the fire and had not seen or heard train No. 11 approach.

Conductor Weaver stated that he had signalled Engineer Hattaway to back up so that the coach might be placed on track 7 while the remainder of the train was taken to the coal chute track. Seeing the headlight of train No. 11, he had called to the brakeman to get out of the way. There was no fog.

Flagman Hodge said that he had been trying to uncouple the coach when he had heard the whistle of train No. 11, of which he knew nothing, and just had time to get from under the cars. The weather had been clear.

Brakeman Warren and Shelton said they had known nothing of train No. 22. Warren said there had been no fog and Shelton that there had not been very much.

Engineer Porter stated that the crossing had appeared clear when he first saw it and he had reduced speed to 1 or 2 miles an hour, but not stopped. Then he had opened the throttle wide to get over the hill beyond the crossing. Initially he said he had seen nothing of the Louisiana train before the collision, which had knocked him out. According to him the fog had been very thick, so that he had been unable to see the caution board or other landmarks. Later he said that when the fog had parted close to the crossing he had seen something which looked like a white cow, had applied the brakes in emergency and reversed his engine.

Fireman Gunn of train No. 11 said that he had seen the cars of the Louisiana train on the crossing when his train was about 150 yards from it and travelling at 15 or 20 miles an hour. At that point, his engineer had applied the brakes in emergency and reversed his engine. At one point he stated that it had been only a little foggy, at another that he thought that the fog had been the cause of his engineer failing to see the cars on the crossing. His engineer usually stopped short of the crossing but had not done so on the day of the accident.

Conductor Berkley said the train brakes had been tested at Monroe. Approaching Gibbsland he had heard the station whistle, but his train had not stopped before the crossing. He had noticed no fog. Afterwards his engineer had told him, "I could not hold them." He was of the opinion that the engineer had seen the box cars on either side of the crossing and not noticed the lower gondola car across it. He insisted that engineers always stopped short of crossings as per the rules.

Flagman Griggs and Train Porter Chocklin thought that the train had been running unusually fast and said that it had not stopped at the stop board. There had been no fog in the vicinity. Train Baggagemaster Patton made a similar statement.

Louisiana and North West train extra No. 24 was stopped at Gibbsland at the time of the accident. Engineer Cobb of this train could not see from his position if train No. 11 had stopped before the crossing but he heard its engine working steam immediately before there was an emergency brake application. The weather had been clear.

Fireman Lyons, Conductor Cain and Flagman Jones of train No. 24 all agreed that train No. 11 had not stopped and that there had been no fog. The fireman and the conductor put the speed of the train as high as 35 or 40 miles an hour and added that its headlight had been shining on the gondola car.

The ICC concluded that the accident had resulted from the failure of Engineer Porter to bring his train to a stand short of the crossing as per the rules. The statements made by the engineer were conflicting, but other testimony showed clearly that train No. 11 had not stopped short of the crossing while the damage done showed that it must have been travelling at a comparatively high speed. The preponderance of evidence was that the weather was clear. In view of the fact that the headlight and brakes of the train were working properly it was impossible to say why Engineer Porter had failed to see the cars in time to stop.

All the railwaymen interviewed had insisted that trains normally stopped on the approach to crossings. However, since the accident an ICC inspector who had been a passenger on a Vicksburg train had noted that it had not stopped at the crossing at Gibbsland, just reduced speed to 7 or 8 miles an hour and whistled. A similar procedure was adopted at an unprotected crossing with the Louisiana and Arkansas Railroad at Sibley, Louisiana, suggesting that the rule as to stopping at crossings was frequently violated.

The commission emphasised that measures should at once be taken to prevent similar accidents by the installation of interlocking plants, signals and derails, gates or other protective devices, and the rules regarding trains stopping at crossings strictly enforced. [1460]

Sunday 27th February 1921 saw a disastrous collision at Porter, Indiana, United States, involving two express trains.

Eastbound Michigan Central Railroad passenger train No. 20, known as *The Canadian*, from Chicago, Illinois, to Detroit, Michigan, consisted of engine No. 8306, a

---

[1460] Ibid.

baggage car, a smoking car, a coach, two sleeper cars, a parlor car, a dining car, a parlor car and a coach, in charge of Conductor Foote and Engineer William Long and Fireman George Block.

Westbound New York Central Railroad passenger train No. 151, known as *The Interstate Express*, from Buffalo, New York, to Chicago, Illinois, consisted of engine No. 4828, an Arms palace horse car, a combination baggage car and coach, two coaches, a dining car and two parlor cars, in charge of Conductor Roy with Engineer Claude Johnson and Fireman George Deland.

The Michigan Central train had left Chicago at 5.05 p.m., the New York Central train left Elkhart, Indiana, at 4.55 p.m.

The two trains approached Porter, where the main lines of the two companies crossed one another, almost at the same time. The crossing was protected by a mechanical interlocking plant (signal tower) with a 54-lever Saxby and Farmer frame, fixed signals, and derails (trap points). The New York Central train was the first to operate the buzzer announcing its approach, so the operator decided to give it priority and cleared his signals for it, the signals thus being locked against the Michigan Central train, and its derail open. Despite this, train No. 20 passed the distant signal on its line, No. 2412, at caution, passed the home signal, No, 35, at danger, was derailed on the derail, continued because of its speed and momentum onto the crossing, where the engine rerailed itself, and its engine finally came to a stand 400 feet beyond the crossing. The coach, the third vehicle in the Michigan Central train, came to a stand directly across the westbound New York Central track. Almost at once this car was struck by the New York Central train and demolished. All of the deaths and fatalities occurred in this coach, apart from Engineer Johnson and Fireman Deland, who were killed when their engine was derailed and partly buried in the ground. Thirty-five passengers died and 11 passengers, 7 other persons and two railwaymen were injured.

Operator Charles Whitehead told the Interstate Commerce Commission (ICC) inquiry that he had been on duty in the tower on the evening of the disaster, assisted by Acting Leverman Joe Cook, who was an assistant signal maintainer but was working as a leverman so that the regular leverman, named Piering, could have a day off. Cook had asked him which train to give priority to and he had replied, "Whichever train hits the current first."

The New York Central one had done so, and Cook had set up the route for it. Shortly afterwards he (Whitehead) had seen the Michigan Central train approaching and he and Cook had remarked upon its speed. Then (according to a newspaper report), Cook had cried, "My God, Charley, the M.C. is coming through the block." [1461] By then it was too late to throw the signal against the New York Central train; the Michigan Central train overshot the crossing and the collision occurred almost at once. Leverman Cook corroborated Whitehead's testimony.

Engineer Long stated that his train had been travelling at 60 miles an hour at the distant signal. Seeing the distant at caution he had made a service application of the brakes and had reduced speed to 10 miles an hour. Then Fireman Block had called, "All the way," indicating that the home signal had been cleared. He then looked for the home signal himself and thought that he saw a green light over a red one. When his train got

---

[1461] *The Kokomo Daily Tribune*, Kokomo, Indiana, United States, 1st March 1921 closer he looked again but smoke and steam obscured his view. The only thing that he could think of to account for his mistake was that he had confused the train order board for the home signal. He insisted that he had not relied entirely upon his fireman and had never previously had difficulty observing signals at that point.

Fireman Block spoke of seeing the distant signal at caution and, from about a mile away, the train order and home signals, two green lights and a red light. He admitted having called out, “All the way,” to his engineer and the latter replying, “All right.” He had then begun to work on the fire.

Conductor Foote said that he had not seen the signals.

Engineer Curtis, of Michigan Central freight extra No. 7489, said that he had been on a siding taking water when train No. 20 approached. The two blades of the home signal had both been in the stop position; the train order signal was showing proceed. Train No. 20 had passed him at 30 or 40 miles an hour. He had never confused the home and train order signals. Fireman Arthur of the freight train corroborated his engineer’s statement.

Brakemen Kubbernuss and Wise of extra No. 7489 agreed that the home signals had been at stop and stated that when they had seen train No. 20 coming at high speed they had given “Ease off” signals with their hand lamps, then when they were not acknowledged, violent “Stop” signals, but there was still no response.

The ICC concluded that the cause of the accident had been the failure of Engineer Long to obey the indication of the home signal. A contributory cause had been the failure of Fireman Block to correctly identify the signal. Having seen that the distant signal was at caution Engineer Long should have exercised particular care to observe the home signal; he had also failed to see the warning signals of the brakemen. It seemed clear that Long had been misled by his fireman but that did not relieve him of responsibility. Fireman Block was responsible to a lesser degree.

The accident, the ICC continued, clearly showed the need for an automatic train control device to be installed in connection with the signal system. The location of the eastbound home signal and train order board should be improved so as to prevent their being obscured by smoke and steam. As the derail had not prevented the accident the maximum permitted speed of trains should be reduced or the derail re-sited further from the crossing. The Michigan Central coach which had been struck was of wooden construction, but an all-steel car would probably have fared little better given the tremendous force of the impact. [1462]

The inquest at Valparaiso upon the victims of the disaster found a verdict of Involuntary Manslaughter against Engineer Long and Fireman Block. [1463]

A year after the disaster, in February 1922, the charges against the engineer and fireman were dismissed on the motion of the prosecutor. [1464]

The Indianapolis Columbus and Southern Traction interurban train known as the *Dixie Flyer*, en route from Indianapolis, Indiana, to Louisville, Kentucky, and consisting of a

---

[1462] Interstate Commerce Commission report.

[1463] *La Crosse Sentinel*, La Crosse, Indiana, United States, 10th March 1921

[1464] *The Cynthiana Argus*, Cynthiana, Indiana, United States, 24th February 1922 motor car and trailer, was in a collision with a Pennsylvania Railroad freight train on the crossing of the two lines at Edinburg, Indiana, United States, on the evening of Monday 30th May 1921. Both the motor car and trailer were filled with passengers who had been to the Speedway in Indianapolis and 57 of them were injured, two it was feared at the time fatally. The motor car was overturned and thrown against a building but the coupling between it and the trailer was broken, and the trailer did not overturn. The train was a special run in connection with the Speedway event which had left Indianapolis at 9 p.m. It was in charge of Conductor Lawrence Bennett with Motorman Will Snyder.

According to a statement made by the motorman after the accident he had stopped short of the crossing and the conductor had got down and walked forward to the crossing. The conductor had signalled for him to come on and he had set off over the crossing when he had seen the Pennsylvania freight train close at hand. He had applied his brakes, but they had not seemed to work, and he had dived back into the smoking compartment. He had not been hurt and had climbed out through a window. [1465]

Trains of the Illinois Central and Toledo St Louis and Western (Clover Leaf) Railroads were involved in a side collision at Lerna, Illinois, United States, on Saturday 13th August 1921. The crossing was protected by stop boards and state law required all trains to come to a stand before entering the crossing. Buildings and trees limited visibility at the crossing on each line.

Northbound Illinois Central Railroad passenger train No. 222 consisted of engine No. 2033, a combination mail and express car, a baggage car and two coaches in charge of Conductor Trott and Engineer Quiett. It stopped short of the crossing, then proceeded towards the depot immediately north of the crossing at a speed estimated as 5 and 14 miles an hour and had almost cleared the crossing when it was run into by a Clover Leaf freight train.

Eastbound Clover Leaf freight train second No. 46 consisted of engine No. 194, 32 cars and a caboose, in charge of Conductor Callahan and Engineer Smith. It had passed Neoga, 11 miles from Lerna, 2¼ hours late, failed to stop at the stop board at Lerna and hit the passenger train at a speed estimated at between 1½ and 15 miles an hour. Train No. 222 was broken in two between the second and third cars and the engine, tender, and leading car of train second No. 16 were derailed. Off-duty Conductor J.H. Gladhart was killed [1466] and four passengers and two mail clerks were injured.

Engineer Quiett told the Interstate Commerce Commission inquiry that having stopped at the stop board he had checked with his fireman that no train was coming from the fireman's side, then whistled and proceeded.

Fireman Scherrer said that he had seen the freight train approaching when his own train was within a few feet of the crossing, too late for it to pull up. He could not tell how fast the freight train was moving but had assumed that it would stop short of the crossing. The crew of train No. 222 estimated its speed at about 5 miles an hour as it was negotiating the crossing. Conductor Trott put the speed of the freight train at 15 miles an hour immediately before the collision.

[1465] *The Seymour Daily Tribune*, Seymour, Indiana, United States, 31st May 1921

[1466] *Quincy Whig-Journal*, Quincy, Illinois, United States, 14th August 1921

Engineer Smith of train second No. 46 stated that he had shut off steam 75 or 80 car lengths from the crossing and drifted towards it, only making a brake application when 35 or 40 feet from the crossing, when he had also reversed his engine upon seeing Train No. 222. The brakes seemed to take hold, but the speed of his train did not decrease, and he had collided with the other train at a speed he put at 1½ or 2 miles an hour. He admitted he had not attempted to stop before the crossing. He said that the brakes had been tested before leaving Madison, but he had not enquired how many were operative.

Fireman Bennett said that his engineer had closed the throttle 300 or 400 feet from the crossing, when they were travelling at 10 or 15 miles an hour. His engineer had then made a service brake application, closely followed by an emergency one.

Head Brakeman Knauss stated that the engineer had shut off steam half a mile from the crossing and 400 yards from it had placed the brake valve in the lap position but had not applied the brakes when he got off the engine immediately before the collision, when the speed had been reduced to three miles an hour.

Conductor Callahan put the speed of the train at 10 miles an hour when he felt the emergency brake application.

The ICC commented that a test of the brakes of train No. 46 after the accident had shown that only 77% of them were operative, below the minimum legal proportion of 85%.

The ICC concluded that the accident had resulted from a failure to stop train second No. 46 short of the crossing, for which Engineer Smith was responsible. Although there was a considerable conflict between members of the crew of the freight train as to how it was operated approaching the crossing all agreed that it had not been brought to a stop, Engineer Smith himself stating that he had not done so because the crossing appeared to be clear.

Had there been an interlocking plant at the crossing the accident would not have occurred, the report concluded. [1467]

Southern Pacific and International-Great Northern Railroad trains disputed a crossing at Houston, Texas, United States, on Wednesday 13th September 1922. The crossing was protected only by a caution board on the Southern Pacific line and a stop board on the International and Great Northern. State law and the operating rules of the two roads required all trains to stop before crossing. Buildings blocked the view of the International and Great Northern line from westbound Southern Pacific trains until they were close to the crossing.

Eastbound Southern Pacific passenger train No. 678 consisted of engine No. 678, a caboose and three coaches, all of wooden construction, in charge of Conductor Patrick and Engineer Harveson. It stopped for the crossing, gave two blasts on the whistle, and set off at a speed estimated at between 10 and 18 miles an hour over the crossing. There the front of the second coach was struck by the International and Great Northern train.

The International and Great Northern train was a northbound freight transfer working consisting of engine No. 147 running tender first and 44 cars, in charge of Engine

[1467] Interstate Commerce Commission report.

Foreman Horrocks with Engineer Trelford. It left the yard south of the crossing at 4.30 p.m. with only the engine brakes operative, the air brakes to the cars not having been coupled up. A stop was made for water, it reduced speed but did not stop at the stop board, gave two whistles and travelling at between 2 and 5 miles an hour collided with the passenger train. All but the engine of the passenger train were derailed and the rear of the second coach was practically demolished. Twenty-seven passengers and one employee were hurt.

Engineer Harveson told the Interstate Commerce Commission inquiry that when he had first seen the transfer working he had assumed that it would stop. Seeing two railwaymen giving violent stop signals he had then realised there was a problem and had increased speed to get his train out of the way.

Fireman McEnroe of the Southern Pacific train also thought that the transfer train would stop. He thought the stop signals were directed at the transfer train. He did not notice the transfer train before his engineer despite it being on his side. He could not explain why.

Engine Foreman Horrocks stated that he had been riding on the rear of the tender of the transfer engine when he heard the whistle of the Southern Pacific train.

Engineer Trelford testified that he had released the air brakes at the stop board and was working steam to climb the ascending gradient when his engine foreman had signalled him to stop. He had then applied the independent engine brake, opened the sanders, and asked Fireman Deason if anything was coming from his side. The fireman had replied first, "No," and then, "Yes." The train was then going too fast to stop, and he and his fireman had jumped from the engine. Engine Foreman Horrocks and Engineer Trelford stated that it had been customary only to reduce speed at the stop board with heavy northbound trains due to the difficulty of ascending the gradient. Terminal Superintendent Brown and Engine Foreman Horrocks stated that since 1st July it had been the practice not to couple up the air brakes for such transfer movements. The engine foreman thought that the transfer train could have stopped had the air brakes been coupled up or the train stopped at the stop board. Engineer Trelford, however, thought that even if a stop had been made the train could not have been pulled up with only the engine brakes.

Engine Foreman Horrocks and Switchman Bell said that there violent stop signals had been directed at the Southern Pacific train, not, as Fireman McEnroe thought, at the engine crew of the transfer train.

A test made after the accident with a similarly made up train showed that if it stopped at the stop board, restarted, and was then signalled to stop as on the occasion of the collision it could be pulled up with only the engine brake 79 feet south of the crossing.

The ICC concluded that the accident had resulted from the air brakes of the transfer train not having been coupled up, in violation of the safety appliance law, plus the failure of Engineer Trelford to stop at the stop board. Engineer Trelford was criticised for not having exercised all possible caution approaching the crossing and not stopping at the stop board even though he knew that the air brakes were not coupled up. The higher officials of the International and Great Northern were, the ICC declared, open to severe censure for permitting trains to run over the crossing without stopping at the stop board and for allowing trains to be worked with their air brakes out of use. [1468]

Passengers on a Chicago South Shore and South Bend Railroad train bound for Chicago had a narrow escape at Calumet Park, Illinois, United States at about 9.25 a.m. on the morning of Wednesday 27th September 1922. The rear coach of their train was just passing over the crossing there when an Indiana Harbor Belt Railroad engine struck it. The steps were torn off the car and the cowcatcher of the engine was demolished. It was assumed that the South Shore conductor, who had signalled to his motorman to proceed, had assumed that as the gates were open it was safe to cross. 'The crossing is considered by authorities as the safest point in the state,' the newspaper report on the accident commented. What exactly the protection of the crossing consisted of was not stated. [1469]

Engineer Matthew Dolthausen and Fireman Otto Strutz of the Minneapolis St Paul and Sault Ste Marie (Soo Line) Railroad were injured badly enough to be taken to hospital when their freight train (No. 27) was run into by a Chicago and North Western Railway train at the crossing of the two lines at Medina Junction, Wisconsin, United States, on the afternoon of Thursday 5th October 1922. A third casualty was a contractor's employee, William Singsime, who was riding on the tender of their engine. He was also taken to hospital. Both Mogul engines were derailed, and the tracks were torn up, blocking both routes. In charge of the Soo train was Conductor Henry Gilboy. The newspaper report on the accident implied that the crew of the North Western train had been at fault. [1470]

On Wednesday 28th March 1923 Great Northern and Illinois Central Railroad trains collided at the intersection of the two lines at Hills, Minnesota, United States. The crossing was another of those protected only by stop boards. State law and the operating rules of the two companies required all trains to stop before negotiating the crossing.

Westbound Illinois Central passenger train No. 711 consisted of engine No. 2032, a combination mail and baggage car, a coach, a dining car, and a Pullman sleeping car, all of all-steel construction, in charge of Conductor Harrington and Engineer Flickinger. It left Hills station at 10.28 a.m., 38 minutes late, and travelling at an estimated 4 miles an hour collided with the Great Northern train on the crossing.

Southbound Great Northern passenger train No. 161 consisted of engine No. 928, a combination mail and baggage car and two coaches, all of wooden construction, in charge of Conductor Butler and Engineer Church. It arrived at Hills station at 10.20 a.m., and after completion of the station work took water from a tank 1,800 feet from the crossing, then continued towards the crossing at an estimated 8 miles an hour and collided with the Illinois Central train. Both engines were derailed in the collision and the Illinois Central one was thrown down an embankment, fatally injuring Patrick Eubank, the Illinois Central fireman. [1471]

1468 Ibid.

1469 *The Times*, Hammond, Indiana, United States, 27th September 1922

1470 *Stevens Point Daily Journal*, Stevens Point, Wisconsin, United States, 7th September 1922

1471 *The Evening Republican*, Mitchell, South Dakota, United States, 29th March 1923

Engineer Flickinger told the Interstate Commerce Commission (ICC) inquiry that on the completion of the station work he had received a proceed signal from his conductor, whistled twice and proceeded. He had first seen the Great Northern train about 1,300 feet north of the crossing, drifting towards its stop board. When his own train was about 75 feet from the crossing, moving at 15 miles an hour, he had seen the Great Northern train working steam and made an emergency brake application, reducing his speed to 4 miles an hour when the trains met. He was positive the Great Northern train did not stop at all.

Conductor Harrington of train No. 711 said that after leaving the station he had been busy collecting transportation and had not seen the other train at all. He considered that the station stop at Hills complied with the rules and said that a second stop was never made.

The statements of Baggagemaster Matzen and Flagman Welker agreed with those of the engineer and conductor.

Engineer Church of the Great Northern train stated that following the station stop he had taken water, then proceeded and had stopped at a point 100 or 150 feet north of the stop board and 550 feet north of the crossing. He could see no Illinois Central train from his side of the engine and his fireman, whom he considered reliable, said that he could not see one from his side. He (the engineer) had then released the brakes, whistled, and started forward, working steam, and was moving at about 8 miles an hour at the point of collision. He thought the other train had run into his.

Fireman Johnson of the Great Northern train said that after his train had stopped he had been busy firing. He was sure that when they had restarted towards the crossing the other train had still been stopped at its station.

The testimonies of Conductor R.A. Butler and Baggagemaster R.R. Butler of train No. 161 added nothing of importance.

The ICC concluded that the collision was the fault of both train crews. Much testimony had concerned which train had reached the crossing first, but the commissioners regarded this as of little importance as they had arrived there at about the same time and if either engine crew had taken sufficient care the accident would have been avoided. According to his own testimony Engineer Flickinger had made no attempt to stop until within 75 feet of the crossing, while Engineer Church had been working steam knowing that his fireman was not keeping a lookout. With neither train having the right of way, both moving at low speed and in plain sight of one another it should have been a simple matter to avoid a collision.

The ICC did not consider that the density of traffic over the crossing warranted an interlocking plant but recommended the provision of crossing gates or of manually operated semaphore signals which would require a trainman to proceed to the crossing and place the signal at danger against the other line before crossing. [1472]

A side collision at Danville, Illinois, United States, on Monday 25th June 1923, injured nine passengers, four persons carried under contract and seven railwaymen. The crossing of the double-track lines of the Chicago and Eastern Illinois and Wabash Railroads was

[1472] Interstate Commerce Commission report.

protected by a ball target signal. This was described as consisting of a 32 feet high mast, to the top of which was attached a 14½ feet long horizontal cross arm. On each end of the cross arm was attached a sheet iron cylinder 2½ feet in diameter and 6 feet high. Two metal balls, 24 inches in diameter, were suspended from the cylinders by metal chains. One ball was painted red and the other white. Lanterns were attached to the balls to provide night indications. A target man on the ground raised and lowered the balls.

Rule No. 5 of the Chicago and Eastern Illinois stated in part:

DANVILLE Wabash Ry., Red signal, crossing clear for C. & E.I., White signal, clear for Wabash road, Red and white signal both displayed, block both roads.

Northbound Chicago and Eastern Illinois passenger train No. 4 consisted of engine No. 1011, two baggage cars and two coaches, in charge of Conductor Forshier and Engineer Fletcher. It left Danville station at 7.50 a.m. and after an emergency brake application stopped with its engine just foul of the crossing where it was struck by a Wabash train.

Westbound Wabash passenger train No. 9 consisted of engine No. 675, a mail car, an express car, three coaches, five Pullman sleeping cars and a dining car, in charge of Conductor Hurlburt and Engineer George Maloney. The train stopped 200 feet east of the crossing, proceeded upon the exhibition of a white ball at the target signal, and was moving at 15 miles an hour when it struck engine No. 2011. Both locomotives were derailed in the collision, No. 1011 to the left and No. 675 to the right. Engineer Maloney jumped before the collision but his fireman, Samuel Shinn, was caught in the gangway and badly scalded. [1473]

Engineer Fletcher alleged at the Interstate Commerce Commission inquiry that when his train had been stopped at the platform of Danville station he had seen the red target ball indicating that he had the road. Leaving the station he saw the target board at the yard office north of the crossing indicating that there were orders for his train. Instead of whistling for the crossing he had sounded the signal for a 19 order (one delivered to the train by a hoop as it was passing) and told his fireman that the target was clear. He had then seen the white target signal exhibited for a Wabash train, seen the engine of that train approaching working steam and applied his brakes in emergency, but could not stop his engine before it fouled the crossing.

Fireman Baldwin agreed that the white target signal had been exhibited when they were at the station and put the speed of his train at 10 miles an hour approaching the crossing.

Targetman Keefe said that the target had been at stop for both trains when he had come on duty. The only movement between then and the accident had been a Chicago and Eastern Illinois light engine at 7.30 a.m., after which both targets had shown stop until he had exhibited the white ball for the Wabash train. He had seen the Chicago and Eastern Illinois train approaching at low speed and had assumed that it would stop short of the crossing.

Fireman Hall, Street Crossing Flagman McFerrin and Switchmen Hannapel and Thiede, all Chicago and Eastern Illinois employees, were unanimous that the white ball had been shown for the Wabash train.

[1473] *Brazil Daily Times*, Brazil, Indiana, United States, 26th June 1923

Tests made after the accident showed that the crossing target could not be seen from the point where Engineer Fletcher (and his fireman) claimed to have seen it at Danville station.

The ICC concluded that the failure of Engineer Fletcher to observe and obey the target signal had caused the collision. He could not have seen the target at the station as he had claimed and there was ample evidence that the white ball had been displayed for the Wabash train. He had clearly been distracted by the train order signal and failed to see the target until too late. His allegation that the air brakes on the engine had not been in first-class condition had no significance in view of the low speed of the train. Had he properly observed the target the brakes on the four train cars should have been enough to stop the train. [1474]

Two New York Chicago and St Louis (Nickel Plate) Railroad trains collided at Continental, Ohio, United States, at 6 a.m. on Thursday 10th July 1924. (Newspaper reports refer to one train as belonging to the Toledo St Louis and Western (Clover Leaf) company, but the Nickel Plate had absorbed the Clover Leaf the previous year, it becoming the Clover Leaf division of the larger company.) The results of the collision were quite serious, but the Interstate Commerce Commission did not investigate it.

An eastbound meat train on the Fort Wayne division of the company reportedly had priority over other freight trains but collided on the crossing with train No. 41 of the Toledo division of the Clover Leaf district (that is, ex Clover Leaf). This train had left Toledo at 3.10 a.m. for Delphos. It was in charge of Conductor R.C. Boham with Engineer A.C. Hutchinson (name unclear in newspaper), Fireman Carl Bumgardner and Brakemen Soffel and Becker. The meat train was in charge of Conductor Compton with Engineer Louis Kores or Koers. Eleven cars of the Toledo division train, some steel gondolas and others steel and wooden box cars were almost totally destroyed, and the engine and four cars of the meat train jumped the rails. Engineer Kores was scalded about his face and body. What was referred to as the shanty of Operator L.E. Wagner was demolished in the accident. He was trapped in the wreckage for a while and at first it was reported that his life was in the balance, but he was back at work the following day. The damage was estimated at $150,000. [1475]

A newspaper reported that 'a freak of fate alone had avoided a slaughter' in a collision between a Baltimore and Ohio Chicago Terminal train (exercising trackage rights on Chicago Rock Island and Pacific Railroad tracks) and a Chicago and Western Illinois Railroad train at Oakdale, Illinois, United States on 7th January 1925. The Baltimore and Ohio train reportedly had a green signal when its engineer, George Campbell, saw the Western Illinois train speeding towards the crossing. He had opened the throttle in an attempt to beat the other train over the crossing, but the Western Illinois train struck the tender of his engine and threw the mail and baggage cars into the 'flimsy suburban depot.' Operator Leo McCann and Mail Clerks L.E. Ruppert and Stanley Day were buried in the wreckage. Operator McCann continued to work his telegraph key and ordered a wrecking crew. Engineer James Mayes and Fireman (?) Albert Grabe were

[1474] Interstate Commerce Commission report.
[1475] *The Lima News*, Lima, Ohio, United States, 10th and 11th July 1924

'being held for questioning.'

It was initially reported that 100 people had been killed (!) and Engineer Campbell reportedly commented that had the Western Indiana train hit his a second later 75 passengers in the diner would have been killed or maimed. [1476]

On an unspecified date in October or November 1925 a side collision occurred on a rail/rail crossing in England. It was between 2 and 3 a.m. on a very frosty morning when the accident happened at Murrow at the crossing of the London and North Eastern Railway (LNE) (ex-Great Eastern & Great Northern Joint) and London Midland and Scottish & London and North Eastern Joint Railway (LMS & LNE Jt) (ex-Midland & Great Northern Joint) lines. The relief signalman at Murrow Crossing box had accepted an up goods train from Turf Fen cabin on the up LMS & LNE Jt line and another up goods train from French Drove cabin on the LNE line, an operation which was in order as he had an outer home signal on the latter line. He had cleared his signals for the former train, intending to hold the latter at his up main outer home signal. What he had failed to notice was that due to the intense cold and snow on the signal wires and bell cranks the up LNE outer and inner home signals had both frozen off. The driver of the LNE train, thinking that he had clear signals, had not stopped, and half of his train was over the crossing when the LMS & LNE Jt train, running under clear signals, cut through it. Many waggons and the engine of the colliding train were derailed but no one was hurt. Breakdown cranes from Peterborough and March cleared up the mess. Local poultry owners were well pleased when they discovered that one waggon loaded with sacks of barley had been thrown into the adjoining fen, scattering barley from burst bags. [1477]

An Erie Railroad passenger train from Hornell and a New York Chicago and St Louis (Nickel Plate) Railroad freight train collided on GB Crossing, Blasdell, New York, United States, just south of Buffalo, on Sunday 4th July 1926. The engine of the passenger train and several cars of the freight train were derailed but none of the passengers (or presumably the train crews) were seriously enough hurt to require medical attention. The Nickel Plate engineer said that he had been unable to stop when he had found the crossing signal against him. [1478]

Beverley Hills, Illinois, United States, on the southern outskirts of Chicago, was the scene of a serious collision involving Pennsylvania Railroad and Chicago Rock Island and Pacific Railroad trains late on the evening of Sunday 26th August 1926. Protection at this crossing was provided by fixed signals worked from a tower with a 12-level Saxby and Farmer mechanical lever frame and there were also derails with detector bars such that when the wheel of a train was on the detector bar the respective home signal protecting the crossing was locked at danger.

Northbound Rock Island passenger train No. 182 consisted of engine No. 1337 and

---

[1476] *The Brazil Daily Times*, Brazil, Indiana, United States, 7th January 1925

[1477] A Short History of the Midland and Great Northern Joint Railway by Ronald H. Clark, pg. 158. Midland and Great Northern Circle Bulletin No. 723

[1478] *San Antonio Express*, San Antonio, Texas, United States, 5th July 1926 three all-steel suburban coaches, in charge of Conductor Postwailer with Engineer Stewart Kauffman and Fireman Joseph Risberk. [1479] It was stopping at 91st Street station when it was run into by the Pennsylvania train.

Eastbound Pennsylvania freight train extra No. 7082 consisted of engine No. 7082 and four cars, in charge of Conductor Gaumer with Engineer Schreckenghaust. It was en route from the 71st Street Yard with four cars to deliver to a train which had stalled due to a disabled engine, and to change engines. It stopped at a block office near the crossing at 10.52 or 10.53 p.m., was given orders, passed the home signal at danger, ran over the derail, which failed to work, entered the crossing, and was struck a glancing blow by the Rock Island train. Both engines were derailed but the Rock Island one suffered worst and Engineer Kauffman and Fireman Risbeck were both killed. In addition 18 passengers and 4 railroad employees were hurt.

Engineer Schreckenghaust and Fireman Kline told the Interstate Commerce Commission (ICC) inquiry that the home signal at the crossing had been clear when they had stopped at the block office for a train order and clearance card, and when they had restarted a minute or so later. They had then been discussing with the conductor the work to be done and they were 20 or 25 feet from the crossing, moving at 10 or 12 miles an hour, when the engineer had noticed the Rock Island train and applied the brakes in emergency. Engineer Schreckenghaust admitted that he had not observed the home signal as he was passing it, though it was not obscured by smoke, fog, or steam.

The statements of Conductor Gaumer, Flagman Murphy (who was riding on the top of the rear car as there was no caboose) and Head Brakeman Matthews brought out no new facts.

Conductor Postwailer of the Rock Island train stated that Engineer Kauffman had told him before he died that the signals had been clear for his train and Brakeman Ditmars of that train said that the route had been properly lined for them.

Towerman Brown stated that he had cleared the route for the Pennsylvania train initially but when it had stopped at the block office he had replaced the home signal, while he thought the train was still stationary, and given priority to the Rock Island train. He could not understand why the derail had not done its job.

Superintendent of Signals and Telegraph Lantz stated that a post-accident inspection had showed damage to the derail indicating that it had been in the normal (derailing) position when the train had run over it.

The ICC concluded that the accident had been caused by Engineer Schreckenghaust overrunning signals. Had he kept a proper lookout the accident would probably have been avoided.

The rules of the two companies set out that once a route had been cleared for a train it must not be altered for another train until the first train had stopped. Towerman Brown maintained that the Pennsylvania train was stopped when he changed the route. Whether or not this was true the fact remained that he had changed the route practically in front of the train. Had he not done so there would have been no collision.

Inspection showed that the derail had been in the correct position, but it had failed to

---

[1479] *The Kokomo Daily Tribune*, Indiana, United States, 30th August 1926 perform its function, probably because of the combined weight of the engine and tender (591,900 pounds). This implied that the type of derail in use was not adequate for the purpose. Had it worked properly there would have been no accident. [1480]

At Bethlehem, Pennsylvania, United States, the quadruple-track main line of the Lehigh Valley Railroad, running along the south bank of the Lehigh River, was crossed by a single-track branch of the Central of New Jersey Railroad. The crossing was protected by signals controlled by Bethlehem tower and, on the Central of New Jersey (CNJ) line, from JU tower. Trains coming from the CNJ line first encountered signal No. 47, close to JU tower, as they entered the single-line connection, and then signal No. 16, controlled from Bethlehem tower, protecting the crossing itself. Despite the signal protection a serious collision occurred on the crossing on Monday 27th September 1926.

On the day of the accident eastbound Lehigh Valley train No. 6 consisted of engine No. 2092, two coaches, a club car and eight Pullman sleeping cars, in charge of Conductor Kromer with Engineer Donlin and Fireman Markan. It left Allentown at 5.40 a.m. and was approaching Bethlehem station under clear signals at 15 miles an hour on track No. 2 when it was struck by CNJ train No. 306.

Eastbound CNJ train No. 306 consisted of engine No. 825, two baggage cars, a mail car, a coach, four Pullman sleeping cars and a combination car, in charge of Conductor Terry with Engineer Henry Schmidt [1481] and Fireman Engler. This train had left Allentown at 5.38 a.m., passed JU tower at 5.47 a.m. and travelling at 12 or 15 miles an hour struck the Lehigh Valley train near the rear end of its first car. The second car of the Lehigh Valley train and engine No. 825 overturned. Eight passengers, all in the second car of the Lehigh Valley train, died and 38 passengers and two railwaymen were injured.

Towerman Reilly told the Interstate Commerce Commission inquiry that he had cleared his signals for train No. 6 when it had reached the annunciator circuit. Signal No. 16 was at danger and train No. 306 was not in sight. The two trains had then come into view and as train No. 306 was approaching signal No. 16 he had noted that the signal was still at danger and that the CNJ train did not look as if it were stopping. He had so remarked to Car Inspector Confer, who had been in the tower. Signal No. 16 had last been cleared for a paper train about 3.50 a.m.

Car Inspector Confer corroborated the statement of Towerman Reilly and added that the towerman had not operated any levers prior to clearing the signal for train No. 6.

Engineer Donlin of train No. 6 testified that signals 892, 1 and 2 had been clear for his train. He had known nothing of the approach of train No. 306 on the fireman's side before the collision.

Fireman Markan said that he had had to go back into the tender as his engine approached the crossing under clear signals. He had seen the CNJ train but had not thought that there would be a collision until the last moment.

Engineer Schmidt stated that the signal at JU tower (No. 47) had been clear for him. He had entered the connection over the bridge across the Lehigh River at a speed of 15 or

[1480] Interstate Commerce Commission report.

[1481] *New Castle News*, New Castle, Pennsylvania, United States, 28th September 1926

18 miles an hour prepared to stop at signal No. 16 but saw it exhibiting a green or proceed indication with the semaphore arm down. He had looked at the water glass and at his fireman, then at the signal again, which was still clear, as was the signal for his train on the far side of the crossing. As his engine reached signal No. 16 he saw that it was in the stop position, which he had assumed was because his engine was on the track circuit. Then he had seen train No. 6 and realising that it was on a conflicting route had applied the brakes in emergency. His fireman had not called the signal and he was positive that he had not misread it.

Fireman Engler stated that he had been cleaning the deck of his engine and putting on some clothing as is train had approached the crossing and he had seen neither the indication of signal No. 16 nor the Lehigh Valley train.

Conductor Terry said that Engineer Schmidt had told him after the accident that signal No. 16 had been clear.

Lehigh Valley train No. 4 was standing at Bethlehem at the time of the collision. Conductor Welch, Baggagemaster Main, Brakeman Folleran and Flagman Kane all spoke of having seen the signals clear for train No. 6, as did Assistant Baggage Agent Osman and Janitor Cunningham, while the latter had also seen signal No. 16 in the stop position.

The ICC concluded that the collision had been caused by the failure of Engineer Schmidt to observe and obey signals. Although he insisted that signal No. 16 had been clear for him no one else supported his assertion, while Towerman Reilly and Janitor Cunningham had seen the signal in the stop position and ten witnesses had seen the signals clear for the Lehigh Valley train. Post-accident tests confirmed that it was impossible for both signals to have been clear at the same time.

It noted that a derail at signal No. 16 had been removed in October 1922 because it was considered that it was a menace rather than a safeguard (presumably because of the risk that a train derailed at that point would end up in the Lehigh River). It did not seem advisable to reinstate the derail but on the other hand signal No. 16 was not well located. The commission recommended that all trains should come to a full stop before using the crossing, or alternatively that the layout should be rearranged so that a train using the crossing would not get a clear signal at JU tower until the entire route over the crossing was set up. [1482]

Four New York Central Railroad carpenters died in a crossing collision at Gibson, Indiana, United States, at noon on Thursday 7th October 1926 when the speeder (gasoline motor car) on which they were riding was struck by a Michigan Central Railroad train at the crossing of the two roads. As at Beverley Hills a tower was located at the crossing, with fixed signals and derails.

Westbound Michigan Central passenger train No. 41 consisted of engine No. 8484 and seven coaches. The signals for it at the crossing had been clear and it had been running at 40 miles an hour when it struck the speeder.

Six carpenters had been laying a plank walk at Indiana Harbor, 4 miles north of Gibson, and at 11.10 a.m. they had set off for the latter place for dinner on motor car No.

[1482] Interstate Commerce Commission report

651. Finding their route blocked by a freight train on the southbound track they had simply lifted the car onto the northbound track and continued against the current of traffic. At Gibson they had found the derail set against them, so they had likewise lifted the car over that and continued onto the crossing, ignoring the signals set against them, and the car, moving at about 12 miles an hour, was struck by the Michigan Central train. Four of the men, Merrill Knoll, Walter Legge, Amil Orr and Maurice Reed were killed outright and a fifth, Arthur Breu, was seriously injured. [1483]

Merrill Turner, the sixth carpenter, described the journey from Indiana Harbor at the Interstate Commerce Commission (ICC) investigation and explained that looking eastwards he had glimpsed the Michigan Central train and jumped. The other men, including the operator of the car, had been looking straight ahead because they were still on the northbound track, though the operator of the car would have had a better view of the other train had he looked. The car's brakes had been in order.

Carpenter Foreman W.L. Turner, who was at Gibson south of the crossing, stated that he had assigned Walter Legge to drive the car and had warned him to take no chances with trains.

Engineer Allen of train No. 41 said that he had known nothing of the car until it had suddenly appeared from behind the tower, the collision following almost immediately.

Fireman Lute added that the engine whistle had been sounding for the Kennedy Avenue road crossing, 500 feet further on, and the engine bell had been ringing.

Towerman Lambert at Gibson said that when he had first seen the speeder 200 feet away he had thrown up his hands to stop it, knowing the passenger train was close at hand. The carpenters could have seen the train for half a mile. The speeder made no attempt to stop in response to his hand signals, which he supposed were mistaken either for a friendly wave or a proceed signal. He said that motor cars frequently used the crossing, carrying signalmen, carpenters, water service men, telephone men and Western Union employees. Apart from the latter they never waited more than a short time at the derail if it was set against them, then lifted their cars over it and proceeding without waiting for a signal from the tower, while looking out for their own safety.

The ICC concluded that the accident had been due to the motor car crossing in front of the passenger train, and it would not have happened had the carpenters kept a proper lookout. [1484]

A crossing near Painesville, Ohio, United States, was the scene of a crossing collision on Tuesday 30th August 1927. The single tracks of the New York Chicago and St Louis (Nickel Plate) and Baltimore and Ohio Railroads crossed one another here at right-angles. Protection at the crossing was rudimentary, being provided by a board signal on a mast controlled by the operator in the adjoining Nickel Plate office. A horizontal board permitted Baltimore and Ohio trains to proceed, a vertical board gave the road to Nickel Plate trains and a diagonal board (its theoretical normal position) ordered all trains to stop. Red lights gave night-time indications. There were also stop boards on each line.

[1483] *The Lake County Times*, Hammond, Indiana, United States, 7th October 1926.
[1484] Interstate Commerce Commission report.

Westbound Baltimore and Ohio passenger train No. 49 consisted of gas-electric motor car No. 6008 and a trailer, in charge of Conductor Bradley and Pilot Wesley Elwell, with Student Engineer Louis Ludt. It stopped briefly short of the crossing at 12.53 p.m., restarted and stopped again with its front end foul of the crossing, where it was struck by the Nickel Plate train a minute later.

Eastbound Nickel Plate stock train second No. 58 consisted of engine No. 669, 67 cars and a caboose in charge of Conductor Terwilliger and Engineer Koelliker. It came to a stand some distance short of the crossing, restarted and the board being vertical continued to the crossing at 6 to 10 miles an hour and collided with the rail motor car. The motor car was overturned by the impact, trapping Elwell and Ludt beneath it. Elwell was killed outright and Ludt died in hospital of his injuries. Three passengers, one of them Ludt's wife, and one railwayman were injured. [1485]

The first witness at the Interstate Commerce Commission inquiry was Inspector Arthur Lee of the Brill Motor Car Company, the makers of the gasoline motor car. He said that after stopping the train very briefly Ludt had apparently looked at the crossing signal and he had then opened the throttle. The car was going towards the crossing at 4 or 5 miles an hour when Elwell had alerted him (Lee) to the Nickel Plate train, shouted to Ludt to stop and run into the engine room. Either Ludt or Ewell had applied the brakes in emergency. Lee said he saw the signal vertical 15 minutes after the accident.

Conductor Bradley said he saw the Nickel Plate train approaching as his car neared the crossing and pulled the whistle cord, he being in the centre of the vehicle away from the conductor's valve. He saw Ewell rush to the engine room but could not say if he had previously applied the brake. He saw the crossing board vertical seven or eight minutes after the accident. He added that Ludt was making his last run as a student engineer.

Brakeman Eddy and Baggagemaster Broughton could not add much of importance except that Broughton saw the board signal vertical at about the time of the collision.

Engineer Koelliker said that train No. 58 had stopped further back from the crossing than he had intended due to a faulty triple valve. Seeing the board signal vertical he had then whistled and proceeded. When he saw the Baltimore and Ohio train first it appeared to be drifting and he had assumed that it would stop close to the crossing. When it did not he had applied his brakes in emergency.

Fireman Hinkle more or less corroborated the statement of his engineer. Conductor Terwilliger and Brakemen Holland and Clements could add little more except that Clements saw the board signal vertical just before the collision.

Operator Hulse said that the board signal had been vertical when the Nickel Plate train had first stopped, before the Baltimore and Ohio train had appeared. Train No. 49 then appeared and whistled for the crossing, but he had decided that it was too late to give it preference as the Nickel Plate train had by then restarted. He saw train No. 49 stop, restart and drift towards the crossing and come to a stand foul of it. He added that as there were considerably more Nickel Plate than Baltimore and Ohio trains and due to the amount of office work the board signal was usually left in the vertical position.

The ICC found that the accident was primarily due to the failure of Student Engineer Ludt to obey the indication of the board signal and the contributory cause was the failure

---

[1485] *The Commercial Tribune*, Cincinnati, Ohio, United States, 31st August 1927 of Pilot Elwell to properly supervise him. It was impossible to determine why Ludt had failed to obey the signal. Pilot Elwell had been in the smoking compartment instead of the engine room where he could have supervised the student engineer. Ludt had over 29 years' experience with engine and was on his final run learning gasoline motor cars. As he was not yet properly qualified, however, it was incumbent on Pilot Elwell to be in a position to take immediate charge of the train, especially at danger points such as junctions and crossings.

The Nickel Plate train should not have stopped as far away as it apparently did from the crossing. [1486]

On Sunday 18th December 1927 there was a minor side collision at Cory Tower, Danville, Illinois, United States, where the lines of the Chicago and Eastern Illinois and Cleveland Cincinnati Chicago and St Louis (Big Four) Railroads intersected. A special passenger train on the Chicago and Eastern Illinois running from New Orleans to Chicago was run into by a Big Four freight train. The locomotive of the freight train and two passengers coaches were derailed. 'The lives of eighty passengers were jeopardized,' a newspaper wrote, but in fact only one man was badly enough hurt to require medical attention. A frozen air line on the freight train was blamed for the collision. [1487]

[1486] Interstate Commerce Commission report

[1487] *The Decatur Review*, Decatur, Illinois, United States, 19th December 1927

CHAPTER 14

FIRES

Fires often broke out on trains following derailments or collisions. While wooden coaches were still in use in passenger trains an overturned stove used for heating or spilt engine coals from a steam locomotive might set them ablaze. Later the substitution of gas for oil for train lighting proved to be another hazard, and a number of accidents were made much more severe by fires started by leaking or damaged gas cylinders until its dangers led to its replacement by electricity. This chapter will confine itself, however, to what might be termed primary fires, that is accidents where there was no precipitating derailment or collision but merely a fire. Serious primary fires were considerably rarer than what might be termed secondary fires, but they did occur from time to time from a variety of causes and more minor fires were fairly common.

On the evening of Saturday 10th March 1917 a vehicle loaded with merchandise and benzine (a mixture of hydrocarbons used for cleaning and dyeing, not to be confused with benzene) attached to the Burke mail train caught fire between Coolabah and Byrock, New South Wales, Australia. The train was brought to a stand between the two stations and adjoining a water tank. When water was played on the burning truck there was an explosion, but the fire was eventually extinguished after the whole of its load had been consumed. It was Sunday morning before the train reached its destination, six hours late. [1488]

A down goods train from London (exact origin not stated) on the London and North Western Railway was found to be on fire at Wilmslow, England, south of Manchester on about Monday 27th July 1917. A large stock of chemical goods was destroyed, one wagon being burnt out and many damaged. [1489]

A munitions train caught fire on the London Brighton and South Coast Railway near

[1488] *Wagga Wagga Express*, Wagga Wagga, New South Wales, Australia, 15th March 1917
[1489] *The Times*, London, England, 28th July 1917

Earlswood, England, on Saturday 22nd September 1917. Two waggons loaded with fuses and other explosives were detached from the train. One was burnt out; the other was moved alongside a brook and the Redhill Fire Brigade were able to put out the blaze. [1490]

A dispatch from Wien, Austria, to the *Nieuwe Rotterdamsche Courant* newspaper in the neutral Netherlands, spoke of a fire on a passenger train near Sambor, Galicia, on or about Monday 31st December 1917. The fire prevented, it was said, the emergency brakes from operating so that the train could presumably not be quickly stopped. An unspecified number of passengers lost their lives. [1491]

Fire broke out on a Berlin express between Lohof and Schleißheim, Germany, on Friday 25th January 1918. A member of an airship crew had brought onto the train as luggage a container filled with petrol. The container turned out to be faulty, the petrol leaked, it was set alight, perhaps by a cigarette, and suddenly the whole of the interior of the carriage was ablaze. It would seem that a gas cylinder then exploded, and before the fire brigade and ambulances had reached the scene of the accident the fire had spread to a second carriage. Eight people died, among them women and soldiers, and 40 were injured, some seriously. [1492]

Five train crew members on the Houston and Texas Central Railway were badly burnt in a fire near Sutton, Texas, United States, on Thursday 24th January 1918 when fire broke out in a caboose hauled by engine No. 504, reportedly due to a gasoline explosion. The train ran wild to Hearne, where the fire was extinguished, and the men, who included Engineer Murphy, were taken to Houston for treatment. [1493]

Crowded Österreichische Staatsbanen passenger train No. 317 from Lemberg (now Lvov, Ukraine) to Czernowitz (now Chernivtsi, Ukraine) caught fire at midnight on Saturday 9th February 1918 between Jezupol (now Yezupil, Ukraine) and Stanislau (now Ivano-Frankivsk, Ukraine) Galicia. The train was brought to a stand on a bridge over the River Dniester and several of its passengers jumped from their carriages and drowned in the river, while others were burnt to death, two carriages being burnt out. In all seven people died, 11 were seriously and eight slightly injured[1494] [1495]

Preußisch Staatsbahnen passenger train No. 511 from Saarbrücken to Trier caught fire at 6.15 a.m. on Monday 4th November 1918 between Völklingen and Bous, Germany. The emergency alarm was pulled, and the train was brought to a stand but by that time the fire was too advanced to be brought under control. Fourteen passengers died and 13 others were

---

[1490] Ibid, 24th September 1917
[1491] Ibid, 2nd January 1918
[1492] *Historische Eisenbahn Katastrophen* by Bernhard Püschel, pg. 115 (in German)
[1493] *Bryan Daily Eagle*, Bryan, Texas, United States, 24th January 1918
[1494] *The Racine Journal-News*, Racine, Wisconsin, United States, 15th February 1918
[1495] *Eisenbahnünfalle Bind I* by Ludwig Stockert, pg. 103 (in German)

more or less seriously injured. (Another source gives 18 fatalities.) The fire was caused by an accident with a container of petrol being illegally carried on the train. [1496]

On Tuesday 17th December 1918, the Vice President of the Canadian Pacific Railway issued a statement regarding a fire which had occurred early that day on one of his company's trains:

"At 6.57 a.m. today the second of train No. 4 stopped for orders at Bonheur station {Ontario, Canada}, 120 miles west of Fort William. Conductor Begg, in charge, walking to his train, saw flames bursting from the front end of tourist car No. 628. At once it was evident that the car would be rapidly destroyed.

"The train crew and engine crew immediately burst into the coach and rescued as many of the passengers as they could reach. We regret to state that as nearly as can be ascertained at the moment that thirteen passengers are missing, although it is just possible that several of these escaped into other coaches and have not reported themselves. Eight passengers who were rescued are slightly injured.

"It would appear that the fire started in the curtains of a berth occupied by a man, his wife and baby, probably through trying to heat the food for the child with a spirit lamp, or that they had struck a light for some reason. When the flames reached the curtains of the berth, it would be impossible to control them." [1497]

A returned soldier who had been a passenger in the coach next to the tourist car told a reporter, "I saw a sudden flash through the doorway. I jumped up and ran to the door. The far end of the doomed coach was then a mass of seething flames, where people were madly fighting and trampling over one another to gain an exit, while the train crew heroically attempted to rescue the survivors. The flames swept along the corridor in a great blast, filling the aisle with suffocating smoke.

"The panic and jam in the aisle must have been the cause of many of the deaths. The coach I was in was of steel construction or it would have gone too."

Among the passengers on the train was Mr Douglas Cameron, former lieutenant-governor of Manitoba, who said, "It all happened in a flash. In fact we in the front part of the train had no knowledge of the tragedy that was taking place behind us and were wondering what was the cause of the delay at the station in the bush. There was a general rush to assist in saving life when the terrible news was learned but most of the people in the doomed coach were then beyond human aid." [1498]

A sleeping passenger, Mr Joseph Mitchell, was reportedly awakened by a Mrs Cochrane who dragged him to the doorway before being herself overcome by the fumes. Rescued then became rescuer and he carried Mrs Cochrane and her baby to safety.

The origin of the fire remained a mystery and with the tourist coach being burnt down its chassis it was thought that it would remain so. Mr Mitchell believed that the fire had begun in the women's lavatory. The final number of dead was probably 14, with 12

[1496] Ibid, pg. 104 (in German)

[1497] *The Brandon Daily Sun*, Brandon, Manitoba, Canada, 17th December 1918

[1498] *Manitoba Free Press*, Winnipeg, Manitoba, Canada, 18th December 1918 injured. [1499]

A fire broke out on a munitions train at Kazan station, Russia, in the Volga region on about 20th May 1918. Russia had made peace with the Axis powers at the treaty of Brest-Litovsk on 3rd March and a newspaper report states that the train was 'destined for the southern front' presumably meaning that the munitions were bound for Armenia or Georgia, who were still fighting the Ottoman Empire. The initial fire was followed by a series of explosions and the fire lasted all day. Seven warehouses full of merchandise and 350 waggons were consumed. Forty people were killed or injured, and the material damage amounted to £3,000,000. [1500]

The fire in Nanteuil Saacy Tunnel, France, on Wednesday 5th February 1919 has been described in the Tunnel Accidents chapter.

Fire broke out in the railcar which had left Baden for Wien, Austria, on the local electrified line between the two places, at 08:10 on Friday 11th July 1919. Twelve passengers suffered serious burns and others lost consciousness. Many saved themselves by jumping from the train. [1501]

A fire which was probably expensive in compensation for the Caledonian Railway company occurred on their main line between Stirling and Larbert, Scotland, on Saturday 16th October 1920. A bull named *Collynie Master Lavender* and a cow named *Boquhan Violet* had been purchased for the Prince of Wales at a livestock sale at Kippen, Stirlingshire, and they were being taken several hundred miles southwards to the Prince's estate at Stoke Climsland, Cornwall, England. The animals were being conveyed in a horse-box, via the North British Railway to Stirling and thence over the Caledonian.

From some undetermined cause the horse box caught fire after the train had left Stirling and the man in charge of them, who had been riding in the groom's compartment, was forced to climb out of the vehicle and make his way to the footboard of another coach. When he was rescued at Larbert he was in a state of collapse. The horse box was quickly detached there, and efforts were made to rescue the beasts, but the flames had by then completely enveloped the sides and roof of the box so that rescue was impossible, and they perished.

The animals had been the property of the late Mr Stephen Mitchell of Boquahan, Kippen. The bull was valued at 3,000 guineas and the cow at 1,500 guineas. [1502]

Benzine again featured in a train fire on Monday 8th November 1920, this time near Te Kuiti, New Zealand. A spark from the locomotive ignited one or more of 300 cases of the substance. A patent fire extinguisher proved inadequate to cope with the blaze and the

---

[1499] *The Lethbridge Daily Herald*, Lethbridge, Alberta, Canada, 18th December 1918
[1500] *The Times*, London, England, 30th May 1918
[1501] *Die Lokomotive*, No. 7, July 1919, pg. 97.
[1502] Ibid, 22nd October 1920

fire brigade had to deal with it. [1503]

The mail van of a Flushing – Berlin express, carrying British mail, caught fire on Sunday 18th December 1921 at an unstated location in the province of Noord Brabant in the Netherlands. Of 150 mail bags only 70 could be saved and some of those were damaged. [1504]

What a newspaper report described as 'an exciting incident,' though that was probably not how the passengers concerned saw it, occurred between Marhoura and Deliniquin, New South Wales, Australia, on the evening of Tuesday 23rd May 1922. A spark from the wood-burning engine alighted on the roof of a first-class carriage and fanned by the breeze set the roof on fire. The passengers in the leading compartment of the carriage noticed that its wall was becoming uncomfortably hot, and when smoke began to enter the compartment through cracks and joints in the woodwork they naturally became alarmed. The carriage doors were locked, and it appears that even at this late date at least some of the passenger trains in New South Wales were not provided with communication cords so the passengers were unable to alert the guard or driver in this way. Fortunately one of them possessed a whistle and shrill blasts on this, accompanied by the waving of newspapers and coats out of the windows, alerted the guard and the train was stopped. A few bucketfuls of water extinguished the fire before too much damage was done but as the newspaper noted had the blaze obtained a firmer hold the results might have been serious. [1505]

The passengers on the 11.30 a.m. London and North Eastern Railway train from Doncaster to Goole, England, on Tuesday 27th February 1923, were better placed than their Australian counterparts the year before when those in the middle, first-class coach, noticed smoke entering their compartments from the corridor. They had a communication cord to pull. The train was stopped, one end of the carriage was found to be burning fiercely, and the passengers hastily abandoned it. Fire extinguishers could not extinguish the blaze and the coach was later taken three miles to sidings at Goole where the station staff put the fire out. Two compartments and a lavatory of the coach were burnt out and other carriages were damaged. [1506]

The rear van of a London and North Eastern Railway express from London Kings Cross to Edinburgh Waverley which had left Carlisle at 4.30 a.m. on Wednesday 2nd May 1923 was found to be on fire as the train was passing through Scotch Dyke station, England, about 12 miles from Carlisle. The van was destroyed together with its contents, which comprised letters, parcels and newspapers posted in the Midlands and Newcastle for delivery in the Border area. The cause of the fire was unknown. [1507]

1503 *The Daily Telegraph*, Sydney, New South Wales, Australia, 9th November 1920
1504 *The Times*, London, England, 20th December 1921
1505 *The Argus*, Melbourne, Victoria, Australia, 26th May 1922
1506 *The Times*, London, England, 28th February 1923
1507 Ibid, 3rd May 1923

A party of Shriners were on their way from Muscatine to a Kaaba temple ceremonial at Davenport on the afternoon of Tuesday 18th March 1924, taking the 2 p.m. interurban train of the United Power and Light Company from Muscatine. Their journey proved more exciting than expected when just on the Muscatine side of the station at Drumm, Iowa, United States, the floor of the car caught fire, accompanied by a loud report. The Shriners and other passengers rushed for the door leading to the baggage room but in their excitement were unable to open it except by breaking the glass. The fire was soon extinguished using a fire extinguisher carried on the train and it then proceeded to Blue Grass under its own power. It was then found that its brakes were not working, and another car had to be sent out from Davenport to pull it there, though it arrived only 29 minutes late. The fire was blamed on a short circuit. One passenger was slightly hurt. [1508]

A goods waggon carrying oil was burnt out in Grünhainichen station, Saxony, Germany, on Thursday 8th May 1924, and one person was injured. [1509]

A further fire occurred in Saxony, Germany, on Tuesday 12th August 1924, near Glaubitz, when a waggon of train G 9009 carrying cellulose waste was involved. The fire spread to four further waggons. A conductor suffered severe burns which proved fatal. [1510]

News of major catastrophe in the U.S.S.R. on about Wednesday 15th October 1924 was allegedly hushed up by the Soviet authorities but a few details of it leaked out to the West via a despatch from Warsaw to Vienna, and it was reported upon in several American newspapers and in the *Railway Gazette* magazine. A passenger train from Ivanovo to Moscow was effectively a mixed train as it included three tank waggons loaded with petroleum. In some manner these caught fire near Moscow, there was an explosion and the whole train was set ablaze. Of the 200 people on the train only 27 reportedly survived. The train passengers included 30 children. [1511] [1512]

Letters, parcels, and newspapers were again the items damaged or destroyed when a van on the 11 p.m. London Midland and Scottish Railway down train from London Euston to Scotland caught fire early on Thursday 11th February 1926. The train was stopped at Newbold signal box, England, probably because the signalman had received the 'Stop and Examine' signal from the previous box. The guard found the front part of the vehicle ablaze. He entered it and was able to save most of the mail bags by throwing them out onto the line. Assistance was summoned from Rugby and an attempt was made to extinguish the fire using water from the locomotive, but the blaze had obtained too strong a hold and after two and a half hours only the wheels and steel framework of the

---

1508 *Muscatine Journal and News Tribune*, Muscatine, Iowa, United States, 19th March 1924
1509 *Eisenbahnunfälle in Sachsen* by Claus Schlegel, pg. 133 (in German)
1510 Ibid.
1511 *The Railway Gazette*, Vol 41, pg. 495
1512 *Rock Valley Bee*, Rock Valley, Iowa, 17th October 1924

vehicle remained. The cause of the fire was being investigated. [1513]

Five passengers died in a train fire on the eastbound Continental Limited of the Union Pacific Railroad from Los Angeles, California, to Salt Lake City, Utah, near Crystal Siding, Nevada, United States, on Wednesday 16th June 1926. It might be argued that as in the Dingle train fire in England in December 1901 at least one and possibly all would have survived the fire but for their own foolishness, though that is doubtless a harsh judgement.

According to the testimony at the inquest upon the victims given by Conductor L.R. Atkinson of the train a small fire was discovered in the roof of a tourist sleeping car at a point 4 miles short of Crystal Siding. The train was stopped, the passengers were ordered out of the burning car, the blaze was attacked with a fire extinguisher and apparently brought under control. He said that he had then checked that no one remained in the car and the train had then continued to Crystal Siding. When they reached there the front six feet of the car were ablaze. The car was uncoupled and at that point he saw a passenger in the car. This it turned out was Mr. J.G. King, a railroad clerk of Drummond, Montana. Mr and Mrs King had only been married nine days and were on their honeymoon. They initially escaped from the car, but Mrs King was very upset by the loss of their belongings, so Mr King re-entered the burning car to retrieve them, at which point he was seen by Conductor Atkinson. Trapped by the flames, Mr King had to jump through a window to escape. Not realising that he had done so, Mrs King then re-entered the car, was overcome by smoke and perished. The conductor insisted that all the passengers had been out of the car when the train reached Crystal Siding and he was supported by other railwaymen who gave evidence at the inquest. The problem was that when the fire had finally been extinguished the bodies of five and possibly six victims were found. (Physicians believed they had discovered the remains of six victims, two men, two women and two children, but according to the passenger list apart from Mrs King only four passengers were missing: Ellen Green and Miss Lydia Hall of Redlands, California; Mr C.H. Gutches of San Diego, California, en route to Minneapolis, Minnesota; and Mr W.H. McCreary, of Riverside, California.) [1514] Either therefore the conductor had failed to check that everyone had left the car or others had also returned for their belongings.

A somewhat different account of the events was given at the inquest by a passenger, Mr George Wilson, of Philadelphia, Pennsylvania, who also attributed the fire to an electrical fault, saying that the train lights had been flickering all night. According to him he had been unable to sleep. He said that the train had not been stopped to extinguish the flames but had run fast to Crystal Siding, and the speed of the train on the ten-minute journey had fanned the flames over the remainder of the car. As far as he knew none of the passengers in the burning car were aroused or taken into other coaches until the train arrived at Crystal Siding, where some of the passengers had had to jump through windows in their night clothes to escape the fire. They had remained at the siding for three hours, then continued to Salt Lake City. No check was made whether all the passengers had been accounted for before the train resumed its journey and it was only later that a telegram was received advising of the finding of the other bodies in the car,

[1513] *The Times*, London, England, 12th February 1926

[1514] *The Salt Lake Tribune*, Salt Lake City, Utah, United States, 18th June 1926 which had been burnt out while they were stopped at Crystal Siding. [1515]

Another newspaper wrote, 'The fire was discovered west of Crystal, whose lone structure is a tottering warehouse, about 5.13 o'clock in the morning. C.E. Drummond, of Glendive, Mont., saw the flames leap forward to consume the highly finished wooden interior of the front end of the sleeper. He notified Conductor Atkinson and before great damage had been done or the occupants of the car alarmed, members of the train crew and volunteers had the fire apparently under control. Only two fire extinguishers were used in the work, according to George J. Wilson, of Philadelphia, Pa., who operated one. Inasmuch as the flames were just gaining headway at the time, it appeared that the fire would soon be overcome.

'It was evident, however, according to Mr Wilson, that the fire had crept under the car roof, and as the smoke issued from the top end of the small compartments on that end of the car, a brakeman seized a pick axe and prepared to cut through the roof of the car to the conflagration. It was at this juncture that Conductor Atkinson, who was in charge of the train, apparently failed to comprehend the seriousness of the situation, according to Mr Wilson, for he ordered the brakeman to cease his activities and the engineer to proceed to a siding which was a few miles ahead. With the doors open the train sped forward at forty miles an hour and the flames were whipped to a raging mass.

'Within two minutes the flames had spread from the front platform down the aisles of the car, eating up with hot tongues the dry draperies and leather cushions which form such an important part of the interior accessories. Now thoroughly aroused from their slumbers, the frantic passengers made valiant attempts to find the exits. Blocked off in front by the blazing heap, they turned to the rear. But even then five were unable to escape before the flames had overcome them. As the five bodies were found in the same general vicinity, it is believed that they were burned to death in the narrow aisle leading to the rear of the car.' [1516]

The inquest took evidence not only from the conductor and passengers but from Engineer T.D. Fanata, Fireman Frank Newlin, Flagman L. Moody, Brakeman L.P. Barker, Porter R.L. Knox, Pullman Conductor J.H. O'Brian and various others. At the close of the inquest, held before Coroner Breeze in Las Vegas, the jury found that "the train crew did not take sufficient precautions in removing the passengers from the burning car or preventing their return" but did not find anyone criminally negligent.

The cause of the fire was attributed by the railroad company to defective wiring. [1517] Surprisingly, in view of its severity, the Interstate Commerce Commission did not investigate the accident.

An overnight Great Western Railway down West of England express from London Paddington, presumably the one nicknamed *The Owl*, was stopped at Ruscombe Siding, near Twyford, England, just before 1 a.m. on Friday 23rd July 1926, either by a passenger pulling the communication cord or because the signalman had received the "Stop and Examine" signal, probably the latter. Fire had broken out in a luggage van and brake van

[1515] *The Evening Tribune*, Albert Lea, Minnesota, United States, 19th June 1926
[1516] *The Mineral Independent*, Superior, Wisconsin, United States, 24th June 1926
[1517] *The Anaconda Standard*, Anaconda, Montana, United States, 19th June 1926

in the middle of the train. The train was divided on each side of the burning vehicles. Its front portion was sent on to Twyford and the passengers in the rear portion were transferred to the 12.15 a.m. train from Paddington which was stopped alongside on the down relief line, and rejoined their own train at Twyford. The two vehicles were found to be well alight and efforts to extinguish the fire were unsuccessful. No passengers or railwaymen were hurt but newspapers and passengers' luggage was destroyed. [1518]

When a train arrived in København from Odense, Denmark (or possibly Odense from København as the reports are conflicting) on Wednesday 15th September 1926 a horse box was found to be on fire. A racehorse in the box was burnt to death and a man lying on the floor of the box was suffering from burns and was taken to hospital. He was identified as a Mr Stapleton from Surrey, England, a trainer of trotting horses. He died later from his burns. The cause of the fire was not stated. [1519]

A York – Newcastle express on the London and North Eastern Railway was in trouble between Bradbury and Ferryhill, England, on Tuesday 28th September 1926 when a fire broke out in its dining car, which was fortunately unoccupied by passengers at the time. Restaurant car attendants George Burns and Edward Taylor attempted to tackle the fire with extinguishers and the train was stopped and divided on each side of the dining car. It took three quarters of an hour to subdue the fire. The kitchen portion of the vehicle was practically burnt out and the two attendants were badly burned and were taken to Newcastle Infirmary. [1520]

Spontaneous combustion was believed to be the cause of a fire in two trucks of chaff on a goods train from Toowoomba to Brisbane on 7th October 1926. At about 5 a.m. when the train was two miles on the Ipswich side of Walloon, Queensland, Australia, the guard noticed dense smoke emanating from the middle of the train. He stopped his train; he and the train crew removed the tarpaulins from the waggon and the chaff was tipped off down a small embankment. The woodwork of the waggon where the fire had evidently begun was destroyed; a second waggon was less badly damaged. Only 100 bags of chaff from the two waggons could be saved. The train was delayed two hours, but no other trains were involved. [1521]

Fire broke out in the saloon carriage of a train running from Leningrad to Murmansk, U.S.S.R., on Saturday 4th February 1928. To escape the fire members of the Railway Commission who were passengers threw themselves from the train, three of them being seriously hurt as a result. [1522]

[1518] *The Times*, London, England, 24th July 1926
[1519] Ibid, 17th and 18th September 1926
[1520] Ibid, 29th September 1926
[1521] *The Queensland Times*, Ipswich, Queensland, Australia, 8th October 1926
[1522] *The Times*, London, England, 6th February 1928

A party of rich American tourists who had been exploring Africa and were on their way home had a lucky escape in a fire on a *train deluxe* from Luxor to Cairo, Egypt, probably on Tuesday 13th March 1928. An American newspaper reported the accident in detail. 'George Eastman, camera manufacturer of Rochester, N. Y., and the Eastman hunting party, enroute home from the heart of Africa, were among passengers on a crowded deluxe train who narrowly escaped when fire swept two coaches including their Pullmans. All the films which Eastman had made on his trip were destroyed, adding to the fury of the flames which threatened the car's occupants.

'Eastman himself just escaped in his pajamas and members of his party similarly lost everything. When the camera manufacturer left the Cairo station this morning for Alexandria, his face was grim. On his arrival in Cairo aboard the train which had come from Luxor, Eastman was wearing a cap, a blue pajamas coat, trousers which apparently had been borrowed and slippers. The other members of his party were similarly attired, although some of them had been able to get into borrowed suits. All looked shaken and hurriedly departed in order to catch the delayed deluxe special for Alexandria to catch the White Star liner *Doric.*

'Most of the passengers on the train were American tourists. The train was crowded and temporary coaches {sic – should read couchettes} were placed in an ordinary first-class carriage. It was here that the fire originated and then spread rapidly to the adjoining sleeper in which the Eastman party was travelling. The engineer of the train stopped it, and this enabled the passengers to beat a hasty retreat in their sleeping garments through the windows.

'An American woman passenger said that the flames enveloped the corridor when the alarm was given and that windows offered the only exit. The train was on a fairly steep embankment and the passengers bundled one another through the windows. In their precipitate flight some of them cut hands, barked shins, sprained ankles and bruised themselves but these were the only mishaps recorded.

'So short was the warning given of the flames that there was not time to save anything or to don suitable garments. Baggage and valuables were abandoned, a prey to the flames, and everything was destroyed in the two coaches. The losses are believed considerable.

'Charles Prescott, British artist, who had been accommodated in the first-class coach, told of being awakened by smoke, opening his door and finding flames raging outside. He said that he had some fearful moments in trying to pull down his window shutter and finally dived through the window in his pajamas.' [1523]

It was theorised that the fire had been caused by a 'hot-box' on the first-class carriage but officially its cause was unknown. [1524]

Forty-seven sacks of mail were destroyed in a train fire near Bennet, Nebraska, United States, at about 5 a.m. on Thursday 6th December 1926. Chicago Burlington and Quincy Railroad passenger and mail train No. 136 between Lincoln and Nebraska City had left Lincoln at 4 a.m. The train crew discovered the blaze when the train was about two miles

[1523] *The Kingston Daily Freeman*, City of Kingston, New York, United States, 14th March 1928
[1524] *The Times*, London, England, 15th March 1926

outside of Bennet and was brought to a stand near the water tower there. Unfortunately, efforts to get water from the tower were unsuccessful and afraid that the burning train car would set the wooden water tower alight the vehicle was moved down the track, detached and allowed to burn itself out. Leaving Lincoln the train had had six pouches of first-class mail and 44 bags of letters, parcels, and newspapers. One pouch and two sacks thrown off at Cheney were the only ones saved. The destroyed mail had been destined for Bennet, Palmyra, Unadilla, and especially Syracuse.

Railroad officials declared that there had been no stoves or lamps in the compartment near the mail sacks and attributed the fire to spontaneous combustion. [1525]

On about Monday 7th May 1928 more mail was lost in a train fire which occurred about six miles from Port Said, Egypt, on a train from Alexandria. Mail from continental Europe for Egypt, East Africa and Australia was destroyed in the blaze. [1526]

What originated as a track fire spread to encompass a train on the Hudson and Manhattan Railway near Christopher Street, New York, United States, on the evening of Tuesday 19th February 1929 and caused a panic in which 150 passengers were injured. A seven-car subway train in charge of Conductor Alfred Horan with Motorman Henry von Zandt was filled with over 1,000 commuters bound for their homes in Jersey City and Newark, New Jersey. It had just left the station and was proceeding towards the tunnel beneath the Hudson River when its motorman noticed a 'lake of fire' composed of burning oil and grease on the line as it dipped to reach the tunnel. He attempted to run the electric train through it but there was a short circuit, the train stalled, and its main lights went out, leaving only dim emergency lights. Flames from the track were reflected in the middle cars of the train, causing frightened passengers there to move towards the front. Women began to cry out, the excitement increased, and people began to push and shove one another. As the cars began to fill with smoke the passengers placed handkerchiefs in their mouths. Then the insulation on the upper part of the cars began to burn and the smoke to become acrid and choking. The fire attacked the wooden battery boxes and even the emergency lights went out. Fear and panic then affected the passengers, and they groped their way forward in the darkness.

Motorman van Zandt called for the passengers to be quiet and told them that they must go back through the car on fire to the rear of the train. It was in the rush to do so that most of them were hurt. Meanwhile a second train had come up behind the first and the passengers were able to make their way back through that one, their panic lessening as the smoke decreased. At the end of the rear train they climbed down to the track or slid down on seats taken from the train. When a policeman saw a woman stumble out of Christopher Street station with her face blackened and her eyes streaked and red he rang the fire alarm. Over two dozen men and women fainted from exhaustion and strain as they reached the station and rescuers had to carry several from the train who had fainted or been overcome by smoke. Over 150 passengers were taken by ambulances and taxis to hospitals where they were treated for smoke inhalation. Many others were treated at drug stores, improvised first-aid stations and private homes.

1525 *Evening State Journal*, Lincoln, Nebraska, 6th December 1928

1526 *The Times*, London, England, 15th May 1928

Investigations into the accident were opened by the transit commission, the city authorities, and the railroad. [1527]

At about 13:30 on Sunday 10th March 1929 a third-class bogie carriage of No. 1 Up Rangoon – Mandalay express caught fire between the 5th and 6th mileposts from Rangoon, Burma. The bodywork of the coach was destroyed, and 70 passengers were injured, five seriously, from jumping out of the train before it had stopped. The damage to the rolling stock was estimated at about 6,250 rupees. From the suddenness of the fire, its location, and the rapidity with which it had spread it was deduced that the fire had resulted from the accidental ignition of a highly inflammable substance, presumably something carried by one of the passengers. A deliberately set fire – that is, arson – was apparently not suspected. [1528]

A major disaster in the U.S.S.R. on Wednesday 16th April 1930 (date given in several American newspapers as 16th May) only became known in the West after the arrest of the railwaymen involved for negligence. A passenger spilt some methylated spirits in a train near Domodedovo and another passenger dropped a lighted match into the liquid. The carriage was soon ablaze from end to end, and the passengers were trapped by luggage piled up in front of the doors. Forty-five of them were burnt, suffocated, or trampled to death before the train stopped and 23 others were injured. [1529] 'Peasants in other coaches prevented the fire from spreading.'

The arrested railwaymen were charged with not having acted quickly enough when confronted with the fire and were to be subjected to a public trial as an object lesson to others. [1530]

Passengers in the coach compartment next to a van of the 11.10 p.m. train from Brighton to London Bridge noticed smoke as the train was passing through Mertsham Tunnel, England, just before 1 a.m. on Saturday 2nd April 1930. One of them pulled the communication cord and the train came to a stand near the Selsdon Road bridge, between Purley Oaks and South Croydon stations. It was then found that the van was on fire. The driver and fireman, at some risk to themselves, uncoupled the vehicle and the Croydon fire brigade was called. After considerable damage had been done to the van and parcels being carried in it the fire was extinguished and the train continued to its destination minus the damaged van. [1531]

[1527] *The Bee*, Danville, Virginia, United States, 20th February 1929
[1528] Report by the Railway Board on Indian Railways for 1928-29 Volume 1
[1529] *Northern Star*, Lismore, New South Wales, Australia, 31st May 1930
[1530] *The Vidette Messenger*, Valparaiso, Indiana, United States, 29th May 1930
[1531] *The Times*, London, England, 4th August 1930

CHAPTER 15

CHEMICAL EXPLOSIONS

Railway accidents involving explosions were of two types; those which might be defined as chemical explosion and locomotive boiler explosions. Accidental chemical explosions, such as those due mishaps with blasting powder or munitions, were rare, whereas in the early days in particular locomotive boiler explosions were much more common, though by the early twentieth century they were becoming less common, while chemical explosions were becoming a little more common, with several very bad ones involving munitions trains around the end of the First World War. This short chapter will deal with chemical explosions.

A goods train bound for Csáktornya (now Čakovec in Croatia) on Tuesday 10th July 1917 via the Südbahn included a waggon laden with naval mines. As the train was departing from Pragerhof, Galicia (now Pragersko in Slovenia) the waggon exploded. Standing on an adjoining line was a leave train, which was showered with numerous pieces of iron and burning waggon parts. On a third line was an armoured train, which also immediately caught fire. A particular danger was posed here because this train included a waggon laden with Ekrasit, a type of high explosive, but through a feat of great bravery this waggon was uncoupled and moved out of danger.

The initial explosion resulted in a crater 9 metres in diameter and 5 metres deep, at the base of which lay the undercarriage of the waggon. The permanent way of eight neighbouring tracks was disturbed over a length of 420 metres and four locomotives and 96 waggons were burnt or otherwise badly damaged. The charred remains of a goods waggon lay on the overturned engine of the armoured train. Worse was the death in the disaster of 42 military personnel, mostly presumably from the leave train, and one railwayman. In addition 55 people were seriously and 20 slightly hurt. [1532]

A munitions train stabled on a siding of the Heeresbahn Nord at Rudnik, near Lublin, Galicia (now in Poland) exploded on Tuesday 17th July 1917 with disastrous results. Of

[1532] *Eisenbahnünfalle Bind I* by Ludwig Stockert, pg. 110 (in German)

the 18 waggons of the train 7 were destroyed and 11 badly damaged. A loading ramp was blown to pieces and an office in the station buildings 150 metres away was demolished and all its occupants were killed. Also wrecked was a coach parked on an adjoining siding which was used as accommodation for railway staff, and again everyone in it lost their life. In all a total of 59 people, military and civilian, succumbed. The munitions train apparently caught fire before it exploded, but the origin of the fire could not be determined. [1533]

As a munitions train which included nine waggonloads of shells, packed in cases, and weighing in total 90,000 kg was passing through the station at Loitsch, Austria, (now Logatec in Slovenia) on the Südbahn on Monday 1st October 1917 an explosion occurred. This was followed by a series of further explosions over the next three days. Thirty-four military personnel died, 55 were seriously injured and 20 slightly hurt. As might be expected there was also great material damage. Fifty-three goods waggons were destroyed or badly damaged, 17 were slightly damaged and the rails and permanent way also suffered greatly. [1534]

According to a memorial in Chocques cemetery, France, 29 soldiers of the 4th Battalion of the Kings Liverpool Regiment were killed by an explosion in a troop train near there on Thursday 11th April 1918. [1535] It seems probable that the train had in fact been struck by a German shell, as a diary entry made by a soldier of the 7th Northamptonshire Regiment on 18th April speaks of 37 members of the Kings Liverpool Regiment – the contents of a cattle truck – being killed in this manner near Béthune, their bodies being buried by the lineside, the graves being marked by 'bloody equipment and shrapnel helmets.' The entry adds that horses in the next waggon were also killed. [1536]

On Friday 24th May 1918 a munitions train exploded between Gutach and Hornberg, Germany, on the Schwarzwaldbahn. There was much material damage, including to lineside buildings, but no one was hurt. [1537]

A petroleum train exploded on Thursday 15th August 1918 at Erlau station, Hungary, 'causing great damage' and interrupting telephone and telegraph communication between Budapest and the north of the country. [1538] There was no report of any casualties.

A telegram to Berne, Switzerland, from Berlin, Germany, on 22nd November 1918 announced the terrible explosion of a munitions train at Hamont, Belgium. Four hospital trains, it continued, had been parked close to the munitions train and were severely

---

[1533] Ibid, pg. 102
[1534] Ibid, 110-11.
[1535] Wikipedia – List of Rail Accidents (1910-1919)
[1536] https://www.greatwarforum.org/topic/19888-4th-kings-liverpool-regiment-in-france/
[1537] Eisenbahn-Katastrophen in Deutschland by Hans Joachim Ritzau, pg. 65
[1538] *The Times*, London, England, 22nd August 1918

damaged. Over 1,000 people had died, mostly Germans. Hamont station was completely in ruins. Dutch doctors were on their way to Belgium to care for the survivors. It blamed the accident upon the local population who had gathered in large numbers around the German soldiers.

The same day reports of the disaster appeared in Dutch newspapers. *Le Telegraaf* in Amsterdam gave a probably more accurate figure of 800 dead and stated that some were Dutch soldiers and Belgian civilians. It added that houses near the site of the explosion had collapsed, burying their occupants in the ruins. In a later report the same day the paper added that munitions and food trains stationed at Hamont had been frequently looted in the days leading up to the catastrophe. The latest incident of looting had occurred on the Monday evening {18th November}. During the looting a fire broke out and quickly spread to a waggon loaded with shells which exploded with a tremendous roar.

Le Handelsblad wrote that shell fragments had fallen on the hamlet of Schoot, near Budel, over the frontier in the Netherlands, where the roofs of houses had collapsed and that men, women and children had fled from Hamont to Budel. Hamont station was a lake of fire from which moans and cries escaped. Three ambulance trains carrying wounded German soldiers caught fire and 18 of the wounded burnt to death. The burning carriages were heroically detached and moved away. One hundred and fifty injured had arrived in Budel. Around Hamont station everything was in ruins. Many houses had collapsed, and the number of dead must be very high.

The Eindhoven correspondent of the *Nieuws van den Dag* stated that the initial explosion in the munitions waggon had been caused by a firework. [1539] Another source blamed the fire on children playing with stolen 'luminous ammunition' (incendiary shells?). [1540]

The newspaper *La Libre Belgique* announced in Brussels on Saturday 11th February 1919 that a waggon of benzine (translated in *The Times* as petrol) had exploded at Ath station, Belgium, alongside of a train returning French deportees to their home country. Six British soldiers had been burnt to death and about 50 others injured. A number of Frenchmen had also been killed and a large number injured. [1541] No further report on this accident seems to have appeared in either British or French newspapers.

A waggon of a train stabled overnight on a siding at Kambove, Belgian Congo, loaded with several tons of dynamite, exploded on about Sunday 16th February 1919. Everyone on the train was killed – seven Europeans and about 20 'natives.' The Europeans included a Belgian non-commissioned officer who had just returned to the colony after spending four years on the Western Front. No trace remained after the explosion of a ganger's house which had stood nearby. [1542] [1543] [1544]

[1539] *Le Petit Journal*, Paris, France, 23rd November 1918

[1540] https://www.reddit.com/r/europe/comments/a15ovk/in_november_1918_the_worst_train_explosion_in_history_occurred.

[1541] *Le Petit Parisien*, Paris, France, 14th January 1919

[1542] *The Times*, London, England, 20th February 1919

The civil war between the Red (Bolshevik) and White forces in Russia was still in progress at the end of 1919. In Siberia, the White forces under Admiral Kolchak had been in retreat from mid-October, the 2nd and 3rd White armies moving eastwards along each side of the Trans-Siberian Railway. Many died on the retreat from attacks by the Bolsheviks and peasant partisans, from the intense cold of the Russian winter, and from typhus. On 17th December 1919 there also occurred what was very possibly the worst railway accident in Russian history, though in the absence of any information on Russian accidents between the mid-1930s and the mid-1980s due to the total censorship of such news items it is impossible to be certain. On the date given a munitions train exploded in Achinsk station with the reported loss of 1,400 [1545] or 1,500 lives. A letter from a witness to the aftermath of the explosion stated that bodies were laid out in a row for many days with parts of women and children scattered in every direction. [1546] The dead included Russians and Poles. Several trains were apparently involved. [1547]

A train loaded with German shells salvaged from the Western Front which were being forwarded to England blew up in the Loubet basin at Boulogne-sur-Mer, France, on Saturday 31st July 1920. A British soldier was seriously injured, and nine trucks were destroyed. [1548]

A mixed train running between Luga on the Gulf of Finland and Novgorod, RSFSR, included a consignment of what was termed benzine but was probably petrol, which was to be bartered for grain. The load blew up, wrecking the train and setting it ablaze. Sixty-eight passengers were burnt to death. [1549] [1550] The accident happened on about Sunday 16th January 1921.

Chemin de fer Fédéraux Suisses train No. 567a, which was due to arrive at Lucerne at 08:00 on Tuesday 3rd April 1923 was in trouble between Lavorgo and Faldo, Switzerland, on the Gothard line, when its electric engine suffered an explosion in its transformer. The engine was one of the first supplied by the factory at Oerlikon and the accident was the first to have occurred to a locomotive of its type. The cause of the explosion was not known but it was believed that it had originated in the formation of gas which had suddenly ignited. The violence of the explosion tore off the door of the engine and the driver was thrown to the ground. He suffered a broken arm and remained unconscious at the time of the report. His assistant, named Bellencini, suffered worse still. He was also

1543 *L'Echo d'Alger*, Paris, France, 19th February 1919
1544 *Railway Gazette*, Vol 30, pg. 348
1545 Historical Dictionary of the Russian Civil Wars by Johnathan D. Smale, pg. 475
1546 *The Lincoln Daily Star*, Lincoln, Nebraska, United States, 29th March 1920
1547 *The Times*, London, England, 25th February 1920
1548 Ibid, 2nd August 1920
1549 *The Capital Times*, Madison, Wisconsin, United States, 18th January 1921
1550 The Railway Gazette, Vol 34, pg. 79

thrown from the locomotive, sustained a fractured skull and died. [1551]

A Missouri Kansas Texas of Texas Railroad freight train of 50 cars suffered a series of explosions four miles south of Taylor, Texas, United States, on the afternoon of Sunday 12th July 1925. The fire that followed the explosions resulted in flames which shot several hundred feet into the air and the noise of the explosions was clearly held in Taylor. Eighteen oil tank cars and two box cars were destroyed, the material loss amounting to hundreds of thousands of dollars. None of the train crew was hurt but the badly burned bodies of three men were found in the wreckage. They could not be identified but it was declared that they were not negroes but were either whites or Mexicans. It was assumed that they had belonged to a group of tramps who were known to have boarded the train north of Taylor. It was theorised that one of the riders had been smoking and a discarded match or cigarette end had ignited some leaking gasoline and set off the explosion. An alternative theory was that the fire had originated from an overheated axle box ('hot-box').

Passenger trains were diverted via Austin over the International Great Northern route while the line was blocked. [1552]

The bodies of three more tramps were later recovered, increasing the total number of dead to six. [1553]

At about 14:10 on Sunday 7th October 1928 an explosion occurred in the seventh vehicle from the engine of No. 198 express of the Great Indian Peninsula Railway between Hisvahal and Manmad, India. Three passengers were killed and nine injured, one fatally. The roof and woodwork of one carriage and part of the woodwork of another were flung 100 yards into an adjoining field and only the steel framework of the carriages remained. The damage done amounted to 5,400 rupees. Two men were subsequently arrested and convicted under the Arms, Explosives, and Indian Railways Acts; each being sentenced to seven years' rigorous imprisonment. [1554] The explosion, which involved both a bomb and dynamite, was accidental, it being theorised that the bomb, which was being carried in a box packed with sand, had by the jolting of the carriage come into contact with the side of the box. It was believed that the bombs were being brought to Bombay (now Mumbai) in connection with the visit of the Simon Commission [1555] (a group of seven British M.P.s sent to India to study constitutional reform) [1556] presumably with the aim of carrying out a terrorist attack.

What was apparently a bomb exploded on the crack Gulf Colorado and Santa Fe Railway train known as *The Ranger* at 2.02 a.m. on Saturday 6th April 1929. Twelve men riding in the car were somehow unhurt but for minor cuts from flying glass, but the

[1551] *Gazette de Lausanne*, Lausanne, Switzerland, 4th April 1923.
[1552] *The Brownsville Herald*, Brownsville, Texas, United States, 14th July 1925
[1553] *Denton Record-Chronicle*, Denton, Texas, United States, 13th July 1925
[1554] Report by the Railway Board on Indian Railways for 1928-29
[1555] *The Times*, London, England, 9th and 12th October 1928
[1556] Wikipedia – Simon Commission

interior of the vehicle was badly damaged, and it was detached at Temple. The explosive had clearly been planted in the lavatory, which was divided into a white smoking section and a section for negroes. This part of the coach had been transformed into an unrecognisable mass of wreckage. It was stated that it had only been the steel construction of the car which had saved it from being derailed causing a serious wreck. A man who had boarded the train at Clifton for Cameron had been seen to be acting suspiciously and it was believed that it was he who had planted the bomb. He could not be found afterwards. [1557] One report stated that the man had been blown to pieces. [1558]

Conductor Maxey Callaway said that it was a miracle that a number of lives had not been lost while Santa Fe Special Officer Rooney theorised that a small bottle of nitro-glycerine might have been left in the lavatory and a jolt had thrown to the floor, causing it to explode. [1559] The reason for the setting off the bomb, if a bomb it was, could not be determined.

The explosion of a steam locomotive near Oita, Kyushu prefecture, Japan, on Sunday 6th April 1930 killed 17 passengers on the train it was pulling and injured another seven. Officially it was attributed to a stick of dynamite which had got mixed up with the coal [1560] and the accident is therefore included in this chapter. Such explosions of steam engines were, however, frequently attributed to some type of explosive such as dynamite when they really were boiler explosions. Such was possibly the case here.

1557 *Waxahachie Daily Light*, Waxahachie, Texas, United States, 6th April 1929
1558 *Del Rio Evening News*, Del Rio, Texas, United States, 6th April 1929
1559 *Corsicana Daily Sun*, Corsicana, Texas, United States, 6th April 1929
1560 *Thomasville Times-Enterprise*, Thomasville, Georgia, United States, 7th April 1930

CHAPTER 16

BOILER EXPLOSIONS

The railways of the early twentieth century were still powered largely using steam engines, though electric traction was making inroads, mostly on suburban and underground railway systems and by the end of the 1920s gasoline or diesel powered locomotives were also coming into use to some degree. Locomotive boiler explosions were therefore more common than chemical explosions. Although the explosion of a locomotive boiler could often prove fatal to the engine crew it was rare for train passengers to suffer and probably because of this, information regarding boiler explosions in countries outside of Britain and the United States has been hard to obtain. By the period dealt with in this book there were very few explosions in Britain so almost all of those dealt with below occurred, as will be seen, in the United States.

Westbound St Louis-San Francisco Railway passenger train No. 9 from St Louis Missouri to Oklahoma City, Oklahoma, was only moving at about ten miles an hour as it climbed a steep gradient four miles west of Sapulpa, Oklahoma, United States, on the morning of Sunday 16th January 1916 when the boiler of its engine exploded. Engineer Harry Smith and Fireman Floyd Ames Conwell were fatally scalded.

The train was a hundred yards beyond a trestle when its passengers saw a large cloud of smoke, fire and steam burst from the locomotive, accompanied by a muffled report and the train stopped suddenly. Passengers and railwaymen left the coaches and went forward and as the cloud of smoke and steam cleared they saw the engine boiler lying beside the right of way a few feet ahead of its bogies which remained on the rails. [1561] [1562]

Brakeman Dola Bryck was killed instantly, Engineer C.Y. Fuller and Fireman A.W. Warren were fatally injured, and two other men were scalded when the boiler of a locomotive gave way at Hoffman, Oklahoma, United States, on the Missouri Oklahoma and Gulf Railroad on Saturday 15th April 1915. The force of the blast blew off the roof of

---

[1561] *Muskogee Times-Democrat*, Muskogee, Oklahoma, United States, 17th January 1916
[1562] *Waco Morning News*, Waco, Texas, United States, 17th January 1916

the station. The cause of the accident was unknown. [1563]

On Saturday 12th August 1916, the engine boiler of an Atchison Topeka and Santa Fe Railway through freight train exploded two and a half miles east of La Rose, Illinois, United States. Engineer M.C. Billington and Fireman Virgil Ross or Roth were killed, and Brakeman Schwartz and Conductor Eastman were injured. The recovery of the brakeman was thought doubtful. All that remained of the locomotive were its trucks (bogies). [1564]

When an eastbound Delaware Lackawanna and Western Railroad freight train of 70 cars, assisted by a pusher engine (banker), was nearing Hainesburg, New Jersey, United States, on Tuesday 14th November 1916 the boiler of the front engine exploded, killing Engineer Patrick Gallagher, his fireman and the head brakeman outright. A fourth man who was badly injured was placed on the pusher engine which made for Stroudsburg Hospital, but he expired before the engine reached Slate Ford, Pennsylvania. [1565]

The boiler of the locomotive of a St Louis-San Francisco Railway freight train gave way as it was climbing a bank just south of Grandview, Missouri, United States, on Friday 12th January 1917. Engineer W.R. Hall, Fireman H.L. Hartley, and Brakeman Hollie Eib were killed. The body of the engineer was found in a field about a hundred feet from the track. The boiler was lifted almost entirely from the bogies of the engine and thrown against the side of the adjoining cutting. The air brakes came on automatically when the explosion occurred and brought the train to a stand. The noise of the blast was heard many miles away. [1566]

Engineer Frank Nesbit and Fireman M.L. Hill were killed in the explosion of a boiler at McLaughlin, North Dakota, United States, on Sunday 25th March 1917 on the Chicago Milwaukee and St Paul Railway. Conductor C.A. Schneider of the same train was very severely injured. [1567]

Only a few hours after the North Dakota accident another boiler explosion occurred in the neighbouring state of Minnesota. A Chicago and North Western Railway passenger train had taken on a second engine at Waseon, Minnesota. At 12.30 a.m. on Monday 26th March 1917 it arrived at Janesville, Minnesota, United States, and while it was standing at the station there, discharging and taking on passengers, the boiler of the train engine let go, throwing Engineer John McDermott and Fireman Roy England off the footplate, one on one side and the other on the other. Neither the pilot engine ahead nor the baggage car behind were damaged. The fireman was, it was feared, fatally scalded while the engineer

[1563] *The Daily Ardmoreite*, Ardmore, Oklahoma, United States, 16th April 1916
[1564] *The Topeka State Journal*, Topeka, Kansas, United States, 12th August 1916
[1565] *Harrisburg Telegraph*, Harrisburg, Pennsylvania, United States, 14th November 1916
[1566] *Topeka State Journal*, Topeka, Kansas, United States, 12th January 1917
[1567] *The Winona Daily-Republican Herald*, Winona, Minnesota, United States, 27th March 1917

was less badly if painfully hurt. Dr Mahler of Janesville, who had been about to board the train, attended to the fireman and after dressing his wounds accompanied him to Mankato hospital. Engineer McDermott stated that the boiler had been well filled with water, there had been the correct steam pressure and he could not account for the explosion. [1568]

Five railroad employees died in a boiler explosion on the Pennsylvania Railroad at Philadelphia, Pennsylvania, United States, on 1st May 1917. A shunting engine, No. 1008, arrived at the yard at 44th and Thompson Streets from Greenwich Point at about midnight and blew up shortly afterwards with a report that could be heard a mile away. Engineer Joseph Coyle, Fireman R.J. Hagy, Conductor W.F. Kelly and Brakemen C.S. Hormickel and Thomas McGorman lost their lives. Flagman J.C. Oakes was slightly hurt.

Superintendent John Lakins of the City Bureau of Boiler Inspections examined the wreckage in the company of Coroner Knight. The inspector concluded that the explosion had been caused by the water in the boiler being allowed to get too low and the opening of the injector then forcing a quantity of cold water onto the red-hot crown sheet, precipitating the explosion. [1569]

Westbound Atchison Topeka and Santa Fe Railway passenger and mail train No. 7 left Holbrook, Arizona, United States, on time at 5.25 p.m. on the evening of Wednesday 2nd May 1917. It consisted of two engines, a mail car, two express cars, a Pullman car and a combination coach and was travelling at some speed two and a half miles west of Holbrook when the boiler of the pilot engine exploded. Engineer Baily and Fireman Alsimino on that engine were blown from the footplate, the fireman being badly hurt, the engineer less so. On the train engine Fireman Moody was fatally scalded and Engineer Wilhelm injured to an unknown degree. No one else on the train was much hurt.

Walter McLaws witnessed the accident from his ranch south of the Little Colorado River, jumped in his automobile and made a record run to Holbrook station with the news of the accident. [1570]

Engineer William Wagoner and Fireman Rudy Harger were killed instantly when the boiler of their engine exploded just west of Uniondale, Indiana, United States, at 5 a.m. on Monday 11th June 1917. A third man, Jerome Wilson, was also on the engine. He was badly bruised and burned and it was feared that he was also internally injured but 'emphatically declared that he would live.' Six of the 34 cars of the Erie Railroad freight train were piled up and both tracks were blocked. [1571]

Three trainmen died and a fourth was injured when an Atchison Topeka and Santa Fe Railway engine suffered a boiler explosion at Orsa block house, 28 miles south of Denver, Colorado, United States, on Tuesday 16th October 1917. Several cars of the

---

1568 *New Ulm Review*, New Ulm, Minnesota, 28th March 1917

1569 *Evening Illustrated Ledger*, Philadelphia, Pennsylvania, United States, 2nd May 1917

1570 *The Holbrook News*, Holbrook, Arizona, United States, 4th May 1917

1571 *Wabash Plain Dealer*, Wabash, Indiana, United States, 11th June 1917 freight train that it had been pulling were derailed and the track was badly torn up. [1572]

Engineer Edward Cummings, Fireman David Kendall and Brakeman Alfred Waddell lost their lives in a boiler explosion a few miles west of Laramie, Wyoming, United States, on about Wednesday 24th October 1917. A westbound Union Pacific Railroad freight train was involved. [1573]

Engine No. 10, drawing the 17 vehicles of mixed train No. 4, which had left Medina del Campo for Salamanca at 2 p.m. on Wednesday 9th January 1918, had just arrived at the station at El Pedroso de la Armuña, Spain, when its boiler gave way at about 2.50 p.m. [1574] The explosion blew the engine tender ten metres, wrecked the leading van, two first and second class carriages and one third-class carriage and damaged a mail van. Driver Eladio Martinez, Fireman Edmundo Garcia and 11 passengers were reportedly killed, and 15 more passengers were injured. Passengers standing on the platform escaped unhurt. The injured were taken to hospital in Medina del Campo.

There had been more than 70 soldiers of the Albuera regiment from the Medina del Campo garrison, on the train. Another train was made up to take the passengers on to Madrid. The uninjured passengers alighted from the train in fear because they assumed that the accident was worse than it had been. A footnote in a newspaper report added that there had been another boiler explosion at Pedroso , with casualties, two nights earlier, which if true seems a strange coincidence. [1575]

Rutland Railroad night express No. 520, which had Montreal, Quebec, Canada at 7 p.m. the evening before and was due at Boston, Massachusetts at 7.10 a.m., was wrecked early on the morning of Tuesday 22nd January 1918 between Beldens and Middlebury, Vermont, United States. The boiler of its engine exploded, and several cars were derailed. Fireman N.S. McCauley was killed and Engineer F.W. Fuller was injured, possibly fatally. Several passengers were hurt, it was thought not seriously. [1576]

The boiler of what was described as 'a mammoth new engine' blew up on the 17 mile grade of the Baltimore and Ohio Railroad at Rodemer, West Virginia, United States, on Tuesday 9th July 1918. Engineer William King, Fireman W. Oster, and Brakeman W.E. Grannan were killed. The explosion tore the boiler from the frame of the engine, and it was thrown 35 or 40 feet up against the mountainside and it then rolled back down to the track in a reversed position. The body of the engineer was thrown 50 yards and it was four hours before it was found in some bushes. Four cars of a freight train passing on the opposite line were knocked off the track and a truckload of lumber next to the engine caught fire. Failure of the crown sheet due to low water was the supposed cause of the

---

[1572] *Topeka State Journal*, Topeka, Kansas, United States, 16th October 1917

[1573] *The Encampment Record*, Encampment, Wyoming, United States, 25th October 1917

[1574] Federación Castellano Manchega de Amigos del Ferrocarril: Accidentes Ferroviarios en España

[1575] *La Vanguardia*, Barcelona, Spain, 11th January 1918

[1576] *The North Adams Transcript*, North Adams, Massachusetts, United States, 22nd January 1918 explosion. [1577] [1578]

At 6.05 a.m. on the morning of Tuesday 1st October 1918 the engine of a Philadelphia and Reading Railroad freight train stopped at the train order board at Front Street, Lebanon, Pennsylvania, United States, for orders. Engineer Franklin Smith stepped to the rear of the cab to receive them and as he did so the boiler exploded, killing Fireman Donald Hazen and Conductor John Bohanon and injuring Engineer Smith, Brakeman Charles Hoke, and Flagman Stephen Bolinskift. All five had been riding on the engine. Hot coals from the engine set an overhead bridge and the copper-shop of the Bethlehem Steel Company alight but employees soon extinguished the fire. Operator Victor Allwein and others then went to the assistance of Engineer Smith who trapped in the wreckage. The brakeman and flagman did not at first appear much hurt but were taken in ambulances with the engineer to the Good Samaritan Hospital. The latter two were thought likely to recover but the engineer was critical.

Local residents, accustomed to explosions in the local stone quarries and industrial plants, were not much alarmed by the explosion. The Harrisburg wrecking crew arrived at 7.30 a.m. and the remains of the boiler were removed and the line cleared.

As to the cause of the accident, a local newspaper wrote, 'Although the locomotive wrecked today was today indicated to be an old one, there were engineers around the scene who declared that a lack of water may have been responsible for the explosion. While the tank on the tender was well supplied with water, and while scalding water ran from the boiler after the explosion, spectators pointed to a wide strip of the torn crown sheet and called attention to the fact that it was a bluish black while the rest of the torn sheet was whitish in color. They declared that the other portion was evidently burned, and though that the boiler was not filled with water. The cause of the accident is being investigated however, by competent railroad experts.' [1579]

Cold water rushing into a dry overheated boiler was blamed for an explosion at Cresco, Pennsylvania, United States, on Sunday 21st March 1919. Delaware Lackawanna and Western Railroad engine No. 746 was one of two hauling 70 freight waggons up the Pocono Mountains. The superstructure of the engine clear from its bogies and two waggons back in the train were derailed, while telephone and telegraph were demolished. Engineer Russell Teets, Fireman Harry Dennis and Trainman A.H. Williams were killed. [1580]

Seaboard Air Line Railroad engine No. 511 blew up at Method, North Carolina, United States, at 1.50 p.m. on Tuesday 13th May 1919, on the Seaboard Air Line and Southern Railway joint line. A negro preacher who witnessed the accident said that the three men on the locomotive were blown 50 feet into the air and parts of their clothing were found hanging from telegraph wires. All were of course killed. They were Engineer

---

[1577] *The Washington Times*, Washington, D.C., United States, 10th July 1918

[1578] *Alexandria Gazette*, Alexandria, Virginia, United States, 10th July 1918

[1579] *Lebanon Daily News*, Lebanon, Pennsylvania, United States, 1st October 1918

[1580] *The Philadelphia Inquirer*, Philadelphia, Pennsylvania, United States, 22nd March 1919

Monroe, his unidentified negro fireman and First Brakeman C. D. Buie. Their bodies were thrown a hundred yards and pieces of the locomotive were thrown 300 yards. The engine had been pulling a freight train. [1581]

Late on Sunday 25th May 1919 a westbound Pennsylvania Railroad freight train of 5-empty cars went into a siding between East Palestine and New Waterford, Ohio, United States, to allow a troop train carrying soldiers of the 83rd Division bound for Chicago to pass. While it was standing in the siding its boiler exploded. A mechanic who was riding on the locomotive named J.W. Crawford was killed instantly, Engineer Samuel Todd was badly burned and died on the way to hospital and it was thought that Fireman George Buchanan was fatally hurt. The track was badly torn up and traffic was delayed for several hours. [1582]

The boiler of the engine of a heavy West Shore Railroad freight train exploded while the train was travelling at 30 miles an hour near Kingston, New York, United States, on the evening of Monday 14th July 1919. Engineer Eugene Hankinson, Fireman Popowich and Brakeman Earl were killed. The train had taken water about half an hour before. The most widely accepted theory for the accident was that the feed pipe had been defective and was not supplying the boiler with water. The coroner stated the next morning that he was not sure if he would hold an inquest upon the victims. [1583]

Union Pacific Railroad officials expressed the opinion that the explosion of engine No. 5008 at Curvo, near Castle Rock, Utah, United States, at about 4.15 p.m. on Monday 21st July 1919 was due to low water. The force of the explosion was enough to lift the boiler and upper parts of the engine from its trucks and throw them 220 feet. The train had been climbing a gradient at 15 miles an hour at the time. Needless to say Engineer Thomas Carroll, Fireman A.H. Woods and Head Brakeman Charles Hobbs, who had also been riding on the engine, were all killed instantly. [1584]

A rack railway on the Portuguese island of Madeira, mainly intended for the use of tourists but also used by locals and owned by the Companhia do Caminho de Ferro do Monte, suffered a boiler explosion on Wednesday 10th September 1919. A train climbing to Monte was involved, the explosion occurring between Pombal and Terreiro da Luta stations. Four people died and a number were injured. There had been 56 passengers on the train. The line was out of use after the accident until February 1920. [1585]

A double-headed Paris – Quimper express on the Chemin de fer de Paris à Orléans was derailed 700 metres from the station at Drefféac, France, at 06:00 on Sunday 14th

[1581] *The Daily Times*, Wilson, North Carolina, United States, 13th May 1919
[1582] *The Evening Review*, East Liverpool, Ohio, United States, 26th May 1919
[1583] *The Kingston Daily Freeman*, Kingston, New York, United States, 15th July 1919
[1584] *The Ogden Standard*, Ogden, Utah, United States, 22nd July 1919
[1585] Wikipedia – Caminho de Ferro do Monte.

September 1919 after the boiler of the train engine had exploded. The engine and 13 carriages left the rails. Driver Salmon and Fireman Leroux were killed; several passengers were slightly hurt. [1586]

As a Chicago Burlington and Quincy Railroad freight train was proceeding westwards near Danville, Iowa, United States, on the afternoon of Monday 20th January 1920 the boiler of its engine exploded. Fireman Henry Engle was killed outright, Engineer F.L. Sylves was fatally injured and Brakeman R.E. Totten was very seriously injured.

A newspaper report stated, 'A farm sale was in progress a mile away when the accident occurred, and hundreds of farmers rushed to the scene and dug the already dead fireman from under the tons of coal and rushed the engineer to a hotel in Danville where he died shortly after. The brakeman partially recovered consciousness and although scalded from head to foot managed to remove part of the heavy automatic stoker which had crushed his arm into the ground and getting to his feet stumbled to a farm house 200 yards distant across a field.'

Burlington Division Superintendent Ustick gave low water in the boiler as the cause of the explosion. [1587]

Two men were killed, and another was probably fatally injured when the boiler of a Chicago Rock Island and Pacific Railroad freight train exploded a mile east of Paxico, Kansas, United States, at noon on Sunday 13th June 1920. Fireman E.D. Hall and Brakeman August Miller were killed, and Engineer Michael Norton was badly hurt. The engine was hurled 50 feet and several cars of the train were derailed. A large refrigerator car was torn in two, the track ripped up and nearby trees and crops blasted. Pieces of the locomotive were scattered over an area of 600 feet. A relief train from Topeka was sent to the scene.

Clogging of the boiler by mud (?) was blamed for the explosion. The train had been running late and Engineer Norton had been anxious to get to McFarland, the next station beyond Paxico, on time, which he could not have done had a stop been made to clean the boiler. [1588]

It was believed that in climbing the heavy grade into Kellogg, Minnesota, United States, early on the morning of Saturday 17th July 1920 the water in the boiler was allowed to get too low. When more water was taken there at 2.45 a.m. the boiler of the Chicago Milwaukee and St Paul Railway engine exploded. Engineer N. Schroedser and Fireman George Hamala were killed outright, and Brakeman Wells Weston died in a Wabasha hospital five hours after the accident. However, two men from Toledo beating their way in a box car immediately behind the engine escaped unhurt. [1589]

1586 *Le Petit Parisien*, Paris, France, 15th September 1919

1587 *Fort Madison Weekly Democrat*, Fort Madison, Iowa, United States, 16th January 1920

1588 *The Topeka State Journal*, Topeka, Kansas, United States, 14th June 1920

1589 *The Evening Tribune*, Albert Lea, Minnesota, United States, 17th July 1920

The narrow gauge Nesttun-Osbanen railway in Norway suffered a boiler explosion on Tuesday 7th September 1920. The afternoon train to Os, consisting of the engine Bjørnen, five passenger carriages and two goods waggons, was ready to start from Nesttun when its boiler suddenly exploded. Fireman Martinius Jordal was just looking back for the departure signal when the boiler gave way. Parts of the engine flew high in the air and the dome of the boiler landed behind Birkeland church, 200 metres away. Passed Fireman Andreas Bertelsen escaped almost unhurt and was able to assist in the clearing away of the wreckage, but Fireman Jordal suffered a broken arm, a crushed elbow, and a head injury. The carriages were only slightly damaged and none of the passengers was hurt.

An investigation disclosed that the boiler, which was 27 years old, had a longitudinal crack in its cylinder. [1590]

The boiler of engine No. 5169 of the New York Central Railroad exploded near Matoon, New York, United States, at about 4.16 a.m. on Tuesday 21st September 1920. The engine had been hauling an extra freight train from Watertown to Malone Junction. Brakeman W.J. Cook died on the way to Watertown hospital, Fireman Leo Moore died soon after reaching the hospital and Engineer William Corey died there the same afternoon. Conductor H.R. Shoen and other crew members escaped injury. Parts of the engine were thrown hundreds of feet in different directions and 18 of the 35 heavy freight cars were thrown on either side of the track or blown to pieces. Over 200 feet of rail were torn up and some twisted and a large hole was formed beneath the track. Telephone and telegraph wires belonging to the railroad company and the Western Union Telegraph Company were torn down but had been repaired by shortly after 8 a.m. Fire also broke out in the wreckage and many wooden box cars and their loads were totally or partly destroyed.

As to how the accident came about a local newspaper wrote, 'All three of the men were experienced railroad men whose reputation for carelessness had been so far tested that it was not believed the accident was due to carelessness. The theory offered is that the injector that forced the water into the boilers was not working as it should and the supply of water in the big boiler had run very low, allowing it to get very hot so that when a fresh supply of water was injected into the boiler the quick generation of steam burst the walls of the boiler like a sheet of parchment. It is not impossible that the water in the indicator did not register the lowness of the water in the boiler, deceiving the engineer. As none of the men could make a statement, on account of severe injuries, it is probable that the real cause of the explosion will never be known.' [1591]

Four Baltimore and Ohio Railroad employees lost their lives when the boiler of an engine exploded without warning as it was pulling a heavy freight train up the grade to Broad Tree Tunnel, near Littleton, West Virginia, United States, early on Saturday 29th January 1921. The men were Road Foreman Matt Walsh, Engineer C.J. Carpenter, Fireman P. Crogan and Brakeman J. Wilfong. [1592]

[1590] Nesttun – Osbanen by Per Ivar Tautra, pg. 77

[1591] *Canton Commercial Advertiser*, Canton, New York, United States, 28th September 1920

[1592] *Cumberland Evening Times*, Cumberland, Maryland, United States, 29th January 1921

The explosion of a Lehigh Valley Railroad switch engine in the yards at Jersey City, New Jersey, United States late on Thursday 24th February 1921 killed four men. Conductor James McGee and Brakeman William McCauley and Engineer George Wells and Fireman George Arnold died later of their injuries. The old theory that the explosion had not been due to steam but to a stick of dynamite was again trotted out, with the variation that it was suggested this time that the explosive had not found its way into the coal accidentally during mining operations but had been placed there deliberately by strikers who had not been reinstated after being involved in industrial action the previous year. As far as the author is aware there has been no substantiated case of a boiler explosion caused by dynamite anywhere at any time though such a cause is often suggested. [1593]

On the evening of Saturday 23rd April 1921 a Kansas City Southern Railroad freight train had stopped for water at Sandel station, Louisiana, United States. As it was pulling out again Fireman Sam Cox turned on the injector and went to the rear of the tender to put the cap on the water tank, when the boiler of the engine, No. 494, gave way. He was thrown 40 feet from the track and badly scalded, but it was thought that he would survive. Engineer Tom Dailey and Brakeman Walter Stevens were killed instantly. [1594]

Very strange circumstances attended a boiler explosion on a westbound Southern Pacific Railroad freight train as it was climbing the bank to Paisano, Texas, United States, on Friday 8th July 1921. The events began when the train suddenly began to lose speed. Brakeman Earl Stirman further back the train started to go forward to find out why and had got about four car lengths from the engine when its boiler exploded. The boiler was thrown 30 feet and the engine cab 600 feet. The body of Engineer William Bohlman was found in the wreckage; he was badly burned and gashed, but he also had a bullet wound in his head. The fireman, Charles Robertson, could not be found until a flagman went back down the line towards Alpine. He was found lying unconscious besides the track four miles back.

When he recovered, Fireman Robertson said that the last thing that he remembered was sanding the line on the heavy grade near Toronto flag station. He had rolled a cigarette and was about to sit down on his seat when he had lost consciousness. He had an ugly gash at the back of his head and severe bruising from falling or having been pushed off the train. He told the coroner that he and the engineer had several times after leaving Alpine seen a man riding the cars a short distance back along the train but thinking that he was just another tramp they had taken no more notice of him.

Brakeman Stirman spoke of seeing a "tall, slender man in black," jump from the train and run off a moment before the boiler explosion. The inquest on the engineer held at Maria, Texas, returned a verdict that he had died from the gunshot wound to his head inflicted by some person unknown, rather than from the boiler explosion.

Railroad officials believed that the murderer must be someone familiar with railroad

[1593] *Syracuse Herald*, Syracuse, New York, United States, 25th February 1921
[1594] *The San Antonio Light*, San Antonio, Texas, United States, 25th April 1921

operation. He had caused the boiler explosion by cutting off the injectors supplying water to the boiler. The motive for the murder of the engineer, supposed assault on the fireman and destruction of the engine was unknown. [1595]

The 12.50 a.m. London and North Western Railway goods train from Buxton to Oldham, England, was just starting its journey on Friday 11th November 1921 with 0-8-0 engine No. 134, 34 loaded waggons and a tender when its boiler exploded, killing Driver W. Holmes and Fireman W. Fletcher. Guard J. Hannah suffered from shock. The boiler was completely wrecked, and the remainder of the engine was damaged.

The explosion was the subject of a formal Ministry of Transport inquiry, conducted by Major G.L. Hall, and as the evidence was of a highly technical nature he was assisted by Mr H. Cranwell, of the Mercantile Marine Department. Much evidence was presented in the report, the conclusion of which was that the explosion had been due to the safety valves failing to open. As an aside it was again suggested, as it had been before and would be elsewhere again (see 1930 Oita explosion in Japan in preceding chapter) that the explosion was due to the presence of some substance such a dynamite, a theory which the inspecting officer dismissed out of hand. [1596]

Because of the war the railway companies had got very behind with maintenance and had had to resort to sending their locomotives for repair to outside engineering firms. Thus in 1920 engine No. 134 had been sent to Messrs Beardmore's Works in Glasgow. The safety valves were of gunmetal but rested in cast-iron columns. Gunmetal has twice the coefficient of expansion of cast iron and the day of the accident had been exceptionally cold. The expansion of the cast iron had thus been reduced even further and the valves, which had been tight when assembled, had become tighter still, so that the excessive steam pressure had been insufficient to force them open. C.H. Hewison in his book on locomotive boiler explosions points out that this problem should have been known to Hewitt Beames, the Chief Mechanical Engineer of the company, as a paper outlining the problem had been read at the Institution of Mechanical Engineers back in 1877. [1597]

A secondary cause of the explosion was that engine sheds were not provided with facilities to test steam gauges. Between the return of the engine from repair in July and its exploding in November drivers had several times reported that the steam gauges were faulty, and they had been changed, the last time the day before the accident, but no one had suspected that in fact the steam gauges were not faulty at all but that it was the safety valve which was the problem.

On the morning of Friday 16th December 1921 a heavy westbound Atchison Topeka and Santa Fe Railway freight train was made up at Marceline, Missouri, with engine No. 3185. The engine had been giving problems so Travelling Engineer E. Kernen accompanied the train crew to try and determine the cause of the trouble. Also riding on the engine was Head Brakeman Malon Dupree. All went right until the train was a

[1595] *The Bisbee Daily Review*, Bisbee, Arizona, United States, 9th July 1921
[1596] Board of Trade report
[1597] *Locomotive Boiler Explosions* by C.H. Hewison, pgs. 119-20

quarter mile west of the station at Standish, Missouri, United States. There the boiler of the engine exploded violently, the engine trucks being reduced to 'a mass of tangled rods and iron,' while the main boiler was lying on the north side of the track, 'a mass of sheet iron and steel scarcely recognizable.' Farmers who witnessed the accident reported that the boiler was thrown 150 feet into the air. Parts of the engine were scattered over a distance of 150 yards, the track was torn up over a long distance, at least half a dozen box cars were smashed, and others turned crossways. The accident happened at about 11.25 a.m.

Engineer George Hunt was killed instantly, Fireman H.L. Gray died in a few minutes, Head Brakeman Dupree was not expected at the time of the report to live more than a few hours and Travelling Engineer Kernen was seriously injured but expected to recover. [1598]

Mobile and Ohio Railroad freight train No. 39, travelling at about 15 miles an hour, was wrecked about 300 yards south of the station at Egypt, Mississippi, United States, at about 10 a.m. on Wednesday 15th March 1922. Of the four men who were on the engine at the time Engineer D.O. Davies was the most seriously hurt, being badly burned, and he died the same afternoon before his family could reach him. Fireman Abe Johnson, Head Brakeman W.T. Enteriken and an un-named negro hobo or extra train man were all hurt to a greater or lesser degree, but all were expected to recover. It was stated that Engineer Davies had several times reported that the boiler of the engine was weak over the fire-box and that the engine was unfit for service. His complaints had been ignored and an engine inspector had passed the engine as fit to run the day before its boiler gave way. [1599]

What was described as a powerful Baltimore and Ohio Railroad locomotive, hauling a fast freight train from Cumberland, Maryland, to Pittsburgh, Pennsylvania, blew up without warning at Fairhope, Pennsylvania, United States, on Tuesday 2nd May 1922. Engineer Timothy Conway, Fireman O.E. Newcomer and Brakeman H.L. Parker were all killed. The front ten cars of the train were thrown from the track, caught fire, and were destroyed. [1600]

A westbound West Shore Railroad freight train from Ravena to Syracuse was labouring up a gradient near Harbor, New York, United States when its boiler exploded at 1.40 p.m. on Friday 29th September 1922. A motorist and a farmer in a field witnessed the accident and ran to help but Engineer John Caldwell and Brakeman Howard Baxter were beyond aid and Fireman George Morrell died in a few minutes. Conductor Curtis Dygert, further back in the train, escaped with a shaking. The three leading cars of the 100 in the train piled up and fell over, snapping off telegraph poles along the right-of-way. The cause of the explosion was tentatively put down to low water in the boiler. [1601]

A southbound New York Central Railroad freight train, was being assisted up Dresden

[1598] *The Daily Constitution*, Chillicothe, Missouri, United States, 17th December 1921
[1599] *Okolona Messenger*, Okolona, Mississippi, United States, 16th March 1922
[1600] *The News*, Frederick, Maryland, United States, 2nd May 1922
[1601] *Syracuse Herald*, Syracuse, New York, United States, 30th September 1922

Hill with a helper (pilot) engine coupled in front of the train engine when the boiler of the train engine exploded at Moreland, New York, United States, at 11 p.m. on Saturday 11th November 1922. The boiler of the engine was thrown forward 400 feet and landed on the track where it was run into by the helper engine. Frank Ferner, Frank Harding, C.W. Hostrander and W.C. Thompson were killed, and F.L. Bennett, Archie Erwin and H. Glover were injured. These must have been the engineers, firemen, brakemen and so on of the train but none of the various newspaper reports states which of them held which position.

Another New York Central Railroad engine which had only returned to service after an overhaul the day before suffered a boiler explosion near Astor Tunnel, two miles south of Barrytown, New York, United States, on Friday 24th November 1922, while pulling a southbound freight train. Fireman Emil Peterson and Brakeman Thomas Haber were killed outright while Engineer George Dugan was very seriously injured. [1602]

The Mikado-type locomotive hauling Southern Railway freight train No. 55 suffered a boiler explosion at Inlet station, Virginia, United States, on the afternoon of Sunday 21st January 1923. Engineer Samuel Harrington and Fireman John Curtin were killed instantly. Brakeman George Allensworth was rushed to the University Hospital in Charlottesville with grievous injuries and died at 7 the same evening. The explosion was attributed to low water. [1603]

The boiler of the engine of St Louis-San Francisco Railway extra freight train No. 28 from Birmingham, Alabama, to Amory, Mississippi, exploded near Glen Allen, Alabama, United States, on the morning of Saturday 24th February 1923, killing Engineer C.A. Curtis, Fireman George Scott and Brakeman Gus Conwell. The cars of the train piled up behind the engine, blocking the line and bringing down the telegraph wires. [1604]

The boiler of the engine of a passenger train exploded between Landau and Insheim, Germany, on Thursday 29th May 1923, and the train was derailed. The casualties, if any, are unknown. Because of the French occupation of the Rhineland the train was being manned by French military personnel, and the explosion was the result of their inexperience in operating the locomotive. [1605] [1606]

A Gulf Colorado and Santa Fe Railway extra freight train left Galveston, Texas, at 8.15 p.m. on the evening of Friday 17th August 1923 for Belleville, pulled by engine No. 1969, in charge of Conductor M. Cannon with Engineer T.A. Abrahams and Fireman Huling Parker. It left Algoa, Texas, United States at 11.40 p.m. and shortly afterwards the

---

[1602] *Olean Evening Herald*, Olean, New York, United States, 25th November 1922
[1603] *Daily News-Record Harrisonburg*, Virginia, United States, 23rd January 1923
[1604] *The Anniston Star*, Anniston, Alabama, United States, 24th February 1923
[1605] Eisenbahn-Katastrophen in Deutschland by Hans Joachim Ritzau, pg. 73.
[1606] Wikipedia – Kesselzerknall (in German)

boiler of the engine exploded. The force of the explosion tore the boiler from the chassis of the engine, and it hit the ground, making a large hole, and then bounded away. The men on the engine were also hurled to the ground. The engineer was killed outright, and the conductor and fireman were fatally injured, dying the following afternoon in St Mary's Infirmary, Galveston. Two other railwaymen, W.R. Vierick and Brakeman S.L. Tucker were injured, Tucker only slightly. [1607]

Only briefly noted in the Press was the explosion of the boiler of a Pittsburgh and Lake Erie Railroad engine at Aliquippa, Pennsylvania, United States, on Wednesday 23rd January 1924. Engineer George Clark, Fireman Urban Kuechler and Road Foreman Jesse Meade. The engine was pulling a passenger train bound from Pittsburgh, Pennsylvania, to Cleveland, Ohio and was reportedly travelling at 40 miles an hour when the boiler gave way. None of the train passengers was hurt but they were reportedly panic stricken. [1608]

One accident led to another on the Wednesday 13th February 1924. The first happened at about 8 a.m. two miles east of Lariat, Texas, United States, when the boiler of the engine pulling Gulf Colorado and Santa Fe Railway passenger train No. 91, bound for California, exploded. Engineer O. E. Plumee and Fireman Tim Cronin were killed, and the track was blocked for seven hours.

When Superintendent F.J. Anton was notified of the accident at Slaton, the divisional point of the railroad, he arranged for a special train consisting of an engine and his private car to rush to the scene with physicians and nurses. This led indirectly to the second accident. Roadmaster E.R. Foster was making his daily track inspection on a track motor car. He had reached the east end of the switch at Burrus, a flag station, when the special, travelling at 55 to 60 miles an hour, caught him up. Either the noise of the motor of his car drowned out the whistle of the special or he heard the whistle but assumed it was a slow freight train. At any rate he failed to get clear of the main line before the special caught him up, struck his car and he was killed. [1609]

Westbound Southern Pacific Railroad train No. 981 from San Antonio had reached Noonan station, Texas, United States, when the boiler of its engine gave way at 7.43 p.m. on Thursday 28th February 1924. Engineer A. Juraschek, Fireman J. R. Gilchrist and Brakeman C. C. Tuttle were killed. The explosion brought down the railroad and Western Union telegraph wires so there was some delay in getting out the wreck train, but it arrived at 11.20 p.m.

Late on the afternoon of Saturday 9th January 1926 the whole town of Hurricane, West Virginia, United States, was rocked by the explosion of a large Mallet engine of the Chesapeake and Ohio Railroad. The locomotive and its tender were demolished and store fronts and business premises in the main street of Hurricane, which ran parallel to the railroad line, were wrecked. Engineer J.D. Rose and Brakeman E.P. Henry were killed

---

[1607] *The Galveston Daily News*, Galveston, Texas, United States, 19th August 1923
[1608] *The Altoona Mirror*, Altoona, Pennsylvania, United States, 23rd January 1924
[1609] *Lubbock Morning Avalanche*, Lubbock, Texas, United States, 14th February 1924

outright and Fireman D. Chattin suffered what were thought to be fatal burns. Mrs H.B. Taylor and her little son were injured by a piece of flying steel. The engine had been pulling a long string of cars and had stopped at Hurricane to take water. [1610]

Early on the morning of Saturday 26th June 1926 westbound New York New Haven and Hartford Railroad freight train No. UN-1 was en route from New Bedford, Massachusetts, to New Haven, Connecticut. It consisted of engine No. 3346, 98 cars and a caboose, carrying oil and cotton cloth, in charge of Conductor Silver with Engineer William Moulton and Fireman John Mulvaney. It passed Kingston, Rhode Island, United States, at 2.33 a.m. and had covered another three miles when at 2.40 a.m. its boiler exploded, killing Engineer Moulton, Fireman Mulvaney and Head Brakeman George Hathaway. The locomotive and leading 11 cars of the train were derailed and seven cars towards the end of the train also left the track.

Approaching the accident site on the opposite track at the time was eastbound passenger train No. 1364, *The Cape Codder*, consisting of engine No. 1364 and nine Pullman cars, in charge of Conductor Lynch with Engineer Chester Frazier and Fireman Andrew Bennett. After the boiler explosion Flagman Ricketts went back to protect the rear of his train and Conductor Silver went forward. He heard the passenger train coming and attempted to flag it but had not even reached the front of his own train before the passenger train collided with the wreckage, but a slight fog and the wreckage of his own train prevented his signals being observed. The engine and first four sleeping cars of the passenger train were derailed in turn, the leading sleeping car being thrown into the Great Swamp alongside the line. Engineer Frazier and Fireman Bennett were seriously injured, and several passengers were slightly hurt. [1611]

The Interstate Commerce Commission issued a fairly short report on the accident, at the end of which it concluded that the secondary collision might have been avoided had the line been continuously track circuited with automatic signals. [1612]

Chicago and Eastern Illinois Railroad engine No. 1928 was pulling freight train No. 41 when its boiler exploded two miles north of Salem, Illinois, United States, at 12.40 a.m. on Tuesday 22nd March 1927. The air brakes were applied as a result of the explosion and the train came to a halt. Conductor I.B. Long and Brakeman Harry Barnett hurried to the front of their train to find that Engineer C.M. Robinson, Fireman T.E. Bruce, and Brakeman C.M. Farlow had all been killed. The body of the engineer had been blown into a coal car and those of the fireman and brakeman were some distance from the locomotive. The engine itself was described as having been practically destroyed; the boiler had been blown across the right of way and was lying against a fence. The remainder of the train was undamaged. As was often the case the cause of the explosion was a mystery. The regular fireman of the engine, John Hodge, chanced to be off sick with a minor complaint and thus probably saved his life. [1613]

[1610] *The Commercial Tribune*, Cincinnati, Ohio, United States, 10th January 1926

[1611] *The North Adams Transcript*, North Adams, Massachusetts, 26th June 1926

[1612] Interstate Commerce Commission report

[1613] *Decatur Evening Herald*, Decatur, Illinois, United States, 22nd March 1927

Engine No. 311, pulling Texas and Pacific Railway mixed train No. 45 from Texarkana to Shreveport, blew up three miles north of Shreveport Louisiana, United States, at 6.30 a.m. on Friday 19th August 1927. After the accident Conductor P.J. Hackett stopped a passing motorist and got a lift into Shreveport where he reported the accident and ordered out the wrecking train. Engineer Charles Millis, Fireman W.C. Hughes and Brakeman C. Burns were all killed. [1614]

Engineer William Carrier, Fireman Richard Criddle and Brakeman J.E. Bell lost their lives when the boiler of Erie Railroad engine No. 3119 hauling fast eastbound freight train No. 78 (*Mayfield*) exploded at about 3.15 p.m. on Sunday 19th August 1928 a mile east of Cameron, New York, United States. Fifteen cars piled up behind the wrecked engine and the line was expected to be blocked for a week. [1615] [1616]

Express passenger train No. 15 from Paris Austerlitz to Bordeaux of the Chemins de fer de Paris à Orléans suffered a problem with its engine on Saturday 25th May 1929. A replacement engine was attached, and its journey resumed but between Rouffec and Salles-Moussac, France, the boiler of the new engine exploded. The driver and fireman could not at first be found after the accident. The body of the fireman, dreadfully mutilated, was then discovered 800 metres back the line, and that of the driver, who was alive but badly hurt, was nearby. None of the passengers was hurt, and after a long delay the train resumed its journey to Boreaux. [1617]

A fatigue fracture in a boiler joint due to defective construction resulted in the failure of a boiler at Reinsfeld, Germany, between Trier and Türkismühle, on Saturday 11th January 1930. Two people were injured. [1618] [1619]

A Toledo and Ohio Central Railroad freight train of 69 cars of coal and a caboose was climbing a bank at about 5 miles an hour and approaching state route 42 south of the station at Arnold, Ohio, United States, on the morning of Tuesday 28th January 1930, when its boiler gave way. The only witness to the explosion was the Arnold operator, S.C. Doherty, who said, "All I saw was a big black cloud of smoke, and when the smoke had cleared away, I saw the engine boiler partly buried in the roadbed." Engineer Roy Taylor, Fireman William Berry and Head Brakeman James Dennison were all killed. Conductor James Wiltshire, Flagman Wesley Clark and Marysville operator William Lucas, riding in the caboose at the end of the train, were unhurt. The line was blocked for a while but a sidetrack could be used and the line was re-opened at 6.30 p.m. Telephone

---

[1614] *Port Arthur News*, Port Arthur, Texas, United States, 19th August 1927
[1615] *Olean Evening Times*, Olean, New York, United States, 22nd August 1928
[1616] *Wellsboro Agitator*, Wellsboro, Pennsylvania, United States, 22nd August 1928
[1617] *La Liberté*, Paris, France, 26th May 1929
[1618] Eisenbahn-Katastrophen in Deutschland by Hans Joachim Ritzau, pg. 73.
[1619] Wikipedia – Kesselzerknall (in German)

communication was also cut for a while. [1620] [1621]

A Louisville and Nashville Railroad through freight train from Nashville, Tennessee to Evansville, Indiana, was wrecked near Springfield, Tennessee, United States, by a boiler explosion early on Tuesday 18th March 1930. Fireman H. L. Russell and Brakeman Aaron Moss were killed outright, and Engineer Joe Crandall was probably fatally injured. Springfield Ticket Agent R.H. Dowden, who witnessed the explosion, spoke of the men being "blown 100 feet in the air." Reports of a hobo also having been killed were later dismissed. Eighteen cars of the train were derailed and set on fire but fortunately a gasoline tank car did not ignite. The line was of course blocked, and the *Dixie Flier* and Dixie Mail Train were diverted via Bowling Green, Kentucky. [1622]

The explosion of a locomotive near Oita, Japan, on 6th April 1930 was officially attributed to an undiscovered stick of dynamite but was more likely an ordinary boiler explosion (see Chemical Explosions chapter).

The boiler of the engine of a northbound Seaboard Air Line Railroad express freight train gave way a mile and a half south of Petersburg, Virginia, United States, on 11th July 1930. According to Trainman E.O. Logan, who was riding in the caboose with Conductor S.S. Wheeler, he heard two separate explosions. The first ten cars of the 50 car train piled up behind the engine and caught fire. Engineer Fred Drewry was killed outright, Fireman S.S. Taylor died in hospital the same night from burns, shock and other injuries and the following day Brakeman S.B. Wiggins was said to be in a critical condition.

The Petersburg fire brigade quickly arrived to extinguish the fire but were hampered by lack of water. Tool cars were sent from Richmond and dynamite was reportedly used to clear the tracks, which seems a rather drastic solution. According to newspaper reports the explosion was thought to be the result of 'low water pressure.' [1623]

Engineer A R. Elliott and Fireman Richard Myers died when the boiler of their Canadian Pacific Railway engine exploded near Molson, Manitoba, Canada, on Friday 18th July 1930. The engine was a new one which had been placed on a freight run before being turned over to working passenger trains. A third railwayman, M.J. Ripley, was severely burned, had to have his right foot amputated and was said to be in a serious condition in a Winnipeg hospital the following day. Four other people were injured, two passengers and two trespassers, the latter two being only slightly hurt.

One of the passengers was a Mr Baczum, who was effectively a drover accompanying cattle and horses, which were mangled in the accident. He was travelling from Procter, British Columbia, to Ingolf, Ontario, and sustained painful chest wounds. The other passenger, Walter McMartin, was going from Orkney, Saskatchewan, to Murillo,

[1620] *The Union County Journal*, Marysville, Ohio, United States, 30th January 1930
[1621] *The Sandusky Register*, Sandusky, Ohio, United States, 30th January 1930
[1622] *The Kokomo Tribune*, Kokomo, Indiana, United States, 18th March 1930
[1623] *The Bee*, Danville, Virginia, United States, 12th July 1930

Ontario, with settler's effects. His right arm was broken, and he had leg injuries. [1624]

The inquest on the engineer and fireman later in the month found that the explosion had been caused by low water in the boiler causing the crown sheet to overheat. No mention was made of Mr Ripley, so he presumably recovered from his injuries. [1625]

[1624] *The Lethbridge Herald*, Lethbridge, Alberta, Canada, 18th July 1930
[1625] *Manitoba Free Press*, Winnipeg, Manitoba, Canada, 26th July 1930

CHAPTER 17

VANDALISM

'Accidents' due to vandalism and sabotage have always been a problem for the railways. They are obviously very closely related, in that both represent deliberate attempts to cause trains to crash, as opposed to accidents caused through non-deliberate carelessness, negligence or purely accidentally. Here they will, however, be treated separately, as they are slightly different. This separation is perhaps a little artificial, and if the perpetrator of a given act is not caught or his or her motives are not sufficiently obvious it may be difficult to decide which it was, but it has nevertheless been felt that the distinction is worth making.

Vandalism is usually, though not invariably, carried out by either children or young adults and its results are most commonly, though not always, less serious than those following from acts of sabotage. The motives for train accidents resulting from vandalism are usually either simple curiosity, in the case of younger children – 'I wanted to see what would happen' – or in the case of older children and teenagers a reflection of general anti-social behaviour; so that such individuals who attempt to derail trains also carry out other acts of 'mindless vandalism,' and perhaps also commit burglaries and or carry out arson attacks.

Sabotage on the other hand is usually undertaken by adults, who employ methods more likely to cause death and injury, so that while a vandal might for example just place an object upon the track or jam a stone into a set of points, a saboteur would be more likely to employ methods such as blowing up a section of track using explosives.

At 8.50 p.m. on the evening of Thursday 11th May 1916 a Cleveland Cincinnati Chicago and St Louis (Big Four) Railroad train ran into obstructions at some facing points adjoining the Seventh Street Crossing in Anderson, Indiana, United States. As the train was only moving slowly at the time it kept the rails. The train crew found several large rocks jammed in the points and a short piece of steel placed across the line. The next morning four young men aged between 18 and 20, named Clarence Gardwood, Albert Gutterson, Roy Newman, and Virgil Smith were arrested. They were all from Richmond, Indiana, nearly 50 miles away, and were unable to explain what they were doing in Anderson, while several rocks were found in their pockets. [1626]

On Monday 6th November 1916 at the Children's Court, Douglas, Isle of Man, three boys aged about 15 or 16 were charged with placing an obstruction on the railway line between Port Erin and Douglas. A train from Port Erin ran into the heaps of sand they had placed there, was derailed, and struck the rock face of a cutting. The fireman suffered a broken collarbone. The boys were severely reprimanded by the Stipendiary and their parents were each fined 40s. [1627] Given that it was the Isle of Man it is surprising that they were not sentenced to a birching.

Early on Monday 20th August 1917 Pennsylvania Railroad train No. 605, travelling reportedly at 60 miles an hour collided with a loose tie (sleeper) which had been planted in the track with one end embedded against a fixed tie, at a point ten miles south of Cleveland, Ohio, United States. The impact threw the tie away and it struck a tool shed, making a hole large enough for a man to walk through. Conductor Rellick, in charge of the train, said that only its speed had saved it from being derailed. The pilot of the engine was shattered and small pieces of it were scattered over half a mile of track. [1628]

About Saturday 29th June 1918 a heap of stones placed on the line by children derailed a workman's train near Ebbw Vale, South Wales. No casualties were reported. [1629]

James Olley, driving a train which left London Broad Street station, England, on the North London Railway, on the afternoon of Sunday 1st June 1919, had not travelled very far when he saw a baulk of timber placed across the track at the Brick Road overbridge. He applied the brakes, but his engine struck the obstruction, jumped over the timber, and came down on the rails again. Had it not done so it might have gone down the embankment and into the street, 25 feet below. The timber was left jammed between a pair of the engine wheels.

Two boys, Edward Charles Benstead, aged 12, and Herbert Bowen, age not stated, both of nearby Bow, subsequently found themselves in the Old Street Children's Court, and were found to have been responsible for the act of vandalism. Benstead was ordered to receive six strokes of the birch and Bowen, who was probably a little younger than his companion, five. [1630]

Westbound New York Central Railroad engine No. 5181, running light, was derailed just east of Potsdam, New York, United States, at about 3 p.m. on the afternoon of Friday 17th October 1919. Boys playing on the railroad had placed spikes and bolts on the rails. Trainmen chased the boys. They got away but were presumably recognised as it was

[1626] *The Fort Wayne Journal-Gazette*, Fort Wayne, Indiana, United States, 13th May 1916
[1627] *The Times*, London, England, 7th November 1916
[1628] *The Xenia Daily Gazette*, Xenia, Ohio, United States, 20th August 1917
[1629] *The Times*, London, England, 1st July 1918
[1630] Ibid, 1st June 1919

reported that a number of them would appear in court the following day. The engine was re-railed without the need to call out the wrecking crew. [1631]

'Boys at play' derailed a mineral train at Abercarn, South Wales, on Monday 12th April 1920. The track was torn up for 100 yards. [1632]

That the results of an act of vandalism by a child can sometimes be as serious as those of sabotage by an adult was shown by a derailment at Glenburn, Pennsylvania, United States, on Friday 22nd July 1921. Delaware Lackawanna and Western Railroad western passenger train No. 3 consisted on that day of engine No. 1120, a mail and baggage car, a dining car, a Pullman sleeping car, two Pullman parlor cars and a Pullman observation parlor car, in charge of Conductor Simrell with Engineer Charles Coolbaugh and Fireman Jacob Gleishman. [1633] The train struck an obstruction on the line at Glenburn at 2.25 p.m. Initially only the leading wheels of the front engine bogie left the rails, but the train could not be stopped before it reached a mains crossover about half a mile further on. Here the remainder of the engine and the first four cars of the train were derailed. Engineer Coolbaugh and Fireman Gleishman were both killed and 79 passengers, three mail clerks and two railroad employees were injured.

Examination of the line by Flagman Gibbons after the derailment led to his discovering a nut which had apparently been run over and the Interstate Commerce Commission concluded that this nut had been the cause of the derailment, and that it had been placed there by some unknown party. Other nuts and bolts which had apparently been run over previously were also found in the vicinity. [1634]

The 'unknown party' was soon found to be 14-year-old John Arre, 'the son of a village huckster,' who signed a confession to having repeatedly placed objects upon the rails until he caused the accident, thus achieving his great ambition to wreck a train. He reportedly witnessed the derailment from the top of an apple tree and was apprehended two days later when he was seen placing more objects on the track, arrested, and placed in the House of Detention for Juveniles, awaiting his trial for murder. Reminded that his act had killed the engineer and fireman and injured many others he was apparently unrepentant, replying, "That's too bad." [1635]

On Tuesday 16th May 1922 four boys named Stanley James, Ernest Lacey, Alfred Miller, and Sidney Windley, all from Dorking, found themselves appearing at the Surrey Quarter Sessions at Kingston, charged with endangering the safety of passengers travelling on the London Brighton and South Coast Railway.

It was stated by Mr Horniman, prosecuting, that the four had placed large pieces of wood and iron on the line near Betchworth, England, and then waited to see what would happen. What happened was that they were struck by an express travelling at 50 miles an

---

[1631] *Syracuse Herald*, Syracuse, New York, United States, 18th October 1919
[1632] *The Times*, London, England, 13th April 1920
[1633] *The Philadelphia Inquirer*, Philadelphia, Pennsylvania, United States, 23rd July 1921
[1634] Interstate Commerce Commission report
[1635] *South Bend News-Times*, South Bend, Indiana, United States, 5th August 1921

hour, damaging its engine. The train was not derailed and pulled up at Dorking.

The Chairman of the Magistrates noted that Miller and Windley appeared to be the ringleaders, and they were sent to the reformatory until they were 18. James and Lacey were bound over for 12 months. Windley had apparently remarked that if the train were wrecked he would go down, and if there were any rich people there he would take their money. [1636]

A Great Western Railway London Paddington to Plymouth express collided with an obstruction near Cowley Bridge Junction, England, on Thursday 12th October 1922, in what a newspaper report described as 'an apparent attempt at train-wrecking.' A ganger (foreman platelayer) found that the gate of an occupation crossing adjoining a bridge over the River Exe had been removed. It was found, in a damaged condition, on an embankment near the down line a few yards away. It was believed that two persons must have been involved. The gate had been intact the night before. 'No arrest has yet been made,' the report added. [1637]

Another serious 'accident' resulting from vandalism occurred Andover, Minnesota, United States, on Sunday 26th November 1922. Eastbound Great Northern Railway freight train No. 3068, consisting of engine No. 3068, 59 cars and a caboose, in charge of Conductor Nickerson with Engineer Eugene Searls and Fireman August Furst, was derailed at a facing switch there at 10.53 a.m. Brakeman Orville Euthis was killed almost instantly, Fireman Furst was severely scalded and died shortly after reaching hospital and Engineer Searls was injured. [1638]

Conductor Nickerson told the Interstate Commerce Commission inquiry that after the derailment he had found the points set for the main track with the switch lock unlocked and the switch lever out of his socket. By chance, the conductor had been in charge of the last train to use the switch, a westbound working at 5 p.m. the night before. He had watched Flagman Stuart operate the switch then and he was sure that the flagman had properly closed and locked the switch.

Upon questioning an 11-year-old boy admitted that in the morning he had found the switch unlocked, had lifted the lever from the socket but had lacked the strength to replace it. He had then left it and gone away without apparently telling anyone what he had done. [1639]

The Great Western Railway suffered from more vandalism on Wednesday 10th January 1923 when an express train ran into pieces of fence laid on the track between Bathampton Junction and Freshford, England, close to the village of Claverton. 'Village lads are suspected.' [1640]

[1636] *The Times*, London, England, 17th May 1922
[1637] Ibid, 13th October 1922
[1638] *The Commercial Tribune*, Cincinnati, Ohio, United States, 22nd November 1922
[1639] Interstate Commerce Commission report
[1640] *The Times*, London, England, 11th January 1923

The engine of *The Mohawk*, one of the crack fliers of the New York Central Railroad, was derailed just after it had left the station at Rochester, New York, United States, on the evening of Thursday 22nd March 1923, when it struck a bolt wedged into a switch. After bumping over the ties for 200 feet the engine overturned and the following baggage and club cars were also derailed but did not overturn. Engineer Michael McCarthy and Fireman James Hedden were bruised and shaken but remained on the engine to draw out its fire. None of the 250 passengers was hurt. [1641]

Three spikes places on the rails derailed the engine of a Rossland – Nelson passenger train at an unstated location in British Columbia, Canada, on Monday 28th May 1923. The locomotive overturned but the injuries were limited to the fireman suffering a sprained ankle. [1642]

Chicago Burlington and Quincy Railroad train No. 28, in charge of Conductor Harry Carson with Engineer Charles Fowler, ran into a piece of iron two feet long and eight inches wide which had been laid on the track, and then into a railroad tie, between Peru and Nebraska City, Nebraska, United States, on the evening of Wednesday 14th November 1923. The engine pushed the piece of iron to one side and cut the timber in two. It had just got dark, but the engineer saw the obstructions by the engine headlight. When the train stopped its crew spotted three men who ran to a boat and rowed away across the adjacent Missouri River. If adults were involved this possibly comes closer to sabotage than vandalism though the 'men' might conceivably have been teenagers. The motive for the act was unclear. [1643] [1644]

When Western Maryland Railroad passenger train No. 1 struck a tie placed across the track near Union Bridge, Maryland, United States, on Friday 23rd November 1923, there were fears that it was an attempt on the life of Matthew C. Byers, the president of the company, who was riding on the train in his private car. The keen eyesight of Engineer C.E. Smith was credited in preventing a serious accident as he saw the obstruction in time to reduce speed considerably despite the poor visibility in the early dawn and a fog and the only damage caused was a broken pilot on the engine. It transpired, however, that it was an act of vandalism, committed by 16-year-old Filmore Hartsock, who was arrested and confessed to placing the obstruction. He said that when he had been on his way to work at the Union Bridge cement plant he had decided to place the tie across the rails, "to see if the engine would push it off." Detectives who questioned him believed that it had been a boyish prank, done without malice or any design on the life of President Byers.

The report on the accident added that, 'Hartsock, according to reports from Union Bridge, has for some time borne a bad reputation, and has several times been charged with malicious destruction of property. He is alleged to have placed a cross-tie on the

---

[1641] *Syracuse Herald*, Syracuse, New York, United States, 23rd March 1923
[1642] *The Winnipeg Evening Tribune*, Winnipeg, Manitoba, Canada, 29th May 1923
[1643] *The Lincoln Star*, Lincoln, Nebraska, United States, 16th November 1923
[1644] *The Lincoln State Journal*, Nebraska, United States, 16th November 1923

tracks of the railroad several days ago, but the obstruction was discovered and removed before any damage was done. Hartsock lives at the home of his father, a farmer, on a small farm close to the railroad tracks and not far from the station at Union Bridge.' [1645]

Grand Trunk Western Railroad train No. 17, *The International Limited*, was wrecked at South Bend, Indiana, United States, on the evening of Sunday 8th June 1924, when it struck an obstruction on the line. The engine reportedly plunged 20 feet from a trestle and buried its nose in the ground, but the only casualty was a slight injury to Engineer Ray Eagy, the 400 passengers being unharmed. Edward McDonald, aged only 5, showed police where he had placed an iron bolt on the rail. [1646] [1647]

Engineer Harry Sargent, described as a veteran employee, lost his life in the derailment of southbound New York Central Railroad passenger train No. 6 at 5.15 p.m. on Wednesday 27th August 1924 at Millfield, Ohio, United States. The train consisted of engine No. 4751, a baggage car, a coach, a combination coach, and café car and a parlor car in charge of Conductor Carruthers.

Fireman Peter Klein, who was seriously injured, told the Interstate Commerce Commission inquiry that Engineer Sargent had made a service brake application as the train approached a curve, released them, then made an emergency brake application as the train hit the obstruction.

Conductor Carruthers said that he had been unable to find the obstruction after the derailment but Baggageman Deardorff said that just south of the first mark of derailment he had found two pieces of board which had obviously originally been one piece, and later a local inhabitant had told him that it was the same board that he had removed from the track a little earlier in the day. [1648]

The perpetrator was found to be a boy named Ralph Roberts, aged 9 or 10 according to different reports, who told Coroner L.W. Sanders that, "I just wanted to see what would happen. I tried it once before two weeks ago, but somebody took it off the track before the train came." He had also made a previous attempt on the day of the accident. 'Young Roberts was seen tying the timber Wednesday afternoon. 10 minutes before the train was due, by Fred Russell, who took the wood from the track and threw it in the ditch. The boy then returned and wired the wood to the track with fine wire. Ralph is the son of Earl Roberts, of Toledo, and has been visiting at the home of his grandparents, Mr and Mrs Frank Roberts, in Millfield. The boy is being held by court probate officials.' [1649]

The engine, mail, and baggage cars of the *Kansas City – Florida Special* of the St Louis-San Francisco Railway left the rails near Greenfield, Missouri, United States, at 10 p.m. on Wednesday, 18th March 1925. The engine overturned but no one was seriously hurt. The next morning a local boy, 15-year-old Gilbert Alee, was taken into custody by

[1645] *The Gettysburg Times*, Gettysburg, Pennsylvania, United States, 24th November 1923
[1646] *The Kokomo Daily Tribune*, Kokomo, Indiana, United States, 9th June 1924
[1647] *The Indianapolis Times*, Indianapolis, Indiana, United States, 9th June 1924
[1648] Interstate Commerce Commission report
[1649] *The Athens Messenger*, Athens, Ohio, United States, 28th August 1924

Frisco special officers. He admitted that he had placed a 20-foot piece of iron on the track because, "he wanted to see the wrecker pick up the train." [1650]

On the evening of Monday 12th July 1926 two railroad employees found some stones placed on the rails near Sang Hollow, Pennsylvania, United States. They removed them and a small boy and girl were seen to run off into nearby woods. A few minutes later Pennsylvania Railroad train No. 511, the *Lehigh Express*, consisting of an engine and seven coaches and carrying about 600 passengers, arrived at the spot on westbound track No. 3, ran over some obstructions on the track and the pony truck of its engine was derailed. Half a mile of track was ripped up before it could be stopped. A little later a freight train travelling in the opposite direction left the rails at the same place but was stopped before much damage was done. The Conemaugh wrecking train had to come out to re-rail the two engines and there was considerable disruption to traffic. Spikes which had been run over were found at the points of derailment.

An investigation led to the arrest of nine-year-old James Sushak from nearby Oakhurst. The boy confessed to having placed stones and spikes on Nos. 1 and 3 track while playing with little Anna Katinsky, also of Oakhurst, whom he exonerated from all blame. When the first ones that he had laid had been removed he had returned and laid more on both tracks. Railroad officials stated that they did not think that the obstructions had been laid with malicious intent but that the occurrence would be a lesson for parents. [1651]

The existence in the United States of large numbers of facing points which were still, in the 1920s, worked by levers on the spot, rather than being worked from a central signal box as in Great Britain, meant that there was much more potential for them to be manipulated by saboteurs or vandals with serious consequences. Thus there was another derailment with fatalities due to vandalism in the form of interfering with a set of facing points near Dryden, New York, United States on Sunday 12th June 1927. On the afternoon of the day westbound Lehigh Valley Railroad passenger train No. 293 consisted of engine No. 1153, a coach, a combination baggage and smoking car, and a milk car, in charge of Conductor West, with Engineer George Barnes and Fireman Andrew Perkins, was derailed at the facing points leading to the Borden Milk Station siding at 5.55 p.m. The entire train was derailed, the engineer overturned, and Engineer Barnes and Fireman Perkins were killed.

Conductor West told the Interstate Commerce Commission inquiry that he had felt the air brakes applied about 400 feet short of the switch. He had examined the switch after the accident and had found the points fully open for the siding and the switch target showing red. The switch target was the only warning to an approaching trainman of the wrong position of the points; it would only be visible from a relatively short distance away. The train, he said, had been travelling at 35 or 40 miles an hour.

Conductor Moore of eastbound train No. 286 told the inquiry that his train had picked up two milk cars from the siding several hours before the accident. He had seen

---

[1650] *Joplin News Herald*, Joplin, Missouri, United States, 19th March 1925
[1651] *Indiana Evening Gazette*, Indiana, Pennsylvania, United States, 16th July 1926

Brakeman Bowe close the switch afterwards and was positive that he had locked it. Trainman Corcoran agreed with Coroner Moore's statement except that he said that he had been too far away to verify that the switch had been locked. Brakeman Bowe himself was positive that he had properly closed and locked the points.

The crews of eastbound train No. 1140 and westbound extra freight No. 814, which had passed the spot between the time that train No. 286 had called at the siding and the accident, stated that they had noticed nothing wrong with the switch. [1652]

It subsequently transpired that the switch had been opened by 12-year-old Harold Brown of Dryden. According to him he had found the switch lock open and hanging by its chain from the guard lever. He had kicked the guard lever down and operated the lever to open the points about one third of the way. Having done so he had then gone on to the home of another boy with whom he had gone fishing. He said that he had had no intention of causing a wreck but had merely been curious to see how the switch worked. [1653]

On Wednesday 7th September 1927, a Chemnitz – Altendorf – Wüstenbrand goods train was derailed at an unstated location in Saxony, Germany, because a block of wood had been fastened to the rails. A reward of 700 marks was offered for information as to the perpetrator. Then at about 7 p.m. on Tuesday 11th October 1927 an Eilgüterzug (express goods train) ran into a tar barrel near Lehndorf, Saxony, Germany. The obstruction was pushed 200 metres, damaging the engine and the brake equipment of the five following trucks. A 300 marks reward was again offered and eventually a 16-year-old youth was apprehended, who had rolled the barrel onto the line. [1654]

Perhaps because acts of railway vandalism in most countries result only in relatively minor 'accidents' reports of them in countries outside of Great Britain, where the author resides, and the United States, have proved hard to discover. However, incidents such as the two just detailed in Saxony above, have come to light, together with one in France on 19th October 1927 and one in Australia the following year. Three small children, who lived near the station at Donnemarie-en-Montois, conceived the idea of derailing a train for amusement. They were Robert Lelièvre, aged 7; his sister Simone, aged 8, and André Gobinot, aged 5½. Some minutes before the arrival at Donnemarie at 16:17 of the train from Bray-sur-Seine, Robert operated the points at the entrance to the station, while his sister and his friend placed stones in the gap between the tongue of the points and the stock rail. Then, hearing the whistle of the train, they ran away without being seen. The locomotive of the train and two coaches were derailed. Fortunately, no one was hurt but the damage to material was considerable.

The police investigating of the affair soon uncovered the facts from a friend of the three children in whom they had confided. [1655]

[1652] Interstate Commerce Commission report
[1653] *Syracuse Herald*, Syracuse, New York, United States, 17th June 1927
[1654] *Eisenbahnunfälle in Sachsen* by Claus Schlegel, pg. 143 (in German)
[1655] *Excelsior*, Paris, France, 20th October 1927

When the 5.40 p.m. goods train from Goulburn to Hardin was between Cullerin and Fish River, New South Wales, Australia, at about 7 p.m. on Sunday 15th April 1928, its engine ran into an obstruction on the line and its leading wheels were derailed. An inspection showed that the obstruction was a heavy piece of timber which had apparently been placed on the track deliberately. A breakdown train was called out from Goulburn but before it arrived the engine crew had been able to re-rail the engine using jacks. There was no settlement nearby, but a large camp of men engaged in road construction was close at hand. [1656]

The following day another goods train ran into an obstruction on the Broken Hill-Sydney line, presumably not far away from the first incident. This time the obstructions were two concrete telephone blocks, 18 inches square and 21 inches thick at one point, and a little further on a 5-foot-long iron telephone pole. This time the train was heavy enough to smash through the obstructions without being derailed. The police were investigating the matter. [1657]

The London and North Eastern Railway down passenger train carrying 200 passengers which had left London Marylebone station at 10.38 a.m. on Wednesday 24th July 1929 collided with a piece of timber and a drum of paint which had been placed on the line at Northolt Junction, England. The driver managed to stop his train before it was derailed but paint from the drum reportedly interfered with the working of the engine. Had there not been a check rail on the curve where the obstruction was placed it was thought that it might have been wrecked. Half a mile down the line a large coil of wire had been placed across the track.

The same night the police arrested a 12-year-old boy who was charged at Greenford police station with endangering the lives of passengers and was due to appear at Uxbridge Juvenile Court tor following Monday. [1658]

Northbound St Louis-San Francisco Railway passenger train No. 118 from Sherman, Texas, to Tulsa, Oklahoma, consisted on Sunday 18th August 1929 of engine No. 1056, a baggage car, a combination mail and baggage car, a coach, a chair car and three Pullman sleeper cars, in charge of Conductor McCaughan with Engineer Pete Wolfe and Fireman H.A. Bryan. [1659] The cars were of all-steel construction except the third and fourth, which had steel underframes. Travelling at between 50 and 60 miles an hour the train reached Henryetta, Oklahoma, United States, at 5.59 a.m., and there was derailed on a partly open set of facing points. The entire train was derailed apart from the last car. The engine was badly damaged, the baggage car mounted it, the combination mail and baggage car went off to the east and the coach, marshalled third in the train, came to rest against the rear end of the boiler, directly opposite the broken safety valves, from which live steam was escaping under pressure. This was the Jim Crow car, in which black passengers were compelled to travel, and it was here that all the fatalities occurred, 12 passengers being scalded to death. Engineer Wolfe and Fireman Bryan also lost their lives, and 27

1656 *Goulburn Evening Penny Post*, Goulburn, New South Wales, Australia, 16th April 1928
1657 *Evening News*, Sydney, New South Wales, Australia, 17th April 1928
1658 *The Times*, London, England, 25th July 1929
1659 *The Salt Lake Tribune*, Salt Lake City, Utah, United States, 19th August 1929

passengers, two railwaymen, two mail clerks and two express messengers were injured.

Conductor McCaughan told the Interstate Commerce Commission inquiry that the first he had known of the derailment was when it happened, without any prior brake application, and the car had begun to fill with steam. He had climbed out of a window, called for assistance from a nearby telephone and then made his way to the switch to find what had happened. The switch he had found unlocked with the switch lock missing and the lever was raised and moved about two inches out of position, so that the switch was open about an inch. He did not notice the switch lamp. The switch target was showing almost full on to the engineer, but he thought that it had probably been moved further as a result of the accident and would not have been visible to the engineer.

Flagman Andre corroborated the statement of the conductor.

Porter Stoner believed that there would have been no deaths but for the escaping steam.

Assistant Superintendent Hutchinson and Divisional Engineer Busch confirmed the state of the switch and that there had been no indication of dragging equipment. Divisional Engineer Busch said that the switch lamp had been out, he had found the wick completely turned down with the appearance of it having been tampered with. Assistant Superintendent Hutchinson said that during the afternoon some small boys had shown him where they had found the switch lock thrown into some weeds.

The switch had last been used two days before the accident. Brakeman Adams was positive that he had locked the switch afterwards. There was evidence given that mischievous boys had previously interfered with switches in the yard, though not on the main line.

The ICC concluded that the accident had been due to a partly opened switch due to malicious tampering. The switch had been inspected since it was last used, with nothing wrong being found, and since its last inspection trains had passed over it in both directions. After the derailment it had been found to be open and tests showed that it could be opened by giving a few hard jerks on the chain. It also appeared that the switch had not been open far enough for the target to be obvious to the engineer.

It was noted that the company intended to install the automatic block system over the subdivision where the derailment occurred, whereby an open switch would cause the signals to return to danger. [1660]

The driver of a Southern Railway train from West Croydon on Sunday 5th January 1930 reported at Mitcham Junction, England, that his engine had struck an obstruction just outside of the station. The stationmaster and another employee examined the line and found an iron skid-pan, such as used as a brake on cart wheels and weighing about 13lb, fixed to a rail with strong steel wire. They then questioned youths on nearby Mitcham Common, as did the police. 'Last night Scotland Yard officers were making inquiries,' a reporter added. [1661]

[1660] Interstate Commerce Commission report

[1661] *The Times*, London, England, 6th January 1930

The fatal derailments in the United States resulting from apparent acts of vandalism continued on Sunday 11th May 1930 with one at Dean's Spur, two miles south of Harriston, Mississippi, United States, on the Yazoo and Mississippi Valley Railroad. Southbound passenger train No. 15, known as *The Planter*, consisted of engine No. 1118, a baggage and mail car, a baggage and express car, three coaches, and three Pullman sleeping cars, all of steel construction, in charge of Conductor C.R. Day, Engineer Emery Maddox and Fireman C.W. Baker. Travelling at 30 to 35 miles an hour the train the engine and first two cars of the train left the rails. Engineer Maddox and Fireman Baker were killed, and two passengers were injured.

Conductor Day stated at the Interstate Commerce Commission inquiry that going back after the derailment he had found the crushed remains of a piece of terra cotta pipe on the east rail and directly opposite to it on the west rail a piece of timber which had been cut in two. Flagman Hattaway and Train Porter Shepherd corroborated the statement of the conductor.

Track Supervisor Brevard also confirmed the existence of the obstructions and added that the objects had been taken from an old mill site nearby. Section Foremen Aldridge and Farr agreed. [1662]

A very sad feature of the accident was that the wife of Fireman Baker had been a widow whose previous husband, an engineer, had died in a wreck the previous year. Now she was widowed again.

Bloodhounds used to track the offender picked up a trail but lost it in the rain. The derailment was attributed to vandals who had reportedly caused several other derailments in previous weeks. [1663] A reward of $10,000 was offered for the apprehension of the culprits. It is possible that this came closer to an act of sabotage rather than vandalism as at one point suspicion attached to 35-year-old J.C. Oversby, a former section foreman, but he was later cleared of involvement. [1664] [1665]

A Denver and Rio Grande Railroad freight train was derailed on Tuesday night, 13th May 1930 a mile south of Spanish River, Utah, United States. The engine and four double-decker livestock cars left the rails. They were carrying 800 sheep – ewes, lambs, and yearlings on their way from winter to summer range. [1666] Nearly 200 of the animals were killed but Engineer H.J. Smith and the other three men on the train escaped unhurt.

Initially the derailment was attributed to a faulty switch, but it was later discovered that two local boys, aged 7 and 9, had chanced to find a switch key the previous Saturday, had used it to unlock the switch and had opened it and reclosed it. Repeating the trick Tuesday they had then found that they were not strong enough to restore it and frightened and afraid to admit that they had been fiddling with the points, they had just run away. Their names were withheld. [1667]

[1662] Interstate Commerce Commission report
[1663] *The Daily Herald*, Biloxi, Mississippi, United States, 12th May 1930
[1664] *Monroe News Star*, Monroe, Louisiana, United States, 16th May 1930
[1665] *Hattiesburg American*, Hattiesburg, Mississippi, United States, 17th May 1930
[1666] *The Ogden Standard-Examiner*, Ogden, Utah, United States, 14th May 1930
[1667] *The Salt Lake Tribune*, Salt Lake City, Utah, United States, 16th May 1930

CHAPTER 18

SABOTAGE

Four distinct motives for sabotaging trains may be distinguished (after Robert B. Shaw):

a. As an aid to intended robbery, either of the train passengers or to steal for example bullion being carried by the train
b. To satisfy a personal grudge against the railway company, such as by an ex-employee who feels that he has been wrongly dismissed
c. To hinder the operation of the railway, as for example during a strike
d. As an act of guerrilla warfare or terrorism, perhaps against an occupying enemy power. Sabotage against the railways has been a feature of many conflicts from the American Civil War onwards.

The period covered by this volume saw many acts of sabotage in a good number of different countries, but especially in Mexico, where a prolonged civil war continued until 1920, in Palestine against the Turks towards the end of the First World War and a little later in Ireland during the struggle for independence from British rule and the civil war which followed. Other acts of sabotage for political reasons occurred in countries such as Russia and China and as will be seen there were more isolated acts of sabotage from the first three causes listed above.

Missing from the 1881 – 1900 volume of this series because it has only just come to light is an act of apparent sabotage at Bubentsch, Bavaria, on Thursday 27th August 1891. A passenger train and a goods train collided, killing 12 people, and injuring a good many more. The collision was allegedly caused by the deliberate act of a Czech switchman, who had recently had an argument with the driver of the passenger train and had decided to engineer a collision in revenge, regardless, as the newspaper report pointed out, of the death and suffering he was to cause to so many innocent people in pursuit of his private revenge. He was arrested and only strong police protection prevented his being lynched by angry locals and passengers. [1668]

[1668] *Winnipeg Tribune*, Winnipeg, Manitoba, Canada, 27th August 1891

Robert Shaw comments in his book that reports of railway accidents in Mexico, despite its geographical proximity to the United States, have been as vague as if they had occurred in Outer Mongolia. [1669] Writing in the third decade of the twenty-first century with access to thousands of newspaper archives online, the situation is somewhat easier, but the frequent acts of sabotage committed in Mexico during its civil war are still hard to disentangle, especially as they tended to happen at times close together in different parts of the country, and to be only reported in American newspapers days or weeks later. However, an attempt will be made, as in the previous volume, to describe them all in chronological order.

On Monday 3rd January 1916, a military train conducted by Colonel Berdusco fell from a fire-damaged trestle near Monclova, Mexico, according to a telegram received by the American consul in Eagle Pass, Texas, United States. Twenty people, including women and children, lost their lives. [1670]Although the brief reports do not specifically say so, as Mexico was involved in a long-running civil war at the time and a military train was involved, it seems probable that the trestle had been deliberately set alight.

An American metallurgical engineer named Andrew Trumbo who arrived in Douglas, Arizona, United States on 22nd January 1916 from Mexico spoke of various atrocities in the latter country, such as the hanging of eight American miners. He also mentioned that the train which had preceded the one on which he and his wife had been riding had been dynamited between Oaxaca and Pueblo and 23 Mexicans had been killed. He added that their own train had been fired upon, but no Americans had been injured.

Separate reports received in Douglas on the same day mentioned that a troop train filled with Carranza troops had been dynamited between Vera Cruz and Jalapa, Mexico, and several hundred soldiers had been killed. [1671]

In about mid-February 1916 or possibly a little earlier Russian guerrillas destroyed a considerable stretch of the railway between Libau (now Liepāja in Latvia) and Poneviezh. As a result a German troop and munitions train carrying men, shells, and guns to the Novo Alexandrovsk positions near Dvinsk (now Daugavpils in Latvia) was wrecked, and the munitions exploded. Over 100 soldiers were killed and injured. [1672]

A troop train on the Pacific Coast extension of the Ferrocarril Central Mexicano was derailed by sabotage at Sayula, Mexico, on Tuesday 21st March 1916. A rail had been loosened or removed and all fifteen cars of the train, loaded with men and cavalry horses, went over a precipice, was wrecked, caught fire, and was burnt out. Fifty troops were

---

[1669] A History of Railroad Accidents, Safety Precautions and Operating Practices by Robert B. Shaw, pg. 408

[1670] *Mt Pleasant Daily News*, Mt Pleasant, Iowa, 4th January 1916

[1671] *The Commercial Tribune*, Cincinnati, Ohio, United States, 23rd January 1916

[1672] *The Times*, London, England, 23rd February 1916

reportedly killed and 100 injured. [1673]

Seventeen bandits derailed a passenger train twenty miles south of Durango at La Punta, Mexico, on Thursday 30th March 1916. They then set fire to the train and threw those who attempted to escape from the burning carriages, who were mostly women and children, back into the fire so that all the passengers and train crew perished. Two hundred Carranza soldiers arrived at the scene three hours later to find only ashes and parts of bodies. Seventy-eight people lost their lives according to a report confirmed by Mr Graham, the former British Consul in Durango, Mexico, who had fled to El Paso, Texas, across the border in the United States, for his own safety. There had been no valuables on the train. [1674]

In late April 1916, the Easter Rising (or the Eastern Rebellion as the British termed it) began in Ireland, all of which at this date was still effectively a British colony. The rising was an attempt to establish an Irish republic independent of British rule, and although unsuccessful it led to the partition of Ireland and the independence of the southern counties in 1922. The Easter Rising was planned to begin in Dublin on Easter Monday, 24th April 1916 but the very first action was an act of sabotage directed against the Great Southern and Western Railway at Colt Wood, Clonadadoran, Ireland, between Portlaoise and Abbeyleix, at 7 p.m. on Sunday 23rd. Telegraph lines and rails were cut down and rails and sleepers were removed. The object was to hinder the transport of British troops who might be sent to Ireland from reaching Dublin by rail. The first shot fired in the uprising was a warning shot over the head of a railwayman, William Dalton, who had been sent out to investigate a signal failure resulting from the wire cutting. A train was derailed on the damaged track and its driver and fireman were thrown from their engine, but no one was much hurt. [1675] Further derailments due to sabotage followed on the Midland Great Western Railway at Blanchardstown and Liffey Junction.

A passenger train left Mexico City at 5 p.m. on Thursday 11th May 1916. Attached to its rear was the private car of Judge Charles A. Douglas, the United States representative in Mexico City. At 2.30 a.m. the next morning, Friday 12th May, between Queretaro and Palma Gonzales, Mexico, its sleeping passengers were awakened by a sudden jar and a terrific detonation, followed by a volley of rifle shots. Outside a party of bandits was shouting, "Viva Villa," and firing repeatedly to frighten the passengers.

The bandits had burned a trestle, derailing the engine and four leading cars of the train, and killing instantly the engineer, fireman, helper and, according to some reports, five other persons riding on the engine, which if true must have made its footplate very cramped. Entering the second-class car at the front of the train the bandits had worked their way along it, robbing the passengers and striking and abusing some of them. Two army paymasters were robbed of $20,000 and $16,000 of the new Carranza currency, a

---

1673 *The Burlington Hawk-Eye*, Burlington, Iowa, United States, 22nd March 1916

1674 *The Commercial Tribune*, Cincinnati, Ohio, United States, 23rd April 1916

1675 www.historyireland.com/volume-24/issue-5-septemberoctober-2016/colt-wood-railway-incident/

paymaster of the El Oro mining company lost $35,000 in the same currency and an American named Hill lost $3,000 in gold. Jewellery and personal belongings were also taken, together with any clothing not actually being worn by the passengers.

Two Carranza lieutenant colonels and two paymasters avoided recognition by the bandits, having removed their uniforms and insignia. Eight soldiers guarding two prisoners were abused and their arms taken from them, while the prisoners were released and left with the bandits. None of the occupants of the train dared put up any resistance for fear of being killed. Either because they were in a hurry to get away or more likely perhaps because they were ignorant of its contents the bandits left a million pesos of the new currency in the express car. The passengers remained in their cars after the attack until 8 a.m., afraid that the bandits would return. They were then rescued by a party from Palma Gonzales, 4 miles away. A strange feature of the affair was that the private car of Judge Douglas was not molested. It was believed that the bandits were Zapatistas, using the rallying cry of the Villaists to conceal their true identity. [1676]

Much more serious than the above attack, but only briefly noted by the American press, was the derailing of another train on or about 16th May 1916 at Ayuzco, Mexico, near Mexico City. This train, it was reported, was carrying 1,000 passengers, 150 of whom were burnt to death in one car, while the train's armed guard was wiped out. [1677]

Political strife in Kwangtung province, China, in 1916, meant that the authorities had lost control of robber bands which were plundering and making trouble in many parts. On Thursday 15th June, the robbers ripped up the track, derailed a train on the Kowloon – Canton (now Guangzhou) line near Fungtong and the leading two third-class carriages were overturned. After the derailment the robbers, who reportedly numbered about 100, calmly went through the train and robbed the passengers of money and valuables worth about $10,000 at 1916 prices. Two or three people died in the derailment and another 22 were seriously injured. One or two of the dead was Chinese, the other was an American missionary of the Hong Kong Pentecostal mission named Elmer B. Hammond. Those who lost their lives had been in the leading coach which was smashed against the engine. [1678] [1679]

Twenty-five soldiers and 11 civilian passengers died when supporters of Carrero Torres derailed a Constitutionalist train between San Luis Potosi and Tampico, Mexico, 'below Cardenas' on Thursday 31st August 1916. The bandits butchered the train guards and carried away everything movable. [1680]

Bandits wrecked a train on Saturday 23rd September 1916 between San Luis Potosi and Gonzales Junction, Mexico, by tearing up the track. Several lives were lost in the

---

[1676] *Laredo Weekly Times*, Laredo, Texas, United States, 21st May 1916

[1677] *The Daily Gate City*, Keokuk, Iowa, United States, 16th May 1916

[1678] *Beatrice Daily Sun*, Beatrice, Nebraska, United States, 6th August 1916

[1679] Forgotten Souls: A Social History of the Hong Kong Cemetery by Patricia Lim, pg. 481

[1680] *Oakland Tribune*, Oakland, California, United States, 4th September 1916

initial derailment, and a few minutes later a southbound mixed train ran into the wreckage. The bandits then allegedly set fire to the wreckage and many more of the passengers were probably burned to death. Information on the attack came from Mr T.B. Hohler, British charge in Mexico City, who arrived in Laredo, Texas, on 27th September, having had to travel via Aguascalientes to reach San Luis Potosi and the United States because of the wreck. [1681]

A Georgia Railroad passenger train was derailed at Union Point, Georgia, United States, at 9.27 a.m. on Friday 27th October 1916. Engineer Dave Printup, Fireman John Curry and Machinist W.H. Whiteside all died in the wreck [1682] and it was the subject of a full Interstate Commerce Commission (ICC) inquiry.

Westbound passenger train No. 1 from Augusta to Atlanta, consisted of engine No. 202, an express car, a combination mail, and baggage car and three coaches, all of wooden construction, though the express car had a steel underframe. It was in charge of Conductor Clary with Engineer Printup and Fireman Curry.

Conductor Clary told the ICC inquiry that his train had been travelling at 30 to 35 miles an hour as it approached the east switch at Union Point. He had felt the brakes being applied and the train had become derailed at 9.27 a.m. Going back to the switch (facing points) he had found it set for the passing track and as the wheels of a derailed vehicle of his train were standing on the points the position of the switch could not have been changed after the derailment.

Engineer Brisendine, of extra No. 57, a westbound light engine which was the last train to pass the location before the accident, at 8.22 a.m., stated that he had been travelling at 15 or 17 miles an hour. The target, visible from half a mile away, had then shown a white indication and the switch had been closed. There had been no one near the switch.

Track Foreman Caldwell said that he and his men had been working half a mile from the switch from 6.30 a.m. He had a key to the switch, which had not left his possession that morning, but none of his men had one.

Chief Law and Special Agent Yarborough stated that there had been a similar accident on 10th January at Swords, Georgia, and suspicion then had fallen on a discharged coloured employee who had a switch key in his possession. The man had subsequently disappeared. The switch stand was of the same type as that at Union Point.

The ICC concluded that the derailment was due to the switch having been opened by some unauthorised person with malicious intent. It commented adversely on the fact that Air Brake Inspector Whiteside had been riding on the locomotive, though his annual pass read "Good on Freight Trains and Locomotives." It was suggested that the presence of Whiteside on the engine might have caused Engineer Printup to neglect his duties, as otherwise he could have brought his train under control while travelling through yards limits, as required by Rule No. 93, and seen the red switch target in time to pull up short of the points. [1683]

[1681] *San Antonio Light*, San Antonio, Texas, United States, 27th September 1916
[1682] *The Lakeland Evening Telegram*. Lakeland , Florida, United States, 28th October 1916
[1683] Interstate Commerce Commission report

Villa supporters dynamited a train carrying Constitutionalist soldiers at a point near Carneros, Mexico, 47 miles south of Saltillo, on Thursday 21st December 1916, killing an unspecified number of troops. Those soldiers who survived the initial derailment were massacred by the attackers. [1684]

The First World War in Palestine, in which British forces, aided by the Arabs, fought the Ottoman Empire, was characterised by a number of acts of sabotage against the Ottoman (Turkish) Hedjaz Railway. The first attack on the railway had been conducted by the Arabs at the beginning of the Arab Revolt against Ottoman rule in June 1916, when they had ripped up a few rails, but the revolt had stalled when they had run out of ammunition, and the Turks had the upper hand, until the arrival of Colonel T.E. Lawrence, who re-organised the Arabs forces and persuaded them to carry the sort of guerrilla actions to which they were more suited than stand-up fights. Thus on Monday 12th February 1917, a party of Arabs led by Major Garland mined a bridge on the Hedjaz Railway south of Toweira, Palestine, and a section of track further south. A train was derailed, the engine going down the embankment. [1685]

An American passenger who arrived in Laredo, Texas, on Thursday 17th February 1917 brought news of attacks on no less than four trains in three hours a few days earlier. The attacks all occurred at or near Orizaba, Mexico, and were carried out against passenger and troop trains by Zapatista forces, who placed dynamite charges under bridges and on the track. In one case a passenger train standing on the track near Orizaba station was dynamited and 'practically blown to atoms.' Reportedly killing more than 50 persons, including an English woman and her daughter. About an hour later another train carrying civilian passengers and about a hundred Carranza soldiers arrived, ran over a mine, and was wrecked. The other two trains were blown up as they were leaving Orizaba for Mexico City. After the attacks, the Zapatistas robbed the bodies of the victims of clothing, jewellery and valuables and then disappeared into the mountains. A special train carrying a 1,000 Carranza soldiers was sent out from Mexico City in pursuit of the bandits, but it was not thought that they would be caught. [1686]

Colonel T.E. Lawrence made his first attack on the Hedjaz Railway on Monday 26th March 1917, when he mined the railway north and south of Aba El Na'am station to seal a train in the station. The train moved south, the driver reversed his engine when he saw the charge and only the front of the engine was damaged. The crew repaired the damage, and the train continued its journey. The next attack was more successful, and a train was derailed. Soon the attacks became almost a weekly occurrence. [1687]

One or possibly two attacks were carried out on trains in Mexico on Sunday 6th May

---

[1684] *Logansport Daily Tribune*, Logansport, Indiana, United States, 24th December 1916
[1685] *Hedjaz Railway* by R. Tourret, pg. 69
[1686] *Laredo Weekly Times*, Laredo, Texas, United States, 18th February 1917
[1687] *Hedjaz Railway* by R. Tourret, pgs. 70-71

1917. A train carrying soldiers, their wives and children was derailed just north of Queretaro. Fourteen died and several score were injured in the crowded train. The line was of course blocked and the ordinary passenger train, which was following, and which was carrying the passenger who brought the news of the derailment to Laredo, Texas, was held up until the following day. Another source attributes this derailment, however, to track defects rather than sabotage and it is described more fully in the first chapter of this book.

The other derailment on Sunday 6th May 1917, either before or after the Queretaro one, was certainly an act of sabotage and had as its object a train carrying $50,000 in its express car, with an escort of 300 soldiers. The train was dynamited near Carneros, Mexico, 30 soldiers being killed and forty injured. No civilian passengers were killed in this attack, but they were reportedly marooned for two days among the dead and dying Carranzistas before a relief train arrived with medical aid. Among the passengers was reportedly the wife of General Murguia, who had been in command of the campaign against Villa in Chihuahua. [1688]

A major strike was in progress in Spain in August 1917, with a general strike in Madrid, and at 5 p.m. on Monday 13th August 1917 the Barcelona – Bilbao mail train was derailed at Fuentelapeña, about three kilometres outside of Bilbao, rocks having been placed on the track. The consequences were serious; the engine, tender and two coaches fell down an embankment, six people died and 18 were injured. There was, however, some debate in the months following as to whether the derailment was in fact connected with the industrial action or whether it was a coincidental act of vandalism or sabotage. [1689] [1690]

On Monday 17th September 1917 T.E. Lawrence carried out his first act of railway sabotage using an improved method, substituting direct firing of a charge by electricity for the less reliable automatic mines used previously. The attack was carried out at km 587 of the Hedjaz Railway, just south of Mudawara. The charge was detonated under the second engine of a double-headed northbound freight train of ten covered waggons as it was crossing a two-arched bridge. The structure collapsed and the waggons fell into the gap or piled up on the line. The second engine was badly damaged and the first derailed and half overturned. Lawrence put a charge of gun-cotton in its outside cylinders, blowing off its front axle. [1691] This is probably the attack detailed in a War Office announcement the following week, in which it was stated that 68 Turks and two German officers had been killed and 80 Turks taken prisoner. [1692]

On about Friday 5th October 1917 Lawrence and his Arabs attacked the railway south of Ma'an, looking first at km 475 but it was guarded by blockhouses (as used by the

---

[1688] *San Antonio Light*, San Antonio, Texas, United States, 9th May 1917
[1689] *The Times*, London, England, 14th and 17th August 1917
[1690] Webpage: Federación Castellano Manchega de Amigos del Ferrocarril – Accidentes Ferroviarios en España
[1691] *Hedjaz Railway* by R. Tourret, pg. 73
[1692] *The Times*, London, England, 27th September 1917

British in the Second Boer War) so they moved south to km 489, where the line ran on a high embankment with culverts. A new automatic Lyddite mine failed to explode under a water train, whereupon Lawrence laid an electric mine upon the Lyddite one, and both exploded beneath a train of 12 loaded goods waggons and the engine was wrecked. [1693]

On about Wednesday 10th October 1917 70 bandits derailed a mail train on the Vladikavkaz – Rostoff – Tiflis line near Vladikavkaz, Russia, attacked the passengers with rifles and robbed them. Eight passengers died and 50 were wounded. [1694] (Another report spoke of 200 brigands and 100 passengers killed or wounded.) [1695]

A Transportzug (troop train?) was derailed near the station at Gjunis, Serbia, on Thursday 11th October 1917, the rails having been damaged by explosives. The engine, tender and all the vehicles of the train were thrown down an embankment and there were 'numerous' dead and injured. [1696]

Villa followers reportedly blew up using dynamite six cars carrying automobiles at a point south of San Luis Potosi, Mexico, on Monday 29th October 1917. No casualties were reported. [1697]

On Sunday 4th November 1917 Villa followers carried out a more deadly attack than that near San Luis Potosi when they dynamited a train at Armendariz station, Mexico, on the Ferrocarril Central Mexicano. Robbery was clearly part of the motive for the attack as the train was carrying 110 bars of silver bullion belonging to the American Mining Company, which was taken, together with passengers' valuables. Sixty Federal train guards were killed and several passengers, including Americans, were executed according to reports, with others wounded. Pancho Villa himself, the leader of the Villa revolutionaries, was said to have been seen near the scene. The Federal major commanding the escort reached Chihuahua, badly wounded, the next day. He stated that he had been forced to witness the execution of several of his men and that the attack had been made under the command of Silvester Quevedomarty Lopez, a former Villa officer. [1698]

On Sunday 11th November 1917, a mine planted on a rebuilt culvert at km 172 at Minifir, Palestine, exploded under the pilot engine of a double-headed train of 12 passenger coaches which chanced to be carrying Jemal Pasha, Commander of the Turkish Eighth Army Corps, who was hurrying down to defend Jerusalem against General Allenby. The pilot engine was destroyed, its boiler blown off, the train engine fell upon it into the gap and the whole train was derailed. [1699] Jemal escaped, but his aide-de-camp

---

1693 *Hedjaz Railway* by R. Tourret 73-4
1694 *The Times*, London, England, 12th October 1917
1695 *Chicago Examiner*, Chicago, Illinois, United States, 21st October 1917
1696 *Eisenbahnünfalle Bind I* by Ludwig Stockert, pg. 118 (in German)
1697 *San Antonio Light*, San Antonio, Texas, United States, 30th October 1917
1698 *The Alaska Daily Empire*, Juneau, Alaska, United States, 6th November 1917
1699 *Hedjaz Railway* by R. Tourret, pg. 74

and Staff Officers were killed. The Turks reportedly suffered 120 casualties, the Arabs 7 dead and 4 wounded. [1700]

On about Friday 16th November another party blew up the Hedjaz Railway between Amman and Ma'an, Palestine, and a double-headed northbound train with 60 men was wrecked, while a stone bridge near El Hasa station was blown up and its guard taken prisoner.

On about Sunday 16th December 1917 an Arab force under Sharif Faisal destroyed a troop train on the Hedjaz Railway south of Tebuk, Palestine. The whole of the Turkish detachment on the train were killed or captured, those killed including Suleiman Pasha Rifada, Paramount Chief of the Bili tribe, 'who, foolishly throwing in his lot with the Turks, was travelling with them.' Three hundred rifles, a large quantity of ammunition and '£24,000 Turkish in gold' were captured. [1701]

A train on the Southern Pacific of Mexico Railroad, which had begun its journey southwards at Nogales the day before, was derailed late on Wednesday 2nd January 1918 by Yaqui Indians, who had torn up the rails on a curve at Pitahaya, Mexico. The Indians then attacked the rear of the train, massacring the mostly Mexican passengers. After looting the baggage and express cars they rode away. The train had an escort of between 50 and 75 Federal soldiers, but the troops were greatly outnumbered by their attackers. A first report stated that between 20 and 30 people had been killed, including the conductor and express messenger, while a further report spoke of 200 dead, including seven Americans from Tucson, Arizona, and an unknown number of wounded. It added that dozens of women had been captured and driven into captivity. [1702]

A third report stated that the Indians had burned a bridge and that 107 men, women and children had been killed and 24 wounded. [1703]

The indigenous Yaqui Indians had been in conflict with the Spanish colonists and later the Mexican state for four centuries. In the opening years of the twentieth century they were greatly oppressed by the Diaz regime and on the outbreak of the Mexican Revolution in 1910 they had joined the armies of all the rebel factions. [1704]

On Saturday 9th February 1918 what was described as the worst attack on railway property up to that time in the country occurred at San Martin, Argentina, on the Ferrocarril Central Argentino. A goods train was derailed and 23 waggons, loaded mostly with wheat, oil and cylinders of carbolic acid were set on fire. The railwaymen were on strike and the object of the sabotage, in a Buenos Aires suburb, was to stop the movement of goods to the port. In this the strikers apparently succeeded, and it was feared that attempts would be made to burn the grain which had accumulated at outlying stations. The railway company had requested additional police protection but only eight officers

---

1700 *The Times*, London, England, 28th December 1917

1701 Ibid, 22nd and 28th December 1917

1702 *The Daily Gazette*, Stirling, Illinois, United States, 3rd January 1918

1703 *Greensburg Daily News*, Greensburg, Indiana, United States, 4th January 1918

1704 Wikipedia – Yaqui Wars

had been available to cover the whole suburb. [1705]

Arabs mined and derailed a train consisting of an engine and ten waggons near Sir Jedid, Palestine, on the Hedjaz Railway, on Wednesday 20th March 1918. Fire and explosives then destroyed the train. Twenty Turks were killed and ten taken prisoner. A large quantity of clothing and rifles was also seized. [1706]

Arab forces wrecked a train near Toweira, Palestine, on the Hedjaz Railway, on or about Saturday 8th June 1918. [1707]

The Finnish Civil War briefly mentioned in the Single-Line Collisions chapter ended with the defeat of the Red Guards in mid-May 1918, but an act of sabotage aimed at members of the Landtag, the Finnish parliament, was carried out on the night of Thursday 18th July 1918. The screw securing the rails were withdrawn and the rails loosened so that the engine and several carriages of a train bound for Helsingfors (Helsinki) were derailed at an unstated location and rolled down an embankment. Two people were killed and five badly wounded. It was supposed that the attack had been perpetrated by members of the defeated Red Guards. [1708]

Two trains were attacked in Mexico by Villa forces on Saturday and Sunday 10th and 11th August 1918. In the first attack a heavy charge of dynamite was exploded under a passenger train at Consuela station, killing a number of women and children passengers. The train guard attempted to resist but were killed. The engine crew escaped and took their locomotive to a nearby station and a relief force drove off the bandits after they had robbed the passengers of their clothes and belongings. Forty soldiers and 26 civilians were reportedly killed and 70 wounded. Martin Lopez's band carried out the attack, but Francisco (Pancho) Villa was also present, and he recognised among the passengers General Justino Catero, one of the Feral commanders. Catero was ordered to be executed and his aged mother was also killed, while his daughter was carried away by the bandits.

In the Sunday attack Lopez's band dynamited a train on the Ferrocarril Central Mexicano in the Bachimba Cañón. The train was a special carrying ammunition to the Santa Rosalla garrison which was captured by the attackers. The rolled down an embankment following the derailment. An initial report stated that no one had died in this attack, but a later report stated that 55 Federal soldiers had been killed and wounded, some as a result of the initial derailment and others shot by the Villa followers. [1709] [1710]

A few days later a man named Jesus Prado was arrested and confessed to having supplied the dynamite for the two attacks. He was tried by court martial, found guilty and

[1705] *The Times*, London, England, 11th February 1918
[1706] Ibid, 10th April 1918
[1707] Ibid, 11th June 1918
[1708] Ibid, 22nd July 1918
[1709] *The Bisbee Daily Review*, Bisbee, Arizona, United States, 13th August 1918
[1710] *Morning Sun*, Yuma, Arizona, United States, 16th August 1918

executed in the Chihuahua penitentiary plaza on Thursday 15th. [1711]

Two Americans who arrived in El Paso, Texas, from Mexico, confirmed reports of an attack on a passenger train at Canatlan, Mexico, on the Durango – Santa Catarina de Tepehuanes branch, about midway between the two places, on Monday 19th August 1918. A bridge was destroyed, the train fell into it, and its occupants, who included women and children, were then shot and bayoneted. The attackers, led by two Durango commanders, Amoza and Gallova, killed the train's escort and captured much ammunition. Ninety soldiers and passengers reportedly died and 60 were wounded. [1712]

The final wrecking of a train by T.E. Lawrence took place on about Tuesday 3rd September 1918 on the Hedjaz Railway near Ma'an, Palestine. Lawrence was anxious to try a new method involving a so-called 'tulip.' Thirty ounces of gun-cotton were planted beneath the centre of the central sleeper of each ten-metre section of track. The charges bent the steel sleepers two feet into the air. The rails were pulled up three inches and inwards six inches and the chairs gripped the bottom flanges, warping them. The damaged track was then made unrepairable. An engine and six waggons were duly derailed and buried in the soft sand. The train crew surrendered after a short fight. [1713]

Mexican newspapers reported a revival of attacks on trains in Vera Cruz state in March and early April 1919, beginning with an attack on a passenger train between Morella and Acambaro on 19th March when the passengers were robbed.

On 21st March, a train on the Ferrocarril Interoceánico de México was dynamited between Mexico City and Vera Cruz. The fireman was badly injured, and the engine and tender were destroyed. The same day on the same line Zapatistas derailed a train and robbed its passengers between Mexico City and Puebla. The government escort was overpowered, three were killed and the others fled to the hills. The engineer, a mail clerk and a passenger were wounded.

A mixed train on the Vera Cruz – Isthmus line was dynamited between Vera Cruz and Tierra Blanca, Mexico on 22nd March 1919. The express car was robbed and four or five cars of freight which the bandits could not carry off was burned. Thirteen government soldiers of the train escort were killed, seven passengers were wounded, and 10,000 pesos were taken from a paymaster. He and several other passengers were then kidnapped for ransom.

A train from Vera Cruz to Alvarado, Mexico, was attacked on 30th March by a body of rebels under Felicia Chieftain Gabay and its entire guard of 50 men was killed.

On 1st April 1919, a double-headed freight train which lacked a government escort was dynamited between Jalapa and Pueblo, Mexico, and a large amount of freight stolen or destroyed. One engine having been dynamited, one of the firemen was then ordered to take the leader of the rebels, some of their party and their prisoners some distance up the

---

[1711] *The Waterloo Times-Tribune*, Waterloo, Iowa, United States, 17th August 1918
[1712] *Morning Sun*, Yuma, Arizona, United States, 28th August 1918
[1713] *Hedjaz Railway* by R. Tourret, pgs. 78-9

line. The leader then being engaged elsewhere the fireman escaped with the engine, two cars of oil and three of sugar. The rebels had reportedly numbered over 500 mounted men. [1714]

On Friday 5th April 1919, a passenger train running from Tampico to Monterey, Mexico, was attacked by bandits at an unstated location 140 kilometres from Tampico. Eight passengers and seven soldiers of the escort died, and the other soldiers were driven off. Some of the attackers were also killed and an American named J.T. Monett was shot in the shoulder. The passengers were robbed of their money, valuables, and clothing, while to complete their work the bandits set the train alight. The then marooned passengers were rescued later. [1715]

In a further attack on Tuesday 9th April 1919 a train carrying General Andre Ozuna and his brother Gregorio, military commander at Tamaulipas, who were returning from a conference with Governor William Hobby of Texas, Governor Zambrano of Coahulla and Governor Mirales of Neuvo Leon, at Nuevo Laredo, was stopped by dynamiting the track at Carizos, Mexico, between Linares and Victoria. According to some reports the two Ozunas had been kidnapped and held for ransom by the bandits, but this was denied by the government, who declared that they were safe. [1716]

On Wednesday 11th June 1919, a troop train carrying reinforcements under the command of General Rentoria Luviana to the Carranza forces near Juarez from Mexico City via Chihuahua was derailed and rolled down an embankment near Aguascalientes, Mexico. The wreckage caught fire and many soldiers perished. It was estimated that between 150 and 200 died. It was not clear whether the derailment was due to track defects or sabotage. [1717] [1718]

A major catastrophe occurred in Russia on Thursday 10th July 1919 between Bakhmatch and Pliski stations, east of Kieff. Bandits removed a rail and derailed a train. It was stated that there were 1,000 victims of whom 600 were burnt alive. [1719]

No railway sabotage occurred in Ireland after the 1916 Eastern Uprising until early on Saturday 16th August 1919, when a Great Northern (Ireland) Railway goods train was derailed between Beauparc and Navan, Ireland. A rail had been removed using tools taken from a platelayer's hut. The engine left the rails, then re-railed itself but 41 of the following 57 waggons of the train were derailed, 30 of which were completely wrecked. That morning a Navan railwayman found a notice written in red ink in his letter box: "Beware! Northern line torn up two miles from Navan. Tell station-master before nine."

[1714] *The Morning Republican*, Mitchell, South Dakota, United States, 10th April 1919
[1715] *Evening State Journal*, Lincoln, Nebraska, United States, 10th April 1919
[1716] *Cheyenne State Leader*, Cheyenne, Wyoming, United States, 11th April 1919
[1717] *Fort Madison Evening Democrat*, Fort Madison, Iowa, United States, 13th June 1919
[1718] *The Daily Gate City*, Keokuk, Iowa, United States, 13th June 1919
[1719] *The Times*, London, England, 27th July 1919

He did so but by then the stationmaster already knew of the derailment. [1720] As was suspected by the locals the sabotage was carried out in support of a strike by farm workers. It was undertaken by the Navan Company of the Irish Volunteers [1721] (the predecessor of the Irish Republican Army).

A Carranza troop train carrying 150 men, machine guns, artillery and provisions was derailed near Espinal de Morelos, Mexico, in Chiapas state on about Monday 25th August 1919. All the Federal troops were either killed or taken prisoner, the prisoners including several generals and a colonel. A quantity of artillery was also taken. There were about 200 attackers; it was not known to which faction they belonged. The attack in a remote, mountainous part of the country came as a complete surprise. [1722]

In a second act of sabotage on Tuesday 26th August 1919 a Federal troop train was derailed at an unstated location 70 miles south of Nuevo Laredo, Mexico. No further details were given. [1723]

Striking railwaymen derailed a train between Oliváes and Sacavém, Portugal, on Thursday 28th August 1919. No casualties were reported. [1724]

A passenger train from Rodez on the Chemins de fer du Midi was derailed near the station at Laissac-Lugans, France, at 20:00 on Tuesday 9th September 1919. Four coaches were thrown down an embankment to the meadow below. A cattle dealer was killed, and two other passengers were seriously injured. An act of sabotage was apparently responsible, a rail having been damaged by an explosive device. [1725] [1726]

Thirty people died in an attack on an armoured train at an unstated location in the Ukrainian Soviet Socialist Republic by 'insurgents' on or about Monday 27th October 1919. [1727]

A train from Guadalajara to Mexico City was 'attacked,' presumably derailed, by bandits, near Celaya, Mexico, early on Saturday, 1st November 1919. Twenty-nine of the 40 train guards were killed, many of the passengers were killed and all were robbed. The bandits were commanded by a General Malagon, 'apparently a newcomer in the bandit field.' [1728]

[1720] Ibid, 18th August 1919
[1721] www.navanhistory.ie/index.php?+railwaystations
[1722] *Oakland Tribune*, Oakland, California, United States, 28th August 1919
[1723] *The Evening Gazette*, Xenia, Ohio, United States, 29th August 1919
[1724] *La Vanguardia*, Barcelona, Spain, 29th August 1919
[1725] *Le Populaire*, Paris, France, 12th September 1919
[1726] *Excelsior*, Paris, France, 11th September 1919
[1727] Source unfortunately mislaid.
[1728] *The Boston Post*, Boston, Massachusetts, United States, 5th November 1919

In 1919 the north-west frontier of India, then a British colony, had long been of importance as the British feared a possible invasion of India by Russia through the Khyber and Bolan Passes. At the same time it was occupied by fiercely independent Muslim tribesmen, who greatly resented British rule. Following the Third Afghan War of 1919 which ended in August of that year the British had been forced to recognise Afghanistan as an independent county. This encouraged the tribesmen of the neighbouring area of Waziristan, bordering Afghanistan on the Indian side, and led to the Waziristan Campaign of 1919-20. Troops were sent to quell the rebellion and on Monday 1st December 1919 a troop train was derailed by what were described by one source as Nomad Waziri tribesmen, 'assisted by a number of Khostwal ruffians,' [1729] though another report described them as Afridis, a neighbouring tribe. [1730] Possibly members of both tribes were involved. One report stated that 21 troops had been killed and 40 wounded, the other that there 36 dead and 50 wounded. In addition all the rifles and two Lewis guns were 'lost.' The train was attacked near Thal, India, by about 500 tribesmen. The report on the accident in the London *Times* concluded with the statement that, 'Steps are being taken for retaliatory measures and the provision of additional protection for the railway.' [1731] No further attacks on troop trains seem to have occurred, partly due, probably, to the fact in some cases that the British substituted using aircraft to attack the tribesmen. [1732]

In early December 1919, a newspaper reported that travelling through northern Mexico had become extremely hazardous due to the activities of the rebels and the inclination of the Federal troops to remain in the cities and towns rather than take the field. It added that a passenger train had been dynamited at Empalme in the north east of the country (on about Tuesday 2nd December) with loss of life. In another attack on about the same date the Cedillos brothers and their forces had captured a train at Auga Nueva, on the main line between Saltillo and San Luis Potosi. The rebels killed or captured the entire force of Federal troops guarding the train. The passengers were left unmolested but had difficulty in completing their journey because the train crew refused (perhaps not surprisingly) to work another train. [1733]

On 22nd December 1919 it was reported that there had been five attacks on trains on the line between San Luis Potosi and Tampico, Mexico, in the previous five weeks.

On Saturday 6th December 1919 28 people were reportedly killed in an attack on that line at an unstated location. [1734]

On Sunday 7th December 1919, a Ferrocarriles Nacionales de México train was

[1729] *The Times*, London, England, 10th December 1919
[1730] *The Daily Mail*, Brisbane, Queensland, Australia, 4th December 1919
[1731] *The Times*, London, England, 10th December 1919
[1732] Wikipedia – Waziristan campaign (1919-20)
[1733] *The Nebraska State Journal*, Lincoln, Nebraska, United States, 8th December 1919
[1734] *The Newark Daily Advocate*, Newark, Ohio, United States, 22nd December 1919

attacked between Colima, the capital of the state of the same name, and Manzanillo, its seaport. The bandits robbed the passengers of money and valuables and looted the express and mail car. Among the passengers robbed was an American, Mr James Cowan, of Fort Worth, Texas, from whom $1,700 was stolen. [1735] A Carranza paymaster was captured and the bandits, who were Zamoristas, set fire to the coaches of the train.

The attacks continued on or about Wednesday 10th December 1919 with the dynamiting of a train on the Ferrocarriles Nacionales de México Tampico line at San Dieguito. Bandits under the command of Castillo killed the Carranza military escort and robbed the 200 passengers of money and baggage. [1736]

Unspecified rebels dynamited a passenger train near Perrole, Vera Cruz, on about Thursday 11th December 1919. One car was derailed but no one was hurt and this time the military escort were able to drive the rebels off. [1737]

Bandits dynamited two passenger coaches and a gondola car carrying the military escort on the Cardenas division between San Luis Potosi and Tampico on Wednesday 17th December 1919. The timely arrival of reinforcements put the rebels to flight and prevented the annihilation of the small escort party. Ten passengers and soldiers reportedly died. [1738]

A Luxor – Cairo express was derailed by armed brigands about 100 kilometres south of Cairo, Egypt, on the evening of Saturday 17th January 1920. They then stole about £2,000 from the postal van. A postal official was wounded but none of the passengers were hurt. [1739]

A prison escapee, together with two Russian prisoners, used a stolen spanner on Tuesday 20th January 1920 to loosen track fittings between Schönlanke and Schneidemühl, Germany. A goods train was thereby derailed, and its wreckage fouled the opposite running line. There was insufficient time after the initial derailment to properly protect the obstruction with detonators and an express (D-Zug) running in the opposite direction collided with the derailed waggons. Eighteen people died and 20 were injured, eight seriously, as a result. Herr Förster, the German Governor of Danzig, was rumoured to have been a passenger on the train. [1740]

The saboteurs were later arrested, tried, and sentenced to death. [1741]

[1735] *The Daily Morning Oasis*, Nogales, Arizona, United States, 10th December 1919
[1736] Ibid, 18th December 1919
[1737] *Nevada State Journal*, Reno, Nevada, United States, 18th December 1919
[1738] *Galveston Daily News*, Galveston, Texas, United States, 21st December 1919
[1739] *The Times*, London, England, 19th January 1920
[1740] Ibid, 21st January 1920
[1741] *Historische Eisenbahn Katastrophen* by Bernhard Püschel, pg. 121 (in German)

On about Saturday 31st January 1920 three bombs exploded under a train of the Ferrocarril Central Norte in the vicinity of Tucuman, Argentina. A number of railway workers (probably scabs) and a soldier were injured, and it was presumed by the authorities that the act was the work of strikers from the railway workshops. Two coaches were partially wrecked, but it was believed that had the timing of the explosion been more precise the entire train would have been destroyed. [1742]

Mexican newspapers reported that the bravery of an engineer had prevented a potential major disaster in about January 1920 near San Isidro in Zacatecas state. A bandit chieftain named Jazal reportedly sent a light engine along the line, loaded with dynamite and other explosives, in the expectation that it would collide with an oncoming passenger train. The heroic engineer, who refused to give his name, uncoupled the engine from his own train and set off in pursuit of the runaway. When his engine had caught it up he left his fireman in charge of their own machine and was able to crawl over the tender of the "wild" locomotive, get to its cab and bring it to a stand. [1743]

An isolated case of sabotage occurred late on Monday 24th May 1920 on the Baghdad – Shergat line in Mesopotamia (Iraq from late August 1921). A band of Arabs, estimated at 200 strong, and paid by Syrian officers of the Sharifian Army, overpowered a permanent way gang between Ain Dibs and Tulul Baq stations, then tore up a number of rails, and in addition laid sleepers and other obstructions on the track. An almost empty goods train arrived at the scene just before midnight and was duly derailed. The few Indian passengers riding on the train were little hurt in the derailment but were afterwards attacked by the Arabs. Six of them were killed or fatally injured and seven others wounded. The British train driver, named King, was also fatally wounded, and died afterwards in hospital. A relief train was sent to the scene and the track repaired, but subsequently trains were only scheduled during daylight hours. [1744]

The sabotage occurred a month before the Great Iraqi Revolution of 1920. Before the First World War Mesopotamia had been a part of the Ottoman Empire. This had ended with the defeat of Turkey and the Treaty of Versailles had granted Britain a mandate for Mesopotamia. Iraqis feared that they were to become part of the British Empire instead of the Ottoman one and were fighting for independence. [1745] News of the attack was censored and only appeared in the British Press nearly two months later.

A Mexican armoured train was blown up by bandits on Tuesday 29th June 1920. The attack, in Chihuahua state, derailed the water car and armoured car of the train but its engine was detached and hurried to Poncho for help. The occupants of the rear car were able to defend themselves against the bandits until government troops arrived and the bandits were driven off. The train then continued to Jiminez. [1746]

[1742] *Biddeford Daily Journal*, Biddeford, Maine, United States, 2nd February 1920
[1743] *The Capital Times*, Madison, Wisconsin, United States, 31st January 1920
[1744] *The Times*, London, England, 16th July 1920
[1745] Wikipedia – Iraqi revolt of 1920
[1746] *The Athens Messenger*, Athens, Ohio, United States, 1st July 1920

The Ferrocarriles Nacionales de México line to Nuevo Laredo was out of use in July 1920, rebels having destroyed a number of bridges. The last of these structures having been rebuilt, the train service from Monterey was resumed but as a freight train heading for Nuevo Laredo was crossing a bridge near Hulsachita, Mexico, on Sunday 25th July 1920, the structure collapsed. No one was killed but nine carloads of Mexican products, valued at $20,000, were lost in the ensuing fire. According to at least some of the reports on the affair the collapse of the structure itself was the result of the rebels having set it on fire in a further act of sabotage. A further temporary bridge was then constructed and the passenger train from Mexico City and Monterey, due in Laredo, Texas, at 6 p.m. on 25th, finally reached there at 3 p.m. the next afternoon. [1747] [1748]

The Madras – Bombay mail train escaped derailment on Wednesday 6th October 1920 because it was preceded by a light engine which reached the spot where a rail had been removed first and was derailed instead. The next day, Thursday 7th, the Madras – Bangalore mail was less fortunate. It was derailed near Arkonam (now Arakkonam) and between there and Minpal stations. This train was carrying five first-class, 32 second-class and 'a large number' of third-class passengers and there were heavy casualties, 13 people being killed and 15 wounded. As on the previous day, a rail had been removed. The two acts of sabotage were blamed on disgruntled platelayers who had been dismissed after a previous strike. [1749] [1750] What an English newspaper termed a strike epidemic was in progress in Bombay at this time with postal and tramway workers also taking industrial action. The strikes were closely linked to a growing demand for Indian independence and protests against the 1919 Rowlatt Act, which had authorised arbitrary arrest, indefinite detention, press censorship and so on, in a bid to curb Indian nationalism. [1751]

An 'optional' goods train from Souk Haras to Tebessa was derailed by bandits at 21:00 on Saturday 4th December 1920 at kilometre 60, on the approach to Clairfontaine, Algeria. The bandits, 30 in number, had torn up the rails over several metres. After the train had left the rails it was fired upon, and grenades were thrown at it. The conductor, brakeman and a gendarme were seriously injured in fighting off the bandits. The arrival of a following passenger train caused the bandits to withdraw. [1752]

A troop train from Derry (Londonderry) to Burtonport, with Driver Turner and Fireman Crawford, was ambushed near Kincasslagh Road, Ireland, on the Londonderry and Lough Swilly Railway, on the evening of Wednesday 12th January 1921. Large boulders placed across the line derailed it and a heavy fire was then opened upon it. The carriages were riddled with bullets and all the windows were broken but no casualties

---

[1747] *San Antonio Evening News*, San Antonio, Texas, United States, 26th July 1920
[1748] *The Evening Gazette*, Cedar Rapids, Iowa, United States, 27th July 1920
[1749] *The Times*, London, England, 11th October 1920
[1750] *The Telegraph*, Brisbane, Queensland, Australia, 11th December 1920
[1751] Wikipedia – Rowlatt Act
[1752] *L'Homme Libre*, Paris, France, 6th December 1920

were reported. [1753]

In January 1921, the Bolsheviks in Russia were still at war with Poland, though a truce had been declared in October 1920. Parts of what is now Belarus were incorporated into the Byelorussian Soviet Socialist Republic, while the eastern provinces, including Vitebsk (now Viciebsk in Belarus), became a part of the Russian Soviet Federative Socialist Republic. [1754] At an unspecified date in early January 1921, discontent among the Red troops in Vitebsk who had been ordered to demobilise led to a mutiny in which they hanged their officers. In response the Bolsheviks in Moscow sent a trainload of Chinese mercenaries to Vitebsk, but their train was derailed on about Saturday 15th January 1921 and all those who did not flee were massacred. [1755]

The Londonderry and Lough Swilly Railway, whose headquarters in Londonderry were in what in July 1921 was to become Northern Ireland, while most of its mileage lay in County Donegal in the future Irish Republic, was in an especially difficult position. Matters were not helped by its lines serving British naval facilities and shore batteries. The struggle for Irish independence and attempts to make a British military occupation impossible meant that train crews were intimidated into not working troop trains and there were further acts of sabotage on 7th and 11th February 1921 at Kincasslagh Road and on 11th February near Cresslough. [1756] Also on 11th February rebels opened fire on a passenger train carrying troops near Mallow, Ireland, on the Great Southern and Western Railway, killing a sergeant. Again on Tuesday 15th February 1921 a passenger train on the Cork Bandon and Coast Railway, carrying 40 soldiers, was attacked by about 50 rebels between Upton and Innishannon stations, Ireland. In the exchange of fire six civilian passengers and two of the attackers were killed. [1757]

Between Merseburg and Ammendorf, Germany, just south of Halle, the fastenings of both tracks were removed on the night of Wednesday 16th March 1921, so that a goods train in each direction was derailed and wrecked, one close upon the other. Three people died and six were hurt. [1758]

An express was derailed at Gröbers, Germany, between Leipzig and Halle, on Monday 16th May 1921, possibly by the same person who had wrecked the two goods trains. This time five people died, and many were injured. [1759]

Well-known engineer James Jackson and his fireman Frank Bizet were fatally injured when Southern Railway passenger train No. 42 from New Orleans, Louisiana, to

---

[1753] *The Times*, London, England, 13th January 1921
[1754] Wikipedia – Byelorussian Soviet Socialist Republic
[1755] *The Evening Gazette*, Xenia, Ohio, United States, 17th January 1921
[1756] *The Lough Swilly Railway* by Edward Patterson, pgs. 84-6
[1757] *The Times*, London, England, 16th February 1921
[1758] Katastrophen der deutschen Bahnen Teil II, by H. J. Ritzau, pg. 49 (in German)
[1759] Ibid.

Cincinnati, Ohio, was derailed four and a half miles south of Hattiesburg, Mississippi, United States, at 11.15 p.m. on Tuesday 14th June 1921. The engineer was able to walk from the wreckage but was severely scalded and died later in hospital, while his fireman was trapped in the wreckage. Conductor Will White, Baggagemaster Frank Haug, passenger Johnny Brezeale, and Philip Cooper, a negro hobo, were injured. The engine and tender, baggage mail and express cars and one coach rolled into a ditch. An investigation disclosed that the spikes had been removed from two lengths of rail on a curve. Bloodhounds were called to the spot but were unable to pick up a trace of a scent. Robbery was ruled out as a motive and suspicion fell on several negro railroad employees who had recently been discharged and had disappeared. 'Arrests are expected before night, however,' a newspaper reported. [1760]

The Irish War of Independence which had begun in January 1919 continued into 1921 with a peak of violence between December 1920 and July 1921, including sporadic acts of sabotage against the railways. The most serious act involving the railway during this period occurred near Adavoyle on the Great Northern (Ireland) Railway. Irish Republican Army (IRA) members planted a mine on the track, their target being a troop train carrying men of the 10th Hussars. One coach of the train and over a dozen trucks loaded with horses plunged down an embankment. Sergeant Bowson and Trooper Harper were killed outright, and Trooper Telford died before reaching hospital. Guard Francis Gallagher also lost his life, and Great Northern Inspector Berkley and four soldiers were seriously injured. A number of badly wounded horses had to be shot while others which had escaped injury ran about the adjoining fields. The troops had been returning to the South after forming part of an escort for a Royal visit to Belfast. The telegraph wires had been cut and signal fires lit on neighbouring mountains had given warning of the approach of the train. [1761]

Further acts of railway sabotage, including arson, followed on 2nd, 5th, 6th, 8th and 11th July. A truce ended the war on the latter date. [1762]

It was suspected that the derailment of train No. 170, an international Amsterdam – Paris express at Hennuyères, Belgium at 22:40 on Wednesday 6th July 1921 was the result of sabotage. Its two locomotives, a postal van and five coaches were wrecked, and six postal workers were killed and Conductor Cardinal, Fireman Van Beethoven of the train engine and four postal officials were seriously injured. A few passengers were severely bruised. Bolts removed from the rails added to the suspicion of sabotage, as did the fact that the train was carrying many valuables, including several million francs destined for the French government. If the derailment was an act of sabotage carried out with the intention of robbing the train the saboteurs did not achieve their objective, for all the items were recovered from the wreckage of the postal van, though some had been torn to pieces in the derailment. They were placed under guard in an annex to the station building. The Hennuyères signalman was interrogated and said that his levers had been

---

1760 *Laurel Daily Leader*, Laurel, Mississippi, United States, 15th June 1921
1761 *The Times*, London, England, 25th June 1921
1762 Wikipedia – Irish War of Independence

working normally. [1763] [1764]

The Entente Powers sent troops to the eastern most part of Siberia in 1918 to support the White Russians against the Red Russians (Bolsheviks). The forces withdrew again in 1920 except for the Japanese, who remained until 1922, having sent a much larger force to Siberia than any other of the Allied powers. As well as having an antipathy towards Communism and socialism the Japanese hoped to regain historical losses and create a buffer state against Russia. [1765] In July 1921 three acts of sabotage directed at Japanese trains were recorded. In the first of these on 20th July 1921 an armoured train was dynamited at a bridge by 'partisans' (presumably Bolshevik guerrillas) near Nicolaevsk, Russia. Twenty officers and men were killed or wounded, one of the casualties being Colonel Miwa, the commanding officer, who died later of his wounds. [1766] Two further attacks on trains occurred near Nicolaevsk on about 26th July 1921. One of the trains was carrying the body of the late Colonel Miwa, guarded by Japanese troops, and there was also a party of Japanese students on board. The report on these attacks noted that 'The only passengers killed in the attacks were Russians.' [1767] On 4th August 1921 the 'partisans' supposedly engineered a collision between a passenger train and a freight train near Nikolsk, Russia (not the same place as Nicolaevsk, unless the name was misspelt in the newspaper report), in which 16 people died and 18, including a few Japanese, were injured. [1768] The Japanese withdrew from Siberia in 1922 after the defeat of the White Russian forces.

Two acts of sabotage occurred in India in August 1921. At 03:35 on Saturday 13th August 1921 No. 7 up mixed train on the Assam-Bengal Railway was derailed about a mile south of Gangasagar station. The engine overturned and went down an embankment and 14 other vehicles were derailed. Two passengers and four railwaymen were injured. Damage to rolling stock and the permanent way was estimated at 30,886 rupees. A pair of rails had been removed. Then at 21:10 on Wednesday August 17th, 1921, also on the Assam-Bengal Railway, No. 3 up passenger train was derailed at mile 4-20, between Pahartali and Bhatiyali. The engine overturned and four coaches were derailed and telescoped. This time one passenger died and three were injured. Damage to permanent way and rolling stock amounted to 53,450 rupees. A pair of fishplates had been removed. [1769]

In 1921 a severe famine was raging in the Volga and Ural River regions of Russia and in order to deal with it food trains were organised from the fertile areas of the Ukraine. On about 14th September 1921 one of these trains was wrecked at Fastoff (now Fastiv) in the Ukrainian Soviet Socialist Republic, near Kieff (now Kiev). Twenty-seven people

---

1763 *Le Gaulois*, Paris, France, 7th July 1921

1764 *Le Journal*, Paris, France, 7th July 1921

1765 Wikipedia – Japanese intervention in Siberia

1766 *The Times*, London, England, 3rd August 1921

1767 Ibid, 6th August 1921

1768 *The Great Falls Tribune*, Great Falls, Montana, United States, 5th August 1921

1769 Administration Report of Indian Railways in 1921-22

died in the accident and 30 were injured. The derailment was blamed by the Soviet authorities on sabotage by followers of Symon Petliura, a Ukrainian politician who fought for that country's independence. Two locomotives and 61 grain waggons were reportedly totally destroyed. The Soviet authorities used the derailment as propaganda, accusing Petliura's followers of being the hirelings of Polish landowners and capitalists. An article in the official *Investiya* read: 'While shiploads of foodstuffs arrive from America, England, Norway and Germany in order to bring some slight relief from the famine, bandits sent out from Poland and Rumania burn our stores and wreck our food trains. This we cannot tolerate; we must put an end to it.' [1770]

In what was probably a different accident, though the reports are difficult to interpret, a passenger train from Kiev were wrecked by 'counter-revolutionists' near the station at Teterevo, Ukrainian Soviet Socialist Republic, probably on about Sunday 9th October 1921. Thirty passengers were said to have been killed and many injured in this attack. The survivors were fired upon by the bandits, who later captured them and took them to a forest, where they were all robbed, and 15 Commissars were slain. [1771] [1772]

At the close of the First World War the last Austrian Emperor, Charles I, who reigned in Hungary as Charles IV, had made a statement relinquishing his right to take part in the politics of the two nations. However in 1921, encouraged by right-wing monarchists in Hungary (legitimists) he made two attempts to return as King of Hungary. The first attempted coup in March failed and he returned to exile in Switzerland. In October, a second attempt was made, Charles flying from Switzerland to western Hungary on 21st October and thence proceeding by train towards Budapest. This train was derailed by anti-monarchists at Cedenburg, Hungary, on the evening of Saturday 22nd October 1921, [1773] without apparently serious consequences. The second attempted coup also failed as the Entente powers would not permit a restoration of the Hapsburg monarchy. Charles was forced into exile in Madeira and died of pneumonia the following year, aged only 34. [1774]

On Wednesday 9th November 1921, an Algarve – Lisbon train from Alentejo was derailed between Aljustrel and Figueirinha, Portugal, rails having been placed across the track and nailed to the sleepers. The results were disastrous; by the following day there were known to be 24 dead and 200 injured. All the injured, it was reported, were in a satisfactory condition, including a little girl named Dinorah, whose condition earlier had been giving concern. [1775] This was apparently the fourth act of sabotage carried out on Portuguese railways in 15 days, the driver of another train, probably on the same day, having been able to just avoid a derailment at Evora. Railwaymen launched a protest against these criminal acts. [1776] [1777] [1778]

[1770] *The Times*, London, England, 22nd September 1921
[1771] *The Richmond Palladium*, Richmond, Indiana, United States, 11th October 1921
[1772] *The Times*, London, England, 1st November 1921
[1773] *Oakland Tribune*, Oakland, California, United States, 23rd October 1921
[1774] Wikipedia – Charles IV of Hungary's attempts to retake the throne.
[1775] *Diário de Lisboa*, Portugal, 11th November 1921
[1776] *Paris-Midi*, Paris, France, 11th November 1921

In 1921 Egypt was a British protectorate but as in India, Iraq and elsewhere there was a growing independence movement. In 1919 there had been anticolonial riots (the 1919 Revolution). In 1920 there was an attempt to replace the protectorate with an alliance, but this failed because of the British wish to maintain control over the Suez Canal. In December 1921, the British imposed martial law in Cairo and this time the demonstrations included attempts to wreck trains, one of which was successful on the Heulan railway on Christmas Day, 25th December 1921. [1779] Britain granted Egypt limited independence at the end of February 1922. [1780]

The Rand Rebellion in South Africa began at the end of 1921 with a strike of white gold miners following a drop in the price of gold and ensuing cut in wages. It developed into an open rebellion in early 1922. On about Tuesday 7th February 1922 the Rhodesian mail train was derailed 80 miles from Johannesburg, South Africa, it was thought due to a rail having been removed. Two coaches were derailed but there were no casualties. It was stated at the time that the incident was not connected with the miners' strike. [1781] However, the dynamiting of the main line south of Dover Junction, South Africa, 'about 30 miles from the Transvaal border,' on Monday 13th March 1922, was linked to the rebellion. A pilot engine was derailed, without casualties. [1782] The rebellion was crushed at the cost of 200 lives. [1783]

The July 1921 truce in Ireland was followed by the Anglo-Irish treaty in December 1921 which provided for the establishment of the Irish Free State within a year, but with Northern Ireland able to opt out (which it did). More radical republicans refused to accept the treaty and the fight for a totally independent Ireland continued with attacks on the Royal Irish Constabulary and the British army, including acts of railway sabotage. Thus on Wednesday 29th March 1922 a Great Northern (Ireland) Railway goods train on the Dublin – Belfast main line was derailed north of Adavoyle station, Ireland, by the removal of a rail and loosening of fishplates, its engine and leading vehicles overturning. About 100 men then looted the train, sprinkled it with petrol and set it alight. No one was reported hurt. [1784]

The employees of the East Indian Railway had been on strike since the beginning of February 1922, the strike being accompanied by some acts of violence. At 01:00 on Tuesday 4th April 1922 the up Calcutta – Punjab mail was derailed between Mooduncatta and Joramow Block Hut, India, eight miles from Madhupur. The engine, tender and six vehicles left the track and went down an embankment and the driver, two firemen, two

---

[1777] *La Voz*, Madrid, Spain, 10th November 1921
[1778] *La Epocha*, Madrid, Spain, 14th November 1921
[1779] *Nebraska State Journal*, Lincoln, Nebraska, United States, 28th December 1921
[1780] Wikipedia – History of Egypt under the British
[1781] *The Times*, London, England, 9th February 1922
[1782] Ibid, 14th March 1922
[1783] Wikipedia – Rand Rebellion
[1784] *The Times*, London, England, 30th March 1922

passengers and a Railway Mail Service peon were killed outright, and eight passengers and a railwayman were seriously injured, one of the injured passengers dying of his wounds. There was considerable damage to rolling stock and the permanent way. [1785] Sabotage was clearly the cause of the derailment, a rail having been fixed across the line and a fishplate loosened. There were 177 passengers on the train. A worse accident was avoided because the corresponding down mail was half an hour late. [1786]

On about Thursday 25th May 1922 another relief train carrying maize for famine relief, this time from the American Relief Association, was completely wrecked near Vladikavkaz, Russia, bandits having removed the rails. No casualties were mentioned. [1787]

On or about Tuesday 13th June 1922 a train carrying a large number of Esthonians who were repatriated from Siberia was derailed at an unstated location in that region of Russia. The bandits responsible then robbed the passengers of all their belongings. [1788]

The Mexican Revolution or Civil War ended in 1920, but sporadic acts of sabotage directed at the railways continued for the next decade and beyond. On Monday 5th June 1922, 25 Federal soldiers died when a train on which they were riding was blown up between Tommelin and Santa Cararina, Mexico, supposedly by bandits. [1789]

Paraguay was in the midst of a civil war in the summer of 1922, involving the government and rebel forces commanded by a Colonel Adolfo Chirife. Having failed to capture Asunción, the capital of Paraguay, the rebels were forced to retreat and on Wednesday 14th June 1922 they were forced back towards Paraguari. [1790] Supporting the government by independent guerrilla actions were a number of local leaders (cadillos) including a Gregorio Flecha, whose men destroyed a long wooden railway bridge over the River Tebicuari, between Itape and Villa Rica, Paraguay, to delay the rebel retreat. Flecha's men were still busy sawing the bridge supports when they saw a rebel troop train approaching. Some of the troops had been instructed by Colonel Chirife to guard the bridge while the remainder of his forces withdrew over it. Meanwhile Flecha's men had hidden themselves on the river bank to await developments. The train drew up at the bridge and some officers made a cursory inspection, failed to discover any signs of damage, and concluded that the structure was safe. The train started slowly over the bridge, which then began to collapse. Efforts made to stop the train were too late and bridge and train fell into the river. The train driver and many soldiers died, mostly by drowning. The remainder saved themselves by swimming to the bank but were captured by Flecha's men. The main force of the rebel army was obliged to cross the river in small

[1785] Administration Report of Indian Railways in 1922-23
[1786] *The Times*, London, England, 5th and 6th April 1922
[1787] Ibid, 27th May 1922
[1788] Ibid, 15th June 1922
[1789] *The Cordova Daily Times*, Cordova, Tennessee, United States, 5th June 1922
[1790] www.histarmar.com.ar/ArmadasExtranjeras/Paraguay/LaGuerraCivil1922-23.htm

boats. [1791]

Conflict between the provisional Irish government (which became the Irish Free State on 6th December 1922) which accepted the Anglo-Irish Treaty as the best compromise which could be achieved, and the Irish Republican Army, which opposed the treaty, led to the Irish Civil War between 28th June 1922 and 24th May 1923. During this period there were many acts of railway sabotage in Ireland and later the Irish Free State, far too many for all to be described in detail, but their dates and locations will be given below with brief details if appropriate.

On Monday 10th July 1922 a train was derailed at Killurin. On Thursday 20th July, a train was derailed at Killester and a bridge blown up at Malahide. [1792] A train carrying rebel prisoners was fired on at Killurin on Wednesday 26th July in an attempt to rescue them [1793] and the same day a train was derailed at Howth Junction. A train was run into the buffers at Tralee on Tuesday 15th August [1794] and on that date a train fell from a damaged viaduct near Killurin Tunnel. After a gap of somewhat less than two months the sabotage then resumed with a train running into an obstruction between Palace East and Rathgarogue on Thursday 5th October. A train was derailed at Ballyanne on Saturday 14th October, another ran into an obstruction at Killurin on Wednesday 1st November, and one was derailed between Castlebellingham and Dromin on Friday 3rd November. On Tuesday 7th November, a train was deliberately set off down the track into collision with another between Liffey Junction and North Wall, Dublin. A train was derailed near Killurin on Saturday 11th November, another between Cafir and Clonmel on Tuesday 14th November and a third at Sparrowsland on 20th November. The railway near Killurin was the location again on Thursday 23rd November when part of the train ran down the embankment into the River Slaney. [1795] Another train was derailed at Killurin on Saturday 25th November, and one at Clonadrum Crossing on the same day. Killurin was clearly a favourite spot for sabotage as yet another train was derailed there on Thursday 7th December in what was now the Irish Free State. The 'rebel outrages,' as the London Times described them, moved to Kildare on Monday 11th December, one engine being run into the turntable buffers, another sent down the line and a third run into it. [1796] An engine was derailed near Ballymote on Thursday 14th December, a train was set on fire and derailed between Durrow and Waterford the next day and another ran into an obstruction near Dundalk the day after. [1797]

On Wednesday 20th December there occurred what a newspaper described as the 'Train wrecker's worst outrage,' though no lives were lost. The mail train from Belfast was stopped at Castlebellingham, set alight and sent along the line towards a goods train from Dublin which had been derailed. [1798] The day before, seven railwaymen who had

---

[1791] *The Times*, London, England, 8th August 1922
[1792] Ibid, 21st July 1922
[1793] Ibid, 27th July 1922
[1794] Ibid, 16th August 1922
[1795] Ibid, 25th November 1922
[1796] Ibid, 12th December 1922
[1797] Ibid, 18th December 1922
[1798] Ibid, 22nd December 1922

been involved in the Kildare sabotage had been executed. [1799] On about Friday 29th December a train was derailed at Ballyhale, on Saturday 6th January 1923 a train was derailed at Killester, on Monday 8th January a train was derailed at Kilkenny and an engine was sent into collision with some waggons at Limerick Junction, [1800] and on Tuesday 9th January there was a derailment at Mountpleasant. Ballyanne was the scene of an engineered collision on Wednesday 10th January, and trains were derailed at Streamstown Junction on Saturday 13th January, near Rathgaroue on Tuesday 16th January, between Ferns and Enniscorthy on Thursday 18th January and on the same day near Ardfert. In this last attack Driver Patrick O'Riordan and Fireman Daniel Crowley lost their lives when their engine went down an embankment. [1801] On Saturday 20th January, a train was derailed at Palace East, an unsuccessful attempt was made to blow up Malahide viaduct and 11 rebels were executed in the Irish Free State. [1802] Three more rebels were executed in Dundalk on Monday 22nd January and the same day an engine was run off a bridge which had been blown up near Kiltoom. [1803] A train was set alight at Drumree on Tuesday 23rd January. [1804] At Kilmacow on Friday 26th January one train was derailed and another run into it. At Macmine Junction, on Saturday 27th January, a train was derailed and on the same day another train was fired upon near Doonberg. A train was set alight near Killiney on Monday 29th January, one was derailed near Macmine Junction on Wednesday 31st January, trains were fired upon near Milltown and between Doonberg and Shragh on Friday 2nd February and at Killala on 3rd February a train was set alight and its engine detached and run off into the sea.

Commenting on the attacks *The Times* in London wrote, 'Dublin itself has lately become once more the centre of the campaign of destruction. Travelling even on the suburban lines has become somewhat of an adventure; for the Irregulars have for a month or more been busy "holding up" trains, whenever occasion offered, burning the carriages, and sending the engines careering down the line under full steam. It is small wonder that the Dublin and South Eastern Railway has announced that it will no longer take any responsibility for the safety of passengers or goods. Elsewhere, in many cases, the railways have given up the struggle, and over large distances there are no longer any communications; markets cannot be held; trade is dying or dead; and whole communities, cut off from their sources of supply, are threatened with starvation.' [1805]

The attacks continued. A train was fired upon and set alight near Castleconnel on Monday 12th February 1923. On Saturday 17th February the mail train from Dublin to Galway was run off a bridge which had been blown up near Streamstown Junction. On the same day, another train was run off a destroyed viaduct near Dungarvan. [1806] A train was derailed near Palace East on Friday 23rd February and there was another derailment there on Thursday 1st March, while the Galway – Dublin mail train was derailed at Kiltoom on the same day. [1807] On Monday 5th March, a train was stopped and set alight at

[1799] Ibid, 20th December 1922
[1800] Ibid, 9th January 1923
[1801] Ibid, 19th January 1923
[1802] Ibid, 22nd January 1923
[1803] Ibid, 23rd January 1923
[1804] Ibid, 25th January 1923
[1805] Ibid, 8th February 1923
[1806] Ibid, 19th February 1923
[1807] Ibid, 3rd March 1923

Barnagh and its engine run off down the line to Newcastle West, while a bridge was blown up between Danagher and Tullamore, another train stopped and set alight and run off the destroyed bridge. [1808] On Sunday 11th March, a train was derailed near Maryborough. On Wednesday 21st March a train was set alight near Foynes. [1809] An engine was derailed in Galway on Saturday 24th March [1810] and a train was derailed between Manorhamilton and Glenfarne on Tuesday 3rd April. By this time, the Republican forces were on the verge of defeat and the last episode of sabotage occurred at Dublin on Saturday 28th April 1923 when a large bomb planted in a signal box at Amiens Street exploded, damaging a passing train. [1811]

Wartime inflation, the employment of non-union labour and a 7% wages cut imposed on repair and maintenance workers led to the Great Railroad Strike of 1922 in the United States. The strike was characterised by violence and several acts of sabotage, especially on Sunday 20th August 1922, when train No. 29, a Michigan Central Railroad New York – Chicago express, travelling at over 50 miles an hour, was derailed, 27 spikes having been removed from one of the rails near Gary, Indiana. Engineer Edward Coy and Fireman Frank Lubbs were killed, and two people were injured. Interviewed by detectives afterwards one of the strikers allegedly remarked, "It's a shame they weren't all killed." [1812] A few days later five men were arrested and charged with murder. [1813] Other acts of sabotage were carried out near Elwood, New York, on 17th August; near Buffalo, New York and near Milton Junction, Wisconsin, on 18th August; near Lima, Ohio, on 24th August; near Spring Lake, Iowa, on about 25th August; at Green Bay, Wisconsin, on 1st September and at Louisville, Kentucky, on 4th September. The strike was supressed in September.

A Burma Railway up mixed train bound for Moulmein was derailed at mile 82 between Mokpalin and Kyaikkatha stations, Burma, on the Pegu – Moulmein branch, on Thursday 8th March 1923. The engine and tender capsized, nine goods waggons piled up and two coaches were telescoped. Two passengers died and 21 were injured. The driver was fatally injured. Two or three joints had been opened up and 30 feet of rail removed. [1814] [1815]

As already noted in an earlier chapter, following the First World War and the Treaty of Versailles Germany was expected to pay a huge sum in reparations, mainly to France. When she defaulted on the payments the French, who also feared for the sanctity of the remainder of the treaty, occupied the Ruhr on 11th January 1923. [1816] Opposition to the

---

1808 Ibid, 7th March 1923
1809 Ibid, 23rd March 1923
1810 Ibid, 26th March 1923
1811 Ibid, 30th April 1923
1812 *The News Sentinel*, Fort Wayne, Indiana, United States, 21st August 1922
1813 *The Lake County Times*, Hammond, Indiana, United States, 29th August 1922.
1814 *The Times*, London, England, 10th March 1923
1815 Administration Report of Indian Railways in 1922-23
1816 Wikipedia – Occupation of the Ruhr

occupation included many acts of railway sabotage. The earliest appears to have been on Friday 30th January 1923, when a turntable was damaged at Mainz Bischofsheim so that two engines fell into the pit. [1817] On about Monday 2nd April the Germans despatched nine driverless trains from Friedrichsfeld, in unoccupied Germany, towards Wesel, 'imitating the methods of the Irish Republicans,' the reporter noted. The eighth train ran into the seventh on the Lippe bridge and 70 waggons were derailed. The object was to bottle up Wesel station, a rail centre. Three persons were arrested on suspicion. [1818] A bomb exploded beneath train No. 7078 between Gerolstein and Oberbettingen on Sunday 15th April, damaging the engine and track and injuring the driver. On Tuesday 17th April, a bomb exploded under a train supposedly carrying French officials in France near the German border. The tender of the engine was derailed but no one was hurt. Rails were also torn up ahead of a train carrying the French Minister of Public Works, Yves de Troquer and the Belgian Minister of Defence, Albert Devèze. [1819] [1820]On Wednesday 25th April a bomb exploded under train A-4 at Viersen- Helenabrunn, damaging the track and numerous vehicles. The next day interference with a set of points derailed at train at Bobenheim. Train No. 503 was derailed by the explosion of a mine between Kierberg and Liblar on Tuesday 1st May, and five waggons were derailed. Two waggons of a coal train were derailed by a mine between Wedau and Lintfort on Wednesday 2nd May. Then on Saturday 5th May a mine was exploded on the railway between Düren and Neuss, Germany, creating a large crater, and train D 414 was derailed, and its driver injured. [1821] A bomb exploded beneath train C 124 at Bobenheim on Monday 7th May, derailing the engine and 13 waggons. A bomb exploded under train C E 51 at Düren on Saturday 12th May, injuring the engine crew and train No. 412 was derailed by a bomb at Föhren on Sunday 13th May and its driver injured. [1822]

Bandits derailed a passenger train near the station at Lincheng, China, on the main line linking Peking and Shanghai, at 2.30 a.m. on Sunday 6th May 1923. The engine was not derailed, only some of the carriages. Following the derailment the bandits, who numbered about 1,000 or 5,000, according to different estimates, entered the train, seized money and jewellery from the passengers and on a second visit rifled suitcases. The passengers were then taken out in their nightclothes and marched off, about 200 Chinese and about 24 foreigners being kidnapped. One of the foreigners, an Englishman named Joseph Rothwell, a Shanghai merchant, was killed, whether as a result of the initial derailment or afterwards by the bandits was not stated. At least two of the passengers escaped through a window, hid until the coast was clear and made their way to Tientsin.

The report on the accident added that, 'There is little doubt that the present outrage was committed by ex-soldiers; in fact, a great part of the Northern armies is recruited in this region, and it is a partial explanation of the difficulty in dealing with bandits in China that the regular forces and the bandits are so akin. The soldiers can seldom be induced to deal faithfully with the bandits. It is a feature of the present situation in this country.

[1817] Wikipedia – Lists von Anschlägen im Schienenverkehr
[1818] *The Evening Republican*, Mitchell, South Dakota, United States, 4th April 1923
[1819] *The Evening World*, Bloomington, Indiana, United States, 17th April 1923
[1820] Wikipedia – Lists von Anschlägen im Schienenverkehr
[1821] *The Times*, London, England, 7th May 1923
[1822] Wikipedia – Lists von Anschlägen im Schienenverkehr

Although the regular armies number nearly one and a half millions, brigandage is more widespread than at any time in recent history. The bandits loot, burn, kidnap, and murder right under the noses of the Tuchuns (Military Governors) and are rarely molested.' [1823]

The hostages were gradually released as the Chinese government gave in to the bandits' demands, which reportedly were not for money but that they be enrolled in the regular army. The last hostages were freed in early July. [1824]

A rebellion apparently was in progress in Brazil in May 1923, though information regarding it has been hard to find. At any rate, a brief report from Buenos Aires, Argentina, stated that the revolutionists had derailed a government armoured train on about Thursday 10th May 1923 at an unstated location, killing 50 soldiers. [1825]

The derailment of a French train near Coblenz (Koblenz from 1926), Germany, on Wednesday 16th May 1923 [1826] was possibly due to sabotage and the next day a bomb exploded under train C.H. 21 between München-Gladbach and Crefeld, Germany. A bomb exploded beneath train No. 136 at Angermund on Tuesday 22nd May, injuring the driver. A bomb derailed an engine at Weidenthal on Tuesday 29th May. On Wednesday 30th May the train which left Strasbourg at 18:57 for Minz was derailed between Insheim and Landau, Germany. The engine and three coaches left the line. There were no serious injuries but both lines were damaged. A German suspected of causing the derailment was arrested.[1827] A train was derailed and wrecked by unknown persons at Ratingen, near Essen, Germany, on Saturday 2nd June and an engine and 13 waggons were derailed and smashed. [1828] The French stated on 12th June that sabotage appeared to be on the increase, and that it had done so since the execution on 26th May of Albert Schlageter, a member of the right-wing Freikorps executed by the French military for railway sabotage. To emphasise the point bombs on the track derailed trains on 12th June between Liblar and Euskirchen and the next day near Bad Godesberg while on 12th June another German was sentenced to death for planting explosives on the line near Ludwigshafen. [1829] Sabotage on Friday 15th June in Germany included the explosion of a bomb beneath a passenger train between Prüm and Sechtem, which derailed the engine but caused no injuries; the exploding of a bomb beneath the fourth carriage of a passenger train between Bingerbrück and Mainz, which slightly injured six passengers; and the explosion of a bomb beneath a train at Bingen, which injured 14 people, two seriously. In retaliation the French arrested officials, prohibited motor and cycle traffic and imposed a curfew. On Monday 18th June train C E 59 was derailed at Elsdorf and another train was derailed between Wickrath and Rheydt. A French Nachschub (replenishment) train was derailed at Wanne on Wednesday 18th June 1923 by the removal of a rail.

The French outlined three different methods employed by the saboteurs. The first was

---

[1823] Ibid, 8th May 1923

[1824] *Mitchell Evening Republican*, Mitchell, South Dakota, United States, 13th June 1923

[1825] *The Daily Ardmoreite*, Ardmore, Oklahoma, United States, 11th May 1923

[1826] *The Times*, London, England, 18th May 1823

[1827] Ibid, 1st June 1923

[1828] *The Daily Silver Belt*, Miami, Arizona, 6th June 1923

[1829] *The Times*, London, England, 14th June 1923

the burying of a high explosive charge with a slow fuse beneath the metals. The second was the planting of bombs beside or beneath the rails which were exploded by percussion when a train or pilot engine arrived. The third and most dangerous method was the use of a delayed-action fuse which would be primed by the pilot engine and was designed to explode some minutes later. 'It is believed by the French that the saboteurs are mainly ex-officers, members of nationalist societies, who, in desperation at French successes in the Ruhr, bring in bombs from unoccupied territory and go to the scene of action by motor-car.' [1830]

A Cologne – Mainz passenger train was 'wrecked' at Landstuhl, Germany, on about Saturday 23rd June 1923, three coaches being damaged (location also given as Boppard). No one was hurt. [1831] The same day train A 10 was derailed by a bomb between Wörth and Jockgrim. Twenty-four waggons were derailed. Some fell down an embankment and a railwayman was seriously injured. Both lines were blocked. A bomb exploded in the leading coach of train No. 626 at Meckenheim. The two leading coaches, both unoccupied, were destroyed. On about Saturday 30th June a bomb exploded under a locomotive between Düren and Buer, Germany. Also on 30th June the most serious act of sabotage during the occupation occurred when a bomb exploded in a compartment of a Belgian leave train between Duisburg and Friemersheim. The French Press Bureau stated that one soldier had been killed and 25 wounded, but a Belgian report stated that there were ten dead and over 30 wounded. [1832] Finally, a train was derailed by wooded blocks wedged between the running and check rails at Stromberg on Monday 6th August 1923. The Ruhr occupation continued until August 1925, but this appears to have been the last definite act of sabotage.

Passenger train P 1009 was derailed at Friedberg, Germany, on Friday 28th September 1923, two passengers being killed and 16 injured, due to a set of facing points having been operated prematurely, but it is unclear to the author if this was due to a deliberate act of sabotage or was due merely to negligence.

A head-on collision occurred early on Thursday 24th May 1923 at Dothan, Texas, United States, on the Texas and Pacific Railway. Eastbound passenger train No. 16 consisted of engine No. 707, a baggage car, a mail car, two baggage cars, two coaches and three Pullman sleeping cars, in charge of Conductor Moss with Engineer H. H. Bailey and Fireman J. N. Gray. At Baird it received train order No. 5, Form 31, reading:

'No. 16, Eng. 707 meet No. 15 Eng. 706 at Dothan. No. 16 take siding.'

Train No. 16 proceeded to Dothan, entered the siding, and was standing there when it was struck by westbound train No. 15 at 2.40 a.m. The latter train consisted of engine No. 706, four baggage cars, a mail car, two coaches, three Pullman sleeping cars and a Texas and Pacific business car, in charge of Conductor Witherspoon with Engineer William Simpson and Fireman J.M. Osborne. [1833] It received the same train order at Ranger and arriving at Dothan, travelling at 35 miles an hour, it was diverted into the siding and collision with train No. 16. Both engines were considerably damaged, the leading

---

[1830] Ibid, 16th June 1923
[1831] Ibid, 25th June 1923
[1832] Ibid, 2nd July 1923
[1833] San Antonio Express, San Antonio, Texas, 25th May 1923

wooden baggage car of train No. 16 was crushed between the tender and the second vehicle, and the front three cars of train No. 15 were destroyed in a post-collision fire. Porter Charley Baker, from train No. 16, who had gone forward to operate the switch once train No. 15 had arrived, was killed, as were six trespassers who had been riding blind baggage on one of the trains, probably No. 16.

Conductor Moss told the Interstate Commerce Commission (ICC) inquiry that with Conductor Witherspoon he had examined the east switch after the collision and found it set for the siding. The switch lock was missing, and they had found a broken piece of the lock hasp nearby. The switch light was on the ground with two pieces of wire attached which appeared to have been used to hold the light on the stand to make it show a false indication.

Engineer Bailey spoke of his train entering the siding and of his flashing his headlight to train No. 15 to indicate that (as he believed) his train was in clear. Fireman Gray corroborated his statement.

Engineer Simpson testified that he had been able to see the green light of the east switch from about a mile distant and he recalled the engineer of train No. 16 flashing his headlight. Near the switch he had braked to stop for water when he had seen that the switch target was red, although the switch light was still green. He had then applied his brakes in emergency, but too late to avoid the collision. Fireman Osborne gave similar evidence.

The ICC concluded that the derailment was due to an open switch due to malicious tampering, probably using a clinker hook which was found discarded nearby to pry off the lock of the switch. [1834]

The railroad company offered a $5,000 reward for the arrest and conviction of the perpetrators. [1835]

A motor trolley carrying a party of railway officials was derailed by 'bandits' near Kalkilieth, Palestine, on the evening of Tuesday 10th July 1923. Shots were then fired, killing the Egyptian driver and a Greek employee, and wounding the Divisional Engineer, Mr Wright, and a Mr Taylor. [1836]

A train carrying a company of 100 soldiers of an engineering party was attacked on Tuesday 23rd October 1923, by rebels who had concealed themselves in a ravine near Tétouan, Spanish Morocco. Six soldiers were killed and 11 wounded. [1837]

A freight train on the Chemins de fer fédéraux suisses was derailed near Porrentruy, Switzerland, probably late on Tuesday 11th December 1923. An unknown person who appeared to have considerable knowledge of the railway had loosened the rails. Driver Reichen, Fireman Wuithier and Conductor Bregnard were all injured, and eight of the 52

[1834] Interstate Commerce Commission report
[1835] *The Victoria Advocate*, Victoria, Texas, United States, 27th May 1923
[1836] *The Times*, London, England, 12th July 1923
[1837] *Paris-Midi*, Paris, France, 24th October 1923

waggons of the train were demolished, the damage totalling 100,000 francs. It was believed that the person responsible had not intended to cause death or injury but merely harm to the railway company. 'It is thought that the perpetrator of the outrage is a man who had been discharged by the Railways Board.' [1838] There had been a previous sabotage attempt on 14th November.[1839]

Spikes were removed and a fast limited train between El Paso, Texas, and Mexico City was derailed near Celaya, Mexico, on Sunday 23rd December 1923. Several coaches overturned and over a score of passengers, among them Americans, were said to have died, some who might otherwise have survived, because of a lack of medical assistance. The rebels responsible were thought to have believed, wrongly, that the train was carrying Federal troops for the protection of the Mexican capital. [1840]

The engine of Cuba Railway passenger train No. 5 was derailed, apparently by a small dynamite charge, on Thursday 3rd January 1924, near Camaguey, Cuba. No one was hurt. Railway employees had been on strike since December. [1841] Another short-lived strike broke out on Sunday 24th February 1924 and a train as derailed on the Northern line near Matanzas, Cuba, without casualties. [1842]

Rebels under Feliciano Garcia, previously police inspector at Tampico, derailed and looted a freight train at Garza Valdez, Mexico, on about Wednesday, 9th January 1924. [1843] Then on about Thursday 24th January 1924 rebel forces commanded by General Salvador Alvarado derailed a northbound train at Pedrito, Mexico, putting the military escort to flight and seizing the armoured cars guarding the train. [1844]

On the Orloff – Vitebsk railway anti-communist forces partially derailed a train near Lazarevka, USSR, on about Sunday 27th January 1924 and made off with a trainload of arms and ammunition. [1845]

An attempt at sabotage was made in Japan on Thursday 31st January 1924. The target was a train carrying 25 members of the Diet who were travelling from Tokyo to a political meeting at Osaka, the passengers including Baron Takahashi, the ex-Premier, Mr Inukia, leader of the Nationalist Party, and a Mr Ozaki, a prominent member of the government. The efforts of the saboteurs were frustrated by the alert train driver, who

---

[1838] *The Times*, London, England, 13th December 1923
[1839] *Journal de Genève*, Geneva, Switzerland, 14th December 1923.
[1840] *Oakland Tribune*, Oakland, California, 25th December 1923
[1841] *The Commercial Tribune*, Cincinnati, Ohio, United States, 4th January 1924
[1842] *Daily Times-Enterprise*, Thomasville, Georgia, United States, 25th February 1924
[1843] *The Anaconda Standard*, Anaconda, Montana, United States, 11th January 1924
[1844] *The Bee*, Danville, Virginia, United States, 25th January 1924
[1845] *The Times*, London, England, 31st January 1924

saw heavy timbers and rocks piled up on the track in time to pull up. [1846]

A 'scout' train and the passenger train following it were both dynamited near Soledad, Mexico, on Friday 22nd February 1924 by rebels under the command of Pedro Gonzales. The scout train was little damaged and escaped. Despite a valiant defence the 50 strong escort of the passenger train was overwhelmed by the rebels, who robbed the passengers and looted and burned the express and mail cars. Some passengers were reported killed and wounded. The rebels had hoped to capture Governor Tejeda of Vera Cruz, but he had remained in Cordoba.[1847] Rebels under the command of Guadalupe Sanchez again attempted to blow up a train near Soledad, this one bound from Vera Cruz to Mexico City, on Sunday 24th February. The mines exploded as they were being planted, killing a large number of rebels. [1848]

A new Russian train was planned to make its inaugural trip from Odessa to Moscow on 1st May 1924 to coincide with the May Day celebrations. Known as the "May Day Train" its engine was surmounted by a huge portrait of Lenin haranguing a large crowd with inscriptions dedicating the train to him "whose spirit lives though his body is dead." (Lenin had died in January following a series of strokes). When the train was 100 miles from Odessa, early on Friday 2nd May 1924, it was derailed, the engine and tender, luggage van and six coaches plunging down an embankment. The train contained important delegates from Moscow, the line showed signs of tampering and the tool used was found alongside the track. It was officially stated that it was a case of sabotage and that many had been killed and injured, but the numbers of casualties or who the delegates were was not stated. [1849] [1850]

A Bucharest-Jassy express was derailed outside the station at Barnova, Romania, on the evening of Sunday 27th July 1924, the track having been interfered with. The train was going slowly and only the engine and a luggage van were overturned, though a passenger coach and a restaurant car left the rails. The driver was the only casualty, being badly burnt. Who was responsible was not known. [1851]

Rebels under the command of General Guadalupe Sanchez, numbering about 100, derailed a Vera Cruz train near San Salvador, Mexico, on Friday 15th August 1924. Obstacles placed on the track threw off the engine and two coaches. A deadly fire was then opened up on the train. About 20 passengers and two of the armed escort were believed killed, including some Germans, and about the same number, including three Germans, were wounded. The passengers having all been killed, wounded, or disarmed

---

1846 *The Daily Telegraph*, Launceston, Tasmania, Australia, 2nd February 1924
1847 *The Commercial Tribune*, Cincinnati, Ohio, United States, 24th February 1924
1848 *The Billings Gazette*, Billings, Montana, United States, 27th February 1924
1849 *The Times*, London, England, 10th May 1924
1850 *The Argus*, Melbourne, Victoria, Australia, 12th May 1924
1851 *The Times*, London, England, 29th July 1924

the rebels robbed them of money and valuables, searching even women and children. [1852]

Forty passengers died and a large number were injured when an express from Odessa to Moscow was derailed 200 miles from Odessa, USSR, early on Thursday 28th August 1924, bandits having torn up the track. The train was travelling at high speed at the time. Following the derailment the engine overturned, its boiler exploded, and the resulting fire spread to the coaches. Relief trains were sent from the nearest station and from Odessa. [1853]

In a repeat of the August attack on Friday 26th September 1924 a Moscow to Odessa express was derailed by bandits about 300 miles from Moscow, USSR. Again, the train was travelling at high speed and disconnected rails caused the engine, baggage car and ten coaches to leave the track, the rear three coaches being spared. Over 300 yards of track were torn up. The driver, fireman and conductor were killed instantly and in all 15 persons were killed, 25 were said to be mortally wounded and 43 seriously hurt. A special train hastened to the scene and brought the injured back to Moscow. The bandits escaped. [1854]

Bandits commanded by Jose Lagunas, Pedro Gonzales and Cayetano Acar – a Turk, what was he doing with bandits in far-off Mexico? – derailed and attacked a passenger train on the isthmus line between Joachim and Piedras Negras, Mexico, on Sunday 5th October 1924. The engine was overturned and five soldiers and two women, one the mother of four children, were killed. The remaining 18 soldiers were taken prisoner. Forcing the passengers to strip to their underclothes the bandits took 16,000 pesos from the express car and kidnapped three young women from among the passengers. Four hours elapsed before the passengers reached Piedras Negras and said what had happened, when Colonel Zuniga set off with government troops in pursuit of the bandits. [1855] Five days later, on Friday 10th October 1924, bandits committed another atrocity near Candelaria, Mexico. About noon a train left the mine of the Eruption Mining Company. Not long afterwards it was derailed by the removal of a rail on a curve, the engine and a dynamite car leaving the track. The entire train crew, two road masters, a labourer and an American, B.F. Barker, the auditor of the company, nine men in all, were lined up on the embankment by the five bandits and shot dead. The only armed man on the train, a customs guard, was missing. The object of the sabotage was a $15,000 payroll. When news of the attack emerged a posse was sent out in pursuit of the bandits. [1856]

Canadian Pacific Railway train No. 11, which had left Nelson late the night before, was wrecked by an explosion in a first-class coach at 1 a.m. on Wednesday 29th October 1924, two miles west of the station at Farron, British Columbia, Canada. Five passengers were killed, among them Peter Veregin, the uncrowned "king" of the Doukhobor religious sect in Canada (the Doukhobors are a Protestant religious sect who originated in

---

[1852] *The Urbana Daily Courier*, Urbana, Illinois, United States, 16th August 1924
[1853] *Medicine Hat News*, Medicine Hat, Alberta, Canada, 28th August 1924
[1854] *The Lancaster Daily Eagle*, Lancaster, Ohio, United States, 26th September 1924
[1855] *The Lowell Sun*, Lowell, Massachusetts, United States, 6th October 1924
[1856] *New Castle News*, New Castle, Pennsylvania, United States, 11th October 1924

Russia). Initially the explosion was attributed to the rupture of an acetylene gas cylinder beneath the coach, a theory which was shown to be untenable when examination revealed the cylinder was intact after the explosion. It then became evident that the explosion must have been due to a bomb, and that it was likely that Peter Veregin had been the target, the death of the other passengers being presumably regarded by the perpetrators as 'collateral damage.' The explosion was believed to have stemmed from a bitter feud between two factions of the Doukhobor colony. Fearful of inciting further violence the Doukhobors refused to co-operate with the government investigation into the explosion and the culprits were never found. [1857] [1858]

An attempt was made to wreck the Calcutta – Darjeeling mail train between Ishurdi and Gopalpur, India, at 1 a.m. on Saturday 15th November 1924. Fishplates, dog-spikes, and bolts had been removed from a section of rail and the engine and seven coaches left the track, but no one was hurt. A previous attempt a few days earlier between Aranghata and Chuadanga had been frustrated by railway staff. [1859]

Bandits under the command of Emilio Espino, a former follower of the late Pancho Villa, captured and wrecked a freight train at Conejos, Mexico, on Friday 21st November 1924. The engine was run into a ravine, several cars were burned and two were sent running down the line in the style of Irish Republicans. Troops were sent in pursuit of the bandits. [1860]

A Belfort express of the Chemins de fer de l'Est left Paris Est at 19:15 on Wednesday 25th March 1925 and had a normal journey. It was followed 40 minutes later by the Vienna express, which took the same route for the first part of its journey. The Vienna train consisted of an engine, a baggage van, four very sturdy *Wagons-Lit* coaches and a restaurant car. Its locomotive was in charge of Driver Rabiet, a man well-known for his vigilance and care, with Fireman Pinsot, and was soon running at 92 kilometre an hour. At 21:00 this train passed Maisons-Rouge station and it was between there and Longueville, France, when suddenly there was a violent shock, and the engine left the track. Rather than applying the brake hard and telescoping the train, Driver Rabiet applied the air brakes gently to bring is train to a stand upright and in line. Shortly after the train came to a stand its passengers began alighting from their coaches, with overcoats thrown over their pyjamas. None of the 27 of them was hurt, only the conductor was injured when he was thrown against the wall of his van. An omnibus train eventually took the passengers to Loungueville and they continued their journeys on the Basle express.

The cause of the derailment was quickly obvious. Six metres of the outer rail had been carefully removed on a curve. At the point of the derailment there was a 20-metre-deep ravine alongside of the line into which the train could have gone. It was clear that the rail

[1857] *The Winnipeg Evening Tribune*, Winnipeg, Manitoba, Canada, 29th October 1924
[1858] Wikipedia – Doukhobors
[1859] *The Times*, London, England, 17th November 1924
[1860] *Brownwood Bulletin*, Brownwood, Texas, United States, 24th November 1924

must have been removed between the passage of the two trains. The tool which had been used had been taken from a trackman's hut a kilometre distant. There was considerable damage to the rolling stock and track, but a temporary line was laid to enable traffic to be resumed.

Two quite well dressed strangers who had been seen in the area were suspected to be the ones responsible for the sabotage. The motive for the attack was not, it would seem, apparent. [1861]

IF the motive for the Longueville sabotage was unclear the reason behind the derailment of northbound Atchison Topeka and Santa Fe Railway train No. 6, running between Oklahoma City and Kansas City at 4.15 a.m. on Tuesday 2nd June 1925 near Oklahoma City, Oklahoma, United States, was apparently robbery. The engine and seven coaches of the train left the track and rolled down a 25 feet embankment. The passengers were severely shaken up but not otherwise hurt. Spikes had been loosened from the ties allowing the rails to spread. Every available officer from Oklahoma City joined a posse to search a dense wood near the accident scene. [1862]

Robbery was also the motive for the derailment of a Louisville and Nashville Railroad train at Knob Lick, Kentucky, United States, on the night of Saturday 5th September 1925. Three Kentucky men, Herbert Hale of Stanford and Elvie and Oscar Dishon of Shepherdsville allegedly plotted to wreck the train by opening a switch and then rob it. Engineer John Peddarar saw the open switch and shouted to Fireman O.F. Walling, "Jump, quick, I'll stay with it!" Engineer Pedderar was killed, and 15 passengers were injured. After the derailment, a man identified as Hale entered the mail car to rob it but confronted by an armed mail clerk he fled and was later captured. [1863] Hale made at least two confessions, in the second of which he exonerated the Dishon Brothers and named another man, Clayton Middleton, as his accomplice. [1864] Hale also later implicated another man, Sam Atkins. [1865]

In 1926 Syria was a French Protectorate, having been, like Iraq, part of the Ottoman Empire before the First World War. In August 1925 Sultan Pasha al-Altrash had launched the Great Syrian Revolt with the intention of driving the French out of Syria. [1866] Probably related to the revolt were several derailments, the first on about Saturday 14th February 1926 of a train near Damascus, Syria. Two passengers on the Damascus – Rayak train were killed and one wounded. The passengers were completely stripped by the 'brigands.' The report on the affair added that loyal (to the French) Hauranis had beaten off a large party of Druse and Beduin raiders, killing 20 of them. [1867] Then on Wednesday 18th February the Hejaz Railway was cut five miles from Damascus. A train

[1861] *Le Gaulois*, Paris, France, 27th March 1925
[1862] *The Brainerd Daily Despatch*, Brainerd, Minnesota, United States, 2nd June 1925
[1863] *The Commercial Tribune*, Cincinnati, Ohio, United States, 8th September 1925
[1864] *The Monroe News-Star*, Monroe., Louisiana, United States, 17th September 1925
[1865] *Port Arthur News*, Port Arthur, Texas, United States, 27th September 1925
[1866] Wikipedia – Great Syrian Revolt
[1867] *The Times*, London, England, 16th February 1926

from Palestine was derailed, three carriages burned, and seven Syrians were killed. However, 'all the European passengers, including the British Vice-Consul, reached Damascus in safety.' The line between Damascus and Rayak was also cut. [1868] Finally, on about 6th April 1926 Druse tribesmen attacked a train south of Damascus and captured the railwaymen. Troops were sent to the scene. [1869]

Acts of railway sabotage were rare in Great Britain, but on Monday 10th May 1926 the country was in the seventh day of the General Strike, which stemmed from an attempt by coal miners to fight a reduction in their already low wages, coupled with an increase in their hours. Winston Churchill, then Chancellor of the Exchequer, organised a strike-breaking movement called the Organisation for the Maintenance of Supplies, to recruit scabs to take over the work of the strikers, and also threatened to call out the troops. Angry and desperate miners determined to stop coal trains running and removed a rail from the London and North Eastern Railway up main line at Cramlington, England, intending to derail a coal train. Instead the train which first arrived was the up *Flying Scotsman*, consisting of Pacific-class engine No. 2565, *Merry Hampton*, a van and ten corridor coaches, in charge of Guards Brown and Fletcher, with Driver Thomas Wedderburn and volunteer (scab) Firemen Aitken and Hird. [1870] The train was duly derailed, the *Merry Hampton* overturned into the ditch and the five leading vehicles were derailed. A large number of passengers were slightly hurt but no one was seriously injured, and no one died. [1871]

The General Strike ended two days later when the Trades Union Council caved in to the government. Ten miners were arrested at the beginning of June and eight stood trial. All were found guilty by a jury which had no working-class members and were sentenced to up to eight years' imprisonment. [1872] The last of the miners to be released – those who had been given eight-year sentences – were Robert Harbottle, Thomas Roberts, and Arthur Wilson, who were freed from Maidstone Prison on 22nd December 1929. [1873]

Up Great Indian *Peninsula Express* No. 604 was derailed between Harauni and Jaitipur stations, India, on the Lucknow – Cawnpore branch of the East Indian Railway on Monday 26th July 1926. Four bogie carriages and a four-wheeled vehicle were derailed and capsized, and 100 yards of track were torn up, but no one was hurt. Dog-spikes and fishplates had been removed from the track. A sub permanent way inspector and two trolleymen were charged with causing the derailment but were acquitted by the Lucknow Sessions Judge. [1874]

On Thursday 19th August 1926 Berlin – Hannover – Köln D-Zug No. 8 was derailed between Leiferde and Meinersen, Germany. The engine and two coaches fell down a 1.5-

---

[1868] Ibid, 22nd February 1926
[1869] *East Liverpool Review-Tribune*, East Liverpool, Ohio, United States, 7th April 1926
[1870] *Let No Wheels Turn* by Margaret Hutcherson, pg. 1
[1871] Ibid, pgs. 16-18
[1872] Ibid, pgs. 49-63
[1873] *The Times*, London, England, 23rd December 1929
[1874] Administration Report of Indian Railways in 1926-27

metre-high embankment and the following coach was telescoped into them. The passengers trapped in the telescoped carriage were unable to release themselves. It was a dark night, and they had a prolonged ordeal before relief trains from Lehrte, Isenbüttel and Oebisfelde finally arrived. It was quickly realised that an act of sabotage had taken place because bolts and fishplates which had been unscrewed were found lying beside the track. The Berlin criminal police were at once alerted and General Director Dorpmüller, the State Advocate and representatives of the Transport Ministry were soon on the scene. The line was dead straight at the point where derailment had occurred, with excellent permanent way and trains were permitted to travel over it at more than 100 kilometres an hour. The various tools which had been used to damage the track were found some 10 metres from the point of derailment. The local track gang did not consider that they were theirs. Four express trains had passed over the spot that night in 50 minutes; the last had derailed because of its higher speed.

The following morning the police searched the whole area with bloodhounds and found that the perpetrators had probably reached the point where they had carried out their attack on bicycles from a nearby level crossing. After a comprehensive investigation two young men, named Otto Schlesinger and Willi Weber, were arrested in a Berlin homeless shelter on 6th September. [1875] They were taken to court, tried, found guilty of carrying out the attack, and sentenced to death, the punishment later being commuted to one of life imprisonment. Their attack had cost 24 lives and seriously injured many more. [1876]

A series of attacks in western Germany at the end of August 1926 included stones being thrown at them, bullets being fired on them, the placing of rocks and rubble on the track near Weiden (for which a man was apprehended) and the derailing of the engine and two carriages of an electric train near Pletenberg on about Sunday 29th, again by the placing of obstructions on the line. No motive was suggested for the attacks. [1877]

Chinese Communist Party inspired activities in China in late 1926 included damaging a portion of the track of the Shanghai-Nanking (now Nanjing) railway on the Shanghai side of Chekiang, resulting in the total wreck of a northbound train on about Saturday 23rd October 1926. There were said to have been comparative few casualties. [1878]

The train on which Major Oliveira Tavares, former police commander of Mozambique, was travelling to Chimoio on the Beira – Salisbury (now Harare) line to take up his new duties there, was derailed on about Saturday 6th November 1926. Details of the incident were lacking but it was believed to be related to a recent railwayman's strike. [1879]

[1875] *The Times*, London, England, 8th September 1926

[1876] *Historische Eisenbahn Katastrophen* by Bernhard Püschel, pg. 128 (in German)

[1877] *The Times*, London, England, 31st August 1926

[1878] Ibid, 26th October 1926

[1879] Ibid, 8th November 1926

Bandits derailed a train at Santiago Papasquiaro, Mexico, on Friday 3rd December 1926 but were driven off by the military escort after a brisk fight. [1880]

Several hundred armed men attacked a Southern Pacific of Mexico Railroad passenger train 15 kilometres south of Acaponeta, Mexico, on Tuesday 23rd August 1927. Forty shots were fired at the train and 15 passengers were killed or wounded. An American woman named Florence Anderson was seriously wounded and died the following day. [1881] [1882]

On 12th April 1927, many Chinese Communist members of the Kuomintang (Chinese Nationalist Party) were purged in a series of arrests and executions (the Shanghai Massacre) on the orders of General Bai Chongxi. A civil war between the Communists and the Nationalists then began on 1st August 1927. [1883] On Friday 26th August 1927, the track was removed at Anting, China, on the Shanghai – Nanking railway, a train was derailed, and 20 people were killed. It was presumed that a train carrying General Bai Chongxi was the target, but another train suffered. [1884]

A train approaching the station at Houtovo, Yugoslavia, at 2 a.m. on Tuesday 20th September 1927 was completely wrecked apart from its engine by the explosion of seven bombs placed on the track. The identity and number of the victims of the attack was not stated. It was supposed that the sabotage had been carried out by Bulgarian Komitadjis, who had recently blown up a hotel at Guevgheli. [1885] The Komitajis were rebel bands organised by the pro-Bulgarian International Macedonian Revolutionary Organisation. [1886]

Six people died, several others were seriously injured and members of the family of President Calles of Mexico had a narrow escape when a passenger train was dynamited by 60 armed men between Villa Reyes and Jural del Berrio, Mexico, on Tuesday 18th October 1927. The dynamite exploded beneath a second-class carriage near the front of the train, the locomotive of which was derailed. The Calles family members, whose identities were not revealed, had been riding in a special car attached to the train from Laredo to Mexico City. The wounded passengers were taken to Mexico City and Federal troops sent to hunt the attackers. [1887]

A passenger train was derailed near Dniepropetrovsk (until 1926 Ekaterinoslav and now Dnipro in the Ukraine), USSR, on Saturday 3rd December 1927. Seven people died and between 20 and 30 were seriously injured. Sabotage was suspected and the OGPU

[1880] *Helena Independent*, Helena, Montana, United States, 5th December 1926
[1881] *Lawrence Daily Journal-World*, Lawrence, Kansas, United States, 24th August 1927
[1882] *The Weekly Ledger-Journal*, Fairfield, Iowa, United States, 25th August 1927
[1883] Wikipedia – Chinese Civil War
[1884] *The Brainerd Daily Despatch*, Brainerd, Minnesota, United States, 26th August 1927
[1885] *The Times*, London, England, 21st September 1927
[1886] Wikipedia – Komitadji
[1887] *The Southwest Times*, Pulaski, Virginia, United States, 19th October 1927

(Soviet secret police) sent a commission to investigate the matter. [1888]

At about 03:12 on Saturday 28th January 1928 Burma Railway train No. 4, the down mail, travelling at 30 miles an hour, was derailed 24 feet short of bridge No. 161 at mile 110/1, near Kyauktaga, Burma. [1889] Fishplates and bolts had been removed from the track. The engine tender and three third-class passenger carriages capsized into the mullah bed and were entirely wrecked. The two brake vans following the tender got across the gap and were partially destroyed, the remaining vehicles were pulled up just short of the bridge. The engine crew and 23 third-class passengers were killed outright, and 14 more passengers died in hospital from extensive scalds and shock. The *Times* correspondent in Rangoon wrote, 'Passengers gave pitiful accounts of the dying and the dead scattered about the river bed, some with arms or legs pinned beneath the heavy wreckage, which could not be removed. An Indian had both legs smashed in this manner, and his wife and children were killed. He begged those present to recover from the bed of the river his box containing 16,000 rupees (£1,200), with which he had hoped to make his family comfortable in India. The injured are said to include two Italian cloth-sellers.' [1890] Both lines were blocked and one lakh (100,000) of rupees damage was caused. After the wounded passengers had been evacuated the up line was cleared at 17:55 the same day but the down line remained blocked until 21st February. [1891]

Robbery was believed to be the motive behind the derailing of a mixed train between Amatlan and Guadalupe in Vera Cruz state, Mexico, on about Tuesday 28th February 1928. Rails piled across the tracks caused the engine and two express cars to derail. The engineer was crushed beneath his machine and killed and a fireman, a brakeman, and two express messengers were injured. If robbery was the motive for the attack it was frustrated by the deployment of the train's armed escort. [1892]

Chang Tso-lin (or Zhang Zuolin) was at the beginning of 1928 a powerful Chinese warlord who ruled Manchuria, supported by Japan, who had economic and political interests in Manchuria. By mid-1927, however, Japan felt that theirs and Chang Tso-lin's interests were diverging and planned to replace him as leader. Junior Japanese army officers, acting without the approval of Tokyo, proceeded to assassinate Tso-lin early on Monday 4th June 1928. On Sunday 3rd June Tso-lin began a journey from Peking (now Beijing) to Mukden (now Shenyang) over the South Manchuria Railway, which was guarded by Japanese troops. When his train reached a point east of Hwangkutun (now Huanggutun) station, Manchuria, at 5.23 a.m. there was an explosion caused by what was described in a contemporary newspaper as an 'infernal machine.' The bomb had been planted by Japanese Sapper First Lieutenant Sadatoshi Fujii. The bomb blew up beneath the eleventh of the 22-saloon coach of the train, destroyed two carriages and set others alight. Several of Tso-lin's companions were killed outright but the first report stated that

---

[1888] *The Times*, London, England, 5th December 1927
[1889] Ibid, 30th January 1928
[1890] Ibid, 31st January 1928
[1891] Administration Report of Indian Railways in 1927-28
[1892] *Laredo Daily Times*, Laredo, Texas, United States, 29th February 1928

he had only suffered a cut nose. In fact he was fatally injured and died some hours later. [1893] The unauthorised assassination delayed Japan's invasion of Manchuria for several years. [1894]

Three different trains were attacked by insurgents in western Mexico on Wednesday 6th June 1928. One train was derailed near Poso (Porfirio Diaz) station in Guanajuato state and fired on. The military escort of the train was reinforced by Federal troops who fought off the attackers. They killed 25 insurgents, including their leader, Francisco Vargas. Three soldiers and a passenger were wounded. In another attack a freight train from Guadalajara to the port of Manzanillo was derailed at Coquimatlàn in Colima state and its crew carried off into the mountains. The third attack occurred near Obregon station when a fast train was stopped, and its crew captured. The attackers uncoupled the engine and started it down the line with the intention of engineering a collision with a passenger train, but the engine jumped the track. Elsewhere three men were apprehended trying to blow up a bridge on the Colima-Guadalajara line. They were executed. [1895]

An American, Mr James Wedgewood, who was an auditor of the Pacific Railroad of Nicaragua, was killed, and General Carmen Diaz, General Manager of the railway and brother of the Nicaraguan President, was badly bruised, when the track motor car on which they were riding was derailed by a spike on the rails near Managua, Nicaragua on Friday 8th June 1928. [1896] The derailment followed several previous attacks on the railway. An American newspaper wrote that the sabotage was the work of a 'political fanatic,' [1897] in reality the attacks were doubtless part of the guerrilla campaign organised by Augusto Sandino against the American occupation of the country. [1898]

Several soldiers were killed when bandits attacked a military train near Lagos De Moreno, Mexico, on about Thursday 21st June 1928, wrecking it by raising the tracks. After the initial confusion, the troops recovered themselves and the attackers were routed. [1899] Then, on about Tuesday 26th June 1928, Engineer Andrew Robinson, who had worked for the Ferrocarriles Nacionales de México for 35 years, was killed when rebels derailed his train near Viborillas, Mexico, on. The train had come from Laredo. Two other railwaymen were wounded before the armed guard drove off the attackers. All the cars of the train except the Pullmans were derailed but none of the passengers was hurt. The opposing, Laredo bound train, was also targeted. Two exploded bombs were found on the track with a third unexploded one nearby. The caution of the engineer prevented an accident. [1900]

[1893] *The Times*, London, England, 5th, 6th and 7th June 1928
[1894] Wikipedia – Huanggutun incident
[1895] *The Daily Silver Belt*, Miami, Arizona, 7th June 1928
[1896] *The Times*, London, England, 7th June 1928
[1897] *Brownwood Bulletin*, Brownwood, Texas, United States, 8th June 1928
[1898] Wikipedia – History of Nicaragua
[1899] *Port Arthur News*, Port Arthur, Texas, United States, 22nd June 1928
[1900] *Denton Record-Chronicle*, Denton, Texas, United States, 27th June 1928

Bandits attacked and presumably derailed a fast train on the Chinese Eastern Railway near Pogranichnaya, USSR, on about Wednesday 4th July 1928. Four passengers were killed and seven wounded, while 40, including 25 women, were dragged off to the mountains to be exchanged for ransom after the train had been looted. [1901] The Russian government was trying to determine the nationality of the passengers. While the apparent motive for the attack was robbery it was theorised that the true motive might be political, that the bandit leader was opposed to the Chinese Nationalists and that he hoped to stimulate the Japanese to increase their forces in Manchuria and thereby check the power of the Nationalists. [1902]

What was described as the worst railway disaster in Bengal, India, for many years occurred at about 22:42 on Sunday 8th July 1928 between Belur and Dankuni stations on the East Indian Railway. The engine and leading five vehicles of No. 17 up passenger train left the rails and plunged down an embankment. Eighteen passengers and Driver Belletti were killed outright, and eight passengers and the fireman were seriously injured, two of the injured passengers dying in hospital. Twenty-four passengers and a second fireman were slightly hurt, and 101,333 rupees damage was done to rolling stock and 1,654 rupees damage was caused to the permanent way. All the passengers killed, and wounded were in the third-class carriages. [1903] A rail had been removed and another twisted out of place. The sabotage was connected with strikes on the railway at Lillooah and Asansol. There had been several previous unsuccessful attempts to wreck trains. [1904]

Insurgents under Ladislao Medina attacked a passenger train between Patzcuaro and Arlo, Mexico, on about Tuesday 10th July 1928. Carrasco Sanduval, a Federal deputy, his brother Carolos, and the mayor of the town of Tacambaro were taken from the train and hanged from lineside telegraph poles. [1905]

A Laredo – Mexico City passenger train was derailed near Rincon, Mexico, on Tuesday 17th July 1928, by 200 bandits, supposedly led by a woman. All but two of the nine coaches of the train were derailed and only a water tower saved the coach in which the American passengers were sleeping from going into an arroyo. The armed guard of the train fought off the first attack of the bandits and in the second attack three of them were captured, the remainder again being driven off. Seven passengers were wounded, and the captured bandits were executed. [1906]

All but one of an escort of about ten soldiers were massacred by bandits who held up a passenger train near Yautepec, between Cuautla and Puente De Ixtla, Mexico, on the night of Thursday 30th August 1928. The engine was derailed and ditched, and the bandits

[1901] *Salt Lake Tribune*, Salt Lake City, Utah, United States, 9th July 1928
[1902] *Olean Evening Times*, Olean, New York, United States, 9th July 1928
[1903] Administration Report of Indian Railways in 1928-29
[1904] *The Times*, London, England, 10th July 1928
[1905] *Moline Daily Despatch*, Moline, Illinois, United States, 10th July 1928
[1906] *Oelwein Daily Register*, Oelwein, Iowa, United States, 19th July 1928

set the train on fire. Several of the train crew were hurt but none of the handful of passengers was harmed, though they were robbed, and the express car was looted. Federal troops from the Guernavaca Garrison set off in pursuit of the bandits, who numbered about 175. [1907] [1908]

Bandits tore up the track near Aguascalientes, Mexico, on or about Monday 10th September 1928, derailing a train and injuring the engineer, fireman and several soldiers of the armed escort. [1909] Then on Sunday 16th September an engineer and fireman died, and several passengers and an express messenger were injured when bandits tore up the track near Moroleon, in Guanajuato state, Mexico, derailing a train. The engine, tender, express car and second-class coaches overturned. A relief train as sent from Guadalajara. [1910] On about Sunday 23rd September another attempt was made to wreck a train between Encarnacion and Tigre, Mexico, the bandits capturing a band of platelayers and forcing them to remove a rail. Their plan was frustrated when the engineer of the targeted train noticed the damage to the track and pulled up his train in time to avoid derailment. Reluctant to attack the intact train and its armed guard the bandits burnt a nearby ranch house, cut telegraph communication and escaped. [1911] Counter measures against the bandits by Federal troops seem to have had some success in the next few weeks for on the 27th of October it was reported from Mexico City that more than 30 rebels had been killed in a serious of engagements over the previous few days, including some who had attacked a Ciudad Juarez to Mexico City passenger train on Tuesday 18th September. [1912]

On about Sunday 16th December 1928 bandits dynamited a Guadalajara to Colima train near Quemado, in Jalisco state, Mexico. The locomotive, express and mail cars were derailed and overturned, and one soldier was injured. The armed escort drove off the attackers. The same day 70 bandits fired on a train near Maravatio in Michoacán state. One soldier and one bandit were killed. [1913]

The special train carrying President-elect Emilio Portes Gil to Mexico City from Tamaulipas, where he had handed over his duties as governor to his successor, was dynamited in the eastern part of Guanajuato state, between Commonfort and Rincon, Mexico, on Sunday 10th February 1929. A pilot train having been allowed to pass unscathed, a bomb was exploded under the locomotive of the Presidential train as it passed through a desolate area. The engine was partly destroyed, and the two leading coaches were thrown from the tracks. Fireman Augustine Martinez, who had been standing on the tender, was the only fatality. The president had been sent an anonymous warning, written on satin, on the day before his journey. It read, “Slave, commend your soul to God. Thou shalt not reach Mexico City.” Saturday was also the day that Jose de Leon Toral, the assassin of President-elect Obregon, was executed. [1914]

A train from Reyes to Yurecuaro and thence to Mexico City was dynamited between

---

[1907] *New Britain Herald*, New Britain, Connecticut, United States, 31st August 1928
[1908] *East Liverpool Review*, East Liverpool, Ohio, United States, 31st August 1928
[1909] *Janesville Daily Gazette*, Janesville, Wisconsin, United States, 11th September 1928
[1910] *The Commercial Tribune*, Cincinnati, Ohio, United States, 17th September 1928
[1911] *Reno Evening Gazette*, Reno, Nevada, United States, 24th September 1928
[1912] *The Commercial Tribune*, Cincinnati, Ohio, United States, 28th October 1928
[1913] *The Brainerd Daily Despatch*, Brainerd, Minnesota, United States, 18th December 1928
[1914] *Wichita Daily Times*, Wichita, Kansas, United States, 11th February 1929

Moreno and Guarachita stations, Michoacán state, Mexico, probably on Friday 15th February 1929. Survivors said that between 25 and 50 might have died as the entire military escort of the train was wiped out, while another report stated simply that 'some of the passengers, some of the crew and some of the soldiers had perished.' The engineer and fireman died in the explosion, but the engine at least was apparently not derailed, for other railwaymen uncoupled it from the coaches and proceeded with some of the passengers to Amora and thence to Yurecuaro. The whole garrison from Zamora was sent in pursuit of the bandits, and increased vigilance was ordered along the railway lines, and all trains must carry armoured cars such as were previously limited to passenger trains. [1915] The escort had fought the bandits until it ran out of ammunition. The bandits set fire to the train. [1916] The attack was followed by an attempt on Sunday 17th February to dynamite a train between Guadalajara and Colima, frustrated when the charge went off prematurely. [1917] Four priests were arrested.

Bandits attacked a Laredo – Mexico City passenger train early on the morning of Monday 18th March 1929 at San Miguel Allende, Mexico, in Guanajuato state. They tore up the tracks, derailing the engine and several coaches. There were no casualties. [1918]

An attempt to dynamite a Mexico City – Laredo passenger train near Queretaro, Mexico, on the night of Thursday 4th May 1929 failed when it was preceded by a freight train which set off the bomb instead. There was no great damage. The passenger train usually carried a good number of Americans. [1919]

Rebels derailed a train between Palmira and Troncosa, Mexico, on Friday 7th May 1929. Federal Cavalry pursued the rebels and attacked them at a ranch on Tuesday 11th, killing four. The remaining 20 were captured, taken to the scene of the wreck, and executed. [1920]

Many soldiers died when a train from Guadalajara to Manzanillo as derailed and attacked by bandits at Fernandez, Mexico, in Colima state, on the night of Thursday 9th May 1929. The military escort and the rebels fought a four hour gun battle with heavy casualties on both sides. None of the passengers was hurt.

Rebels tore up the tracks at Cortes, Jalisco state, Mexico, on Monday 27th May 1929, throwing the whole train into a ditch. As soon as the train stopped they opened fire on it from positions alongside the track. One passenger was wounded before the rebels were driven off by the train's military escort. [1921] This was the last act of railway sabotage in

---

[1915] *The Bismark Tribune*, Bismark, North Dakota, United States, 16th February 1929
[1916] *Freeport Journal-Standard*, Freeport, Illinois, United States, 16th February 1929
[1917] *The Bismark Tribune*, Bismark, North Dakota, United States, 18th February 1929
[1918] *The Amarillo Daily News*, Amarillo, Texas, United States, 19th March 1929
[1919] *The Bismark Tribune*, Bismark, North Dakota, United States, 4th May 1929
[1920] *The Vidette Messenger*, Valparaiso, Indiana, United States, 14th May 1929
[1921] *Reno Evening Gazette*, Reno, Nevada, United States, 28th May 1929

Mexico during the period covered by this volume.

Two locomotives attached to trains on the Chinese Eastern Railway were attacked at unstated locations in China on about Friday 16th August 1929, the bombs, presumably planted on the track, killing the drivers and firemen. No passengers were hurt. [1922] The sabotage came on the eve of the Sino-Soviet Conflict over control of the railway. In a further attack on Friday 1st November 1929 a Harbin to Pogranichnaya passenger train was blown up near Taipingling, China, killing two passengers. [1923]

At 7.45 p.m. the evening of Sunday 10th November 1929 train No. 59, the crack *West Coast Flyer* of the Southern Pacific Railroad was derailed between Saugus and Bouquet Junction, California. The locomotive, baggage, and mail cars and two coaches left the rails and Engineer Richard Ball was badly scalded, though Fireman R.C. Fowler escaped injury. A single bandit then boarded the train dressed as a railroad employee, persuaded the passengers to move into an undamaged coach and robbed them. Conductor Otis French was also robbed. First reports spoke of there having been two robbers, but in fact there was only one. The track had been interfered with using clawbars stolen from a lineside tool box. [1924] The bandit escaped with less than $400.

On Monday 25th November 1929, train No. 17, the westbound *Portland Limited* express of the Union Pacific Railroad was derailed near Cheyenne, Wyoming, United States. The engine of the train managed to negotiate the point where the track had been loosened by the removal of spikes without leaving the rails, but the seven following seven day coaches and Pullman cars left the rails and bumped along the ties. As soon as the train came to a stand a 'young desperado,' as he was described, moved quickly through the train, relieving passengers of their valuables, and then disappeared. To some of the women passengers he remarked, "I've got a wife and two kids at home and the railroad won't give me work." A possible worse accident was averted by Brakemen H.E. Lowell and Russell Grandy, who respectively went west and east along the double track and flagged down an eastbound freight train and the westbound *Gold Coast Limited.* The saboteur had been careful to leave the rail in place to prevent the automatic block signals from going to danger. [1925]

The same person was responsible for both wrecks. He was soon apprehended and turned out to be Tom Vernon, an ex-cowboy who had already served five prison sentences in Pennsylvania, Ohio, and California. He readily confessed to both offences. At his trial in Los Angeles on 18th December District Attorney Burton Fitts argued for the death penalty on the grounds of the great danger that he had caused to the train passengers and crew and because he had used a deadly weapon. The defence argued that no one had actually been killed, Vernon had readily confessed and had helped to clear up details of the Saugus derailment. Vernon was given two life sentences, one for causing

---

[1922] *The Times*, London, England, Wednesday 21st August 1929
[1923] Ibid, 2nd November 1929
[1924] *The Bakersfield Californian*, Bakersfield, California, United States, 11th November 1929
[1925] *Xenia Evening Gazette*, Xenia, Ohio, United States, 26th November 1929

the Saugus derailment and the other for being a habitual criminal. He was not charged with the Wyoming sabotage. He had been released from prison by 1957 and died of tuberculosis. [1926]

An attack was made on the eastbound *Orient Express* on the evening of Thursday 21st November 1929 near Tsaribrod, Yugoslavia (now Dmitrovgrad in Serbia), close to the Yugoslav-Bulgaria frontier. Shortly before the arrival of the train Yugoslav border guards saw eight or ten brigands planting bombs on the track and fired on them. The brigands returned fire. The train then arrived at 7.40 p.m., and as it passed one of the bombs went off, damaging its engine and throwing a postal van from the track. The engine was able to continue to Tsaribrod to summon help. None of the train passengers was hurt and they were able to continue their journey to Sofia, Bulgaria, after a delay of 2½ hours. Frontier guards and troops pursued the brigands, who escaped over the frontier. [1927]

Another attack on a train in Yugoslavia occurred on about Wednesday 18th December 1929 when a bomb exploded on a Pirot – Belgrade passenger train. Three passengers were seriously injured. [1928]

On about Friday 20th December 1929 an attempt was made to wreck a Chisinau – Bucharest train at Brazi, Romania, by cutting through the rails at two points about a metre and a half apart. Because the train was running at full speed it was able to negotiate the damaged portion of track safely and it was a following goods train, moving more slowly, which was derailed. There was no loss of life. [1929] No suggested motive for the attack was suggested in the report on it.

A Vice Regal Special train on the North Western Railway carrying His Excellency the Viceroy from Bombay to New Delhi on Monday 23rd December 1929 was between Hazrat Nizamuddin and New Delhi, India, when a bomb planted on the track at mile 925/6 exploded at about 07:40. The explosion blew off the end verandah of a saloon carriage and caused slight injury to one servant travelling on the train. [1930]

Two bombs exploded on the Great Indian Peninsula Railway on Saturday 12th April 1930. One was in the third-class waiting room at Byculla, but the other was on an early morning local train just outside the Bombay Victoria terminus (now Mumbai Chhatrapati Shivaji Maharaj Terminus), India. One man in the compartment where the explosion occurred was wounded. A connection with an on-going strike was not thought likely. [1931]

No. 82 down parcels express on the East Indian Railway was derailed at 04:05 on

---

[1926] www.scvhistory.com/scvhistory/tomvernon
[1927] *The Times*, London, England, 22nd and 23rd November 1929
[1928] Ibid, 19th December 1929
[1929] Ibid, 23rd December 1929
[1930] Administration Report of Indian Railways in 1929-30
[1931] *The Times*, London, England, 14th April 1930

Sunday 25th May 1930 between Maitha and Bhanpur, India. Fishplates, nuts, and bolts had been removed from one rail and the rail turned on its side. No one was hurt. There was slight damage to the rolling stock and permanent way. [1932]

Chemins de fer de Paris à Lyon et à la Méditerranée express train No. 27, which had left Paris Lyon at 21:20 on Saturday 31st May 1930, was wrecked near the station at Montereau, France, at 22:30. Head Signalman Laclaute, in his box 200 metres from the station, heard a terrible noise as the express was passing his box. He rushed to a window. The night was not very dark, and he could see smoke rising from the ballast, 'like the agonised breath of the massive engine,' as a French newspaper wrote. He seized his horn and sounded a long blast towards the station. Goods train No. 102, bound for Paris, was stopped in time before it could collide with the wreckage fouling track No. 2.

The express consisted of seven well-filled first- and second-class coaches, in charge of Conductor Henri Poulain with Driver Arsène Chausse and Fireman Robert Schlik. It was travelling at 70 kilometres an hour as it approached Montereau. "All of a sudden," Driver Chausse related afterwards, "I felt that we were on the ballast. I applied the brakes and my engine overturned onto its right side. I was underneath, as was my fireman, of course. It seemed to us that a coach was on top of the engine. Schlik and I got out… We looked at one another for a few seconds. We were alive…"

"Yes, it was a close shave," Schlik added.

The leading luggage van of the train was overturned and wrecked, and the four leading coaches were derailed. They remained more or less upright but were badly telescoped. In the leading van was Conductor Poulain, who suffered two broken legs, but his first concern, like a good railwayman, was that his train had been protected. When he was assured that this had been done he asked if his broken limbs could be ligatured. [1933]

Seven passengers died in the accident and about 30 were injured. The rescue operation was hampered by the darkness but proceeded after some delay with the aid of torches and lanterns. A number of passengers could only be released with great difficulty and in one case a whole carriage roof had to be removed to get at the victims. Mlle Rey, whose legs were badly crushed, was imprisoned for three hours with the bodies of her father, mother and two brothers around her and was finally released only with the aid of oxy-acetylene cutting equipment. The injured were taken to hospital in Montereau and by 1 a.m. the last body had been removed from the wreckage. [1934]

When the initial inquiry into the accident opened it had been assumed that the derailment had been due to some defect in the train or the track, but it quickly became evident that it was no accident. The express had collided with an inspection trolley which had been placed upon the running line in the ten-minute interval which had elapsed since the passing of the previous train. Three trollies had been left by the side of the line, two of which were still there after the derailment. A reconstruction of the affair showed that it would have taken one man considerable difficulty to lift the trolley onto the track, and the noise made in doing so would have alerted Signalman Laclaute. It was thus evident that

---

[1932] Administration Report of Indian Railways in 1930-31

[1933] *Le Petit Parisien*, Paris, France, 2nd June 1930

[1934] *The Times*, London, England, 2nd June 1930

several persons must have been involved. There was talk that the sabotage had been carried out by militant members of the Communist Party, which was active in the area, but there was no proof that such was the case. The railway company offered a reward for the apprehension of the perpetrators. [1935]

All the vehicles except the engine of Great Indian Peninsula Railway No. 16 up passenger train were derailed at 15:15 on Thursday 24th July 1930 between Narayanpeth and Yadgiri, India, on the Poona (now Pune) – Raichir section of the railway. Three passengers and two railway servants were injured. Wooden keys had been removed from the track and an obstruction placed on the rails. The damage to rolling stock and the engine amounted to 2,000 rupees; to the permanent way it was 1,584 rupees. [1936]

Eastern Bengal Railway train No. 7, the up Dacca mail, was derailed at 00:52 on Sunday 7th September 1930 between Banpur and Darsana, India, on the Ranaghat – Poradah section of the railway. The driver of the train felt the wheels of his engine leave the track beneath his cab. He immediately applied the brakes, and the train was pulled up in its own length, but the remainder of the train was derailed, and its third vehicle struck a bridge parapet and overturned into a culvert. Four people died and 56 were injured. Examination showed that a rail had been stripped of its fastenings, the bolts, nuts, and fishplates being found loose and undamaged opposite the ends of the rail. The damage to the rolling stock amounted to 55,000 rupees and to the permanent way to 3,000 rupees. [1937] No evidence was forthcoming as to who had committed the sabotage, or why. [1938] The annual report on railways in India stated that there had been a great increase in cases of train wrecking and attempted train wrecking recently, amounting to 145 instances in 1930-31 compared to only 58 instances in 1929-30.

On Wednesday 17th September 1930 No. 302 down goods train on the East Indian Railway was derailed between Karchana and Bheerpur, India, on the Mugalsarai – Allahabad section of the railway. The engine overturned, the leading 12 waggons were derailed, and the driver, fireman and guard were slightly hurt. One rail had been completely removed and laid in the cess. Damage to rolling stock and engine was put at 13,000 rupees and to the permanent way at 500 rupees. [1939]

In October 1930 Brazil was in the throes of a revolution. Unconfirmed reports reaching the United States stated that a train crowded with 1,000 Federal troops and travelling from São Paulo to Santos over a hazardous mountain route had derailed on a curve and fallen 4,800 feet down the mountainside on about Friday 10th October 1930. The casualties in this apparent catastrophe and whether it was due to sabotage were

---

[1935] Ibid, 3rd and 4th June 1930

[1936] Administration Report of Indian Railways in 1930-31

[1937] Ibid.

[1938] *The Times*, London, England, 8th, 9th and 12th September 1930

[1939] Administration Report of Indian Railways in 1930-31

unknown. [1940] Some light on the affair was shed by the report of an Associated Press correspondent who visited the area at the end of October. He found the wreck of a locomotive lying in a creek near the station at Inginiero Schamer, Brazil. It had apparently been wrecked following the blowing up of a bridge by the Federal forces. Station officials told the reporter that several people had been killed. Possibly this was the source of the story of the wrecking of the troop train. [1941]

Burma Railway No. 4 down mail train was derailed between Sibintha and Nyaungchidauk, Burma, on the Rangoon – Mandalay section of the railway at about 01:50 on Tuesday 28th October 1930 (location also given as 'near Toungoo'). [1942] The engine, tender and six rear bogies of the train remained on the line but a four-wheeled luggage van, bogie brake and seven bogie passenger coaches were derailed and four capsized. Two passengers were killed and seven injured, one of whom subsequently died. Damage to rolling stock amounted to 65,000 rupees and to the permanent way to 400 rupees. A joint and 12 spikes in a short rail length south of a bridge at mile 144-17 had been removed. [1943]

A passenger train was derailed 26 miles from Shanghai, China, on the night of Thursday 20th November 1930, and many persons were injured. Rails had been removed by miscreants. The train was derailed 30 minutes after the passage of an express from Nanking bringing back to Shanghai delegates to the Kuomintang conference at Nanking. Possibly it was this train which was the intended target of the saboteurs. [1944]

On about Tuesday 9th December 1930 bandits derailed a passenger train on the Chinese Eastern Railway near Szepingkai, Manchuria. Thirty-six passengers were wounded, and six Japanese were briefly captured but a rescue train arrived, fought with the bandits, and released the captives. Eight police were wounded in the battle. [1945]

In a much more serious act of sabotage on the Peking – Mukden Railway near Peipiao, Manchuria, on Saturday 27th December 1930 bandits removed a rail, derailing a passenger train. The engine boiler then exploded, the carriages were set on fire and many third-class passengers were burnt to death. Two firemen and two officials also died, and the death toll was variously put at 80 and 100. Many others were wounded. While the flames were raging the bandits carried off 90 merchants for ransom but did not take other passengers whom they discovered were poor. The mails were also taken. [1946] [1947]

[1940] The Chillicothe Constitution-Tribune, 11th October 1930
[1941] *The Register*, Sandusky, Ohio, United States, 29th October 1930
[1942] *The Times*, London, England, 30th October 1930
[1943] Administration Report of Indian Railways in 1930-31
[1944] *The Times*, London, England, 22nd November 1930
[1945] *The Daily Examiner*, Grafton, New South Wales, Australia, 12th December 1930
[1946] *The Times*, London, England, 31st December 1930
[1947] *Brisbane Courier*, Brisbane, Queensland, Australia, 31st December 1930

CHAPTER 19

MISCELLANEOUS ACCIDENTS

Most railway accidents fall into one of a small number of categories such as derailments, collisions, bridge failures and so on. A few, though, do not easily fall into any obvious group, hence this short chapter on miscellaneous accidents. Some of the accidents listed in this chapter were fairly minor ones; they have been included because of their unusual features.

Great Northern Railway train No. 25, the *Cascade Limited*, was stopped near Corea station, Washington, United States, on the morning of Saturday 22nd January 1916, by a small snow slide which was blocking the line. Workmen had just finished clearing the track and the train was about to proceed at 7.15 a.m. when a much larger avalanche came down the mountainside, struck the dining car and day coach and carried them 300 feet down the slope and the dining car caught fire and was destroyed. By the evening, the bodies of Mr W.F. Carter, Mr Bert Kirkman, Mr Edward Batterman and his baby had been recovered, and a ten-year-old son of Mr Batterman and another passenger were believed dead. Mrs Batterman had been in the lavatory at the time, warming a bottle of milk for the baby. The accident happened not far from the site of the 28th of February 1910 disaster. [1948] The final number of dead varied between six and ten according to different reports but was probably eight. [1949]

At about 11 p.m. on Tuesday 21st March 1916 Toledo St Louis and Western (Clover Leaf) Railroad passenger train No. 5, the *Commercial Traveller*, was blown from the track four miles east of Marion, Indiana, United States, during a cyclone. The train was reportedly moving at high speed at the time and was overturned and wrecked. Four of the five cars of the train were completely overturned, and the remaining vehicle left tilted at an angle of 25 degrees. The engine and tender remained on the track and continued to

---

[1948] *The Salt Lake Tribune*, Salt Lake City, Utah, United States, 23rd January 1916
[1949] *The Athena Press*, Athena, Oregon, United States, 28th January 1916

Marion to raise the alarm. About 25 people were injured. [1950]

On Friday 23rd June 1916, an Italian goods train on the east coast was shelled by an Austrian torpedo boat near Giulio Nova, Italy. The engine and four waggons were destroyed, and several waggons were damaged. [1951]

Meanwhile on the Western Front air attacks on trains continued. German attacks on Allied trains doubtless happened but were not publicised; French or Allied attacks on German trains took place for example between Douai and Cambrai, France, on Saturday 1st July 1916; at Amage Lucquy, France, on Sunday 2nd July and near Bapume, France, on Friday 6th July. [1952] Many more would follow before the end of the war. Naturally the details of most of them have not come to light but a German official communiqué from Berlin on Monday 6th July 1916 stated that a stationary German hospital train had been bombed from a low altitude by an enemy (Allied) aircraft at a point south-west of Cambrai, France. Six of the patients on the train had been killed. [1953]

A Rhaetian Railway train was struck by an avalanche near Davos, Switzerland, on the morning of Sunday 29th April 1917. The two rearmost carriages were thrown more than 50 metres from the track and the remainder of the train was buried beneath 10 metres of snow. After the accident 200 men of the Grisons infantry battalion worked through the night to clear the snow. There had been 30 people on the train and by the end of Monday the body of a tenth victim had been recovered. Those who lost their lives included Conductor Schmid. The German military authorities granted 2,000 marks to the families of the victims in recognition of Swiss hospitality. [1954] [1955]

Although the aerial bombing of London and other English cities is normally associated with the Second World War there were limited bombing attacks during the First World War, at least on London, by both aircraft and Zeppelin airships. A communique issued by the Commander-in-Chief, Home Forces, mentioned an air attack on the East End of London on Wednesday 13th June 1917. According to one report a bomb fell on an unspecified station, hitting an incoming train, killing seven people, and injuring 17. Fifteen German aeroplanes took part in the midday attack. [1956]

Nine cars of chlorine were wrecked in a derailment between Camelot and New Hamburg, New York, United States, on the New York Central Railroad, on the morning of Thursday 31st October 1918. Gas spread over a wide area of what was fortunately an

---

[1950] *Norwich Bulletin*, Norwich, Connecticut, United States, 22nd March 1916
[1951] *The Times*, London, England, 26th June 1916
[1952] Ibid, 3rd, 4th and 7th July 1916
[1953] Ibid, 7th July 1916
[1954] Ibid, 1st May 1917
[1955] *Gazette de Lausanne*, Lausanne, Switzerland, 1st May 1917
[1956] *The Times*, London, England, 14th June 1917

unpopulated area adjoining the River Hudson. [1957]

Presumably because the cause was obvious and because a proportion of the victims were only tramps, the Interstate Commerce Commission did not bother to investigate an accident near Debeque, Colorado, United States, early on Sunday 12th September 1920, in which at least five and probably six people died. Denver and Rio Grande Railroad passenger train No. 1 was struck by a landslide at Nigger Hill, described as 'a great shale mountain.' The engine, tender, baggage, and mail cars and two coaches were thrown from the track. Initial reports stated that Engineer George Hardenburg and Fireman Charles Schweendenman had been killed and a tramp, Russell Hager, of Dayton, Ohio, had been fatally injured, while the mail clerk, express messenger and several passengers were less badly hurt. The slide began moving as the train was passing and hundreds of tons of rock and earth were piled upon the track. At the end of the day the slide was still active, with boulders weighing many tons crashing down the mountainside, bringing much loose dirt with them. Railroad officials hoped in the evening that the slide would soon cease, when workmen could begin clearing the track, but warned that it would take a week or ten days to do so. [1958]

During the clearance operations the bodies of three more tramps were discovered: brothers Floyd, 17, Ralph, 19, and Lawrence Brundle, 21, of Glenwood Springs, Colorado. [1959]

The Hokuriku main line of the Japanese State Railways ran along the side of the Hida Mountains between Ichiburi and Oyashirazu stations, with the land sloping steeply towards the Japanese Sea and several tunnels. On Friday 3rd February 1922 after several days of snow the temperature rose, and the snow turned to rain. The sudden thaw precipitated an avalanche which blocked the line between the two stations. In view of the importance of the railway, especially to the military, the blockage had to be quickly cleared, and a train (No. 65) was made up to take about 200 labourers to the scene. The labourers were reservists and recruits from local youth organisations, clad in *minokasa* (capes and straw hats) and with shoes which were not waterproof. The arduous clearance work had to cease when darkness set in and the train, which consisted of engine 2296 and 11 vehicles, set off back towards Itoigawa. At about 7.50 p.m., in heavy rain, it was between Fukaya and Katsuyama Tunnels when it was struck by another avalanche. The engine and leading coach were forced into Katsuyama Tunnel, the second coach was badly damaged and the wooden superstructures of the third and fourth were destroyed. The fourth coach was also telescoped by the fifth.

Eighty-eight people died at the scene and two more later; another 42 were injured. One of those killed was a railwayman, one an ordinary passenger who for some reason was on the train and the other 88 were labourers.

There was a dispute over compensation for the families of the victims, who were initially offered nothing by the Ministry of Railways on the grounds that they were not

---

1957 *The Rock Island Argus*, Rock Island, Illinois, United States, 31st October 1918
1958 *The Nebraska State Journal*, Lincoln, Nebraska, United States, 13th September 1920
1959 *The Commercial Tribune*, Cincinnati, Ohio, United States, 14th September 1920

employed by the railway, then 200 yen and finally, after representations in parliament, they received 2,200 yen (a labourer would earn about 1 yen a day). [1960]

A Burma Railway mixed train on the Pegu – Moulmein branch was struck by a tornado between Naunggala and Naungbo, Burma, on Sunday 31st July 1922. The last nine vehicles of the train, mostly bogie passenger carriages, were blown off the line and landed on their sides or on their roofs on the flooded area alongside of the embankment. Four passengers lost their lives and 32 were injured. [1961]

What at first appeared to be just another level crossing accident on Monday 18th September 1922 had occurred according to the survivor of the collision under sensational circumstances. The crash happened at King's Crossing, near Chippewa Falls, Wisconsin, United States. Chicago St Paul Minneapolis and Omaha Railroad northbound passenger train No. 96 crashed into a Ford sedan owned by a Mr Peter Kelly of Chippewa Falls, killing him and apparently seriously injuring his passenger, Mrs Lulu Poston, the wife of a Minneapolis St Paul and Sault Ste Marie section hand. It had been supposed that rain on the windscreen of the car had prevented the car driver from seeing the train as the view at the crossing was unobstructed. However, Mrs Posten, who suffered a fractured skull, told her husband the next day that Kelly had been trying to abduct her. She said that she had been at home when someone had telephoned her to say that her eight-year-old son had been hurt and to come at once to the Omaha station. As she had set off there Kelly had driven up and had offered her a lift, but instead of going there the car had headed for the country. She had told Kelly to stop the car and let her out, there had been a struggle and he had hit her over the head with something. At that point she had reached down and shut off the ignition to stop the car and it had stopped directly on the crossing. She remembered nothing more. [1962]

Situated, as it is, on the Pacific Ring of Fire, Japan has always been liable to earthquakes. The worst in the twentieth century was the Great Kantō Earthquake on Saturday 1st September 1923. An estimated 142,800 people died in the catastrophe. Details of accidents to trains are necessarily vague in the face of such an overwhelming event. There were reports of 600 people perishing in the collapse of a tunnel at Sasako, of a train on the Tokyo-Sendai railway falling into a river when crossing a bridge damaged by the earthquake and of 26 passengers dying in the wreck of a train between Hitatsuka and Oiso stations on the Tokaido line. Another nine trains were reportedly derailed by the earthquake on Honshu island. [1963] A Japanese source, however, mentions only one major train accident resulting from the earthquake, when the ground collapsed beneath a train entering the station at Nebukawa, Japan. A train overturned and rolled down a cliff into the sea, killing 112 people and injuring 13. [1964]

Two trains were in fact involved in the Nebukawa accident, Nos. 109 and 116. Train

---

1960 Wikipedia – Eisenbahnunfall von Oyashirazu

1961 Administration Report of Indian Railways in 1922-23

1962 *The Eau Claire Leader*, Eau Claire, Wisconsin, United States, 19th September 1922

1963 *The Times*, London, England, 3rd, 4th and 7th September 1923

1964 Webpage – Japanese Railway Safety and the Technology of the Day

No. 109 consisted of engine No. 960 and eight coaches and had left Tokyo at 9.05 a.m. for Manazuru. It was to cross train No. 116 at Nebukawa. The earthquake occurred just as it was pulling into the station there, derailing it. Four minutes later there was a violent aftershock and a landslide in the mountainside over the station carried the train, together with the station buildings, the platform and two water towers over a headland and into the sea.

Train No. 116 was still approaching Nebukawa and avoided being caught in the main landslide but ran into a mudslide as it was leaving the 360 metre Sainome Tunnel, two railwaymen being killed, and several passengers injured. Surviving railwaymen then tried to lead the passengers from this train back through the tunnel to Manazuru but they were caught by another landslide and four railwaymen, and several passengers were killed.

The total of 112 dead included seven railwaymen. The conductor and driver of train No. 109 survived, as did the stationmaster and his wife. About 30 of the passengers were able to climb up the slope from the remains of the train and 13 others were rescued by boats. [1965]

Another earthquake occurred in Japan at 05:45 on Tuesday 15th January 1924. It was much less serious than the Great Kantō one, but six trains were derailed between Gotemba and Tokyo and a seventh was precipitated into a river at Banyugawa. No casualties were reported. [1966] [1967]

On about Saturday 19th January 1923 the British garrison at Khartoum, the capital of the Sudan, carried out an artillery practice using live ammunition. Unnoticed by either the firing party or the crew of an observing aeroplane one of the shells broke a rail on the Khartoum-El Obeid railway line. A goods train was derailed but 'No serious personal injury was caused.' [1968]

On the evening of Friday 14th March 1924 a cyclone caused a disaster in India. At about 19:00 up Rohilkund and Kuamon Railway mixed train No. 7 was crossing the Ramganga Bridge, near Bareilly, in the United Provinces, when a sudden and unexpected storm struck it, overturning the last five carriages, third-class bogie coaches filled with the Indian passengers. Four of the vehicles fell roof downwards into the water and were completely smashed. Two coaches were submerged. Twenty-seven passengers were killed and 40 injured. [1969] [1970]

In Australia in the 1920s evening commuter trains on the Main North line from Sydney to Hornsby, New South Wales, were invariably very crowded. Because of the packed condition of the coaches passengers who could find a place standing near a window so that they could get some fresh air often felt themselves privileged. Such was

---

[1965] Wikipedia – Eisenbahnunfall von Nebukawa

[1966] *The Times*, London, England, 16th January 1924

[1967] *The Northern Champion*, Taree, New South Wales, Australia, 19th January 1924

[1968] *The Times*, London, England, 24th January 1924

[1969] Report by the Railway Board on Indian Railways for 1923-24

[1970] *The Times*, London, England, 17th March 1924

the case on the 6.17 p.m. train from Sydney Central on Thursday 15th April 1924, and a tragedy would result from passengers leaning with their heads and shoulders out of the window, when an empty train passed in the opposite direction near Rhodes signal box. The door of a carriage of the incoming empty train from Ryde had come open and was swinging wildly to and fro. As the two trains passed, it struck the outgoing train a series of blows and twin brothers named James and John Gilbert were hit by it and received fatal head injuries. A third passenger, Mr Frederick Weeks, suffered a compound fracture to his arm and shock. [1971] [1972]

As a passenger train on the Burtonport Extension of the Londonderry and Lough Swilly Railway, consisting of an engine and two coaches, a van and a waggon, was crossing a viaduct over the Owencarrow valley, between Kilmacrennan and Creeslough, in the Irish Free State, on Friday 30th January 1925 when it was struck by the full force of a gale blowing down the valley. The leading coach was blown from the track and against the bridge parapet, part of which broke off. The second coach was pulled off the rails in turn and had the couplings not held the two would have fallen into the valley. Despite this, 15 passengers fell through the roof of the first coach and the side of the second. Guard Neil Boyle and Fireman Con Hannigan hurried off to Cresslough to get help, while Driver Robert McGuiness made his way cautiously down the embankment where he found the passengers lying in a huddled heap. The unlucky ones had fallen onto rocks, the more fortunate landed on soft ground. Mr Phillip Boyle, his wife Sarah Boyle, Neal Doogan and Mrs Una Mulligan were killed, and all the other passengers were more or less hurt except a girl named Campbell who was driven feet first into soft sand. Assistance in the form of a doctor and others arrived from Cresslough within an hour. The Boyles had been bringing their son home from Letterkenny Hospital; injured in the accident, he ended up back there, now an orphan with ten younger brothers and sisters. [1973]

At the inquests on the victims the juries suggested that the installation of a protective rail along the length of the viaduct would have prevented the accident. Instead anemometers were provided at Dunfanaghy Road, where wind speeds were similar to those at Owencarrow. In their first few weeks of use winds speeds of 80 miles an hour were recorded three times, and trains were suspended. [1974]

A train from Verkhneudinsk (now Ulan-Ude) was struck by a large avalanche in Siberia near Irkutsk, USSR, on Saturday 28th March 1925. 'After several hours of exhausting efforts' 16 dead and 30 injured were recovered from the wreckage. How many more victims there were, if any, was not known. The tracks were blocked over a considerable distance. [1975]

A courier arriving in Sofia, Bulgaria, from Greece reported that an earthquake in the Peloponnesus region of the latter country had caused a landslide which had struck a train

---

[1971] *The Maitland Daily Mercury*, Maitland. New South Wales, Australia, 16th April 1924
[1972] *Truth*, Sydney, New South Wales, Australia, 20th April 1924
[1973] *The Times*, London, England, 2nd and 7th February 1925
[1974] www.donegalheritage.wordpress.com/2020/01/31/the-owencarrow-viaduct-disaster/
[1975] *The Davenport Democrat and Leader*, Davenport, Iowa, United States, 29th March 1925

on about Tuesday 2nd March 1926, causing a heavy loss of life. Three passenger coaches were stated to have been crushed by the landslide coming down a mountainside. [1976]

A rare triple collision occurred on the Pennsylvania Railroad, near Marsh Run, Pennsylvania, United States, at about 8.12 a.m. on Friday 16th March 1928. The line here was quadruple track, with tracks 1 and 2 for eastbound and westbound passenger trains respectively and tracks 3 and 4 for 3 and 4 for eastbound and westbound freight trains respectively. Trains were controlled by timetable train orders and a block system, automatic on the passenger lines and manual on the freight lines.

Westbound freight train No. P-9 on track 4 consisted of engines Nos. 3098 and 1498 and 67 cars, in charge of Conductor Rote with Engineers Everhart and Shaffer and Firemen Kauffman and Uleau, was travelling at 10 or 15 miles an hour when it became divided between the 50 and 51 cars. It stopped and was shortly run into from behind by westbound freight train second No. TH-1, consisting of engine No. 120, 32 cars and a caboose, in charge of Conductor Davidson and Engineer Levi Simmers and Fireman Luther Williams. Eastbound freight train No. P-2-12-14, on track 3, consisting of engine No. 4626, 69 cars and a caboose, in charge of Conductor Early with Engineer Diffenderfer and Fireman Bickle, was struck by the wreckage of the first two trains. Finally eastbound passenger train No. 992, on track 1, consisting of engine No. 3333, a combination mail and express car and two coaches, in charge of Conductor E.L. Reichard with Engineer J.G. Norris and Fireman Rembould, ran into the wreckage of train No. P-2-12-14. Fireman Williams was fatally scalded, and three railwaymen and 17 passengers were injured. The wreckage blocked all four lines and caught fire. [1977]

The engineers and firemen of train No. P-9 could not tell the Interstate Commerce Commission (ICC) inquiry anything regarding the initial collision. Conductor Rote said that as soon as his train had stopped he had sent his flagman back to protect and seeing train second No. TH-1 coming had remarked to the flagman that if it was on their track there would be an accident.

Flagman Skunk of train No. P-9 stated that he had thrown off a lighted fusee before his train stopped and had run back, passing the fusee, which was still burning, and waved a flag at train No. second TH-1 until he had had to jump into a ditch out of the way. There had been no time to lay torpedoes.

Brakeman Singer of train No. P-9, riding with Flagman Skunk in the caboose, put the speed of train second No. TH-1 at 30 to 40 miles an hour at the moment of collision and said that its engine had been still working steam.

Engineer Simmers of train No. P-1 was not in a fit state to give evidence at the ICC inquiry. Conductor Davidson of that train put its speed at 25 miles an hour approaching Marsh Run, when the brakes had been applied in emergency. After the accident he had spoken to his engineer but had received no reply while the still-conscious fireman had said that he did not know what had happened.

The ICC concluded that the initial collision had been due to the failure of Engineer Simmers to operate his train under proper control after entering an occupied block section

[1976] *The Winona Republican-Herald*, Winona, Minnesota, United States, 4th March 1926
[1977] *Warren Tribune*, Warren, Pennsylvania, United States, 16th March 1928

for reasons which could not be ascertained. The following collisions had been unavoidable. An adequate train stop or train-control device on track 4 would have prevented the accident. [1978]

On Sunday 6th January 1929, a light lattice girder bridge was being erected across the London Midland and Scottish Railway line between Bispham and Blackpool Talbot Road, England, to carry a belt conveyor to a brickworks. An occupation of the line was arranged between 7.10 a.m. and 8.35 a.m. and between 8.50 a.m. and 9.35 a.m. on the morning of the accident for the erection of the two girder spans across the line, with further periods being granted if need be. The work of erecting the floorings, struts, bracings and so on carried on continuously except when trains were actually passing. Flagmen were stationed on either side of the site with a third flagman at the site itself, the signalmen on either side were instructed not to let any train approach except under the express instructions of Traffic Inspector Seward and all trains were restricted to ten miles an hour past the site.

In addition to the two earlier possessions of the line, after consulting with Traffic Inspector Seward the contractors had taken a third possession at 10.10 a.m. to continue with their work, which was given up on the approach of the 8.02. a.m. passenger train from Manchester Victoria to Blackpool. Inspector Seward had checked with the foreman erector that all was safe, then instructed the train, which had just stopped at Bispham station, to proceed. It did so at low speed and was just passing beneath the girders when, without warning, both beams collapsed onto the train at 10.40 a.m. No one on the train was hurt but three of the contractor's men were killed and four others were injured.

The accident was the subject of an exhaustive Ministry of Transport inquiry conducted by Colonel A. C. Trench. He concluded that the probable cause of the collapse of the girders was the movement of the contractor's men upon that on the Bispham side, the girder being in an unsafe condition after the removal of its top bracings. He considered that the contractors, Harry Peers and Co., while experienced in the general manufacture and erection of steelwork, had not appreciated the extreme lateral instability of the girders while they were being placed in their final positions. As for the railway company, he considered that it would have been better had they appointed a responsible engineer to supervise the operation, though the inspecting officer conceded that the engineer would have been in a difficult position unless he had actually taken charge of the work. There could be no criticism of the railway company in terms of the protection to trains which it had provided with flagmen, a speed restriction and so on. [1979]

A passenger train was buried by a landslide at Yamaguchi, Japan, on Saturday 28th June 1930. Eight hundred workmen laboured to dig in the debris and free the occupants of the train. Five bodies were recovered, and the remaining 50 passengers were rescued alive, though many were on the point of exhaustion. [1980]

[1978] Interstate Commerce Commission report

[1979] Ministry of Transport report

[1980] *The Times*, London, England, 30th June 1930

INDEX

TABLE OF MAJOR RAILWAY ACCIDENTS 1901-1915

Date	Location	Railway Company	Description	Passenger Fatalities
3 Jan 1916	Near Monclova, Mexico		Fire-damaged trestle collapsed under train, sabotage?	20
c16 Jan 1916	Between Oaxaca and Pueblo, Mexico		Derailment, sabotage	23
c16 Jan 1916	Between Vera Cruz and Jalapa, Mexico		Derailment, sabotage	?
22 Jan 1916	Near Corea, WA, USA	Great Northern	Train struck by avalanche, fire	8
1 Feb 1916	Saint-Denis, France	Chemins de fer du Nord	Collision with waggon, fire	13
c5 Feb 1916	Near Vitebsk, Russia		Collision	16
15 Feb 1916	Near Campania, TS, Australia		Derailment, excessive speed	6
c16 Feb 1916	Between Libau and Poneviezh, Russia		Derailment, sabotage	?
20 Feb 1916	South Cheney, WA, USA	Spokane Portland and Seattle	Rear collision in fog	5
22 Feb 1916	Milford, CT, USA	New York New Haven and Hartford	Double collision boiler explosion	7
26 Feb 1916	Cortona, Italy		Derailment	17
29 Feb 1916	Maisons-Alfort, France	Chemins de fer de Paris à Lyon et à la Méditerranée	Collision with waggon, fire	8
4 Mar 1916	Pedaso, Italy		Head-on collision	6
11 Mar 1916	La Loupe, France	Chemins de fer de l'État	Rear collision in fog	7

21 Mar 1916	Sayula, Mexico	Ferrocarril Central Mexicano	Derailment, sabotage, fire	50
29 Mar 1916	Near Amherst, OH, USA	New York Central	Rear collision in fog	27
30 Mar 1916	La Punta, Mexico		Derailment, fire, sabotage and arson	78
17 Apr 1916	Bradford, RI, USA	New York New Haven and Hartford	Rear collision. fire	5
11 May 1916	Between Lawoczne and Slawsko, Austria	Kaiserliche-Königliche-Österreiche-Staatsbahnen	Head-on collision, single line	8
12 May 1916	Between Queretaro and Palma Gonzales Mexico		Derailment, sabotage	5
c12 May 1916	Ayuzeo, Mexico		Trestle collapsed under train, fire, sabotage, and arson	150
2 Jun 1916	Flood Creek Bridge, near Packard, IA, USA	Chicago Rock Island and Pacific	Bridge collapsed under train	17
8 Jun 1916	Near Saltillo, Mexico		Derailment, broken rail	10
6 Jul 1916	'South west of Cambrai,' France		German train bombed by Allied aeroplane	6
12 Aug 1916	Between Brookvale and Echo, PA, USA	Southern Cambria	Runaway train, collision	25
30 Aug 1916	Kalanour, India	North Western	Rear collision	9
31 Aug 1916	Near Gripp, France	Tramway de la Bigorre	Runaway train, collision	6
31 Aug 1916	Between San Luis Potosi and Tampico, Mexico		Derailment, sabotage	36
10 Sep 1916	Near Ortiz, Mexico		Derailment, track defects	22
23 Sep 1916	Between San Luis Potosi and Gonzales Junction, Mexico		Derailment, sabotage, collision	40
26 Sep 1916	Between Medzilaborce and Lupków, Hungary		Collision, fire	140
8 Oct 1916	Jansfelde, near Zantoch, Germany		Double collision	12
15 Oct 1916	Smithfield, NE, USA	Chicago Burlington and Quincy	Rear collision	11

5 Nov 1916	Near Celaya, Mexico		Collision	40
12 Nov 1916	Between Dehosa and San Miguel, Mexico	Ferrocarril Interoceánico de Mexico	Derailment, excessive speed	159
29 Nov 1916	Between Shimoda and Furumagi, Japan		Head-on collision, single line	20
1 Dec 1916	Herezehalom, Hungary	Magyar Állumvasutak	Sidelong collision	68
21 Dec 1916	Near Carneros, Mexico		Derailment, sabotage	?
27 Dec 1916	De Beaujeu, QC, Canada	Canadian Pacific	Rear collision	6
30 Dec 1916	Near Hamburg, Germany		Collision	7
3 Jan 1917	Queensferry Junction, Scotland	North British	Head-on collision	12
13 Jan 1917	Ciurea, Romania		Runaway train, derailment, fire, explosion	374
16 Jan 1917	Between Trifail and Sagor, Austria	Südbahn	Derailment, train ran into landslide, then struck by further landslide	11
17 Jan 1917	Near Massy-Palaiseau, France		Collision between portions of divided train	14
27 Jan 1917	Châteauneuf-sur-Cher, France	Chemin de fer de Paris à Orléans	Collision with shunting goods	11
27 Jan 1917	Near Pumping Station, OH, USA	Cleveland South Western and Columbus	Head-on collision, single line, fire	5
3 Feb 1917	Finkenrath, Germany		Collision	7
5 Feb 1917	Near Glenwood, Newfoundland	Reid-Newfoundland	Derailment, fire	8
c12 Feb 1917	Near Orizaba, Mexico		Derailment, sabotage	50
13 Feb 1917	Roncherolles-en-Bray, France		Rear collision	26
26 Feb 1917	Near Holmsveden, Sweden		Train diverted onto wrong line; buffer stop collision	11
27 Feb 1917	Mount Union, PA, USA	Pennsylvania	Rear collision in fog	19

8 Apr 1917	Divaca, Austria	Südbahn	Rear collision	10
18 Apr 1917	Nannhofen, Germany	Bayerisch Staatsbahnen	Collision with shunting train in blizzard	33
29 Apr 1917	Near Davos, Switzerland	Rhaetian	Train struck by avalanche	9
5 May 1917	Mkata, German East Africa	Central	Bridge collapsed under train	20
6 May 1917	Near Queretaro, Mexico	Ferrocarriles Nacionales de México	Derailment, track defects?	14
6 May 1917	Oarneros, Mexico		Derailment, sabotage	30
13 Jun 1917	East London		Train bombed by German aeroplane	7
10 Jul 1917	Pragerhof, Galicia	Südbahn	Explosion	42
23 Jul 1917	Between Ghunsore and Pindrai, India	Bengal-Nagpur	Collision	20
5 Aug 1917	Milwaukee, WI, USA	Chicago Milwaukee and St Paul	Head-on collision	5
7 Aug 1917	Arquata Scrivia, Italy		Derailment, excessive speed	38
13 Aug 1917	Between Petrograd and Moscow, Russia		Rear collision?	60
13 Aug 1917	Yemaras, India	Great Indian Peninsula Railway	Head-on collision	5
13 Aug 1917	Fuentelapeña, Spain		Derailment, sabotage by strikers?	5
13 Aug 1917	North Branford, CT, USA	Shore Line Electric	Head-on collision, single line	19
6 Sep 1917	Earlville, IL, USA	Chicago Burlington and Quincy	Rear collision	7
17 Sep 1917	'90 miles south east of Beersheba,' Palestine	Hejaz	Derailment, sabotage	70
17 Sep 1917	Willenrod, Germany		Head-on collision, single line	6
17 Sep 1917	Großneukirch, Germany		Collision	6

23 Sep 1917	Between Matapozuelos and Pozaldez, Spain		Rear collision	13
24 Sep 1917	Between Rakek and Planina, Austria	Südbahn	Rear collision	29
28 Sep 1917	Kellyville, OK, USA	St Louis-San Francisco	Head-on collision, single line	32
10 Oct 1917	Near Vladikavkaz, Russia		Derailment, sabotage	8
11 Oct 1917	Gjunis, Serbia		Derailment, sabotage	11
16 Oct 1917	Schönhausen (Elbe), Germany		Rear collision	25
c25 Oct 1917	Between Petrograd and Moscow, Russia		Collision	16
2 Nov 1917	'In the Caucasus,' Russia	Vladikavkaz	Head-on collision	25
4 Nov 1917	Armendariz, Mexico	Ferrocarril Central Mexicano	Derailment, sabotage	?
11 Nov 1917	Minifir, Palestine	Hejaz	Derailment, sabotage	?
12 Nov 1917	Near Cotopaxi, CO, USA	Denver and Rio Grande	Rear collision	5
16 Nov 1917	Between Amman and Ma'an, Palestine	Hejaz	Derailment, sabotage	?
3 Dec 1917	Heeßen, Germany	Preußisch Staatsbahnen	Double collision	36
11 Dec 1917	Düren, Germany		Rear collision	18
12 Dec 1917	Saint-Michel-de-Maurienne, France	Chemins de fer de Paris à Lyon et à la Méditerranée	Runaway train, derailment, fire, explosions	675
16 Dec 1917	'South of Tabuk,' Palestine	Hejaz	Derailment, sabotage	?
20 Dec 1917	Shepherdsville, KY, USA	Louisville and Nashville	Rear collision	44
21 Dec 1917	Sporniak, Galicia	Heeresbahn Nord	Sidelong collision	17
31 Dec 1917	Near Sambor, Galicia		Fire	?
2 Jan 1918	Near Pitahaya, Mexico	Southern Pacific of Mexico	Derailment, sabotage	?

4 Jan 1918	Dorval, QC, Canada	Canadian Pacific	Rear collision	7
c5 Jan 1918	Near Changsa, China		Head-on collision, single line	300
c7 Jan 1918	Karpovo, Ukraine		Rear collision	10
7 Jan 1918	Bruchmühlbach, Germany	Pfälzer Eisenbahn	Rear collision in blizzard?	12
9 Jan 1918	El Pedroso de la Armuña, Spain		Boiler explosion, derailment	11
14 Jan 1918	Trcziana, Galicia	Kaiserliche-Königliche-Österreiche-Staatsbahnen	Collision	14
14 Jan 1918	Hammond, TX, USA	Houston and Texas Central	Derailment, defective points	17
16 Jan 1918	Hochstetten (Kirn), Germany		Derailment, track subsidence	38
16 Jan 1918	Ölingen, Germany		Collision	31
17 Jan 1918	Between Benapur and Khargpur, India	Bengal-Nagpur	Collision with brake van	16
18 Jan 1918	Between Argeningkem and Pamletten, Germany		Collision	32
19 Jan 1918	Between Little Salkeld and Lazonby, England	Midland	Derailment, train ran into cutting slip	7
25 Jan 1918	Between Lohof and Schleißheim, Germany		Fire, explosion	8
29 Jan 1918	Cordoba, Argentina		Derailment	5
31 Jan 1918	Sedro Woolley, WA, USA	Northern Pacific/Great Northern	Side collision on rail/rail level crossing	5
3 Feb 1918	Near Kemi, Finland		Head-on collision, single line	?
7 Feb 1918	Between Güsten and Aschersleben,	Preußisch Staatsbahnen	Runaway vehicles, collision	15
8 Feb 1918	Köhn-Ehrenfeld, Germany	Preußisch Staatsbahnen	Rear collision	6
9 Feb 1918	Between Jezupol and Stanislau, Galicia	Österreiche-Staatsbahnen	Fire	7
16 Feb 1918	Between Tamarite de Litera and Binefar, Spain		Head-on collision, single line	21

19 Feb 1918	Pizzo Calabro, Italy		Derailment	42
25 Feb 1918	Near Frost, SC, USA	Southern	Rear collision	13
20 Mar 1918	Near Sir Jedid, Palestine	Hejaz	Derailment, sabotage	20
11 Apr 1918	Near Béthune, France		Explosion	?
10 May 1918	Near Camp Jackson, SC, USA	United States Government	Derailment, track defects, part of train fell from trestle	9
16 May 1918	Raphèle, France	Chemins de fer de Paris à Lyon et à la Méditerranée	Collision	8
22 Jun 1918	Ivanhoe, IN, USA	Michigan Central	Rear collision, fire	85
28 Jun 1918	Between Mircesti and Roman, Romania		Collision	45
30 Jun 1918	Hrastnigg, Austria	Südbahn	Rear collision, fire	6
9 Jul 1918	Near Nashville, TN, USA	Nashville Chattanooga and St Louis	Head-on collision, single line, fire	87
16 Jul 1918	Between Theillay and Salbris, France		Derailment	15
20 Jul 1918	Near Chelsea, MI, USA	Detroit Jackson and Chicago	Head-on collision, single line	15
30 Jul 1918	Between Gurkow and Zantoch, Germany		Derailment, engine defect, collision, fire	42
3 Aug 1918	Oswiecim, Galicia	Kaiserliche-Königliche-Österreiche-Staatsbahnen	Rear collision, train fell from bridge	11
10 Aug 1918	Near Solers, France	Chemin de fer de l'Est	Double collision, part of train fell from bridge	24
10 Aug 1918	Consuelo, Mexico		Derailment, sabotage	66
11 Aug 1918	Bachimba Cañón		Derailment, sabotage	?
14 Aug 1918	Uggowitz, Austria	Kaiserliche-Königliche-Österreiche-Staatsbahnen	Train diverted onto wrong line, collision	11

16 Aug 1918	Between Hoenningen and Dümpelfeld, Germany		Head-on collision during temporary single line working, fire, explosions	29
19 Aug 1918	Canatlan, Mexico		Bridge collapsed under train, sabotage	90
10 Sep 1918	Birdsell, NE, USA	Chicago Burlington and Quincy	Collision	11
11 Sep 1918	Schneidemühl, Germany	Preußisch Staatsbahnen	Rear collision	34
13 Sep 1918	Weesp, the Netherlands	Hollandsche Ijzeren	Derailment, track subsidence, fire	41
17 Sep 1918	Marshfield, MO, USA	St Louis-San Francisco	Head-on collision, single line	12
19 Sep 1918	Pacy Tunnel, near Tonnerre, France	Chemins de fer de Paris à Lyon et à la Méditerranée,	Rear collision in tunnel	28
22 Sep 1918	Dresden-Neustadt, Germany	Sächsichen Staatsbahnen	Rear collision, fire	38
1 Oct 1918	Getä, Sweden		Derailment, track subsidence, fire	40
1 Oct 1918	Sames-Guiche, France		Collision	8
9 Oct 1918	Piatra, Romania	Căile Ferate Române	Derailment, culvert washed away, boiler explosion, fire	100
9 Oct 1918	Jünkenrath, Germany		Collision	16
12 Oct 1918	Celrá, Spain		Collision with shunting goods	7
18 Oct 1918	Uerdingen, Germany		Collision	7
21 Oct 1918	Kapfenberg, Austria	Südbahn	Double collision in fog, fire	13
26 Oct 1918	Gael, France		Rear collision	38
1 Nov 1918	Between Jacobsdorf and Briesen, Germany		Rear collision in fog	19
1 Nov 1918	Malbone Street Tunnel, NY, USA	Brooklyn Rapid Transit	Derailment, excessive speed	92

4 Nov 1918	Rakos, Hungary	Ungarish Staatsbahnen	Derailment, 'hot-box,' broken axle	60
4 Nov 1918	Between Völkingen and Bous, Germany		Fire, explosion	18
5 Nov 1918	Between Sternthal and Pettau, Austria	Südbahn	Derailment, 'hot-box,' broken axle	63
13 Nov 1918	Malmedy, Germany		Collision	8
18 Nov 1918	Near Sterzing, Italy		Collision	21
22 Nov 1918	Hamont, Belgium		Explosion	?
5 Dec 1918	Meung-sur-Loire, France	Chemin de fer de Paris à Orléans	Rear collision	20
7 Dec 1918	Lothiers, France		Rear collision in fog	73
10 Dec 1918	Herzele, Belgium		Collision	18
13 Dec 1918	Syrau, Germany		Collision between portions of divided train	8
16 Dec 1918	Belleville, France		Rear collision in fog	9
17 Dec 1918	Bonheur, ON, Canada	Canadian Pacific	Fire	14
25 Dec 1918	Norge, OK, USA	St Louis-San Francisco	Rear collision	5
10 Jan 1919	Ath, Belgium		Explosion	?
12 Jan 1919	South Byron, NY, USA	New York Central	Rear collision	21
13 Jan 1919	Near Fort Washington, PA, USA	Philadelphia and Reading	Rear collision	13
13 Jan 1919	London, OH, USA	Ohio Electric	Derailment, excessive speed	5
20 Jan 1919	Near Mauvages, France	Chemins de fer de L'Est	Rear collision, fire	13
Jan 1919	Between Tiflis and Batum, Georgia		Derailment?	200
2 Feb 1919	Montieramey, France		Collision with light engines	7

4 Feb 1919	Cousances-au-Bois, France	Chemins de fer de L'Est	Rear collision	9
5 Feb 1919	Nanteuil-Saacy Tunnel, France	Chemins de fer de L'Est	Fire in tunnel	10
16 Feb 1919	Kambove, Belgian Congo		Explosion	27
17 Feb 1919	Between Trieste and Nabresina, Italy		Runaway waggons, collision	16
22 Feb 1919	Laughlin Junction, PA, USA	Baltimore and Ohio	Sidelong collision	8
5 Mar 1919	Between Famechon and Namps, France		Runaway coaches, collision	18
22 Mar 1919	Between Vera Cruz and Tierra Blanca, Mexico		Derailment, sabotage	13
30 Mar 1919	Near Vera Cruz, Mexico		Derailment, sabotage	?
5 Apr 1919	'140 km from Tampico,' Mexico		Derailment, sabotage	15
17 Apr 1919	Crissé, France	Chemins de fer de l'État	Rear collision	33
11 May 1919	Between Rettmer and Melbeck, Germany		Derailment, track defects?	9
25 May 1919	Near Persan-Beaumont, France		Collision	6
29 May 1919	Between Marchienne-au-Pont and Luttre, Belgium		Derailment	7
11 Jun 1919	Near Aguascalientes, Mexico		Derailment, sabotage?, fire	?
19 Jun 1919	Near Firozabad, India	East Indian	Head-on collision, single line, fire	100
29 Jun 1919	Kinana, Punjab, India	North Western	Collision	41
1 Jul 1919	Dunkirk, NY, USA	New York Central	Runaway train, collision, fire	9
10 Jul 1919	Between Bakhmatch and Pliski, RSFSR		Derailment, sabotage, fire	600

12 Jul 1919	Lembecq, Belgium		Derailment on points	14
14 Aug 1919	Near Blainville-la-Grange, France		Rear collision	8
23 Aug 1919	Artix, France		Derailment, faulty train composition, traction shock	7
25 Aug 1919	Espinal de Moreles, Mexico		Derailment, sabotage	?
1 Sep 1919	Between Uchhri and Chhab, India	North Western	Derailment, train ran into landslide	19
4 Sep 1919	Between Castelnau-D'Estretefonds and St Jory, France		Rear collision	19
26 Sep 1919	Tarascon, France	Chemins de fer de Paris à Lyon et à la Méditerranée	Sidelong collision?	7
4 Oct 1919	Dunkirk, NY, USA	Buffalo and Lake Erie Traction	Rear collision	6
5 Oct 1919	Near Yurecuaro, Mexico		Derailment	60
18 Oct 1919	Between Ujda and Oran, Algeria		Derailment	12
24 Oct 1919	Kranowitz, Germany		Sidelong collision, fire	40
27 Oct 1919	Ukraine		Armoured train attacked by 'insurgents.'	30
1 Nov 1919	Vigerslev, Denmark	Danske Statsbaner	Rear collision	38
1 Nov 1919	Near Celaya, Mexico		Derailment, sabotage	?
3 Nov 1919	Near Pont-sur-Yonne, France	Chemins de fer de Paris à Lyon et à la Méditerranée	Rear collision	17
10 Nov 1919	Neumühl-Katzdorf, Germany		Collision	45
18 Nov 1919	Töllschütz, Germany		Head-on collision, runaway train, derailment, part of train fell from bridge	5
24 Nov 1919	Engo Forest, Belgian Congo		Runaway waggon, collision	6

1 Dec 1916	Near Thal, India		Derailment, sabotage	36
2 Dec 1919	Empalme, Mexico		Derailment, sabotage	?
6 Dec 1919	Between San Luis Potosi and Tampico, Mexico		Derailment, sabotage	28
10 Dec 1919	Turkey		Sidelong collision?	?
17 Dec 1919	Achinsk, RSFSR		Explosion	?
17 Dec 1919	Near Cardenas, Mexico		Derailment, sabotage	10
18 Dec 1919	Walton, VA, USA	Norfolk and Western	Rear collision	5
20 Dec 1919	Near Onawa, ME, USA	Canadian Pacific	Head-on collision, single line	21
22 Dec 1919	Near Douai, France		Sidelong collision	12
10 Jan 1920	Caen, France	Société Métallurgique de Normandie de Colombelles	Derailment on points	7
20 Jan 1920	Between Schönlake and Schneidemühl, Germany		Derailment, collision, sabotage	18
23 Jan 1920	Tunnel No. 1, near Balanzona, Spain		Runaway train, derailment in tunnel	8
25 Jan 1920	Near Corbeil, ON, Canada	Canadian Pacific	Rear collision	7
28 Jan 1920	Czersk, Germany		Collision	6
3 Feb 1920	Clayton, LA, USA	Missouri Pacific	Train fell from open drawbridge	5
4 Feb 1920	Near Perrigny, France	Chemins de fer de Paris à Lyon et à la Méditerranée	Rear collision	18
14 Mar 1920	Near Riverside, VT USA	Rutland	Head-on collision, single line	5
27 Mar 1920	Between Pontebba and Dogna, Italy		Runaway waggons, collision, part of train fell from viaduct	13
26 Apr 1920	Mewa Newada, India	Oudh and Rohilkhand	Rear collision, fire	45

30 Apr 1920	Near Edgewick, WA, USA	North Bend Lumber	Runaway train, derailment,	5
9 May 1920	Near Bertha, OR, USA	Southern Pacific	Head-on collision, single line	8
17 May 1920	Jakhwada, India	Bombay Baroda and Central India	Train diverted onto wrong line, collision	26
20 May 1920	Near Yaroslav, Poland		Head-on collision	7
9 Jun 1920	Schenectady, NY, USA	New York Central	Rear collision	14
21 Jun 1920	Near Sabadell, Spain		Runaway coaches, collision, fire	5
27 Jun 1920	Boemish Bod, Czechoslovakia		Collision	5
3 Jul 1920	Near Livermore, IA, USA	Minneapolis and St Louis	Derailment, track defects, part of train fell from bridge	8
3 Jul 1920	Near South Pittston, PA, USA	Lackawanna and Wyoming Valley	Double collision	16
4 Jul 1920	Atoka, OK, USA	Missouri Kansas Texas	Collision with coaches, fire	7
5 Jul 1920	Aubrais, France		Derailment, points moved under train	6
6 Sep 1920	Near Globeville, CO, USA	Denver and Interurban	Head-on collision, single line	11
30 Sep 1920	Piedmont, MT, USA	Chicago Milwaukee and St Paul	Runaway train, collision	7
7 Oct 1920	Near Arkonam, India	Madras and Southern Mahratta	Derailment, sabotage	13
8 Oct 1920	Venice, Italy	Fellovie dello Stato	Rear collision	25
9 Oct 1920	Houilles, France		Double collision	45
20 Oct 1920	Near Erie, PA, USA	New York Central	Sidelong collision	8
23 Oct 1920	Pogranichnaya, RSFSR		Derailment?	100
27 Oct 1920	Lufany, Romania		Collision	50

6 Nov 1920	Between Mornington Mills and Wokalup, WA, Australia	Millar's Timber and Trading	Runaway train, derailment	7
Nov 1920	Between Panbult and Piet Retief, South Africa		Derailment	13
12 Dec 1920	Near Armentières, France	Chemins de fer du Nord	Sidelong collision in blizzard	10
14 Dec 1920	Bommidi, India	South Indian	Collision with shunting goods, fire	52
15 Dec 1920	Kreuzwald, France		Collision with shunting goods	17
16 Dec 1920	Between Hüpstedt and Silberhausen, Germany		Runaway train, buffer stop collision	17
4 Jan 1921	Near Matadi, Belgian Congo		Bridge collapsed under train	9
c15 Jan 1921	Near Vitebsk, RSFRS		Derailment, sabotage	?
16 Jan 1921	Between Luga and Novogorod, RSFRS		Explosion, fire	68
26 Jan 1921	Between Abermule and Newtown, Wales	Cambrian	Head-on collision, single line	14
30 Jan 1921	Huachana, near La Oroya, Peru		Collision	7
2 Feb 1921	Bardys, France		Sidelong collision	10
6 Feb 1921	Felixdorf, Austria	Südbahn	Head-on collision	7
15 Feb 1921	Upton and Inishannon, Ireland	Cork Bandon and South Coast	Train fired on by the IRA	6
27 Feb 1921	Porter, IN, USA	New York Central/Michigan Central	Side collision on rail/rail level crossing	35
21 Mar 1921	Niigata, Japan		Derailment in tunnel, fire	10
c3 Apr 1921	Villa Garta, Mexico	Ferrocarriles Nacionales de México	Collision	33
6 Apr 1921	Near New River, TN, USA	Southern	Derailment, track defects	5

2 May 1921	Near Welch, WV, USA	Norfolk and Western	Train diverted onto wrong line, collision	6
11 May 1921	Matundah, India	Great Indian Peninsula	Collision, fire	11
15 May 1921	Near La Bachellerie, France		Derailment, excessive speed	8
16 May 1921	Gröbers, Germany		Derailment, sabotage	5
23 May 1921	Between Artigues and Gripp	Tramway de la Bigorre	Runaway train, derailment	9
11 Jun 1921	Villaverde, Spain		Head-on collision	11
18 Jun 1921	Warsaw, MI, USA	Detroit Jackson and Chicago	Rear collision	5
25 Jun 1921	Near Amroha, India	Oudh and Rohilkhand	Bridge collapsed under train	40
25 Jun 1921	Beaumont Hamel, France	Chemins de fer du Nord	Derailment, track distorted by the heat	29
6 Jul 1921	Hennuyères, Belgium		Derailment, sabotage?	6
7 Jul 1921	Las Cososos, Mexico		Derailment, track washed away	11
20 Jul 1921	Near Nicolaevsk, RSFSR		Armoured train attacked by 'partisans.'	?
21 Jul 1921	Near Carneros, Mexico		Collision	9
25 Jul 1921	Between Tawwi and Peinzalok, Burma	Burma	Head-on collision, single line	100
26 July 1921	Near Nicolaevsk, RSFSR		Train attacked by 'partisans.'	?
26 July 1921	Near Nicolaevsk, RSFSR		Train attacked by 'partisans.'	?
4 Aug 1921	Nicolsk, RSFSR		Collision, sabotage?	16
27 Aug 1921	Magliana, Italy	Fellovie dello Stato	Collision	23
28 Aug 1921	Near Baranowicze, Poland		Derailment	13
4 Sep 1921	Iserlohn, Germany		Sidelong collision	5

10 Sep 1921	Les Eschets, France	Chemins de fer de Paris à Lyon et à la Méditerranée	Derailment, excessive speed through connection	38
14 Sep 1921	Fastoff, Ukraine		Derailment, sabotage	27
18 Sep 1921	Nidareid, Norway		Head-on collision, single line	6
1 Oct 1921	Near Ragusa, Yugoslavia		Derailment?	?
5 Oct 1921	Batignolles Tunnel, France	Chemins de fer de l'État	Rear collision in tunnel, fire	28
c9 Oct 1921	Near Teterevo, Ukrainian Soviet Socialist Republic		Derailment, sabotage	30
12 Oct 1921	Villagonzalo, Spain		Derailment, obstruction (sand) on line	12
9 Nov 1921	Between Aljustrel and Figueirinha, Portugal		Derailment, sabotage	24
28 Nov 1921	Duffel, Belgium		Rear collision	5
1 Dec 1921	Near Celilo, OR, USA	Oregon Washington Railroad and Navigation	Head-on collision	8
5 Dec 1921	Near Woodmont, PA, USA	Philadelphia and Reading	Head-on collision, single line, fire	23
8 Jan 1922	Between Košice and Odenberg, Czechoslovakia		Derailment	5
3 Feb 1922	Oyashirazu, Japan		Train struck by avalanche	89
12 Mar 1922	Ben Hill, GA, USA	Atlanta Birmingham and Atlantic	Derailment, broken wheel, part of train fell from trestle	7
2 May 1922	Rilly Semuy, France		Collision with shunting train	6
5 Jun 1922	Between Tommelin and Santa Cararina, Mexico		Derailment, sabotage	25
15 Jun 1922	Tebicuari River Bridge, Paraguay		Bridge collapsed under train, sabotage	?

11 Jul 1922	Paredes de Nava, Spain		Head-on collision	33
21 Jul 1922	'South of Baghdad,' Iraq		Collision	7
22 Jul 1922	Logan, MO, USA	St Louis-San Francisco	Head-on collision	5
1 Aug 1922	Near Miélan, France		Runaway train, collision	33
5 Aug 1922	Near Sulphur Springs, MO, USA	Missouri Pacific	Rear collision, part of train fell from bridge	33
12 Aug 1922	Annandale, MN, USA	Minneapolis St Paul and Sault Ste Marie	Collision with petrol tanker on open level crossing, derailment	6
19 Sep 1922	Near Empalma Gonzales, Mexico	Ferrocarriles Nacionales de México	Derailment, track washed away, collision	14
21 Oct 1922	Rouffach, France		Train diverted onto wrong line, collision	12
c23 Oct 1922	Near Costanza, Romania		Collision, driver intoxicated	30
29 Oct 1922	Châtelaudren, France		Collision with shunting goods in blizzard	15
13 Dec 1922	Humble, TX, USA	Houston East and West Texas	Sidelong collision	20
22 Dec 1922	Onteniente, Spain		Runaway train, collision	12
Dec 1922	Near Smyrna, Turkey		Bridge collapsed under train	?
Dec 1922	Near Smyrna, Turkey		Collision	?
26 Jan 1923	Near Anuradhapura, Ceylon		Derailment, track washed away	39
30 Jan 1923	Near Mexico City, Mexico		Derailment	6
18 Feb 1923	Port-à-Binson, France	Chemins de fer de L'Est	Collision with shunting goods	13
17 Mar 1923	Friemersheim, Germany		Collision with light engine	40

2 May 1923	Grassy, UT, USA	Denver and Rio Grande Western	Derailment, high speed with unstable engine over poor track	5
6 May 1923	Near Canasi, Cuba	Hershey Electric	Head-on collision, single line	24
9 May 1923	St Goar, Germany		Derailment, train fell into river	29
c10 May 1923	Brazil		Derailment, sabotage	50
20 May 1923	Braga, Portugal	Bom Jesus do Monte	Runaway coach?, derailment (funicular railway)	6
24 May 1923	Dothan, TX, USA	Texas and Pacific	Train diverted onto wrong line, collision, fire, sabotage	6
25 Jun 1923	Brooklyn, NY, USA	Brooklyn Manhattan Transit	Derailment	7
28 Jun 1923	Between Wanne and Gelsenkirchen, Germany		Collision	5
30 Jun 1923	Between Duisberg and Friemersheim, Germany		Bomb exploded on train	8
1 Jul 1923	Near Băile Felix, Romania		Collision	7
2 Jul 1923	Vinty-Leanca, Romania		Train diverted onto wrong line, collision	63
5 Jul 1923	Fairdale, IL, USA	Chicago Milwaukee and St Paul	Derailment, broken rail	5
6 Jul 1923	Ongarue, New Zealand		Derailment, train ran into landslide, fire	17
23 Jul 1923	Near Plevna, Bulgaria		Collision	200
31 Jul 1923	Kreiensen, Germany		Rear collision	48
23 Aug 1923	Near Rozhanka, Poland		Derailment	50
1 Sep 1923	Between Hitatsuka and Oiso, Japan		Derailment, track damaged by earthquake	26

1 Sep 1923	Nebukawa, Japan		Derailment, track damaged by earthquake	105
1 Sep 1923	Sasako Tunnel, Japan		Tunnel roof fell onto train during earthquake	?
6 Sep 1923	Lohnde, Germany		Rear collision	18
8 Sep 1923	Omsk, USSR		Derailment	82
27 Sep 1923	Cole Creek, Near Lockett, WY, USA	Chicago Burlington and Quincy	Bridge washed away or collapsed under train	33
8 Oct 1923	Between Pilkhani and Sarsawa, India	North Western	Runaway waggons, collision	21
17 Oct 1923	Near Frydek, Czechoslovakia		Collision	9
23 Oct 1923	Near Tétouan, Spanish Morocco		Derailment, sabotage	6
15 Nov 1923	Stuttgart-Unterturkheim, Germany		Head-on collision	12
9 Dec 1923	Near Forsyth, NY, USA	New York Central	Rear collision	9
12 Dec 1923	Near Tarki, India		Head-on collision, single line, fire	7
23 Dec 1923	Near Celaya, Mexico		Derailment, sabotage	?
2 Feb 1924	Ingalls, IN, USA	Union Traction	Head-on collision, single line, fire	16
14 Mar 1924	Ramganga Bridge, India	Oudh and Rohilkhand	Train blown from bridge by cyclone	27
24 Mar 1924	Near Bénestroff, France		Train diverted onto wrong line, collision	6
23 Apr 1924	Near Bellinzona, Switzerland		Head-on collision, fire	17
26 Apr 1924	Near London Euston, England	London Midland and Scottish	Rear collision	5
2 May 1924	Between Odessa and Moscow, USSR		Derailment, sabotage?	?
1 Jun 1924	Williamsport, IN, USA	Wabash	Derailment, points moved under train	12

3 Jun 1924	Near Dorpat, Estonia		Derailment	12
30 Jun 1924	Near Buda, IL, USA	Chicago Burlington and Quincy	Rear collision	6
12 Jul 1924	Vitry-la-Ville, France	Chemins de fer de l'Est	Collision with shunting goods	5
28 Jul 1924	Haymarket, Scotland	London and North Eastern	Rear collision	5
14 Aug 1924	Near Entrocamento, Portugal		Runaway engine, collision, fire	12
15 Aug 1924	Near San Salvador, Mexico		Derailment, sabotage	?
19 Aug 1924	Belem, Portugal		Collision	5
19 Aug 1924	Paporah Bridge, French West Africa		Bridge collapsed under train	29
28 Aug 1924	200 miles from Odessa, USSR		Derailment, sabotage, boiler explosion, fire	40
29 Aug 1924	Between Harappa Road and Mirdad Mirafi, India	North Western	Head-on collision, single line	107
9 Sep 1924	'Near the Lithuanian frontier,' Poland		Derailment	9
12 Sep 1924	Near Kaiki, Japan		Derailment, train ran into landslide	9
26 Sep 1924	300 miles from Moscow on the Odessa line, USSR		Derailment, sabotage	30
1 Oct 1924	Between Mainz and Mainz Süd, Germany		Rear collision in tunnel	14
c15 Oct 1924	Moscow, USSR		Fire	173
29 Oct 1924	Near Farron, BC, USA	Canadian Pacific	Bomb exploded on train	9
3 Nov 1924	Near Lytham, England	London Midland and Scottish	Derailment, broken tyre, fire	14
4 Dec 1924	Oakland, CA, USA	Key Transit System	Rear collision	8
20 Dec 1924	Chippewa Falls, WI, USA	Minneapolis St Paul and Sault Ste Marie	Derailment, defective points, part of train fell from bridge	8

13 Jan 1925	Herne, Germany	Deutsche Reichsbahn	Rear collision in fog	24
17 Jan 1925	Magny Saint Médard, France		Derailment, excessive speed	5
6 Mar 1925	Mogosaia, Romania		Head-on collision	8
20 Mar 1925	Near Davis, IL, USA	Illinois Traction	Head-on collision, single line	6
22 Mar 1925	Ricohoc, LA, USA		Head-on collision, single line in fog	6
25 Mar 1925	Near Poitiers, France		Derailment, excessive speed over line under repair (broken rail), part of train fell from bridge	5
28 Mar 1925	Irkutsk, USSR		Train ran into or struck by avalanche	16
30 Mar 1925	Near Kazansky, USSR		Head-on collision	50
9 Apr 1925	Between Los Planas and Sarrià, Spain		Derailment, excessive speed and traction shock	24
1 May 1925	Swaroschin and Preussich Stargard, Poland		Derailment, high speed over poor track?, sabotage?	29
12 May 1925	Between Rumiantsevo and Volokolamsk, USSR		Derailment, track defects?	13
c26 May 1925	Near Gorna Banya, Bulgaria		Derailment	10
16 June 1925	Rockport Crossing, Hackettstown, NJ, USA	Delaware Lackawanna and Western	Derailment, obstruction (soil) on line	45
29 Jun 1925	Taiga, USSR		Derailment	9
12 Jul 1925	Near Taylor, TX, USA	Missouri Kansas Texas of Texas	Explosion, fire	6
29 Jul 1925	Saint Antoine-de-Rocher, France	Chemin de fer de Paris à Orléans	Derailment, excessive speed through crossover	17
13 Aug 1925	Amiens, France	Chemins de fer du Nord	Derailment, excessive speed through crossover	13

22 Sep 1925	Estonia		Collision	5
9 Oct 1925	Between Catanzaro and Badolato, Italy		Bridge washed away; train fell into river	13
15 Oct 1925	'on the Pleskau line, USSR		Derailment	6
16 Oct 1925	Halsa Junction, India	Eastern Bengal	Collision	11
16 Oct 1925	Bressana, Italy		Rear collision	12
27 Oct 1925	Near Victoria, MS, USA	St Louis-San Francisco	Derailment, broken rail	20
12 Nov 1925	Near Monmouth Junction, NJ, USA	Pennsylvania	Rear collision in fog	9
11 Dec 1925	Between Pattukkottai and Adirampatnam, India	South Indian	Derailment, track washed away	9
23 Dec 1925	Vallenar, Chile	Lonjitudinal Chileno	Derailment	7
25 Dec 1925	Arad, Romania		Collision	10
4 Feb 1926	Moscow Oktyabrsky, USSR		Buffer stop collision	5
18 Feb 1926	Damascus, Syria	Hejaz	Derailment, sabotage, fire	7
c2 Mar 1926	Peloponnesus, Greece		Train struck by landslide	?
14 Mar 1926	Virillá River Bridge, Costa Rica	Northern	Derailment, 'faulty loading' and weak track, part of train fell from bridge	248
17 Mar 1926	Sicamous Lake, BC, Canada	Canadian Pacific	Derailment, track subsidence	6
22 Mar 1926	Wülknitz, Germany		Train diverted onto wrong line, collision	6
27 Mar 1926	Near Pedro Grande, Brazil		Derailment	5
29 Mar 1926	El Corso, Columbia		Rear collision	10
30 Mar 1926	Edgewater, AL, USA	Birmingham Southern	Collision	7
29 Apr 1926	Between Hervás and Baños de Montemayor, Spain		Derailment, broken rail	5

12 May 1926	Between Peking and Hankow, China		Derailment	50
16 May 1926	Near Ixmiquilpan, Mexico		Derailment	10
20 May 1926	München Ost, Germany	Deutsche Reichsbahn	Collision with train standing at platform	30
9 Jun 1926	Salt River, South Africa		Derailment, defective vehicle	16
10 Jun 1926	Between Kayuga and Aberdeen, NSW, Australia		Derailment, track defects, part of train fell from viaduct	5
16 Jun 1926	Near Gray, PA, USA	Pennsylvania	Rear collision	12
16 Jun 1926	Crystal Siding, NV, USA	Union Pacific	Fire	5
20 Jun 1926	Between Noizay and Vernou, France	Chemin de fer de Paris à Orléans	Derailment, broken axle	7
3 Jul 1926	Des Ambassadeurs, Near Achères, France	Chemins de fer de l'Ouest	Derailment, excessive speed	18
c15 Jul 1926	Ploesti, Romania		Derailment?	?
21 Jul 1926	Seba Ayoun, Morocco		Train diverted onto wrong line, collision	7
13 Aug 1926	Langenbach, Germany		Derailment, excessive speed over track under repair	12
13 Aug 1926	Calverton, NY, USA	Long Island	Derailment, defective points	5
19 Aug 1926	Between Leiferde and Meinersen, Germany	Deutsche Reichsbahn	Derailment, sabotage	24
22 Aug 1926	Wyanet, IL, USA	Chicago Burlington and Quincy	Derailment, defective vehicle	7
1 Sep 1926	Between Kalgan and Hankow, China		Runaway train, derailment	76
1 Sep 1926	Between Ametlia and Ampolla, Spain		Derailment, train ran into landslide	21
2 Sep 1926	Near Monroe, MI, USA	Detroit United	Head-on collision, single line	9

5 Sep 1926	Near Waco, CO, USA	Denver and Rio Grande Western	Derailment, excessive speed	27
6 Sep 1926	Clybourn, IL, USA	Chicago and North Western	Rear collision	5
13 Sep 1926	Between Murulla and Blandford, NSW, Australia	New South Wales State	Runaway waggons, collision, fire	27
13 Sep 1926	Contesci, Romania		Sidelong collision	8
23 Sep 1926	Between Aki-nakano and Kaidaichi, Japan		Derailment, track washed away	34
23 Sep 1926	Vulaines-sur-Seine, France	Chemin de fer de Paris á Lyon et á la Méditerranée	Rear collision in fog	8
27 Sep 1926	Bethlehem, PA, USA	Central of New Jersey/ Lehigh Valley	Side collision on rail/rail level crossing	8
2 Oct 1926	Cotzoreni, Romania		Collision	5
13 Oct 1926	Near Vila Anastácio, Brazil		Collision	8
4 Nov 1926	Near Farlane, ON, Canada	Canadian National	Rear collision, fire	5
16 Nov 1926	Near Tiflis, USSR		Runaway train, derailment	7
19 Nov 1926	Near Parkgate and Rawmarsh, England	London Midland and Scottish	Collision with derailed waggons	9
8 Dec 1926	Near Machungo, Manchuria	South Manchurian	Collision	25
16 Dec 1926	Fabrica, The Philippines	Insular Lumber	Derailment	8
23 Dec 1926	Rockmart, GA, USA	Southern	Head-on collision	12
28 Dec 1926	Near Lota, Chile		Collision with 'animal' in tunnel, collision	15
8 Jan 1927	Arsaki, USSR		Collision	16
14 Feb 1927	West Parade, near Hull Paragon, England	London and North Eastern	Head-on collision	12

24 Feb 1927	Between Chocoy and Manuel, Mexico		Derailment, broken axle	40
9 Mar 1927	Crivina, Romania		Collision	5
13 Mar 1927	Bhadrak, India	Bengal-Nagpur	Train diverted onto wrong line, collision	14
25 May 1927	Near Pulpi, Spain		Runaway train, collision	17
3 Jun 1927	Near Bessay-sur-Allier, France		Derailment, defective vehicle, collision	7
18 Jun 1927	Near Baku, USSR		Collision	5
25 Jun 1927	Near Dayton, OH, USA	Dayton and Troy Electric	Head-on collision, single line	5
7 Jul 1927	Near Alpacatal, Argentina	Buenos Aires and Pacific	Head-on collision in fog, fire	12
27 Jul 1927	Hiedelberg, South Africa		Head-on collision, single line	22
24 Aug 1927	Near Sevenoaks, England	Southern	Derailment, high speed with unstable engine over poor track	13
25 Aug 1927	Montenvers, Switzerland	Chemin de fer du Montenvers	Runaway train, derailment, part of train fell from viaduct	21
26 Aug 1927	Anting, China		Derailment, sabotage	20
20 Sep 1927	Near Houtovo, Yugoslavia		Explosions, sabotage	?
18 Oct 1927	Between Villa Reyes and Jaral Del Jerrito, Mexico		Derailment, sabotage	6
30 Oct 1927	Triggiano, Italy		Rear collision	9
27 Nov 1927	Near Dockrill Siding, ON, Canada	Canadian Pacific	Head-on collision, fire	5
3 Dec 1927	Near Dniepropetrovsk, USSR		Derailment, sabotage?	7
22 Jan 1928	Between Haiyaghat and Kishenpur, India	Bengal and North Western	Derailment, faulty train composition?	8
24 Jan 1928	Fishhoek South Africa		Derailment	5

28 Jan 1928	Near Kyauktaga, Burma	Burma	Derailment, sabotage, part of train fell from bridge	37
12 Mar 1928	Near Katukurunda, Ceylon		Head-on collision, single line	28
11 Apr 1928	Near Paris Nord	Chemins de fer du Nord	Head-on collision	14
27 Apr 1928	Hex River Pass, South Africa		Derailment, fire	7
4 May 1928	Csóka, Yugoslavia		Derailment on points	5
9 May 1928	Between San Benetto and Grottamare, Italy		Derailment, train ran into landslide	11
31 May 1928	Near Delong Siding, IL, USA	Illinois Traction	Head-on collision, single line	6
10 Jun 1928	Siegelsdorf, Germany	Deutsche Reichsbahn	Derailment, high speed over poor track	25
22 Jun 1928	Länninge, between Bollnäs and Granbo, Sweden,		Head-on collision, single line, fire	16
27 Jun 1928	Darlington Bank Top, England	London and North Eastern	Head-on collision	25
8 Jul 1928	Between Belur and Dankuni, India	East Indian	Derailment, sabotage	20
15 Jul 1928	München Hbf, Germany		Rear collision, fire	10
18 Jul 1928	Grónjec, Poland		Collision	10
31 Jul 1928	Dinkelsherben, Germany		Train diverted onto wrong line, collision	16
2 Aug 1928	Le Mans, France	Chemins de fer de l'État	Derailment, excessive speed through crossover	5
6 Aug 1928	Mounds, IL, USA	Illinois Central	Derailment, obstruction on line (faulty loading)	6
24 Aug 1928	Times Square, NY, USA	Interborough Rapid Transit	Derailment on points	17
10 Sep 1928	Zaječí, Czechoslovakia		Head-on collision	21
15 Sep 1928	Near Cadiz, The Philippines	Negros Lumber	Runaway train?, derailment	16

18 Sep 1928	Between Sessano and Carpinone, Italy		Runaway waggon, collision	9
29 Sep 1928	Near Las Madrigueras, Spain		Head-on collision, single line	9
13 Oct 1928	Charfield, England	London Midland and Scottish	Double collision, fire	16
26 Oct 1928	Resca, Romania		Train diverted onto wrong line, collision, fire	34
15 Jan 1929	Near Prusnila, Romania		Derailment, collision, fire	14
15 Feb 1929	Between Moreno and Guarachito, Mexico		Derailment, sabotage, fire, arson	?
20 Feb 1929	Near Bismark Siding, IL, USA	Peoria Terminal	Derailment, track defects, fire	6
20 Mar 1929	Drocourt, ON, Canada	Canadian National	Head-on collision, single line, fire	18
4 Apr 1929	Boboc, Romania		Derailment on points	20
17 Apr 1929	Hal, Belgium		Sidelong collision in fog, fire	8
18 Apr 1929	Between Iquique and Antofagasta, Chile		Derailment, engine defect?	14
23 Apr 1929	Between Irkutsk and Chita, USSR		Derailment, train ran into landslide	6
9 May 1929	Fernandez, Mexico		Derailment, sabotage	?
19 Jun 1929	Viane-Moerbeke, Belgium		Derailment, track subsidence, collision	9
3 Jul 1929	Plaszow, Poland		Collision	19
18 Jul 1929	Near Stratton, CO, USA	Chicago Rock Island and Pacific	Bridge collapsed under train	8
18 Aug 1929	Henryetta, OK, USA	St Louis-San Francisco	Derailment, vandalism	12
22 Aug 1929	Between Schwarzach-St Veit and Loifarn, Austria,		Head-on collision, single line	5
25 Aug 1929	Buir, Germany		Derailment, excessive	14

			speed through crossover	
12 Sep 1929	Between Prieto and Lena, Mexico		Derailment, track washed away	5
23 Sep 1929	Near Zuevka, USSR		Derailment	30
9 Oct 1929	Near Deblin, Poland		Head-on collision, single line	5
21 Oct 1929	Near Briansk, USSR		Collision	?
1 Dec 1929	Onley, VA, USA	Pennsylvania	Derailment, broken rail	8
10 Dec 1929	Namur, Belgium		Runaway train, derailment, boiler explosion, fire	10
6 Jan 1930	Between Nador and Petit, Algeria		Bridge collapsed under train	18
10 Jan 1930	Clutterbuckganj, India	East Indian	Head-on collision	12
23 Feb 1930	Burlington Road Crossing, near Kenosha, WI, USA	Chicago North Shore and Milwaukee	Collision with car on open level crossing, derailment, collision, fire	12
9 Mar 1930	Between Alto and Plumby, Brazil		Runaway train, derailment (rack railway)	6
6 Apr 1930	Near Oita, Japan		Boiler(?) explosion	17
11 Apr 1930	Near Laissey, France		Derailment, inadequate protection of PW work	6
16 Apr 1930	Domodedovo, USSR		Fire	45
19 May 1930	Tcherni, USSR		Collision	28
31 May 1930	Near Montereau, France	Chemins de fer de Paris à Lyon et à la Méditerranée	Collision with platelayer's trolley, derailment, sabotage	7
26 Jun 1930	Vermilion River, ON, Canada	Canadian National	Derailment, track washed away	5
27 Jun 1930	Near Crerar, ON, Canada	Canadian National	Derailment, track washed away	8
28 Jun 1930	Yamaguchi Ken, Japan		Train struck by landslide	5
29 Jun 1930	Cherepok, USSR		Derailment	22

3 Jul 1930	Sasso, Italy		Train diverted onto wrong line, collision	18
16 Jul 1930	Between Petrova and Viseu Bistra, Romania		Collision	?
20 Jul 1930	Canaan, NB, Canada	Canadian National	Derailment, defective vehicle	5
14 Aug 1930	Seceleanu, Romania		Head-on collision	8
18 Aug 1930	La Houve, France		Collision	6
c10 Oct 1930	Near Inginiero Schamer, Brazi		Derailment, sabotage?	?
29 Oct 1930	Between Négrondes and Ligueux, France		Derailment, track subsidence	13
26 Nov 1930	Guadalupe, Mexico		Collision during shunting operations, explosion, fire	12
26 Nov 1930	Fermo Santa Lucia, Italy		Derailment	7
17 Dec 1930	Near Cuadros, Spain		Head-on collision, single line in blizzard	8
27 December 1930	Near Peipiao, Manchuria		Derailment, sabotage, boiler explosion, fire	80